# The Orbitofrontal Cortex

# The Orbitofrontal Cortex

Edited by

David H. Zald
Assistant Professor
Department of Psychology
Vanderbilt University, USA

and

Scott L. Rauch
Associate Chief of Psychiatry for Neuroscience Research
Massachusetts General Hospital;
Associate Professor of Psychiatry
Harvard Medical School, USA

# OXFORD
UNIVERSITY PRESS

Great Clarendon Street, Oxford OX2 6DP

Oxford University Press is a department of the University of Oxford.
It furthers the University's objective of excellence in research, scholarship,
and education by publishing worldwide in

Oxford  New York

Auckland  Cape Town  Dar es Salaam  Hong Kong  Karachi
Kuala Lumpur  Madrid  Melbourne  Mexico City  Nairobi
New Delhi  Shanghai  Taipei  Toronto

With offices in

Argentina  Austria  Brazil  Chile  Czech Republic  France  Greece
Guatemala  Hungary  Italy  Japan  Poland  Portugal  Singapore
South Korea  Switzerland  Thailand  Turkey  Ukraine  Vietnam

Oxford is a registered trade mark of Oxford University Press
in the UK and in certain other countries

Published in the United States
by Oxford University Press Inc., New York

© Oxford University Press 2006

The moral rights of the authors have been asserted
Database right Oxford University Press (maker)

First published 2006

All rights reserved. No part of this publication may be reproduced,
stored in a retrieval system, or transmitted, in any form or by any means,
without the prior permission in writing of Oxford University Press,
or as expressly permitted by law, or under terms agreed with the appropriate
reprographics rights organization. Enquiries concerning reproduction
outside the scope of the above should be sent to the Rights Department,
Oxford University Press, at the address above

You must not circulate this book in any other binding or cover
and you must impose the same condition on any acquirer

British Library Cataloguing in Publication Data

Data available

Library of Congress Cataloging in Publication Data

Data available

Typeset by Newgen Imaging Systems (P) Ltd., Chennai, India
Printed in Great Britain
on acid-free paper by
Biddles Ltd., King's Lynn

ISBN 0–19–856574–7 (Hbk.)    978–0–19–856574–1 (Hbk.)

10 9 8 7 6 5 4 3 2 1

# Foreword

The exploration of the human frontal lobes started with a bang, not with a whimper. At around 4:30 p.m. on September 13, 1848, an accidental dynamite explosion at a railroad construction site in Cavendish, Vermont, U.S., hurled an iron tamping bar through the head of a 25-year-old foreman named Phineas Gage. The reconstruction of the trajectory, as described by Dr John Harlow (Harlow 1848), showed that the bar must have torn through the frontal lobes, including a substantial part of the orbitofrontal cortex (OFC). The nearly miraculous medical recovery of Gage, the dramatic shift of his personality toward the irresponsible and profane, his chaotic career as a curiosity at Barnum's, stagecoach driver in Chile, and drifter in San Francisco, and the exhumation of his remains for the recovery of the skull are now part of the neurological legend.

With such a headstart, one would have guessed that the clarification of frontal lobe function would have led all other developments in behavioral neurology. Nonetheless, over a hundred years later in 1949, when Carney Landis, Professor of Psychology at Columbia University was called upon to summarize another dramatic chapter in the story of the frontal lobes, namely the effects of surgical prefrontal ablations and lobotomies, he concluded that, 'No existing theory or hypothesis dealing with the psychological significance of the human frontal lobes is tenable. There is no point in either propping up or demolishing established theories. They add only to the confusion of thought and had best be forgotten' (Landis 1949).

Did Professor Landis forget Gage, or did he forget that brain is responsible not only for cognition, but also for comportment? Landis was not alone in being baffled by the elusive consequences of frontal lobe lesions. As so many clinicians have observed, prefrontal lesions do not give rise to salient apraxia, agnosia, or aphasia, deficits that are easily documented by standard tests and bedside examinations. Instead, prefrontal lesions give rise to context-dependent impairments of judgement, insight, inference, decision-making, prioritization, perspective shifting, mental flexibility, and the ability to harmonize feeling with thought. The one common theme that I have been able to find among the numerous behavioral affiliations attributed to the frontal lobes is the transcending of what might be called a 'default mode' of neural procesing (Mesulam 2002).

Prefrontal lesions promote the emergence of this default mode, a neural mode characterized by a tendency for stimulus-bound responses and perseverative behaviours. The default errs more on the side of immediate gratification; it cannot distinguish appearance from significance; it is dominated by the attraction of the here-and-now rather than concerns about future consequences. How does one test for the resurgence of the default mode, especially since it tends to thrive in unconstrained naturalistic settings much more than in the structured environment of the medical office? No wonder that the characterization of prefrontal function and dysfunction remains so challenging, even 150 years after Gage.

Among all prefrontal components, the orbitofrontal region is also the most complex. Part of this complexity reflects the membership of the OFC in the paralimbic belt of the hemispheres, a ring of cortical areas that provides a neuronal bridge linking association neocortex with the limbic system. Paralimbic areas have heterogeneous inputs, heterogeneous outputs, heterogeneous functions. They would not provide much comfort to a casual phrenologist. Paralimbic areas most closely interconnected with the amygdala and olfactory cortex (e.g., the OFC, temporal pole and insula) tend to be associated with emotion, drive, endocrine balance and autonomic function, whereas those more closely interconnected with the hippocampal complex (e.g., the parahippocampal, retrosplenial and posterior cingulate cortices) tend to be associated with memory and learning. These distinctions are relative rather than absolute.

The OFC has no strict boundaries. Anteriorly, it blends into lateral prefrontal association cortex. Posteriorly, it merges with two other members of the paralimbic belt: the insula (laterally) and the cingulate complex (medially). The posterior orbital surface also contains the olfactory nuclei and the piriform cortex, making it a pivotal part of the olfactory system. The posterior and lateral parts of the OFC remain within the amygdaloid sphere of influence, whereas its medial part also displays a close affiliation with the hippocampal formation. The anthropomorphic question of what a brain region 'does', never easy to answer outside of primary sensory-motor areas, becomes particularly difficult to address in the case of the OFC. In a figurative sense, the OFC provides a convergence site for numerous processing streams that collectively transform sensory reality into a subjective reality shaped by private recollections, expectations, and feelings.

Few regions of the brain have attracted as much drama and controversy as the frontal lobes, and few regions have remained as mysterious as the paralimbic areas. The OFC, the topic of this thoughtfully edited volume, sits at the intersection of these two crosscurrents in behavioral neurology. David Zald and Scott Rauch deserve to be congratulated for having had the foresight and good judgement to decide that the time was ripe for a comprehensive volume on this subject. As I look at the list of contributors, a veritable who-is-who in this field, I am convinced that this book will remain the principal source of reference for the OFC for many years to come.

Marsel Mesulam, MD<br>Chicago

**Harlow, J. M.** (1848). Passage of an iron rod through the head. *Boston Medical and Surgical Journal* **39**:389–393.

**Landis, C.** (1949). Psychology. In: Mettler, F. A. (ed.). *Selective Partial Ablation of the Frontal Cortex*. New York: Paul B. Hoeber.

**Mesulam, M-M.** (2002). The human frontal lobes: Transcending the default mode through contingent encoding. In: Stuss, D. and Knight, R. (eds.) *Principles of Frontal Lobe Function*. New York: Oxford University Press.

# Preface

For much of the last century, research on the orbitofrontal cortex (OFC) has lagged behind that on other brain regions. Methodological difficulties in accessing the region, complexities regarding inferring homology across species, inconsistency in conceptualizing its sub-territories, and uncertainty regarding how to approach its functions have fostered this neglect. Other than noting OFC involvement in olfaction and emotion, many textbooks skirt the area in favor of discussions of other brain regions. Even books on frontal lobe functions have typically only provided limited coverage of the OFC, choosing instead to focus on more dorsal frontal regions. This lack of detailed coverage has been evident despite emerging theoretical constructs and the progressive accrual of evidence implicating the OFC in a wide range of psychiatric and neurological disorders.

Over the last decade, there has been an enormous increase in data addressing the structure, connections and functions of the OFC. In 1996, one of the editors of this volume (DHZ) published a pair of reviews in the *Journal of Neuropsychiatry and Clinical Neuroscience* on the anatomy and function of the OFC (Zald and Kim 1996a,b). The reviews included reference to approximately 230 citations, which represented a substantial portion of the published literature at the time. Since then, there has been an explosion of research on this region, with dramatic advances in knowledge regarding its structure, neurocircuitry, and functional contributions to sensation, emotion, social processing, decision-making and memory. Additionally, the OFC's involvement in neuropsychiatric disorders and personality has become increasingly apparent. Because of this groundswell of knowledge and interest, a comprehensive exploration of the OFC can no longer be dealt with in a chapter or two, but instead warrants a full volume.

A major challenge in putting together a book of this sort lies in where to draw the boundaries of coverage. While the OFC proper is readily delineated based on topography (i.e., the cortex lying on the ventral surface of the frontal lobe), a functional parcellation proves much more difficult, especially since much of the data on the subject derives from rodent, and non-human primate models in which the homologies are often uncertain and the topographical boundaries differ from those in humans. Based both on work in monkeys and humans, researchers frequently refer to the ventrolateral prefrontal cortex. However, definitions of this region often encompass the lateral OFC in addition to areas that are more formally on the lateral surface of the prefrontal cortex. Similarly, on the medial surface, researchers frequently refer to the ventromedial prefrontal cortex, which although centered on the medial wall of the frontal lobe (i.e., subgenual and pregenual anterior cingulate) typically extends into the gyrus rectus and neighboring medial orbital areas. The distinctions between these areas and the OFC proper are not always functionally clear, and both clinical and experimental studies frequently involve patients or animals whose lesions do not constrain themselves to the topographical boundaries between these regions. Indeed, in many cases the relative contributions of the orbital and extra-orbital territories have yet to be clarified. Our approach to this problem has been to give contributors full freedom to discuss issues regarding the ventrolateral and ventromedial prefrontal regions, but at the same time, we have largely avoided chapters in which the primary focus is restricted to the extra-orbital components of ventrolateral and ventromedial cortices. We believe that this approach has struck a successful balance, but also acknowledge that there are important aspects of the ventrolateral and medial wall structure and functioning that have simply been omitted from this text due to our focus on the OFC.

We were fortunate to assemble an outstanding collection of experts on different aspects of the OFC to author chapters for this volume. **Part 1** focuses on the anatomy of the OFC. These chapters detail the cytoarchitecture, topography and neurocircuitry of the OFC in both human and non-human primates. **Part 2** broadly covers the functions of the OFC, and includes studies ranging from rodent neurophysiological studies to lesion studies in monkeys to functional neuroimaging in humans. This section also includes a detailed chapter on methodological issues in applying functional MRI to the OFC; given the emergence of functional MRI as one of the major tools for examining OFC functions in humans, and the substantial methodological hurdles related to using functional MRI in the OFC, we felt that it was essential to address this topic. Finally **Part 3** focuses on clinical issues, particularly as they relate to the involvement of the OFC in neuropsychiatric disorders.

We anticipate that while some readers may consume the volume from start to finish, others may use it as a reference, reading one or more chapters at a time, and not necessarily in the prescribed order. Consequently, we encouraged a structure whereby each chapter could stand alone, while minimizing undue redundancies across the volume. Thus, common behavioral tasks are discussed in more than one place in the volume, with the emphasis varying depending upon the specific chapter in question. Similarly, while the classic case of Phineas Gage is mentioned in several chapters, the discussions are restricted to the most relevant features for the given chapter. We thank our contributors for their willingness to sculpt their chapters accordingly. (Readers looking for a more detailed historical account of Phineas Gage may see Macmillan (2000).)

Several individuals deserve special thanks for helping us bring this book to fruition. They include Martin Baum at OUP who showed support and enthusiasm for the project from the very outset, and Amy Cooter who provided administrative support for DHZ. We also thank a number of individuals who helped with the process of reviewing chapters. They include many of the volume's authors as well as other individuals, notably Brian Knutson, Doug Bremner, Jeffrey Schall, Jon Kaas, Edythe London, Thilo Deckersbach, Larry Wald, Chris Wright, Robert Zatorre, Lesley Fellows, and Stephen Smith

## References

Zald, D.H. and Kim, S.W. (1996a). Anatomy and function of the orbital frontal cortex, I: anatomy, neurocircuitry; and obsessive-compulsive disorder. *Journal of Neuropsychiatry and Clinical Neurosciences* **8**:125–138.

Zald, D.H. and Kim, S.W. (1996b). Anatomy and function of the orbital frontal cortex, II: Function and relevance to obsessive-compulsive disorder. *Journal of Neuropsychiatry and Clinical Neurosciences* **8**:249–261.

Macmillan, M. (2000). *An Odd Kind of Fame: Stories of Phineas Gage*. Cambridge, MA: The MIT Press.

## Dedications

I thank my parents, who instilled in me a love of ideas and learning, and my wife, who put up with me despite having in no uncertain terms told me that I shouldn't edit a book at this moment in my career.

DHZ

I acknowledge the profound debt I owe to my family and close friends for their support for this work. I wish to dedicate this volume especially to my father—a model for me both as a man and as a scientist.

SLR

# Contents

Contributors  xv

## Part 1 **Anatomy**

1. Architectonic structure of the orbital and medial prefrontal cortex  3
   Joseph L. Price

2. The orbitofrontal cortex: sulcal and gyral morphology and architecture  19
   Michael Petrides and Scott Mackey

3. Connections of orbital cortex  39
   Joseph L. Price

4. Sequential and parallel circuits for emotional processing in primate orbitofrontal cortex  57
   Helen Barbas and Basilis Zikopoulos

## Part 2 **Functions and methods**

5. The neurophysiology and functions of the orbitofrontal cortex  95
   Edmund T. Rolls

6. The chemical senses  125
   Jay A. Gottfried, Dana M. Small, and David H. Zald

7. Involvement of primate orbitofrontal neurons in reward, uncertainty, and learning  173
   Wolfram Schultz and Leon Tremblay

8. From associations to expectancies: orbitofrontal cortex as gateway between the limbic system and representational memory  199
   Matthew Roesch and Geoffrey Schoenbaum

9. A componential analysis of the functions of primate orbitofrontal cortex  237
   Angela C. Roberts and John Parkinson

10. The role of human orbitofrontal cortex in reward prediction and behavioral choice: insights from neuroimaging  265
    John P. O'Doherty and Raymond J. Dolan

11. Memory processes and the orbitofrontal cortex  285
    Matthias Brand and Hans J. Markowitsch

12. The role of lateral orbitofrontal cortex in the inhibitory control of emotion  307
    Christine I. Hooker and Robert T. Knight

13 Visceral and decision-making functions of the ventromedial prefrontal cortex  *325*
   Nasir Naqvi, Daniel Tranel, and Antoine Bechara

14 Intracranial electrophysiology of the human orbitofrontal cortex  *355*
   Ralph Adolphs, Hiroto Kawasaki, Hiroyuki Oya, and Matthew A. Howard

15 Orbitofrontal cortex activation during functional neuroimaging studies of emotion induction in humans  *377*
   Darin D. Dougherty, Lisa M. Shin, and Scott L. Rauch

16 Neurochemical modulation of orbitofrontal cortex function  *393*
   Trevor W. Robbins, Luke Clark, Hannah Clarke, and Angela C. Roberts

17 Technical considerations for BOLD fMRI of the orbitofrontal cortex  *423*
   V. Andrew Stenger

## Part 3 **Neuropsychiatry**

18 Neuropsychological assessment of the orbitofrontal cortex  *449*
   David H. Zald

19 The orbitofrontal cortex in drug addiction  *481*
   Rita Z. Goldstein, Nelly Alia-Klein, Lisa A. Cottone, and Nora D. Volkow

20 The orbitofrontal cortex and anxiety disorders  *523*
   Mohammed R. Milad and Scott L. Rauch

21 The role of the ventral prefrontal cortex in mood disorders  *545*
   Carolyn A. Fredericks, Jessica H. Kalmar, and Hilary P. Blumberg

22 Effect of orbitofrontal lesions on mood and aggression  *579*
   Pamela Blake and Jordan Grafman

23 Pseudopsychopathy: a perspective from cognitive neuroscience  *597*
   Michael Koenigs and Daniel Tranel

24 Frontotemporal dementia and the orbitofrontal cortex  *621*
   Po H. Lu, Negar Khanlou, and Jeffrey L. Cummings

Index  *643*

# Contributors

**Ralph Adolphs**
Department of Neurosurgery,
The University of Iowa,
Iowa City, IA, USA

Division of Humanities and
Social Sciences,
California Institute of Technology,
Pasadena, CA, USA

**Nelly Alia-Klein**
Medical Department,
Neuroimaging Group,
Brookhaven National Laboratory,
Upton, NY, USA

**Helen Barbas**
Department of Health Sciences,
Boston University,
Boston, MA, USA

Department of Anatomy and
Neurobiology,
Boston University School of Medicine,
Boston, MA, USA

NEPRC,
Harvard Medical School,
Southborough, MA, USA

**Antoine Bechara**
Department of Neurology,
Division of Cognitive Neuroscience and
Neuroscience Training Program,
University of Iowa,
Iowa City, IA, USA

**Pamela Blake**
Cognitive Neuroscience Section,
National Institute of Neurological
Disorders and Stroke,
National Institutes of Health,
Bethesda, MD, USA

Department of Neurology,
Georgetown University Hospital,
Washington, DC, USA

**Hilary P. Blumberg**
Departments of Psychiatry and
Diagnostic Radiology,
Yale University School of Medicine,
Department of Veterans' Affairs,
New Haven, CT, USA

**Matthias Brand**
Department of Physiological Psychology,
University of Bielefeld
Bielefeld, Germany

**Hannah Clark**
MRC-Wellcome Trust Behavioural and
Clinical Neurosciences Institute,
Department of Experimental Psychology,
University of Cambridge,
Cambridge, UK

**Luke Clark**
MRC-Wellcome Trust Behavioural and
Clinical Neurosciences Institute,
Department of Experimental Psychology,
University of Cambridge,
Cambridge, UK

**Lisa A. Cottone**
Medical Department,

Neuroimaging Group,
Brookhaven National Laboratory,
Upton, NY, USA

**Jeffrey L. Cummings**
Department of Neurology,
David Geffen School of Medicine at UCLA,
Los Angeles, CA, USA
Department of Psychiatry and Biobehavioral Sciences,
David Geffen School of Medicine at UCLA,
Los Angeles, CA, USA.

**Raymond J. Dolan**
Wellcome Department of Imaging Neuroscience,
University College London,
London, UK

**Darin D. Dougherty**
Department of Psychiatry,
Massachusetts General Hospital,
Boston, MA, USA

**Carolyn A. Fredericks**
Departments of Psychiatry,
Yale University School of Medicine,
New Haven, CT, USA

**Rita Z. Goldstein**
Medical Department,
Neuroimaging Group,
Brookhaven National Laboratory,
Upton, NY, USA

**Jay A. Gottfried**
Department of Neurology,
Northwestern University Feinberg School of Medicine,
Chicago, IL, USA

**Jordan Grafman**
Cognitive Neuroscience Section,
National Institute of Neurological Disorders and Stroke,
National Institutes of Health,
Bethesda, MD, USA
Department of Neurology,
Georgetown University Hospital,
Washington, DC, USA

**Christine I. Hooker**
Helen Wills Neuroscience Institute,
University of California at Berkeley,
CA, USA

**Matthew A. Howard**
Department of Neurosurgery,
The University of Iowa,
Iowa City, IA, USA
Division of Humanities and Social Sciences,
California Institute of Technology,
Pasadena, CA, USA

**Jessica H. Kalmar**
Department of Psychiatry,
Yale University School of Medicine,
New Haven, CT, USA

**Hiroto Kawasaki,**
Department of Neurosurgery,
The University of Iowa,
Iowa City, IA, USA
Division of Humanities and Social Sciences,
California Institute of Technology,
Pasadena, CA, USA

**Negar Khanlou,**
Department of Pathology and Laboratory Medicine,

David Geffen School of Medicine at UCLA,
Los Angeles, CA, USA

**Robert T. Knight**
Helen Wills Neuroscience Institute,
Department of Psychology,
University of California at Berkeley,
Berkeley, CA, USA

**Michael Koenigs**
Department of Neurology,
Division of Cognitive Neuroscience and
Neuroscience Training Program,
University of Iowa,
Iowa City, IA, USA

**Po H. Lu**
Department of Neurology,
David Geffen School of Medicine at UCLA,
Los Angeles, CA, USA

**Scott Mackey**
Montreal Neurological Institute,
McGill University,
Montreal, Quebec,
Canada

**Hans J. Markowitsch**
Department of Physiological
Psychology,
University of Bielefeld,
Bielefeld, Germany

**Mohammed R. Milad**
Department of Psychiatry,
Massachusetts General Hospital and
Harvard Medical School,
Charlestown, MA, USA

**Nasir Naqvi**
Department of Neurology,
Division of Cognitive Neuroscience and
Neuroscience Training Program,
University of Iowa,
Iowa City, IA, USA

**John P. O'Doherty**
Wellcome Department of Imaging
Neuroscience,
University College London,
London, UK

**Hiroyuki Oya**
Department of Neurosurgery,
The University of Iowa,
Iowa City, IA, USA

Division of Humanities and
Social Sciences,
California Institute of Technology,
Pasadena, CA, USA

**John Parkinson**
School of Psychology,
University of Wales,
Bangor,
Gwynedd, UK

**Michael Petrides**
Montreal Neurological Institute and
Department of Psychology,
McGill University,
Montreal, Quebec,
Canada

**Joseph L. Price**
Department of Anatomy and
Neurobiology,
Washington University School of Medicine,
St Louis, MO, USA

**Scott L. Rauch**
Department of Psychiatry,
Massachusetts General Hospital,
Boston, MA, USA

**Trevor W. Robbins**
MRC-Wellcome Trust Behavioural and
Clinical Neurosciences Institute,
Department of Experimental Psychology,
University of Cambridge,
Cambridge, UK

**Angela C. Roberts**
MRC-Wellcome Trust Behavioral and
Clinical Neurosciences Institute,
Department of Experimental Psychology,
University of Cambridge,
Cambridge, UK

**Matthew Roesch**
University of Maryland School of
Medicine,
Department of Anatomy and
Neurobiology,
Baltimore, MD, USA

**Edmund T. Rolls**
Department of Experimental Psychology,
University of Oxford,
Oxford, UK

**Geoffrey Schoenbaum**
University of Maryland School of Medicine
Departments of Anatomy, Neurobiology,
and Psychiatry,
Baltimore, MD, USA

**Wolfram Schultz**
Department of Anatomy,
University of Cambridge,
Cambridge, UK

**Lisa M. Shin**
Department of Psychiatry,
Massachusetts General Hospital,
Boston, MA, USA

Department of Psychology,
Tufts University,
Medford, MA, USA

**Dana M. Small**
J. B. Pierce Laboratory and Department of
Psychology,
Yale University,
New Haven, CT, USA

**V. Andrew Stenger**
John A. Burns School of Medicine
University of Hawaii
Honolulu, HI, USA

**Daniel Tranel**
Department of Neurology,
Division of Cognitive Neuroscience and
Neuroscience Training Program,
University of Iowa,
Iowa City, IA, USA

**Leon Tremblay**
Neurologie et Thérapeutique
Expérimentale (INSERM U289),
Hôpital de la Salpêtrière,
Paris, France

**Nora D. Volkow**
National Institute on Drug Abuse,
Bethesda, MD, USA

**David H. Zald**
Department of Psychology,
Vanderbilt University,
Nashville, TN, USA

**Basilis Zikopoulos**
Department of Health Sciences,
Boston University,
Boston, MA, USA

# Part 1

# Anatomy

Chapter 1

# Architectonic structure of the orbital and medial prefrontal cortex

Joseph L. Price

The orbital and medial prefrontal cortex (OMPFC) is a large and heterogeneous region. Although it makes up a substantial fraction of the cortex, which increases markedly in non-human primates, and even more in humans, our understanding of its organization and function has lagged behind that of most other cortical regions. In the past 15 years, however, anatomical, physiological, and imaging studies have greatly expanded our knowledge. A central part of this advance has been linked to a better understanding of the regional variation in the structure of the cortex, and the subdivision of the cortex into distinct architectonic areas based on these variations. By delineating a set of areas that can be recognized from animal to animal and across species, a framework can be established for a detailed analysis of connections and functions. Further, by transferring such architectonic maps from monkeys to humans, experimental data from animals can be correlated with observations on the human brain.

In primates, including humans, the cortex on both the orbital and medial frontal surfaces varies from an agranular region caudally at the junction with the insular cortex and septal nuclei, to a dysgranular zone in the central region, and then a granular region near the frontal pole (Barbas and Pandya 1989). These broad cortical zones are, however, not homogeneous, and each of them can be divided into several architectonic areas that have distinct connections, and presumably distinct functions. Although there is considerable variation in the prefrontal cortex across species, especially in the amount of granular vs. agranular cortex, similarities in the position and connections of orbital and medial areas indicate that this part of the prefrontal cortex is at least relatively comparable across species (Uylings and Van Eden 1990; Petrides and Pandya 1994). Such comparative cytoarchitectonic studies of the prefrontal cortex may allow a further extension of the correlation from extensively studied animal species to humans, where experimental data are scarce. This is of particular relevance to human neuroimaging studies, where knowledge of the anatomy and physiology of brain areas is crucial for sound interpretation of the data.

This account will begin with architectonic maps in monkeys, followed by their correlation with equivalent maps in rats and humans. While it will focus on the orbital or orbitofrontal cortex, this region is closely linked to the areas on the medial surface, the anterior portion of the insula, and the ventrolateral convexity of the frontal lobe, so those areas will also be briefly described.

## 1.1. Monkey

All the areas at the caudal edge of the orbital and medial prefrontal cortex have a five-layered agranular cortical structure, without a granular layer IV. The central orbital areas have an incipient layer IV and may be termed dysgranular. Finally, the most rostral orbital and medial areas are fully granular and sharply laminated, similar to most areas in the dorsolateral prefrontal cortex.

On the medial wall of the prefrontal cortex of monkeys (*Cercopithecus*), Brodmann (1909) and Garey (1994) delineated three agranular areas—area 24, just dorsal to the genu of the corpus callosum, area 25 in the subgenual region, and area 32, rostral to area 25 (see Fig. 1.1a). Brodmann's granular area 10 occupies the frontal pole. He did not carry out a detailed study of the orbital cortex, and the later map by Walker (1940) has formed the basis for most maps of this region in monkeys. Walker recognized area 10 on the frontal pole and area 11 on the rostral orbital surface. More caudally, his areas 12, 13, and 14 occupy the lateral, central, and medial orbital surface, respectively (see Fig. 1.1b). Walker (1940) also described the medial prefrontal cortex, but his numbering of areas does not correspond to the locations in Brodmann's maps.

Many later studies have essentially adapted Walker's map of the orbital cortex, and Brodmann's map of the medial prefrontal cortex, and used them to map connectional data onto the cortex (Porrino *et al.* 1981; Amaral and Price 1984; Preuss and Goldman-Rakic 1991; Petrides and Pandya 1994, 2002). Carmichael and Price (1994) used Walker's and Brodmann's maps as a starting point for a more detailed map based on several histochemical and immunohistochemical stains.

Another approach to analysis of cortical architectonic structure, based on a proposed phylogenetic progression, has been used in some studies of the orbital cortex (Sanides 1970; Mesulam 2000). According to this scheme, the cortex developed from the primitive allocortex, consisting of the olfactory cortex (paleocortex) and hippocampus (archicortex). From these limbic areas, the hypothesis is that successive rings of cortex form during evolution, producing first an intermediate paralimbic or periallocortex around the allocortex, and then a larger isocortex (or neocortex). The evolutionary basis of this scheme is difficult to assess because of the absence of a fossil record of the brain structure; even in primitive reptiles, a dorsolateral 'general' cortex (neocortical primordium) is present between the lateral cortex (olfactory primordium) and dorsomedial cortex (hippocampal primordium). When applied to the orbital and the medial prefrontal cortex, a periallocortical zone has been recognized caudally, near the olfactory cortex, surrounded by proisocortical and isocortical zones further rostrally. Barbas, Pandya, and their colleagues have published and used a map of the OMPFC that combines recognition of small periallocortical and proisocortical areas in the caudal orbital cortex, combined with Walker's areas in the central and rostral orbital cortex and Brodmann's areas on the medial wall (Barbas and Pandya 1989; Dombrowski *et al.* 2001).

The description in this chapter will be based primarily on the architectonic analysis by Carmichael and Price (1994). Using nine different staining methods, we applied

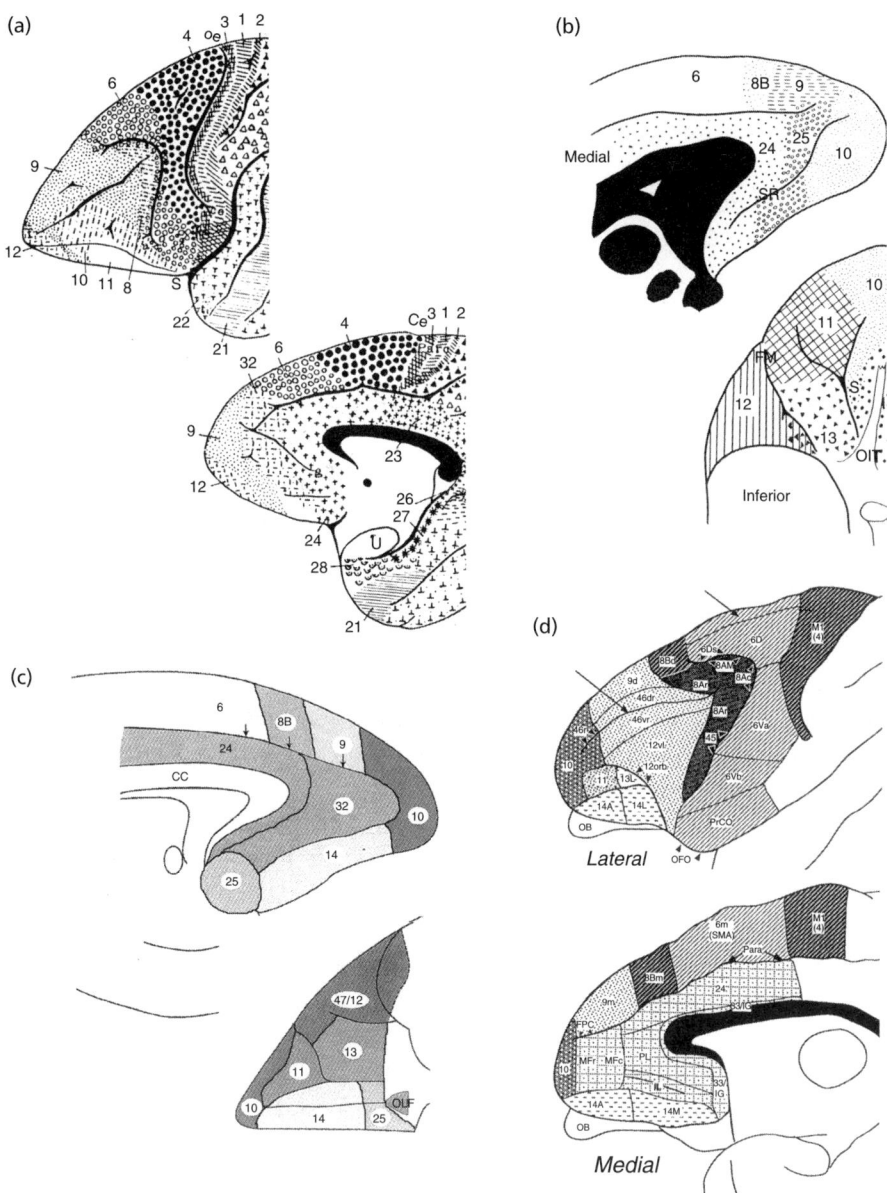

**Fig. 1.1** Architectonic maps of the prefrontal cortex of monkeys, taken from four different studies. (a) The classical map of the guenon brain by Brodmann (1909), showing the medial and lateral prefrontal cortex; Brodmann did not illustrate the orbital cortex as such. (b) The map by Walker (1940), which divided the orbital cortex into areas 10, 11, 12, 13, and 14. Note that area 25 on the medial surface is in the location usually occupied by area 32 in other maps. (c) A recent map published by Petrides and Pandya (1994). This map is very similar to that of Walker, except for areas 32 and 25 on the medial wall. The cortex on the ventrolateral convexity is designated area 47/12 instead of simply area 12, for better correspondence with Brodmann's human map. (d) A map of the lateral and medial surfaces by Preuss and Goldman-Rakic (1991).

a working definition that a local region could be recognized as a distinct area if it exhibited characteristic staining patterns in a consistent location with at least three stains, had distinct connections, and showed relatively little variation across brains. This method has the advantage that multiple criteria can be used to distinguish areas. While the traditional cell and myelin stains provide a good distinction of some areas, they are not very useful for others; by using additional markers, such as acetylcholinesterase and parvalbumin immunoreactivity, the other boundaries often become apparent. Further, this method is purely observational, and not based on any pre-existing hypothetical scheme. It is less likely, therefore, to be led astray by theoretical considerations.

The analysis by Carmichael and Price (1994) identified over 20 areas in the OMPFC, many of which are subdivisions of previously recognized areas (Fig. 1.2a). For example, Walker's rostral granular areas 10 and 11 were subdivided into areas 10m, 10o, 11m and 11l. Of these, areas 10o, 11m and 11l are in the rostral orbital surface, while area 10m is situated primarily on the medial surface of the hemisphere. In the central part of the orbital surface, Walker's areas 12, 13 and 14 were subdivided into areas 12r, 12l, 12m, 12o, 13a, 13b, 13m, 13l, 14r and 14c. Finally, five areas (Iam, Iai, Ial, Iapm and Iapl) were recognized in the posterior orbital cortex, which represent a rostral extension of the agranular insular cortex onto the orbital surface. Each of these areas represents a cortical module characterized by consistent staining patterns and specific input/output relations, similar to the large number of areas described within the visual-processing stream. Studies of the anatomical connections of these individual areas have shown that they have distinct interactions with many other cortical and subcortical areas (Carmichael and Price 1995a, b, 1996; An *et al.* 1998; Öngür *et al.* 1998; Ferry *et al.* 2000; Kondo *et al.* 2003).

## Agranular insula

Although the insula is well known as the cortical structure in the depth of the lateral sulcus, it is better defined by the presence of claustrum deep into the cortex. As Rose (1928) found, this allows areas to be delineated in animals, such as rats, that are equivalent to at least part of the insular cortex. By this criterion, there is a substantial extension of the insula onto the caudal orbital surface. All of these areas are agranular, as are the adjacent areas in the rostroventral part of the insula proper. In their description of the insular cortex, Mufson and Mesulam (1982) recognized this continuity from the insula to the orbital cortex, and applied the term 'agranular, periallocortical orbitofrontal cortex' to the orbital part, and 'agranular, periallocortical insular cortex' to the insular part. The orbito-insular areas also correspond approximately to the 'orbital periallocortex' (OPall) and 'orbital proisocortex' (OPro) areas of Barbas and Pandya (1989). Of the areas recognized by Carmichael and Price (1994), areas Iam and Iapm (medial and posteromedial agranular insula) are located immediately adjacent to the primary olfactory cortex (anterior olfactory nucleus and piriform cortex). Both of these areas have relatively poorly developed lamination, such that layers II and III, and also layers

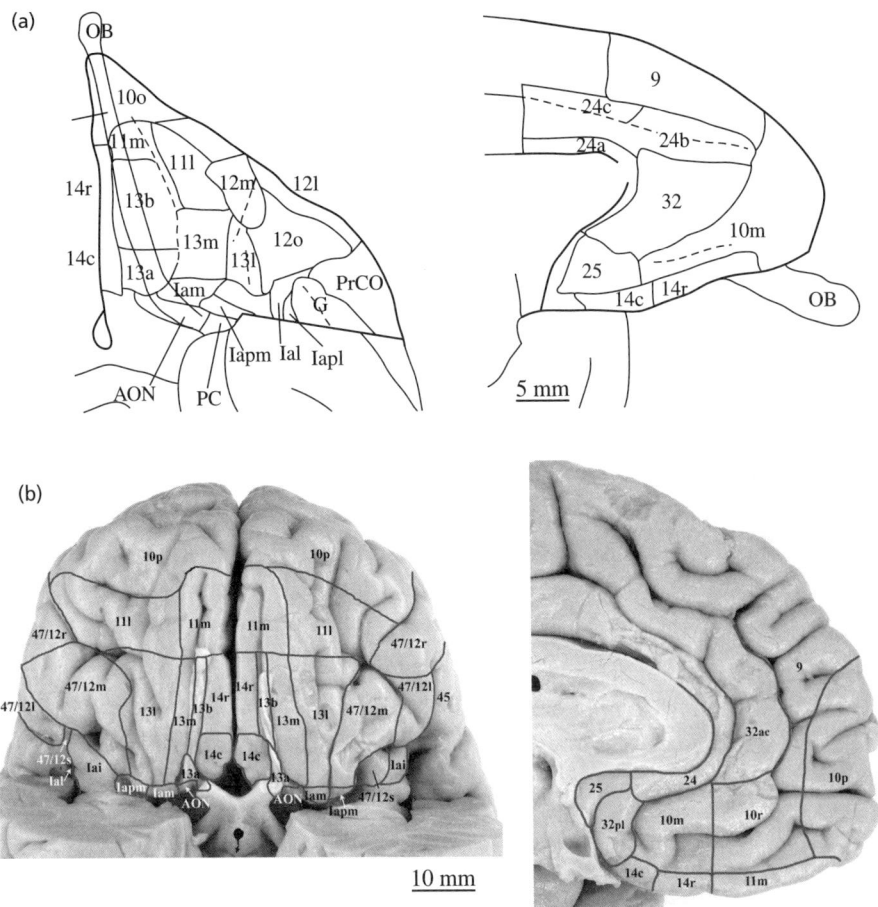

**Fig. 1.2** Architectonic maps of the orbital and medial prefrontal cortex (a) in monkeys, by Carmichael and Price (1994), and (b) in humans, by Öngür et al. (2003). In both studies, patterns seen with several different staining methods were correlated to produce a more detailed and reliable subdivision. In monkeys, these patterns were also correlated with experimentally defined connections. The map of the orbital cortex was based on the map by Walker (1940), although each of his areas was subdivided into two to four different areas. The map of the medial cortex was based on Brodmann's (1909) description. As in the map by Petrides and Pandya (1994), the ventrolateral convexity in humans was designated area 47/12.

V and VI, are combined into single layers. In area Iam, these layers are pushed together, but in area Iapm there is a cell-sparse layer separating them. Areas Ial and Iapl (lateral and posterolateral agranular insula) have well-defined layers II, III, V, and VI, with layer V especially sharp, and subdivided into Va and Vb. Area Iai (intermediate agranular insula) is in the caudal part of the lateral orbital sulcus, and is somewhat intermediate in structure. It has a more developed lamination than areas Iam and Iapm, but the layers are not as sharp and distinct as in Ial and Iapl.

### Central orbital areas

Areas 12, 13, and 14 of Walker (1940), in the central part of the orbital cortex have a slight granular layer IV, and for the most part can be considered to be dysgranular. Both were subdivided by Carmichael and Price (1994) into four separate areas, which have somewhat different structures and connections. The subdivisions of area 12 (12o, 12m, 12l, 12r) are located lateral to the lateral orbital sulcus. Area 12o, on the lateral bank of the sulcus, is best characterized by its weak myelin staining pattern. Area 12l is relatively granular, and located at and lateral to the ventrolateral convexity; it is characterized by vertical bundles of myelinated fibres. Area 12m, rostral to area 12l, is marked by horizontal plexuses of myelinated fibres, especially in layers III and V (bands of Baillarger). Area 12r occupies the rostral part of the ventrolateral convexity. Areas 13a and 13b are located on the medial side of the orbital surface, on the medial bank of the medial orbital sulcus, and dorsal to the course of the olfactory peduncle. Area 13a is a small area that, like area Iam, is agranular and poorly laminated. The more rostral area 13b is dysgranular, and is characterized by a combination of radial and horizontal striations. Areas 13m and 13l are dysgranular with well-developed lamination; area 13m occupies the lateral bank of the medial orbital sulcus, and most of the cortex between it and the lateral orbital sulcus. Area 13l is situated in the lateral orbital sulcus; in optimally stained sections, it is characterized by a strong plexus of parvalbumin positive fibres. Area 14 occupies the medial convexity between the orbital cortex and the medial prefrontal cortex. It can be divided into a caudal agranular area 14c and a more rostral dysgranular area 14r.

### Rostral orbital areas

Areas 11 and 10 have a fully formed granular layer IV, and occupy the rostral orbital cortex and the frontal pole, respectively. Area 11 can be divided into area 11m, in and medial to the medial orbital sulcus, and area 11l more laterally. These areas are difficult to distinguish on purely architectonic grounds, but have different connections to several other parts of the brain. Area 10 is subdivided into area 10o on the orbital surface and frontal pole, and area 10 m in the medial prefrontal cortex. Area 10o is an especially well-developed granular cortex, the most granular part of the orbital cortex.

### Medial prefrontal cortex

The cortex on the medial wall also varies from the agranular cortex caudally (areas 25, 32, 24a/b) to the granular cortex rostrally (area 10 m). Area 25 is located ventral to the genu of the corpus callosum; it is poorly laminated, with conjoined layer II/III superficially, and V/VI deep. Area 24 is located dorsal to the corpus callosum, and extends rostrally ventral to the cingulate sulcus. It is agranular, and is characterized by marked radial striations. Area 32, rostral to the corpus callosum, is also agranular; it has prominent layers III and V that are marked by horizontal striations. Area 10 m occupies the medial side of the frontal pole, and extends caudally between areas 32 and 14r. It is distinguished by mixed horizontal and radial striations, and a granular layer IV, which increases in size and prominence from caudal to rostral. This gradient in granularity suggests that rostral and caudal parts of area 10m may be recognized.

## 1.2. Rat

The most prominent cytoarchitectonic feature of the rat prefrontal cortex is that it is composed exclusively of agranular cortical areas. Brodmann (1909) and Garey (1994) did not use the term 'prefrontal cortex', but in primates and carnivores Brodmann defined the granular cortex anterior to the motor strip as the 'frontal region'. He did not map the rat brain, but from his analysis of other related animals (flying fox, ground squirrel, and rabbit) he indicated that rodents do not have a comparable frontal cortex, since they lack a rostral granular cortex. In an effort to address this question, Rose and Woolsey (1948a) proposed that equivalent areas could be recognized in different species on the basis of similar connections; they specifically proposed that the 'orbitofrontal cortex' of rabbits and cats was similar to the primate prefrontal cortex due to its connections with the mediodorsal thalamic nucleus (MD).

More recent studies have shown that MD is reciprocally connected with all parts of the prefrontal cortex in both rats and monkeys, including both orbital and anterior cingulate regions as well as dorsolateral prefrontal areas (Krettek and Price 1977; Goldman-Rakic et al. 1985; Giguere and Goldman-Rakic 1988; Groenewegen 1988; Uylings and Van Eden 1990; Ray and Price 1992, 1993). Based on this criterion, the 'prefrontal cortex' in rats includes the medial frontal cortex (around and rostral to the genu of the corpus callosum), the cortex at the dorsomedial corner of the hemisphere, and the 'orbital' cortex in the dorsal bank of the rhinal sulcus. Despite the usefulness of this criterion, the problem that Brodmann raised has not been completely solved. The absence of a granular zone in the rat prefrontal cortex still makes it difficult to establish which areas, if any, in the rat are homologous to the rostral or dorsolateral prefrontal cortex in primates (see also Preuss 1995).

On topological grounds, it would be expected that the orbital cortex in rats would be situated in the dorsal bank of the rostral end of the rhinal sulcus. Rose (1929, 1931) included most of this area in the agranular insular region in mice, but in rabbits he also identified the orbital cortex in the region rostral to the junction of the frontal cortex with the olfactory peduncle. Krettek and Price (1977) subdivided the orbital region in rats into medial, ventrolateral, and lateral orbital areas (MO, VLO, LO). Area VLO continues caudally in the depth of the rhinal sulcus, but the ventral and dorsal agranular insular areas (AIv, AId) occupy the dorsal bank and lip of the rhinal sulcus (Krettek and Price 1977; Ray and Price 1992), (Fig. 1.3). In the initial description, area AId extended rostrally, lateral to the orbital areas (Krettek and Price 1977), but subsequently the part of area AId rostral to the claustrum was recognized as the dorsolateral orbital area (DLO; Ray and Price 1992). More rostrally, Ray and Price (1992) also delineated lateral and medial frontal pole areas (FPl and FPm).

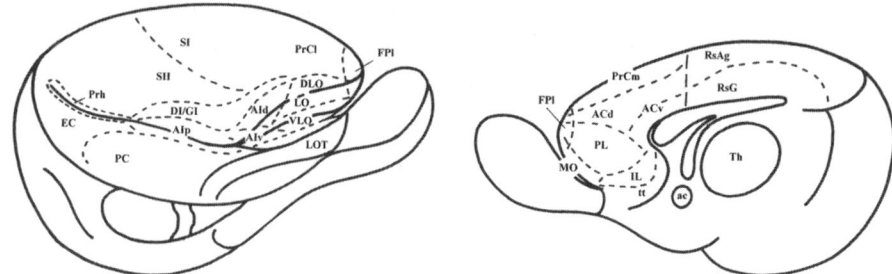

**Fig. 1.3** Architectonic map of the agranular insular, orbital, and medial prefrontal areas in rats, from Krettek and Price (1977), with slight modifications by Ray and Price (1992). See text for correlation between the areas in rats and those in monkeys and humans.

The orbital areas in rats appear to correspond to several of the orbital areas in monkeys although it is difficult to establish conclusive homology. Areas AIv and AId are comparable to areas Iam and Iai, respectively, based on structure and connections (see Price, other chapter). Similarly, area VLO is topologically similar to areas 13a and 13b. Area LO appears to correspond to the central orbital areas 13m and 13l, while area DLO may correspond to area 12o in monkeys. The correspondence, if any, between the frontal polar areas in rats and the more rostral orbital areas in monkeys is uncertain.

Brodmann mapped the medial prefrontal cortex of a variety of animals, including ground squirrels and hedgehogs (Brodmann 1909; Garey 1994). Beginning with these maps, subsequent analyses in rats have divided the medial prefrontal cortex from the olfactory areas ventrally to the dorsomedial corner into infralimbic (IL), prelimbic (PL), ventral and dorsal anterior cingulate (ACv and ACd), and medial precentral (PrCm) areas (Fig. 1.3) (Krettek and Price 1977; Ray and Price 1992). Zilles (1995) and Paxinos and Watson (1997) recognized very similar areas, but termed area PrCm as the frontal area 1 (Fr1), and renamed the dorsal and ventral anterior cingulate areas and the prelimbic area as cingulate areas Cg1, Cg2, and Cg3, respectively. This difference is almost entirely terminological, and most investigators agree on the boundaries of prefrontal cortex areas. The medial prefrontal areas in rats and monkeys are relatively comparable, based on several studies going back to Brodmann. Areas IL and PL in rats are very similar to areas 25 and 32 in monkeys, while areas ACv and ACd can be compared to area 24. It is uncertain whether rats have an area that is equivalent to area 10 m in monkeys.

## 1.3. Humans

Brodmann's human map generally resembles his monkey map (Fig. 1.1A), though it is clear that they are not identical (Brodmann 1909; Garey 1994). Thus, in the human map the medial prefrontal cortex consists of areas 10, 11, 24, 25, and 32, while in his monkey (guenon) map there is no area 10 or 25, and area 12 occupies the ventral frontal pole. Further, Brodmann specifically stated that area 32 in the human map is not homologous to area 32 in the monkey map. Another terminological problem is related to area 12. In one version of his human map that has been republished in several subsequent reviews (e.g. Petrides and Pandya 1994), Brodmann included an area 12 on the medial wall anterior to the corpus callosum, but this is not found in the map published in his monograph (Brodmann 1909), or in his preceding journal articles (Brodmann 1905, 1908). Where illustrated, this medial area 12 is clearly different from area 12 in the lateral orbital cortex in Walker's (1940) and the subsequent maps of monkeys (see above). The lateral orbital area that is usually denoted as area 12 in monkeys appears to correspond to area 47 in Brodmann's human map. Petrides and Pandya (1994) have made an attempt to resolve this by using the term area 47/12 to refer to the lateral orbital areas in both humans and monkeys.

Brodmann did not illustrate the rest of the orbital surface, but most of it appears to have been denoted as area 11 in the map of the human brain. Several other investigators have proposed delineations that are in relatively closer accord with the map made by Walker (1940) in monkeys. Beck (1949) recognized the cortex of the human gyrus rectus as area 14, and medial part of the orbital cortex as area 13 (caudally) and area 11 (rostrally). Semendeferi *et al.* (1998) also recognized area 13 on the posterior orbital surface of several different primate species, including those of macaque monkeys and humans. Using several staining methods, Hof *et al.* (1995) divided the orbital cortex into eight areas, although they denoted the areas by letters that indicate the position of each area, instead of with numbers; it is not clear how this subdivision can be correlated with other recent maps. Petrides and Pandya (1994; 2002) have published very comparable maps in monkeys and humans using Walker's map of the orbital cortex and Brodmann's map of the medial prefrontal cortex.

Recently, Öngür *et al.* (2003) did a more detailed analysis that attempted to translate the monkey map of the OMPFC by Carmichael and Price (1994) onto the human brain. Using five different histological and immunohistochemical stains, twenty three distinct areas were recognized in humans, all of which are correlated with a specific area in monkeys. Each cortical area was distinct with at least two different stains and was found in a similar location in different brains. As in monkeys, many of these areas are subdivisions of previously recognized zones from Walker's (1940) map (Beck 1949; Petrides and Pandya 1994, 2002). To a remarkably high degree, the human cortical areas were defined by staining characteristics similar to the corresponding areas in monkeys and were located in similar locations (see Figs. 1.2b and 1.4).

## Agranular insula

Beginning with the work of von Economo and Koskinas (1925), the cortex on the posterior orbital surface of human brains has been recognized to be a continuation of the insular cortex. The 'precentral insular' area of von Economo and Koskinas (1925) extended into the caudolateral orbital region, and their 'frontoinsular' area occupied the central part of the caudal orbital cortex. Brockhaus (1940) used the term 'claustrocortex' to describe this region. Petrides and Pandya (1994, 2002) labeled this region as 'periallocortex' and 'proisocortex'.

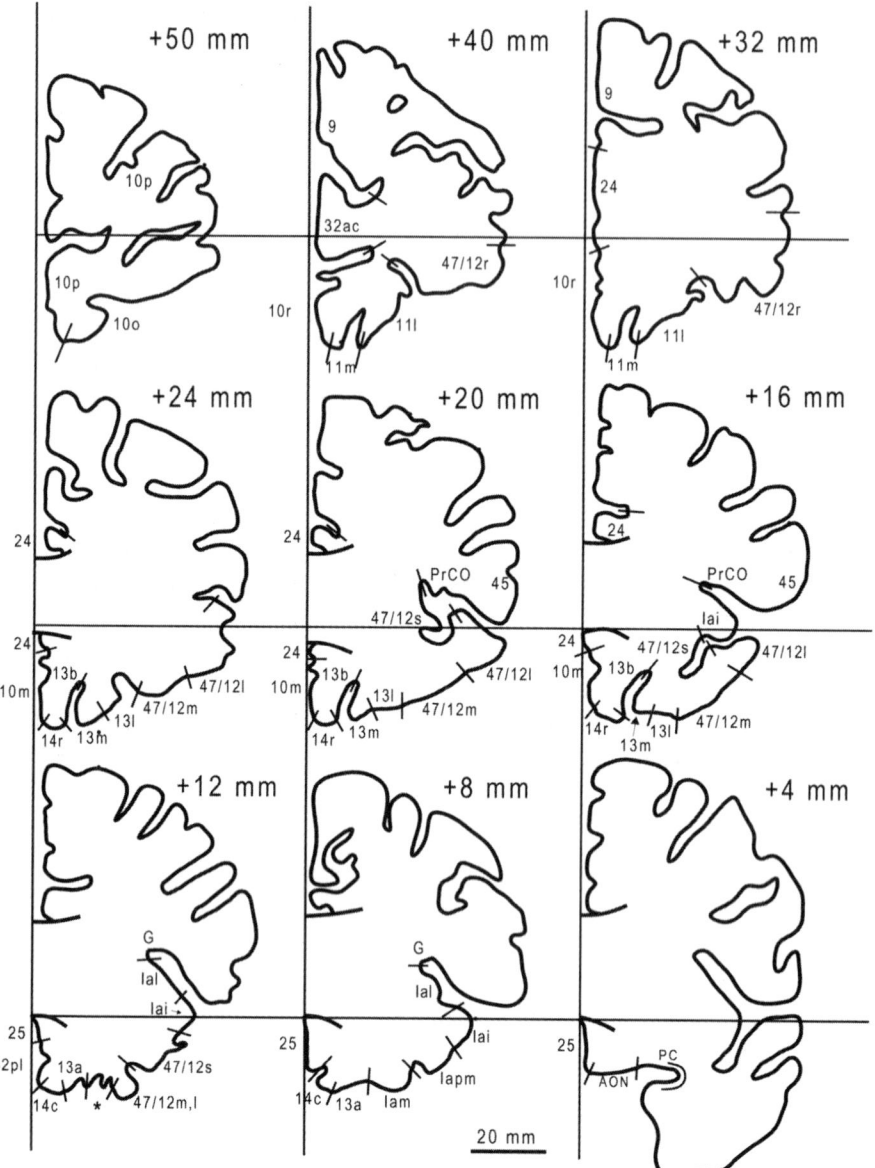

**Fig. 1.4** The architectonic areas of the OMPFC in humans, drawn onto sections from the atlas by Tailairch and Tournoux (1988). The numbering of slices reflects the number of mm anterior to the anterior commissure.

Öngür et al. (2003) recognized four subdivisions of the agranular insula on or adjacent to the caudal orbital surface (areas Iam, Iapm, Iai, Ial). Areas Iam and Iapm possess a relatively thin cortex and lie immediately lateral to the olfactory cortex (anterior olfactory nucleus and piriform cortex). Lateral and rostral to these areas, area Iai is a thicker and more laminated cortex, which extends rostrally into the horizontal ramus of the lateral sulcus (the rostral continuation of the lateral sulcus into the frontal lobe). This area is particularly interesting, because it has different connections than the areas around it (see Chapter 3). Area Ial is a well-laminated area lateral and caudal to area Iai, which is distinguished by a prominent layer Va and a prominent myelin plexus in layer III (outer line of Baillarger). It adjoins the gustatory cortex in the fundus of the superior limiting sulcus.

## Lateral orbital cortex

The lateral orbital cortex is recognized in this study as area 47/12, following the suggestion of Petrides and Pandya (1994), in order to highlight the similarity of this region on the ventrolateral frontal convexity to the monkey area 12. In parallel with Carmichael and Price's (1994) analysis in monkeys, area 47/12 was subdivided into four areas in the human brain. Taken together, these largely correspond to Petrides and Pandya's area 47/12 (1994, 2002) and to Hof et al.'s area PL (1995).

Area 47/12s of the human OMPFC is structurally similar to area 12o in monkeys. The apparent discrepancy between the topological relationships that these two areas have with their neighbors is resolved when the changes in folding patterns and the growth of nearby areas are taken into consideration. The growth of areas 47/12m and 47/12l has pushed area 47/12s onto the dorsal bank of the lateral orbital gyrus within the horizontal ramus of the lateral sulcus. Area 47/12s is characterized by its very sparse myelination, while area 47/12l has marked radial bundles of myelinated fibres, and area 47/12m has prominent myelin plexuses in layers III and V (inner and outer bands of Baillarger).

## Central and rostral orbital cortex

The location, morphology and staining characteristics of the four subdivisions of area 13 in humans (areas 13a, 13b, 13m, 13l) are similar to those in the monkey, and are among the most consistent in the human orbital cortex. Petrides and Pandya (1994) did not subdivide area 13, but their area 13 corresponds approximately to areas 13m and 13l. The medial bank of the olfactory sulcus, which is occupied by areas 13b and 13a, has sometimes been included in area 14 by previous investigators (Beck 1949; Petrides and Pandya 1994, 2002). Because of marked structural differences between areas 13a/13b and 14c/14r, however, the terminology previously used by Carmichael and Price (1994) in monkeys may be preferred.

Area 13a is a small, agranular area, deep to the point at which the olfactory peduncle is linked onto the orbital cortex, with relatively simple laminar structure. Area 13b is a larger, dysgranular region on the medial bank of the olfactory sulcus, which is characterized by both radial and tangential striations in Nissl stained sections. Areas 13m and 13l are fully developed dysgranular cortexes that extend from the fundus of the olfactory sulcus to the fundus of the lateral orbital sulcus. They are relatively similar to each other, but can be distinguished by several architectonic features, including stronger staining for acetylcholinesterase in area 13l.

Area 14c is a small area at the ventromedial corner of the hemisphere, medial to area 13a. Area 14c is thicker than 13a, but it also has a relatively simple laminar structure. Area 14r is a longer, more developed cortical region, with a sub-laminated layer V, and two dense fiber bands in parvalbumin stained sections.

## Caudal medial wall

The differential growth of frontal pole regions in humans has been accompanied by a caudoventral shift of the agranular areas on the medial wall, as compared to monkeys. Öngür et al. (2003) identified agranular areas 25 and 32pl on the caudal medial wall ventral to the corpus callosum. The comparable infralimbic and prelimbic areas in rats (Krettek and Price 1977), and areas 25 and 32 in monkeys (Carmichael and Price 1994), are located more dorsally and rostrally within the medial wall, ventral to area 24a/b and the cingulate sulcus. In humans, the anterior cingulate areas 24a and 24b wrap around the genu of the corpus callosum and form the subgenual part of area 24. Thus, areas 25 and 32pl come to lie ventral to area 24b in the human brain as well.

The agranular region that Brodmann (1909) identified in the human brain as area 32 lies anterior and dorsal to the cingulate area 24 on the paracingulate gyrus (Paus et al., 1996). Brodmann himself wrote that his human area 32 (which he called the 'dorsal anterior cingulate area'; see p.148 of Brodmann (1909), or p.127 of Garey (1994) is not homologous to the area 32 that he identified in monkeys (which he called the 'prelimbic area'; see p.159 of Brodmann (1909), or p.138 of Garey (1994). Öngür et al. (2003), therefore, defined area 32ac as the area Brodmann designated the anterior cingulate area 32 in humans, and area 32pl as the area that is comparable to the prelimbic area 32 in monkeys. Area 32ac bears many similarities to areas 24b and 24c, but it is distant and distinct from visceromotor areas 25 and 32pl on the ventral medial wall. The parcellation and staining characteristics of areas 24a, 24b and 25 reported here are largely consistent with the previously published data (Vogt et al. 1995; Nimchinsky et al. 1997). For example, Nimchinsky et al. (1997) observed the same double-plexus arrangement of parvalbumin-immunoreactive fibres in the middle and deep layers of cortex in areas 24a and 24b in a region dorsal to the corpus callosum. They also point out the presence of SMI-32 positive pyramidal cells in layer III in this region. This is a feature not seen in the anterior part of areas 24a and 24b in monkeys (Hof and Nimchinsky 1992; Carmichael and Price 1994). In the present study, the subgenual part of area 24 was found to be devoid of layer III pyramidal cells stained with the SMI-32 antibody, suggesting that the anterior part of areas 24a and 24b in monkeys is comparable to the subgenual part of areas 24a and 24b in humans.

## Frontal pole

The most striking change in the OMPFC between monkeys and humans is the expansion of the granular cortex at the frontal pole. Most investigators have recognized the rostral orbital cortex and frontal pole as areas 11 and 10, respectively (Brodmann 1909; Beck 1949; Petrides and Pandya 1994). Öngür et al. (2003) subdivided area 11 into 11m and

11l, and area 10 into 10m, 10r, and 10p. Areas 11m and 11l, in the rostral part of the orbital cortex, correspond to the same areas in monkeys (Carmichael and Price 1994). The monkey area 10m was divided into areas 10m and 10r, reflecting the further elaboration of this region; the monkey area 10o was renamed 10p, in recognition of its much larger size and its position at the frontal pole.

All of these areas are granular, and become more so nearer to the frontal pole. Area 11m occupies the medial and lateral banks of the olfactory sulcus, rostral to areas 14r, 13b, and 13m. It is granular, but not strongly so. Area 11l is a thicker, more granular cortex. Areas 10m and 10r are located on the ventral part of the medial wall, ventral to area 24, while area 10p is a large area that covers the entire frontal pole. The major distinction between areas 10m, 10r, and 10p is an increasing granularity, with area 10p being thick, eulaminate cortex, with all layers fully represented, including a thick, granular layer IV.

## References

Amaral, D. G. and Price, J. L. (1984) Amygdalo-cortical projections in the monkey (*Macaca fascicularis*). *J Comp Neurol* **230**:465–496.

An, X., Bandler, R., Öngür, D., and Price, J. L. (1998) Prefrontal cortical projections to longitudinal columns in the midbrain periaqueductal gray in macaque monkeys. *J Comp Neurol* **401**:455–479.

Barbas, H. and Pandya, D. N. (1989) Architecture and intrinsic connections of the prefrontal cortex in the Rhesus monkey. *J Comp Neurol* **286**:353–375.

Beck, E. (1949) A cytoarchitectural investigation into the boundaries of cortical areas 13 and 14 in the human brain. *J Anat* **83**: 147–157.

Brockhaus, H. (1940) Die Cyto- und Myeloarchitektonik des Cortex claustralis und der Claustrum beim Menschen. *J Pyschol Neurol* **49**:249–348.

Brodmann, K. (1905) Beitraege zur histologischen Lokalisation der Grosshirnrinde. III Mitteilung. Die Rindenfelder der niederen Affen. *J Psychol Neurol* **4**:177–226.

Brodmann, K. (1908) Beitraege zur histologischen Lokalisation der Grosshirnrinde. VI Mitteilung. Die Cortexgliederung des Menschen. *J Psychol Neurol* **10**:231–246.

Brodmann, K. (1909) *Vergleichende Lokalisationslehre der Grosshirnrinde in ihren Prinzipien dargestellt auf Grund des Zellenbaues*. Leipzig: JA Barth.

Carmichael, S. T. and Price, J. L. (1994) Architectonic subdivision of the orbital and medial prefrontal cortex in the macaque monkey. *J Comp Neurol* **346**:366–402.

Carmichael, S. T. and Price, J. L. (1995a) Limbic connections of the orbital and medial prefrontal cortex in macaque monkeys. *J Comp Neurol* **363**:615–641.

Carmichael, S. T. and Price, J. L. (1995b) Sensory and premotor connections of the orbital and medial prefrontal cortex. *J Comp Neurol* **363**:642–664.

Carmichael, S. T. and Price, J. L. (1996) Connectional networks within the orbital and medial prefrontal cortex of macaque monkeys. *J Comp Neurol* **371**:179–207.

Carmichael, S. T., Clugnet, M. C., and Price, J. L. (1994) Central olfactory connections in the macaque monkey. *J Comp Neurol* **346**:403–434.

Dombrowski, S. M., Hilgetag, C. C., and Barbas, H. (2001) Quantitative architecture distinguishes prefrontal cortical systems in the rhesus monkey. *Cereb Cortex* **11**:975–988.

Ferry, A. T., Ongur, D., An, X., and Price, J. L. (2000) Prefrontal cortical projections to the striatum in macaque monkeys: evidence for an organization related to prefrontal networks. *J Comp Neurol* **425**:447–470.

Garey, L. J. (1994) *Brodmann's 'Localisation in the Cerebral Cortex'*. London: Smith-Gordon and Co.

Giguere, M. and Goldman-Rakic, P. S. (1988) Mediodorsal nucleus: areal, laminar, and tangential distribution of afferents and efferents in the frontal lobe of rhesus monkeys. *J Comp Neurol* **277**:195–213.

Groenewegen, H. J. (1988) Organization of the afferent connections of the mediodorsal thalamic nucleus in the rat, related to mediodorsal-prefrontal topography. *Neuroscience* **24**:379–431.

Hof, P. R. and Nimchinsky, E. A. (1992) Regional distribution of neurofilament and calcium-binding proteins in the cingulate cortex of the macaque monkey. *Cereb Cortex* **2**:456–467.

Hof, P. R., Mufson, E. J., and Morrison, J. H. (1995) Human orbitofrontal cortex: Cytoarchitecture and quantitative immunohistochemical parcellation. *J Comp Neurol* **359**:48–68.

Kondo, H, Saleem, K. S., and Price, J. L. (2003) Differential connections of the temporal pole with the orbital and medial prefrontal networks in macaque monkeys. *J Comp Neurol* **465**:499–523.

Krettek, J. E. and Price, J. L. (1977) The cortical projections of the mediodorsal nucleus and adjacent thalamic nuclei in the rat. *J Comp Neurol* **171**:157–191.

Mesulam, M. M. (2000) *Principles of Behavioural and Cognitive Neurology*. Oxford University Press.

Mesulam, M. M. and Mufson, E. J. (1982) Insula of the old world monkey. III: Efferent cortical output and comments on function. *J Comp Neurol* **212**:38–52.

Nimchinsky, E. A., Vogt, B. A., Morrison, J. H., and Hof, P. R. (1997) Neurofilament and calcium-binding proteins in the human cingulate cortex. *J Comp Neurol* **384**:597–620.

Öngür, D., Ferry, A. T., and Price J. L. (2003) Architectonic subdivision of the human orbital and medial prefrontal cortex. *J Comp Neurol* **460**:425–449.

Paus, T., Tomaiuolo, F., Otaky, N., MacDonald, D., Petrides, M., Atlas, J., Morris, R., and Evans, A. C. (1996) Human cingulate and paracingulate sulci: pattern, variability, asymmetry, and probabilistic map. *Cereb Cortex* **6**:207–214.

Paxinos, G. and Watson, C. (1997) *The Rat Brain in Stereotaxic Coordinates*. San Diego: Academic Press.

Petrides, M. and Pandya, D. N. (1994) *Comparative Cytoarchitectonic Analysis of the Human and the Macaque Frontal Cortex*. In: Handbook of Neuropsychology, Vol. 9. (Boller, F. and Grafman, J. ed.), pp. 17–58, Amsterdam: Elsevier Science B.V.

Petrides, M. and Pandya, D. N. (2002) Comparative cytoarchitectonic analysis of the human and the macaque ventrolateral prefrontal Cortex and corticocortical connection patterns in the monkey. *Eur J Neurosci* **16**:291–310.

Porrino, L. J., Crane, A. M., and Goldman-Rakic, P. S. (1981) Direct and indirect pathways from the amygdala to the frontal lobe in rhesus monkeys. *J Comp Neurol* **198**:121–136.

Preuss, T. M. (1995) Do rats have prefrontal cortex? The Rose-Woolsey-Akert program reconsidered. *J Cog Neurosci* **7**:1–24.

Preuss, T. M. and Goldman-Rakic, P. S. (1991) Myelo- and cytoarchitecture of the granular frontal cortex and surrounding regions in the strepsirhine primate Galago and the anthropoid primate Macaca.
*J Comp Neurol* **198**:121–136.

Ray, J. P. and Price, J. L. (1992) The organization of the thalamocortical connections of the mediodorsal thalamic nucleus in the rat, related to the ventral forebrain-prefrontal cortex topography. *J Comp Neurol* **323**:167–197.

Ray, J. P. and Price J. L. (1993) The organization of projections from the mediodorsal nucleus of the thalamus to orbital and medial prefrontal cortex in macaque monkeys. *J Comp Neurol* **337**:1–31.

Rose, M. (1928) Die Inselrinde des Menschen und der Tiere. *J Psychol Neurol* **37**:467–624.

Rose, M. (1929) Cytoarchitektonischen Atlas der Grosshirnrinde der Maus. *J Psychol Neurol* **40**:1–51.

Rose, J. E. and Woolsey, C. N. (1948a) The orbitofrontal cortex and its connections with the mediodorsal nucleus in rabbit, sheep and cat. *Res Publ Ass Nerv Ment Dis* **27**:210–232.

Rose, J. E. and Woolsey C. N. (1948b) Structure and relations of limbic cortex and anterior thalamic nuclei in rabbit and cat. *J Comp Neurol* **77**:61–130.

Sanides, F. (1970) *Functional architecture of motor and sensory cortices in primates in the light of a new concept of neocortex evolution.* In The Primate Brain: Advances in Primatology. (Noback, C.R. and Montagna, W., ed), pp.137–205. New York: Appleton-Century-Crofts.

Semendeferi, K., Armstrong, E., Schleicher, A, Zilles, K, van Hoesen, G. W. (1998) Limbic frontal cortex in hominoids: a comparative study of area 13. *Am J Phys Anthropol* **106**:129–155.

Talairach, J. and Tournoux, P. (1988) *Co-planar stereotaxic atlas of the human brain: 3-dimensional proportional system: An approach to cerebral imaging.* translated Mark Rayport, G. Thieme, New York.

Uylings, H. B. M. and van Eden, C. G. (1990) Qualitative and quantitative comparison of the prefrontal cortex in rat and in primates, including humans. *Prog Brain Res* **85**:31–62.

Vogt, B. A., Nimchinsky, E. A., Vogt, L. J., and Hof, P. R. (1995) Human cingulate cortex: surface features, flat maps, and cytoarchitecture. *J Comp Neurol* **359**:490–506.

von Bonin, G. and Bailey, P. (1947) *The Neocortex of Macaca Mulatta.* Urbana: University of Illinois Press.

von Economo, C. and Koskinas, G. N. (1925) *Die Cytoarchitektonik der Hirnrinde des Erwachsenen Menschen.* Springer, Vienna.

Walker, A. E. (1940) A cytoarchitectural study of the prefrontal area of the macaque monkey. *J Comp Neurol* **73**:59–86.

Zilles, K. and Wree, A. (1995) *Cortex: areal and laminar structure in The Rat Nervous System.* (Paxinos, G., ed), pp.649–685 San Diego: Academic Press.

Chapter 2

# The orbitofrontal cortex: sulcal and gyral morphology and architecture

Michael Petrides and Scott Mackey

## 2.1. Introduction

The morphology of the sulci and gyri of the human cerebral cortex was intensively investigated during the late 19th century (e.g., Weisbach 1870; Eberstaller 1890; Cunningham 1892; Retzius 1896) and continued to be the focus of research during the first half of the 20th century (e.g., Landau 1911, 1914; Shellshear 1926, 1937; Connoly 1950). During the last 50 years, there have been few investigations of the sulcal patterns of the human brain, with the notable exception of Ono *et al.* (1990). The development of functional neuroimaging, with its demonstration of specific foci of cortical activation, has given rise to the need for a more detailed description of the sulcal and gyral patterns of the human cerebral cortex, and their variations, than has typically been provided in standard textbooks of neuroanatomy and neuroradiology. The interpretation of functional neuroimaging studies aimed at the orbitofrontal region would be greatly facilitated by the provision of quantitative information on the location and variability of the major sulci of this region of the brain in the standard proportional stereotaxic space utilized by the neuroimaging community, namely the Talairach space (Talairach and Tournoux 1988), and its recent evolution into the Montreal Neurological Institute (MNI) space (Evans *et al.* 1994). Consequently, the first part of this chapter will provide a description of the sulcal and gyral patterns of the orbital surface of the human frontal lobe, as well as that of the macaque monkey, which is the nonhuman primate most often used in architectonic and experimental anatomical connectivity studies. The description provided is based on the classical studies, as well as recent investigations in our laboratory (e.g., Chiavaras and Petrides 2000; Chiavaras *et al.* 2001). The second part of this chapter will provide a general description of the cytoarchitectonic organization of the human prefrontal cortex with comments on its relation to the sulcal and gyral patterns.

## 2.2. Sulcal and gyral morphology of the orbitofrontal cortical region

### Human brain

The ventral surface of the frontal lobe of the primate brain, namely the orbitofrontal region, extends from the frontal pole to the insula, posteriorly. At the medial edge of the

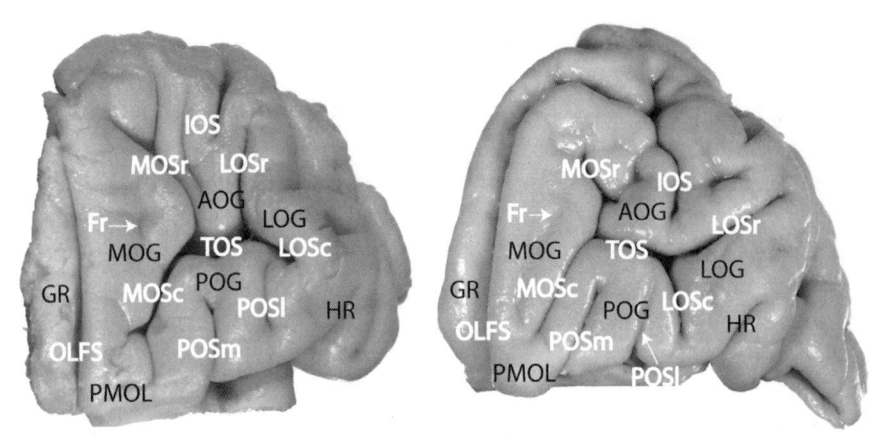

**Fig. 2.1** Photographs of the orbitofrontal surface of two human brains to illustrate the typical sulci and gyri of this region. The names of sulci are in white and those of the gyri in black. Abbreviations: AOG, anterior orbital gyrus; HR, horizontal ramus of the Sylvian fissure; GR, gyrus rectus; Fr, sulcus fragmentosus; IOS, intermediate orbital sulci; MOG, medial orbital gyrus; MOSc, caudal ramus of the medial orbital sulcus; MOSr, rostral ramus of the medial orbital sulcus; LOG, lateral orbital gyrus; LOSc, caudal ramus of the lateral orbital sulcus; LOSr, rostral ramus of the lateral orbital sulcus; OLFS, olfactory sulcus; PMOL, posteromedial orbital lobule; POG, posterior orbital gyrus; POS, posterior orbital sulci; TOS, transverse orbital sulcus.

orbital surface of the frontal lobe, the olfactory sulcus, which runs in a rostro-caudal direction, delimits the narrow gyrus rectus (Fig. 2.1). Typically, the posterior end of the olfactory sulcus lies more laterally than its anterior end, which comes very close to the medial margin. As a result, the gyrus rectus is very narrow at its anterior part, but much wider posteriorly. In the human brain, lateral to the olfactory sulcus, there is a complex of orbital sulci that give the impression of an H- or a K-pattern. This complex of sulci which dominates the morphological appearance of the orbital frontal region was first referred to as the 'scissure en H' by Gratiolet (1854). Figure 2.1 shows the orbitofrontal region of two human cerebral hemispheres that display the typical H-pattern. Weisbach (1870), Eberstaller (1890), and Economo and Koskinas (1925) referred to the entire posterior arch-like part of the H-pattern as the sulcus orbitalis transversus. The anterior limbs of the medial and lateral sections of the H-pattern were referred to as the sulcus orbitalis medius and externus by Weisbach (1870), as the ramus medialis and ramus lateralis of the sulcus orbitalis by Eberstaller (1890), and as the sulcus orbitalis medialis and sulcus orbitalis lateralis by Economo and Koskinas (1925). Subsequently, Bailey and Bonin (1951) expanded the terms medial and lateral orbital sulci to refer to both the anterior and posterior limbs of the medial and lateral longitudinal components of the H-pattern, and referred to the horizontal connecting link as the arcuate orbital sulcus. The horizontal connecting link in the H-pattern is now referred to as the transverse orbital sulcus (Ono *et al.* 1990; Chiavaras and Petrides 2000). Consistent with our previous work, we refer to the anterior extensions of the H-pattern as the rostral branches of the medial and lateral orbital sulci, and the posterior extensions of the H-pattern as the caudal branches of the medial and lateral orbital sulci (Fig. 2.1). A major departure from the classical nomenclature is that by Duvernoy (1991), who uses the term medial orbital sulcus as a synonym for the olfactory sulcus, and lateral orbital sulcus for a sulcus that lies lateral to the lateral component of the H-pattern.

The classic H-pattern in which the rostral and caudal branches of the medial and lateral orbital sulci appear continuous and are joined by the transverse orbital sulcus is observed in about 30% of the cases, and is slightly more frequent in the left, as compared with the right hemisphere. The remaining cases appear as variations on this classic pattern. In the most frequent variation (56% of hemispheres), the rostral branch of the medial orbital sulcus is separated from the transverse orbital sulcus lying slightly anterior to it, while in a minority of cases (14%) the rostral branches of both the medial and lateral orbital sulci are separated from the transverse orbital sulcus (Chiavaras and Petrides 2000).

Four gyri are defined by the H-pattern of orbital sulci: 1) the medial orbital gyrus, which is bordered by the olfactory sulcus and the medial orbital sulcus; 2) the anterior orbital gyrus, which lies anterior to the transverse orbital sulcus between the rostral branches of the medial and lateral orbital sulci; 3) the posterior orbital gyrus, which lies posterior to the transverse orbital sulcus between the caudal branches of the medial and lateral orbital sulci; and 4) the lateral orbital gyrus, which lies lateral to the lateral orbital sulcus (Fig. 2.1). Since there is no sharp boundary between the orbitofrontal region and the ventral part of the lateral surface of the frontal lobe, the lateral orbital gyrus can be thought of as part of the orbital extension of the ventrolateral prefrontal region. Indeed, in terms of function and architecture, there is considerable evidence that the cortex that lies just lateral to the lateral orbital sulcus belongs more to the ventrolateral prefrontal cortex rather than to the orbital frontal cortex (see Petrides 2005).

The most posterior part of the orbitofrontal region is closely related to the anterior part of the insula, both in terms of gross morphology as well as architecture (Barbas and Pandya 1989; Carmichael and Price 1994; Öngür et al. 2003; Price, Chapter 1 this volume). Just behind the caudalmost part of the medial orbital sulcus, the medial orbital gyrus and the posterior orbital gyrus merge to form the posteromedial orbital lobule, which is continuous with the transverse insular gyrus, a short narrow transitional gyrus that links the orbitofrontal cortex (OFC) with the anterior margin of the pole of the insula, i.e., the ventralmost agranular insula (Fig. 2.1).

Apart from the main sulci described above, which define the gross morphology of the orbitofrontal region of the human brain, one or two longitudinal sulci—the intermediate orbital sulci—are often encountered within the anterior orbital gyrus (i.e., between the rostral parts of the medial and lateral orbital sulci: see Figs. 2.1 and 2.2a). When only one intermediate sulcus is present (81% of cases), it usually joins, posteriorly, the transverse orbital sulcus, though in some cases it may stop just in front of it, or merge with the medial or lateral orbital sulci (Fig. 2.2a). When there are two intermediate sulci (19% of cases), the medial one is usually the deeper one, and tends to join the transverse orbital sulcus or stop just anterior to it. Within the posterior orbital gyrus

**(a)** Intermediate orbital sulci

**(b)** Posterior orbital sulci

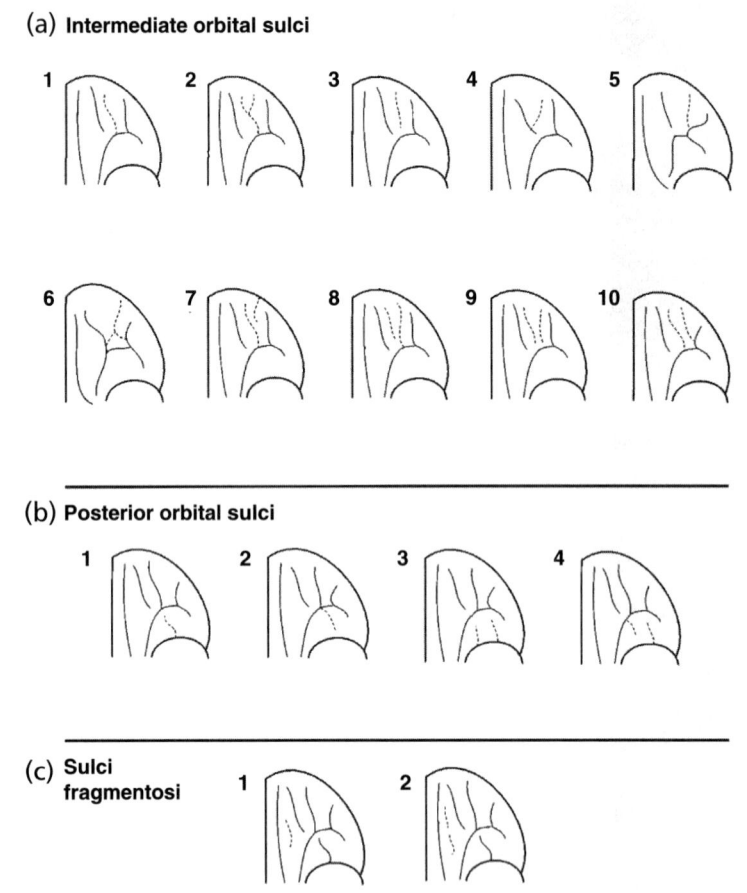

**(c)** Sulci fragmentosi

**Fig. 2.2** Schematic drawings of the human orbitofrontal region to illustrate the variations in appearance of the intermediate orbital sulci (a), the posterior orbital sulci (b), and the sulci fragmentosi (c). The sulci of interest are represented by a dotted line while the olfactory sulcus and the sulci that form the H-pattern are represented by solid lines. Reproduced with permission from Chiavaras and Petrides (2000).

(i.e., behind the transverse orbital sulcus and between the caudal parts of the medial and lateral orbital sulci), there may be one or two short sulci that are referred to as the posterior orbital sulci (Fig. 2.2b). In some brains, a short, shallow, longitudinally arranged sulcus can be observed between the olfactory sulcus and the medial orbital sulcus (Fig. 2.1). This sulcus is known as the sulcus fragmentosus and is observed twice as often in the left hemisphere than in the right hemisphere. The sulcus fragmentosus may appear as two shallow branches in some brains (Fig. 2.2c).

Figure 2.3 presents a series of coronal and sagittal sections of the orbitofrontal region as seen in magnetic resonance images. The coordinates provided are in the MNI space. Coronal sections provide an excellent definition of all orbitofrontal sulci, except for the transverse orbital sulcus that is best visualized in sagittal sections. The transverse orbital

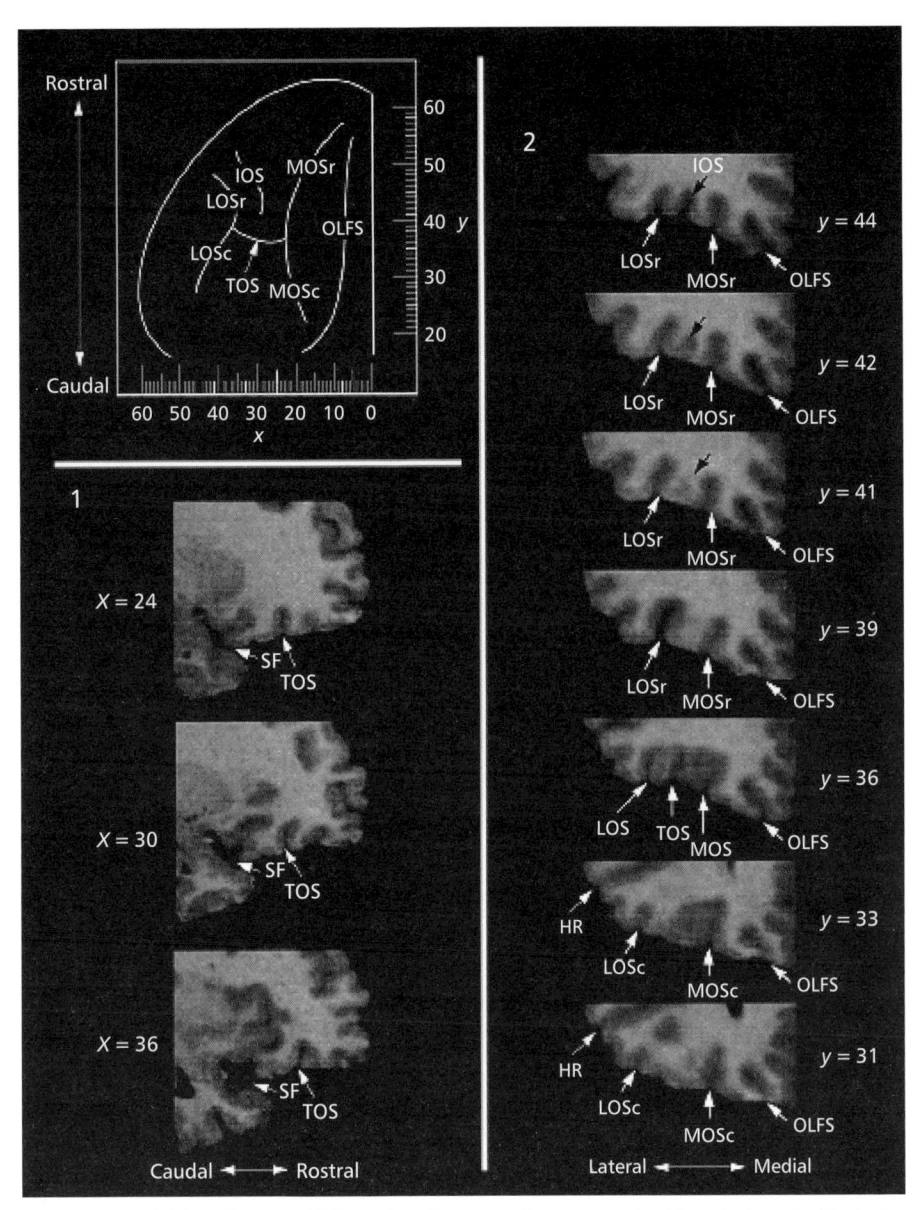

**Fig. 2.3** Sagittal (1) and coronal (2) sections from one human cerebral hemisphere to illustrate the typical sulcal patterns of the orbitofrontal region as seen in magnetic resonance images. MR scans were performed on a Phillips Gyroscan 1.5 T system. A 3-D fast-field echo sequence was used to acquire 160 contiguous 1-mm $T_1$-weighted images (TR = 18 ms; TE = 10 ms; flip angle 30°) in the sagittal plane. Each MR volume was then transformed into the Montreal Neurological Institute (MNI) space using an automatic registration program. The coordinates provided are in MNI space, which is based on the standard proportional stereotaxic space of Talairach ($x$ = millimeters lateral to the midline, $y$ = millimeters anterior to the anterior commissure). Abbreviations are the same as in Figure 2.1, with the following addition: SF, Sylvian fissure.

sulcus in sagittal sections demonstrates clearly the anterior orbital gyrus in front of it and the posterior orbital gyrus that lies behind it. Note that the gyrus rectus is much more ventrally located than the rest of the OFC and that, lateral to the olfactory sulcus, the cortex forming the medial orbital gyrus slopes upwards at approximately a 45° angle. At the end of this slope, the medial orbital sulcus is encountered: a deep and long sulcus that runs rostro-caudally and parallel to the olfactory sulcus. The medial orbital sulcus has a curved appearance creating a concavity facing the gyrus rectus. Table 2.1 provides

**Table 2.1** Bounding box coordinates for the orbitofrontal sulci in MNI space

|   |   | OLFS | MOS | LOS | TOS |
|---|---|---|---|---|---|
| x | Right | +3 to +24<br>r: +8 to +11<br>($P > 0.9$)<br>c: +10 to +16<br>($P > 0.9$) | +11 to +30<br>r: +19 to +24<br>($P = 0.65$)<br>c: +21 to +23<br>($P > 0.9$) | +30 to +49<br>+36 to +43<br>($P = 0.65$) | +21 to +39<br>+27 to +33<br>($P = 0.65$) |
|   | Left | −3 to −23<br>r: −5 to −10<br>($P > 0.9$)<br>c: −8 to −16<br>($P > 0.9$) | −10 to −30<br>r: −18 to −23<br>($P = 0.65$)<br>c: −20 to −25<br>($P = 0.8$) | −30 to −49<br>−37 to −41<br>($P = 0.65$) | −20 to −38<br>−27 to −34<br>($P = 0.65$) |
| y | Right | +7 to +61<br>+11 to +46<br>($P > 0.9$) | +14 to +61<br>+21 to +46<br>($P = 0.65$) | +24 to +55<br>+30 to +46<br>($P = 0.5$) | +26 to +42<br>+35 to +37<br>($P = 0.8$) |
|   | Left | +6 to +60<br>+11 to +49<br>($P > 0.9$) | +10 to +60<br>+18 to +47<br>($P = 0.65$) | +20 to +55<br>+26 to +47<br>($P = 0.5$) | +26 to +41<br>+34 to +36<br>($P = 0.8$) |
| z | Right | −10 to −31<br>r: −22 to −24<br>($P > 0.9$)<br>c: −15 to −25<br>($P > 0.9$) | −9 to −25<br>r: −14 to −16<br>($P = 0.65$)<br>c: −15 to −21<br>($P = 0.8$) | −4 to −21<br>−11 to −18<br>($P = 0.65$) | −5 to −21<br>−10 to −16<br>($P = 0.8$) |
|   | Left | −10 to −32<br>r: −21 to −26<br>($P > 0.9$)<br>c: −15 to −24<br>($P > 0.9$) | −8 to −26<br>r: −14 to −19<br>($P = 0.65$)<br>c: −15 to −21<br>($P = 0.8$) | −6 to −21<br>−13 to −17<br>($P = 0.65$) | −5 to −21<br>−11 to −17<br>($P = 0.8$) |

Reproduced with permission from Chiavaras et al. (2001).

*Note:* Bounding box coordinates for the orbitofrontal sulci. The values in bold represent the boundaries of the x, y, and z coordinates containing a particular sulcus in at least 10% of subjects. The italicized values represent the region of greatest probability for a particular sulcus. For the olfactory and medial orbital sulci, values are presented separately for the rostral (r) and caudal (c) portions of the sulci.

bounding box coordinates for the major orbitofrontal sulci which can be used to interpret the location of activation foci from functional neuroimaging studies.

## Macaque monkey brain

The pattern of the sulci in the OFC of the macaque monkey, the most commonly used primate in experimental anatomical and physiological studies, has recently been examined in detail and compared to that of the human brain (Chiavaras and Petrides 2000). This work showed that the basic pattern of sulci and gyri encountered in the orbitofrontal region of the human brain can also be identified in the brains of macaque monkeys. As in the human brain, the olfactory sulcus in the macaque monkey lies below the olfactory tract and runs in an anteroposterior direction delimiting a narrow gyrus, the gyrus rectus that forms the medial edge of the orbitofrontal region. Also as in the human brain, the gyrus rectus of the monkey is much more ventrally located than the rest of the OFC and the medial orbital gyrus slopes at approximately 45° towards the medial orbital sulcus (Fig. 2.4). The olfactory sulcus in the macaque monkey brain is only a shallow groove in comparison to the deep olfactory sulcus of the human brain. The medial orbital sulcus is a long sulcus that is consistently found in different macaque monkeys. In sharp contrast, the lateral orbital sulcus, which lies in the posterior-lateral edge of the OFC, is small and is less consistently expressed than the medial orbital sulcus. In some cases, the orbitofrontal sulci of the macaque monkey form the H-pattern, which is often observed in the human brain (Fig. 2.4a). There may be a clearly identifiable

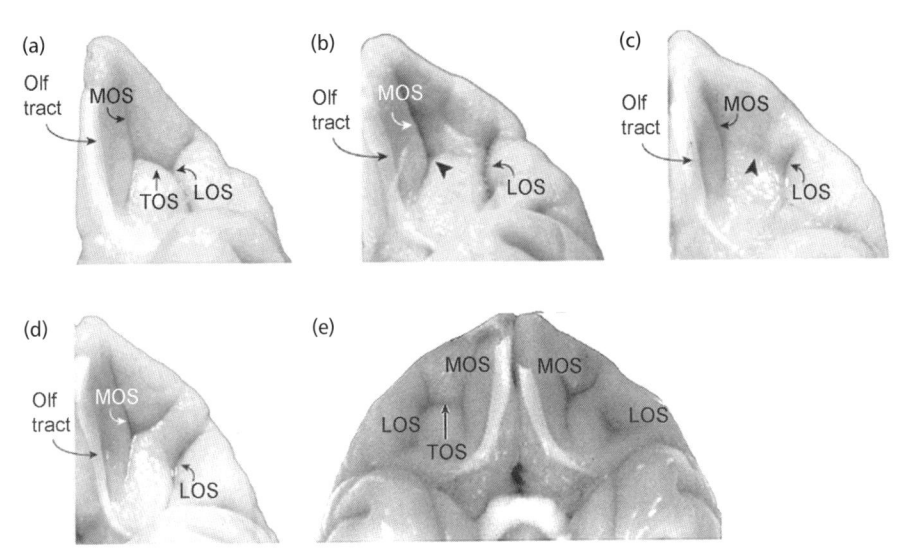

**Fig. 2.4** Photographs of the orbital frontal surface of four rhesus monkey (*Macaca mulatta*) hemispheres (a–d) and one cynomolgus monkey (*Macaca fascicularis*) brain (e) to illustrate the typical orbitofrontal sulcal patterns in macaque monkey brains. The olfactory sulcus is a shallow groove that runs underneath the olfactory tract. Abbreviations: MOS, medial orbital sulcus; LOS, lateral orbital sulcus; Olf tract, olfactory tract; TOS, transverse orbital sulcus.

transverse orbital sulcus, but in many cases there is nothing more than a slight depression or a dimple separating the cortex between the medial and lateral orbital sulci into an anterior and a posterior component, as in the human brain (Fig. 2.4b and c). In many macaque monkey brains, a short sulcus emerges from approximately the middle part of the medial orbital sulcus and is directed rostrolaterally (Fig. 2.4d) (Chiavaras and Petrides 2000). This Y-pattern occurred in 64 of 100 hemispheres of rhesus monkeys (macaca mulatta) that we examined, but it should be noted that it may not necessarily be the most common pattern in other macaque monkeys (unpublished observations). Figure 2.4e shows a photograph of the OFC of both hemispheres in a cynomolgus monkey (*Macaca fascicularis*) in which the H-pattern of orbitofrontal sulci is clearly seen in the right hemisphere and the Y-pattern in the left hemisphere. In conclusion, the above observations show that there are fundamental similarities in the basic sulcal patterns in the orbitofrontal region of the human and the macaque monkey brains, though the sulcal patterns in the monkey brain are much simpler.

## 2.3. **Architectonic organization**

While topographic similarities between the human and nonhuman primate OFC may encourage cross-species functional comparisons, knowledge of the location of comparable cytoarchitectonic areas is necessary in order to compare directly monkey and human data. Unfortunately, defining the relationship between the topography and cytoarchitecture of the human OFC has not been resolved because the nomenclature and architectonic criteria employed by various investigators of the OFC, though often similar, have not been applied consistently. A short review of the classic studies will first be provided to help explain the current issues, our parcellation of the architectonic areas of the OFC, and the numerical nomenclature used. The numerical designations of areas of the OFC that are widely used today and, to a large extent, the uncertainty in their application originate in Brodmann's original studies.

In his classic work on the cytoarchitecture of the human cerebral cortex, Brodmann (1908, 1909) divided the OFC into two large regions that he designated as areas 47 and 11 (Fig. 2.5a, b). He acknowledged, however, that these two large regions are architectonically heterogeneous, and that they can be further subdivided. He also made a reference to the myeloarchitectonic studies of Vogt, published a little later in 1910, which also suggested that these large regions of the OFC are not homogeneous. For instance, Brodmann (1909) pointed out that certain structural differences permit a division of his area 11 into an area recta and an internal orbital area, although he did not show this subdivision in his map. With regard to area 47, Brodmann (1909) indicated that it lies around the posterior branches of the orbital sulcus and that it extends onto the pars orbitalis of the inferior frontal gyrus. He also pointed out that area 45, which occupies the pars triangularis of the inferior frontal gyrus, may on occasion extend partially on the pars orbitalis of the inferior frontal gyrus. Thus, Brodmann (1909) was unclear about the border between area 45 and area 47. It should be noted here that Brodmann never published photomicrographs of these areas that might have helped subsequent

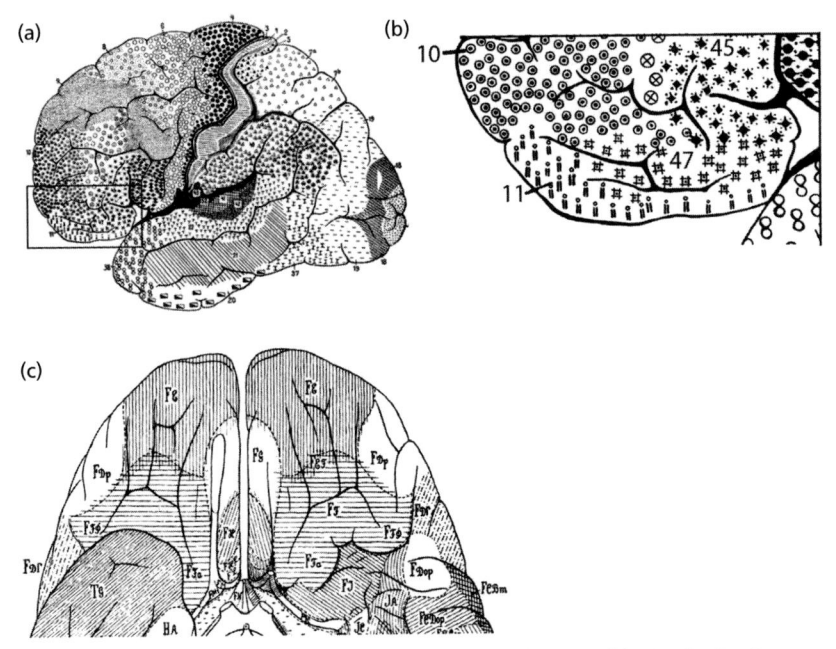

**Fig. 2.5** Cytoarchitectonic maps of the human orbital frontal cortex. (a) Map by Brodmann (1909), (b) magnification of outlined region in (a), (c) map by Economo and Koskinas (1925). Note that Brodmann illustrated the orbital frontal surface (b) at the most ventral part of the map of the lateral surface of the brain (a). Note also that Economo and Koskinas identified many more areas on the orbitofrontal surface of the brain than Brodmann.

investigations of the OFC. In the cytoarchitectonic atlas that Economo and Koskinas published in 1925, Brodmann's area 47 was referred to as area FF (Fig. 2.5c). Economo and Koskinas also acknowledged that this is a large and heterogeneous region, ranging from an almost agranular part most caudally and medially (FFa) to a more granular part anteriorly and laterally. Economo and Koskinas stated that the most lateral part of this region that curves onto the ventrolateral frontal convexity is so different in structure that it can hardly be included with the rest of area FF. Later cytoarchitectonic studies by Sarkissov *et al.* (1955) similarly confirmed the early work of Vogt (1910) that the large region designated as area 47 by Brodmann is heterogeneous and can be divided into distinct areas.

Another major problem with Brodmann's subdivision of the orbitofrontal region is the lack of correspondence with the architectonic areas of the OFC of the macaque monkey. In fact, Brodmann had published a map of the monkey cortex in 1905 before his map of the human cortex appeared in 1908. However, there were such gross discrepancies in the numerical designations employed by Brodmann in his maps of the monkey and human cortices for even obviously comparable areas of the frontal cortex that Brodmann himself repeatedly warned in his 1909 book that the numbers he used did not always denote comparable areas in the two species. For this reason Brodmann's map of the monkey cortex was not widely used. Later, Walker (1940) attempted to correct these problems, namely the lack of correspondence in the nomenclature used for the human and macaque frontal lobes in the Brodmann maps and the lack of detail in the parcellation of the macaque prefrontal cortex. In his map of the macaque monkey cortex, Walker used a numerical scheme similar to that used by Brodmann for the human cortex and, in addition, introduced new subdivisions (Fig. 2.6a, b). Walker's map has been the basis of all subsequent investigations of the architecture of the frontal cortex of the monkey (e.g., Barbas and Pandya 1989; Preuss and Goldman-Rakic 1991; Carmichael and Price 1994; Semendeferi et al. 1998).

Walker divided the OFC into five basic areas: the gyrus rectus area 14, the caudal orbitofrontal area 13, the lateral orbitofrontal area 12 that extends onto the ventrolateral prefrontal cortex, area 11 in the rostral part of the OFC, and the orbital extension of frontopolar area 10 (Fig. 2.6). Since Walker did not carry out a comparison of the human and monkey cerebral cortices, the correspondence between the areas he identified in the monkey and those in the human OFC remained unclear. In order to address this issue, Petrides and Pandya (1994) explicitly compared the human prefrontal cortex with that of the rhesus monkey (macaca mulatta) and identified comparable areas in the two species. More recently, Öngür et al. (2003) published an architectonic subdivision of the human orbital and medial prefrontal cortex that is based on their earlier work on the macaque monkey (Carmichael and Price 1994).

The description of the architectonic organization of the human OFC that is presented here is based on investigations in our laboratory, as well as on the available evidence from classical and recent studies. Although there is a great deal of agreement between our analysis and the map presented by Price in Chapter 1, there are some differences between our map and that produced by Price's group regarding the sulcal borders of these areas. Our work is based on an examination of 36 μ thick sections from a human brain cut serially in the coronal plane and 12 μ thick sections from blocks cut at angles optimal for cytoarchitectonic analysis of specific parts of the OFC. All sections were histologically prepared with cresyl violet for cytoarchitectonic analysis. Cytoarchitectonic analysis involved examination of the sections under the microscope, and was supplemented by quantitative analysis (Mackey and Petrides 2005; see also supplementary methods in Petrides et al, 2005). Briefly, measurements were obtained from digital images of

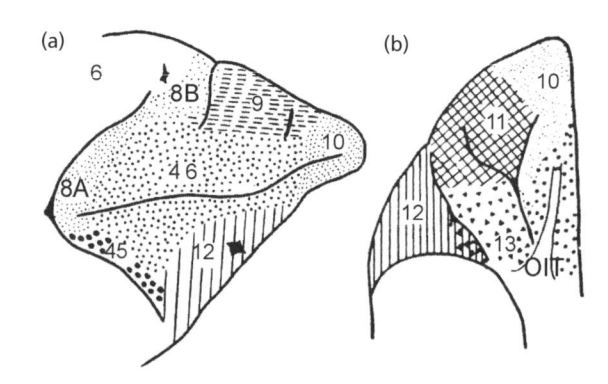

**Fig. 2.6** Cytoarchitectonic map of the lateral (a) and orbital (b) surfaces of the frontal lobe of the macaque monkey as parcellated by Walker (1940).

the specimen acquired under low power magnification. Cells were segmented from the cellular matrix to eliminate the influence of inhomogeneous staining by convolving the images with two local adaptive filters followed by thresholding. For each histological section, the locations of cells within the size range typical of granule cells were identified by specialized software, and coded in a separate *granularity* image. The thresholded and granularity images were then processed with various blurring filters in order to accentuate the laminar organization of the cortex. Specifically, in the illustrative data presented here, the granularity image was smoothed by gaussian blurring before sampling (Fig. 2.7, second column). The granularity image was also subtracted from the thresholded image to provide an image of the cortex without granule cells, then processed with a median filter to remove cells smaller than granule cells and smoothed by gaussian blurring before sampling (Fig. 2.7, third column). This method separates the contribution of each individual cortical layer to neuronal density. The smoothed granularity images select the inner and outer granule layers (i.e., layers II and IV). The inverse images select the pyramidal layers III, V and VI. Sampling was guided by equidistant transverse lines placed across the width of the cortex from layer II to the internal cortical boundary with the underlying white matter (Schleicher *et al.* 2000; Petrides *et al.* 2005). The pixel values lying under the transverse lines were sampled serially from the outer to the inner cortical contour. The sample at any one transverse line represents a profile of cortical density at that location. All profiles were standardized to a common length by linear interpolation, and the neuronal density of various layers was quantified by calculating the mean value of the region on each profile, corresponding to the individual locations of the cortical layers that were traced manually on the original digital images of the specimens.

There is general agreement that the most caudal part of the OFC at the interface with the insula is composed of agranular cortex, i.e., cortex that lacks layer IV. This cortex has been referred to as orbital proisocortex by Barbas and Pandya (1989) and as agranular insula by Carmichael and Price (1994) and Öngür *et al.* (2003), because it lacks layer IV and lies next to the claustrum. Just in front of this agranular cortex, in the posterior part of the orbital region, Walker (1940) identified an area with a faint layer IV that he designated as area 13. In our comparative studies of the monkey and human frontal cortical areas, we similarly labeled the posterior part of the human OFC that has a faint layer IV and which lies in front of the agranular insular cortex as area 13 (see Figs. 2.7 and 2.8).

In Walker's map, area 13 is replaced medially by area 14 on the gyrus rectus (Fig. 2.6b). As pointed out above, Brodmann (1908, 1909) in his map of the human cerebral cortex included the gyrus rectus within the large region that he labeled as area 11 (Fig. 2.5a), although he wrote that the architecture of the gyrus rectus is distinct from that of the rest of this region. In the monkey, Walker (1940) designated the cortex of the gyrus rectus as area 14 and retained the term area 11 for the granular anterolateral orbital region, thus implementing the distinction that Brodmann alluded to (but did not implement) in the human cortex. The architecture of the cortex of the gyrus rectus has been studied by

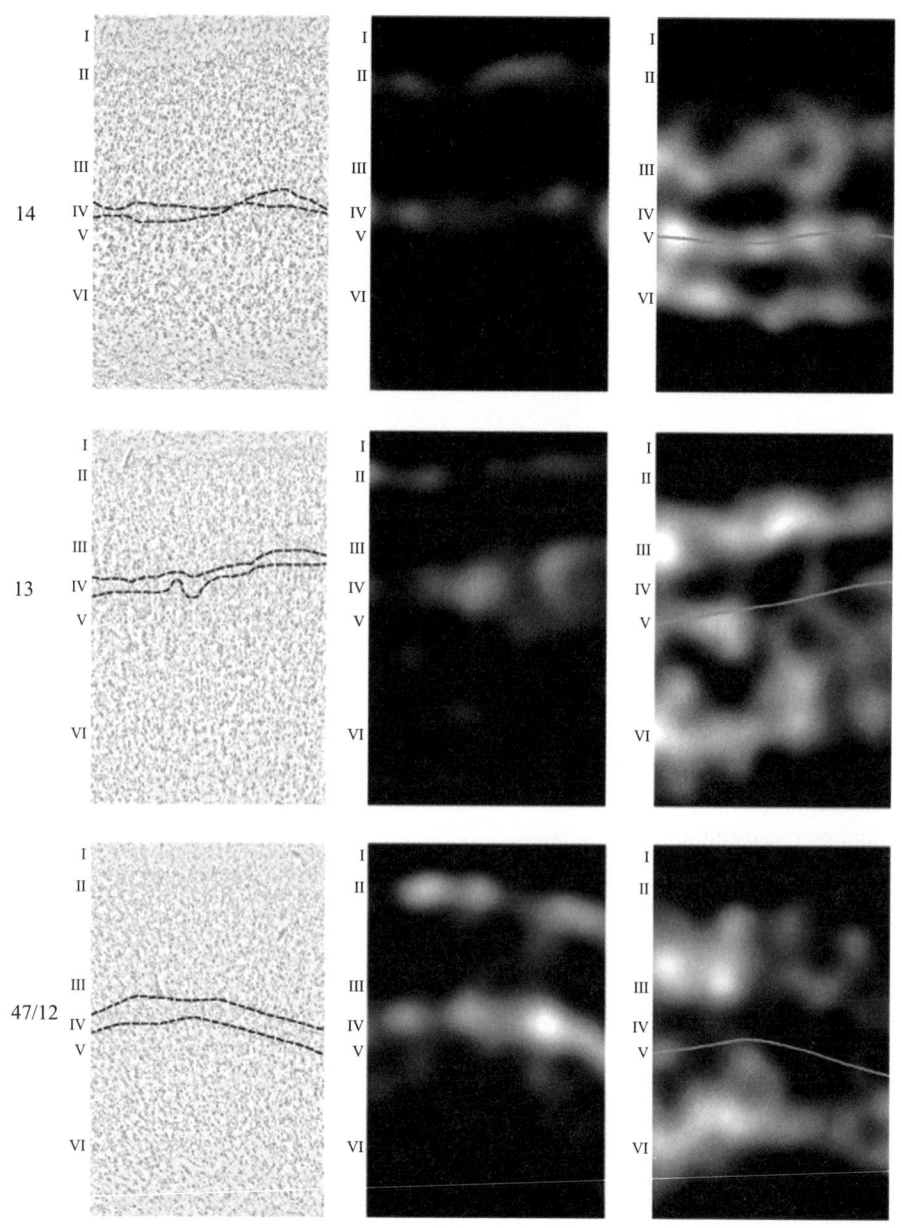

**Fig. 2.7** Illustrative images of areas 14, 13, and 47/12: (first column) digital photomicrographs of sections stained with cresyl violet, (second and third columns) images that have been digitally processed to highlight the granule and pyramidal layers, respectively. The layers are labeled by Roman numerals. The boundaries of layer IV are highlighted by dashed lines in the first column and the center of layer Va is highlighted by a thin white line in the third column.

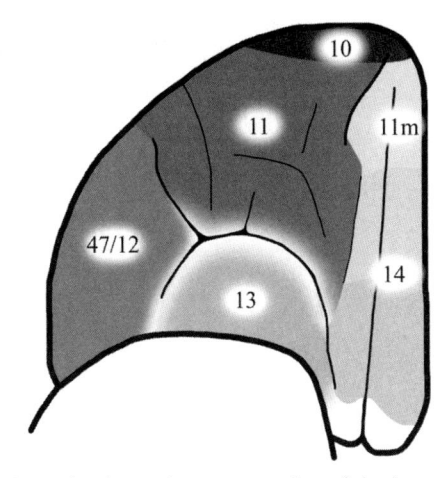

**Fig. 2.8** (Also see Color Plate 1.) Schematic reconstruction of the human orbital surface with architectonic areas shown in different colors. Note area 14 is shown extending into the medial part of the medial orbital gyrus. A graded color scheme was selected for the gyrus rectus and adjacent cortex reflecting the increasing granularity as one moves forward from the caudal gyrus rectus to the more rostral region that we have labeled area 11m in keeping with nomenclature proposed by Öngür et al. (2003). The area in white around the transverse orbital sulcus and posterior arm of the lateral orbital sulcus reflects a transitional zone between area 13 and 11 rostrally and between area 13 and 47/12 laterally. The precise boundaries between these areas remain to be determined in studies with larger number of subjects capable of addressing variability in topography relative to cytoarchitectual features.

many investigators in both the monkey (e.g., Walker 1940; Barbas and Pandya 1989; Preuss and Goldman-Rakic 1991; Morecraft et al. 1992; Carmichael and Price 1994) and the human (e.g., Economo and Koskinas 1925; Beck 1949; Petrides and Pandya 1994; Hof et al. 1995; Öngür et al. 2003) brain. In our comparative architectonic study of the human and the rhesus monkey prefrontal cortex (Petrides and Pandya 1994), we used Walker's designation of area 14 for an area on the gyrus rectus that was comparable in both species. The cortex on the gyrus rectus, which is designated as area 14 in both the human and the monkey brain, is a narrow cortex that has a faint but distinct layer IV and a dense layer Va that stands out clearly against a less dense layer Vb. In the posterior part of this cortical area, layer IV becomes even less distinct and, therefore, a caudal dysgranular subdivision might be distinguished from a rostral subdivision, that has a more clearly defined (albeit narrow) layer IV (Figs. 2.7, 2.8, 2.9). Figure 2.10 presents a quantitative comparison of the relative granularity of layer IV in areas 14, 13, and 47/12 sampled from a posterior coronal section. The granularity in areas 13 and 14 is significantly lower, statistically, than the granularity in area 47/12.

We have observed that the architectonic features that characterize area 14 (i.e., the faint layer IV and the dense, band-like layer Va) can be observed both on the gyrus rectus and within both banks of the adjacent olfactory sulcus, extending for some distance onto the

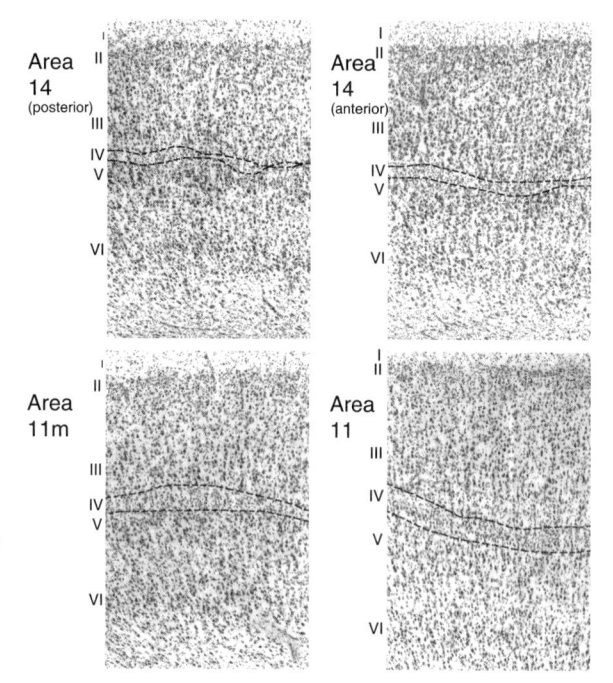

**Fig. 2.9** Digital photomicrographs of area 14 from a posterior and an anterior part of the gyrus rectus, and from areas 11m and 11 to illustrate differences in the relative granularity of layer IV and the density of the upper part of layer V (i.e., Va).

medial orbital gyrus (Fig. 2.8). This observation is consistent with the earlier studies of Economo and Koskinas (1925) and Beck (1949), who also noted that the gyrus rectus area extends within the banks of the adjacent olfactory sulcus and partly onto the adjacent medial orbital gyrus. In contrast, Öngür et al. (2003) consider the medial bank of the olfactory sulcus next to area 14 as area 13b, and the lateral bank and adjacent lip of the medial orbital gyrus as area 13m. Although the structure of the cortex changes somewhat within the sulcus as compared to the crown of the gyrus rectus, we were not able to assure ourselves that these minor modifications reflected different architectonic areas rather than differences in the relative width of the individual layers caused by cortical folding, or an impression resulting from the angle of sectioning. We have, therefore, elected to maintain the term, area 14, for both the gyrus rectus and the cortex within the adjacent olfactory sulcus and medial orbital gyrus (Fig. 2.8). It should be noted that any problems in the interpretation of architectonic features due to cortical folding or non-optimal sectioning will be reflected in material with different stains and, therefore, cannot be resolved by the use of multiple stains.

In both the human and the monkey brain, the rostral part of the orbitofrontal region consists of the granular cortex belonging to area 11 and the orbital extension of the frontopolar area 10 (Figs. 2.5, 2.6, 2.8). Area 11 has a well-defined inner granular layer IV, and a layer III that shows a greater sublamination compared to areas 14 and 13, which lie caudal and medial to it (Fig. 2.9). Interestingly, the most rostral part of the gyrus rectus also has a well-developed granular layer IV and, therefore, it could be

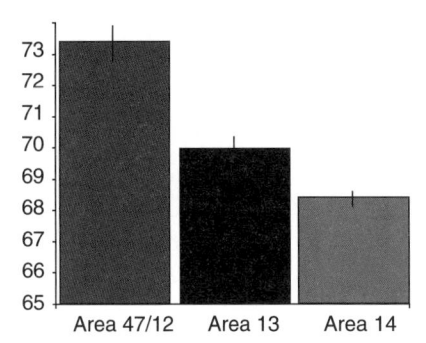

**Fig. 2.10** Histogram of the average granularity of layer IV in areas 47/12, 13, and 14 in one example section. The units on the *y*-axis are an index of granule cell density on a scale of 0–255, where higher values represent greater density. Bar lines indicate standard error. An analysis of variance indicated significant differences ($F$ (2, 294) = 9.17, $p < 0.001$). Post-hoc comparison with the Newman-Keuls test showed that the granularity in area 13 and in area 14 was less than that in area 47/12 ($p < 0.01$).

considered to be part of area 11. Indeed, this part of the gyrus rectus has been referred to as area 11m by Öngür *et al.* (2003) to distinguish it from the more laterally lying part of area 11. It should be noted that this medial part of area 11 has some characteristics typical of the more posteriorly lying area 14. Thus, area 11m could be thought of as possessing a mixture of characteristics of area 14 and area 11 and could be referred to as area 11/14.

On the lateral part of the orbital gyrus of the monkey cortex and extending onto the ventrolateral prefrontal convexity, Walker (1940) identified an area that he designated as area 12 (Fig. 2.6a, b). In the classical maps of the human cerebral cortex (Brodmann 1908, 1909; Beck 1949; Sarkissov *et al.* 1955), the cortical region occupying the most ventral part of the inferior frontal gyrus (pars orbitalis), and extending onto the caudal half of the OFC, was included within the heterogeneous area 47. In our comparative architectonic examination, we observed that the part of this large region that lies rostral and ventral to area 45 and extends as far as the lateral orbital sulcus has architectonic features comparable to those of Walker's area 12 in the monkey (Petrides and Pandya 1994, 2002a). We therefore labeled this part as area 47/12 to denote the fact that this particular portion of the previously labeled area 47 is comparable in architecture to Walker's ventrolateral area 12 in the monkey brain (Fig. 2.8). The transition from area 45 (which occupies the pars triangularis of the inferior frontal gyrus) to area 47/12 can be clearly defined and lies close to the horizontal ramus of the Sylvian fissure. The conspicuously large and deeply stained pyramidal neurons in the deeper part of layer III, which are the defining feature of area 45, are replaced in area 47/12 by smaller neurons. Thus, in area 47/12, layer III contains small and medium pyramidal cells in its upper part, and medium and somewhat larger pyramidal neurons in its lower part. In area 47/12, layer IV is not as broad as in area 45.

Öngür *et al.* (2003), in their map of the human OFC, adopted the term area 47/12 that was introduced by Petrides and Pandya (1994) to describe the human ventrolateral and adjacent orbital region that was considered to correspond to Walker's area 12 in the monkey. However, in their map, granular area 47/12 is shown to extend much more medially and posteriorly than our studies would suggest. We have found that the caudal part of the posterior orbital gyrus, in front of the agranular cortex at the interface with the insula, consists of dysgranular cortex comparable to area 13 in the monkey, while the most anterior and anterolateral part of the posterior orbital gyrus near the transverse orbital sulcus is more granular (i.e. the cortex shown in white in figure Fig. 2.8). The borders that mark the transition between areas 47/12 and 13 and areas 11 and 13 remain to be determined in studies in which many different brains are used so that variations in the sulcal and gyral morphological features can be correlated with architectonic borders. Such studies are under way in our laboratory.

## 2.4. Conclusion

Considerable improvements in functional neuroimaging during the last 20 years have yielded several findings of relatively specific activation patterns in the human OFC in relation to cognitive, affective, and motivational states. The location and interpretation of the cortical areas within which modulations of brain activation occurred has most often been interpreted in terms of Brodmann's architectonic areas, as provided in the Talairach and Tournoux (1988) atlas. Architectonic areas in the Talairach and Tournoux atlas were not the result of architectonic analysis, but rather the result of simple projection of the Brodmann (1908, 1909) map onto the brain presented in the atlas. This is particularly problematic for the OFC because Brodmann (1908, 1909) subdivided this part of the cerebral hemisphere into only two large regions (areas 11 and 47) that were, by his own admission, no more than broad territories encompassing multiple architectonic areas. Furthermore, he expressed considerable uncertainty about the boundaries of these regions. For example, the interpretation of three different activation foci, one located within the dysgranular cortex in the caudal part of the gyrus rectus, the other in the more granular cortex in the middle section of the gyrus rectus, and the third in the granular cortex that lies anterolaterally in the orbitofrontal region, as all occurring within area 11 can only impede progress in appreciating the meaning of such diverse activation patterns. Furthermore, such interpretation of the locus of activation does not encourage appropriate integration of functional neuroimaging findings on human subjects with the wealth of experimental physiological and anatomical data obtained on macaque monkeys. The evidence presented in this article revealed a similar architectonic organization scheme in the human orbitofrontal region and that of the macaque monkey. Many more architectonic areas can be recognized than the two broad territories shown in the Brodmann (1908, 1909) map and its projection into the Talairach and Tournoux (1988) atlas, which has provided the basis for the interpretation of activation patterns in the human orbitofrontal region until now. Since the new subdivision of the human orbitofrontal region is similar to that of the macaque monkey, insights resulting from experimental anatomical (e.g., Barbas 1988; Morecraft *et al.* 1992; Carmichael and Price 1995a, b; Petrides and Pandya 2002a, b) and neuronal recording (e.g., Rolls and Bayliss 1994; Rolls *et al.* 1996, 1999) studies in the monkey can more easily be integrated with functional neuroimaging findings on human subjects.

## Acknowledgments

This work was supported by grant MOP-14620 from the Canadian Institutes of Health Research. We wish to thank Ms. Veronika Zlatkina for help with the preparation of the figures.

## References

Bailey, P., and Bonin, G. (1951) *The Isocortex of Man*. University of Illinois Press, Urbana.

Barbas, H. (1988) Anatomic organization of basoventral and mediodorsal visual recipient prefrontal regions in the rhesus monkey. *J Comp Neurol* **276**:313–342.

Barbas, H., and Pandya, D. N. (1989) Architecture and intrinsic connections of the prefrontal cortex in the rhesus monkey. *J Comp Neurol* **286**:353–375.

Beck, E. (1949) A cytoarchitectural investigation into the boundaries of cortical areas 13 and 14 in the human brain. *J Anat* **83**:147–157.

Brodmann, K. (1908) Beitraege zur histologischen Lokalisation der Grosshirnrinde. VI. Mitteilung: Die Cortexgliederung des Menschen. *J Psychol Neurol* (Lzp.) **10**:231–246.

Brodmann, K. (1909) *Vergleichende Lokalisationslehre der Grosshirnrinde in ihren Prinzipien dargestellt auf Grund des Zellenbaues*. Barth, Leipzig.

Chiavaras, M. M., and Petrides, M. (2000) Orbitofrontal sulci of the human and macaque monkey brain. *J Comp Neurol* **422**:35–54.

Carmichael, S. T., and Price, J. L. (1994) Architectonic subdivision of the orbital and medial prefrontal cortex in the macaque monkey. *J Comp Neurol* **346**:366–402.

Carmichael, S. T., and Price, J. L. (1995a) Sensory and premotor connections of the orbital and medial prefrontal cortex of macaque monkeys. *J Comp Neurol* **363**:642–664.

Carmichael, S. T., and Price, J. L. (1995b) Limbic connections of the orbital and medial prefrontal cortex in macaque monkeys. *J Comp Neurol* **346**:403–434.

Chiavaras, M. M., LeGoualher, G., Evans, A., and Petrides, M. (2001) Three-dimensional probabilistic atlas of the human orbitofrontal sulci in standardized stereotaxic space. *Neuroimage* **13**:479–496.

Connolly, C. J. (1950) *External morphology of the primate brain*. Charles C. Thomas, Springfield, Illinois.

Cunningham, D. J. (1892) *Contribution to the surface anatomy of the cerebral hemispheres*. Royal Irish Academy of Science. Cunningham Memoirs. No. VII., Academy House, Dublin.

Duvernoy H. (1991) *The Human Brain: Surface, Three-Dimensional Sectional Anatomy and MRI*. Springer-Verlag, Wien.

Eberstaller, O. (1890) *Das Stirnhirn*. Urban & Schwarzenberg, Wien.

Economo, C., and Koskinas, G. N. (1925) *Die Cytoarchitektonik der Hirnrinde des erwachsenen Menschen*. Springer, Wien.

Evans, A. C., Kamber, M., Collins, D. L., and MacDonald, D. (1994) *An MRI-based probabilistic atlas of neuroanatomy*. In: Magnetic resonance scanning and epilepsy (Shorvon, S. D., ed.), pp. 263–274. Plenum Press, New York.

Gratiolet L. P. (1854) *Memoire sur les plis cérébraux de l' homme and des primates*. A. Bertrand, Paris.

Hof, P.R., Mufson, E. J., and Morrison, J. H. (1995) Human orbitofrontal cortex: Cytoarchitecture and quantitative immunohistochemical parcellation. *J Comp Neurol* **359**:48–68.

Landau, E. (1911) Ueber die Grosshirnfurchen am basalen Teile des temporooccipitalen Feldes bei den Esten. *Ztschr F Morphol u Anthropol* **13**:423–438.

Landau, E. (1914) Ueber die Furchen an der Lateralflaeche des Grosshirns bei den Esten. Ztschr F Morphol u Anthropol **16**:239–279.

Mackey, S., and Petrides, M. (2005) *Quantitative analysis of the cortical architecture of the posterior orbitofrontal gyrus in the human brain*. Program No. 410.3. 2005 Abstract Viewer/Itinerary Planner, Washington, D.C.: Society for Neuroscience 2005, Online.

Morecraft, R. J., Geula, C., and Mesulam, M.-M. (1992) Cytoarchitecture and neural afferents of orbitofrontal cortex in the brain of the monkey. *J Comp Neurol* **323**:341–358.

Öngür, D., Ferry, A. T., and Price, J. L. (2003) Architectonic subdivision of the human orbital and medial prefrontal cortex. *J Comp Neurol* **460**:425–449.

Ono, M., Kubik, S., and Abernathey, C. D. (1990) *Atlas of the Cerebral Sulci*. Thieme, Stuttgart.

Petrides, M. (2005) Lateral prefrontal cortex: architectonic and functional organization. *Phil. Trans. R. Soc. B* **360**:781–795.

Petrides, M., and Pandya, D. N. (1994) *Comparative architectonic analysis of the human and the macaque frontal cortex*. In:Handbook of Neuropsychology (F. Boller, F., and Grafman, J., eds), Vol. 9, pp. 17–58, Elsevier, Amsterdam.

Petrides, M., and Pandya, D. N. (2002a) Comparative architectonic analysis of the human and the macaque ventrolateral prefrontal cortex and corticocortical connection patterns in the monkey. *Eur J Neurosci* **16**:291–310.

Petrides, M., and Pandya, D. N. (2002b) *Association pathways of the prefrontal cortex and functional observations*. In Principles of Frontal Lobe Function (Stuss, D.T., and Knight, R.T., eds) pp. 31–50, Oxford University Press, New York.

Petrides, M., Cadoret, G., and Mackey, S. (2005) Orofacial somatomotor responses in the macaque monkey homologue of Broca's area. *Nature* **435**:1235–1238.

Preuss, T. M., and Goldman-Rakic, P. S. (1991) Myelo- and cyto-architecture of the granular frontal cortex and surrounding regions in the strepsirhine primate galago and the anthropoid primate macaca. *J Comp Neurol* **310**:429–474.

Retzius, G. (1896) *Das Menschenhirn*. P. A. Norstedt und Söner, Stockholm.

Rolls, E. T., Critchley, H. D., Mason, R., and Wakeman, E. A. (1996) Orbitofrontal cortex neurons: role in olfactory and visual association learning. *J Neurophysiol* **75**:1970–1981.

Rolls, E. T., Critchley, H. D., Browning, A. S., Hernadi, A., and Lenard, L. (1999) Responses to the sensory properties of neurons in the primate orbitofrontal cortex. *J Neurosci* **19**:1532–1540.

Schleicher, A., Amunts, K., Geyer, S., Kowalski, T., Schormann, T., Palmero-Gallagher, P., and Zilles, K. (2000) A stereological approach to human cortical architecture: identification and delineation of cortical areas. *J Chem Neuroanatomy* **20**:31–47.

Semendeferi, K., Armstrong, E., Schleicher, A., Zilles, K., and Van Hoesen, G. W. (1998) Limbic frontal cortex in hominoids: a comparative study of area 13. *American Journal of Anthropology* **106**:129–155.

Shellshear, J. L. (1926) The occipital lobe in the brain of the Chinese with special reference to the sulcus lunatus. *J Anat* **61**:1–13.

Shellshear, J. L. (1937) The brain of the Australian aboriginal. A study in cerebral morphology. *Phil Trans Roy Soc Lond B* **227**:293–409.

Talairach, J., and Tournoux, P. (1988) *Co-planar stereotaxic atlas of the human brain*. Thieme, New York.

**Vogt, O.** (1910) Die myeloarchitektonische Felderung des menschlichen Strinhirns. *J Psychol Neurol* **15**:221–232.

**Walker, A. E.** (1940) A cytoarchitectural study of the prefrontal area of the macaque monkey. *J Comp Neurol* **73**:59–86.

**Weisbach, A.** (1870) Die Supraorbitalwindungen des menschlichen Gehirns. *Weiner medicinische Jahrbucher* **19**:88–162.

Chapter 3

# Connections of orbital cortex

Joseph L. Price

The orbitofrontal cortex (OFC) is a complex, heterogeneous cortical region. There have been several studies of the connections of this region in the past 10–15 years that have greatly expanded our understanding of its relations with other cortical and subcortical structures (e.g., Amaral and Price 1984; Barbas 1988, 1993; Barbas and Pandya 1989; Barbas and De Olmos 1990; Morecraft et al. 1992; Ray and Price 1993; Barbas and Blatt 1995; Carmichael and Price 1994, 1995a, 1995b, 1996; Carmichael et al. 1995; An et al. 1998; Öngür et al. 1998; Rempel-Clower and Barbas 1998; Barbas et al. 1999; Cavada et al. 2000; Ferry et al. 2000; Kondo et al. 2003, 2005). Together, these studies have shown that the OFC has extensive connections with other cortical areas, especially in the temporal and insular cortices, with the amygdala, hippocampus and other limbic structures, and with several subcortical structures, including the medial thalamus, ventromedial striatum, hypothalamus and midbrain. In addition, some of the studies have shown that there is an intrinsic organization within the OFC and the adjacent medial prefrontal cortex, which appears to be closely related to the organization of orbital connections with other parts of the brain.

The studies described in this chapter were based on an architectonic analysis in which the OFC was divided into 18 architectonic areas (see Chapter 1, this volume). This architectonic analysis, together with small injections of anterograde and retrograde axonal tracers that were restricted to individual architectonic areas, allowed distinct patterns of cortico-cortical connections to be defined within the OFC, and with adjacent medial and ventrolateral prefrontal areas. Analyses of these patterns of connections indicate that at least two functional systems can be recognized in the orbital and medial prefrontal cortex (Carmichael and Price 1996) (Fig. 3.1). These have been referred to as the 'orbital' and 'medial' prefrontal networks, although the medial network also includes several areas on the orbital surface. More recent experiments have shown that these two networks also have distinct connections with other parts of the brain, and appear to be associated with different, though related, functional roles (An et al. 1998; Öngür et al. 1998; Ferry et al. 2000; Öngür and Price 2000; Kondo et al. 2003, 2005).

## 3.1. Intrinsic connections

Experiments in which anterograde or retrograde axonal tracers are injected into the agranular insular areas (Iam, Iapm, Ial, Iapl) at the caudal edge of the orbital surface label

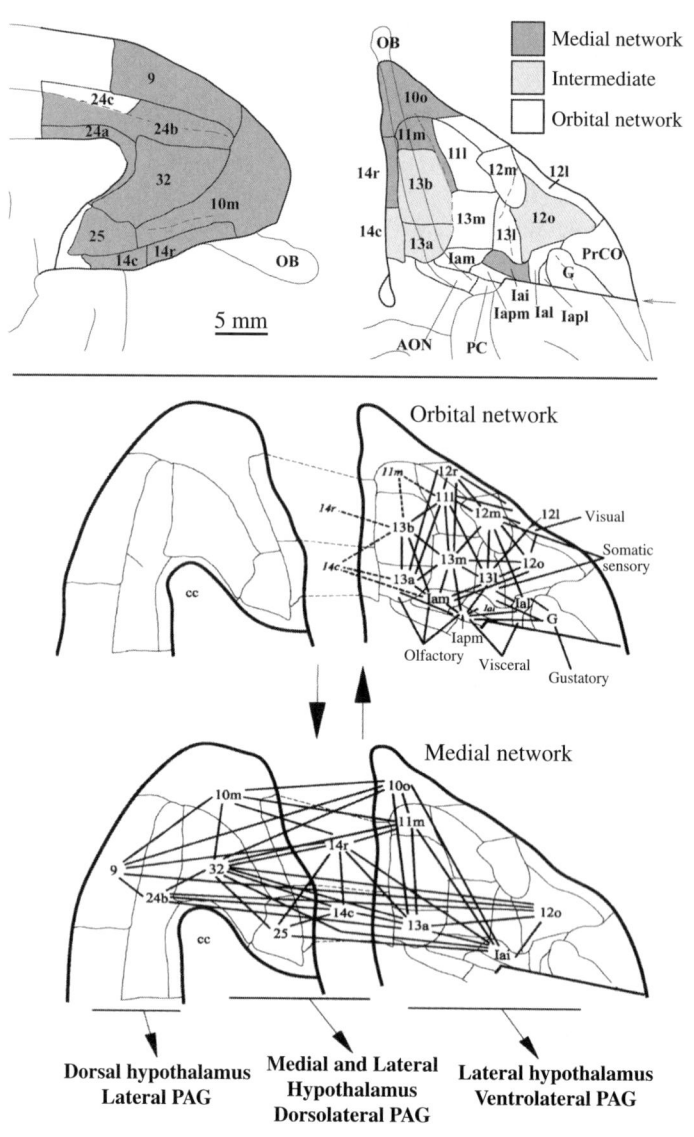

**Fig. 3.1** Diagrammatic illustrations of the 'medial' and 'orbital' prefrontal networks, and the architectonic areas that comprise them. In the upper part, the medial network areas are shaded dark gray and the orbital network areas are unshaded. Light gray shading indicates intermediate areas that have connections with both networks. In the lower part of the figure, the intrinsic cortico-cortical connections that define the two networks are indicated. Note that the orbital network receives several sensory inputs, while the medial network has outputs to visceral control systems in several parts of the hypothalamus and the periaqueductal gray (PAG). Modified with permission from Carmichael and Price (1996).

substantial connections to areas 12l, 12m, 13l, and 13m in the central OFC (Carmichael and Price 1996) (Fig. 3.1). Those areas, in turn, are substantially connected with more rostral and lateral areas 11l and 12r. Some of the connections, especially from the lateral orbital areas, extend around the ventrolateral convexity of the frontal lobe, and also involve area 45 and the ventral part of area 46. The labeled axons or cells are often arranged in patches or columns; in general, these extend across all layers of the cortex.

The areas in the medial prefrontal cortex (areas 24, 25, 32, and 10m) have a largely complementary pattern of interconnections (Carmichael and Price 1996) (Fig. 3.1). These areas are substantially connected with other medial prefrontal areas, but have few connections with most of the OFC, especially the areas in the central OFC. The exceptions to this rule are the areas along the medial edge of the OFC (areas 11m, 13a, 13b, 14c, 14r), and two areas in the caudolateral part of the OFC (areas 12o and Iai). Many of these areas are connected to both the orbital and the medial network areas, but three of them, areas 11m, 14r, and Iai, are primarily connected to the medial areas.

Based on these observations, the areas of the orbital and the medial prefrontal cortex have been divided into 'orbital,' 'medial' and prefrontal networks (Carmichael and Price 1996) (Fig. 3.1). The areas on the caudal and central parts of the orbital surface are included in the orbital network, but areas 11m, 14r, and Iai are part of the 'medial' network. The most medial orbital areas, 13a, 13b, 14c, and lateral area 12o, also have prominent connections with the medial network, but they also have connections to the orbital network, so they can be considered intermediate areas which may serve as interfaces between the two networks. Because the intermediate areas are most closely associated with the medial network, for the purposes of this chapter, they will be considered together with the medial network.

As discussed below, the orbital network receives inputs from several sensory areas, and appears to integrate them, particularly in relation to the assessment of food and reward. In contrast to this, the medial network receives few sensory inputs, but it has outputs to the hypothalamus and other visceral control areas, and appears to be involved in visceral modulation and emotion (Öngür and Price 2000).

Other groups have also reported an organization within the orbital and the medial prefrontal cortex. Although they do not match the two networks described by Carmichael and Price (1996), there are some similarities. Most strikingly, Barbas and Pandya (1989) proposed that the entire prefrontal cortex could be separated into two architectonic 'trend'. These extend from supposedly less differentiated 'periallocortical' areas at the caudal edge of the orbital or medial cortex around to parts of area 8, which were suggested to represent the most differentiated cortex. Their 'basoventral trend' includes all of the OFC and the ventral part of the lateral prefrontal cortex, while the 'mediodorsal trends' includes the medial prefrontal cortex and the dorsal part of the lateral prefrontal cortex. The mediodorsal trend, therefore, includes most of the areas of the medial prefrontal network of Carmichael and Price (1996), while the basoventral trend includes the areas of the 'orbital prefrontal network', though the correlation between the trends and the networks is not close. In particular, the mediodorsal trend does not include the areas along the medial side of the OFC, or areas Iai and 12o in the lateral OFC, all of which have important associations with the medial prefrontal network.

In another study, Cavada et al. (2000) reported differences between the connections of the medial and lateral parts of the OFC. Although this study used relatively large injections that do not allow detailed analysis, the medial orbital region appears to correspond with the areas along the medial edge of the OFC that are largely associated with the medial prefrontal network of Carmichael and Price (1996). Similarly, the lateral orbital region of Cavada et al. (2000) corresponds to the core of the orbital prefrontal network.

## Sensory inputs

A chief characteristic of the orbital network is that it receives inputs from the cortical areas associated with most of the sensory systems, including olfaction, taste/visceral afferents, somatic sensation, and vision (Fig. 3.2). As discussed below, inputs from the auditory system are still uncertain. Taken together, the constellation of sensory inputs

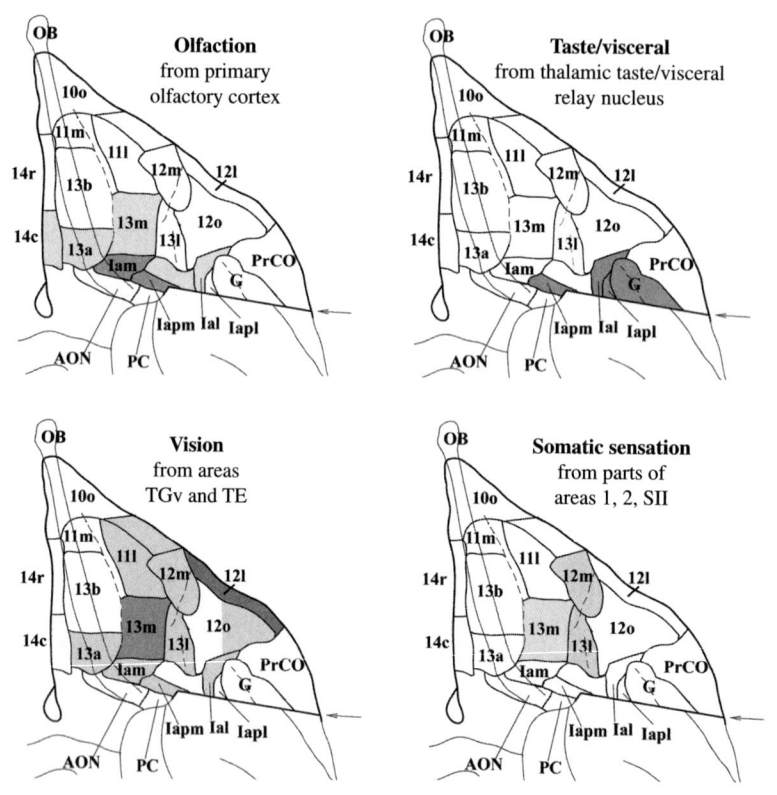

**Fig. 3.2** Areas of the OFC that receive direct inputs from the olfactory, taste/visceral, visual, and somatic sensory systems. Darker shading indicates areas that appear to receive stronger inputs. Note that while different sensory modalities project into somewhat different areas, the interconnections that define the orbital prefrontal network would serve to integrate these together. Based on data from Carmichael and Price (1995b) and Kondo et al. (2003).

suggests that the orbital network is particularly involved in assessment of food. This suggestion is supported by physiological recording studies, which have shown that orbital neurons respond to multisensory, food-related stimuli in a way that appears to code for the flavor, appearance and texture of food (e.g., Rolls 2005). In addition, the studies also indicate that the orbital neurons respond to affective characteristics (reward vs. aversion) as well as sensory characteristics of the stimuli (e.g., Rolls 2000; Schultz *et al.* 2000). It may be noted that the orbital network is connected with the ventrolateral prefrontal cortex, which also receives multiple sensory inputs, and also may be involved in sensory object assessment (e.g., Petrides 2005).

## Olfaction

The primary olfactory cortex is a paleocortical region situated immediately caudal to the OFC, contiguous with the agranular insular areas. It consists of several areas, including the anterior olfactory nucleus, the piriform cortex, and parts of the periamygdaloid cortex and entorhinal cortex. The primary olfactory cortex receives direct projections from the olfactory bulb, and is the first cortical region involved in the integration and analysis of olfactory sensory information (Price 1973; Carmichael *et al.* 1995; Price 2003). During the past decades both anatomical and physiological data from experiments on rats and monkeys have shown that olfactory sensory information is projected from the primary olfactory cortex to the posterior part of the OFC, including the agranular insular areas and adjacent orbital area 13m (e.g., Tanabe *et al.* 1975; Price 1985; Carmichael *et al.* 1995) (Fig. 3.2). The earliest indications were that this pathway involves a transthalamic relay, similar to that in other sensory systems (Yarita *et al.* 1980). Thus, the olfactory cortex projects to the medial part of the mediodorsal nucleus of the thalamus, which in turn projects to the OFC (Heimer 1972; Krettek and Price 1977a; Russchen *et al.* 1987; Ray and Price 1992, 1993). However, the thalamic projection from the olfactory cortex is through a relatively small population of large neurons at the deep edge of the cortex (the ventral endopiriform nucleus) (Krettek and Price 1977b; Price and Slotnick 1983; Russchen *et al.* 1987). It is likely that olfactory sensory information is carried by the more numerous fiber pathways that run directly from the olfactory cortex to the posterior OFC (Price 1985; Carmichael *et al.* 1995).

Axons to the OFC arise in several components of the olfactory cortex, including the anterior olfactory nucleus, piriform cortex, and several parts of the periamygdaloid cortex. They terminate most strongly in areas Iam and Iapm, which are immediately adjacent to the olfactory cortex, but fibers also extend to other agranular insular areas, and to areas 13a and 13m (Fig. 3.2). This projection is very different than other sensory inputs to the neocortex. Not only do they reach the cortex without a thalamic relay, they also end predominantly in layer I. Only in areas Iam and Iapm do the fibers extend significantly into the deeper cortical layers.

It should be noted that olfactory sensory information is not restricted to posterior orbital areas. Olfaction-related activity is further processed, and integrated with other sensory modalities, in more rostral orbital areas, through the connections of the orbital prefrontal network (Fig. 3.1). Recordings in more rostral orbital areas have shown that neurons there respond to multisensory stimuli, especially those related to eating such as the flavor (combined taste and olfaction), texture, and sight of food (e.g., Rolls 2005).

## Taste/visceral afferents

Taste and visceral afferent information is relayed from the nucleus of the solitary tract to the medial, parvicellular part of the ventroposterior medial nucleus of the thalamus (Beckstead *et al.* 1980). In turn, this taste/visceral nucleus projects to the cortex in the anterior insula. The major projection from the dorsal part of the nucleus, is to the primary taste cortex at the dorsal edge of the insula, within the fundus of the superior limiting sulcus. Cells in the ventral lamina of the taste/visceral thalamic nucleus, however, project to the agranular insular areas, including Ial and Iapm (Carmichael and Price 1995b; Fig. 3.2). This ventral part of the taste/visceral thalamic nucleus appears to receive input from the caudal part of the nucleus of the solitary tract that is related to vagal afferents, and may function as a general visceral relay (Beckstead *et al.* 1980). In contrast, the more anterior, taste-related part of the solitary tract nucleus projects to the dorsal lamina of the taste/visceral nucleus. A similar organization has been described in rats (Cechetto and Saper 1987; Allen *et al.* 1991).

It should be noted that area Iapm receives convergent olfactory input from the olfactory cortex, and visceral afferent input from the thalamus (Fig 3.2). Further, the primary taste cortex, and areas Ial and Iapm, all project into the orbital network, with special convergence on area 13l (Carmichael and Price 1995b, 1996; Fig. 3.1). This area also receives input from visual and somatic sensory cortical areas, and approximately coincides with the 'second taste cortex' of Rolls and his colleagues (e.g., Rolls 2005). Neurons in this region respond to multi-modal, food-related stimuli.

## Vision

Sensory information about the visual form and color of objects reaches the OFC from the ventral visual association stream in the inferior temporal cortex (Fig. 3.2). The strongest input is from the ventral temporal pole (TGv) and the anterior ventral part of area TE (TEav), but there are also inputs from as far caudal as area TEO (Webster *et al.* 1994; Carmichael and Price 1995b; Kondo *et al.* 2003). The more caudal areas project mainly to areas 12l and 45, on the ventrolateral convexity, both of which have connections into the orbital network (Carmichael and Price 1966). The more rostral inferior temporal areas (areas TEav and TGv) are connected to all of the areas in the orbital network (but not those of the medial prefrontal network: Kondo *et al.* 2003; Fig. 3.2). The visual inputs to the orbital network contribute to the multimodal nature of this region, and would be important in the assessment and selection of food based on appearance as well as flavor.

The visual input to areas 45 and 12l has also been considered in relation to the object memory functions of the ventrolateral prefrontal cortex (Wilson *et al.* 1993; Romanski 2004), though language functions are also found in this region in humans

(Buckner *et al.* 1999). In addition, other aspects of higher level sensory processing appear to be found in the ventrolateral prefrontal cortex, including object selection, comparison, and judgment (Passingham *et al.* 2000; Petrides 1996, 2005).

## Somatic sensation

There are inputs to the OFC from several cortical areas related to somatic sensation, including the ventral, opercular parts of areas 1–2, and parts of areas SII and 7b (Carmichael and Price 1995b) (Fig. 3.2). The parts of SII, at least those that project to the OFC, appear to be related to sensation on the face, digits, or forelimb which would be expected to be involved in feeding. Within the OFC, the connections are primarily with area 12m, though they also extend to areas 13l, 13m, and 11l. As with the other sensory inputs, the somatic sensory inputs would be integrated with other modalities through the connections that make up the orbital network.

## Audition

It is uncertain whether there are direct auditory inputs to the OFC, and to the extent that auditory information enters the OFC, it appears to be associated more with the medial rather than the orbital network. Auditory connections have been reported from the parabelt auditory association areas to the ventrolateral prefrontal cortex (e.g., area 45 or ventral area 46; Hackett *et al.* 1999; Romanski *et al.* 1999), and recordings of neuronal responses to auditory stimuli have been made in the ventrolateral prefrontal cortex (Romanski and Goldman-Rakic 2002). There has been little published physiological evidence for auditory input to the OFC proper, however, and most of the superior temporal connections with the OFC come from the rostral part of the superior temporal gyrus, just anterior and lateral to the rostral parabelt area. When demonstrated with retrograde tracer injections into the medial network areas, the origin of the projection is restricted to the cortex in the dorsal part of the temporal pole (TGd), the rostral part of the superior temporal gyrus (STGr) and the dorsal bank of the superior temporal sulcus (STSd) (Carmichael and Price 1995b; Kondo *et al.* 2003; Barbas *et al.* 2005). This region is adjacent to, but separate from, the auditory/belt/parabelt areas, which appear to lie dorsal and then medial to the TGd/STGr/STSd region (Hackett *et al.* 1998, 2001). Further, in sharp contrast to other sensory inputs that preferentially target the orbital network, these connections are primarily into areas associated with the medial network, including the medial prefrontal areas 25, 32, and 10m, areas 12o and Iai in the caudolateral OFC, and areas along the medial edge of the OFC (Carmichael and Price 1995b; Romanski *et al.* 1999; Kondo *et al.* 2003; Barbas *et al.* 2005). Further studies may be needed to resolve this issue.

It is of course possible that auditory sensory information reaches the orbital network through the ventrolateral prefrontal cortex. In this region, most of the auditory-responsive neurons respond to species-specific vocalizations, suggesting that the ventrolateral prefrontal cortex is particularly involved in assessing distinctive acoustic features (Romanski *et al.* 2005). There is still little information on the functional role of the TGd/STGr/STSd region. It is possible that it receives multimodal sensory information, but this has not yet been defined.

## Limbic connections

In addition to the sensory inputs, the OFC as a whole is also connected with a number of limbic structures, including the amygdala, hippocampus, entorhinal cortex, and parahippocampal gyrus. Most of these connections are distributed to the areas of the medial prefrontal network that are located along the medial edge and caudolateral portions of the OFC. The agranular insular areas in the caudal part of the orbital network also have limbic connections, but there are few limbic connections to other areas in the orbital network. The only exception to this pattern is the connection with the perirhinal cortex (areas 35 and 36), which is primarily connected with the orbital network (see below).

The amygdaloid nuclei have substantial reciprocal connections with the orbital and the medial prefrontal cortex (Amaral and Price 1984; Barbas and de Olmos 1990; Carmichael and Price 1995a; Cavada et al. 2000). Within the OFC, the amygdaloid fibers terminate around the junction of cortical layers I and II, and in layer V; they are distributed primarily to the areas along the medial edge (11m, 13a, 13b, 14c, and 14r), and to the caudolateral areas Iai and 12o (Amaral and Price 1984; Carmichael and Price 1995a; Fig. 3.3), all of which are associated with the medial network or are intermediate between the two networks (Carmichael and Price 1996). There are also substantial amygdaloid connections with areas 25, 32, and 24 in the medial prefrontal cortex. Other areas at the caudal edge of the OFC, such as Iam, Iapm, and Ial, have also been reported to be connected to the amygdala, but the orbital network areas in the central and rostral parts of the OFC have far fewer amygdaloid connections (Amaral and Price 1984; Carmichael and Price 1995a).

Most of the amygdalo-cortical connections arise or terminate in the basal and accessory basal nuclei of the amygdala, though there are also some connections with the lateral nucleus. Although tracer injections in different cortical areas label distinct patterns of cells or axons in the amygdaloid nuclei, it has been difficult to ascertain a clear, consistent organization within these connections. Barbas and de Olmos (1990) reported that there are dorso-ventral and medio-lateral gradients within the basal amygdaloid nucleus related to connections to different cortical areas. Carmichael and Price (1995b) also noted differences in the cortical connections of the ventrolateral and dorsomedial parts of the basal nucleus. It is difficult to correlate the two sets of observations, or to discern an overall organizing principle in the connections. Further study will be needed to resolve this question.

The connections of the entorhinal and posterior parahippocampal cortex with the orbital and the medial prefrontal cortex are very similar to those with the amygdala (Carmichael and Price 1995a; Kondo et al. 2005; Fig. 3.3, 3.4). In both cases, the frontal projections are distributed in the medial and caudolateral OFC (and in the medial prefrontal cortex), but avoid the central and rostral parts of the OFC. Thus, the connections are primarily with the medial prefrontal network and avoid the areas of the orbital

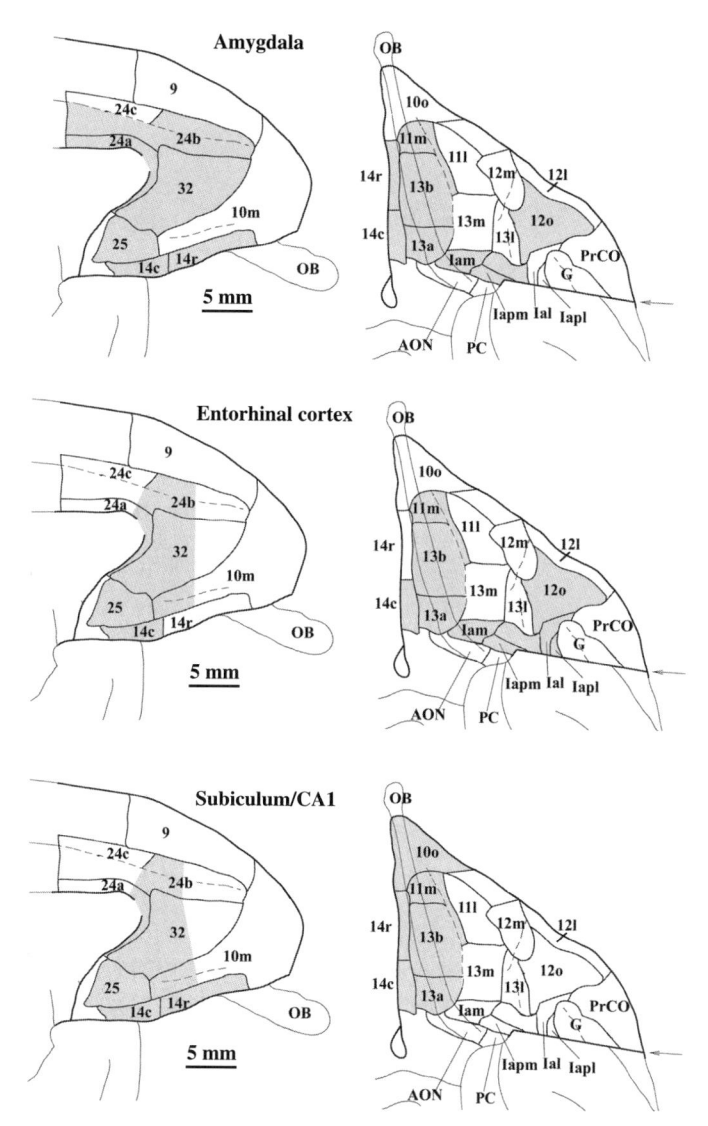

**Fig. 3.3** The shading indicates the areas in the orbital and medial prefrontal cortex that are connected with the amygdala, the entorhinal cortex, and the subiculum/CA1 region of the hippocampus. The amygdaloid and entorhinal connections are bidirectional, but the hippocampal connection is unidirectional, from the hippocampus to the frontal cortex. Note that in all of these cases, the major connections are with areas in the medial prefrontal network. Based on data from Carmichael and Price (1995a).

network. The hippocampus also projects to a very similar area in the medial OFC and medial prefrontal cortex, though this projection is not reciprocated (Barbas and Blatt 1995; Carmichael and Price 1995a; Kondo *et al.* 2005; Fig. 3.3). Within the hippocampus, the projection arises primarily from the region along the junction between the subiculum and CA1.

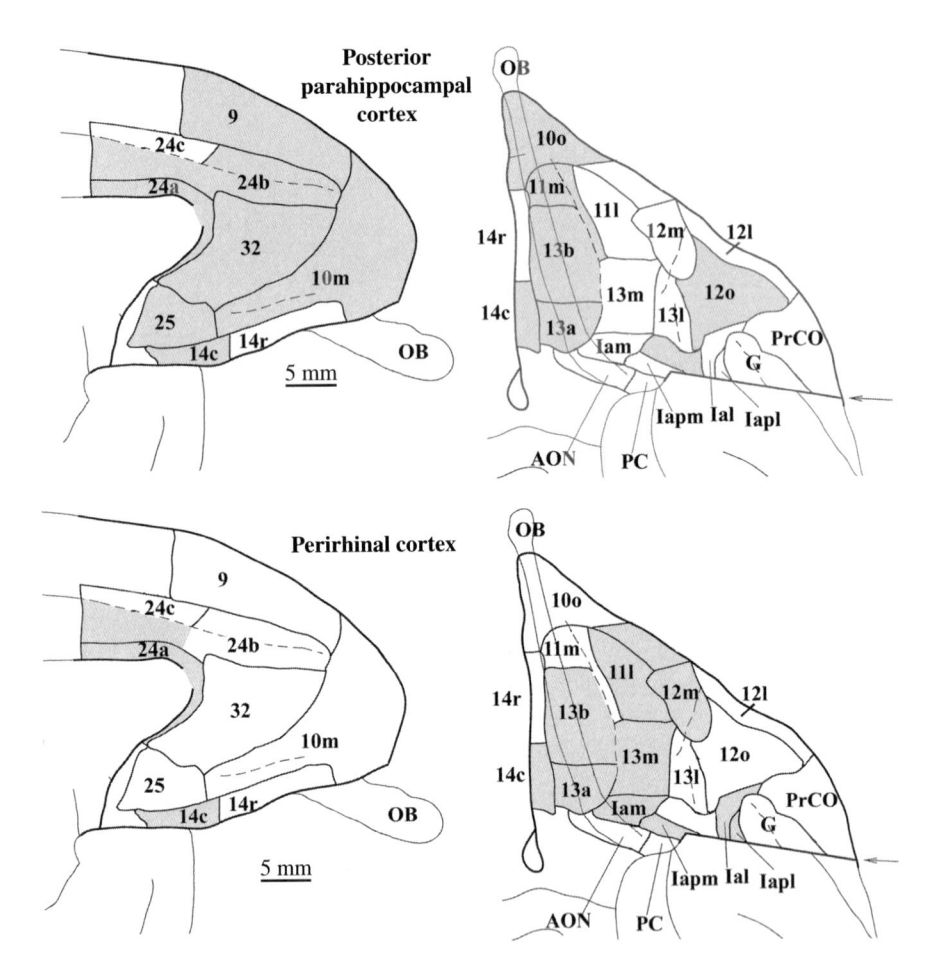

**Fig. 3.4** The shading indicates the areas of the orbital and medial prefrontal cortex that are connected with the posterior parahippocampal cortex (areas TF and TH), and with the perirhinal cortex (areas 35 and 36). Note that although there is some overlap along the medial edge of the OFC, the parahippocampal cortex is primarily connected with the medial prefrontal cortex, while the perirhinal cortex is primarily connected with the orbital network. Based on data from Kondo et al. (2005).

The medial prefrontal network and the parahippocampal cortex show a related pattern of connections in that both regions connect with the posterior cingulate/retrosplenial cortex (Carmichael and Price 1995a; Kobayashi and Amaral 2003; Kondo et al. 2005) (Fig. 3.4). Several areas within the medial network, including especially the medial orbital area 11m as well as areas in the medial frontal wall, have reciprocal connections with retrosplenial areas 29 and 30 and parts of the posterior cingulate area 23. These same areas are also connected to the parahippocampal cortex (Kobayashi and Amaral 2003; Blatt et al. 2003; Kondo et al. 2005).

As mentioned above, the perirhinal cortex provides a contrast to the other limbic connections in that its connections are focused on areas in the orbital network. These include areas Iam, Iapm, Ial, 13m, 13l, and 11l in the central part of the OFC, as well as the intermediate areas 13a, 13b, and 14c (Kondo *et al.* 2005) (Fig. 3.4). It is notable that the perirhinal cortex is also connected with the visual areas in the inferior temporal cortex that also project to the orbital network, and is generally considered to be involved in visual object recognition memory (Squire *et al.* 2004). Lesions of the OFC also produce memory-related deficits in tasks that require the animal to learn and modify stimulus-reward associations (Gaffan and Murray 1990), and in tests of visual recognition memory (Bachevalier and Mishkin 1986; Meunier *et al.* 1997). It is likely that the OFC and the perirhinal cortex play different roles in overall memory function. The OFC may support associations between stimuli and reward, while the perirhinal cortex and other medial temporal structures are more important for recognition memory as such.

## Outputs to hypothalamus and midbrain

Although the medial prefrontal network is not substantially connected with sensory systems, it can be characterized by substantial outputs to visceral control structures in the hypothalamus and midbrain, including the periaqueductal gray (PAG) (An *et al.* 1998; Öngür *et al.* 1998; Rempel-Clower and Barbas 1998; Freedman *et al.* 2000) (Fig. 3.5). Both of these brain regions serve to coordinate several aspects of visceral function, and it is likely that the cortical projections are a major pathway for forebrain modulation of bodily reactions.

Although the strongest projections to the hypothalamus are from areas 25 and 32 in the medial prefrontal cortex, there are also projections from all of the areas of the medial network, including the caudolateral orbital areas Iai and 12o (Öngür *et al.* 1998) (Fig. 3.5). There are some differences between the projections from different parts of the medial network. The fibers from the medial prefrontal cortex are distributed to both medial and lateral hypothalamic nuclei and areas, suggesting that the cortex influences both autonomic and endocrine functions. Fibers from areas 13a, Iai, and 12o, on the other hand, are restricted to the lateral hypothalamus. There are relatively few projections to the hypothalamus from areas of the orbital network, but the agranular insular areas Iam, Iapm, and Ial have light projections to the caudal part of the lateral hypothalamus (Öngür *et al.* 1998).

Many of the projections to the hypothalamus continue caudally into the ventral midbrain and tegmentum, eventually reaching the PAG (An *et al.* 1998; Öngür *et al.* 1998). As in the hypothalamus, the projections to the midbrain arise almost exclusively from the medial prefrontal network, and from adjacent related areas, such as the dorsomedial prefrontal area 9 and the dorsal temporal pole; there are very few, if any, fibers from the orbital network to the midbrain (Fig. 3.5).

The axons from the medial network ramify within the ventral tegmental area and the pars compacta of the substantia nigra, and appear to form a moderate terminal field there before continuing onto the midbrain tegmentum and PAG (An *et al.* 1998; Öngür *et al.* 1998; Freedman *et al.* 2000). Many of the axons also ramify around the serotonergic cells of the median raphe and the cholinergic cells of the peripeduncular pontine tegmental nucleus. A cluster of axons from area 25 has been reported to aggregate around

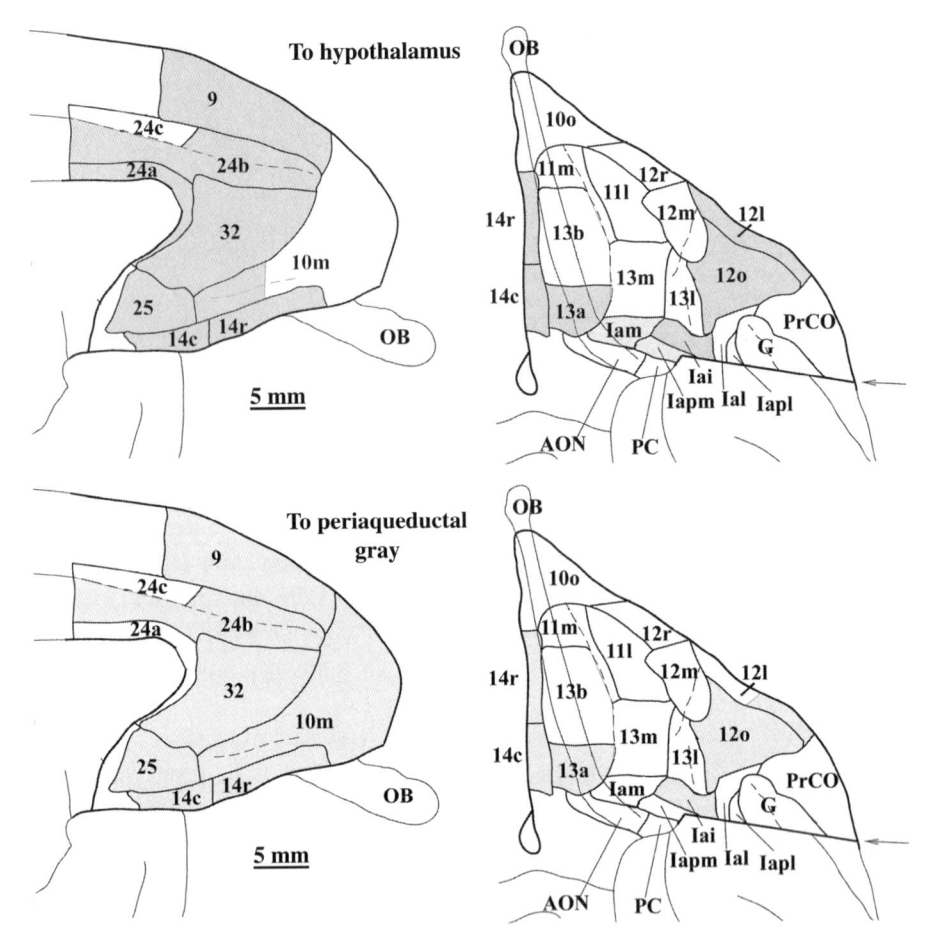

**Fig. 3.5** Areas that project to the hypothalamus and periaqueductal gray, indicated by the shading. Note that in both cases, the outputs to these visceral control structures arises from areas of the medial prefrontal network. Based on data from An et al. (1998) and Öngür et al. (1998).

a group of tyrosine hydroxylase immunoreactive cells in the lateral parabrachial nucleus (Freedman et al. 2000).

Within the PAG, different parts of the medial network are related to different components (or columns). The medial prefrontal areas 25, 32, and 10m project to both the dorsolateral and ventrolateral columns, while the caudolateral orbital areas Iai and 12o project only to the ventrolateral column (An et al. 1998) (Fig. 3.1). Just as the medial and lateral hypothalamus are thought to have different roles in visceral control, the columns in the PAG have been associated with different functions (Bandler et al. 2000). In both cases, the differential projections suggest that different parts of the medial network have different,though related functions.

## Thalamic and striatal connections

In addition to their cortical connections, the orbital cortical areas have specific connections with the thalamus and striatum. The principal thalamic nucleus related to the OFC is the medial, magnocellular part of the mediodorsal nucleus, but the totality of the connections is complex. Large injections of retrograde axonal tracers in the OFC label cells in many nuclei in the medial thalamus, including components of the anterior thalamic group, several midline and intralaminar nuclei, and the medial pulvinar, as well as the mediodorsal nucleus (e.g., Cavada *et al.* 2000). Many of these, including the anteromedial nucleus and many of the midline nuclei, target the areas of the medial network (Carmichael and Price 1995a). The anteromedial nucleus connects with approximately the same zone along the medial edge of the OFC that is related to the subiculum and posterior parahippocampal cortex (Carmichael and Price 1995a; Kondo *et al.* 2005).

Within the mediodorsal nucleus, different zones are connected to the orbital and medial networks (Ray and Price 1993). The medial network is connected to the dorsomedial and dorsocaudal region of the nucleus, while the orbital network is connected to

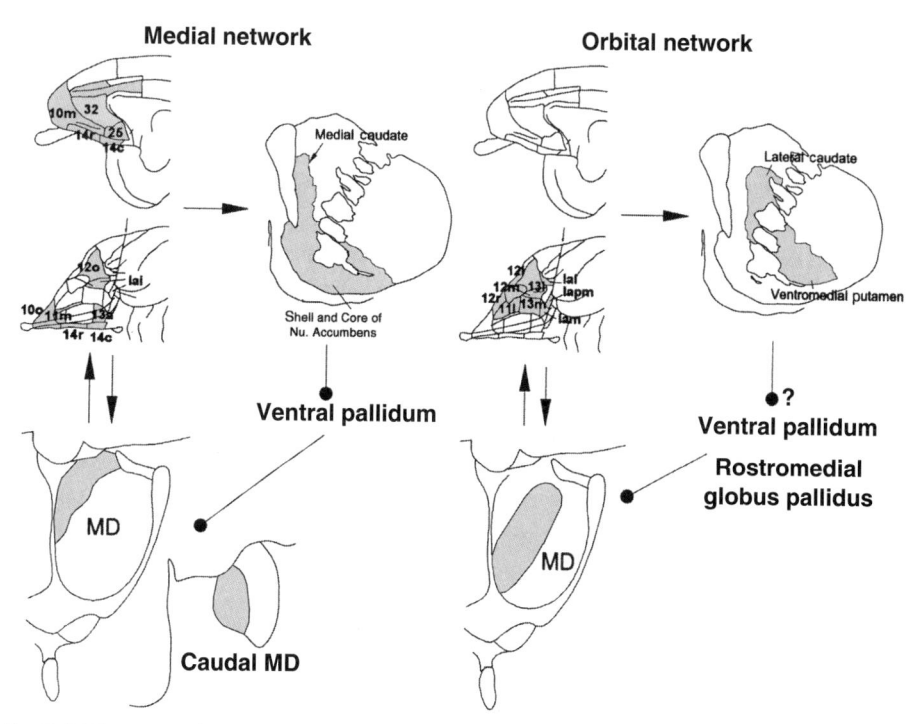

**Fig. 3.6** Diagrammatic summary of the connections of the orbital and medial prefrontal networks with the mediodorsal nucleus of the thalamus and with the striatum. Although all of the details are not clear, it appears that there are two relatively separate cortico-striato-pallido-thalamic loops for the orbital and medial networks. Based on data from Russchen *et al.* (1987), Ray and Price (1993), and Ferry *et al.* (2000).

the ventromedial region of the nucleus. It is still uncertain how much overlap there is between these two regions, but the distinction appears clear. These regions of the mediodorsal nucleus also receive excitatory inputs from the olfactory cortex, amygdala, entorhinal cortex, subiculum and other limbic areas (Ray and Price 1993).

There is a comparable separation in the projections of the orbital and medial networks to the striatum. The areas of the medial network project into the classic ventromedial or 'limbic' striatum, including the nucleus accumbens, rostromedial caudate nucleus, and ventral putamen (Ferry et al. 2000; Fig. 3.6). This striatal region is the same that receives input from the amygdala (Russchen et al. 1985). In comparison, the orbital network areas, especially areas 12l, 12m, 12r, 11l, 13m, and 13l in the central and lateral OFC, project to a more central part of striatum, dorsolateral to the region related to the medial network (Ferry et al. 2000).

These striatal regions project to the ventral pallidum and rostral globus pallidus, which in turn, projects to the medial part of the mediodorsal thalamic nucleus (Russchen et al. 1987; Ray and Price 1992). The details of the pathway have not yet been fully worked out, but it is likely that there are two relatively closed cortico-striato-pallido-thalamic loops, one for the medial network and one for the orbital network (Öngür and Price 2000; Fig. 3.6). As in other striato-pallido-thalamic systems, the synapses at both the pallidum and mediodorsal thalamus are inhibitory (Kuroda and Price 1991).

## 3.2. Conclusion

Taken together, these results indicate that the OFC, and the adjacent medial prefrontal cortex, can be divided into two distinct but inter-related systems, which have been termed the orbital and medial prefrontal networks. These networks were defined on the basis of intrinsic cortico-cortical connections, but they also reflect connections with other parts of the cerebral cortex and with subcortical structures. The two systems are closely related to each other, and can only be recognized by experiments with relatively small, restricted injections of axonal tracers, analyzed with careful attention to architectonic areas.

The orbital network is characterized by its sensory inputs from almost all of the sensory modalities. These inputs link the areas of the orbital network with the sensory association cortex in the inferior temporal cortex and the insula, as well as the primary olfactory and taste/visceral cortical areas. It is uncertain whether there is an auditory input to the orbital network, but there are inputs to related ventrolateral prefrontal cortex. Many of these sensory inputs appear to be related to assessment of food. In addition, to their sensory function, however, orbital neurons code for reward and other affective aspects of stimuli.

The medial network, which includes areas along the gyrus rectus on the medial edge of the OFC and a caudolateral orbital region, does not have obvious sensory inputs. Instead, it is characterized by outputs to the visceral control structures in the hypothalamus and brainstem, and is involved in cortical modulation of visceral function. In addition, this system is involved in mood and emotional behavior. It is substantially interconnected with limbic areas, including the amygdala, entorhinal cortex, and hippocampus, and with a wider cortical system that includes the rostral part of the superior temporal cortex, the parahippocampal cortex, and the posterior cingulate/retrosplenial cortex.

## References

Allen, G. V., Saper, C. B., Hurley, K. M., and Cechetto, D. F. (1991) Organization of visceral and limbic connections in the insular cortex of the rat. *J Comp Neurol* **311**:1–16.

Amaral, D. G. and Price, J. L. (1984) Amygdalo-cortical projections in the monkey (*Macaca fascicularis*). *J Comp Neurol* **230**:465–496.

An, X., Bandler, R., Öngür, D., and Price, J. L. (1998) Prefrontal cortical projections to longitudinal columns in the midbrain periaqueductal gray in macaque monkeys. *J Comp Neurol* **401**:455–479.

Bachevalier, J. and Mishkin, M. (1986) Visual recognition impairment follows ventromedial but not dorsolateral prefrontal lesions in monkeys. *Exp Brain Res* **20**:249–261.

Bandler, R., Keay, K. A., Floyd, N., and Price, J. (2000) Central circuits mediating patterned autonomic activity during active vs. passive emotional coping. *Brain Res Bull* **53**:95–104.

Barbas, H. (1993) Organization of cortical afferent input to orbitofrontal areas in the rhesus monkey. *Neuroscience* **56**:841–864.

Barbas, H. and Pandya, D. N. (1989) Architecture and intrinsic connections of the prefrontal cortex in the rhesus monkey. *J Comp Neurol* **286**:353–375.

Barbas, H. and Blatt, G. J. (1995) Topographically specific hippocampal projections target functionally distinct prefrontal areas in the rhesus monkey. *Hippocampus* **5**:511–533.

Barbas, H. and de Olmos, J. (1990) Projections from the amygdala to basoventral and mediodorsal prefrontal regions in the rhesus monkey. *J Comp Neurol* **300**:549–571.

Barbas, H., Ghashghaei, H., Dombrowski, S.M., and Rempel-Clower, N. L. (1999) Medial prefrontal cortices are unified by common connections with superior temporal cortices and distinguished by input from memory-related areas in the rhesus monkey. *J Comp Neurol* **410**:343–367.

Barbas, H. (1988) Anatomic organization of basoventral and mediodorsal visual recipient prefrontal regions in the rhesus monkey. *J Comp Neurol* Oct 15; **276**(3):313–342.

Beckstead, R. M., Morse, J. R., and Norgren, R. (1980) The nucleus of the solitary tract in the monkey: projections to the thalamus and brain stem nuclei. *J Comp Neurol* **190**:259–282.

Blatt, G. J., Pandya, D. N., and Rosene, D. L. (2003) Parcellation of cortical afferents to three distinct sectors in the parahippocampal gyrus of the rhesus monkey: An anatomical and neurophysiological study. *J Comp Neurol* **466**:161–179.

Buckner, R. L., Kelley, W. M., and Petersen, S. E. (1999) Frontal cortex contributes to human memory formation. *Nat Neurosci* **2**:311–314.

Carmichael, S. T. and Price, J. L. (1994) Architectonic subdivision of the orbital and medial prefrontal cortex in the macaque monkey. *J Comp Neurol* **346**:366–402.

Carmichael, S. T. and Price, J. L. (1995a) Limbic connections of the orbital and medial prefrontal cortex in macaque monkeys. *J Comp Neurol* **363**:615–641.

Carmichael, S. T. and Price, J. L. (1995b) Sensory and premotor connections of the orbital and medial prefrontal cortex. *J Comp Neurol* **363**:642–664.

Carmichael, S. T. and Price, J. L. (1996) Connectional networks within the orbital and medial prefrontal cortex of macaque monkeys. *J Comp Neurol* **371**:179–207.

Carmichael, S. T., Clugnet, M.-C. and Price, J. L. (1994) Central olfactory connections in the macaque monkey. *J Comp Neurol* **346**:403–434.

Cavada, C., Company, T., Tejedor, J., Cruz-Rizzolo, R. J., and Reinoso-Suarez, F. (2000) The anatomical connections of the macaque monkey orbitofrontal cortex. A review. *Cereb Cortex* **10**:220–242.

Cechetto, D. F. and Saper, C. B. (1987) Evidence for a viscerotopic sensory representation in the cortex and thalamus in the rat. *J Comp Neurol* **262**:27–45.

Ferry, A. T., Ongur, D., An, X., and Price, J.L. (2000) Prefrontal cortical projections to the striatum in macaque monkeys: evidence for an organization related to prefrontal networks. *J Comp Neurol* **425**:447–470.

Freedman, L. J., Insel, T. R., and Smith, Y. (2000) Subcortical projections of area 25 (subgenual cortex) of the macaque monkey. *J Comp Neurol* **421**:172–188.

Friedman, D. P., Murray, E. A., O'Neill, J.B., and Mishkin, M. (1986) Cortical connections of the somatosensory fields of the lateral sulcus of macaques: evidence for a corticolimbic pathway for touch. *J Comp Neurol* **252**:323–347.

Gaffan, D. and Murray, E. A. (1990) Amygdalar interaction with the mediodorsal nucleus of the thalamus and the ventromedial prefrontal cortex in stimulus-reward associative learning in the monkey. *J Neurosci* **10**:3479–3493.

Hackett, T. A., Stepniewska, I., and Kaas, J. H. (1998) Subdivisions of auditory cortex and ipsilateral cortical connections of the parabelt auditory cortex in macaque monkeys. *J Comp Neurol* **394**:475–495.

Hackett, T. A., Stepniewska, I., and Kaas, J. H. (1999) Prefrontal connections of the parabelt auditory cortex in macaque monkeys. *Brain Res* **817**:45–58.

Hackett, T. A., Preuss, T. M., and Kaas, J. H. (2001) Architectonic identification of the core region in auditory cortex of macaques, chimpanzees, and humans. *J Comp Neurol.* **441**:197–222.

Heimer, L. (1972) The olfactory connections of the diencephalon in the rat. An experimental light- and electron-microscopic study with special emphasis on the problem of terminal degeneration. *Brain Behav Evol* **6**:484–523.

Kobayashi, Y., and Amaral, D. G. (2003) Macaque monkey retrosplenial cortex: II. Cortical afferents. *J Comp Neurol* **466**:48–79.

Kondo H., Saleem K. S., Price, J. L. (2003). Differential connections of the temporal pole with the orbital and medial prefrontal networks in macaque monkeys. *J Comp Neurol* **465**:499–523.

Kondo, H., Saleem, K. S., and Price, J. L. (2005) Differential Connections of the Perirhinal and Parahippocampal Cortical Areas with the Orbital and Medial Prefrontal Networks in Macaque Monkeys *J Comp Neurol* in press.

Krettek, J. E. and Price, J. L. (1977) The cortical projections of the mediodorsal nucleus and adjacent thalamic nuclei in the rat. *J Comp Neurol* **171**:157–191.

Kuroda, M. and Price, J. L. (1991) Synaptic organization of projections from basal forebrain structures to the mediodorsal thalamic nucleus of the rat. *J Comp Neurol* **303**:513–533.

Morecraft, R. J., Geula, C., and Mesulam, M-M. (1992) Cytoarchitectonic and Neural Afferents of orbitofrontal cortex in the brain of the monkey. *J Comp Neurol* **323**:341–358.

Meunier, M., Bachevalier, J., Mishkin, M. (1997) Effects of orbital and anterior cingulate lesions on object and spatial memory in rhesus monkeys. *Neuropsychologia* **35**:999–1015.

Öngür, D., An, X., and Price, J. L. (1998) Prefrontal cortical projections to the hypothalamus in macaque monkeys. *J Comp Neurol* **401**:480–505.

Öngür, D. and Price, J. L. (2000) The organization of networks within the orbital and medial prefrontal cortex of rats, monkeys and humans. *Cereb Cortex* **10**:206–219.

Passingham, R. E., Toni, I., and Rushworth, M. F. (2000) Specialisation within the prefrontal cortex: the ventral prefrontal cortex and associative learning. *Exp Brain Res* **133**:103–113.

Petrides, M. (1996) Specialized systems for the processing of mnemonic information within the primate frontal cortex. *Philos Trans R Soc Lond B Biol Sci* **351**:1455–1461.

Petrides, M. (2005) Lateral prefrontal cortex: architectonic and functional organization. *Philos Trans R Soc Lond B Biol Sci* **360**:781–795.

Price, J. L. (1973) An autoradiographic study of complementary laminar patterns of termination of afferent fibers to the olfactory cortex. *J Comp Neurol* **150**:87–108.

Price, J. L. (1985) Beyond the olfactory cortex: olfactory-related areas in the neocortex, thalamus and hypothalamus. *Chem Sens* **10**: 239–258.

Price, J. L. (2003) The Olfactory System. In: *The Human Nervous System*, 2nd ed., Paxinos, G. (ed.) San Diego: Academic Press.

Price, J. L. and Slotnick, B. M. (1983) Dual olfactory representation in the rat thalamus: an anatomical and electrophysiological study. *J Comp Neurol* **215**:63–77.

Ray, J. P. and Price, J. L. (1992) The organization of the thalamocortical connections of the mediodorsal thalamic nucleus in the rat, related to the ventral forebrain-prefrontal cortex topography. *J Comp Neurol* **323**:167–197.

Ray, J. P. and Price, J. L. (1993) The organization of projections from the mediodorsal nucleus of the thalamus to orbital and medial prefrontal cortex in macaque monkeys. *J Comp Neurol* **337**:1–31.

Rempel-Clower, N. L. and Barbas, H. (1998) Topographic organization of connections between the hypothalamus and prefrontal cortex in the rhesus monkey. *J Comp Neurol* **398**:393–419.

Rolls, E. T. (2000) The orbitofrontal cortex and reward. *Cereb Cortex* **10**:284–294.

Rolls, E. T. (2005) Taste, olfactory, and food texture processing in the brain, and the control of food intake. *Physiol Behav* **85**:45–56.

Romanski, L. M. (2004) Domain specificity in the primate prefrontal cortex. *Cogn Affect Behav Neurosci* Dec; **4**(4):421–9.

Romanski, L. M. and Goldman-Rakic, P. S. (2002) An auditory domain in primate prefrontal cortex. *Nat Neurosci* **5**:15–16.

Romanski, L. M., Tian, B., Fritz, J., Mishkin, M., Goldman-Rakic, P. S., Rauschecker, J. P. (1999) Dual streams of auditory afferents target multiple domains in the primate prefrontal cortex. *Nat Neurosci* **2**:1131–136.

Romanski, L. M., Averbeck, B. B., Diltz, M. (2005) Neural representation of vocalizations in the primate ventrolateral prefrontal cortex. *J Neurophysiol* **93**(2):734–747.

Russchen, F. T., Bakst, I., Amaral, D. G., and Price, J. L. (1985) The amygdalostriatal projections in the monkey. An anterograde tracing study. *Brain Res* **329**:241–257.

Russchen, F. T., Amaral, D. G., Price, J. L. (1987) The afferent input to the magnocellular division of the mediodorsal thalamic nucleus in the monkey, *Macaca fascicularis*. *J Comp Neurol* **256**:175–210.

Schultz, W., Tremblay, L., and Hollerman, J. R. (2000) Reward processing in primate orbitofrontal cortex and basal ganglia. *Cereb Cortex* **10**:272–284.

Squire, L. R., Stark, C.E.L., and Clark, R. E. (2004) The medial temporal lobe. *Annu Rev Neurosci* **27**:279–306.

Tanabe, T., Yarita, H., Iino, M., Ooshima, Y., and Takagi, S. F. (1975) An olfactory projection area in orbitofrontal cortex of the monkey. *J Neurophysiol* **38**:1269–1283.

Webster, M. J., Bachevalier, J., and Ungerleider, L. G. (1994) Connections of inferior temporal areas TEO and TE with parietal and frontal cortex in macaque monkeys. *Cereb Cortex* **4**:470–483.

Wilson, F. A., Scalaidhe, S. P., and Goldman-Rakic, P. S. (1993) Dissociation of object and spatial processing domains in primate prefrontal cortex. *Science* **260**:1955–1958.

Yarita, H., Iino, M., Tanabe, T., Kogure, S., and Takagi, S. F. (1980) A transthalamic olfactory pathway to orbitofrontal cortex in the monkey. *J Neurophysiol* **43**:69–85.

Chapter 4

# Sequential and parallel circuits for emotional processing in primate orbitofrontal cortex

Helen Barbas and Basilis Zikopoulos

## 4.1. Overview

The prefrontal cortex in primates guides behavior by selecting information through a vast communication network with cortical and subcortical structures (Goldman-Rakic 1988; Fuster 1993; Barbas 1995a; Petrides 1996; Barbas 2000a). In addition to its long-held position as the central node for executive processing associated with lateral prefrontal areas, it is now clear that the orbitofrontal and medial sectors of the prefrontal cortex have rich connections with structures that process emotions (Barbas 1995a; Price et al. 1996; Barbas et al. 2002). The three distinct prefrontal sectors, namely the lateral, orbitofrontal and medial, are interconnected, and provide the basis for interaction of pathways underlying cognitive and emotional processes. The anatomic bond between these functionally distinct areas underscores the biologic significance of the synthesis of cognitive and emotional processes. Disruption of this linkage has profound effects on behavior, as seen in several psychiatric and neurologic diseases affecting the orbitofrontal cortex (OFC).

This chapter focuses on the circuits that underlie the motivational component for the selection of relevant signals in behavior. Decisions and actions are inextricably linked to the emotional significance of events, and the OFC has a key role in this process. This idea is supported by striking deficits in emotional behavior and social interactions in both human and nonhuman primates after orbitofrontal damage (Damasio et al. 1994; Damasio 1994). Several features of the circuits of the OFC, particularly its connections with temporal structures, suggest that it is in an ideal position to convey signals associated with the emotional significance of events. Ultimately, the OFC communicates with lateral and caudal medial prefrontal cortices in synergistic functions in decision and action in behavior. We address below the key features of circuits that likely underlie the role of the OFC in cognitive-emotional interactions.

## 4.2. OFC: location and definitions

The OFC encompasses a large area on the rostral basal surface of the brain (Fig. 4.1). In macaque monkeys it is found directly behind the eye socket, and assumes a concave

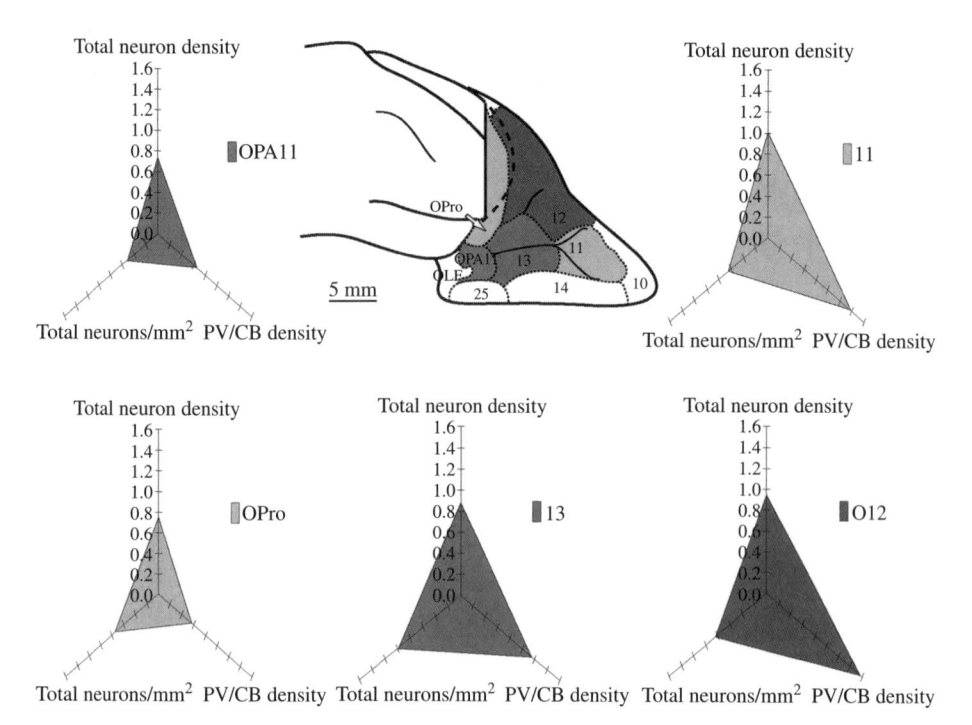

**Fig. 4.1** (Also see Color Plate 2.) Quantitative architectonic profiles of orbitofrontal cortices of the rhesus monkey. Areas are shown on the basal surface of the right hemisphere (top center). Fingerprints of the distinct orbitofrontal areas were constructed quantitatively on the basis of features that discriminate among areas, including density of neurons, total neurons/mm² and the ratio of parvalbumin (PV) to calbindin (CB) neurons. The temporal pole at top center was rendered transparent (thick dotted lines) to show the underlying orbitofrontal cortex. The map is according to Barbas and Pandya (1989). Abbreviations: O12, orbital area 12; OPro, orbital proisocortex (dysgranular cortex); OPAll, orbital periallocortex (agranular cortex).

shape around the orbit. The OFC is bordered caudally by the olfactory areas and the anterior insula, and rostrally it extends to the frontal pole. The most anterior part encompasses area 10, the only prefrontal area to have orbital, medial and lateral components.

This chapter focuses primarily on the caudal part of the OFC, the part with the strongest and most specialized anatomic interactions with temporal lobe structures. References to the OFC here do not include the basal (orbital) part of area 10, whose structural features and connections differ from other orbitofrontal areas. In addition, the discussion does not include the cortex of the medial wall, namely areas 9, 14 and 10 rostrally, or areas 24, 32 and 25 in the anterior cingulate. The areas of the medial wall are not geographically part of the OFC and have a distinct set of connections. In the rhesus monkey, areas 14 and 25 extend over a short distance to the medial edge of the basal surface (Barbas and Pandya 1989). The connections of the orbital parts of areas 14 and 25 are similar to their medial counterparts, and are not included in the descriptions of the OFC.

Areas 14, 25, and 32 share some features with the OFC, but also differ from the OFC in several key connections. We suggest that medial and orbitofrontal cortices have distinct roles in emotional function and cognitive-emotional interactions, as discussed towards the end of this chapter.

The focus of this review is on those orbitofrontal cortices that are situated lateral to areas 14 and 25, as shown in Figure 4.1. The orbitofrontal region is composed of several types of cortices, including an agranular area, which has only three identifiable layers, dysgranular areas, which have four identifiable layers, and granular (eulaminate) areas, which have six distinct layers. Agranular and dysgranular type cortices are collectively called limbic. The distinction of eulaminate from limbic cortices has several implications for the organization of connections and ultimately the function of these structurally distinct cortices (Dombrowski et al. 2001). On the one hand, higher laminar complexity increases local processing power, but, on the other hand, limbic areas have widespread connections and may exercise a tonic influence on the neuraxis (Barbas 1995b). The significance of cortical type is underscored by evidence that the pattern of cortico-cortical connections can be predicted on the basis of the broad laminar features of the interconnected areas (Barbas 1986; Barbas and Rempel-Clower 1997; Barbas 2000b), as discussed under the section titled 'The role of the OFC in emotional memory'.

The rostral extent of the the OFC includes areas 11 and the orbital part of area 12, both of which are eulaminate areas, characterized by six layers, including a distinct granular layer 4. The most caudal extent of the OFC includes the agranular cortex, which lacks a granular layer 4, and has been named area OPAll (Barbas and Pandya 1989), or area OFap (Morecraft et al. 1992), which are coextensive. The agranular cortex has been subdivided into areas13a, Iam, Iapm by Carmichael and Price (1994). The agranular cortex is bordered laterally and rostrally by the dysgranular cortex, which has an incipient granular layer 4 (area OPro in Barbas and Pandya (1989); OFdg in Morecraft et al. (1992); Iai, Ial, Iapl in Carmichael and Price (1994). An adjacent region, area 13, is also dysgranular in type Barbas and Pandya (1989). Area 13 is incorporated into area OFdg in the map of Morecraft et al. (1992), and is largely coextensive with areas 13b, 13m, 13l in the map of Carmichael and Price (1994). References to architectonic areas in this chapter and in Figure 4.1 are according to the map of Barbas and Pandya (1989), which was modified from the classic map of Walker (1940).

There are both structural and connectional features to suggest that the posterior and anterior orbitofrontal cortices (areas OPAll, OPro and 13, O12, 11, respectively) are distinct. From a structural perspective, the identity of individual areas can be determined objectively by analyzing their architectonic features using unbiased quantitative methods (Dombrowski et al. 2001). As shown in Fig. 4.1, areal 'fingerprints' based on highly informative cellular features can distinguish among architectonic areas in the orbitofrontal region. These features include neuronal density, the total number of neurons found under 1 mm$^2$ (which depends on the depth of the cortex), and the distribution of two classes of inhibitory neurons, identified by the expression of the calcium binding proteins parvalbumin (PV) and calbindin (CB). The quantitative architectonic profiles demonstrate the individuality of orbitofrontal areas, but also suggest similarities among areas. Fingerprints based on these features suggest that the posterior orbitofrontal areas OPAll and OPro have similar architectonic profiles, as do areas O12 and 11. Area 13 occupies an intermediate position between the caudally situated areas OPro and OPAll, on one hand, and the rostrally situated areas O12 and 11, on the other. Consistent with the architectonic findings, area 13 shares connectional features with its rostral and caudal neighbors. Moreover, although area 13 and area OPro are both dysgranular and have been considered to be one area (Morecraft et al. 1992), their quantitative architectonic profiles suggest that they are distinct.

We present evidence that connectional features distinguish the posterior (areas OPAll, OPro) from the anterior orbitofrontal cortices. The topography and pattern of connections of the posterior OFC suggest that they are earlier processing areas for emotional information than the anterior orbitofrontal areas. Throughout this chapter we focus on the most striking features of posterior OFC, as they pertain to emotions. Posterior orbitofrontal cortices are connected with anterior orbitofrontal areas, which, in turn, are connected with lateral prefrontal areas, in pathways that likely underlie the flow of information for the synthesis of cognitive and emotional processes.

## 4.3. **The posterior OFC is a global environmental integrator**

### Processing signals from the external environment

Analysis of the features of the environment is prerequisite to emotional processing. The OFC, in general, is distinguished for its multimodal circuitry, comparable in this respect to the perirhinal region in the temporal lobe (Van Hoesen et al. 1972; Van Hoesen 1975; Insausti et al. 1987; Suzuki and Amaral 1994), with which it has strong connectional and functional ties. The orbitofrontal cortices are enriched with projections from visual, auditory, somatosensory, and polymodal cortices (Morecraft et al. 1992; Barbas 1993; Carmichael and Price 1995b). While all orbitofrontal areas receive projections from several unimodal sensory association cortices, there are foci within the OFC that receive more robust projections from one modality over the others (e.g., Morecraft et al. 1992; Barbas 1993; Baylis et al. 1995; Carmichael and Price 1995b; Cavada et al. 2000; Barbas et al. 2002). However, it should be noted that neurophysiological studies show considerable overlap of sensory inputs within the OFC and many neurons respond to stimuli from several modalities (Kringelbach and Rolls 2004).

The posterior orbitofrontal areas (areas OPro, OPAll) have the most multimodal circuits. They are distinguished for their unique connections with olfactory cortices (Barbas 1993; Carmichael et al. 1994), in addition to connections with visual, auditory, somatosensory, and gustatory cortices (Takagi 1986), as shown in an example in Fig. 4.2. The anterior orbitofrontal areas also receive input from visual, auditory, and somatosensory areas. In addition, area 12 and rostral area 13 receive projections from

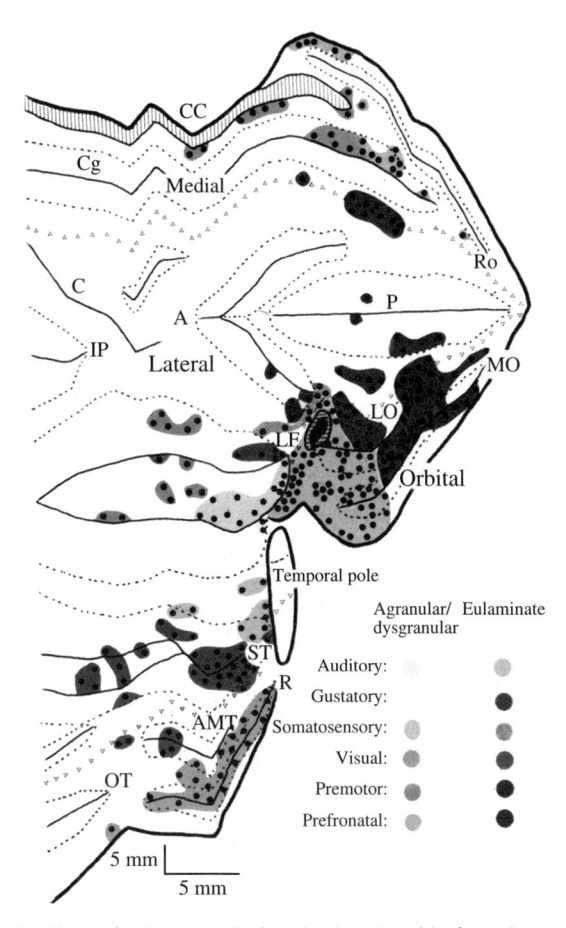

**Fig. 4.2** (Also see Color Plate 3.) Diverse cortical projections in orbitofrontal cortex. Unfolded two-dimensional map of the prefrontal cortex showing the origin of projection neurons (black dots) directed to orbitofrontal area OPro on the basal surface (black area, injection site; surround horizontal lines, halo of injection site). Projection neurons originate in visual association (red overlay), gustatory (purple), somatosensory (green), auditory (yellow), premotor (brown) and prefrontal cortices (blue). Area OPro receives some projections from the olfactory prepiriform cortex as well (not shown). Dark shades show eulaminate areas, light shades agranular/dysgranular (limbic) cortices. Triangles separate the medial (top) from the lateral (center) surfaces, and the lateral from the basal (bottom) surfaces. Vertical stripes (top) show the corpus callosum. A, arcuate sulcus; AMT, anterior middle temporal dimple; C, central sulcus; CC, corpus callosum; Cg, cingulate sulcus; IP, intraparietal sulcus; LF, lateral fissure; LO, lateral orbital sulcus; MO, medial orbital sulcus; OT, occipitotemporal sulcus; P, principal sulcus; R, rhinal sulcus; Ro, rostral sulcus; ST, superior temporal sulcus.

gustatory areas. However, the rostral orbitofrontal areas do not receive direct projections from olfactory areas (Morecraft *et al.* 1992; Barbas 1993; Carmichael *et al.* 1994).

The sensory input to all orbitofrontal cortices originates from late-processing sensory association areas (Barbas 1992; Barbas 1995a). The sole exception to this pattern is the

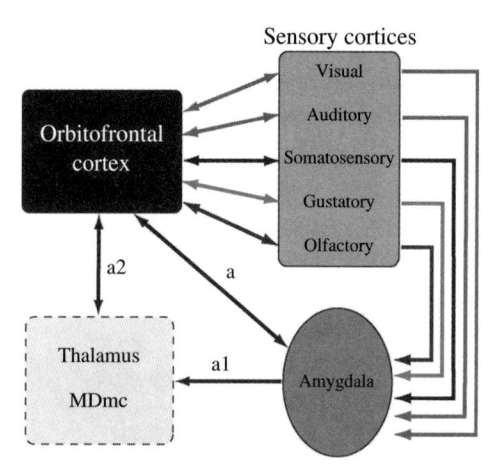

**Fig. 4.3** Direct and indirect sensory input to orbitofrontal cortex. Projections from sensory association cortices reach orbitofrontal cortex as well as the amygdala, which has strong bidirectional connections with orbitofrontal cortex (a). The amygdala also projects to MDmc of the thalamus (a1), which has strong bidirectional connections with orbitofrontal cortex (a2). The connections of the amygdala with sensory areas are presumed to be bidirectional as well (not shown).

olfactory input to the posterior OFC, which emanates from the piriform cortex and the anterior olfactory nucleus (Barbas 1993) that are considered to be a part of the primary olfactory cortex (Price 1990). Unlike the large extent of cortical areas devoted to the visual, auditory and somatosensory modalities, the olfactory cortex is comparatively small in macaque monkeys.

### Processing signals from the internal environment

The above discussion shows that through their connections with sensory association cortices, the OFC can sample the entire sensory periphery (Fig. 4.3). In addition, input from the internal environment, related to drives and motives, is necessary for emotional processing. In this respect, all orbitofrontal areas receive particularly robust projections from cortical limbic areas in the medial temporal lobe and the cingulate cortex (Cavada et al. 2000; Barbas 2000a), which likely provide information about the internal environment. In addition, as discussed below, the OFC has a special relationship with the amygdala, and strong connections with the medial thalamic nuclei that belong to the limbic thalamus. To sum up, the OFC can be viewed as an integrator of the external and internal environments, a feature that may be necessary for signaling the emotional significance of events.

### Specialized bidirectional connections linking OFC with the amygdala may underlie emotional processing

One of the most distinctive features of the OFC is its anatomic and functional linkage with the amygdala, an anterior temporal structure with a key role in emotions (Nishijo *et al.* 1988; Davis 1992; Damasio 1994; LeDoux 1996). The anatomic connections of the OFC with the amygdala are robust and bidirectional (Nauta 1961; Pandya *et al.* 1973; Jacobson and Trojanowski 1975; Aggleton *et al.* 1980; Porrino *et al.* 1981; Van Hoesen 1981; Amaral and Price 1984; Barbas and De Olmos 1990; Morecraft *et al.* 1992; Carmichael and Price 1995a; Chiba *et al.* 2001). The projections from the OFC are organized along a rostrocaudal gradient. The strongest connections in this circuit involve the posterior sector of the OFC and the posterior half of the amygdala (Ghashghaei and Barbas 2002; Barbas and De Olmos 1990; Stefanacci and Amaral 2002). Moreover, the amygdala resembles the OFC in its wealth of circuits with sensory areas, as summarized in Fig. 4.3. Like the OFC, the amygdala receives projections from the sensory association cortices representing each of the sensory modalities (Herzog and Van Hoesen 1976; Turner *et al.* 1980). Like the OFC, the cortical sensory projections to the amygdala originate preferentially in areas representing late-processing sensory cortices, which are specialized in processing the features of stimuli and their memory.

## Direct and indirect sensory input to OFC may be necessary for emotional processing

The above discussion suggests that sensory input reaches the OFC directly by projections from sensory association cortices, and indirectly through the amygdala, which receives projections from the same sensory association cortices, as depicted in the diagram in Figure 4.3. However, previous studies did not directly investigate whether the sectors of the amygdala that receive sensory information are the same sites that are connected with orbitofrontal cortices. A recent study addressed this issue in the same experiments, and provided direct evidence that pathways from anterior temporal visual and auditory association cortices, on one hand, and caudal orbitofrontal areas, on the other, occupy overlapping territories within the basal complex of the amygdala (Ghashghaei and Barbas 2002). This evidence suggests that a closely linked triadic network links the amygdala, orbitofrontal, and anterior temporal cortices. Moreover, this circuit implies that sensory input reaches orbitofrontal cortices directly through cortico-cortical pathways (Barbas 1993; Rempel-Clower and Barbas 2000), and indirectly through the amygdala (Barbas 1995a). This highly organized network may form the basis for processing the emotional significance of events within a behavioral context, forming reward associations in cognitive tasks that are inextricably linked with emotional associations (Malkova *et al.* 1997; Schoenbaum *et al.* 1999; Hikosaka and Watanabe 2000; Baxter *et al.* 2000; Wallis and Miller 2003). There is evidence that orbitofrontal areas have an important role in encoding in long term memory changes in reward contingencies that accompany changes in task demands (Tremblay and Schultz 2000). Neurons in the OFC and the amygdala respond to sensory stimuli when they are significant for behavior, and cease to respond when the stimuli lose their motivational value (Lipton *et al.* 1999; Tremblay and Schultz 1999; Schoenbaum *et al.* 2000). This habituation of activity has also been shown in recent functional imaging studies of the human amygdala during repeated exposure to fearful and neutral faces, suggesting the regulation of neural activity in the brain regions involved in detecting novelty (Fischer et al. 2003). The mechanisms for this habituation are unclear, but it is possible that there is a reduction in the allocation of resources to the stimuli of decreasing salience.

The motivational value of the stimuli mentioned above is in many cases dependent on reward expectancy, which is based on information provided by associated reward-predicting stimuli. Numerous studies have shown that neurons in the OFC and the amygdala respond to reward-predicting cues and are also activated in relation to the expectation and detection of reward (Schoenbaum and Eichenbaum 1995; Schoenbaum et al. 1999; Rolls 2000; Schultz et al. 2000; Tremblay and Schultz 2000; Baxter and Murray 2002; Schoenbaum et al. 2003; Saddoris et al. 2005).

## Relationship of OFC to striatal and brainstem reward centers

The striatum (including the nucleus accumbens) is another brain region that responds similarly to the OFC, but is additionally activated during the preparation, initiation and execution of movements, related to the expected reward (Schultz et al. 2000). As lesion studies have indicated, the nucleus accumbens (NAc) is critical for encoding and using information regarding the learned significance of cues predictive of reward or aversive outcome (Schoenbaum and Setlow 2003). Just like the OFC and the amygdala, the NAc is a site of convergence for numerous sources of motivational information, including the basolateral complex of the amygdala (BLA), the OFC, the medial prefrontal cortex and dopaminergic ventral tegmental area (VTA) neurons (Groenewegen et al. 1990; Lynd-Balta and Haber 1994; Haber et al. 1995; Phillips et al. 2003). Studies in the rat have shown that dopamine terminals from the VTA synapse on PFC pyramidal cells that project to the NAc. Moreover, in rats there seems to be a selective PFC synaptic input to GABA-containing mesoaccumbens and dopamine-containing mesocortical VTA neurons (Carr et al. 1999; Carr and Sesack 2000). A direct path from BLA sends glutamatergic afferents to the NAc that overlap anatomically and functionally with dopaminergic afferents in this nucleus. Indirectly, the central nucleus of the amygdala probably disinhibits dopaminergic neurons in the VTA, which project to the NAc (Phillips et al. 2003). Thus, it appears that information about the value of stimuli is conveyed from the amygdala to the NAc, where it can be potentiated or 'gain-amplified' by dopamine (Cardinal et al. 2002). Although links between the PFC, the amygdala, the ventral striatum and the midbrain dopaminergic neurons exist in the primate brain (Fudge and Haber 2000; Ghashghaei and Barbas 2002), to date there are no detailed studies demonstrating overlaps and other possible interactions of those pathways. This is in contrast to the compelling literature analyzing such interactions in the rat brain and therefore, any cross-species comparisons should be made with caution.

Midbrain dopamine neurons, which respond to unpredicted primary rewards, also rapidly adapt to the information provided by reward-predicting stimuli (Schultz et al. 2000) and through interactions with the NAc, the amygdala and the OFC adjust the gain of reward-related responses participating in mechanisms that select the action associated

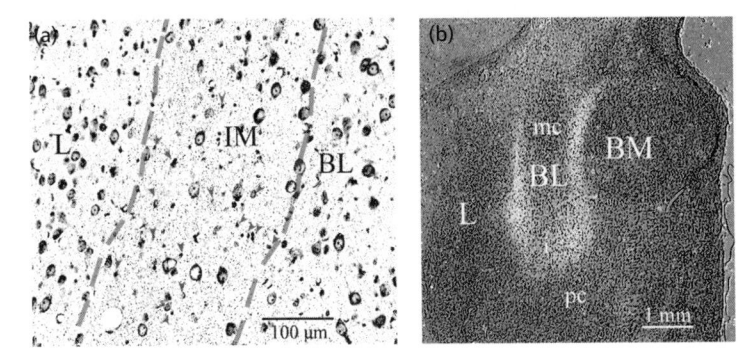

**Fig. 4.4** (Also see Color Plate 4.) Pattern of axonal terminations from posterior orbitofrontal cortex in the amygdala.(a) Brightfield photomicrograph showing the intercalated masses of the amygdala (IM) interposed between the nuclei of the amygdala, which are composed of small neurons and are GABAergic. (b) Darkfield photomicrograph of the section shown in A, showing axons from posterior orbitofrontal area OPro targeting heavily the intercalated masses of the amygdala (white grain), found between the lateral (L), basolateral (BL) and basomedial (BM; also known as accessory basal) nuclei of the amygdala. Adapted from Ghashghaei and Barbas (2002).

with the largest reward (Kalivas and Nakamura 1999; Cardinal *et al.* 2002; Tobler *et al.* 2005). Therefore, the circuitry linking midbrain dopaminergic neurons, NAc, amygdala and OFC is clearly implicated in a variety of reward perception and adaptive behavioral responses. It seems that dopaminergic afferents signal changes in the availability or receipt of rewarding stimuli, the value of which is potentiated in the NAc, whereas the input from the amygdala enables necessary associations with the conditioned stimuli, and afferents from the OFC integrate the information from short-term memory into behavioral responses. Complex interactions between these structures play an important role in forming associations between specific sensory stimuli and biologically significant events that have emotional and motivational valence.

## Distinction of input-output zones in the amygdala linking it with posterior OFC

As discussed above, the connectional studies show that the posterior OFC is robustly connected with the caudal sector of the amygdala, and both structures receive projections from sensory cortices representing each sensory modality. The most striking distinguishing feature of the connections of orbitofrontal areas with the amygdala is in their pattern. Axons from posterior orbitofrontal areas terminate around the borders of the magnocellular basolateral nucleus, forming a dense U-shaped pattern (Ghashghaei and Barbas 2002), as shown in Fig. 4.4. Projection neurons in the amygdala directed to posterior orbitofrontal areas are positioned around the axonal terminations, occupying adjacent sites of the basal complex (Ghashghaei and Barbas 2002). This unique pattern of connectivity suggests a certain degree of segregation of input and output zones in the amygdala that link it with the posterior OFC. No other prefrontal areas show such a remarkable degree of anatomic specialization in the amygdala.

The projections from the posterior OFC are particularly striking for preferentially targeting the outskirts of the basal complex of the amygdala, and terminating most densely not in the basal nuclei themselves, but in the narrow margins interposed between the nuclei of the basal complex (Fig. 4a). The margins include a dense array of small neurons, known as intercalated masses (IM), whose function was unknown in classic studies (De Olmos 1990). It is now evident that neurons in most IM nuclei in macaques are GABAergic (Pitkanen and Amaral 1994), as they are in several other species, and project to the central nucleus of the amygdala, the basal forebrain, and the brainstem (Moga and Gray 1985; Nitecka and Ben Ari 1987; Paré and Smith 1993a, b; Paré and Smith 1994; Pitkanen and Amaral 1994). The central nucleus, in turn, is enriched in inhibitory neurons, and is the main output of the amygdala to hypothalamic and brainstem autonomic structures (Jongen-Relo and Amaral 1998; Saha et al. 2000; Ghashghaei and Barbas 2002).

## Implications of orbitofrontal circuits in behavior

Evidence on the GABAergic nature of IM neurons adds special significance to the pattern of projections from the posterior OFC to the amygdala. Thus, beyond the reciprocity of connections linking orbitofrontal areas with the basal nuclei of the amygdala, caudal orbitofrontal efferents target heavily and unidirectionally the intercalated masses, suggesting a direct influence on an internal system of the amygdala. The characteristic pattern of termination is evident only for axons originating in the posterior OFC. Neither rostral orbitofrontal, nor medial prefrontal areas show this pattern of connection with the amygdala. A previous study includes a figure that shows this pattern (van Hoesen 1981), without reference to the IM, as it predated recent studies on the significance of the IM in inhibitory control within the amygdala (Moga and Gray 1985; Paré and Smith 1993b; Paré and Smith 1994).

On the basis of the above evidence, we summarize the implications of the circuitry from the posterior OFC to the IM of the amygdala in Figure 4.5. This circuitry suggests that orbitofrontal projections have a net effect of suppressing activity in the central nucleus and down-regulating its inhibitory output to hypothalamic and brainstem autonomic centers. The orbitofrontal pathway to the IM thus appears to have a permissive effect on autonomic centers, allowing the activation of hypothalamic and brainstem autonomic centers as the circumstance demands. Activity in hypothalamic and brainstem autonomic centers can be enhanced by direct excitatory projections from orbitofrontal and medial prefrontal areas (Ongur et al. 1998; Rempel-Clower and Barbas 1998; Barbas et al. 2003). Pathways with disinhibitory or excitatory effects on the hypothalamus and brainstem may be recruited when the synergistic activity of the amygdala and the OFC signal an emotionally charged circumstance. This could arise when a potential danger lurks in the environment, such as a loud noise that may suggest a dangerous explosion.

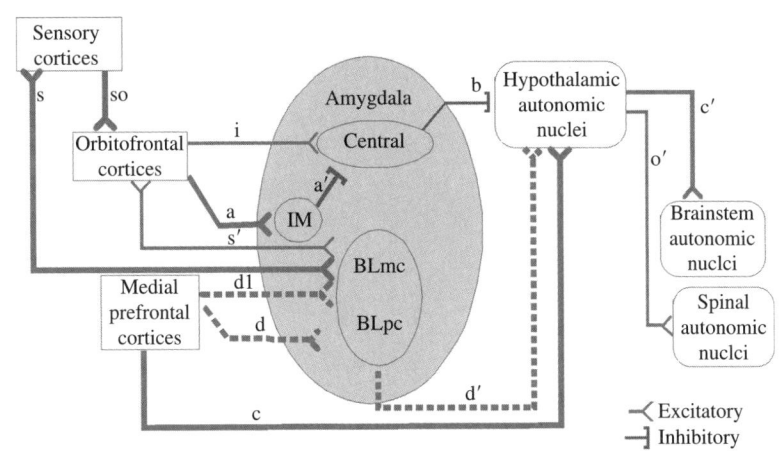

**Fig. 4.5** Summary of pathways linking prefrontal cortex with structures associated with perception and expression of emotions. Line thickness represents the density of projections. The orbitofrontal cortex as sensor of the environment: The orbitofrontal cortex and the amygdala receive projections from every sensory modality through the cortex (pathways so, s), and are interconnected, providing pathways for direct (so), and indirect (s and s') sensory input to orbitofrontal cortex. The orbitofrontal cortex disinhibits hypothalamic autonomic centers: Orbitofrontal axons terminate heavily in the intercalated masses of the amygdala (IM, pathway a), which project to the central nucleus (a'), which projects to hypothalamic autonomic centers (pathway b). Activation of pathways (a, a') leads to disinhibition of hypothalamic autonomic centers, which innervate brainstem and spinal autonomic centers (pathways c', o'). Posterior medial prefrontal (anterior cingulate) areas as effectors for emotional expression: A direct pathway from medial prefrontal cortex innervates hypothalamic autonomic centers (c), forming asymmetric, and presumed excitatory synapses in the lateral and posterior hypothalamic areas (not shown); this pathway also innervates brainstem autonomic centers (not shown). An indirect pathway courses from medial prefrontal cortices to the parvicellular sector of the basolateral nucleus of the amygdala (BLpc, pathway d), which projects to hypothalamic autonomic centers (d'), and is presumed to be excitatory. Activation of the direct or indirect pathways ultimately activates brainstem and spinal autonomic nuclei (pathways c, c', o'), which directly innervate peripheral organs. Reproduced with permission from Barbas et al. 2003.

There is yet another pathway from the caudal OFC, which is lighter than the pathway to IM, and targets directly the central nucleus of the amygdala (Carmichael and Price 1995a; Ghashghaei and Barbas 2002). The activation of this pathway would have an effect opposite to that of the pathway to the IM by inhibiting autonomic centers. This pathway has the potential to suppress the hypothalamic autonomic activity, and may be recruited when further information arrives that the loud noise was from a bursting balloon and was harmless.

The above circuits suggest that a highly specialized set of connections links the OFC with the amygdala. This circuit partially segregates the input and output zones in the amygdala that link it with the posterior OFC. In addition, the circuit has specific effects on the internal systems of the amygdala. A robust pathway from the posterior OFC to the

IM is positioned to increase the autonomic drive, and a smaller direct pathway to the central nucleus of the amygdala is consistent with dampening autonomic drive. No other area shows such a remarkable degree of specificity for processing signals related with emotionally charged events as the posterior OFC. Further, the posterior OFC shares with the sensory association cortices sites of interaction in the basal complex of the amygdala (Ghashghaei and Barbas 2002). On the basis of the highly specific circuits, we suggest that the caudal OFC is the earliest processing cortical area for emotional signals.

The proposed circuits have functional implications for information processing through the amygdala. The orbitofrontal to IM pathway may be akin to the pathway from the cortical areas to the basal ganglia (Alexander et al. 1986). In the cortico-striatal system, cortical axons target inhibitory neurons in the neostriatum, which innervate and inhibit neurons in the internal segment of the globus pallidus (GPi), which, in turn, synapse and inhibit the motor nuclei of the thalamus. These thalamic nuclei, such as the ventral anterior nucleus, have robust and bidirectional connections with the frontal cortex, including premotor and prefrontal cortices (McFarland and Haber 2002; Xiao and Barbas 2004). The mediodorsal nucleus, which is the principal thalamic relay for prefrontal cortices, also receives projections from GPi. The overall effect of the activation of the corticostriatal system is disinhibition downstream, at the level of the thalamus, allowing the activation of other circuits. In the corticostriatal system, the final effect is disinhibition of the motor nuclei of the thalamus and the mediodorsal nucleus, allowing the initiation of executive functions and motor activity through an activation of the reciprocal pathways with the frontal cortex, including prefrontal, premotor and motor areas.

By analogy with the corticostriatal system, in the orbitofrontal pathways to the IM nuclei of the amygdala, the final effect is the disinhibition of hypothalamic autonomic centers. The disinhibited hypothalamus, in turn, can be activated from other sources. Posterior orbitofrontal cortices as well as posterior medial prefrontal cortices in the anterior cingulate project robustly to hypothalamic autonomic centers (Rempel-Clower and Barbas 1998; Ongur et al. 1998). A pathway from medial area 32 forms excitatory synapses with hypothalamic neurons (Barbas et al. 2003), comparable to the highly efficient thalamocortical system (Ahmed et al. 1994). The hypothalamus, in turn, can activate brainstem and spinal autonomic centers (Barbas et al. 2003), the final efferent pathway to autonomic organs, such as the lungs and heart, whose activity increases during an emotional arousal.

## 4.4. Conscious awareness of the significance of the environment may depend on the OFC

The above evidence suggests that the topography and pattern of the connections linking the amygdala with the OFC likely underlie the processing for emotional behavior, but do not necessarily indicate that this interaction is critical for interpreting the emotional significance of events for action. A clue that the interaction between the amygdala and the OFC is necessary for a conscious appreciation of emotions emerged from disparate experimental approaches. For example, there is evidence that activity in the human amygdala can increase even when stimuli are presented tachistoscopically, below the level of conscious awareness (Whalen *et al.* 1998). A subcortical pathway between the right amygdala and the pulvinar and superior colliculus seems more active than pathways between the amygdala and the OFC in similar situations (Morris *et al.* 1999). Similarly, in rats, a short subcortical loop connecting the amygdala with the thalamus can support fear conditioning (Romanski and LeDoux 1992). The wiring of the amygdala appears to allow fast and fairly automatic processing for emotional attention and vigilance through its connections with limbic structures underlying the processing of the internal environment. In this sense the amygdala may act as a supraspinal reflex for the internal milieu, poised to direct attention to an event of significant emotional import.

The above evidence raises the question of the neural structure(s) that may account for awareness of the emotional significance of the environment, which is necessary for taking informed action. Classic studies indicate that the cortex is necessary for conscious perception of emotions (Kennard 1945). We suggest that the direct and indirect connections of orbitofrontal cortices with sensory cortices (discussed above) have a critical role in this function. Direct cortical sensory input may provide a global overview of the external environment, and the indirect sensory input through the amygdala may provide the emotional context necessary to interpret the significance of environmental events (Barbas 1995a). The connectional architecture between the caudal OFC and the amygdala allows several processes to occur. First, the partial segregation of input and output zones that link the amygdala with the posterior OFC may allow local processing within the amygdala, akin to the processing among linked layers within a cortical column. Second, the bidirectional connections between the OFC and neurons of the basal nuclei of the amygdala may allow prolongation of a signal between the OFC and the amygdala, as sensory information is assessed to determine whether autonomic structures must be recruited. This prolongation of activity may be comparable to the enhancement of feature-specific activity in sensory cortices, such as between the primary visual cortex and the lateral geniculate nucleus in nonhuman primates (Sillito *et al.* 1994). In this case, feedback cortical projections to the thalamus lead to the synchronization of 'preferred' or 'relevant' relay cell firing, effectively increasing the gain of the input for feature-linked events detected by the cortex. By analogy with the sensory cortex, the IM nuclei, the basal nuclei, and the central nucleus of the amygdala may be likened to different layers, allowing local control of neural computations. The entire complement of extrinsic and intrinsic connections of the amygdala enables monitoring, updating, and integrating sensory signals, so that an adequate judgment can be made regarding the emotional significance of events. This may be necessary in order to determine to what extent autonomic structures need to be recruited.

## 4.5. The role of the OFC in emotional memory

### Connections with medial temporal cortices

Common experience tells us that people zero in on stimuli that are emotionally salient in the environment, be it a beautiful flower or a colorful bird while walking through a meadow, or a terrible accident while driving on a busy highway. Such motivationally relevant stimuli are more likely to be remembered among the barrage of 'neutral' stimuli impinging upon our senses at any one time. In this respect, the posterior OFC has strong connections with several structures associated with long-term memory that may process emotionally salient information. Some of these connections involve the entorhinal (area 28) and perirhinal (areas 35, 36) cortices (van Hoesen *et al.* 1975; Morecraft *et al.*1992; Barbas 1993; Carmichael and Price 1995a). These medial temporal areas are interconnected, and the entorhinal cortex robustly innervates the hippocampus (Rosene and Van Hoesen 1987; Witter *et al.* 1989; Leonard *et al.* 1995; Nakamura and Kubota 1995; Squire and Zola 1996; Suzuki *et al.* 1997). But what is the nature of the input transmitted from the posterior OFC to the entorhinal cortex? As discussed above, the OFC has widespread connections with sensory areas and with the amygdala, and is in a strategic position to process information about the emotional significance of stimuli, and to associate stimuli with reward (Rolls 1996; Watanabe 1998). Neurons in the OFC respond to stimuli that predict reward (Tremblay and Schultz 1999). This process requires memory for associations between specific stimuli, and that is based largely on the emotional significance of such experiences (Cahill and McGaugh 1998). In humans, the OFC has a specific role in distinguishing between mental representations of the current situation and irrelevant memories (Schnider and Ptak 1999; Schnider *et al.* 2000). The orbitofrontal cortex thus is in a strategic position to issue signals about emotional significance to the entorhinal cortex, which may be used to encode events into long-term memory.

Of special significance in this context is the pattern of connections, as axons from the OFC innervate the middle layers of the entorhinal cortex (Rempel-Clower and Barbas 2000). By analogy with sensory systems, this laminar pattern of projection is associated with 'feedforward' or 'bottom-up' communication, linking earlier processing areas with later processing areas. This pattern is contrasted with projections that terminate mostly in layer I, linking the later processing and earlier processing cortices. This evidence suggests that the orbitofrontal cortical input to the entorhinal cortex reflects feedforward rather than feedback projections.

We have previously linked the laminar pattern of connections to the structural relationship of the linked areas within the prefrontal cortex (Barbas and Rempel-Clower, 1997), a model supported in the connections between prefrontal and temporal areas as well (Barbas 1986; Barbas *et al.* 1999; Rempel-Clower and Barbas 2000; Barbas *et al.* 2005b). The structural model for connections states that the pattern of connections is determined by the structural relationship of the connected areas. In this model, 'forward' or bottom-up connections simply reflect connections proceeding from an area with either more layers or higher cell density than the cortex of destination. According to the rules of the structural model, projections from the dysgranular posterior OFC to the agranular entorhinal cortex are expected to be mostly 'feedforward', which is substantiated by data (Rempel-Clower and Barbas 2000), and shown in Fig. 4.6. By analogy with sensory systems, 'forward' projections reflect 'bottom-up' processing. It is thus possible that information gathered from sensory areas is processed in the OFC and sent as feedforward signals to the entorhinal cortex to encode an emotionally memorable event into long-term mem-

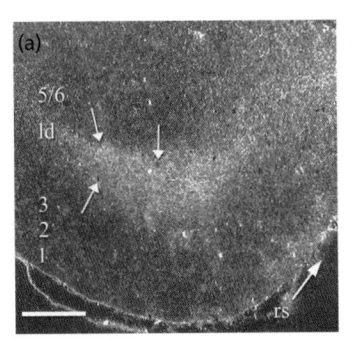

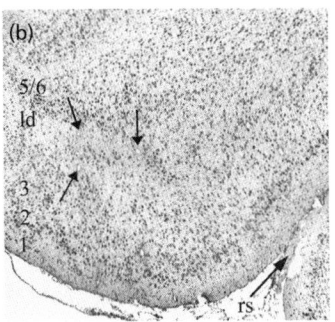

**Fig. 4.6** 'Forward' pattern of termination from orbitofrontal cortex to entorhinal cortex. (a) Pathways from orbitofrontal cortex terminate preferentially in the middle layers of the entorhinal cortex in the temporal lobe (white grain), the lamina dissecans (ld), shown under darkfield illumination. (b) Brightfield photomicrograph of the same section, showing the cell sparse lamina dissecans (ld). Numbers indicate cortical layers. rs, rhinal sulcus. Arrows show matching blood vessels. Scale bar, 500 μm. Adapted with permission from Rempel-Clower and Barbas (2000).

ory through projections to the hippocampus. In turn, the hippocampus sends direct projections to the posterior OFC (Rosene and Van Hoesen 1977; Barbas and Blatt 1995; Insausti and Munoz 2001). A recent study suggests that the hippocampus is part of a flexible, rapid memory system that forms unique associations between the places where rewards are found in the environment (Rolls and Xiang 2005). Therefore, there seems to be a close interaction between the hippocampal learning system and the orbitofrontal cortical system involved in associating objects and rewards. Lesions of the OFC in rhesus monkeys and marmosets impair rapid object-reward learning (Pears *et al.* 2003; Izquierdo *et al.* 2004).

## Common connections of the OFC and the amygdala with the mediodorsal and anterior thalamic nuclei

In addition to having a panoramic view of the entire sensory environment, the OFC and the amygdala have another feature in common: the well-known projection to the magnocellular part of the thalamic mediodorsal (MDmc) nucleus (Aggleton *et al.* 1980; Aggleton and Mishkin 1984; Russchen *et al.* 1987). The significance of this common connection is based on the role of the MDmc nucleus in long-term memory and emotional memory. Specifically, the OFC is connected preferentially with the caudal part of MD (Barbas *et al.* 1991; Dermon and Barbas 1994), whose damage in both human and non-human primates results in severe deficits in long-term memory (Victor *et al.* 1971; Isseroff *et al.* 1982; Zola-Morgan and Squire 1985). The interconnected network linking the amygdala, the MD and the OFC (Fig. 4.3) may underlie the processing of emotionally significant events into long-term memory.

There is yet another common thalamic pathway between the amygdala and the OFC, the anterior medial (AM) nucleus (Xiao and Barbas 2002b). The AM is part of the anterior thalamic limbic nuclei (Armstrong 1990), and links several orbitofrontal areas, as well as anterior cingulate areas with multiple systems underlying the processing of long-term memory and emotions (Xiao and Barbas 2002a; Xiao and Barbas 2002b). Like MDmc, the anterior thalamic nuclei are part of a circuit associated with long-term memory, through direct connections with the hippocampal formation, a circuit seen in several species, including rats, cats and monkeys (Meibach and Siegel 1977; Somogyi et al. 1978; DeVito 1980; Aggleton et al. 1986; Van Groen and Wyss 1990).

The implication of the anterior nuclei in long-term memory has emerged from behavioral studies (Aggleton and Brown 1999). There is evidence that monkeys with lesion of the anterior thalamus, including the AM nucleus and the MD, are impaired in object recognition and in associating objects with reward (Aggleton and Mishkin 1983). In humans, infarction of the AM nucleus results in deficits in episodic and recognition memory (Gaffan and Gaffan 1991; Parkin et al. 1994; Ghika-Schmid and Bogousslavsky 2000; Kishiyama et al. 2001; Nolan et al. 2001). In diseases affecting long-term memory, including Alzheimer's disease, and Korsakoff's syndrome, there are structural abnormalities or neuronal loss in the AM nucleus (Victor et al. 1971; Mair et al. 1979; Braak and Braak 1991; Kopelman 1995; Braak et al. 1996).

Beyond their role in long-term memory, the anterior thalamic nuclei are a focal point in the classic circuit for emotions, as proposed by Papez on the basis of clinical observations (Spiegel and Wycis 1962; Mark et al. 1963; Mark et al. 1970; Clarke et al. 1994; Tasker and Kiss 1995; Young et al. 2000; Ghika-Schmid and Bogousslavsky 2000; George et al. 2001; ). The association of the AM nucleus in emotional behavior is also evident in other species. For example, in rats, lesion of the AM nucleus reduces auditory fear conditioning responses (Celerier et al. 2000).

The initial circuit proposed by Papez did not include the amygdala (Papez 1937). As discussed above, there is strong evidence that the amygdala is an important component in circuits underlying emotional processing. Recent evidence indicates that the amygdala projects to the AM thalamic nucleus as well (Xiao and Barbas 2002b). This circuit, therefore, extends the classic circuit for emotions, and further demonstrates a common link between the OFC and the amygdala, with the MD as well as the AM nucleus. The involvement of both of these thalamic nuclei in long-term memory and emotions suggests that they may have a special role in emotional memory.

The rich variety of the connections of orbitofrontal areas extends to their thalamic connections as well. In contrast to lateral prefrontal cortices, whose thalamic projections originate predominantly in the multiform and parvicellular sectors of the mediodorsal nucleus, orbitofrontal areas receive substantial projections from numerous 'limbic' thalamic nuclei, including the midline and anterior, in addition to projections from MDmc (Barbas et al. 1991; Dermon and Barbas 1994). The intralaminar thalamic nuclei also provide a significant projection to posterior orbitofrontal and posterior medial areas, and to a lesser extent to rostral orbital areas, another feature that differentiates the posterior from the anterior OFC (Dermon and Barbas 1994).

## 4.6. Orbitofrontal-striato-thalamic circuits for emotional processing

The prefrontal cortex is an action-oriented region, a feature generally associated with lateral prefrontal areas. But are all prefrontal areas, including the OFC, associated with action? The prefrontal cortex, in general, has a special relationship with the basal ganglia, because it not only projects to the neostriatal parts of the basal ganglia, like the rest of the cortex, but its thalamic interactions are modulated by signals from the internal segment of the globus pallidus (GPi). Among the thalamic nuclei that project to prefrontal cortex, the MD and ventral anterior (VA) receive the most prominent projections from GPi (Steriade *et al.* 1997). The VA nucleus is connected with prefrontal as well as premotor areas (Kievit and Kuypers 1977; Jacobson *et al.* 1978; Kunzle 1978; Baleydier and Mauguiere 1980; Ilinsky *et al.* 1985; Asanuma *et al.* 1985; Goldman-Rakic and Porrino 1985; Preuss and Goldman-Rakic 1987; Yeterian and Pandya 1988; Barbas *et al.* 1991; Cavada *et al.* 2000; Chiba *et al.* 2001; Middleton and Strick 2002; McFarland and Haber 2002) in circuits classically implicated in motor functions (Goldman-Rakic 1987; Graybiel 2000; Haber and McFarland 2001; Anderson 2001). However, recent evidence suggests that circuits through the basal ganglia have a role beyond motor control, including cognition, reward evaluation, motivated behavior, learning, and memory (Middleton and Strick 1994; Mitchell *et al.* 1999; Hikosaka *et al.* 1999; Schultz *et al.* 2000; Middleton and Strick 2000; Hollerman *et al.* 2000; Graybiel 2000; Sato and Hikosaka 2002; Toni *et al.* 2002) which may be traced to pathways through the prefrontal cortex.

The VA nucleus is connected with all prefrontal cortices. However, connections are particularly robust with only a few areas, including lateral areas 9 and 8, medial area 32, and the posterior OFC (Xiao and Barbas 2004). Among orbitofrontal areas, area OPro stands apart from all others in having very robust and bidirectional connections with the VA nucleus (Dermon and Barbas 1994; Xiao and Barbas 2004). The comparatively denser projections of specific prefrontal areas to the VA may be significant in view of the role of areas 8 and 9 in executive functions. Similarly, the robust connections of medial area 32 and the posterior OFC with the VA may be associated with a different set of actions in emotional expression.

The most striking feature of the connections of the posterior OFC with the VA nucleus lies in their pattern. Bidirectional connections in the VA nucleus that link it with area OPro are organized in distinct clusters, as shown in Fig. 4.7. With the exception of area 8, only the caudal OFC shows a focused, modular pattern in these connections. These patchy connections in the VA may represent functional modules that can be selected for specific actions. Like the MD, the VA receives robust projections from GPi of the basal ganglia in circuits that likely support the central executive functions of the prefrontal cortex. Moreover, the strong connections linking medial and orbitofrontal cortices with the VA suggest a role of this pathway in translating motivational and emotional processing into action.

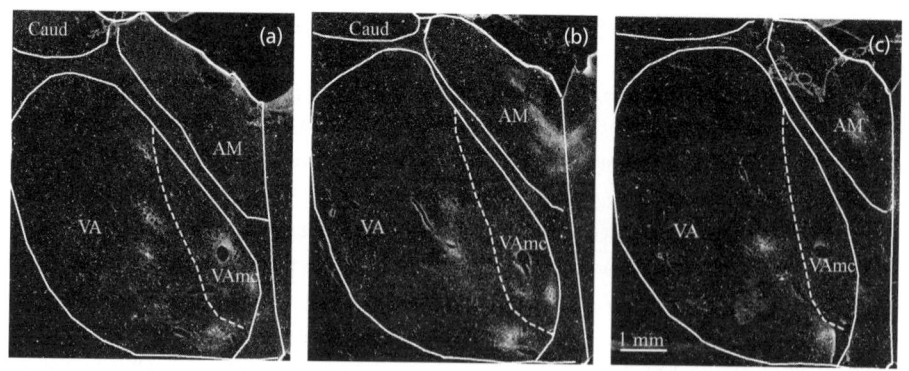

**Fig. 4.7** Modular connections of posterior orbitofrontal cortex with the ventral anterior (VA) nucleus of the thalamus. (a–c) Darkfield photomicrographs from coronal sections from rostral (a) to caudal (c) VA nucleus showing anterograde label in VA (yellow grain) from axonal terminations originating in posterior orbitofrontal area OPro (not shown), labeled with HRP-WGA. AM, anterior medial thalamic nucleus. Caud, caudate nucleus; VA, principal VA nucleus; VAmc, VA magnocellular. Scale bar, 1 mm. Adapted with permission from Xiao and Barbas (2004).

An important feature of the projections from prefrontal cortices to the VA nucleus is their origin in layer 5 as well as layer 6. This pattern is unusual, as the predominant origin of cortical projections to the thalamus is layer 6 in all cortices (Gilbert and Kelly 1975; Lund et al. 1976; Robson and Hall 1975; Jones and Wise 1977; Jones 1985; Steriade et al. 1997), with layer 5 contributing a much lower proportion of projections (~10–20%). In sharp contrast, about half of the projection neurons to the VA originate in layer 5 (Xiao and Barbas 2004). The significance of this laminar pattern of projection is based on evidence from sensory systems, indicating that corticothalamic projections from layer 5 differ markedly from projections originating in layer 6 (Rouiller and Welker 2000). There is some evidence that axons from neurons in cortical sensory areas in layer 6 terminate as small and diffuse terminals in the thalamus, targeting the distal dendrites of thalamic neurons in the nuclei that project focally to cortical layer 4 (Rockland 1996; Rouiller and Welker 2000). On the other hand, projections from layer 5 terminate as large and clustered axonal terminals in the thalamus, targeting the dendrites of thalamocortical projection neurons, which terminate widely, especially in cortical layer 1 (Jones 1985; Steriade et al. 1997; Rouiller and Welker 2000; Haber and McFarland 2001; Jones 2002). Although the pattern of termination in the VA nucleus of prefrontal neurons originating from layer 5 versus layer 6 is not known, there is evidence of substantial projections from the VA to layer 1 of the frontal cortex (Castro-Alamancos and Connors 1997; McFarland and Haber 2002). The patchy distribution of connections in the VA from the posterior OFC (as seen in Fig. 4.7) is consistent with a high distribution of neurons in layer 5 of the OFC that project to the VA (Xiao and Barbas 2004). Axons from the VA to layer 1 of the prefrontal cortex (including the OFC) may initiate activity in the apical dendrites of layer 5 neurons (Castro-Alamancos and Connors 1997), spreading excitation across cortical areas and back to the thalamus, recruiting other cortical areas in behavior. Being one of the few prefrontal areas with robust connections with the VA, and a significant contribution of projections to the VA from layer 5, the OFC may have a key role in recruiting other prefrontal areas in emotional situations.

The AM nucleus is another thalamic nucleus that is connected with the OFC and the basal ganglia in both rats and monkeys (Groenewegen 1988; Xiao and Barbas 2002b). In the rhesus monkey, projections to the AM nucleus originate from the medial part of GPi, and are restricted to its rostral sector (Xiao and Barbas 2002b). The medial tip of GPi appears to represent a distinct output channel linking the basal ganglia with the AM nucleus and with the limbic component of the prefrontal cortex. The GPi site occupied by projections to the AM nucleus appears to be distinct from circuits linking lateral prefrontal, premotor, and motor cortices through the VA and VL nuclei (Hoover and Strick 1993; Middleton and Strick 2002). The pathway from the GPi site to the AM nucleus adds another parallel loop to the classic striato-thalamic circuit. The latter describes basal ganglia feedback projections to their principal cortical targets, the premotor and prefrontal areas, coursing indirectly through the thalamic motor nuclei VA/VL, MD and intralaminar nuclei (Rosvold 1972; Alexander *et al.* 1986; Groenewegen *et al.* 1990; Hoover and Strick 1993; Joel and Weiner 1994; Strick *et al.* 1995; Middleton and Strick 2000; Haber and McFarland 2001; McFarland and Haber 2002). Of added significance in this respect is the association of the AM nucleus with the amygdala, as noted above. These findings extend the classic basal ganglia circuit and give additional support for a role of the basal ganglia in emotional functions through association with nuclei with a demonstrated role in emotional processes.

## 4.7. Transfer of information from OFC to lateral prefrontal cortices for decision and action

The orbitofrontal areas have connections with structures that have a role in the emotional significance of events, but lack direct access to the key systems that would allow decision and action in behavior. One of the limitations is the level of detail of the signals conveyed from sensory association cortices to orbitofrontal cortices. Orbitofrontal areas are distinguished for their connections with cortices associated with each of the sensory modalities, but sensory projections originate from late-processing sensory cortices. While this input may provide an overview of the sensory environment, it lacks the detail that may be necessary for discrimination. Take, for example, the situation where one must decide whether a figure in a path in the forest is a snake or a harmless limb from a tree. The sensory input from late-processing visual areas likely lacks resolution for discrimination, and thus may not be adequate for decision. However, lateral prefrontal areas are connected with a large variety of visual areas, going as far back as V2 (Barbas and Mesulam 1981; Barbas 1988). Lateral area 8, for example, receives a rich variety of visual input, much like other visual association cortices (Barbas 1988; Schall *et al.* 1995).

The pattern of connections among prefrontal areas, on one hand, and that with sensory cortices, on the other, suggest that prefrontal areas have a complementary role in processing information for decision and action in behavior. Posterior orbitofrontal cortices have strong connections with anterior orbitofrontal areas (Barbas and Pandya 1989; Carmichael and Price 1996), which, in turn, are robustly connected with lateral prefrontal areas, as shown in Fig. 4.8. Information about the emotional significance of events thus may be transmitted sequentially from the posterior OFC to the anterior OFC, and then to lateral prefrontal areas. Lateral prefrontal cortices, which are connected with early-processing visual areas, may retrieve information that may form the basis for making a decision about the identity of an ambiguous object in the path in the forest. This information may then be transmitted from lateral prefrontal areas to orbitofrontal cortices, activating one of the two paths from the posterior OFC to the amygdala, as discussed above: a pathway to the IM that would allow mobilization of autonomic structures, if the object in the path is a snake, or a direct path to the central nucleus of the amygdala that would dampen autonomic activity and allow return to autonomic homeo-stasis, if the figure is a stick.

Importantly, the laminar pattern of connections specified by the rules of the structural model (Barbas and Rempel-Clower 1997) is consistent with the flow of information for decision and action in behavior. As discussed above, the relative distribution of connections in cortical layers depends on the structural relationship of the linked areas. When an area with fewer layers or lower cell density projects to an area with more layers or higher cell density, projection neurons originate predominantly in the deep layers, and their axons terminate in the upper layers (layers 1–3a), a pattern akin to 'feedback' connections. In contrast, projections proceeding in the reverse direction originate in the upper layers (2–3), and their axons terminate in the middle to deep layers (deep layer 3 to 6), analogous to 'feedforward' connections. When areas that are close in structure are linked, connections involve most layers. Importantly, when the entire complement of connections is considered within a cortical system (such as the entire prefrontal cortex), the laminar pattern of connections is graded in a pattern that is predictable by the graded differences in the laminar structure of cortical areas.

The structural model for connections can be used to demonstrate a plausible flow of information for emotional processing in the prefrontal cortical system that is predicted by rules relating connections to cortical structure. Caudal orbitofrontal areas have fewer layers and lower overall density than rostral orbitofrontal areas (Dombrowski et al. 2001), though the differences in structure between these areas are not large. Consequently, the posterior OFC issues projections predominantly (though not exclusively) from its deep layers to the anterior OFC, targeting predominantly the superficial, but also substantially the deep layers of anterior orbitofrontal cortices. In turn, anterior orbitofrontal areas issue projections predominantly from their deep layers and target the upper layers (1–3a) of lateral prefrontal areas, including posterior area 46 (Fig. 4.9). The latter receives robust projections from early-processing visual areas, and

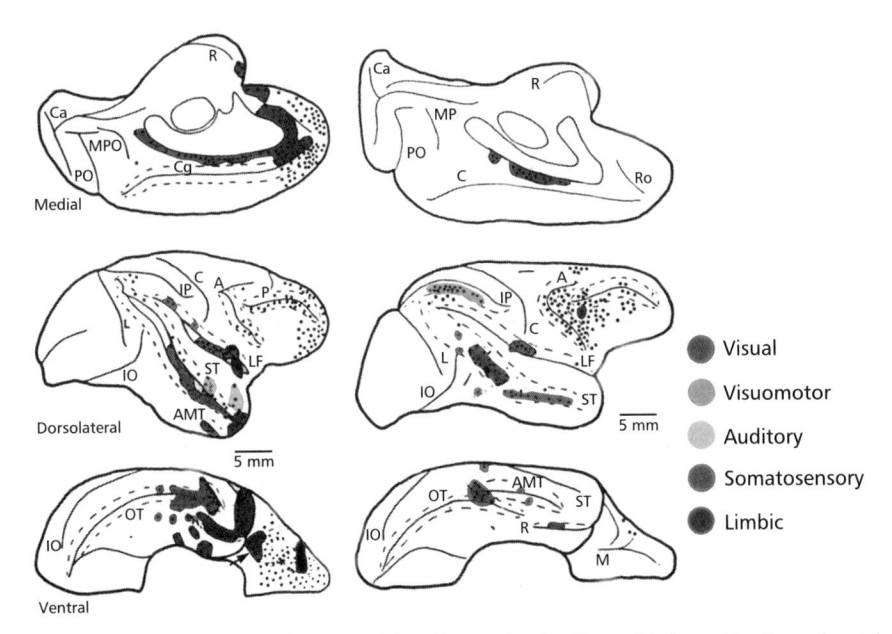

**Fig. 4.8** (Also see Color Plate 5.) Sequential pathways for the flow of information in prefrontal cortices. Comparison of the connections of an orbitofrontal area (left) and a lateral prefrontal area (right). (Left) Projection neurons (dots) directed to orbitofrontal area 11 (black area, bottom) originate from several sensory association cortices, as well as from posterior OFC (arrow), a bidirectional pathway. In turn, area 11 is connected with lateral prefrontal cortices, such as area 46 in the principalis region (P). (Right) Projection neurons directed to the ventral part of lateral area 46 (black area, center) originate predominantly from visual (red overlay) and visuomotor (brown) cortices. Information about the external and internal environments reaching posterior OFC could be conveyed to anterior orbitofrontal cortex and then to lateral prefrontal cortex through these pathways.

may be in a position to extract information pertaining to the finer details of the sensory environment, either directly, or through connections with adjacent area 8 (Barbas and Pandya 1989). Lateral prefrontal areas have more layers and a higher neuronal density than the orbitofrontal areas, and thus issue projections from their upper layers and target primarily the middle to deep layers of the OFC, including layer 5, a pattern akin to 'feedforward' connections. In turn, layer 5 is a major source of projections from prefrontal areas to the amygdala. The laminar relationships of these sequential connections provide a plausible path for the speedy transfer of information necessary for decision and action in behavior, as summarized in Fig. 4.9.

The other factor that suggests a collaborative action of prefrontal areas in decision and action in behavior is their linkage with the motor systems that allow a motor response. Among prefrontal areas, posterior lateral area 8 and caudal area 46, the areas that have connections with early-processing sensory areas, also have the strongest connections

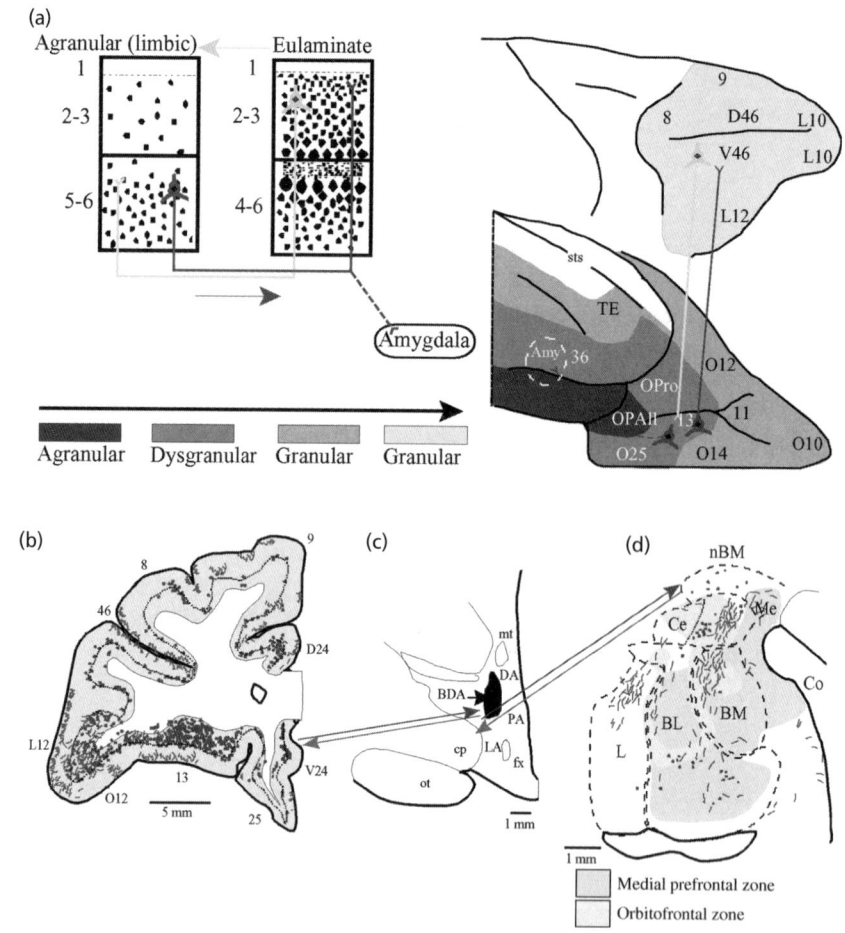

**Fig. 4.9** (Also see Color Plate 7.) Laminar-specific pathways underlying the perception and expression of emotions. (a) Eulaminate prefrontal areas (such as areas 46 and 8 on the lateral surface) receive detailed sensory input from early-processing sensory areas (not shown) and target the middle-deep layers of the agranular/dysgranular (limbic) prefrontal cortices (blue), such as posterior orbitofrontal cortex. Neurons from the same layers, particularly layer 5, issue robust projections to the amygdala, and 'feedback' projections to lateral eulaminate areas (red). The laminar pattern of connections is according to the rules of the structural model for connections (Barbas and Rempel-Clower 1997), and the pattern is consistent with the rapid transmission of sensory signals for emotional processing. (b) Medial and orbitofrontal cortices are bidirectionally connected with hypothalamic autonomic centers (c), which are also connected with the amygdala (d). These pathways form the basis for transmission of signals to hypothalamic autonomic centers which project to brainstem and spinal autonomic centers (not shown); the latter innervate peripheral organs for the expression of emotions, as shown in Fig. 4.5.

with the neighboring premotor areas (Barbas 1992). Thus, information may flow from orbitofrontal to lateral prefrontal areas and to premotor areas for decision and action in behavior.

A more direct involvement of orbitofrontal areas in motor activity appears to lie within the realm of autonomic activation necessary for the expression of emotions. This is based on the direct connections of orbitofrontal as well as medial prefrontal cortices with hypothalamic and brainstem autonomic centers, as discussed above. In addition, the termination of axons from the posterior OFC to the IM of the amygdala has a permissive effect on this process, by disinhibiting the hypothalamus and allowing its activation from other sources, as discussed above. Finally, the interface of orbitofrontal cortices with the thalamic VA nucleus, associated with motor systems, and the AM nucleus, associated with emotions, may underlie pathways necessary for emotional expression, as noted above.

## 4.8. Distinction of medial prefrontal from orbitofrontal areas in emotional processing

The posterior OFC is considered to be part of the limbic component of the prefrontal cortex (Yakovlev 1948; Nauta 1979). Yakovlev's and Nauta's inclusion of the OFC as part of the limbic system followed the classic incorporation of caudal medial prefrontal areas in the anterior cingulate as the limbic system, as suggested by Broca (Broca 1878) and Papez (Papez 1937). The prefrontal cortex, therefore, has two distinct limbic components, characterized not only by their robust connections with other cortical and subcortical limbic structures, but also by their agranular or dysgranular architecture (Barbas 1997).

The two components of the prefrontal limbic system share some connectional features, but are also distinguished at the level of circuits and function in important ways (Barbas 1997). They are similar in their strong connections with other cortical and subcortical limbic structures, and in their widespread connections with the thalamus, and medial temporal memory-related cortices (Barbas et al. 1991; Dermon and Barbas 1994; Barbas et al. 1999).

The orbitofrontal and medial components of the prefrontal limbic system, however, also have important differences in their connections. One of their most distinguishing features is the extent of their connections with unimodal sensory association cortices. Unlike the posterior orbitofrontal areas, the medial prefrontal cortices lack direct connections with sensory association areas. The most notable exception is the strong and bidirectional connection of medial prefrontal cortices with auditory association areas (Barbas et al. 1999). Like the OFC, medial prefrontal areas are bidirectionally linked with the amygdala (Barbas and De Olmos 1990; Carmichael and Price 1995a), but their pattern differs markedly. The connections of the posterior OFC have partly segregated input and output zones in the amygdala, as discussed above. No such segregation is seen for the afferent and efferent connections of medial (cingulate) cortices, which are largely intermingled in the amygdala (Ghashghaei and Barbas 2002). Moreover, the axons from posterior medial prefrontal cortices occupy a wider territory than orbitofrontal axons in the amygdala, extending to the cortical and medial nuclei. The medially situated nuclei of the amygdala have different connections than the basal nuclei, and have a role in gustatory, olfactory and reproductive functions (Ghashghaei and Barbas 2002).

Medial prefrontal cortices have stronger projections to hypothalamic autonomic centers, the spinal cord (Rempel-Clower and Barbas 1998) and brainstem autonomic centers (Ongur et al. 1998) than the OFC has. Several of the medially situated nuclei of the amygdala, which are connected with medial prefrontal cortices (but not orbitofrontal areas), project to the hypothalamus (Petrovich et al. 2001). These circuits suggest that the medial prefrontal cortex has multiple ways to access autonomic centers, and ultimately the emotional motor system. One pathway courses directly from medial prefrontal areas in the anterior cingulate to hypothalamic and brainstem autonomic centers (Rempel-Clower and Barbas 1998; Ongur et al. 1998; Freedman et al. 2000), and indirectly through the amygdala (Ghashghaei and Barbas 2002). The direct pathway from cingulate area 32 targets preferentially excitatory neurons in the hypothalamus, forming asymmetric, and presumed excitatory synapses mostly on spines in the hypothalamus (Barbas et al. 2003), which are enriched on dendrites of excitatory neurons (Peters et al. 1991). An indirect pathway involves a more diffuse projection from medial prefrontal cortices to the basolateral nucleus of the amygdala, which is known to project to hypothalamic and brainstem autonomic centers. There is evidence that the latter pathway has a role in the process of learning the significance of motivationally relevant cues (Petrovich et al. 2002). Another pathway from caudal medial prefrontal areas innervates, preferentially, the extended amygdala, a striatal-related structure in the basal forebrain, which is also involved in autonomic function (Ghashghaei and Barbas 2001). These varied pathways from medial prefrontal cortices may allow direct cortical control of autonomic functions in response to complex emotional situations.

## Orbitofrontal areas as sensors, and medial areas as effectors of emotions

The connectional distinctions suggest that posterior orbitofrontal and posterior medial prefrontal cortices have complementary roles in emotional processing. Based on connectional features, posterior orbitofrontal areas may be viewed as the primary 'sensory' areas for emotional processing, as recipient of information from cortices associated with each and every sensory modality, as well as the internal environment. Orbitofrontal areas may specialize in integrating the external and internal environments. The rich sensory input to the OFC may make it possible to determine the emotional value of events and their conscious appreciation.

In contrast, medial prefrontal areas receive only limited information from sensory cortices, emanating preferentially from auditory association cortices. It is not clear what this strong connection of medial prefrontal areas with auditory association cortices signifies, though it may be related to the role of anterior cingulate areas in emotional communication (Vogt and Barbas 1988; Barbas et al. 2002). On the other hand, medial prefrontal cortices have rich, direct and indirect, and highly efficient networks with central autonomic pathways extending as far as spinal autonomic centers (Barbas et al.

2003). On the basis of these connectional features, medial prefrontal areas may be the chief effectors for autonomic response, appropriately named the emotional motor system (Holstege 1991; Alheid and Heimer 1996; Holstege et al. 1996). The dichotomy within the prefrontal emotional system is reminiscent of the distinction of separate streams for processing features and spatial information in the sensory systems (Ungerleider and Mishkin 1982), or distal and axial body representations in the motor-premotor system (Barbas and Pandya 1987). The strong and bidirectional connections linking the posterior OFC with posterior medial prefrontal cortices may facilitate a continuously updated dialogue between these functionally distinct cortices in behavior. Based on the structural similarity of orbitofrontal and anterior cingulate areas, the laminar pattern of their connections is generally columnar, each affecting the input as well as the output layers of the other. The strong linkage between the OFC and anterior cingulate areas also extends to their callosal connections (Barbas et al. 2005a).

## 4.9. Implications for psychiatric diseases

The OFC has been implicated in a number of psychiatric disorders, including depression, anxiety, phobias and obsessive-compulsive disorder (Zald and Kim 1996; Simpson et al. 2001; Mayberg 2003). Data from unmedicated subjects with depression show that regional cerebral blood flow and glucose metabolism are increased in the amygdala, the OFC and the medial thalamus. Structures implicated in modulating and mediating emotional and stress responses are pathologically activated during major depressive episodes, whereas regions implicated in attention and sensory processing are deactivated (Davidson 2002).

The OFC may have a critical role in monitoring the environment with direct effect on one of two pathways with opposite effects on autonomic activation: The pathway from the OFC to the IM would increase the gain of autonomic centers in emotionally charged circumstances. Or, the pathway from the OFC to the central nucleus of the amygdala would inhibit the hypothalamus and return to autonomic homeostasis. Changes in opposite directions in the activity of these circuits may be at the root of the dramatically different clinical symptomatologies in psychiatric diseases. Disorders characterized by abnormal fear, such as panic disorder, anxiety, phobias and obsessive-compulsive disorder (Zald and Kim 1996; Simpson et al. 2001; Mayberg 2003), may have over activity in the OFC in common, and in particular in the pathway from the caudal OFC to the IM. In contrast, the under activity of the OFC appears to underlie sociopathic personality disorder, characterized by dampening of autonomic responses in emotional situations (Bechara et al. 1996). These contrasting deficits demonstrate at the clinical level the primary role of the posterior OFC in the flow of information for emotions.

## Acknowledgment

We thank our collaborators who participated in the original reports that contributed information for this review. Supported by NIH grants from NINDS and NIMH.

# References

Aggleton, J. P. and Mishkin, M. (1983) Memory impairments following restricted medial thalamic lesions in monkeys. *Exp Brain Res* **52**:199–209.

Aggleton, J. P. and Mishkin, M. (1984) Projections of the amygdala to the thalamus in the cynomolgus monkey. *J Comp Neurol* **222**:56–68.

Aggleton, J. P. and Brown, M. W. (1999) Episodic memory, amnesia, and the hippocampal-anterior thalamic axis. *Behav Brain Sci* **22**:425–444.

Aggleton, J. P, Burton, M. J., and Passingham, R.E. (1980) Cortical and subcortical afferents to the amygdala of the rhesus monkey (*Macaca mulatta*). *Brain Res* **190**:347–368.

Aggleton, J. P., Desimone, R., and Mishkin M (1986) The origin, course, and termination of the hippocampothalamic projections in the macaque. *J Comp Neurol* **243**:409–421.

Ahmed, B., Anderson, J. C., Douglas, R. J., Martin, K. A., and Nelson, J. C. (1994) Polyneuronal innervation of spiny stellate neurons in cat visual cortex. *J Comp Neurol* **341**:39–49.

Alexander, G. E., Delong, M. R., and Strick, P. L. (1986) Parallel organization of functionally segregated circuits linking basal ganglia and cortex. *Ann Rev Neurosci* **9**:357–381.

Alheid, G. F. and Heimer, L. (1996) Theories of basal forebrain organization and the 'emotional motor system'. *Prog Brain Res* **107**:461–484.

Amaral, D. G. and Price, J. L. (1984) Amygdalo-cortical projections in the monkey (*Macaca fascicularis*). *J Comp Neurol* **230**:465–496.

Anderson, M. E. (2001) Pallidal and cortical detriments of thalamic activity. In: *Basal ganglia and thalamus in health and movement disorders*. Kultas-Ilinsky, K. and Ilinsky, I.A., (eds), pp. 93–104. New York: Kluwer Academic/Plenum Publishers.

Armstrong, E. (1990) Limbic thalamus: anterior and mediodorsal nuclei. In: *The Human Nervous System* (Paxinos, G., ed.), pp. 469–481. New York: Academic Press, Inc.

Asanuma, C., Andersen, R. A., and Cowan, W. M. (1985) The thalamic relations of the caudal inferior parietal lobule and the lateral prefrontal cortex in monkeys: Divergent cortical projections from cell clusters in the medial pulvinar nucleus. *J Comp Neurol* **241**:357–381.

Baleydier, C. and Mauguiere, F. (1980) The duality of the cingulate gyrus in monkey. Neuroanatomical study and functional hypothesis. *Brain* **103**:525–554.

Barbas, H. (1986) Pattern in the laminar origin of cortico-cortical connections. *J Comp Neurol* **252**:415–422.

Barbas, H. (1988) Anatomic organization of basoventral and mediodorsal visual recipient prefrontal regions in the rhesus monkey. *J Comp Neurol* **276**:313–342.

Barbas, H. (1992) Architecture and cortical connections of the prefrontal cortex in the rhesus monkey. *Adv Neurol* **57**:91–115.

Barbas, H. (1993) Organization of cortical afferent input to orbitofrontal areas in the rhesus monkey. *Neuroscience* **56**:841–864.

Barbas, H. (1995a) Anatomic basis of cognitive-emotional interactions in the primate prefrontal cortex. *Neurosci Biobehav Rev* **19**:499–510.

Barbas, H. (1995b) Pattern in the cortical distribution of prefrontally directed neurons with divergent axons in the rhesus monkey. *Cereb Cortex* **5**:158–165.

Barbas, H. (1997) Two prefrontal limbic systems: Their common and unique features. In: *The Association cortex: Structure and function* (Sakata, H., Mikami, A., and Fuster, J.M., eds), pp. 99–115. Amsterdam: Harwood Academic Publ.

Barbas, H. (2000a) Complementary role of prefrontal cortical regions in cognition, memory and emotion in primates. *Adv Neurol* **84**:87–110.

Barbas, H. (2000b) Connections underlying the synthesis of cognition, memory, and emotion in primate prefrontal cortices. *Brain Res Bull* **52**:319–330.

Barbas, H. and Blatt, G. J. (1995) Topographically specific hippocampal projections target functionally distinct prefrontal areas in the rhesus monkey. *Hippocampus* **5**:511–533.

Barbas, H. and de Olmos J (1990) Projections from the amygdala to basoventral and mediodorsal prefrontal regions in the rhesus monkey. *J Comp Neurol* **301**:1–23.

Barbas, H. and Mesulam, M. M. (1981) Organization of afferent input to subdivisions of area 8 in the rhesus monkey. *J Comp Neurol* **200**:407–431.

Barbas, H. and Pandya, D. N. (1987) Architecture and frontal cortical connections of the premotor cortex (area 6) in the rhesus monkey. *J Comp Neurol* **256**:211–218.

Barbas, H. and Pandya, D. N. (1989) Architecture and intrinsic connections of the prefrontal cortex in the rhesus monkey. *J Comp Neurol* **286**:353–375.

Barbas, H. and Rempel-Clower, N. (1997) Cortical structure predicts the pattern of cortico-cortical connections. *Cereb Cortex* **7**:635–646.

Barbas, H., Henion, T. H., and Dermon, C. R. (1991) Diverse thalamic projections to the prefrontal cortex in the rhesus monkey. *J Comp Neurol* **313**:65–94.

Barbas, H., Ghashghaei, H., Dombrowski, S. M., and Rempel-Clower, N. L. (1999) Medial prefrontal cortices are unified by common connections with superior temporal cortices and distinguished by input from memory-related areas in the rhesus monkey. *J Comp Neurol* **410**:343–367.

Barbas, H., Ghashghaei, H., Rempel-Clower, N., and Xiao, D. (2002) Anatomic basis of functional specialization in prefrontal cortices in primates. In: *Handbook of Neuropsychology* (Grafman J., (ed)), pp. 1–27. Amsterdam: Elsevier *Science* B.V.

Barbas, H., Saha, S., Rempel-Clower, N., and Ghashghaei, T. (2003) Serial pathways from primate prefrontal cortex to autonomic areas may influence emotional expression. *BMC Neurosci* **4**:25.

Barbas, H., Hilgetag, C. C., Saha, S., Dermon, C. R., and Suski, J. L. (2005a) Parallel organization of contralateral and ipsilateral prefrontal cortical projections in the rhesus monkey. *BMC Neurosci* **6**:32.

Barbas, H., Medalla, M., Alade, O., Suski, J., Zikopoulos, B., and Lera, P. (2005b) Relationship of prefrontal connections to inhibitory systems in superior temporal areas in the rhesus monkey. *Cereb Cortex* **15**:1356–1370.

Baxter, M. G. and Murray, E. A. (2002) The amygdala and reward. *Nat Rev Neurosci* **3**:563–573.

Baxter, M. G., Parker, A., Lindner, C. C., Izquierdo, A. D., and Murray, E. A. (2000) Control of response selection by reinforcer value requires interaction of amygdala and orbital prefrontal cortex. *J Neurosci* **20**:4311–4319.

Baylis, L. L., Rolls, E. T., and Baylis, G. C. (1995) Afferent connections of the caudolateral orbitofrontal cortex taste area of the primate. *Neuroscience* **64**:801–812.

Bechara, A., Tranel, D., Damasio, H., and Damasio, A. R. (1996) Failure to respond autonomically to anticipated future outcomes following damage to prefrontal cortex. *Cereb Cortex* **6**:215–225.

Braak, H., and Braak, E. (1991) Alzheimer's disease affects limbic nuclei of the thalamus. *Acta Neuropathol (Berl)* **81**:261–268.

Braak, H., Braak, E., Yilmazer, D., de Vos, R. A., Jansen, E. N., and Bohl J (1996) Pattern of brain destruction in Parkinson's and Alzheimer's diseases. *J Neural Transm* **103**:455–490.

Broca, P. (1878) Anatomie compareé des enconvolutions cérébrales: Le grand lobe limbique et la scissure limbique dans la serie des mammifères. *Rev Anthropol* **1**:385–498.

Cahill, L. and McGaugh J. L. (1998) Mechanisms of emotional arousal and lasting declarative memory. *Trends Neurosci* **21**:294–299.

Cardinal, R. N., Parkinson, J. A., Hall, J., and Everitt, B. J. (2002) Emotion and motivation: the role of the amygdala, ventral striatum, and prefrontal cortex. *Neuroscience and Biobehavioral Reviews* **26**:321–352.

Carmichael, S. T. and Price, J. L. (1994) Architectonic subdivision of the orbital and medial prefrontal cortex in the macaque monkey. *J Comp Neurol* **346**:366–402.

Carmichael, S. T. and Price, J. L. (1995a) Limbic connections of the orbital and medial prefrontal cortex in macaque monkeys. *J Comp Neurol* **363**:615–641.

Carmichael, S. T. and Price, J. L. (1995b) Sensory and premotor connections of the orbital and medial prefrontal cortex of macaque monkeys. *J Comp Neurol* **363**:642–664.

Carmichael, S. T. and Price, J. L. (1996) Connectional networks within the orbital and medial prefrontal cortex of macaque monkeys. *J Comp Neurol* **371**:179–207.

Carmichael, S. T., Clugnet, M.-C., and Price, J. L. (1994) Central olfactory connections in the macaque monkey. *J Comp Neurol* **346**:403–434.

Carr, D. B., O'Donnell, P., Card, J. P., and Sesack, S. R. (1999) Dopamine terminals in the rat prefrontal cortex synapse on pyramidal cells that project to the nucleus accumbens. *J Neurosci* **19**:11049–11060.

Carr, D. B. and Sesack, S. R. (2000) Projections from the rat prefrontal cortex to the ventral tegmental area: target specificity in the synaptic associations with mesoaccumbens and mesocortical neurons. *J Neurosci* **20**:3864–3873.

Castro-Alamancos, M. A. and Connors, B. W. (1997) Thalamocortical synapses. *Prog Neurobiol* **51**:581–606.

Cavada, C., Company, T., Tejedor, J., Cruz-Rizzolo, R. J., and Reinoso-Suarez, F. (2000) The anatomical connections of the macaque monkey orbitofrontal cortex. A review. *Cereb Cortex* **10**:220–242.

Celerier, A., Ognard, R., Decorte, L., and Beracochea, D. (2000) Deficits of spatial and non-spatial memory and of auditory fear conditioning following anterior thalamic lesions in mice: comparison with chronic alcohol consumption. *Eur J Neurosci* **12**:2575–2584.

Chiba, T., Kayahara, T., and Nakano, K. (2001) Efferent projections of infralimbic and prelimbic areas of the medial prefrontal cortex in the Japanese monkey, *Macaca fuscata*. *Brain Res* **888**:83–101.

Clarke, S., Assal, G., Bogousslavsky, J., Regli, F., Townsend, D. W., Leenders, K. L. *et al.* (1994) Pure amnesia after unilateral left polar thalamic infarct: topographic and sequential neuropsychological and metabolic (PET) correlations. *J Neurol Neurosurg Psychiat* **57**:27–34.

Damasio, A. R. (1994) *Descarte's Error: Emotion, Reason, and the Human Brain*. New York: G. P. Putnam's Sons.

Damasio, H., Grabowski, T., Frank, R., Galaburda, A. M., and Damasio A. R (1994) The return of Phineas Gage: clues about the brain from the skull of a famous patient. *Science* **264**:1102–1105.

Davidson, R. J. (2002) Anxiety and affective style: role of prefrontal cortex and amygdala. *Biol Psychiatry* **51**:68–80.

Davis, M. (1992) The role of the amygdala in fear and anxiety. *Ann Rev Neurosci* **15**:353–375.

de Olmos, J. (1990) Amygdaloid nuclear gray complex. In: *The Human Nervous System* (Paxinos, G., ed.), pp. 583–710. San Diego: Acedemic Press, Inc.

Dermon, C. R. and Barbas, H. (1994) Contralateral thalamic projections predominantly reach transitional cortices in the rhesus monkey. *J Comp Neurol* **344**:508–531.

DeVito, J. L. (1980) Subcortical projections to the hippocampal formation in squirrel monkey (*Saimira sciureus*). *Brain Res Bull* **5**:285–289.

Dombrowski, S. M., Hilgetag, C. C., and Barbas, H. (2001) Quantitative architecture distinguishes prefrontal cortical systems in the rhesus monkey. *Cereb Cortex* **11**:975–988.

Fischer, H., Wright, C. I., Whalen, P. J., McInerney, S. C., Shin, L. M., and Rauch, S. L. (2003) Brain habituation during repeated exposure to fearful and neutral faces: a functional MRI study. *Brain Res Bull* **59**:387–392.

Freedman, L. J., Insel, T. R., and Smith, Y. (2000) Subcortical projections of area 25 (subgenual cortex) of the macaque monkey. *J Comp Neurol* **421**:172–188.

Fudge, J. L. and Haber, S. N. (2000) The central nucleus of the amygdala projection to dopamine subpopulations in primates. *Neuroscience* **97**:479–494.

Fuster, J. M. (1993) Frontal lobes. *Curr Opin Neurobiol* **3**:160–165.

Gaffan, D. and Gaffan, E. A. (1991) Amnesia in man following transection of the fornix. A review. *Brain* **114**:2611–2618.

George, M. S., Anton, R. F., Bloomer, C., Teneback, C., Drobes, D. J., Lorberbaum, J. P., *et al.* (2001) Activation of prefrontal cortex and anterior thalamus in alcoholic subjects on exposure to alcohol-specific cues. *Arch Gen Psychiatry* **58**:345–352.

Ghashghaei, H. T. and Barbas, H. (2001) Neural interaction between the basal forebrain and functionally distinct prefrontal cortices in the rhesus monkey. *Neuroscience* **103**:593–614.

Ghashghaei, H. T. and Barbas, H. (2002) Pathways for emotions: Interactions of prefrontal and anterior temporal pathways in the amygdala of the rhesus monkey. *Neuroscience* **115**:1261–1279.

Ghika-Schmid, F. and Bogousslavsky, J. (2000) The acute behavioral syndrome of anterior thalamic infarction: a prospective study of 12 cases. *Ann Neurol* **48**:220–227.

Gilbert, C. D. and Kelly, J. P. (1975) The projections of cells in different layers of the cat's visual cortex. *J Comp Neurol* **163**:81–105.

Goldman-Rakic, P. S. (1987) Motor control function of the prefrontal cortex. *Ciba Found Symp* **132**:187–200.

Goldman-Rakic, P. S. (1988) Topography of cognition: Parallel distributed networks in primate association cortex. *Ann Rev Neurosci* **11**:137–156.

Goldman-Rakic, P. S. and Porrino, L. J. (1985) The primate mediodorsal (MD) nucleus and its projection to the frontal lobe. *J Comp Neurol* **242**:535–560.

Graybiel, A. M. (2000) The basal ganglia. *Curr Biol* **10**:R509–R511.

Groenewegen, H. J. (1988) Organization of the afferent connections of the mediodorsal thalamic nucleus in the rat, related to the mediodorsal-prefrontal topography. *Neuroscience* **24**:379–431.

Groenewegen, H. J., Berendse, H. W., Wolters, J. G., and Lohman, A. H. (1990) The anatomical relationship of the prefrontal cortex with the striatopallidal system, the thalamus and the amygdala: evidence for a parallel organization. *Prog Brain Res* **85**:95–116.

Haber, S. and McFarland, N. R. (2001) The place of the thalamus in frontal cortical-basal ganglia circuits. *Neuroscientist* **7**:315–324.

Haber, S. N., Kunishio, K., Mizobuchi, M., and Lynd-Balta, E. (1995) The orbital and medial prefrontal circuit through the primate basal ganglia. *J Neurosci* **15**:4851–4867.

Herzog, A. G. and van Hoesen, G. W. (1976) Temporal neocortical afferent connections to the amygdala in the rhesus monkey. *Brain Res* **115**:57–69.

Hikosaka, K. and Watanabe, M. (2000) Delay activity of orbital and lateral prefrontal neurons of the monkey varying with different rewards. *Cereb Cortex* **10**:263–271.

Hikosaka, O., Nakahara, H., Rand, M. K., Sakai, K., Lu, X., Nakamura, K., *et al.* (1999) Parallel neural networks for learning sequential procedures. *Trends Neurosci* **22**:464–471.

Hollerman, J.R., Tremblay, L., and Schultz, W. (2000) Involvement of basal ganglia and orbitofrontal cortex in goal-directed behavior. In: *Progress in Brain Research* (Uylings, H.B.M., van Eden, C.G., de Bruin, J.P.C., Feenstra, M.G.P., and Pennartz, C.M.A., eds), pp. 193–215. Paris: Elsevier Science.

Holstege, G. (1991) Descending motor pathways and the spinal motor system: limbic and non-limbic components. *Prog Brain Res* **87**:307–421.

Holstege, G., Bandler, R., and Saper, C. B. (1996) The emotional motor system. *Prog Brain Res* **107**:3–6.

Hoover, J. E. and Strick, P. L. (1993) Multiple output channels in the basal ganglia. *Science* **259**:819–821.

Ilinsky, I. A., Jouandet, M. L., and Goldman-Rakic, P. S. (1985) Organization of the nigrothalamocortical system in the rhesus monkey. *J Comp Neurol* **236**:315–330.

Insausti, R., Amaral, D. G., and Cowan, W. M. (1987) The entorhinal cortex of the monkey: II. Cortical afferents. *J Comp Neurol* **264**:356–395.

Insausti, R. and Munoz, M. (2001) Cortical projections of the non-entorhinal hippocampal formation in the cynomolgus monkey (Macaca fascicularis). *Eur J Neurosci* **14**:435–451.

Isseroff, A., Rosvold, H. E., Galkin, T. W., and Goldman-Rakic, P. S. (1982) Spatial memory impairments following damage to the mediodorsal nucleus of the thalamus in rhesus monkeys. *Brain Res* **232**:97–113.

Izquierdo, A., Suda, R. K., and Murray, E. A. (2004) Bilateral orbital prefrontal cortex lesions in rhesus monkeys disrupt choices guided by both reward value and reward contingency. *J Neurosci* **24**:7540–7548.

Jacobson, S. and Trojanowski, J. Q. (1975) Amygdaloid projections to prefrontal granular cortex in rhesus monkey demonstrated with horseradish peroxidase. *Brain Res* **100**:132–139.

Jacobson, S., Butters, N., and Tovsky, N. J. (1978) Afferent and efferent subcortical projections of behaviorally defined sectors of prefrontal granular cortex. *Brain Res* **159**:279–296.

Joel, D. and Weiner, I. (1994) The organization of the basal ganglia-thalamocortical circuits: open interconnected rather than closed segregated. *Neuroscience* **63**(2):363–379.

Jones, E. G. (1985) *The Thalamus*. New York (NY): Plenum Press.

Jones, E. G. (2002) Thalamic organization and function after Cajal. *Prog Brain Res* **136**:333–357.

Jones, E. G. and Wise, S. P. (1977) Size, laminar and columnar distribution of efferent cells in the sensory-motor cortex of monkeys. *J Comp Neurol* **175**:391–438.

Jongen-Relo, A. L. and Amaral, D. G. (1998) Evidence for a GABAergic projection from the central nucleus of the amygdala to the brainstem of the macaque monkey: a combined retrograde tracing and in situ hybridization study. *Eur J Neurosci* **10**:2924–2933.

Kalivas, P. W. and Nakamura, M. (1999) Neural systems for behavioral activation and reward. *Curr Opin Neurobiol* **9**:223–227.

Kennard, M. A. (1945) Focal autonomic representation in the cortex and its relation to sham rage. *J Neuropathol Exp Neurol* **4**:295–304.

Kievit, J. and Kuypers, H. G. J. M. (1977) Organization of the thalamocortical connexions to the frontal lobe in the rhesus monkey. *Exp Brain Res* **29**:299–322.

Kishiyama, M. M., Kroll, N. E. A., Yonelinas, A. P., Lazzara, M. M., Jones, E. G., and Jagust, W. J. (2001) The effects of bilateral thalamic lesions on recollection- and familiarity-based recognition memory jedgements: a case study. *Neuroscience Abstracts* **27**, 530.9.

Kopelman, M. D. (1995) The Korsakoff syndrome. *Br J Psychiatry* **166**:154–173.

Kringelbach, M. L. and Rolls, E. T. (2004) The functional neuroanatomy of the human orbitofrontal cortex: evidence from neuroimaging and neuropsychology. *Prog Neurobiol* **72**:341–372.

Kunzle, H. (1978) An autoradiographic analysis of the efferent connections from premotor and adjacent prefrontal regions (Areas 6 and 9) in Macaca fascicularis. *Brain Behav Evol* **15**:185–234.

LeDoux, J. (1996) *The emotional brain*. New York: Simon & Schuster.

Leonard, B. W., Amaral, D. G., Squire, L. R., and Zola-Morgan, S. (1995) Transient memory impairment in monkeys with bilateral lesions of the entorhinal cortex. *J Neurosci* **15**:5637–5659.

Lipton, P. A., Alvarez, P., and Eichenbaum, H. (1999) Crossmodal associative memory representations in rodent orbitofrontal cortex. *Neuron* **2**:349–359.

Lund, J. S., Lund, R. D., Hendrickson, A. E., Hunt, A. B., and Fuchs, A. F. (1976) The origin of efferent pathways from the primary visual cortex, area 17, of the macaque monkey as shown by retrograde transport of horseradish peroxidase. *J Comp Neurol* **164**:287–304.

Lynd-Balta, E. and Haber, S. N. (1994) The organization of midbrain projections to the ventral striatum in the primate. *Neuroscience* 1–15.

Mair, W. G. P., Warrington, E. K., and Weiskrantz, L. (1979) Memory disorder in Korsakoff's psychosis a neuropathological and neuropsychological investigation of two cases. *Brain* **102**:749–783.

Malkova, L., Gaffan, D., and Murray, E. A. (1997) Excitotoxic lesions of the amygdala fail to produce impairment in visual learning for auditory secondary reinforcement but interfere with reinforcer devaluation effects in rhesus monkeys. *J Neurosci* **17**:6011–6020.

Mark, V. H., Barry, H., McLardy, T., and Ervin, F. R. (1970) The destruction of both anterior thalamic nuclei in a patient with intractable agitated depression. *J Nerv Ment Dis* **150**:266–272.

Mark, V. H., Ervin, F. R., and Yakovlev, P. I. (1963) Stereotactic Thalamotomy III. The verification of anatomical lesion sites in the human thalamus. *Arch Neurol* **8**:528–538.

Mayberg, H. S. (2003) Modulating dysfunctional limbic-cortical circuits in depression: towards development of brain-based algorithms for diagnosis and optimised treatment. *Br Med Bull* **65**:193–207.

McFarland, N. R. and Haber, S. N. (2002) Thalamic relay nuclei of the basal ganglia form both reciprocal and nonreciprocal cortical connections, linking multiple frontal cortical areas. *J Neurosci* **22**:8117–8132.

Meibach, R. C. and Siegel, A. (1977) Thalamic projections of the hippocampal formation: evidence for an alternate pathway involving the internal capsule. *Brain Res* **134**:1–12.

Middleton, F. A. and Strick, P. L. (1994) Anatomical evidence for cerebellar and basal ganglia involvement in higher cognitive function. *Science* **266**:458–461.

Middleton F. A, and Strick, P. L. (2000) Basal ganglia and cerebellar loops: motor and cognitive circuits. *Brain Res Brain Res Rev* **31**:236–250.

Middleton, F. A., and Strick, P. L. (2002) Basal-ganglia 'projections' to the prefrontal cortex of the primate. *Cereb Cortex* **12**:926–935.

Mitchell, I. J., Cooper, A. J., and Griffiths, M. R. (1999) The selective vulnerability of striatopallidal neurons. *Prog Neurobiol* **59**:691–719.

Moga, M. M. and Gray, T. S. (1985) Peptidergic efferents from the intercalated nuclei of the amygdala to the parabrachial nucleus in the rat. *Neurosci Lett* **61**:13–18.

Morecraft, R. J., Geula, C., and Mesulam, M.-M. (1992) Cytoarchitecture and neural afferents of orbitofrontal cortex in the brain of the monkey. *J Comp Neurol* **323**:341–358.

Morris, J. S., Ohman, A., and Dolan, R. J. (1999) A subcortical pathway to the right amygdala mediating 'unseen' fear. *Proc Natl Acad Sci USA* **4**:1680–1685.

Nakamura, K. and Kubota, K. (1995) Mnemonic firing of neurons in the monkey temporal pole during a visual recognition memory task. *J Neurophysiol* **74**:162–178.

Nauta, W. J. H. (1961) Fibre degeneration following lesions of the amygdaloid complex in the monkey. *J Anat* **95**:515–531.

Nauta, W. J. H. (1979) Expanding borders of the limbic system concept. In: *Functional Neurosurgery* (Rasmussen, T. and Marino, R., eds), pp. 7–23. New York: Raven Press.

Nishijo, H., Ono, T., and Nishino, H. (1988) Single neuron responses in amygdala of alert monkey during complex sensory stimulation with affective significance. *J Neurosci* **8**:3570–3583.

Nitecka, L. and Ben Ari, Y. (1987) Distribution of GABA-like immunoreactivity in the rat amygdaloid complex. *J Comp Neurol* **266**:45–55.

Nolan, E. C., Yonelinas, A. P., Hopfinger, J. B., Kroll, N. E. A., Baynes, K., Buonocore, M. H., et al. (2001) Functional brain activation during associative and item recognition in a patient with thalamic infarct. *Neuroscience Abstract* **27**, 530.10.

Ongur, D., An, X., and Price, J. L. (1998) Prefrontal cortical projections to the hypothalamus in macaque monkeys. *J Comp Neurol* **401**:480–505.

Pandya, D. N., van Hoesen, G. W., and Domesick, V. B. (1973) A cingulo-amygdaloid projection in the rhesus monkey. *Brain Res* **61**:369–373.

Papez, J. W. (1937) A proposed mechanism of emotion. *AMA Arch Neurol Psychiat* **38**:725–743.

Pare, D. and Smith, Y. (1993a) Distribution of GABA immunoreactivity in the amygdaloid complex of the cat. *Neuroscience* **57**:1061–1076.

Pare, D. and Smith, Y. (1993b) The intercalated cell masses project to the central and medial nuclei of the amygdala in cats. *Neuroscience* **57**:1077–1090.

Pare, D. and Smith, Y. (1994) GABAergic projection from the intercalated cell masses of the amygdala to the basal forebrain in cats. *J Comp Neurol* **344**:33–49.

Parkin, A. J., Rees, J. E., Hunkin, N. M., and Rose, P. E. (1994) Impairment of memory following discrete thalamic infarction. *Neuropsychologia* **32**:39–51.

Pears, A., Parkinson, J. A., Hopewell, L., Everitt, B. J., and Roberts, A. C. (2003) Lesions of the orbitofrontal but not medial prefrontal cortex disrupt conditioned reinforcement in primates. *J Neurosci* **23**:11189–11201.

Peters, A., Palay, S. L., and Webster, H. D. (1991) *The fine structure of the nervous system. Neurons and their supporting cells.* New York: Oxford University Press.

Petrides, M. (1996) Lateral frontal cortical contribution to memory. Seminars in the *Neurosciences* **8**:57–63.

Petrovich, G. D., Canteras, N. S., and Swanson, L. W. (2001) Combinatorial amygdalar inputs to hippocampal domains and hypothalamic behavior systems. *Brain Res Brain Res Rev* **38**:247–289.

Petrovich, G. D., Setlow, B., Holland, P. C., and Gallagher, M. (2002) Amygdalo-hypothalamic circuit allows learned cues to override satiety and promote eating. *J Neurosci* **22**:8748–8753.

Phillips, A. G., Ahn, S., and Howland, J. G. (2003) Amygdalar control of the mesocorticolimbic dopamine system: parallel pathways to motivated behavior. *Neurosci and Biobehav Rev* **27**:543–554.

Pitkanen, A. and Amaral, D. G. (1994) The distribution of GABAergic cells, fibers, and terminals in the monkey amygdaloid complex: an immunohistochemical and in situ hybridization study. *J Neurosci* **14**:2200–2224.

Porrino, L. J., Crane, A. M., and Goldman-Rakic, P. S. (1981) Direct and indirect pathways from the amygdala to the frontal lobe in rhesus monkeys. *J Comp Neurol* **198**:121–136.

Preuss, T. M. and Goldman-Rakic, P. S. (1987) Crossed corticothalamic and thalamocortical connections of macaque prefrontal cortex. *J Comp Neurol* **257**:269–281.

Price, J. L. (1990) Olfactory system. In: *The Human Nervous System* (Paxinos, G., ed.), pp 979–998. San Diego: Academic Press.

Price, J. L., Carmichael, S. T., and Drevets, W. C. (1996) Networks related to the orbital and medial prefrontal cortex; a substrate for emotional behavior? *Prog Brain Res* **107**:523–536.

Rempel-Clower, N. L. and Barbas, H. (1998) Topographic organization of connections between the hypothalamus and prefrontal cortex in the rhesus monkey. *J Comp Neurol* **398**:393–419.

Rempel-Clower, N. L. and Barbas, H. (2000) The laminar pattern of connections between prefrontal and anterior temporal cortices in the rhesus monkey is related to cortical structure and function. *Cereb Cortex* **10**:851–865.

Robson, J. A. and Hall, W. C. (1975) Connections of layer VI in striate cortex of the grey squirrel (*Sciurus carolinensis*). *Brain Res* **93**:133–139.

Rockland, K. S. (1996) Two types of corticopulvinar terminations: round (type 2) and elongate (type1). *J Comp Neurol* **368**:57–87.

Rolls, E. T. (1996) The orbitofrontal cortex. *Philos Trans R Soc Lond B Biol Sci* **351**:1433–143.

Rolls, E. T. (2000) The orbitofrontal cortex and reward. *Cereb Cortex* **10**:284–294.

Rolls, E. T. and Xiang, J. Z. (2005) Reward-spatial view representations and learning in the primate hippocampus. *J Neurosci* **25**:6167–6174.

Romanski, L. M. and LeDoux, J. E. (1992) Equipotentiality of thalamo-amygdala and thalamo-cortico-amygdala circuits in auditory fear conditioning. *J Neurosci* **12**:4501–4509.

Rosene, D. L. and van Hoesen, G. W. (1977) Hippocampal efferents reach widespread areas of cerebral cortex and amygdala in the rhesus monkey. *Science* **198**:315–317.

Rosene, D. L. and van Hoesen, G. W. (1987) *The hippocampal formation of the primate brain. A review of some comparative aspects of cytoarchitecture and connections*. In: Cerebral Cortex, Vol. 6 (Jones, E.G. and Peters, A., eds), pp. 345–455. New York: Plenum Publishing Corporation.

Rosvold, H. E. (1972) The frontal lobe system: cortical-subcortical interrelationships. *Acta Neurobiol Exp (Warsz)* **32**:439–460.

Rouiller, E. M. and Welker, E. (2000) A comparative analysis of the morphology of corticothalamic projections in mammals. *Brain Res Bull* **53**:727–741.

Russchen, F. T., Amaral, D. G., and Price, J. L. (1987) The afferent input to the magnocellular division of the mediodorsal thalamic nucleus in the monkey, *Macaca fascicularis*. *J Comp Neurol* **256**:175–210.

Saddoris, M. P., Gallagher, M., and Schoenbaum, G. (2005) Rapid associative encoding in basolateral amygdala depends on connections with orbitofrontal cortex. *Neuron* **46**:321–331.

Saha, S., Batten, T. F., and Henderson, Z. (2000) A GABAergic projection from the central nucleus of the amygdala to the nucleus of the solitary tract: a combined anterograde tracing and electron microscopic immunohistochemical study. *Neuroscience* **99**:613–626.

Sato, M. and Hikosaka, O. (2002) Role of primate substantia nigra pars reticulata in reward-oriented saccadic eye movement. *J Neurosci* **22**:2363–2373.

Schall, J. D., Morel, A., King, D. J., and Bullier, J. (1995) Topography of visual cortex connections with frontal eye field in macaque: Convergence and segregation of processing streams. *J Neurosci* **15**:4464–4487.

Schnider, A. and Ptak, R. (1999) Spontaneous confabulators fail to suppress currently irrelevant memory traces. *Nat Neurosci* **2**:677–681.

Schnider, A., Treyer, V., and Buck, A. (2000) Selection of currently relevant memories by the human posterior medial orbitofrontal cortex. *J Neurosci* **20**:5880–5884.

Schoenbaum, G., Chiba, A. A., and Gallagher, M. (1999) Neural encoding in orbitofrontal cortex and basolateral amygdala during olfactory discrimination learning. *J Neurosci* **19**:1876–1884.

Schoenbaum, G., Chiba, A. A., and Gallagher, M. (2000) Changes in functional connectivity in orbitofrontal cortex and basolateral amygdala during learning and reversal training. *J Neurosci* **20**:5179–5189.

Schoenbaum, G. and Eichenbaum, H. (1995) Information coding in the rodent prefrontal cortex. I. Single-neuron activity in orbitofrontal cortex compared with that in pyriform cortex. *J Neurophysiol* **74**(2):733–750.

Schoenbaum, G. and Setlow, B. (2003) Lesions of nucleus accumbens disrupt learning about aversive outcomes. *J Neurosci* **23**:9833–9841.

Schoenbaum, G., Setlow, B., Saddoris, M. P., and Gallagher, M. (2003) Encoding predicted outcome and acquired value in orbitofrontal cortex during cue sampling depends upon input from basolateral amygdala. *Neuron* **39**:855–867.

Schultz, W., Tremblay, L., and Hollerman, J. R. (2000) Reward processing in primate orbitofrontal cortex and basal ganglia. *Cereb Cortex* **10**:272–284.

Sillito, A. M., Jones, H. E., Gerstein, G. L., and West, D. C. (1994) Feature-linked synchronization of thalamic relay cell firing induced by feedback from the visual cortex. *Nature* **369**:479–482.

Simpson, J. R., Snyder, A. Z., Gusnard, D. A., and Raichle, M. E. (2001) Emotion-induced changes in human medial prefrontal cortex: I. During cognitive task performance. *Proc Natl Acad Sci USA* **98**:683–687.

Somogyi, G., Hajdu, F., Tombol, T., and Madarasz, M. (1978) Limbic projections to the cat thalamus. A horseradish peroxidase study. *Acta Anat (Basel)* **102**:68–73.

Spiegel, E. A. and Wycis, H. T. (1962) Clinical and physiological applications: thalamotomy. In: *Monographs in biology and medicine Vol. 2: Stereoencephalotomy* (Spiegel, E.A., ed.), pp. 132–135. New York: Grune and Stratton.

Squire, L. R. and Zola, S. M. (1996) Structure and function of declarative and nondeclarative memory systems. *Proc Natl Acad Sci USA* **93**:13515–13522.

Stefanacci, L. and Amaral, D. G. (2002) Some observations on cortical inputs to the macaque monkey amygdala: an anterograde tracing study. *J Comp Neurol* **451**:301–323.

Steriade, M., Jones, E. G., and McCormick, D. A. (1997) *Thalamus—Organization and function*. Oxford: Elsevier Science.

Strick, P. L., Dum, R. P., and Picard, N. (1995) Macro-organization of the circuits connecting the basal ganglia with the cortical motor areas. In: *Models of Information Processing in the Basal Ganglia* (Houk, J. C., Davis, J. L., and Beiser, D. G., eds), pp 117–130. Cambridge, MA: The MIT Press.

Suzuki, W. A. and Amaral, D. G. (1994) Perirhinal and parahippocampal cortices of the macaque monkey: Cortical afferents. *J Comp Neurol* **350**:497–533.

Suzuki, W. A. Miller, E. K., and Desimone, R. (1997) Object and place memory in the macaque entorhinal cortex. *J Neurophysiol* **78**:1062–1081.

Takagi, S. F. (1986) Studies on the olfactory nervous system of the old world monkey. *Prog Neurobiol* **27**:195–250.

Tasker, R. R. and Kiss, Z. H. (1995) The role of the thalamus in functional neurosurgery. *Neurosurg Clin N Am* **6**:73–104.

Tobler, P. N., Fiorillo, C. D., and Schultz, W. (2005) Adaptive coding of reward value by dopamine neurons. *Science* **307**:1642–1645.

Toni, I., Rowe J., Stephan K. E., and Passingham, R. E. (2002) Changes of cortico-striatal effective connectivity during visuomotor learning. *Cereb Cortex* **12**:1040–1047.

Tremblay, L. and Schultz, W. (1999) Relative reward preference in primate orbitofrontal cortex. *Nature* **398**:704–708.

Tremblay, E. and Schultz, W. (2000) Reward-related neuronal activity during go/no-go task performance in primate orbitofrontal cortex. *J Neurophysiol* **83**:1864–1876.

Turner, B. H., Mishkin, M., and Knapp, M. (1980) Organization of the amygdalopetal projections from modality- specific cortical association areas in the monkey. *J Comp Neurol* **191**:515–543.

Ungerleider, L. G. and Mishkin, M. (1982) Two cortical visual systems. In: Ingle, D.J., Goodale, M.A., and Mansfield, R.J.W. (eds) *Analysis of Visual Behavior*, pp 549–586. MIT Press, Cambridge, MA

van Groen, T. and Wyss, J. M. (1990) The connections of presubiculum and parasubiculum in the rat. *Brain Res* **518**:227–243.

van Hoesen, G. W. (1975) Some connections of the entorhinal (area 28) and perirhinal (area 35) cortices of the rhesus monkey. I. Temporal lobe afferents. *Brain Res* **95**:1–24.

van Hoesen, G. W. (1981) The differential distribution, diversity and sprouting of cortical projections to the amygdala of the rhesus monkey. In: *The Amygdaloid complex* (Ben-Ari, Y., ed.), pp 77–90. Amsterdam: Elsevier/North Holland Biomedical Press.

van Hoesen, G. W., Pandya, D. N., and Butters, N. (1972) Cortical afferents to the entorhinal cortex of the rhesus monkey. *Science* **175**:1471–1473.

van Hoesen, G. W., Pandya, D. N., and Butters, N. (1975) Some connections of the entorhinal (area 28) and perirhinal (area 35) cortices of the rhesus monkey. II. Frontal lobe afferents. *Brain Res* **95**:25–38.

Victor, M., Adams, R. D., and Collins, G. H. (1971) *The Wernicke-Korsakoff Syndrome*. Philadelphia: F.A. Davis Company.

Vogt, B. A. and Barbas, H. (1988) *Structure and connections of the cingulate vocalization region in the rhesus monkey*. In: The Physiological control of mammalian vocalization (Newman JD, ed.), pp. 203–225. New York: Plenum Publ. Corp.

Walker, A. E. (1940) A cytoarchitectural study of the prefrontal area of the macaque monkey. *J Comp Neurol* **73**:59–86.

Wallis, J. D. and Miller, E. K. (2003) Neuronal activity in primate dorsolateral and orbital prefrontal cortex during performance of a reward preference task. *Eur J Neurosci* **18**:2069–2081.

Watanabe, M. (1998) Cognitive and motivational operations in primate prefrontal neurons. *Rev Neurosci* **9**:225–241.

Whalen, P. J., Rauch, S. L., Etcoff, N. L., McInerney, S. C., Lee, M. B., and Jenike, M. A. (1998) Masked presentations of emotional facial expressions modulate amygdala activity without explicit knowledge. *J Neurosci* **18**:411–418.

Witter, M. P., van Hoesen, G. W., and Amaral, D. G. (1989) Topographical organization of the entorhinal projection to the dentate gyrus of the monkey. *J Neurosci* **9**:216–228.

Xiao, D. and Barbas, H. (2002a) Pathways for emotions and memory I. input and output zones linking the anterior thalamic nuclei with prefontal cortices in the rhesus monkey. *Thalamus and Related Systems* **2**:21–32.

Xiao, D. and Barbas, H. (2002b) Pathways for emotions and memory II. afferent input to the anterior thalamic nuclei from prefrontal, temporal, hypothalamic areas and the basal ganglia in the rhesus monkey. *Thalamus and Related Systems* **2**:33–48.

Xiao, D. and Barbas, H. (2004) Circuits through prefrontal cortex, basal ganglia, and ventral anterior nucleus map pathways beyond motor control. *Thalamus & Related Systems* **2**:325–343.

Yakovlev, P. I. (1948) Motility, behavior and the brain: Stereodynamic organization and neurocoordinates of behavior. *J Nerv Ment Dis* **107**:313–335.

Yeterian, E. H. and Pandya, D. N. (1988) Corticothalamic connections of paralimbic regions in the rhesus monkey. *J Comp Neurol* **269**:130–146.

Young, K. A., Manaye, K. F., Liang, C., Hicks, P. B., and German, D. C. (2000) Reduced number of mediodorsal and anterior thalamic neurons in schizophrenia. *Biol Psychiatry* **47**:944–953.

Zald, D. H. and Kim, S. W. (1996) Anatomy and function of the orbital frontal cortex, I: anatomy, neurocircuitry; and obsessive-compulsive disorder. *J Neuropsychiatry Clin Neurosci* **8**:125–138.

Zola-Morgan, S. and Squire, L. R. (1985) Amnesia in monkeys after lesions of the mediodorsal nucleus of the thalamus. *Ann Neurol* **17**:558–564.

Part 2

# Functions and methods

## Chapter 5

# The neurophysiology and functions of the orbitofrontal cortex

Edmund T. Rolls

## 5.1. Introduction

The functions of the orbitofrontal cortex (OFC), as shown primarily by analysis of neuronal activity, are considered here. The focus is primarily on recordings made in nonhuman primates, macaques, because there are many topological, cytoarchitectural, and probably connectional similarities between macaques and humans with respect to the OFC (Fig. 5.1) (Carmichael and Price 1994; Petrides and Pandya 1995; Ongur and Price 2000; Kringelbach and Rolls 2004; Price 2006). This brain region is substantially expanded in primates relative to rodents. Moreover, cross-species differences in connections may lead to differential functions in primates relative to rodents. For instance, the OFC receives visual information in primates from the inferior temporal visual cortex, which is a highly developed area for primate vision enabling invariant visual object recognition (Rolls 2000e; Rolls and Deco 2002), and which provides visual inputs used in the primate OFC for one-trial visual object-reward association reversal learning, and for representing face identity and expression. Further, even the taste system of primates and rodents may be different, with obligatory processing from the nucleus of the solitary tract via the thalamus to the cortex in primates, but a subcortical pathway in rodents via a pontine taste area to the amygdala, and differences in where satiety influences taste-responsive neurons in primates and rodents (Norgren 1984; Rolls and Scott 2003; Rolls 2005a). To understand the function of the OFC in humans, the majority of the studies described here were therefore performed with macaques or with humans.

This chapter describes evidence on the types of neuron found in the primate OFC; relates these to its anatomical connections; shows how these and related findings lead to the concept that the primate OFC represents primary (unlearned) reinforcers and implements rapid stimulus-reinforcer association learning; shows how this concept helps to provide a basis for understanding the neural basis of emotion; describes a model of how the OFC may perform these functions; and shows how these discoveries are linked to the understanding of the human OFC.

## 5.2. Connections

Part of the background for understanding neuronal responses in the OFC is the anatomical connections of the OFC, described in Chapter 3, this volume, and by Carmichael and Price

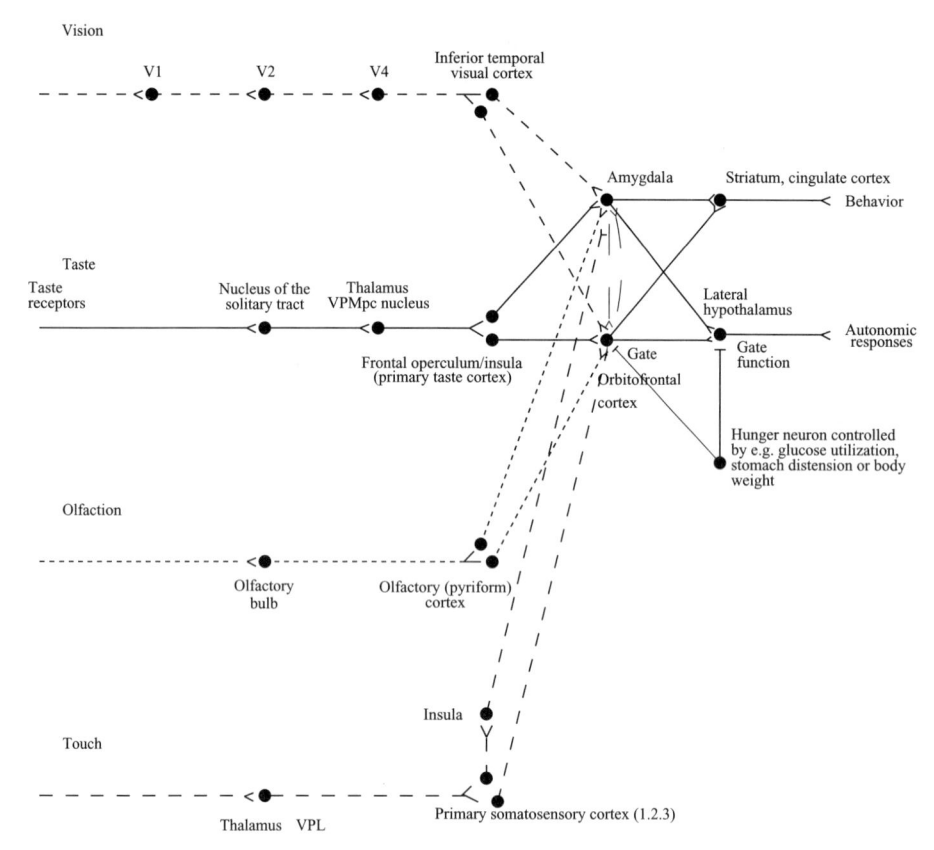

**Fig. 5.1** Schematic diagram showing some of the gustatory, olfactory, visual, and somatosensory pathways to the OFC, and some of the outputs of the OFC, in primates. The secondary taste cortex and the secondary olfactory cortex, are within the OFC. V1—primary visual cortex. V4—visual cortical area V4. Figure based on Rolls (1999b).

(1994; 1995), Petrides and Pandya (1995), Pandya and Yeterian (1996), Barbas (1995), and Öngür and Price (2000). A schematic diagram which helps to show the stage of processing in different sensory streams of the OFC is provided in Fig. 5.1. Rolls *et al.* (1990) discovered a taste area with taste-responsive neurons in the lateral part of the macaque OFC, and showed anatomically that this was a secondary taste cortex in that it receives a major projection from the primary taste cortex (Baylis *et al.* 1995). More medially, there is an area with olfactory neurons (Rolls and Baylis 1994), and anatomically, there are direct connections from the primary olfactory cortex, the piriform cortex, to area 13a of the posterior OFC, which in turn has onward projections to a middle part of the OFC (area 11) (Price *et al.* 1991; Morecraft *et al.* 1992; Barbas 1993; Carmichael *et al.* 1994; ) (see Fig. 5.1). Thorpe *et al.* (1983) found neurons with visual responses in the OFC, and anatomically, visual inputs reach the OFC directly from the inferior temporal cortex, the cortex in the superior temporal sulcus, and the temporal pole (see Barbas 1988; Barbas and

Pandya 1989; Seltzer and Pandya 1989; Morecraft *et al.* 1992; Barbas 1993, 1995; Carmichael and Price 1995). There are corresponding auditory inputs (Barbas 1988, 1993). Some neurons in the OFC respond to oral somatosensory stimuli such as the texture of food (Rolls *et al.* 1999; Rolls, Verhagen *et al.* 2003), and anatomically there are inputs to the OFC from the somatosensory cortical areas 1, 2 and SII in the frontal and pericentral operculum, and from the insula (Barbas 1988; Carmichael and Price 1995). The caudal OFC receives strong inputs from the amygdala (Price *et al.* 1991). The OFC also receives inputs via the mediodorsal nucleus of the thalamus, pars magnocellularis, which itself receives afferents from the temporal lobe structures such as the prepiriform (olfactory) cortex, amygdala, and inferior temporal cortex (see Ongur and Price 2000). These connections provide some routes via which the responses of OFC neurons can be produced.

The OFC projects back to the temporal lobe areas such as the inferior temporal cortex. The OFC also has projections to the entorhinal cortex, cingulate cortex (Insausti *et al.* 1987), the preoptic region, the lateral hypothalamus, the ventral tegmental area (Johnson *et al.* 1968; Nauta 1964), and the head of the caudate nucleus (Kemp and Powell 1970), and these connections provide some routes via which the OFC can influence behavior (Rolls 2005a) and memory (Rolls and Xiang 2005).

## 5.3. **Effects of lesions of the OFC**

Part of the background to investigations of neuronal activity in the OFC is the effect of lesions of the OFC. Macaques with lesions of the OFC are impaired at tasks which involve learning about which stimuli are rewarding and which are not, and especially impaired at altering behavior when reinforcement contingencies change. The monkeys may respond when responses are inappropriate, e.g., no longer rewarded, or may respond to a non-rewarded stimulus. For example, monkeys with orbitofrontal damage are impaired on go/no-go task performance in that they go on the no-go trials (Iversen and Mishkin 1970); in an object reversal task in that they respond to the object which was formerly rewarded with food; and in extinction in that they continue to respond to an object which is no longer rewarded (Butter 1969; Jones and Mishkin 1972; Izquierdo and Murray 2004; Izquierdo *et al.* 2004). There is some evidence for dissociation of function within the OFC, in that lesions to the inferior convexity produce the go/no-go and object reversal deficits, whereas damage to the caudal OFC produces the extinction deficit (Rosenkilde 1979).

Damage to the caudal OFC in the monkey also produces emotional changes (e.g., decreased aggression to humans and to stimuli such as a snake and a doll), and a reduced tendency to reject foods such as meat (Butter *et al.* 1969, 1970; Butter and Snyder 1972) or to display the normal preference ranking for different foods (Baylis and Gaffan 1991). In humans, euphoria, irresponsibility, lack of affect, and impulsiveness can follow frontal lobe damage (Damasio 1994; Rolls 1999b; Kolb and Whishaw 2003), particularly orbitofrontal damage (Rolls *et al.* 1994; Hornak *et al.* 1996; Rolls 1999b; Berlin *et al.* 2004; Hornak *et al.* 2003; Berlin *et al.* 2005; Rolls 2005a).

Lesions more laterally, in for example the inferior convexity, can influence tasks in which objects must be remembered for short periods, e.g., delayed matching to sample tasks and delayed matching to non-sample tasks (Passingham 1975; Mishkin and Manning 1978; Kowalska *et al.* 1991), and neurons in this region may help to implement this visual object short-term memory by holding the representation active during the delay period (Rosenkilde *et al.* 1981; Wilson *et al.* 1993; Rao *et al.* 1997). Whether this inferior convexity area is specifically involved in a short- term object memory (separately from a short-term spatial memory) is not yet clear (Rao *et al.* 1997), and a medial part of the frontal cortex may also contribute to this function (Kowalska *et al.* 1991). It should be noted that this short- term memory system for objects (which receives inputs from the temporal-lobe visual cortical areas in which objects are represented) is different to the short-term memory system in the dorsolateral part of the prefrontal cortex, which is more concerned with spatial short-term memories, consistent with its inputs from the parietal cortex (Rolls and Deco 2002).

## 5.4. Neurophysiology of the OFC

### Taste

One of the recent discoveries, that has helped us to understand the functions of the OFC in behavior, is that it contains a major cortical representation of taste (Fig. 5.1) (Rolls 1989, 1995, 1997; Rolls and Scott 2003). Given that taste can act as a primary reinforcer, that is without learning as a reward or punisher, we now have the start for a fundamental understanding of the function of the OFC in stimulus-reinforcer association learning (Rolls 1999b; Rolls 2004a, 2005a). We know how one class of primary reinforcers reaches and is represented in the OFC. A representation of primary reinforcers is essential for a system that is involved in learning associations between previously neutral stimuli and primary reinforcers, for example, between the sight of an object, and its taste.

The representation (shown by analyzing the responses of single neurons in macaques) of taste in the OFC includes robust representations of the prototypical tastes, sweet, salt, bitter and sour (Rolls *et al.* 1989), but also separate representations of the 'taste' of water (Rolls *et al.* 1990), and of protein or umami as exemplified by monosodium glutamate (Baylis and Rolls 1991; Rolls 2000a) and inosine monophosphate (Rolls *et al.* 1996a; Rolls *et al.* 1998). An example of an OFC neuron with different responses to different taste stimuli is shown in Fig. 5.2b. As will be described below, some neurons have taste-only responses, and others respond to a variety of oral somatosensory stimuli, including for some neurons viscosity (Rolls *et al.* 2003c), and for other neurons astringency, as exemplified by tannic acid (Critchley and Rolls 1996a).

The nature of the representation of taste in the OFC is that for the majority of neurons the reward value of the taste is represented. The evidence for this is that the responses of orbitofrontal taste neurons are modulated by hunger (as is the reward value or palatability of a taste). In particular, it has been shown that OFC taste neurons gradually stop responding to the taste of a food as the monkey is fed to satiety (Rolls *et al.* 1989). For

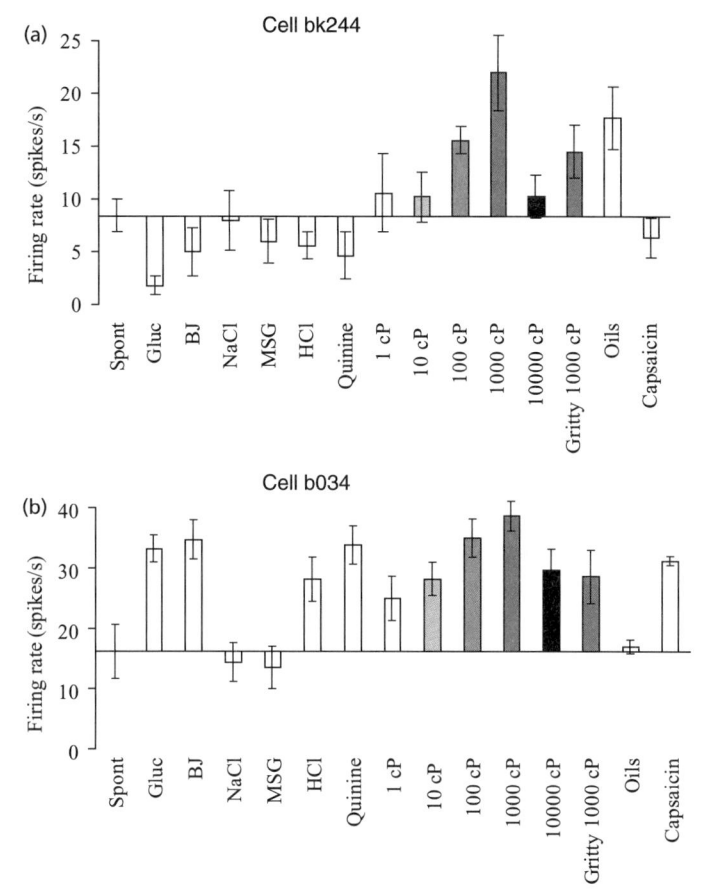

**Fig. 5.2** Oral somatiosensory and taste inputs to OFC neurons. (a) Firing rates (mean ± sem) of viscosity-sensitive neuron bk244, which did not have taste responses, in that it did not respond differentially to the different taste stimuli. The firing rates are shown to the viscosity series, to the gritty stimulus (carboxymethylcellulose with Fillite microspheres), to the taste stimuli 1 M glucose (Gluc), 0.1 M NaCl, 0.1 M MSG, 0.01 M HCl, and 0.001 M QuinineHCl, and to fruit juice (BJ). Spont = spontaneous firing rate. (b) Firing rates (mean ± sem) of viscosity-sensitive neuron bo34, which had no response to the oils (mineral oil, vegetable oil, safflower oil and coconut oil, which have viscosities which are all close to 50 cP). The neuron did not respond to the gritty stimulus in a way that was unexpected given the viscosity of the stimulus, was taste tuned, and did respond to capsaicin. Figure based on Rolls et al. (2003).

example, in Fig. 5.7 (which appears later in the chapter in the section on visual responses) a single neuron with taste, olfactory, and visual responses to food, is shown. The neuronal responses elicited through all these sensory modalities showed a decrease with satiety. The decrease is relatively specific to the food eaten to satiety, and the responses of these neurons are thus very closely related to sensory-specific satiety. Moreover, this type of neuronal responsiveness shows that it is the relative preference for different stimuli that is represented by these neurons, in that the neuronal response decreases in parallel with the decrease in the acceptability or reward value of the food being eaten to satiety, but the neuronal responses remain high (or even sometimes become a little larger) to foods not eaten in the meal, which remain acceptable, with a high reward value. The responses of these OFC neurons thus reflect the relative preferences of the macaque for different sensory stimuli (Rolls et al. 1989; Critchley and Rolls 1996b), as also shown by Tremblay and Schultz (1999). In contrast, the representation of taste in the primary taste cortex (Scott et al. 1986; Yaxley et al. 1990) is not modulated by hunger (Rolls et al. 1988; Yaxley et al. 1988), and thus represents the sensory properties and not the hedonic value of taste. Thus in the primate primary taste cortex, the reward value of taste is not represented, and instead the identity of the taste is represented. Additional evidence that the reward value of food is represented in the OFC is that monkeys work for electrical stimulation of this brain region if they are hungry, but not if they are satiated (Mora et al. 1979; Rolls 2005a). Further, neurons in the OFC are activated from many brain-stimulation reward sites (Mora et al. 1980; Rolls et al. 1980). Thus there is clear evidence that it is the reward value of taste that is represented in the OFC (see Rolls 1999b, 2000d, 2005a), and this is further supported by the finding that feeding to satiety decreases the activation of the human OFC to the food eaten to satiety in a sensory-specific way (Kringelbach et al. 2003).

The secondary taste cortex is in the caudolateral part of the OFC, as defined anatomically (Baylis et al. 1995). This region projects on to other regions in the OFC (Baylis et al. 1995), and neurons with taste responses (in what can be considered as a tertiary gustatory cortical area) can be found in many regions of the OFC (Rolls et al. 1990, 1996b; Rolls and Baylis 1994). Although some taste neurons are found laterally in the OFC (Rolls et al. 1990, 1996; Rolls and Baylis 1994), others are found through the middle and even towards the medial part of the OFC (Critchley and Rolls 1996a; Rolls and Baylis 1994; Rolls 2005a).

Corresponding to the findings in nonhuman primate single neuron neurophysiology, in human functional neuroimaging experiments (e.g., with functional magnetic resonance image, fMRI) it has been shown that there is an OFC area activated by sweet taste (glucose: Francis et al. 1999; Small et al. 1999), and that there are at least partly separate areas activated by the aversive taste of saline (NaCl, 0.1 M) (O'Doherty et al., 2001a), by pleasant touch (Francis et al. 1999; Rolls et al. 2003b), and by pleasant vs. aversive olfactory stimuli (Francis et al. 1999; O'Doherty et al. 2000; Rolls 2000d, 2003). Umami (protein) taste is not only represented by neurons in the primate OFC (Baylis and Rolls 1991; Rolls et al. 1996a), but human fMRI studies also show that umami taste is represented in the OFC, with an anterior part responding supralinearly to a combination of monosodium glutamate and inosine monophosphate (de Araujo et al. 2003a). Some OFC neurons respond to the 'taste' of water in the mouth (Rolls et al. 1990), and their responses occur only when thirsty and not when satiated (Rolls et al. 1989); and correspondingly in humans, the pleasantness of the taste of water in the mouth is represented in the OFC (de Araujo et al. 2003b). Although the representations of pleasant and unpleasant tastes overlap at least partly with each other and with activations produced by these other classes of sensory stimuli as shown by fMRI methodology, the single neuron neurophysiology shows that at the neuronal level there is an exquisitely rich representation of oral stimuli, with different neurons responding unimodally or to different combinations of taste, olfactory, somatosensory and visual stimuli, thus providing an information rich and distributed representation of these sensory properties of stimuli, which allows different combinations to be separated from each other, and for sensory-specific satiety to occur to particular combinations of these sensory stimuli (Rolls and Baylis 1994; Rolls et al. 2003c; Verhagen et al. 2003; Kadohisa et al. 2004; Kadohisa et al. 2005; Rolls 2005a).

## Convergence of taste and olfactory inputs in the OFC: the representation of flavor

In these further parts of the OFC, not only unimodal taste neurons, but also unimodal olfactory neurons are found. In addition some single neurons respond to both gustatory and olfactory stimuli, often with correspondence between the two modalities (Rolls and Baylis 1994) cf. Fig. 5.3. It is probably here, in the OFC of primates including humans, that these two modalities converge to produce the representation of flavor (Rolls and Baylis 1994; de Araujo et al. 2003c), for neurons in the primary taste cortex in the insular/frontal opercular cortex do not respond to olfactory (or visual) stimuli (Verhagen et al. 2004). Evidence will soon be described that indicates that these representations are built by olfactory-gustatory association learning, an example of stimulus-reinforcer association learning.

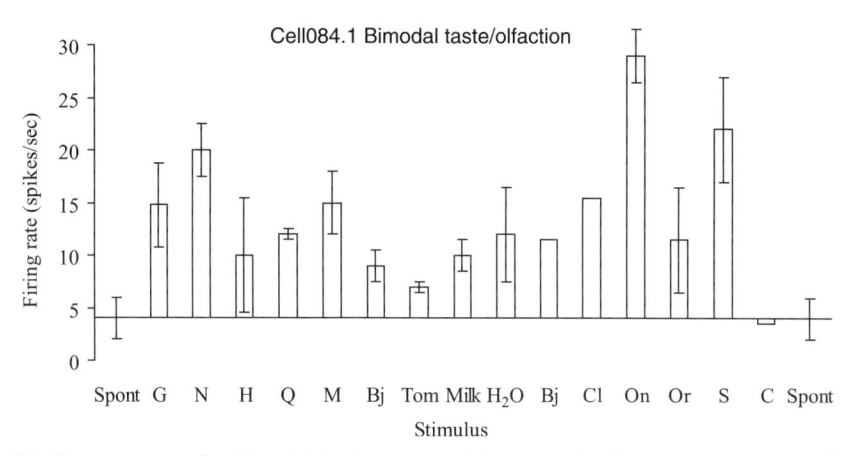

**Fig. 5.3** The responses of a bimodal single neuron with taste and olfactory responses recorded in the caudolateral OFC. G, 1 M glucose; N, 0.1 M NaCl; H, 0.01 M HCl; Q, 0.001 M quinine HCl; M, 0.1 M monosodium glutamate; Bj, 20% blackcurrant juice; Tom, tomato juice; B, banana odor; Cl, clove oil odor; On, onion odor; Or, orange odor; S, salmon odor; C, control no-odor presentation. The mean responses ± sem. are shown. The neuron responded best to the tastes of NaCl and monosodium glutamate and to the odors of onion and salmon. Figure based on Rolls and Baylis (1994).

## An olfactory representation in the OFC

Takagi, Tanabe, and colleagues (Takagi 1991) described single neurons in the macaque OFC that were activated by odors. A ventral frontal region has been implicated in olfactory processing in humans (Jones-Gotman and Zatorre 1988; Zatorre et al. 1992). Rolls and colleagues have analyzed the rules by which orbitofrontal olfactory representations are formed and operate in primates. For 65% of neurons in the orbitofrontal olfactory areas, Critchley and Rolls (1996c) showed that the representation of the olfactory stimulus was independent of its association with taste reward (i.e. these neurons did not categorize the odors depending on whether they were associated with glucose reward or saline punishment in an olfactory-to-taste discrimination task). For the remaining 35% of the neurons, the odors to which a neuron responded were influenced by the taste (glucose or saline) with which the odor was associated. Thus the odor representation for 35% of orbitofrontal neurons appeared to be built by olfactory- to taste-association learning. This possibility was confirmed by reversing the taste with which an odor was associated in the reversal of an olfactory discrimination task. It was found that 68% of the sample of neurons analyzed altered the way in which they responded to odor when the taste reinforcement association of the odor was reversed (Rolls et al. 1996b). (See example in Fig. 5.4, in which 25% showed reversal, and 43% no longer discriminated after the reversal. The olfactory to taste reversal was quite slow, both neurophysiologically and behaviorally, often requiring 20–80 trials, consistent with the need for some stability of flavor representations. The relatively high proportion of olfactory neurons with modification of responsiveness by taste association in the set of neurons in this experiment was probably related to the fact that the neurons were pre-selected to show differential responses to the odors associated with different tastes in the olfactory discrimination task). Thus the rule according to which the orbitofrontal olfactory representation was formed was for some neurons by association learning with taste.

To analyze the nature of the olfactory representation in the OFC, Critchley and Rolls (1996b) measured the responses of olfactory neurons that responded to food while they fed the monkey to satiety. They found that the majority of orbitofrontal olfactory neurons decreased their responses to the odor of the food with which the monkey was fed to satiety (see example in Fig. 5.7). Thus for these neurons, the reward value of the odor is what is represented in the OFC (cf. Rolls and Rolls 1997). In that the neuronal responses decreased to the food with which the monkey is fed to satiety, and may even increase to a food with which the monkey has not been fed, it is the relative reward value of stimuli that is represented by these OFC neurons (as confirmed by Schultz et al. 2000), and this parallels the changes in the relative pleasantness of different foods after a food is eaten to satiety (Rolls, B. J. et al. 1981a, b; Rolls 1997, 1999b, 2000d, 2005a). We do not yet know whether this is the first stage of processing at which reward value is represented in the olfactory system in macaques (although in rodents the influence of reward-association learning appears to be present in some neurons in the piriform cortex—Schoenbaum and Eichenbaum 1995). However, an fMRI investigation in humans showed that whereas in the OFC the pleasantness vs unpleasantness of odors is represented, this was not

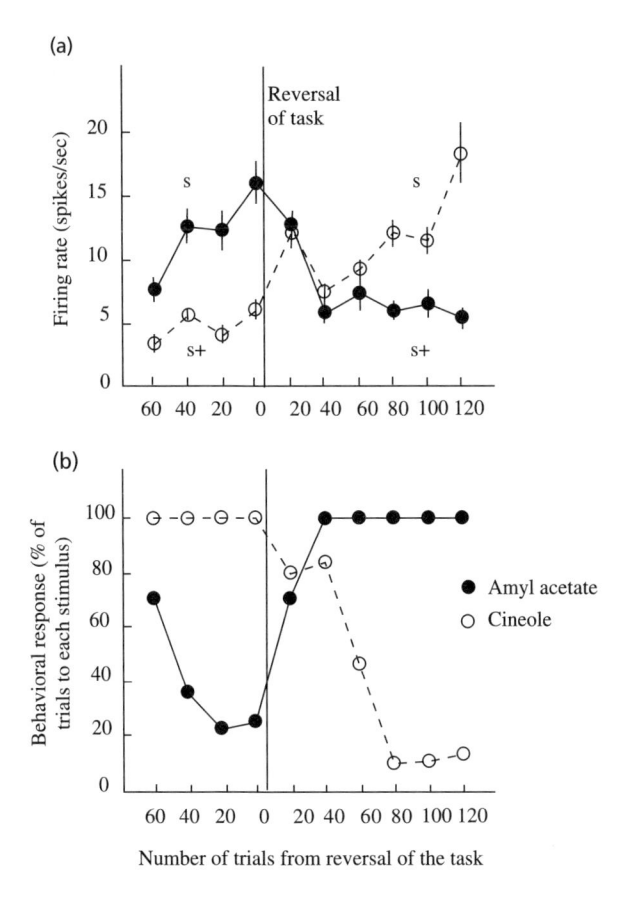

**Fig. 5.4** Olfactory to taste association reversal shown by an OFC neuron. (a) The activity of a single orbitofrontal olfactory neuron during the performance of a two-odor olfactory discrimination task and its reversal is shown. Each point represents the mean poststimulus activity of the neuron in a 500-ms period on approximately 10 trials of the different odorants. The standard errors of these responses are shown. The odorants were amyl acetate (closed circle) (initially S−) and cineole (o) (initially S+). After 80 trials of the task the reward associations of the stimuli were reversed. This neuron reversed its responses to the odorants following the task reversal. (b) The behavioral responses of the monkey during the performance of the olfactory discrimination task. The number of lick responses to each odorant is plotted as a percentage of the number of trials to that odorant in a block of 20 trials of the task. Figure based on Rolls *et al.* (1996).

the case in primary olfactory cortical areas, where instead the activations reflected the intensity of the odors (Rolls *et al.* 2003a).

Although individual neurons do not encode large amounts of information about which of 7–9 odors has been presented (Rolls *et al.* 1996c), we have shown that the information does increase linearly with the number of neurons in the sample. This ensemble encoding does result in useful amounts of information about which odor has been presented being provided by OFC olfactory neurons.

Corresponding to the findings in nonhuman primate single-neuron neurophysiology, in human neuroimaging experiments, it has been shown that there is an OFC area activated by olfactory stimuli (Francis *et al.* 1999; Jones-Gotman and Zatorre 1988; Zatorre *et al.* 1992). Moreover, the pleasantness or reward value of odor is represented in the OFC, in that feeding the humans to satiety decreases the activation found to the odor of that food, and this effect is relatively specific to the food eaten in the meal (Francis *et al.* 1999; cf. O'Doherty *et al.* 2000; Morris and Dolan 2001). Further, the human medial OFC has activation that is related to the subjective pleasantness of a set of odors, and a more lateral area has activation that is related to how unpleasant odors are subjectively (Rolls *et al.* 2003a).

An fMRI study has shown that cognitive effects can reach down into the human OFC and influence activations produced by odors (de Araujo *et al.* 2005). In this study, a standard test odor, isovaleric acid with a small amount of cheese flavor, was delivered through an olfactometer. (The odor alone, like the odor of brie, might have been interpreted as pleasant, or perhaps as unpleasant.) On some trials the test odor was accompanied with the visually presented word label 'cheddar cheese', and on other trials with the word label 'body odor'. It was found that the activation in the medial OFC to the standard test odor was much greater when the word label was cheddar cheese that when it was body odor. Controls with clean air were run to show that the effect could not be accounted for by the word label alone. Moreover, the word labels influenced the subjective pleasantness ratings to the test odor, and the changing pleasantness ratings were correlated with the activations in the human medial OFC. Part of the interest and importance of this finding is that it shows that cognitive influences, originating here purely at the word level, can reach down and modulate activations in the first stage of cortical processing that represents the affective value of sensory stimuli (de Araujo *et al.* 2005; Rolls 2005a).

## Visual inputs to the OFC, error detection neurons, visual stimulus-reinforcement association learning and reversal, and neurons with face-selective responses

We have been able to show that there is a major visual input to many neurons in the OFC, and that what is represented by these neurons is in many cases the reinforcement association of visual stimuli. The visual input is from the ventral, temporal lobe, visual stream concerned with 'what' object is being seen (Rolls 2000e; Rolls and Deco 2002). Many neurons in these temporal cortex visual areas have responses to objects or faces that are invariant with respect to size, position on the retina, and even view, making these neurons ideal as an input to a system that may learn about the reinforcement association properties of objects and faces, for after a single learning trial, the learning then generalizes correctly to other views etc. (Rolls 2000e, 2005a; Rolls and Deco 2002). Using this object-related information, OFC visual neurons frequently respond differentially to objects or images depending on their reward association (Thorpe *et al.* 1983; Rolls *et al.* 1996b). The primary reinforcer that has been used is taste. Many of these neurons (approximately 71%—Rolls *et al.* 1996) show visual-taste reversal in one or a very few trials (see example in Fig. 5.5). (In a visual discrimination task, they will reverse the stimulus to which they respond, e.g., from a triangle to a square, in one trial when the taste delivered for a behavioral response to that stimulus is reversed). This reversal learning probably occurs in the OFC, for it does not occur one synapse earlier in the visual inferior temporal cortex (Rolls *et al.* 1977), and it is in the OFC that there is convergence of visual and taste pathways onto the same neurons (Thorpe *et al.* 1983; Rolls and Baylis 1994; Rolls *et al.* 1996b; ). The probable mechanism for this learning is an associative modification of synapses conveying visual input onto taste-responsive neurons, implementing a pattern association network (Rolls and Treves 1998; Rolls 1999b; Rolls and Deco 2002) (see Fig. 5.9). When the reinforcement association of a visual stimulus is reversed, other OFC neurons stop responding, or stop responding differentially, to the visual discriminanda (Thorpe *et al.* 1983). For example, one neuron in the OFC responded to a blue stimulus when it was rewarded (blue S+) and not to a green stimulus when it was associated with aversive saline (green S−). However, the neuron did not respond after reversal to the blue S− or to the green S+ (see Fig. 5.6). Similar conditional reward neurons were found for olfactory stimuli (Rolls *et al.* 1996b). Such conditional reward neurons convey information about the current reinforcement status of particular stimuli. They may be part of a system that can implement very rapid reversal, by being biased on by rule neurons if that stimulus is currently associated with reward, and being biased off if that stimulus is currently not associated with reward (Deco and Rolls 2005b).

To analyze the nature of the visual representation of food-related stimuli in the OFC, Critchley and Rolls (1996b) measured the responses of neurons that responded to the sight of food while they fed the monkey to satiety. They found that the majority of orbitofrontal visual food-related neurons decreased their responses to the sight of the food with which the monkey was fed to satiety (see example in Fig. 5.7). Thus for these neurons, the reward value of the sight of food is what is represented in the OFC (Rolls and Rolls 1997). In that the neuronal responses decreased to the food with which the monkey is fed to satiety, and may even increase to a food with which the monkey has not been fed, it is the relative reward value of stimuli that is represented by these OFC neurons. At a stage of visual processing one synapse earlier, in the inferior temporal visual cortex, neurons do not show visual discrimination reversal learning, nor are their responses modulated by feeding to satiety (Rolls *et al.* 1977). Thus both these functions are implemented for visual processing in the OFC.

In addition to these neurons that encode the reward association of visual stimuli, other, 'error', neurons in the OFC detect non-reward, in that they respond, for example, when an expected reward is not obtained when a visual discrimination task is reversed (Thorpe *et al.* 1983) (see Fig. 5.8, and Table 5.1, Visual Discrimination Reversal), or when reward is no longer made available in a visual discrimination task (Table 5.1, Visual Discrimination Extinction). Different populations of such neurons respond to other types of non-reward, including the removal of a formerly approaching taste reward

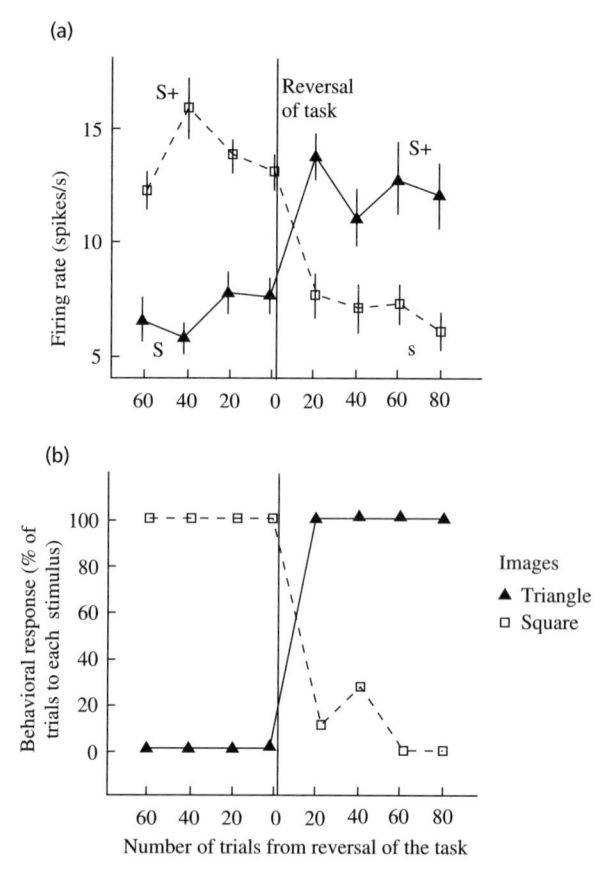

**Fig. 5.5** (a) Visual discrimination reversal of the responses of a single neuron in the macaque OFC when the taste with which the two visual stimuli (a triangle and a square) were associated was reversed. Each point is the mean poststimulus firing rate measured in a 0.5 s period over approximately 10 trials to each of the stimuli. Before reversal, the neuron fired most to the square when it indicated (S+) that the monkey could lick to obtain a taste of glucose. After reversal, the neuron responded most to the triangle when it indicated that the monkey could lick to obtain glucose. The response was low to the stimuli when they indicated (S–) that if the monkey licked then aversive saline would be obtained. (b) The behavioral response to the triangle and the square, and indicates that the monkey reversed rapidly. Figure based on Rolls et al. (1996b).

(Removal in Table 5.1), and the termination of a taste reward in the extinction of ad lib licking for juice (see Table 5.1), or the substitution of juice reward for aversive tasting saline during ad lib licking (Table 5.1, Ad Lib Licking Reversal) (Thorpe et al. 1983). The presence of these neurons is fully consistent with the hypothesis that they are part of the mechanism by which the OFC enables very rapid reversal of behavior by stimulus-reinforcement association relearning when the association of stimuli with reinforcers is

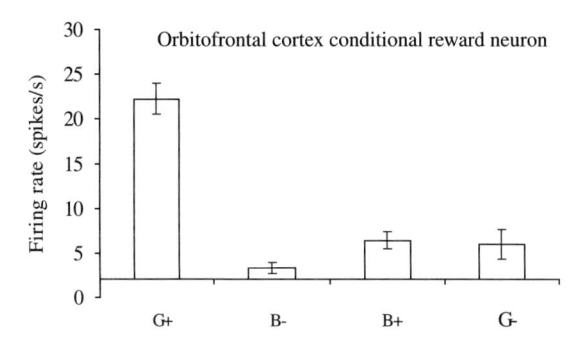

**Fig. 5.6** A conditional reward neuron recorded in the OFC by Thorpe, et al. (1983) which responded only to the Green stimulus when it was associated with reward (G+), and not to the Blue stimulus when it was associated with Reward (B+), or to either stimuli when they were associated with a punisher, the taste of salt (G- and B-).

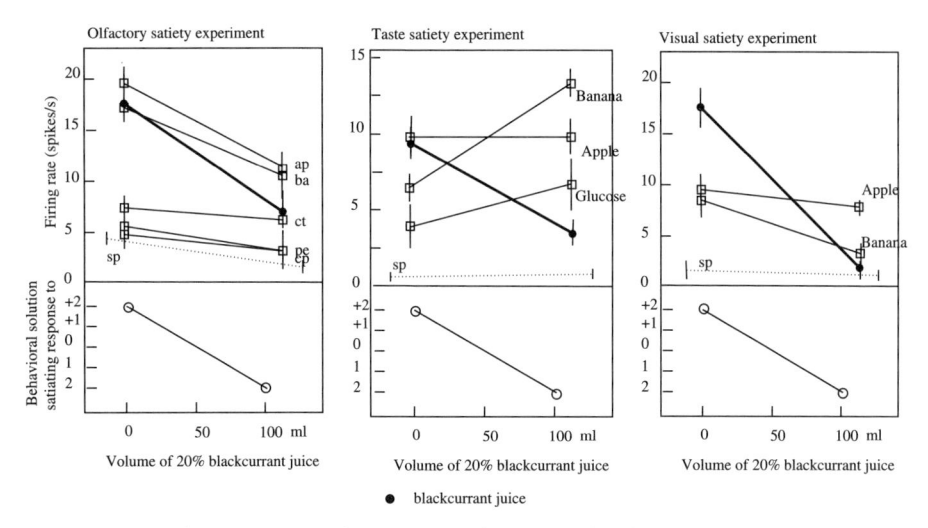

**Fig. 5.7** Multimodal OFC neuron with sensory-specific satiety-related responses to visual, taste and olfactory sensory inputs. The responses are shown before and after feeding to satiety with blackcurrant juice. The solid circles show the responses to blackcurrant juice. The olfactory stimuli included apple (ap), banana (ba), citral (ct), phenylethanol (pe), and caprylic acid (cp). The spontaneous firing rate of the neuron is shown (sp). Figure based on Critchley and Rolls (1996).

altered or reversed (Deco and Rolls 2005b; Rolls 1990, 2005a). The finding that different OFC neurons respond to different types of non-reward (Thorpe et al. 1983) may provide part of the brain's mechanism that enables task- or context-specific reversal to occur. Evidence that there may be similar error neurons in the human OFC was found in

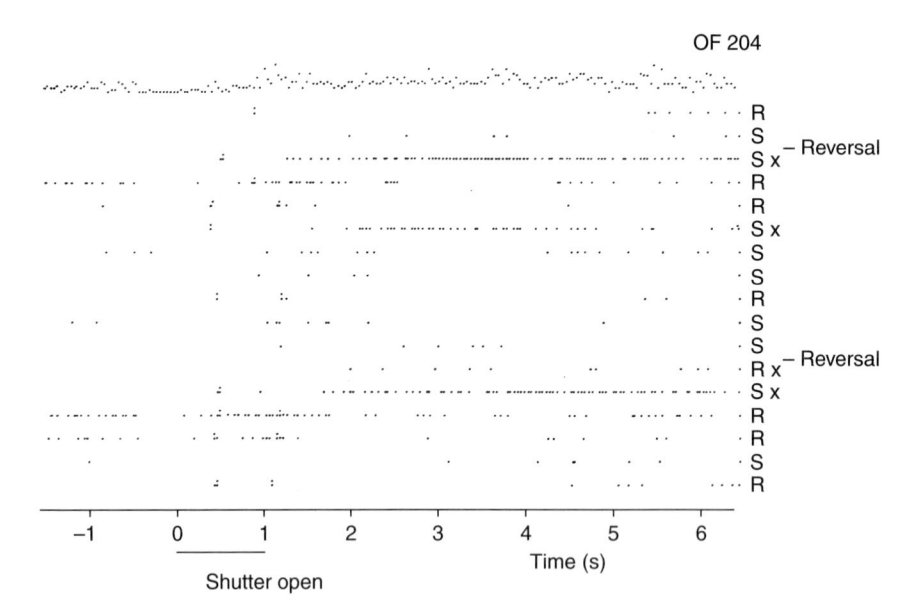

**Fig. 5.8** Error neuron: Responses of an OFC neuron that responded only when the monkey licked to a visual stimulus during reversal, expecting to obtain fruit juice reward, but actually obtaining the taste of aversive saline because it was the first trial of reversal. Each single dot represents an action potential; each vertically arranged double dot represents a lick response. The visual stimulus was shown at time 0 for 1 s. The neuron did not respond on most reward (R) or saline (S) trials, but did respond on the trials marked x, which were the first trials after a reversal of the visual discrimination on which the monkey licked to obtain reward, but actually obtained saline because the task had been reversed. Figure based on Thorpe et al. (1983).

a model of social learning described below, a face discrimination reversal task (Kringelbach and Rolls 2003).

Another type of visual information represented in the OFC is information about faces. There is a population of orbitofrontal neurons which respond in many ways similar to those in the temporal cortical visual areas (Rolls 1984, 1992a, 1996; 2000e; for a description of their properties see Wallis and Rolls 1997; Rolls and Deco 2002). The orbitofrontal face-responsive neurons, first observed by Thorpe et al. (1983), then by Rolls et al. (2006), tend to respond with longer latencies than temporal lobe neurons (140–200 ms typically, compared to 80–100 ms); also convey information about which face is being seen, by having different responses to different faces; and are typically rather harder to activate strongly than temporal cortical face-selective neurons, in that many of them respond much better to real faces than to two-dimensional images of faces on a video monitor (Rolls and Baylis 1986). Some of the OFC face-selective neurons are responsive to face gesture or movement. The findings are consistent with the likelihood that these neurons are activated via the inputs from the temporal cortical visual areas in which face-selective neurons are found (see Fig. 5.1). The significance of the neurons is

**Table 5.1** Different types of non-reward to which OFC neurons respond

| | D 90 | D 127 | D 153 | D 154 | D 195 | D 204 | D 262 | F 466 | B 24 | B 7B | B 37B | B 57B | D 44A | D 48A | D 20 | D 40 | D 61 | D 66 |
|---|---|---|---|---|---|---|---|---|---|---|---|---|---|---|---|---|---|---|
| *Visual discrimination* | | | | | | | | | | | | | | | | | | |
| Reversal | 1 | 0 | 1 | 0 | 0 | 1 | 1 | 0 | | | | | | 0 | | | | |
| Extinction | 1 | | | | | | | | | | | | | | | | | |
| *Ad lib licking* | | | | | | | | | | | | | | | | | | |
| Reversal | 1 | 1 | | 0 | 0 | 0 | | 0 | 1 | | | | | | | | | |
| Extinction | 0 | 0 | | 0 | 0 | 0 | | 0 | 1 | | | | | | | | | |
| Taste of saline | 0 | | 0 | 0 | 0 | 0 | 0 | 0 | 1 | 0 | 0 | 0 | 0 | 0 | 0 | 0 | 0 | 0 |
| Removal | 0 | | 0 | 1 | 1 | 1 | 0 | 1 | 0 | 1 | 1 | 1 | 1 | 1 | 1 | 1 | 1 | 1 |
| Visual arousal | 1 | | 1 | 0 | 0 | 0 | 0 | 0 | 1 | 0 | 0 | 0 | 0 | 0 | 1 | 0 | 0 | 0 |

*Source*: After Thorpe et al. (1983).

*Note*: Tasks (rows) (see text) in which individual neurons (columns) responded (1), did not respond (0), or were not tested (blank).

likely to be related to the fact that faces convey information that is important in social reinforcement in at least two ways that could be implemented by these neurons. The first is that some may encode face expression (cf. Hasselmo et al. 1989), which can indicate reinforcement. The second way is that they encode information about which individual is present, which by stimulus-reinforcement association learning is important in evaluating and utilizing learned reinforcing inputs in social situations, e.g. about the current reinforcement value as decoded by stimulus reinforcement association of a particular individual.

This neurophysiology has led to the following types of discovery in humans. In one example, Kringelbach and Rolls (2003) showed that activation of a part of the human OFC occurs during a face discrimination reversal task. In the task, the faces of two different individuals are shown, and when the correct face is selected, the expression turns into a smile (the expression turns to angry if the wrong face is selected). After a period of correct performance, the contingencies reverse, and the other face must be selected to obtain a smile expression as a reinforcer. It was found that activation of a part of the OFC occurred specifically in relation to the reversal, that is when a formerly correct face was chosen, but an angry face expression was obtained. In a control task, it was shown that the activations were not related just to showing an angry face expression. Thus in humans, there is a part of the OFC that responds selectively in relation to face expression specifically when it indicates that behavior should change, and this activation is error-related (Kringelbach and Rolls 2003) and occurs when the error neurons in the OFC become active (Thorpe et al. 1983).

Also prompted by the neuronal recording evidence of face and auditory neurons in the primate OFC (Rolls et al. 2006) (with a previous paper reporting auditory neuronal responses, Benevento et al. 1977), and the evidence from a PET study showing activation of a mid/lateral part of the OFC by unpleasant auditory information (Frey et al. 2000), it has further been shown that there are impairments in the identification of facial and vocal emotional expression in a group of patients with ventral frontal lobe damage who had socially inappropriate behavior (Hornak et al. 1996). The expression identification impairments could occur independently of perceptual impairments in facial recognition, voice discrimination, or environmental sound recognition. Poor performance on both expression tests was correlated with the degree of alteration of emotional experience reported by the patients. There was also a strong positive correlation between the degree of altered emotional experience and the severity of the behavioral problems (e.g., disinhibition) found in these patients (Hornak et al. 1996). A comparison group of patients with brain damage outside the ventral frontal lobe region, without these behavioral problems, was unimpaired on the face expression identification test, was significantly less impaired at vocal expression identification, and reported little subjective emotional change (Hornak et al. 1996). In recent studies, it has been shown that patients with discrete surgical lesions of the OFC may have face and/or voice expression identification impairments (Hornak et al. 2003).

## Somatosensory inputs to the OFC

Some neurons in the macaque OFC respond to the texture of food in the mouth. Some neurons alter their responses when the texture of a food is modified by adding gelatine or methyl cellulose, or by partially liquefying a solid food such as apple (Critchley et al. 1993). Another population of orbitofrontal neurons responds when a fatty food such as cream is in the mouth. These neurons can also be activated by pure fat such as glyceryl trioleate, and by non-fat substances with a fat-like texture such as paraffin oil (hydrocarbon) and silicone oil ($(Si(CH_3)_2O)_n$). These neurons thus provide information by somatosensory pathways that a fatty food is in the mouth (Rolls et al. 1999). These inputs are perceived as pleasant when hungry, because of the utility of ingestion of foods that are likely to contain essential fatty acids and to have a high calorific value (Rolls 2000d, 2005a). Satiety produced by eating a fatty food, cream, can decrease the responses of OFC neurons to the texture of fat in the mouth (Rolls et al. 1999). We have recently shown that the OFC receives inputs from a number of different oral texture channels, which together provide a rich sensory representation of what is in the mouth. Using a set of stimuli in which viscosity was systematically altered (carboxymethylcellulose with viscosity in the range 10–10,000 centiPoise), we have shown that some OFC neurons encode fat texture independently of viscosity (by a physical parameter that varies with the slickness of fat) (Verhagen et al. 2003); that other OFC neurons encode the viscosity of the texture in the mouth (with some neurons tuned to viscosity, and others showing increasing or decrease firing rates as viscosity increases) ((Rolls et al. 2003c)); and that other neurons have responses that indicate the presence of texture stimuli (such as grittiness and capsaicin) in the mouth independently of viscosity and slickness (Rolls et al. 2003c). A further population of OFC neurons represents the temperature of what is in the mouth (Kadohisa et al. 2004). These single-neuron recording studies thus provide clear evidence on the rich sensory representation of oral stimuli, and of their reward value that is provided in the primate OFC, and how this differs from what is represented in the primary taste cortex and in the amygdala (Kadohisa et al. 2005). In a complementary human functional neuroimaging study, it has been shown that activation of parts of the OFC, the primary taste cortex, and the mid-insular somatosensory region posterior to the insular taste cortex, have activations that are related to the viscosity of what is in the mouth, and that there is in addition a medial prefrontal/cingulate area where the mouth feel of fat is represented (de Araujo and Rolls 2004).

In addition to these oral somatosensory inputs to the OFC, there are also somatosensory inputs from other parts of the body, and indeed an fMRI investigation we have performed in humans indicates that pleasant and painful touch stimuli to the hand produce greater activation of the OFC relative to the somatosensory cortex than do affectively neutral stimuli (Francis et al. 1999; Rolls et al. 2003b); (see below). (In a PET study by Hagen et al.(2002), activation of the OFC by somatosensory stimuli has been confirmed, but no measures of the relative activations of the OFC relative to the somatosensory cortex for pleasant and painful vs neutral stimuli were made.)

### A representation of novel visual stimuli in the OFC

A population of neurons has been discovered in the primate OFC that responds to novel but not familiar visual stimuli, and takes typically a few trials to habituate (Rolls *et al.* 2005). The memories of these neurons last for at least 24 hours. Exactly what role these neurons have in memory is not yet known, but there are connections from the area in which these neurons are recorded to the temporal lobe, and activations in a corresponding area in humans are found when new visual stimuli must be encoded in memory (Frey and Petrides 2002, 2003).

## 5.5. A neurophysiological basis for stimulus-reinforcer association learning and reversal in the OFC

The neurophysiological, imaging, and lesion evidence described above suggests that one function implemented by the OFC is rapid stimulus-reinforcement association learning and the correction of these associations when reinforcement contingencies in the environment change. To implement this, the OFC has the necessary representation of primary (unlearned) reinforcers, including taste and somatosensory stimuli, as described above. In addition, in humans it has been shown in drug-naïve participants that amphetamine activates the OFC (Völlm *et al.* 2004). It also receives information about objects, e.g., visual view-invariant information (Booth and Rolls 1998; Hasselmo *et al.*1989a, b Rolls 2000e; Rolls and Deco 2002), and can associate this at the neuronal level with primary reinforcers such as taste, and reverse these associations very rapidly (Thorpe *et al.* 1983; Rolls, Critchley *et al.* 1996b). Another type of stimulus which can be conditioned in this way in the OFC is olfactory, though the learning here is slower. It is likely that auditory stimuli can be associated with primary reinforcers in the OFC, though there is less direct evidence of this yet. The OFC also has neurons that detect non-reward, which are likely to be used in behavioral extinction and reversal (Thorpe *et al.* 1983). These OFC error neurons may alter behavior not only by helping to reset the reinforcement association of neurons in the OFC (as described below), but also by sending a signal to the striatum which could be routed by the striatum to produce appropriate behaviors for non-reward (Rolls and Johnstone 1992; Rolls 1994; Williams *et al.* 1993; ). Indeed, one output from the OFC through which it may influence behavior is via the striatum (Rolls 1999b, 2005a). Some of the evidence for this is that neurons which reflect these orbitofrontal neuronal responses, are found in the ventral part of the head of the caudate nucleus and the ventral striatum, which receive from the OFC (Rolls *et al.* 1983b; Williams *et al.* 1993); and lesions of the ventral part of the head of the caudate nucleus impair visual discrimination reversal (Divac *et al.* 1967).

The dopamine projections to the prefrontal cortex and other areas are not likely to convey information about reward to the OFC, which instead is likely to be decoded by the neurons in the OFC that represent primary reinforcers, and the OFC neurons that learn associations of other stimuli to the primary reinforcers. Although it has been suggested that the firing of dopamine neurons may reflect the earliest signal in a task that indicates reward and could be use as an error signal during learning (Schultz et al. 2000), it is likely that any error information to which they fire is computed in and originates in the OFC (Rolls 2005a). The anterior cingulate cortex is another region that receives information from the OFC, and which may provide a site for the reinforcing outcomes to influence the actions selected (Cardinal et al. 2002; Rushworth et al. 2004; Rolls 2005a).

Decoding the reinforcement value of stimuli, which involves, for the previously neutral (e.g., visual) stimuli, learning their association with a primary reinforcer, often rapidly, and which may involve not only rapid learning but also rapid relearning and alteration of responses when reinforcement contingencies change, is a proposed function for the OFC (Rolls 1996, 2005a). This way of producing behavioral responses would be important for motivational and emotional behavior. It would be important, for example, in motivational behavior such as feeding and drinking by enabling primates to learn rapidly about the food reinforcement to be expected from visual stimuli (Rolls 1999b; Rolls 2004b, 2005a, b). This is important, for primates frequently eat more than 100 varieties of food; vision by visual-taste association learning can be used to identify when foods are ripe; and during the course of a meal, the pleasantness of the sight of a food eaten in the meal decreases in a sensory-specific way (Rolls et al. 1983a), a function that is probably implemented by the sensory-specific satiety-related responses of orbitofrontal visual neurons (Critchley and Rolls 1996b).

With respect to emotional behavior, decoding and rapidly readjusting the reinforcement value of visual signals is likely to be crucial, for emotions can be described as responses elicited by reinforcing signals[1] (Rolls 1990; Rolls 1999a, b, 2000b; Rolls 2005a). The ability to perform this learning very rapidly is probably very important in social situations in primates, in which reinforcing stimuli are continually being exchanged, and the reinforcement value of stimuli must be continually updated (relearned), based on the actual reinforcers received and given. Although the functions of the OFC in implementing the operation of reinforcers such as taste, smell, tactile and visual stimuli including faces are mostly understood, in humans the rewards processed in the OFC include quite general and abstract rewards such as working for 'points' or for monetary reward, as shown by visual discrimination reversal deficits in patients with OFC lesions working for these rewards (Berlin et al. 2004; Fellows and Farah 2003, 2005; Hornak et al. 2004; Rolls et al. 1994), and the activation of different parts of the human OFC by monetary gain vs. loss (O'Doherty et al., 2001a), and other reinforcers (Kringelbach and Rolls 2004).

Although the amygdala is concerned with some of the same functions as the OFC, and receives similar inputs (see Fig. 5.1), there is evidence that it may function less effectively in the very rapid learning and reversal of stimulus- reinforcement associations, as indicated by the greater difficulty in obtaining reversal from amygdala neurons (Rolls 1992b, 2000f;

---

[1] For the purposes of this chapter, a positive reinforcer or reward can be defined as a stimulus which the animal will work to obtain, and a punisher as a stimulus that will reduce the probability of an action on which it is contingent or that an animal will work to avoid or escape (Rolls, 2005a).

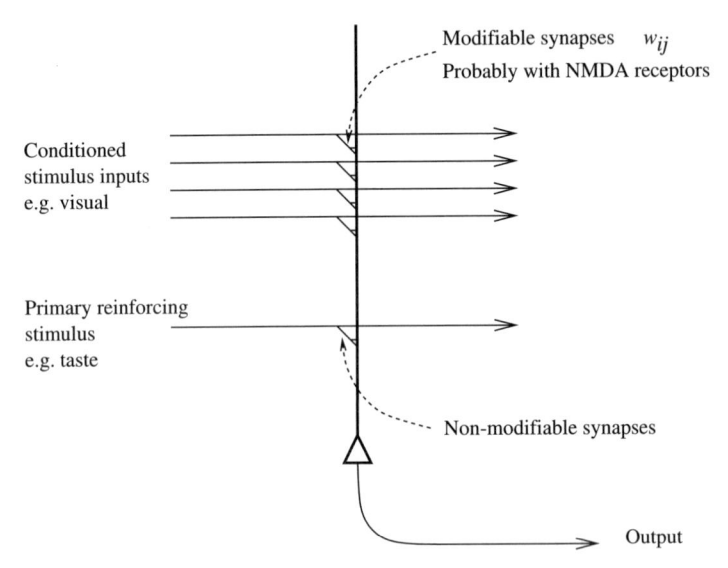

**Fig. 5.9** Pattern association between a primary reinforcer, such as the taste of food, which activates neurons through non-modifiable synapses, and a potential secondary reinforcer, such as the sight of food, which has modifiable synapses on to the same neurons. The associative rule for the synaptic modification is that if there is both presynaptic and post-synaptic firing, then that synapse should increase in strength. Such a mechanism appears to be implemented in the amygdala and OFC. (Homosynaptic) long-term depression (produced by presynaptic firing in the absence of strong postsynaptic firing) in a pattern associator in the amygdala could account for the habituating responses to novel visual stimuli which are not associated with primary reinforcers. For further details, see Rolls (2005a).

Wilson and Rolls 2005), and by the greater effect of orbitofrontal lesions in leading to continuing choice of no-longer rewarded stimuli (Jones and Mishkin 1972). In primates, the necessity for very rapid stimulus-reinforcement re-evaluation, and the development of powerful cortical learning systems, may result in the OFC effectively taking over this aspect of amygdala functions (Rolls 1992b, 1999b, 2005a).

How might this rapid stimulus-reinforcer association learning and reversal be implemented at the neuronal and neuronal network levels? One mechanism could be implemented by the Hebbian modification of synapses conveying the visual input onto taste-responsive neurons, implementing a pattern association network (Rolls and Treves 1998; Rolls 1999b, 2000c; Rolls and Deco 2002). Long-term potentiation would strengthen synapses from active conditioned stimulus neurons onto neurons responding to a primary reinforcer such as a sweet taste, and homosynaptic long-term depression would weaken synapses from the same active visual inputs if the neuron was not responding because an aversive primary reinforcer (e.g., a taste of saline) was being presented (see Fig. 5.9).

As noted above, the conditional reward neurons in the OFC convey information about the current reinforcement status of particular stimuli. In a new theory of how the OFC implements rapid, one-trial, reversal, these neurons play a key part, for particular conditional reward neurons (responding to, for example, 'green is now rewarding'; see example in Fig. 5.6) are biased on by a rule set of neurons if the association is being run direct, and are biased off if the association is being run reversed ('green is now not rewarding') (Deco and Rolls 2005b). A diagram of the neural architecture of the network model is shown in Fig. 5.10. One set of rule neurons in the short-term memory attractor network is active when the rule is direct, and a different set of neurons when the association is reversed. An attractor network is a network with excitatory recurrent collateral connections, that are associatively modified to store a set of patterns of firing, in which some of the neurons have high rates, and others have low rates. Each pattern of firing is associated with itself via the recurrent collateral associative connections, and hence these networks are called autoassociation networks. After learning a set of patterns, presenting one of the patterns can produce continuing firing of the neurons in that learnt pattern. The network thus implements a short-term memory. The term attractor refers to the fact that patterns close to a learnt pattern are attracted into firing that represents the closest learnt pattern, and is then stable (descriptions of the properties of attractor networks are provided elsewhere (Rolls and Deco 2002; Rolls 2005a)).

The state of the rule network is reversed when the error neurons fire in reversal, because the firing of the error neurons quenches the attractor by activating inhibitory neurons, and the opposite set of rule neurons emerges to activity after the quenching, because of some adaptation in the synapses or neurons in the rule attractor that have just been active. The error-detection neurons themselves may be triggered by a mismatch between what was expected when the visual stimulus was shown, and the primary reinforcer that was obtained, both of which are represented in the primate OFC (Thorpe *et al.* 1983). The whole system maps stimuli (such as green and blue) through a biased competition layer of conditional reward neurons in which the mapping is controlled by the biasing input from the rule neurons, to output neurons which fire if a stimulus is being shown which is currently associated with reward (Deco and Rolls 2005b). (See (Deco and Rolls 2005a; Deco and Rolls 2005c; Desimone and Duncan 1995; Rolls and Deco 2002) for a description of biased competition mechanisms.) The model gives an account of the presence of conditional reward and error neurons in the OFC, as well as of neurons that respond to whichever visual stimulus is currently associated with reward, and neurons which signal whether a reward or punishment has just been obtained. The model also suggests that the OFC may be especially appropriate for this rapid-reversal mechanism, because, in contrast to the amygdala, the OFC as a cortical structure has a well-developed system of recurrent collateral synapses between the pyramidal cells which provides an appropriate basis for implementing a working memory to hold the current rule. The model also shows how, when on a reversal trial, a reward is not obtained to a previously rewarded stimulus, on the very next trial when the recently punished stimulus is shown, it is treated as a reward, and it is chosen. This type of behavior is in fact illustrated in Fig. 5.8 (e.g., trials 4 and 14), and cannot be accounted for by a new association of the now-to-be rewarded stimulus with reward, for in its recent past the stimulus has been associated with saline. Thus this type of one-trial rapid reversal cannot be accounted for by direct stimulus-reward association learning, and a rule-based system, such as the type implemented in the model, is needed.

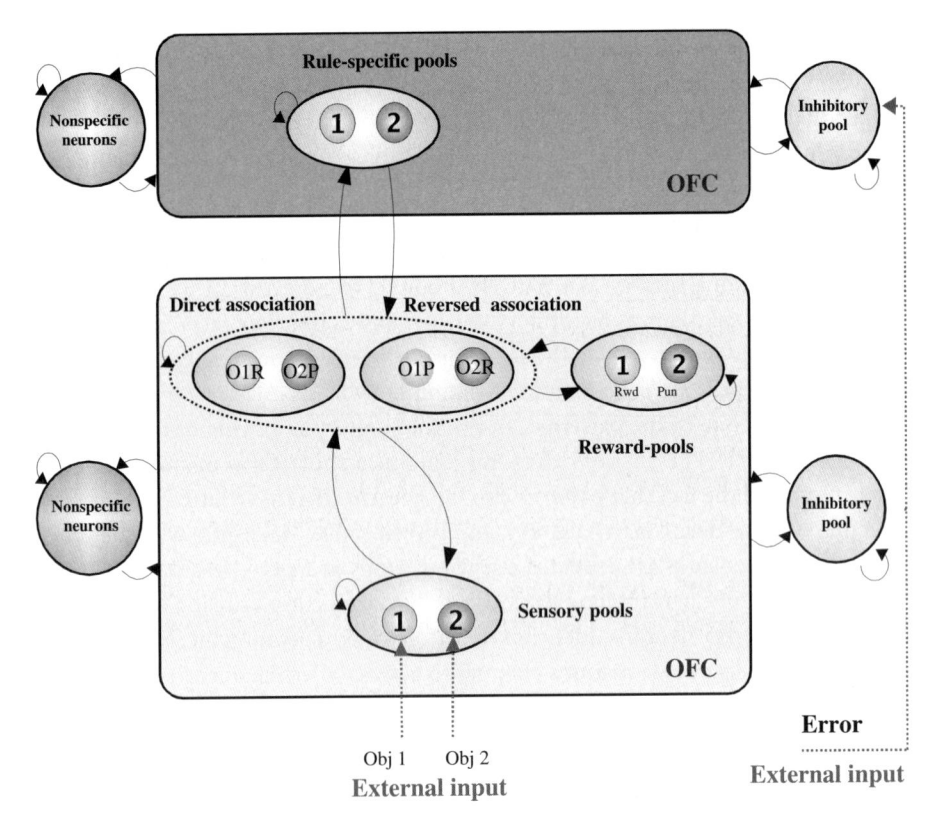

**Fig. 5.10** Cortical architecture of a reward reversal model implemented at the level of integrate-and-fire neurons so that the spiking activity of neurons in the model can be implemented and compared to real neuronal recordings. There is a rule module (top) and a sensory–intermediate neuron-reward module (below). Neurons within each module are fully connected, and form attractor states. The sensory–intermediate neuron-reward module consists of three hierarchically organized levels of attractor network, with stronger synaptic connections in the forward than the backprojection direction. The intermediate level of the sensory–intermediate neuron-reward module contains conditional reward neurons (as recorded in the OFC) that respond to combinations of an object and its association with reward or punishment, for example, object 1–reward (O1R, in the direct association set of pools), and object 1- punishment (O1P in the reversed association set of pools). The rule module acts as a biasing input to bias the competition between the object–reward combination neurons at the intermediate level of the sensory–intermediate neuron–reward module. The synaptic current that flows into the cells is mediated by four different families of receptors. The recurrent excitatory postsynaptic currents are given by two different types of EPSP (excitatory post-synaptic potentials), respectively mediated by AMPA and NMDA receptors. These two glutamatergic excitatory synapses are on the pyramidal cells and interneurons. The external background is mediated by AMPA synapses on pyramidal cells and interneurons. The visual input is also introduced by AMPA synapses on specific pyramidal cells. Inhibitory GABAergic synapses on pyramidal cells and interneurons yield corresponding IPSPs (inhibitory post-synaptic potentials). OFC, OFC. Figure based on Deco and Rolls (2005).

The model has been worked out in detail at the level of integrate-and-fire spiking neurons, and makes predictions about further types of neuron expected to be present in the OFC such as rule neurons (Deco and Rolls 2005b; Rolls 2005a).

In conclusion, the neuronal data from the OFC provides an essential component for understanding how the OFC functions, for it provides the data necessary to understand what is represented in an area, how it is represented, and around which computational and functional models can be built. Modelling at the integrate-and-fire level (Deco and Rolls 2005b) enables the different levels of explanation, from neuronal firing to fMRI signals (Deco et al. 2004) to neuropsychology (Rolls and Deco 2002) to be integrated.

## Acknowledgment

The author has worked on some of the experiments described above with I. Araujo, L.L. Baylis, G.C. Baylis, R. Bowtell, A.D. Browning, H.D. Critchley, S. Francis, M.E. Hasselmo, J. Hornak, M. Kadohisa, M. Kringelbach, C.M. Leonard, F. McGlone, F. Mora, J. O'Doherty, D.I. Perrett, T.R. Scott, S.J. Thorpe, J. Verhagen, E.A. Wakeman and F.A.W. Wilson, and their collaboration is sincerely acknowledged. Some of the research described was supported by the Medical Research Council, PG8513790 and PG9826105.

## References

Barbas, H. (1988) Anatomic organization of basoventral and mediodorsal visual recipient prefrontal regions in the rhesus monkey. *Journal of Comparative Neurology* **276**:313–342.

Barbas, H. (1993) Organization of cortical afferent input to the orbitofrontal area in the rhesus monkey. *Neuroscience* **56**:841–864.

Barbas, H. (1995) Anatomic basis of cognitive-emotional interactions in the primate prefrontal cortex. *Neuroscience and Biobehavioral Reviews* **19**:499–510.

Barbas, H. and Pandya, D. N. (1989) Architecture and intrinsic connections of the prefrontal cortex in the rhesus monkey. *Journal of Computational Neurology* **286**:353–375.

Baylis, L. L. and Gaffan, D. (1991) Amygdalectomy and ventromedial prefrontal ablation produce similar deficits in food choice and in simple object discrimination learning for an unseen reward. *Experimental Brain Research* **86**:617–622.

Baylis, L. L. and Rolls, E.T. (1991) Responses of neurons in the primate taste cortex to glutamate. *Physiology and Behavior* **49**:973–979.

Baylis, L. L., Rolls, E. T., and Baylis, G.C. (1995) Afferent connections of the orbitofrontal cortex taste area of the primate. *Neuroscience* **64**:801–812.

Benevento, L. A., Fallon, J., Davis, B.J., and Rezak, M. (1977) Auditory-visual interaction in single cells in the cortex of the superior temporal sulcus and the orbital frontal cortex of the macaque monkey. *Experimental Neurology* **57**:849–872.

Berlin, H., Rolls, E. T., and Kischka, U. (2004) Impulsivity, time perception, emotion, and reinforcement sensitivity in patients with orbitofrontal cortex lesions. *Brain* **127**:1108–1126.

Berlin, H., Rolls, E.T., and Iversen, S. D. (2005) Borderline Personality Disorder, impulsivity and the orbitofrontal cortex. *American Journal of Psychiatry* **162**:2360–2373.

Booth, M. C. A. and Rolls, E. T. (1998) View-invariant representations of familiar objects by neurons in the inferior temporal visual cortex. *Cerebral Cortex* **8**:510–523.

Butter, C. M. (1969) Perseveration in extinction and in discrimination reversal tasks following selective prefrontal ablations in Macaca mulatta. *Physiology and Behavior* **4**:163–171.

Butter, C. M., McDonald, J. A., and Snyder, D.R. (1969) Orality, preference behavior, and reinforcement value of non-food objects in monkeys with orbital frontal lesions. *Science* **164**:1306–1307.

Butter, C. M., Snyder, D. R., and McDonald, J. A. (1970) Effects of orbitofrontal lesions on aversive and aggressive behaviors in rhesus monkeys. *Journal of Comparative and Physiological Psychology* **72**:132–144.

Butter, C. M. and Snyder, D. R. (1972) Alterations in aversive and aggressive behaviors following orbitofrontal lesions in rhesus monkeys. *Acta Neurobiologiae Experimentalis* **32**:525–565.

Cardinal, N., Parkinson, J. A., Hall, J., and Everitt, B. J. (2002) Emotion and motivation: the role of the amygdala, ventral striatum, and prefrontal cortex. *Neuroscience and Biobehavioral Reviews* **26**:321–352.

Carmichael, S. T. and Price, J. L. (1994) Architectonic subdivision of the orbital and medial prefrontal cortex in the macaque monkey. *Journal of Comparative Neurology* **346**:366–402.

Carmichael, S. T. and Price, J. L. (1995) Sensory and premotor connections of the orbital and medial prefrontal cortex of macaque monkeys. *Journal of Comparative Neurology* **363**:642–664.

Carmichael, S. T., Clugnet, M-C., and Price, J.L. (1994) Central olfactory connections in the macaque monkey. *Journal of Comparative Neurology* **346**:403–434.

Critchley, H. D. and Rolls, E. T. (1996a) Responses of primate taste cortex neurons to the astringent tastant tannic acid. *Chemical Senses* **21**:135–145.

Critchley, H. D. and Rolls, E. T. (1996b) Hunger and satiety modify the responses of olfactory and visual neurons in the primate orbitofrontal cortex. *Journal of Neurophysiology* **75**:1673–1686.

Critchley, H. D. and Rolls, E. T. (1996c) Olfactory neuronal responses in the primate orbitofrontal cortex: analysis in an olfactory discrimination task. *Journal of Neurophysiology* **75**:1659–1672.

Critchley, H. D., Rolls, E. T., and Wakeman, E. A. (1993) Orbitofrontal cortex responses to the texture, taste, smell and sight of food. *Appetite* **21**:170.

Damasio, A. R. (1994) *Descartes' Error*. New York: Putnam

de Araujo, I. E. T., Kringelbach, M. L., and Rolls, E. T. (2003a) Taste-olfactory convergence, and the representation of the pleasantness of flavor, in the human brain. *European Journal of Neuroscience* **18**:2374–2390.

de Araujo, I. E. T., Kringelbach, M. L., Rolls, E. T., and Hobden, P. (2003b) The representation of umami taste in the human brain. *Journal of Neurophysiology* **90**:313–319.

de Araujo, I.E.T., Kringelbach, M.L., Rolls, E.T., and McGlone, F. (2003c) Human cortical responses to water in the mouth, and the effects of thirst. *Journal of Neurophysiology* **90**:1865–1876.

de Araujo, I. E. T. and Rolls, E. T. (2004) The representation in the human brain of food texture and oral fat. *Journal of Neuroscience* **24**:3086–3093.

de Araujo, I. E. T., Rolls, E.T., Velazco, M. I., Margot, C., and Cayeux, I. (2005) Cognitive modulation of olfactory processing. *Neuron* **46**:671–679.

Deco, G. and Rolls, E. T. (2005a) Attention, short-term memory, and action selection: a unifying theory. *Progress in Neurobiology*.

Deco, G. and Rolls, E. T. (2005b). Synaptic and spiking dynamics underlying reward reversal in orbitofrontal cortex. *Cerebral Cortex* **15**:15–30.

Deco, G. and Rolls, E. T. (2005c) Neurodynamics of biased competition and co-operation for attention: a model with spiking neurons. *Journal of Neurophysiology* **94**:295–313.

Deco, G., Rolls, E. T., and Horwitz, B. (2004) 'What' and 'where' in visual working memory: a computational neurodynamical perspective for integrating fMRI and single-neuron data. *J Cog Neurosci* **16**:683–701.

Desimone, R. and Duncan, J. (1995) Neural mechanisms of selective visual attention. *Annual Review of Neuroscience* **18**:193–222.

Divac, I., Rosvold, H. E., and Szwarcbart, M. K. (1967) Behavioral effects of selective ablation of the caudate nucleus. *Journal of Comparative Physiological Psychology* **63**:184–190.

Fellows, L. K. and Farah, M. J. (2003) Ventromedial frontal cortex mediates affective shifting in humans: evidence from a reversal learning paradigm. *Brain* **126**:1830–1837.

Fellows, L. K. and Farah, M. J. (2005) Different underlying impairments in decision-making following ventromedial and dorsolateral frontal lobe damage in humans. *Cerebral Cortex* **15**:58–63.

Francis, S., Rolls, E. T., Bowtell, R. *et al.* (1999) The representation of the pleasant touch in the brain and its relationship with taste and olfactory areas. *Neuroreport* **10**:453–459.

Frey, S., Kostopoulos, P., and Petrides, M. (2000) Orbitofrontal involvement in the processing of unpleasant auditory information. *European Journal of Neuroscience* **12**:3709–3712.

Frey, S. and Petrides, M. (2002) Orbitofrontal cortex and memory formation. *Neuron* **36**:171–176.

Frey, S. and Petrides, M. (2003) Greater orbitofrontal activity predicts better memory for faces. *European Journal of Neuroscience* **17**:2755–2758.

Hagen, M. C., Zald, D. H., Thornton, T. A., and Pardo, J. V.(2002) Somatosensory processing in the human inferior prefrontal cortex. *Journal of Neurophysiology* **88**:1400–1406.

Hasselmo, M. E., Rolls, E. T., and Baylis, G.C. (1989a) The role of expression and identity in the face-selective responses of neurons in the temporal visual cortex of the monkey. *Behavioral Brain Research* **32**:203–218.

Hasselmo, M. E., Rolls, E. T., Baylis, G. C., and Nalwa, V. (1989b) Object-centred encoding by face-selective neurons in the cortex in the superior temporal sulcus of the the monkey. *Experimental Brain Research* **75**:417–429.

Hornak, J., Rolls, E. T., and Wade, D. (1996) Face and voice expression identification in patients with emotional and behavioral changes following ventral frontal lobe damage. *Neuropsychologia* **34**:247–261.

Hornak, J., Bramham, J., Rolls, E.T. *et al.*(2003) Changes in emotion after circumscribed surgical lesions of the orbitofrontal and cingulate cortices. *Brain* **126**:1691–1712.

Hornak, J., O'Doherty, J., Bramham, J., *et al.* (2004) Reward-related reversal learning after surgical excisions in orbitofrontal and dorsolateral prefrontal cortex in humans. *Journal of Cognitive Neuroscience*, **16**:463–478.

Insausti, R., Amaral, D. G., and Cowan, W. M. (1987) The entorhinal cortex of the monkey. II. Cortical afferents. *Journal of Comparative Neurology* **264**:356–395.

Iversen, S. D. and Mishkin, M. (1970) Perseverative interference in monkeys following selective lesions of the inferior prefrontal convexity. *Experimental Brain Research* **11**:376–386.

Izquierdo, A. and Murray, E. A. (2004) Combined unilateral lesions of the amygdala and orbital prefrontal cortex impair affective processing in rhesus monkeys. *Journal of Neurophysiology* **91**:2023–2039.

Izquierdo, A., Suda, R. K., and Murray, E. A. (2004) Bilateral orbital prefrontal cortex lesions in rhesus monkeys disrupt choices guided by both reward value and reward contingency. *Journal of Neuroscience* **24**:7540–7548.

Johnson, T. N., Rosvold, H. E., and Mishkin, M. (1968) Projections from behaviorally defined sectors of the prefrontal cortex to the basal ganglia, septum and diencephalon of the monkey. *Experimental Neurology* **21**:20–34.

Jones, B. and Mishkin, M. (1972) Limbic lesions and the problem of stimulus-reinforcement associations. *Experimental Neurology* **36**:362–377.

Jones-Gotman, M. and Zatorre, R. J. (1988) Olfactory identification in patients with focal cerebral excision. *Neuropsychologia* **26**:387–400.

Kadohisa, M., Rolls, E.T., and Verhagen, J. V. (2004) Orbitofrontal cortex neuronal representation of temperature and capsaicin in the mouth. *Neuroscience* **127**:207–221.

Kadohisa, M., Rolls, E. T., and Verhagen, J.V. (2005) Neuronal representations of stimuli in the mouth: the primate insular taste cortex, orbitofrontal cortex, and amygdala. *Chemical Senses* **30**:401–419.

Kemp, J. M. and Powell, T.P.S. (1970) The cortico-striate projections in the monkey. *Brain* **93**:525–546.

Kolb, B. and Whishaw, I.Q. (2003) *Fundamentals of Human Neuropsychology*. New York:Worth.

Kowalska, D-M., Bachevalier, J., and Mishkin, M. (1991) The role of the inferior prefrontal convexity in performance of delayed nonmatching-to-sample. *Neuropsychologia* **29**:583–600.

Kringelbach, M. L., O'Doherty, J., Rolls, E. T., and Andrews, C. (2003) Activation of the human orbitofrontal cortex to a liquid food stimulus is correlated with its subjective pleasantness. *Cerebral Cortex* **13**:1064–1071.

Kringelbach, M. L. and Rolls, E. T. (2003) Neural correlates of rapid reversal learning in a simple model of human social interaction. *Neuroimage* **20**:1371–1383.

Kringelbach, M. L. and Rolls, E. T. (2004) The functional neuroanatomy of the human orbitofrontal cortex: evidence from neuroimaging and neuropsychology. *Progress in Neurobiology* **72**:341–372.

Mishkin, M. and Manning, F. J. (1978) Non-spatial memory after selective prefrontal lesions in monkeys. *Brain Research* **143**:313–324.

Mora, F., Avrith, D. B., Phillips, A. G., and Rolls, E. T. (1979) Effects of satiety on self-stimulation of the orbitofrontal cortex in the monkey. *Neuroscience Letters* **13**:141–145.

Mora, F., Avrith, D. B., and Rolls, E. T. (1980) An electrophysiological and behavioral study of self-stimulation in the orbitofrontal cortex of the rhesus monkey. *Brain Research Bulletin* **5**:111–115.

Morecraft, R.J., Geula, C., and Mesulam, M-M. (1992) Cytoarchitecture and neural afferents of orbitofrontal cortex in the brain of the monkey. *Journal of Comparative Neurology* **232**:341–358.

Morris, J. S. and Dolan, R. J. (2001) Involvement of human amygdala and orbitofrontal cortex in hunger-enhanced memory for food stimuli. *Journal of Neuroscience* **21**:5304–5310.

Nauta, W. J. H. (1964) Some efferent connections of the prefrontal cortex in the monkey. In: Warren, J.M. and Akert, K., (eds.) *The Frontal Granular Cortex and Behavior*, pp. 397–407. New York: McGraw Hill.

Norgren, R. (1984) Central neural mechanisms of taste. In: Darien-Smith, I., (ed.) *Handbook of Physiology—The Nervous System III. Sensory Processes 1*, pp. 1087–1128. Washington: American Physiological Society.

O'Doherty, J., Rolls, E.T., Francis, S., et al. (2000) Sensory-specific satiety related olfactory activation of the human orbitofrontal cortex. *Neuroreport* **11**:893–897.

O'Doherty, J., Kringelbach, M.L., Rolls, E.T., Hornak, J., and Andrews, C. (2001a) Abstract reward and punishment representations in the human orbitofrontal cortex. *Nature Neuroscience* **4**:95–102.

O'Doherty, J., Rolls, E.T., Francis, S., Bowtell, R., and McGlone, F. (2001b) The representation of pleasant and aversive taste in the human brain. *Journal of Neurophysiology* **85**:1315–1321.

Ongur, D. and Price, J.L. (2000) The organization of networks within the orbital and medial prefrontal cortex of rats, monkeys and humans. *Cerebral Cortex* **10**:206–219.

Pandya, D. N. and Yeterian, E. H. (1996) Comparison of prefrontal architecture and connections. *Philosophical Transactions of the Royal Society of London Series B* **351**:1423–1431.

Passingham, R. (1975) Delayed matching after selective prefrontal lesions in monkeys (Macaca mulatta) *Brain Research* **92**:89–102.

Petrides, M. and Pandya, D.N. (1995) Comparative architectonic analysis of the human and macaque frontal cortex. In Boller, F. and Grafman, J., eds. *Handbook of Neuropsychology*, pp. 17–58. Amsterdam: Elsevier Science.

Price, J. L. (2006) Connections of frontal cortex. In: Zald, D.H. and Rauch, S.L. (eds.) *The Orbitofrontal Cortex*. Oxford: Oxford University Press.

Price, J. L., Carmichael, S.T., Carnes, K. M., Clugnet, M-C., Kuroda, M., and Ray, J. P. (1991) Olfactory input to the prefrontal cortex. In: Davis, J.L. and Eichenbaum, H.(eds.) *Olfaction: A Model System for Computational Neuroscience*, pp. 101–120. Cambridge, Mass.: MIT Press.

Rao, S. C., Rainer, G., and Miller, E.K. (1997) Integration of what and where in the primate prefrontal cortex. *Science* **276**:821–824.

Rolls, E. T. (1984) Neurons in the cortex of the temporal lobe and in the amygdala of the monkey with responses selective for faces. *Human Neurobiology* **3**:209–222.

Rolls, E. T. (1989) Information processing and basal ganglia function. In C Kennard and M Swash, eds. *Hierarchies in Neurology*, pp. 123–142. London:Springer-Verlag.

Rolls, E. T. (1990) A theory of emotion, and its application to understanding the neural basis of emotion. *Cognition and Emotion* **4**:161–190.

Rolls, E. T. (1992a) Neurophysiological mechanisms underlying face processing within and beyond the temporal cortical visual areas. *Philosophical Transactions of the Royal Society of London B* **335**:11–21.

Rolls, E. T. (1992b) Neurophysiology and functions of the primate amygdala. In Aggleton, J.P., (ed.) *The Amygdala*, pp. 143–165. New York: Wiley-Liss.

Rolls, E. T. (1994). Neurophysiology and cognitive functions of the striatum. *Revue Neurologique (Paris)* **150**:648–660.

Rolls, E. T. (1995) Central taste anatomy and neurophysiology. In Doty, R.L., ed. *Handbook of Olfaction and Gustation*, pp. Ch. 24, pp. 549–573. New York: Dekker.

Rolls, E. T. (1996) The orbitofrontal cortex. *Philosophical Transactions of the Royal Society of London* B **351**:1433–1444.

Rolls, E. T. (1997) Taste and olfactory processing in the brain and its relation to the control of eating. *Critical Reviews in Neurobiology* **11**:263–287

Rolls, E. T. (1999a) The functions of the orbitofrontal cortex. *Neurocase* **5**:301–312.

Rolls, E. T. (1999b) *The Brain and Emotion*. Oxford: Oxford University Press.

Rolls, E. T. (2000a) The representation of umami taste in the taste cortex. *Journal of Nutrition* **130**:S960–S965.

Rolls, E. T. (2000b) Précis of The Brain and Emotion. *Behavioral and Brain Sciences* **23**:177–233.

Rolls, E. T. (2000c) Memory systems in the brain. *Annual Review of Psychology* **51**:599–630.

Rolls, E. T. (2000d) Taste, olfactory, visual and somatosensory representations of the sensory properties of foods in the brain, and their relation to the control of food intake. In: Berthoud, H.-R. and Seeley, R.J., (eds). *Neural and Metabolic Control of Macronutrient Intake*, pp. 247–262. Boca-Raton, Florida: CRC Press.

Rolls, E. T. (2000e) Functions of the primate temporal lobe cortical visual areas in invariant visual object and face recognition. *Neuron* **27**:205–218.

Rolls, E. T. (2000f) Neurophysiology and functions of the primate amygdala, and the neural basis of emotion. In Aggleton, J.P. (ed). *The Amygdala: A Functional Analysis*, pp. 447–478. Oxford: Oxford University Press.

Rolls, E. T. (2004a) The functions of the orbitofrontal cortex. *Brain and Cognition* **55**:11–29.

Rolls, E. T. (2004b) Smell, taste, texture and temperature multimodal representations in the brain, and their relevance to the control of appetite. *Nutrition Reviews* **62**:193–204.

Rolls, E. T. (2005a) *Emotion Explained*. Oxford: Oxford University Press.

Rolls, E. T. (2005b) Taste, olfactory, and food texture processing in the brain, and the control of food intake. *Physiology and Behavior* **85**:45–56.

Rolls, E. T. and Rolls, J.H. (1997) Olfactory sensory-specific satiety in humans. *Physiology and Behavior* **61**:461–473.

Rolls, E. T. and Baylis, G.C. (1986) Size and contrast have only small effects on the responses to faces of neurons in the cortex of the superior temporal sulcus of the monkey. *Experimental Brain Research* **65**:38–48.

Rolls, E. T. and Johnstone, S. (1992) Neurophysiological analysis of striatal function. In: Vallar, G. and Wallesch, C.W., eds. *Neuropsychological Disorders Associated with Subcortical Lesions*, pp. 61–97. Oxford: Oxford University Press.

Rolls, E. T. and Baylis, L.L. (1994) Gustatory, olfactory, and visual convergence within the primate orbitofrontal cortex. *Journal of Neuroscience* **14**:5437–5452.

Rolls, E. T. and Treves, A. (1998) *Neural Networks and Brain Function*. Oxford: Oxford University Press.

Rolls, E. T. and Deco, G. (2002) *Computational Neuroscience of Vision*. Oxford: Oxford University Press.

Rolls, E. T. and Scott, T.R. (2003) Central taste anatomy and neurophysiology. In Doty, R.L.(ed.) *Handbook of Olfaction and Gustation*, pp. 679–705. New York: Dekker.

Rolls, E. T. and Xiang, J.-Z. (2005) Reward-spatial view representations and learning in the hippocampus. *Journal of Neuroscience* **25**:6167–6174.

Rolls, E. T., Judge, S.J., and Sanghera, M. (1977) Activity of neurones in the inferotemporal cortex of the alert monkey. *Brain Research* **130**:229–238.

Rolls, E. T., Burton, M.J., and Mora, F. (1980) Neurophysiological analysis of brain-stimulation reward in the monkey. *Brain Research* **194**:339–357.

Rolls, B. J., Rolls, E.T., Rowe, E. A., and Sweeney, K. (1981a) Sensory specific satiety in man. *Physiology and Behavior* **27**:137–142.

Rolls, B. J., Rolls, E.T., Rowe, E. A., and Sweeney, K. (1981b) How sensory properties of foods affect human feeding behavior. *Physiology and Behavior* **29**:409–417.

Rolls, E. T., Rolls, B.J., and Rowe, E.A. (1983a) Sensory-specific and motivation-specific satiety for the sight and taste of food and water in man. *Physiology and Behavior* **30**:185–192.

Rolls, E. T., Thorpe, S.J., and Maddison, S.P. (1983b) Responses of striatal neurons in the behaving monkey. 1: Head of the caudate nucleus. *Behavioral Brain Research* **7**:179–210.

Rolls, E. T., Scott, T.R., Sienkiewicz, Z.J., and Yaxley, S. (1988) The responsiveness of neurones in the frontal opercular gustatory cortex of the macaque monkey is independent of hunger. *Journal of Physiology* **397**:1–12.

Rolls, E. T., Sienkiewicz, Z.J., and Yaxley, S. (1989) Hunger modulates the responses to gustatory stimuli of single neurons in the caudolateral orbitofrontal cortex of the macaque monkey. *European Journal of Neuroscience* **1**:53–60.

Rolls, E. T., Yaxley, S., and Sienkiewicz, Z.J. (1990) Gustatory responses of single neurons in the caudolateral orbitofrontal cortex of the macaque monkey. *Journal of Neurophysiology* **64**:1055–1066.

Rolls, E. T., Hornak, J., Wade, D., and McGrath, J. (1994) Emotion-related learning in patients with social and emotional changes associated with frontal lobe damage. *Journal of Neurology, Neurosurgery and Psychiatry* **57**:1518–1524.

Rolls, E. T., Critchley, H., Wakeman, E.A., and Mason, R. (1996a) Responses of neurons in the primate taste cortex to the glutamate ion and to inosine 5`-monophosphate. *Physiology and Behavior* **59**:991–1000.

Rolls, E. T., Critchley, H.D., Mason, R., and Wakeman, E.A. (1996b) Orbitofrontal cortex neurons: role in olfactory and visual association learning. *Journal of Neurophysiology* **75**:1970–1981.

Rolls, E. T., Critchley, H.D., and Treves, A. (1996c) The representation of olfactory information in the primate orbitofrontal cortex. *Journal of Neurophysiology* **75**:1982–1996.

Rolls, E. T., Critchley, H.D., Browning, A., and Hernadi, I. (1998) The neurophysiology of taste and olfaction in primates, and umami flavor. *Annals of the New York Academy of Sciences* **855**:426–437.

Rolls, E. T., Critchley, H.D., Browning, A.S., Hernadi, A., and Lenard, L. (1999) Responses to the sensory properties of fat of neurons in the primate orbitofrontal cortex. *Journal of Neuroscience* **19**:1532–1540.

Rolls, E. T., Kringelbach, M.L., and de Araujo, I.E.T. (2003a) Different representations of pleasant and unpleasant odors in the human brain. *European Journal of Neuroscience* **18**:695–703.

Rolls, E. T., O'Doherty, J., Kringelbach, M.L., Francis, S., Bowtell, R., and McGlone, F. (2003b) Representations of pleasant and painful touch in the human orbitofrontal and cingulate cortices. *Cerebral Cortex* **13**:308–317.

Rolls, E. T., Verhagen, J.V., and Kadohisa, M. (2003c) Representations of the texture of food in the primate orbitofrontal cortex: neurons responding to viscosity, grittiness and capsaicin. *Journal of Neurophysiology* **90**:3711–3724.

Rolls, E. T., Browning, A.S., Inoue, K., and Hernadi, S. (2005) Novel visual stimuli activate a population of neurons in the primate orbitofrontal cortex. *Neurobiology of Learning and Memory* **84**:111–123.

Rolls, E. T., Critchley, H.D., Browning, A.S., and Inoue, K. (2006) Face-selective and auditory neurons in the primate orbitofrontal cortex. *Experimental Brain Research* **170**:74–87.

Rosenkilde, C. E. (1979) Functional heterogeneity of the prefrontal cortex in the monkey: a review. *Behavioral and Neural Biology* **25**:301–345.

Rosenkilde, C. E., Bauer, R.H., and Fuster, J.M. (1981) Single unit activity in ventral prefrontal cortex in behaving monkeys. *Brain Research* **209**:375–394.

Rushworth, M. F., Walton, M.E., Kennerley, S.W., and Bannerman, D.M. (2004) Action sets and decisions in the medial frontal cortex. *Trends in Cognitive Sciences* **8**:410–417.

Schoenbaum, G. and Eichenbaum, H. (1995) Information encoding in the rodent prefrontal cortex.1. Single-neuron activity in orbitofrontal cortex compared with that in piriform cortex. *Journal of Neurophysiology* **74**:733–750.

Schultz, W., Tremblay, L., and Hollerman, J. R. (2000) Reward processing in primate orbitofrontal cortex and basal ganglia. *Cerebral Cortex* **10**:272–284.

Scott, T. R., Yaxley, S., Sienkiewicz, Z. J., and Rolls, E. T. (1986) Gustatory responses in the frontal opercular cortex of the alert cynomolgus monkey. *Journal of Neurophysiology* **56**:876–890.

Seltzer, B. and Pandya, D.N. (1989) Intrinsic connections and architectonics of the superior temporal sulcus in the rhesus monkey. *Journal of Comparative Neurology* **290**:451–471.

Small, D. M., Zald, D. H., Jones-Gotman, M., *et al.* (1999) Human cortical gustatory areas: a review of functional neuroimaging data. *Neuroreport*, 10, 7–14.

Takagi, S. F. (1991) Olfactory frontal cortex and multiple olfactory processing in primates. In: Peters, A. and Jones, E.G. (eds) *Cerebral Cortex*, pp. 133–152. New York: Plenum Press.

Thorpe, S. J., Rolls, E. T., and Maddison, S. (1983) Neuronal activity in the orbitofrontal cortex of the behaving monkey. *Experimental Brain Research* **49**:93–115.

Tremblay, L. and Schultz, W. (1999) Relative reward preference in primate orbitofrontal cortex. *Nature* **398**:704–708.

Verhagen, J. V., Kadohisa, M., and Rolls, E. T. (2004) The primate insular/opercular taste cortex: neuronal representations of the viscosity, fat texture, grittiness and taste of foods in the mouth. *Journal of Neurophysiology* **92**:1685–1699.

Verhagen, J. V., Rolls, E. T., and Kadohisa, M. (2003) Neurons in the primate orbitofrontal cortex respond to fat texture independently of viscosity. *Journal of Neurophysiology* **90**:1514–1525.

Völlm, B. A., de Araujo, I. E. T., Cowen, P. J. *et al.* (2004) Methamphetamine activates reward circuitry in drug naïve human subjects. *Neuropsychopharmacology* **29**:1715–1722.

Wallis, G. and Rolls, E.T. (1997) Invariant face and object recognition in the visual system. *Progress in Neurobiology* **51**:167–194.

Williams, G. V., Rolls, E. T., Leonard, C. M., and Stern, C. (1993) Neuronal responses in the ventral striatum of the behaving macaque. *Behavioral Brain Research* **55**:243–252.

Wilson, F. A. W., O'Scalaidhe, S. P. O., and Goldman-Rakic, P. S. (1993) Dissociation of object and spatial processing domains in primate prefrontal cortex. *Science* **260**:1955–1958.

Wilson, F. A. W. and Rolls, E. T. (2005) The primate amygdala and reinforcement: a dissociation between rule-based and associatively-mediated memory revealed in amygdala neuronal activity. *Neuroscience* **133**:1061–1072.

Yaxley, S., Rolls, E.T., and Sienkiewicz, Z.J. (1988) The responsiveness of neurons in the insular gustatory cortex of the macaque monkey is independent of hunger. *Physiology and Behavior* **42**:223–229.

Yaxley, S., Rolls, E.T., and Sienkiewicz, Z.J. (1990) Gustatory responses of single neurons in the insula of the macaque monkey. *Journal of Neurophysiology* **63**:689–700.

Zatorre, R.J., Jones-Gotman, M., Evans, A.C., and Meyer, E. (1992) Functional localization of human olfactory cortex. Nature, **360**: 339–340.

# Chapter 6

# The chemical senses

Jay A. Gottfried, Dana M. Small, and David H. Zald

## 6.1. Introduction

While the orbitofrontal cortex (OFC) receives information from all sensory modalities, it is most intimately linked to the chemical senses of smell and taste. Portions of the OFC receive robust projections directly from the primary olfactory and gustatory cortices, and these areas are often referred to as the secondary olfactory and gustatory cortex. Given the prominence of the OFC's involvement in chemosensory processing, examination of the manner in which the OFC codes and uses chemosensory information provides a critical avenue for understanding OFC functions more generally. A review of the animal literature on the OFC suggests the pervasiveness of chemosensory issues even in tasks that are usually interpreted primarily in terms of learning and executive functions. Indeed, most of the basic concepts and theories regarding OFC processing originate from paradigms that utilize or manipulate aspects of food reward or the contingencies necessary to obtain food reward. For instance, animal studies of the OFC's role in learning and emotion have almost always relied on chemosensory stimuli as either reinforcers or associative cues. Many of the non-chemosensory functions of the OFC may actually represent evolutionary extensions or exaptions of processes involved in coding chemosensory information (exaption is used here to refer to a trait evolved for one purpose that is later used for another). If this hypothesis is correct, the chemosensory functions of the OFC provide a critical context for understanding the broad range of topics covered in this book. The present chapter reviews data on the primate and human OFC's involvement in chemosensory processing and feeding behavior.

## 6.2. Olfaction

### Background

For over 100 years it has been well-recognized that the temporal lobe contributes to the human experience of smell (Hughlings-Jackson and Beevor 1890; Hughlings-Jackson and Stewart 1899). By historical comparison, a role for the OFC in olfactory processing was slow to emerge. Throughout the 19th and early 20th centuries, anosmia (smell loss) was frequently documented as a result of post-traumatic head injury, but the inevitable damage to peripheral olfactory structures and olfactory bulb, along with the scarcity of detailed post-mortem studies, generally confounded efforts to relate these smell impairments to frontal lobe pathology (reviewed in Eslinger *et al.* 1982). Research by Allen provided the first demonstration of a critical role for the frontal cortex in olfactory function (Allen 1940; 1943a, b). Bilateral ablation of the frontal lobes in dogs caused a delay in learning an olfactory-conditioned reflex (lifting the foreleg in response to an odor in order to avoid an electric shock), and interrupted the ability to discriminate between positive- and negative-conditioned odors (Allen 1940). In contrast, total ablation of parieto-temporal lobes (sparing piriform areas) or hippocampi had no effect on these responses, indicating that discrimination learning selectively relied on the structural integrity of the prefrontal cortex. Parallel experiments revealed that the prefrontal ablation had no impact on auditory, tactile, or visual conditioning (Allen 1943b), highlighting the olfactory specificity of this effect. In subsequent work, extracellular recordings in unoperated dogs showed that electrical stimulation of the piriform cortex evoked short-latency spike activity in ventrolateral areas of the prefrontal cortex (Allen 1943a), suggesting that this region might have rapid access to olfactory information. These physiological findings were complemented by a series of strychnine neuronography experiments in monkeys, emphasizing the presence of reciprocal connections between the piriform, posterior orbital, and anterior insular areas (Pribram et al. 1950; Pribram and MacLean 1953). However, following these studies, scientific interest in the subject of olfactory neocortex waned, and further data did not become available for another two decades.

## Olfactory input to the OFC and the localization of secondary olfactory cortex

Odor-evoked responses are initially conducted from the first-order olfactory receptor neurons at the nasal mucosa toward the olfactory bulb, where sensory axons make contact with the second-order (mitral and tufted cell) dendrites within discrete glomeruli. Axons of the mitral and tufted cells of each bulb coalesce to form the olfactory tract, one on each side. This structure lies in the olfactory sulcus, just lateral to gyrus rectus, and conveys olfactory information ipsilaterally to a wide number of brain areas within the posterior orbital surface of the frontal lobe and the dorsomedial surface of the temporal lobe (Fig. 6.1). Collectively, the areas receiving direct bulbar input are sometimes referred to as the 'primary olfactory cortex' (POC: Price 1973; de Olmos et al. 1978; Turner et al. 1978). These structures (from rostral to caudal) include the anterior olfactory nucleus, olfactory tubercle, anterior and posterior piriform cortex, amygdala (peri-amygdaloid region, anterior and posterior cortical nuclei, nucleus of the lateral olfactory tract, and medial nucleus), and rostral entorhinal cortex, all of which are substantially interconnected via associational intracortical fiber systems (Carmichael et al. 1994; Haberly 1998). Note that because the piriform cortex is the largest of the central olfactory areas and receives the densest input from the olfactory bulb, the term POC is frequently used to designate just the piriform cortex.[1]

---

[1] Other regions of the basal forebrain, such as the taenia tecta, induseum griseum, anterior hippocampal continuation, and nucleus of the horizontal diagonal band, have been shown to receive direct bulbar input in animal models (Shipley and Adamek 1984; Carmichael et al. 1994), but whether similar connections are preserved in humans is unknown.

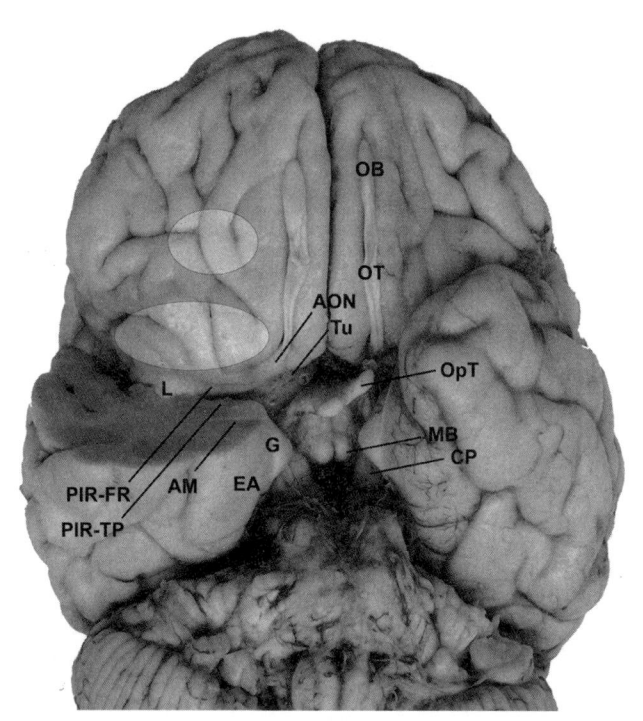

**Fig. 6.1.** (Also see Color Plate 9.) Macroscopic view of the human ventral forebrain and medial temporal lobes, depicting the olfactory tract, its primary projections, and surrounding non-olfactory structures. The right medial temporal lobe has been resected horizontally through the mid-portion of the amygdala (AM) to expose olfactory cortex. The blue oval (posterior) signifies the cytoarchitectural homologues to the monkey olfactory region, whereas the pink oval (anterior) denotes the putative human olfactory region, as defined by neuroimaging findings. AON, anterior olfactory nucleus; CP, cerebral peduncle; EA, entorhinal area; G, gyrus ambiens; L, limen insula; MB, mammillary body; PIR-FR, frontal piriform cortex; OB, olfactory bulb; OpT, optic tract; OT, olfactory tract; Tu, olfactory tubercle; PIR-TP, temporal piriform cortex. Figure prepared with the help of Dr. Eileen H. Bigio, Dept. of Pathology, Northwestern University Feinberg School of Medicine, Chicago, IL.

Higher-order projections arising from each of these primary structures converge on the secondary olfactory regions in the OFC (Fig. 6.1), agranular insula, additional amygdala subnuclei, hypothalamus, mediodorsal thalamus, and hippocampus. Together this complex network of connections provides the basis for odor-guided regulation of behavior, feeding, emotion, autonomic states, and memory. In addition, each region of the primary olfactory cortex (except for the olfactory tubercle) sends feedback projections to the olfactory bulb (Carmichael *et al.* 1994), supplying numerous physiological routes for central or 'top-down' modulation of olfactory processing.

## Nonhuman primates

The first methodical investigation of an olfactory projection area in the primate OFC was carried out by Tanabe and colleagues using electrophysiological techniques (Tanabe *et al.*

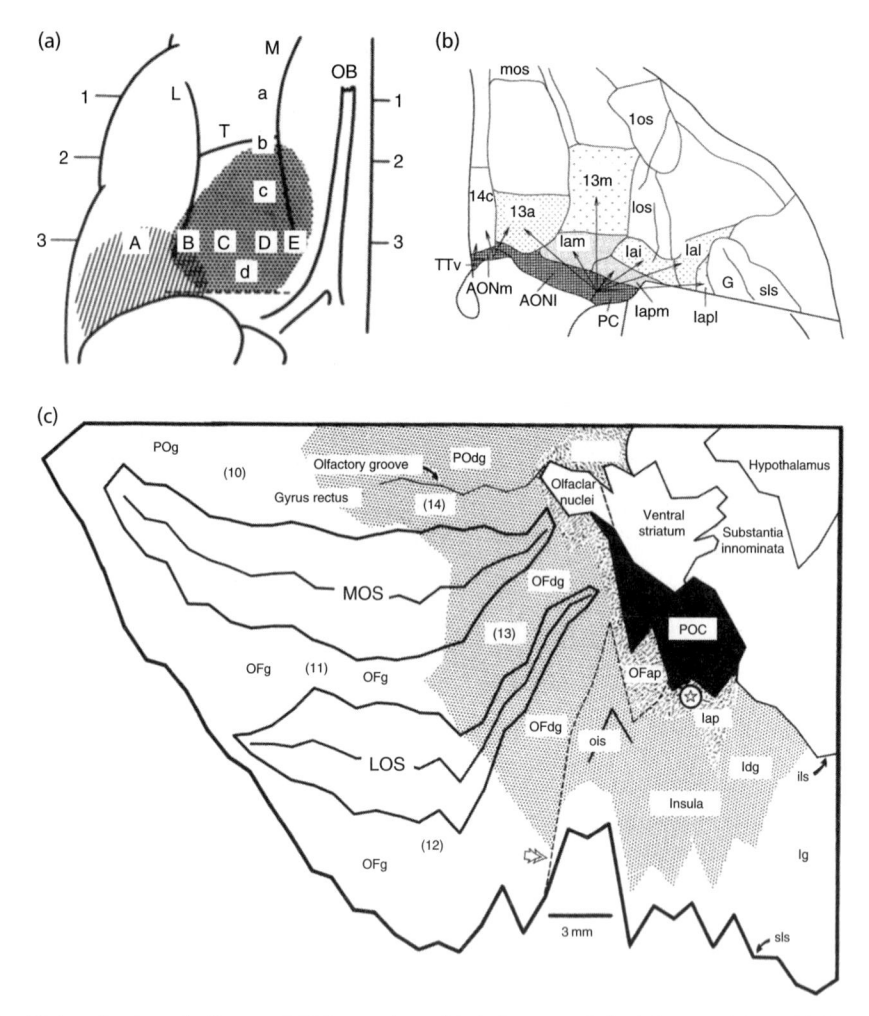

**Fig. 6.2** Localization of olfactory OFC in monkeys. (a) A diagram of the left ventral frontal lobe shows the approximate locations of regions LPOF (hatched lines) and CPOF (stippled area). Note 'H-shaped' orbital sulci comprised of medial (M), lateral (L), and transverse (T) segments, and the olfactory bulb/tract (OB) medially. Numbers and letters refer to electrode stimulation sites within these regions (data not shown). (Modified from Fig. 3 of Yarita et al. (1980). Used with permission of the American Physiological Society, copyright 1980.) Copyright 1992 by Wiley-Liss. Reprinted with permission of Wiley-Liss, a subsidiary of John Wiley and Sons, Inc). (b) A summary of the primary olfactory inputs to monkey OFC on this ventral view of the macaque brain shows that projection density (indicated by the density of dots) is highest in regions adjacent to POC (e.g., Iam, Iapm) and decreases with distance (e.g., 13m). (Adapted from Fig. 22 of: Carmichael et al. (1994). Copyright 1994 by Wiley-Liss. Reprinted with permission of Wiley-Liss, a subsidiary of John Wiley and Sons, Inc.) (c) This two-dimensional flattened cytoarchitectonic map of monkey OFC highlights the olfactory projection areas in posterior agranular and dysgranular regions of OFC and insula and demonstrates the concentric organization of these paralimbic structures.

1974; Tanabe *et al.* 1975a; Tanabe *et al.* 1975b). These researchers demonstrated that in awake, unanesthetized monkeys, electrical stimulation of the olfactory bulb or anterior piriform cortex elicited extracellular potentials in the lateral portions of the posterior OFC, substantiating Allen's earlier findings in the canine prefrontal cortex (Allen 1943a). This area was named the lateral posterior orbitofrontal area, or LPOF, encompassing the posterior part of Walker's area 12 (Walker 1940), the lateroposterior part of area 13, and more posteriorly, the frontotemporal junction and frontal operculum (Fig. 6.2a).

In the same study, damage to either the anterior piriform cortex or the hypothalamus abolished the bulb-evoked potentials in the LPOF, leading to the conclusion that olfactory projections to the OFC were conducted via piriform and hypothalamic relays (Tanabe *et al.* 1975b). In contrast, ablation of the mediodorsal thalamus (MD) had no effect on the evoked potentials in the LPOF, nor did direct MD stimulation elicit potentials within the LPOF region of the OFC. These findings were a surprise at the time, as it was generally assumed that the primate MD was an obligatory checkpoint through which all afferent fibers must pass en route to the orbitofrontal surface (von Bonin and Green 1949; Pribram and MacLean 1953; Nauta 1971). Moreover, the medial subdivision of the MD was already known to receive direct olfactory projections in both rodents (Powell *et al.* 1965) and primates (Benjamin and Jackson 1974), so it seemed plausible that this region should form a node in the projection pathway from the olfactory bulb to the LPOF. The heretical idea, that olfactory information in the OFC could bypass the thalamus, prompted a retrograde tracer study in primates intended to characterize the afferents to the LPOF (Potter and Nauta 1979). Neurons were labeled in many regions, including the MD thalamus and prorhinal (entorhinal) cortex, implying that the LPOF was accessible by dual olfactory fiber systems: a thalamocortical pathway via the piriform cortex/MD, and a cortico-cortical pathway via the piriform/entorhinal cortex.

The same research group that initially characterized the LPOF later identified a second olfactory projection area in the primate OFC, that was in fact mediated by a transthalamic pathway (Yarita *et al.* 1980). This region was situated in between the medial and lateral orbital sulci, within Walker's area 13 (Fig. 6.2a). As this region was positioned medial and slightly anterior to the LPOF, it was labeled the centroposterior orbitofrontal area (CPOF). Both extracellular and intracellular recordings indicated that electrical stimulation of the olfactory bulb or the CPOF evoked spike potentials in the same recording

---

The open double arrow indicates approximate border between OFC and insula. Iap, agranular-periallocortical insula; Idg, dysgranular insula; Ig, granular insula; ils, inferior limiting sulcus; LOS, lateral orbital sulcus; MOS, medial orbital sulcus; OFap, agranular-periallocortical OFC; OFdg, dysgranular OFC; OFg, granular OFC; ois, orbital insular sulcus; POC, primary olfactory cortex; POap, agranular-periallocortical paraolfactory cortex; POdg, dysgranular paraolfactory cortex; POg, granular paraolfactory cortex; sls, superior limiting sulcus. (Reproduced and modified with permission from Fig. 1 of Morecraft *et al.* (1992).

sites within the medial (magnocellular) division of the MD thalamus, suggesting direct participation of this nucleus in an olfactory pathway between the bulb and the OFC.

While the above findings suggested the presence of two discrete olfactory areas (LPOF, CPOF) in the monkey OFC, specific details regarding the underlying projection pathway remained somewhat uncertain, with conflicting evidence for involvement of the hypothalamus, thalamus, entorhinal cortex, substantia innominata, and amygdala (Tanabe *et al.* 1975b; Potter and Nauta 1979; Yarita *et al.* 1980; Naito *et al.* 1984). However, recent advances in cytoarchitectonic and histochemical techniques have made clear that the OFC is no more than two synapses removed from the nasal periphery. Mesulam and Mufson first showed that projections to anteroventral insular regions (agranular and dysgranular) on the monkey posterior orbital surface arise directly from the POC (Mesulam and Mufson 1982a, b; Mesulam and Mufson 1982b; Mufson and Mesulam 1982). The agranular region consists of a relatively undifferentiated two- to three-layer periallocortex, is contiguous laterally with the frontal piriform cortex, and was termed Iap (agranular-periallocortical) (Mesulam and Mufson 1982a). In turn, the dysgranular region is a more highly differentiated five- to six-layered cortex, is located rostral and dorsal to the agranular sector, and was termed Idg (Mesulam and Mufson 1982a). Subsequent work has documented analogous olfactory projection patterns in agranular (OFap) and dysgranular (OFdg) sectors of the caudal OFC (Morecraft *et al.* 1992; Barbas 1993), which themselves represent medial continuations of Iap and Idg. Together these insular and orbital territories comprise concentric rings emanating from an allocortical piriform 'root' (Fig. 6.2b), and appear to provide the substrate for the convergence of olfactory, gustatory, visceral, autonomic, endocrine, and emotional information (Mesulam and Mufson 1982b; Morecraft *et al.* 1992).

In 1994 Price and colleagues conducted further investigations of the POC and its projections to the OFC using anterograde and retrograde tracers and electrophysiological recordings (Carmichael *et al.* 1994). These experiments showed that a total of nine different orbital areas, each with distinct structural characteristics (Carmichael and Price 1994), received direct input from virtually all portions of the POC, including the anterior and the posterior piriform cortex, the anterior olfactory nucleus, the periamygdaloid cortex, and the entorhinal cortex. Olfactory neocortical targets included areas Iam, Iapm, Iapl, Iai, and Ial in the agranular insula, and areas 13a, 13m, 14, and 25 in the posterior and medial OFC, which broadly overlaps the previously identified representations in Iap/OFap and Idg/OFdg (Mufson and Mesulam 1982; Morecraft *et al.* 1992; Barbas 1993). Among the cytoarchitectural subregions identified by Carmichael and Price, areas Iam and Iapm received the highest density projections from olfactory regions, followed by area 13a. All three areas showed action potentials in response to electrical stimulation of the olfactory bulb. Area Iam was notable for the presence of rapid (4–10ms latency) action potentials, which were not characteristic of Iapm or 13a responses. In comparing their findings to the prior physiology data from the Takagi laboratory (Tanabe *et al.* 1975b; Yarita *et al.* 1980), Carmichael *et al.* (1994) suggested that area LPOF roughly corresponded to their projection sites in Iam, Iai, and Ial, and area CPOF corresponded to 13m and portions of Iam and 13a.

Four general principles emerged out of this work. First, despite the broad pattern of distribution, neocortical projections were most heavily concentrated in the regions nearest to the POC (in areas Iam and Iapm), and progressively declined with distance from the POC (Fig. 6.2c). It was concluded that Iam and Iapm are the principal neocortical sites of olfactory information processing, though the role of Iapm may subsume a more general visceral function, given its substantial input from the ventroposteromedial nucleus of the thalamus (Carmichael and Price 1995). Second, there is overall preservation of olfactory topography, such that more anterior-medial portions of the OFC received olfactory inputs from the anterior-medial POC (e.g., anterior olfactory nucleus), whereas posterior-lateral portions of the OFC received inputs from the posterior-lateral POC (e.g., temporal piriform cortex, periamygdaloid cortex). Third, projections between the POC and Iam/Iapm are typically bidirectional, consistent with the previous findings (Mesulam and Mufson 1982b; Mufson and Mesulam 1982), apart from the olfactory tubercle, which lacks input to the OFC (Carmichael *et al.* 1994). Finally, most of the olfactory areas identified in the agranular insula and the posterior OFC are connected to the region of the MD thalamus that receives input from the POC (Russchen *et al.* 1987; Ray and Price 1993), substantiating the presence of both direct and indirect (thalamic) pathways to the olfactory neocortex.

## Humans

Until recently, information regarding the anatomical localization of the human olfactory neocortex has lagged behind the primate (and rodent) data due to methodological limitations. However, the advent of modern functional neuroimaging techniques has permitted a more precise determination of the structural underpinnings of human olfactory processing *in vivo*. In 1992, Zatorre and colleagues published the first imaging (PET) study of human olfaction (Zatorre *et al.* 1992). Healthy volunteers were asked to inhale during birhinal presentation of an odorless cotton wand (first scan) or during presentation of 8 different odorants (second scan). Comparison of odor to no-odor scans revealed significant neural activity in the left and the right piriform cortex and the right OFC. The right OFC activation was located centrally within the orbital cortex, in between the medial and lateral orbital sulci, but more rostral than would be predicted from the monkey data. Odor-evoked OFC activity was also observed on the left side, albeit at reduced statistical threshold. The greater right-hemispheric OFC activity, combined with evidence of greater impairments in olfactory function following right vs. left OFC lesions (discussed below), led to a hypothesis of a right frontal dominance for olfaction.

Following this groundbreaking work, numerous investigators using both PET and fMRI techniques (and a variety of experimental paradigms) have further examined the location of an olfactory representation in the OFC. To document the localization of the human olfactory OFC more definitively, we compiled the OFC voxel coordinates reported in all imaging studies where: (a) the neural characterization of basic olfactory processing was a central aim; (b) odor-evoked activity was not complicated by the use of aversive odorants; and (c) subjects were not asked to make cognitive olfactory judgements (other than odor

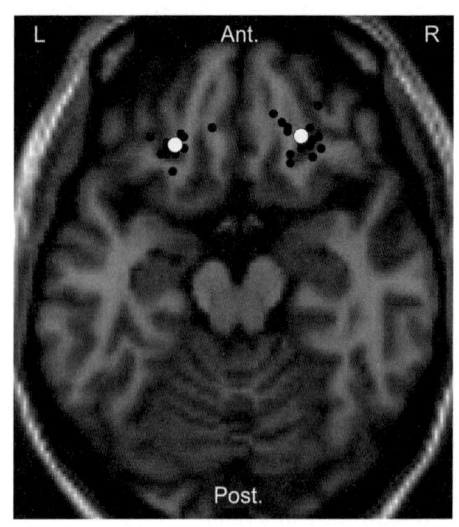

**Fig. 6.3** (Also see Color Plate 10.) Localization of putative olfactory OFC in humans. The peak activation coordinates (23 voxels: dots) from 12 neuroimaging studies of basic olfactory processing are plotted on a canonical T1-weighted anatomical MRI scan in Talairach space. The large circles indicate the mean voxel coordinates for right and left hemisphere. For presentation, coordinates are collapsed across the z-axis (superior-inferior) in order to depict all voxels on a single image. It may be noted that separate foci also emerged around $y = 6$ in three of the studies included in the analysis. These foci are not displayed because they are so far posterior that if mapped to the current slice they would appear to be located in temporal cortex.

detection) during scanning (Gottfried and Zald 2005). Given the data from monkeys, this analysis was broadly constrained to OFC activations posterior to $y = 45$ along the anterior-posterior axis (Talairach coordinate space). Data came from the following 13 studies (5 PET, 8 fMRI): Zatorre *et al.* 1992; Small *et al.* 1997; Sobel *et al.* 1998; Zald and Pardo 1998; Francis *et al.* 1999; O'Doherty *et al.* 2000; Savic and Gulyas 2000; Savic *et al.* 2000; Sobel *et al.* 2000; Poellinger *et al.* 2001; Gottfried *et al.* 2002; Gottfried and Dolan 2003; Kareken *et al.* 2004. To reduce inconsistencies across studies, MNI coordinates were converted to Talairach space. A total of 26 coordinates were identified, including 15 on the right and 11 on the left (Fig. 6.3). Seven of these studies reported significant bilateral activation, five reported significant right-sided activation only, and one reported significant left-sided activation only. While the data generally confirm a right-hemisphere predominance, it is important to note that left hemisphere activations arose in over 60% of the studies.

Examination of the location of activations reveals that olfactory stimulation consistently activates a bilateral area along the medial orbital sulcus (i.e., the medial posterior limb of the 'H'-shaped sulcus), close to the transverse orbital sulcus (i.e., the horizontal limb of the 'H'-shaped sulcus) (Fig. 6.3). Voxel coordinate means (with standard deviations) and medians for this putative olfactory region are shown in Table 6.1.

**Table 6.1** Functional localization of putative human olfactory OFC averaged across 13 neuroimaging studies

| Coordinates[a] | Mean (± S.D.) | | | Median | | |
|---|---|---|---|---|---|---|
| | x | y | z | x | y | z |
| Right OFC | 23.9 | 33.8 | ±12.1 | 24.0 | 33.0 | ±12.0 |
| | (5.4) | (5.6) | (5.3) | | | |
| Left OFC | −21.2 | 30.8 | −15.5 | −21.8 | 30.0 | −17.0 |
| | (5.7) | (4.2) | (4.6) | | | |

[a] Coordinates are in Talairach space in mm, where x = medial/lateral from the midline (+ = right), y = posterior/anterior from the anterior commissure (+ = anterior), and z = inferior/superior from the intercommissural plane (+ = superior).

A comparison of these activations to the monkey (Carmichael and Price 1994) and human (Ongur et al. 2003) architectonic maps delineated by Price and colleagues suggests that the putative human olfactory OFC roughly corresponds to the posterior part of area 11, a location that is strikingly more anterior than one would predict on the basis of the animal data discussed above (Yarita et al. 1980; Morecraft et al. 1992; Carmichael et al. 1994). Regardless of specific cytological subdivisions, it is clear that the putative human region is situated at least 2 cm rostral to the monkey counterpart of the agranular insula. Indeed, among the 13 studies cited above, only three reported activations that could possibly be considered to lie within the agranular insular extension into the posterior OFC (Poellinger et al. 2001; Gottfried, Deichmann et al. 2002; Gottfried and Dolan 2003). We did not include these foci in the above analyses (or Table 6.1 and Fig. 6.3), because they were so far posterior (around $y = 6$) that they were difficult to distinguish from the frontal piriform cortex, and clearly reflected a topographical area different from the dominant location of foci in the OFC. A complete survey of the olfactory imaging literature shows that the agranular insular region is only rarely activated and, if anything, appears to be preferentially activated during the higher-order tasks involving cross-modal integration rather than during passive olfactory stimulation (De Araujo et al. 2003; Small et al. 2004).

It is possible that this discrepancy between predictions from animal studies and human data are the result of methodological differences (see Gottfried and Zald 2005). We note, however, that the difference is unlikely to be simply a result of signal dropout in fMRI, since a similar distribution of responses is observed in the PET studies of basic olfactory processing, which are not susceptible to these technical artifacts (see Stenger, this volume). Another possible source of these differences is that in the human neuroimaging studies, subjects are often actively attending to the odorants even if they are not told to make any specific judgements. Such cognitive or attentional factors may fundamentally alter the underlying activation patterns, such that a more central region is preferentially activated relative to the area that would be predicted from the primate literature. Regardless of the source of this displacement, it seems clear that this more central region plays an important role in olfaction in humans. Given the different cytological and connectional features of this region, its functional properties are almost certainly different from those seen in the studies of the more posterior OFC regions in monkeys, and may reflect behavioral differences in the role that the sense of smell plays in these two species. Of particular note in this regard, the agranular OFC receives strong input from the amygdala, which is almost completely absent in the central OFC (see Price, this volume).

## Response characteristics of secondary olfactory cortex

### Nonhuman primates

**Stimulus specificity:** In their initial characterizations of the LPOF and CPOF, Tanabe and colleagues showed that these olfactory areas exhibited different activity patterns in response to odor stimulation (Tanabe et al. 1974, 1975a,b; Yarita et al. 1980) (Fig. 6.4). Out of 44 neurons recorded in the LPOF, 50% responded to just one of eight different odors, and no

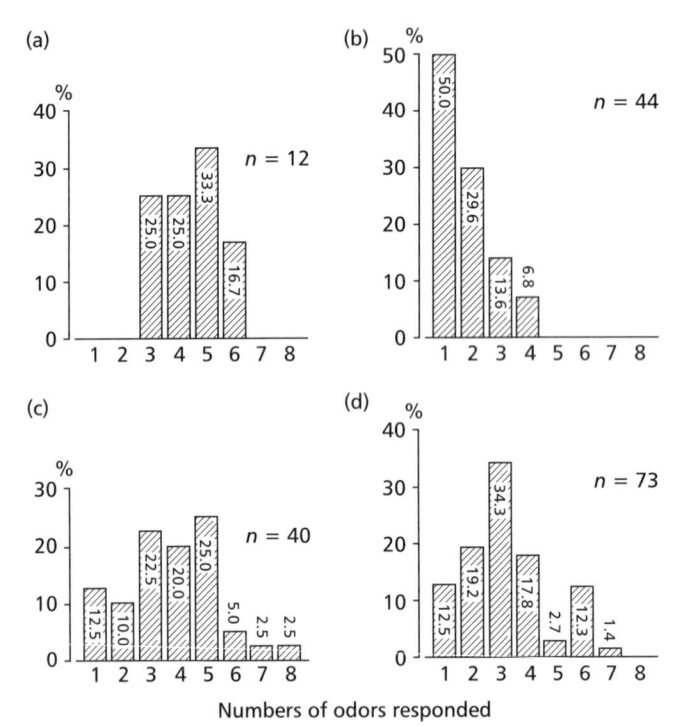

**Fig. 6.4** Response properties of secondary olfactory cortex in monkeys. Neuronal response histograms to eight different kinds of odors depicts the activity profiles in olfactory areas CPOF (a) (broadly tuned) and LPOF (b) (finely tuned), as well as in olfactory bulb (c) and medial amygdala/anterior piriform cortex (d). Bars represent the percentage of cells in each respective area that responded to a given number of odors, as indicated on the abscissa. Adapted from Fig. 9 of Yarita et al. (1980). Used with permission of the American Physiological Society, copyright 1980.

cell responded to more than four odors (Tanabe, Iino *et al.* 1975). In contrast, out of 12 neurons recorded in the CPOF, none responded to less than three of the eight odorants (Yarita *et al.* 1980). Although recordings in the CPOF were limited, the data suggested that odor tuning and discrimination were highly specialized and selective in the LPOF, but much more generalized in the CPOF.

The specificity of the LPOF responses was also notable in relation to other olfactory structures (Fig. 6.4). Whereas only 20% of cells in the LPOF responded to three or more odorants, the majority of recorded cells in olfactory bulb (78%) responded to three or more odorants, and an intermediate response pattern was observed in the anterior piriform cortex and the medial amygdala (69%). Tanabe *et al.* (1975a) additionally noted that LPOF responses were particularly effective in distinguishing odor pairs with overlapping perceptual qualities (in this case, camphoraceous smells) relative to other olfactory regions. In other words, the LPOF was particularly able to resolve subtle differences between odors when they smelled alike, underscoring the idea that this orbitofrontal area supports highly refined olfactory discriminations.

In a complementary attempt to define the role of the LPOF, Tanabe *et al.* (1975b) performed a lesion study of this region. After monkeys were conditioned to associate one of several odors with an unpleasant (bitter) taste, the LPOF was removed bilaterally. This excision resulted in severe impairments of olfactory discrimination. On the surface, these findings accord with their data on odor-evoked responses in the LPOF, as detailed above. However, inspection of Table 1 of Tanabe *et al.* (1975a) reveals that monkeys with the most complete LPOF excisions (subjects M4 and M5) actually responded well *below* chance for the majority of odor discriminations, consistently choosing the wrong odor. This observation suggests that odor discrimination was not abolished *per se* (which would have yielded at-chance performance), but was distorted, raising the possibility of a more complex role for the LPOF. For example, if LPOF activity reflected information about conditioned odor value, odor preference, or odor-response linkages, then LPOF ablation might have disrupted these mechanisms, yielding an aberrant performance pattern such as that observed, without necessarily implicating this region simply in olfactory discrimination.

In contrast to area LPOF, the characterization of the CPOF was less comprehensive, and no lesion data were obtained. Nevertheless, on the basis of the known fiber projections from the limbic areas to the MD thalamus, Yarita *et al.*(1980) proposed a synthetic, integrative function for this 'transthalamic' olfactory area, perhaps as a site of convergence for olfactory and emotional information. However, the suggestion that the olfactory system has two anatomically distinct OFC projection areas, each with different tuning properties and subserving different functions, was called into question by Rolls and colleagues in a series of single-unit recording studies (Rolls and Baylis 1994; Critchley and Rolls 1996a). In an initial investigation (Rolls and Baylis 1994), neuronal responses in the caudal two-thirds of the monkey OFC were tested to nine different olfactory stimuli, representing food and non-food smells, as part of a larger study of multimodal processing. Unimodal olfactory-responsive cells ($n = 15$) were found primarily in medial, but also in lateral, portions of the OFC. Critically, most of these cells were broadly tuned to the odorant set, though each showed clear preferences across the different stimuli. In a subsequent study, both broadly tuned and finely tuned olfactory-responsive cells were identified throughout the posterior orbital surface (Critchley and Rolls 1996a). Taken together, these findings cast some doubt on the notion of functionally heterogeneous olfactory areas in the OFC. However, it is possible that the LPOF may represent a specific functional zone that was under-sampled in the work of Rolls' group. Tanabe *et al.* (1975b) identified 44 of 119 (37%) olfactory-responsive cells in the LPOF, indicating a highly odor-sensitive neuronal population, whereas Rolls and Baylis (1994) identified only 15 of 2000 (0.75%) cells that responded to unimodal olfactory stimulation. This leaves open the possibility that there is a unique olfactory region in the vicinity of posterior-most OFC and agranular insula. Unfortunately, we know of no recent investigations that have specifically tested this hypothesis.

**Odor identity and hedonic value:** Several lines of evidence suggest that both odor identity and hedonic value are separately represented in the monkey OFC. In a study of olfactory discrimination learning, monkeys were trained to associate nine different food and non-food odors with either rewarding (sweet) or aversive (saline) tastes (Critchley and Rolls 1996a). Single-unit recordings were subsequently collected from 1,580 cells throughout the OFC, 48 (3.1%) of which responded to at least one odorant. Characterization of the tuning profiles indicated that the majority of the odorant-responsive cells ($n = 34$; 71%) responded differentially to the odorants, while the remainder ($n = 14$; 29%) did not. Among the 34 OFC neurons showing differential selectivity, the responses in 22 cells (65%) were not based on odor-taste associations. Thus, these neurons appear to encode representations of odor identity, independently of the reinforcing value of the olfactory stimuli. In contrast, the responses in 12 cells (35%) reflected the taste-reward association of the odorants, irrespective of odorant identity. Some neurons preferentially responded to the reward-associated odorants, and others to the saline-associated odorants. Odor-selective, reinforcer-selective, and non-selective cell types were located throughout the medial and the lateral OFC without evidence for functional segregation. Additional support for the presence of OFC cells that respond to odorants on the basis of their reinforcer associations rather than their identity comes from work in which the reward and aversive outcomes are reversed (Rolls *et al.* 1996a), described in detail in Chapter 5 by Rolls (this volume).

Such findings demonstrate that the hedonic or motivational value of an olfactory stimulus is encoded within the OFC, and can be flexibly updated based on changing reinforcement contingencies. Nonetheless, it is important to keep in mind that coding of odor identity, not reinforcement value, is the rule in the majority of olfactory-responsive neurons in the primate OFC. Computational models suggest that even despite the relative sparseness and broad tuning observed in the posterior OFC (as opposed to the specificity observed by Tanabe *et al.*), information regarding odor identity may be robustly supported across local ensembles within the OFC (Rolls 1996b). As discussed below, there is also evidence that odor identity may actually be coded across a network of olfactory regions.

## Humans

There have been relatively few neuroimaging investigations of the response characteristics of the human olfactory OFC, or human POC for that matter. This is partly due to the experimental challenges of dissociating the many psychological factors that comprise odor perception. Intensity, pleasantness, quality/character, pungency (trigeminality), familiarity, edibility, and nameability are all distinctive properties of an olfactory stimulus, but it has been difficult to study these elements in isolation, since any one factor frequently influences the perception of another (Distel *et al.* 1999).

**Odor hedonics:** A number of human imaging studies have centered on odor valence using emotionally salient olfactory stimuli. The first study of this kind was conducted by Zald and Pardo (Zald and Pardo 1997), who determined that highly aversive sulfides evoked activity in the left posterior lateral OFC (and bilateral amygdala), when compared to an odorless control (Fig. 6.5). The OFC activation was situated 10–15 mm lateral to the putative human olfactory area in the OFC, and may therefore represent a region that is selectively engaged by unpleasant odors. This finding was complemented by a regression analysis showing that the neural activity in the left lateral OFC significantly correlated with subjective ratings of odor unpleasantness (but not odor intensity), further supporting the idea that this region participates in hedonic processing of olfactory stimuli. Interestingly, during exposure to weaker and less unpleasant odorants, the peak of the subjects' responses corresponded more closely to the putative human OFC olfactory region in the left hemisphere.

Since this initial report, several additional studies have supported the idea of left lateral OFC sensitivity to aversive odorants. As part of an event-related fMRI study on

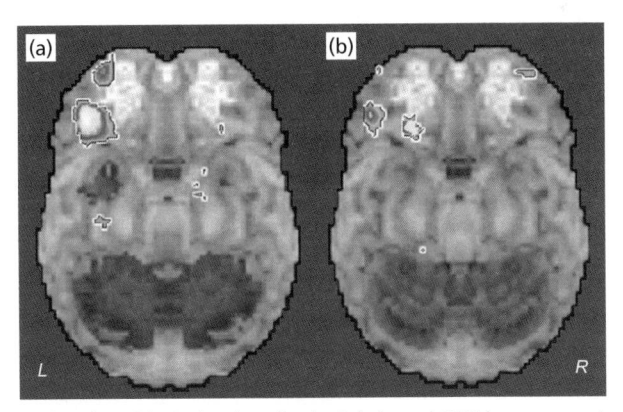

**Fig. 6.5** (Also see Color Plate 8.). Activations in the left lateral OFC in response to aversive odorants. (a) Left lateral OFC activation during exposure to highly aversive sulfides. (b) Left putative olfactory region and weaker left lateral OFC activations in the same subjects during exposure to several more mildly unpleasant (and less intense) odorants from the University of Pennsylvania Smell Identification Test (Doty *et al.* 1984). The PET data derive from Zald and Pardo (1997). Note, the most posterior portions of the OFC are covered by the temporal poles in this surface rendering, but no statistically significant activations were seen in this area.

visual-olfactory associative learning, presentation of an unpleasant odor evoked significant left-sided responses in the lateral posterior OFC, whereas a pleasant odor evoked significant right-sided responses in the medial posterior OFC (Gottfried *et al.* 2002a). Examining the effects of three aversive odorants, Rolls *et al.* (2003) similarly observed specific activation of the left lateral OFC, whereas pleasant odorants did not clearly engage this region. In an elegant study that dissociated intensity and valence within a single fMRI experiment, the right medial OFC was activated to pleasant odor, and the left lateral OFC was activated to unpleasant odor (Anderson *et al.* 2003). Finally, similar medial-lateral dissociations in the human OFC according to valence have been identified in olfactory cross-modal paradigms of episodic memory (Gottfried *et al.* 2004).

To characterize these response patterns more accurately, we plotted the activation peaks for pleasant and unpleasant odor stimuli from the following six studies: Zald and Pardo 1997, 1998; Gottfried *et al.* 2002a; Anderson *et al.* 2003; Rolls *et al.* 2003; and Gottfried *et al.* 2004 (Fig. 6.6). Although further data are clearly needed to make definitive statements, some basic patterns nevertheless emerge. First, there is a strong tendency of aversive odorants to engage the left OFC, whereas pleasant odorants have a more bilateral representation. Second, neural responses evoked by either pleasant or unpleasant odors do not localize to discrete regions within OFC, nor do these activations systematically cluster around the putative olfactory OFC area. Third, while aversive odorants tend to activate more lateral areas of OFC than do pleasant odorants, it is important to point out that the 'aversive' voxels are equally situated in the medial (areas 11 and 13) and lateral (area 47/12) portions of the OFC. The greater appearance of more lateral responses in the left OFC appears consistent with a recent analysis by Kringelbach and Rolls (2004), who emphasize a general medial-positive, lateral-negative distinction in the OFC. However, the presence of several more medially located foci in response to aversive

**Fig. 6.6** (Also see Color Plate 11.) Olfactory hedonics in human OFC. The peak neural responses to pleasant (8 voxels; dark red dots) and unpleasant (9 voxels; light green dots) odors were identified from six imaging studies of olfactory hedonic processing and mapped onto a transverse T1-weighted anatomical scan. Presentation as in Fig. 6.3. For reference, the large yellow circles indicate the putative olfactory OFC areas.

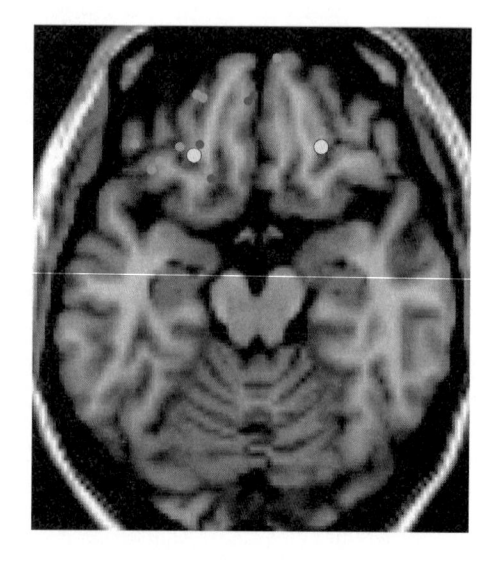

stimuli indicate that the pattern of responses is not fully explained by a medial-positive, lateral-negative model of OFC function.

Neuroimaging studies of explicit hedonic judgements have also been used to explore the emotional underpinnings of olfactory processing. Royet and colleagues (Royet *et al.* 2001, 2003; Royet and Plailly 2004) have emphasized the importance of the left OFC in the conscious hedonic evaluation of odorants. This premise arises out of their initial finding that hedonic judgements produced left OFC activation (at $x = -28$, $y = 16$, $z = -10$) which exceeded that produced by judgements of familiarity, edibility, intensity, or detection. However, two studies suggest that hedonic judgements are not exclusively lateralized to the left OFC. First, in a study by Zatorre *et al.* (2000), both intensity and pleasantness judgements were observed to activate the right OFC, and there was no preferential activation of the left OFC during pleasantness relative to intensity judgements. An additional concern may be raised regarding the earlier Royet studies, in that the hedonic judgement tasks are commonly made in the exclusive presence of emotionally salient odors, whereas other judgements were not. Therefore, any task-related activations could be confounded by mere stimulus properties of the odorants themselves. In an effort to overcome this issue, Royet *et al.* (2003) conducted an fMRI study in which subjects smelled pleasant and unpleasant odors during alternating blocks of either hedonic judgements or a control task (random button press). A comparison of hedonic and control tasks (thereby controlling for stimulus effects) revealed significant activity in the OFC bilaterally, further supporting the possibility of right (as well as left) OFC involvement in hedonic coding.

**Odor quality:** In contrast to the neurophysiology data in monkeys, there has been very little research on the neural substrates of odor quality in the human brain. As used here, the term 'quality' denotes odor-object identity, i.e., the perceptual character of a smell emanating from an odorous object (in contrast to other odor qualities such as intensity or valence). The characterization of odor quality coding faces particular challenges using neuroimaging, due to the limited spatial resolution of the technique. Take for example, as one's hypothetical research aim, the comparison of the smell of 'strawberry' vs. the smell of 'lavender'. Presuming that quality is represented as a distributed ensemble code in the OFC, 'strawberry' will evoke a hemodynamic response that is indistinguishable from 'lavender'. Thus, in the absence of differential activity, one cannot conclude whether the OFC codes odor quality. Moreover, the comparison of 'strawberry' and 'lavender' will be sensitive not only to differences of quality, but also to differences of intensity, pleasantness, familiarity, pungency, and even molecular structure, thereby confounding attempts to delineate the neural components of odor quality.

In recent studies, we (Gottfried 2006) have overcome these problems by combining an olfactory version of fMRI cross-adaptation (Buckner *et al.* 1998; Grill-Spector and Malach 2001; Kourtzi and Kanwisher 2001) with a careful selection of odorant pairs that systematically differ in perceptual quality or physical structure. Critically, this experimental design was fully balanced, controlling for the possibility that the findings could be confounded by variations in intensity, hedonics, or other perceptual dimensions, and ensuring that only quality and structure differed across condition types. Sixteen healthy subjects participated in a task of odor detection (rather than an odor quality judgement task), so that the neural substrates of odor quality would not be confounded by explicit cognitive judgements of odor quality. Across the group, neural activity in the right OFC, the left posterior piriform cortex, and the left hippocampus decreased (cross-adapted) in response to pairs of odorants that were similar (vs. different) in quality, irrespective of the effects of molecular structure (Fig. 6.7). These data are consistent with the idea that representations of odor quality may be distributed across a network of olfactory-related regions, in keeping with prior monkey (Tanabe *et al.* 1975a) and human (Eichenbaum *et al.* 1983; Royet *et al.* 1999;

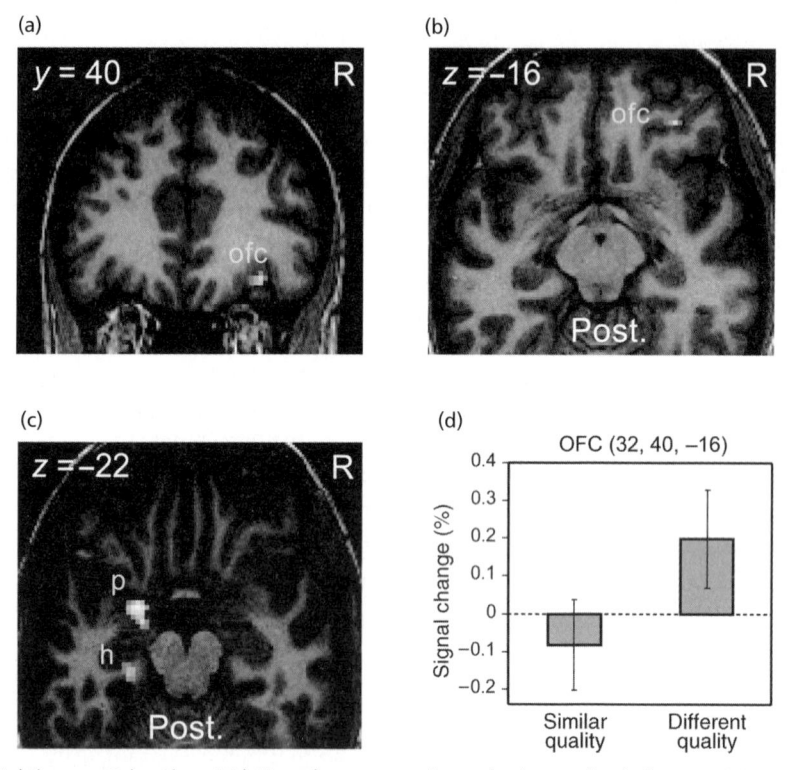

**Fig. 6.7** (Also see Color Plate 12.) Neural representations of odor quality in human OFC. (a, b) A comparison of qualitatively different vs. qualitatively similar odorants revealed significant activity in right OFC. Functional responses are superimposed on coronal (a) and transverse (b) sections of a T1-weighted anatomical scan. (c) Neural representations of odor quality were also detected in piriform cortex ('p') and hippocampus/parahippocampal cortex ('h') (responses overlaid on a transverse T1 section). (d) Group-averaged plots of the peak fMRI signal from right OFC indicate decreased (cross-habituated) responses to qualitatively similar odorants, in comparison to qualitatively dissimilar odorants.

Savic *et al.* 2000; Gottfried and Dolan 2003) studies of olfactory discrimination and semantic processing. We speculate that each of these areas encodes a distinct qualitative element (e.g., odor-object identity, semantic and verbal mnemonic associations, affective and motivational features) that together define an odor percept. It is also possible that the OFC and the hippocampus are directly involved in sculpting odor quality representations in the posterior piriform cortex. This idea is supported by our data on olfactory-visual semantics (Gottfried and Dolan 2003) showing OFC and hippocampal activation to semantically related odor-picture pairs, suggesting that semantic representations in these regions might shape the expression of quality-based codes in the piriform cortex.

## Olfactory cognitive judgements and the human OFC

The ability to code odor quality is essential for making explicit judgements of odor identification. Consistent with the evidence of odor quality representations in the OFC (as described above), partial excisions of the prefrontal lobe for the relief of intractable epilepsy have been associated with impairments of odor identification (Jones-Gotman and Zatorre 1988). In this study, patients were asked to identify common odors using the University of Pennsylvania Smell Identification Test (Doty *et al.* 1984), a 'scratch-and-sniff' test that requires matching each of 40 different odors to different lists of 4 verbal descriptors. Compared to a normal control group, only those prefrontal patients whose excisions involved the OFC were impaired on this task. Interestingly, patients with temporal lobe excisions (frontal sparing) also performed poorly, though they were relatively less impaired than the OFC group. These effects were generally observed in the presence of normal detection thresholds, excluding any primary loss of smell, although there was a significant negative correlation between right-nostril odor detection values and identification performance in the right frontal patients.

Complementary studies on odor quality discrimination have demonstrated similar effects. Potter and Butters (1980) asked patients with prefrontal damage to rate the similarity of odor quality between successive pairs of stimuli. Performance impairment appeared to be specific to the olfactory modality, as visual hue discrimination was left intact. Curiously, odor detection thresholds were lower (more sensitive) in the prefrontal subgroup than in controls, though the overall younger age of the prefrontal patients might have accounted for this finding. This same experiment suggested that the smelling deficit might be specific to the ipsilateral nostril (Potter and Butters 1980), although subsequent work has demonstrated a more complicated picture. In a quality discrimination task, Zatorre and Jones-Gotman (1991) gave monorhinal presentations of odorant pairs to post-surgical epilepsy patients, who judged odor quality similarity for each pair of stimuli. Right OFC patients showed discrimination impairments in both nostrils, whereas right frontal patients with orbital sparing lesions were impaired only in the ipsilateral nostril. Performance was also markedly reduced in the ipsilateral nostril of left OFC patients, though this was non-significant (possibly due to low sample size).

Together with the finding that discrimination scores were higher in control subjects when odorants were delivered to the right (vs. left) nostril, the data were interpreted to suggest a right hemisphere (and right OFC) advantage in odor quality discrimination (Zatorre and Jones-Gotman 1991). This conclusion is in keeping with the greater frequency of right sided activations in basic olfactory neuroimaging studies, and the fMRI adaptation data described above, in which the right (but not the left) OFC was implicated in odor quality coding. However, one concern with this interpretation warrants mention. Inspection of the brain lesion maps delineating extents of surgical excision (cf. Figs. 1 and 2 in Zatorre and Jones-Gotman 1991) suggests that the right frontal groups as a whole may have engendered greater damage to olfactory projection areas in the centroposterior OFC than did the left frontal groups. This might explain why odor detection thresholds correlated with identification performance in right (but not left) frontal patients, and also why detection thresholds were lower in the right (but not the left) nostril of right OFC patients, compared to both left OFC patients and controls (cf. Table 2 in: Zatorre and Jones-Gotman 1991). Thus, although the data are consistent with a right OFC dominance for olfactory identification and discrimination, it remains conceivable that differential hemispheric damage to the secondary olfactory cortex could be contributing to this apparent asymmetry.

A variety of other explicit cognitive tasks influence neural responses in the OFC (reviewed in: Savic 2002; Royet and Plailly 2004). Judgements of intensity (Zatorre *et al.* 2000), familiarity (Royet *et al.* 1999, 2001), edibility (Royet *et al.* 1999, 2001), and pleasantness-unpleasantness, have each been associated with activtion of the OFC in PET studies. Some of these activations are situated near the purported olfactory projection area in the OFC, while others are found more rostrally. As already noted, Royet *et al.* (2001) have suggested that there is a degree of specificity in the engagement of the left OFC, such that the activity induced by hedonic judgements was greater than that seen for other judgements. Specificity for other judgements has been less clear. For instance, in the right OFC, familiarity judgements were observed to produce greater activation than that produced by odor detection or intensity judgements. However, familiarity judgements did not cause activations beyond those caused by judgements of hedonicity or edibility, indicating that the involvement of the right OFC in explicit judgement tasks is not limited to familiarity. Finally, it is worth noting that many judgement-related foci lie outside of the OFC altogether: olfactory judgement tasks typically evoke responses in large portions of the frontal, temporal, parietal, and occipital cortices (Royet *et al.* 1999, 2000, 2003; Savic *et al.* 2000). Although the specific contributions of these regions remain unclear, it appears that several areas that are tertiary to the olfactory system help mediate a higher-level olfactory decision-making. Together, these findings are consistent with both serial hierarchical and parallel distributed modes of olfactory information processing that vary according to task demands (Savic *et al.* 2000; Royet and Plailly 2004). However, it remains unclear specifically how these non-olfactory brain areas interact with the OFC and the POC in mediating these tasks.

## Olfactory memory and learning in the human OFC

Following from their lesion studies on olfactory identification and discrimination, Jones-Gotman and Zatorre (1993) also showed that patients with postsurgical excisions of the right frontal lobe performed poorly on immediate, short-term (20 min), and long-term (24 hr) tests of odor recognition memory. While these data make it evident that right OFC lesions impair performance on odor recognition memory, it remains unclear if this reflects a specific mnemonic problem, or is secondary to deficits in discrimination and identification described above (Zatorre and Jones-Gotman 1991). In subsequent PET studies, olfactory working memory (Dade *et al.* 2001) and short- and long-term recognition memory (Savic *et al.* 2000; Dade *et al.* 2002) have been associated with enhanced right or bilateral OFC activity. Unfortunately, in some of these studies, the control conditions failed to control for olfactory stimulation, again leaving open to question the extent to which OFC involvement in olfactory memory is specific to mnemonic vs. other olfactory processes.

In an effort to characterize odor memory in the absence of odor stimulus confounds, Gottfried *et al.* (2004) conducted an event-related fMRI study on episodic memory retrieval of olfactory and emotional contexts. During an initial study phase, subjects were given combinations of smells varying in affective valence (pleasant, neutral, and unpleasant) and neutral pictures and asked to imagine a link or association between the two stimuli. The aim of this session was to encourage episodic memory encoding of odor-picture pairs. In a subsequent recognition memory test, subjects had to decide whether they were viewing a study (old) or novel (new) picture, in the absence of any odor cues. Comparison of correctly remembered items to correct rejections ('old/new' effect) revealed significant memory-related activity in the piriform cortex (but not in the OFC). Critically, odor absence during the memory session ruled out the possibility that piriform activity was merely being driven by olfactory stimulation. On the other hand, the neural substrates of emotional contextual retrieval were expressed in dissociable subregions of the OFC. The contrast of positive (vs. negative) hits was associated with activity in the medial OFC, while negative (vs. positive) hits elicited activity in the more lateral OFC regions.

While the above findings do not provide compelling evidence that the OFC is directly involved in explicit forms of odor memory, the related work on implicit memory implicates this structure more directly. In fMRI studies of visual-olfactory associative learning (Pavlovian conditioning), a neutral visual stimulus (the conditioned stimulus, or CS+) is repetitively paired with an emotionally charged odor (the unconditioned stimulus, or UCS). By selecting either a pleasant or unpleasant odor as the UCS, both appetitive (reward-based) and aversive learning can be investigated (Gottfried *et al.* 2002b, 2003; Gottfried and Dolan 2004). Presentation of an appetitive or aversive CS+, in the absence of odor, elicits robust neural activity in the OFC (Gottfried *et al.* 2002b), suggesting that this region participates in the establishment of picture-odor contingencies. Moreover, when the current motivational value of an olfactory reinforcer (UCS) is either decreased (via selective satiety) or increased (via odor inflation), cue-evoked OFC responses are modulated in parallel with the behavioral manipulation (Gottfried et al. 2003; Gottfried and Dolan 2004). The implication is that a predictive cue has direct access to internal representations of value in the OFC, and that these representations are flexibly updated according to an individual's motivational state.

Interestingly, a caudal-rostral pattern of specialization in the human OFC has tentatively emerged out of the above imaging data on olfactory memory and learning. As detailed above, odor-evoked neural activity around the putative olfactory region identified in Table 1 is typically associated with low-level aspects of olfactory processing. In contrast, more rostral areas of the OFC (roughly within Walker's area 13) appears more engaged in higher-order olfactory computations, including working memory (Dade et al. 2001), recognition memory (Dade et al. 2002), and associative learning (Gottfried, et al. 2002b, et al. 2003; Gottfried and Dolan 2004) (Fig. 6.8). In accord with such a distinction,

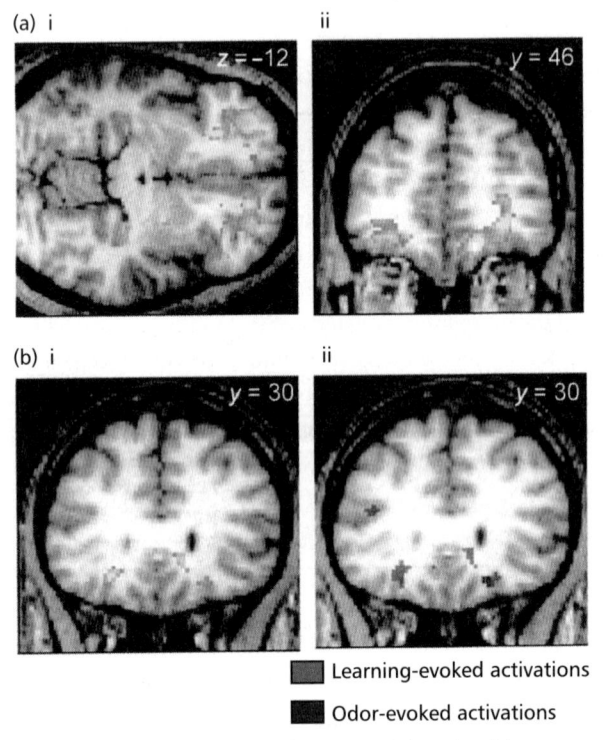

**Fig. 6.8** (Also see Color Plate 14.) Caudal-rostral pattern of functional heterogeneity in human OFC. In an fMRI paradigm of visual-olfactory associative learning, learning-evoked activity was detected across extensive areas of rostral OFC (a, i, a ,ii), but was also observed in caudal OFC (b,i). In panel (b,ii) the learning-related responses in caudal OFC (illustrated in red) overlap odor-evoked responses in blue. The learning-related responses more posteriorly may reflect modulation of secondary olfactory cortex by higher-order (top-down) centers. Conversely, more complex operations pertaining to odor learning would be specifically executed in rostral OFC. Adapted with permission from Fig. 2 of Gottfried et al. (2002b). Copyright 2002 by the Society for Neuroscience.

cross-modal integration of vision and olfaction is associated with more anterior activations than when the stimuli are presented unimodally. (See Part III for more information on multisensory integration involving the chemical senses). Such findings appear consistent with an anatomical hierarchy of the OFC specialization, whereby caudal regions (such as the olfactory OFC) converge on medial and anterior territories to permit more complex information processing in the service of feeding and other goal-directed behaviors (Carmichael and Price 1996; Ongur and Price 2000). Thus, with regard to functional complexity, the architecture of the olfactory OFC appears to be broadly preserved across humans and nonhuman primates.

## 6.3. Gustation

Taste refers to the qualities of sweet, sour, salty, bitter, and savory (umami). These sensations occur when molecules interact with receptors on the tongue. What is colloquially referred to as 'taste' is actually a combination of the physiologically distinct sensory experiences of taste and smell, or flavor. As described below, the OFC plays an important role in both taste and flavor processing, as well as a more general role in feeding behavior and the representation of food reward.

### Localization of OFC gustatory areas

#### Nonhuman primates

In primates, gustatory information travels through cranial nerves XII, IX, and X to the nucleus of the solitary tract (Beckstead *et al.* 1980). The second-order gustatory fibers leave the nucleus of the solitary tract to join the central tegmental tract and project to the parvocellular division of the ventral posterior medial nucleus (VPM) of the thalamus (Beckstead *et al.* 1980). Fibers from the VPM project most densely to the dorsal sections of the anterior insula and adjacent operculum (Pritchard *et al.* 1986). This region is often designated as the 'primary gustatory cortex', although it appears to be sensitive to a wide range of oral sensation in addition to taste (Scott and Plata-Salaman 1999). A secondary projection exists in a region extending from areas in the precentral opercular area near the base of the central sulcus. Both regions are activated by peripheral stimulation of the gustatory nerves (Ogawa *et al.* 1985). Taste-responsive cells are also found in the adjacent regions of the insula (Ogawa 1994; Scott and Plata-Salaman 1999). Since these additional areas do not receive inputs from the VPM they can be considered higher-order gustatory regions (Ogawa 1994).

Neurons that respond unimodally to taste stimulation have also been identified in the OFC. The first gustatory region identified in the OFC is located in a dysgranular, caudolateral portion of the OFC (Rolls *et al.* 1989). In the macaque, this region is just lateral to the caudomedial orbital olfactory area (Price *et al.* 1991) at the lateral point of the transition between the ventral agranular insula and the caudal OFC and corresponds to areas 13l and Ial according to Carmichael and Price (Carmichael and Price 1995). Baylis *et al.* (1995) isolated taste neurons in the gustatory orbital region and injected horseradish peroxidase into this site. They found substantial labelling in the frontal opercular taste area and anterior dorsal insula extending into more ventral regions of the insula. Since no labeling occurred in the VPM, Baylis and colleagues proposed that the caudolateral OFC region represents the secondary taste cortex. However, it has recently been demonstrated that there is at least one additional region in the OFC with prominent unimodal gustatory responses. In a recent study, Pritchard and colleagues identified a taste-responsive region caudal and medial to the so-called secondary taste area described by Rolls and colleagues (Pritchard *et al.* 2005). This region, which is 12 mm² in extent, straddles areas 13m, 13a, 13b and Iam, and is remarkable because 20% of the cells isolated responded to taste stimulation. This represents the highest concentration of gustatory neurons in any cortical region yet described. Based upon anatomy and the characteristics of the evoked taste responses, the authors proposed that it comprises a major gustatory relay that lies anatomically and functionally between the insular and caudolateral taste areas. Gustatory responsive neurons have also been identified in a more anterior medial region of the OFC (Thorpe *et al.* 1983), which is consistent with evidence presented by Barbas (1993), indicating the presence of projections from the insular operculum gustatory area to an extreme rostral portion of area 13. However, the nature of gustatory processing in this region remains poorly defined.

### Humans

The human gustatory pathway is assumed to be equivalent to the monkey's (Scott and Plata-Salaman 1999), and cytoarchitectural studies (Petrides and Pandya 1994) and functional neuroimaging studies (Barry *et al.* 2001; Cerf-Ducastel *et al.* 2001; de Araujo *et al.* 2003; Faurion *et al.* 1998, 1999; Francis *et al.* 1999; Frey and Petrides 1999; Kinomura *et al.* 1994; Kobayakawa *et al.* 1996, 1999; O'Doherty s *et al.* 2001a; O'Doherty *et al.* 2002; Small *et al.* 1997a, 2003b, 1999; Zald *et al.* 1998, 2002) support its location in the anterodorsal insula and the adjacent frontal opercular cortex in and around the cortex of the horizontal ramus of the Sylvian fissure. As in the monkey, there is evidence for some secondary taste representations in the midinsula, the ventral insula, and the parietal and temporal opercula (Faurion *et al.* 1999; Small *et al.* 1999).

The OFC is frequently activated in response to gustatory stimulation (Francis *et al.* 1999; Frey and Petrides 1999; O'Doherty *et al.* 2001b, 2002; Small *et al.* 1997a, b; 2003b; Zald *et al.* 1998, 2002). However, the area of activation is often anterior and medial to the orbital taste region described in monkeys (Small *et al.* 1997a, 1999, 2003b). Small and colleagues used the anatomical descriptions of the orbital taste area in the monkey provided by Rolls and colleagues to identify the anatomically homologous region in the human brain. This is shown in Fig. 6.9. Activation foci generated in response to taste stimulation from all published (and some unpublished) PET studies of human gustation were plotted onto a high-resolution MRI image. Eighteen peaks localized to the OFC. However, of these 18 peaks, the authors noted that only 2 were located in the region anatomically homologous to the monkey caudolateral orbital taste area (Fig. 6.9). The rest were dispersed throughout areas 11, 13, and 14, according to the terminology employed by Petrides and Pandya (1994). This raises the possibility of inter-species differences in the precise location of the orbital gustatory area. The anterior location of OFC taste responses converges with the olfactory literature in demonstrating greater activations in more anterior regions than would be predicted based on a simple examination of cytoarchitectural homology. As with the olfactory system, one may question whether this apparent displacement is due to a methodological artifact of the functional imaging technology. However, the original analysis by Small and colleagues was based exclusively on PET data, where the prominent susceptibility related artifacts and distortion that hamper fMRI imaging of the OFC are not an issue. Alternatively, these anterior activations in both modalities may represent higher-order processing of chemosensory information. If so, this would imply that humans preferentially activate higher-order processing regions relative to the more formally defined secondary taste and smell regions in the more caudal OFC.

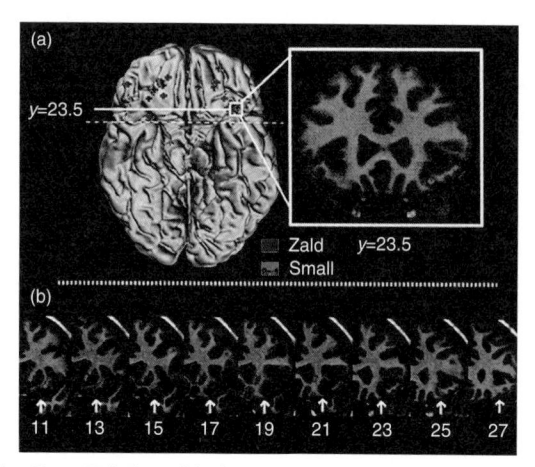

**Fig. 6.9**. (Also see Color Plate 15.) The orbitofrontal gustatory region in the human brain (a) Crosses represent peaks from three gustatory (Small et al. 1997a; Small et al. 1997b; Zald et al. 1998) studies projected onto a surface rendering with anterior temporal lobes removed to expose the underlying caudal orbital surface. This was accomplished with a segmentation technique in which the region of the temporal lobe at the point where it begins to separate from the insula and OFC ($y = 9$ in the left and $y = 8$ in the right hemisphere) is isolated and removed (indicated by the white dotted line). Based on anatomical descriptions in non-human primates (Baylis et al. 1995; Rolls et al. 1990), the proposed gustatory region is at the transition between the ventral agranular insular cortex and the caudal OFC at its lateral point. The two peaks in the small white box fall within a homologous region of the human brain. These peaks are also shown in the coronal plane to illustrate their medial-lateral location. (b) Coronal sections illustrating the transition from agranular insular cortex (at the level where temporal, insular and orbital areas are continuous) to the dysgranular cortex in the caudolateral orbitofrontal region. The numbers represent the y-value at which each section is taken. The white arrows point to agranualr cortex in sections 11–17, dysgranular cortex in sections 19–25, and the disappearance of this cortical region in section 27. The cortical area indicated by the arrows in 19–25 represents an analogous region to the 'secondary gustatory area' described in monkeys. Reproduced with permission from Small et al. (1999).

## Response characteristics of the orbital taste region

### Stimulus identity and specificity

Rolls and colleagues have used single-cell recording techniques to characterize the responsiveness of taste neurons in the macaque caudolateral OFC region (Rolls et al. 1989, 1990; Rolls and Baylis 1994). In one study, they isolated 49 gustatory responsive neurons in this region. Eighty percent of these neurons responded to just a single taste, 12% responded to two of the four basic tastes, and only 4% responded to three or more of the four basic tastes (Rolls et al. 1990). This response profile differs considerably from the profile elicited in the caudomedial orbital area and the insula/opercular gustatory areas, where each neuron responded to more than one of the prototypical stimuli (Scott et al. 1986a; Yaxley et al. 1988, 1990; Pritchard et al. 2005). In the gustatory literature, the specificity of individual neural responses to chemical compounds is often described using a breadth of tuning metric developed by Smith and Travers (Smith and Travers 1979). This coefficient can range from 0.0, representing a complete specificity to one of the stimuli, to 1.0, which indicates an equal response to all tastants. In the primate, peripheral taste nerves have a breadth of tuning coefficient of .54, indicative of a fairly high specificity (Hellekant et al. 1997). Specificity is then lost in the nucleus of the

**Fig. 6.10** Breadth-of-tuning index for taste neurons in the brainstem, insula, operculum and OFC. Breadth-of-tuning index for taste responsive neurons recorded in the nucleus of the solitary tract (NST), operculum, insula and orbitofrontal cortex (OFC) in the primate. This coefficient can range from 0.0, representing a complete specificity to one of the stimuli, to 1.0, which indicates an equal response to all tastants (Smith and Travers, 1979). Means, standard error and the number of neurons in the sample for each area are shown. Adapted with permission from Rolls et al. (1990).

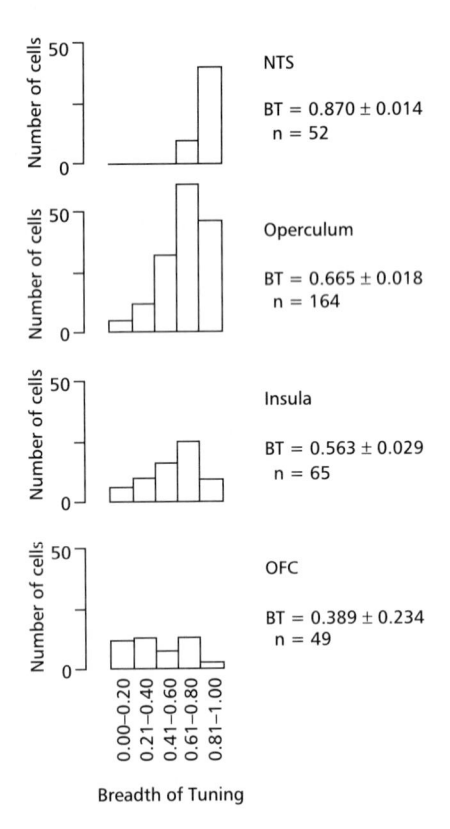

solitary tract (0.87) (Scott *et al.* 1986b) and gradually recovered in the progression from the thalamus (VPM) = 0.73 (Pritchard *et al.* 1989) to the frontal operculum = 0.67 (Scott *et al.* 1986a) and the insula = 0.56 (Smith-Swintosky *et al.* 1991). In the caudomedial OFC the breadth of tuning is 0.79, whereas in the caudolateral OFC the breadth of tuning is 0.39, indicating the highest level of response specificity in the gustatory neuroaxis (Fig. 6.10).

This finding resembles the results from the olfactory system, indicating that high stimulus specificity is a consistent trait across both smell and taste modalities. However, as with the olfactory modality, questions remain as to what extent the OFC processing reflects a necessary component in stimulus quality decoding, vs. the OFC being the recipient of the results of early processing, or participating in a broader ensemble involved in the process. Recent animal studies have highlighted the importance of patterns of cell ensemble activity in taste quality decoding. (Smith and Scott 2003). For example, Katz and colleagues (Katz *et al.* 2002) used multiunit recordings in rodents to isolate transient quality-specific cross-correlations in ensembles of taste neurons within the insular taste region. This ensemble activity in the insula may act to provide the information necessary to allow stimulus specific representations to emerge at the level of single cells in the OFC.

**Hedonic value:** Initial neurophysiological studies of the caudolateral OFC taste region were notable in the relative paucity of neurons responding to aversive stimuli (quinine hydrochloride and hydrochloride) (Rolls *et al.* 1990), which led to speculation that there might be anatomical segregation between areas responsive to pleasant and unpleasant taste stimuli. Consistent with this possibility, neuroimaging studies in humans frequently report valence-specific gustatory responses in the OFC (Zald *et al.* 1998, 2002; O'Doherty *et al.* 2001b, 2002; Small *et al.* 2003a). However, although segregated responses are consistently reported within studies, there is a great deal of variability across studies. Fig 6.11 shows the location of all activation foci in the OFC reported in analyses designed to isolate the response to either pleasant or unpleasant taste in eight analyses from five published studies using pure gustatory stimuli (Zald *et al.* 1998, 2002; O'Doherty, *et al.* 2001b, 2002; Small *et al.* 2003b). The only consistent finding is that the pleasant taste appears to preferentially activate the right OFC. This finding is inconsistent with the conclusions of Kringelbach and Rolls (2004), who argued against the presence of hemispheric lateralization related to valence (based on their meta-analysis of a wide range of studies related to positive and negative stimulus exposures in multiple modalities). The present data also run against a strict model of medial-positive/lateral-negative segregation of responses (Small *et al.* 2001; Kringelbach and Rolls 2004). As already described, analysis of hedonic olfactory data provides partial support for this medial-lateral model of segregation, but does not indicate a complete segregation of responses. These discrepancies lead us to speculate that some of the unique specializations for chemosensory processing in the OFC override or obscure the larger trends of hedonic segregation suggested by the Kringelbach and Rolls analysis. Our speculation is further

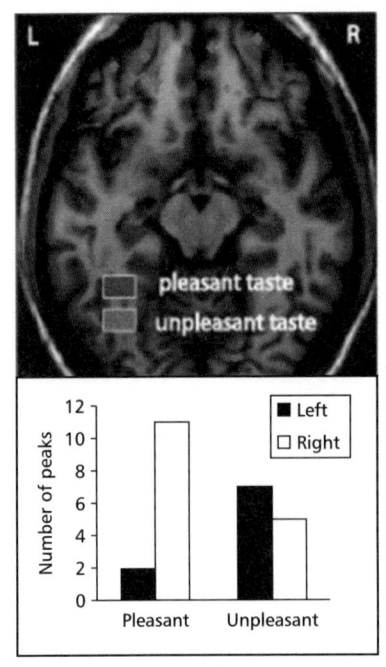

**Fig. 6.11** (Also see Color Plate 13.) Lateralization of responses to pleasant and unpleasant tastes. Top: 25 activation foci from 8 analyses of pleasant taste (either pleasant taste > a neutral taste or a pleasant taste > an unpleasant taste) and 8 analyses of a unpleasant taste (either unpleasant taste > a neutral taste or unpleasant taste > a pleasant taste) collected from 5 published studies (O'Doherty et al. 2001b, 2002; Small et al. 2003; Zald et al. 2002, 1998) and plotted onto an anatomical image. L = left and R = right. To display peaks in a single plane the average level of z (superior to inferior = −14 in MNI coordinates) was calculated and the peaks were plotted onto this plane according to their x (medial to lateral) and y (anterior to posterior) coordinate values. Peaks from the pleasant analyses are in red and peaks from the unpleasant analyses are in green. Bottom: A bar graph showing the number of peaks (y axis) that fell in the left or right OFC for the pleasant and unpleasant analyses.

supported by the observation that taste neurons are equally likely to fire to any one of the taste qualities in the caudomedial taste region, whereas there is clearly a dominant response to sweet taste in the caudolateral OFC (Pritchard et al. 2005).

In considering the apparent right lateralization of responses to pleasant tastes, it is useful to consider that affective value and quality are frequently confounded in taste studies. Sweet generally represents the pleasant taste, and bitter or intensely salty the unpleasant taste. Thus, an alternative interpretation of hedonic lateralization is that the quality of sweetness, or its nutritive value, is preferentially coded in the right OFC. In general, the link between specific taste quality, nutritive value and hedonic value is quite stable and can be seen even in infancy (Steiner et al. 2001). This has led some to argue that the perception of specific tastes is so integrally affective in nature that taste in isolation can be treated as a primary reinforcer. However, the affective value of tastes is not entirely immutable (Moskowitz *et al.* 1975; Hill *et al.* 1986; Hill and Przekop 1988; Krimm and Hill 1999; Pittman and Contreras 2002; Hendricks *et al.* 2004; Shuler *et al.* 2004) indicating that taste quality and hedonic value can at least partially be dissociated.

Irrespective of whether responses are anatomically segregated according to valence, quality or physiological significance, it appears that OFC responses in humans are relatively independent of stimulus intensity (Small *et al.* 2003a). Small and colleagues gave subjects a weak and a strong sweet solution and a weak and a strong bitter solution while recording brain response with fMRI. Importantly, the two sweet solutions were rated as similarly pleasant, the two bitter solutions as similarly unpleasant, the weak solutions as equally weak, and the strong solutions as equally strong. Thus intensity and valence were effectively dissociated. To isolate the regions sensitive to intensity and not valence, a conjunction analysis was performed to localize the regions activating to both strong intensity solutions compared to the two weak solutions. Activation was observed in the insula/operculum, amygdala, cerebellum and brainstem, but not in the OFC. In contrast, preferential activity in the right caudolateral OFC was observed to the pleasant sweet vs. the unpleasant bitter stimuli, irrespective of the intensity, and a region of the left anterior medial OFC was responsive to the unpleasant bitter taste

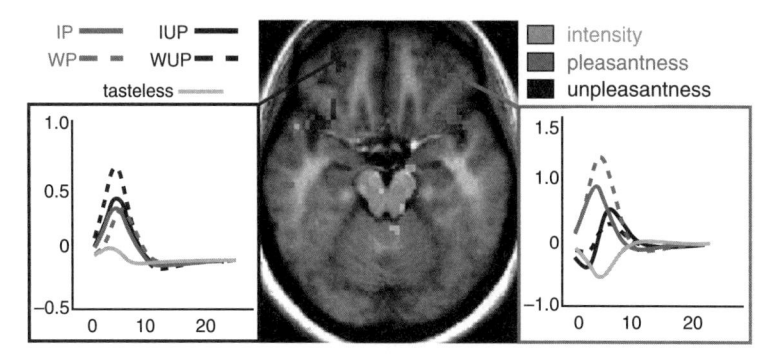

**Fig. 6.12** (Also see Color Plate 16.) Dissociation of taste intensity and affective coding. Results from three random effects contrasts color-coded and superimposed upon an axial section from the averaged anatomical image. Green denotes activity related to taste intensity from the analysis {(intense pleasant + intense unpleasant) > (weak pleasant + weak unpleasant)}; pink denotes activity related to taste pleasantness from the analysis {(intense pleasant + weak pleasant > tasteless}; and blue denotes activity related to taste unpleasantness from the analysis {(intense unpleasant + weak unpleasant) > tasteless}. Note the valence-specific responses and lack of intensity response in the OFC. In contrast, activity related to intensity but not valence is seen in the amygdala. The graphs depict the fitted hemodynamic response function from the unpleasant (left = blue box) and the pleasant (right = pink box) responses in the orbitofrontal cortex. The blue lines represent the unpleasant taste (quinine), the pink lines represent the pleasant taste (sucrose) and the beige line indicates tasteless. Dotted lines represent weak solutions and solid lines indicate strong solutions. Adapted with permission from Small *et al.* (2003).

irrespective of intensity (Fig. 6.12). The observation, that the human OFC does not show a general effect of intensity, is consistent with the single cell literature in monkeys in that the narrow breadth of tuning and suggestions of regional specializations would make it unlikely that single cells in the OFC would have the ability to code for intensity of multiple tastes. Thus, to the extent that single cells in the OFC show concentration response functions that track intensity (Rolls *et al.* 1990), such responses are likely to be stimulus-specific and indeed may reflect changes in pleasantness that occur with increasing concentrations (Engel 1928). The lack of a general effect of intensity in the OFC contrasts with the response in the amygdala, which appears to reflect a complex function of taste intensity, valence and familiarity (Small *et al.* 1997a, 2001a, b, 2003a; Zald 2003).

## Anticipation of taste

Studies in animals on expectancy of reward have typically used tastes as the expected reinforcer, and demonstrated neurons in the OFC that respond during the anticipation of tastes (see Roesch and Schoenbaum, this volume). In humans, O'Doherty, *et al.* (2001b) measured BOLD responses while subjects were presented with one of three arbitrary visual stimuli, which reliably predicted the subsequent delivery of either a pleasant sweet taste, an unpleasant salty taste or a neutral tasteless solution (O'Doherty *et al.* 2002). The right OFC was not only activated more in response to the receipt of a pleasant taste (compared to either the neutral tasteless solution or the aversive saline solution) but also to the anticipation of that pleasant taste, compared to either the anticipation of a neutral tasteless solution or to the anticipation of an unpleasant saline solution (O'Doherty *et al.* 2002). Similarly, anticipation of the unpleasant salty solution activated the left lateral OFC compared to anticipation of the neutral tasteless solution. Both anticipatory and consummatory responses to taste were attenuated over the course of the study, suggesting that processes related to either habituation or satiation influence both anticipatory and consummatory responses in the OFC (see Part IV below for a more detailed discussion of satiety issues).

Whether or not a taste or flavor is anticipated may also alter the OFC response to food reward. As outlined by Schultz (Chapter 7) and Montague and Berns (Montague and Berns 2002), the predictability of a rewarding stimulus is a critical parameter for activation of reward pathways, and OFC-striatal circuits are thought to be critical for ongoing tabulation of reward prediction (Schultz *et al.* 2000 Berns *et al.* 2001; Montague and Berns 2002; McClure *et al.* 2003; O'Doherty *et al.* 2003; ). Data from Berns and colleagues (Berns *et al.* 2001) indicate that the predictability of a sapid stimulus influences the location of the OFC responses. Preferential activation of the medial OFC occurred to the unpredictable arrival of fruit juice, whereas preferential activation of the lateral OFC emerged in response to predictable arrival of the juice. Preliminary findings from one of our laboratories (DMS) suggest that this dissociation also generalizes to pure tastes.

## 6.4. Chemosensory integration and the assembly of flavor

We rarely experience smells or tastes in isolation. For example, taste sensations are normally experienced in the context of eating or drinking when they are combined with other sensory experiences into a unitary flavor percept (Small and Prescott 2005; Delwiche 2004). What is colloquially referred to as 'taste' is therefore the combination of a variety of physiologically distinct sensory experiences associated with the flavor of food, such as its smell, texture and temperature. Flavor perception may also be influenced by the sight of a food (Morrot *et al.* 2001; Zellner and Durlach 2002; Delwiche 2004).

### Primates

In nonhuman primates, neural representations of flavor in the caudal OFC are not limited to taste and smell information, but also encode many of the other sensory constituents that together comprise flavor (Morecraft *et al.* 1992; Carmichael *et al.* 1994; Rolls and Baylis 1994; Carmichael and Price 1996; Rolls *et al.* 2003b). Dispersed throughout this region of the OFC are cells that respond to the sight (Rolls and Baylis 1994), texture (Rolls *et al.* 1999 de Araujo *et al.* 2003;) and temperature (Kadohisa *et al.* 2004) of food. Moreover, cells that respond to various combinations of these sensory experiences are also found in the OFC (Rolls and Baylis 1994; Rolls, *et al.* 2003b; Verhagen *et al.* 2003; Kadohisa *et al.* 2004). In a series of studies Rolls and colleagues examined the response of single cells to multiple sensory inputs (Thorpe *et al.* 1983; Rolls and Baylis 1994; Rolls *et al.* 2003b; Verhagen *et al.* 2003; Kadohisa *et al.* 2004). In one study 2000 cells were sampled for their responsiveness to taste, smell, flavor and visual stimuli (Rolls and Baylis 1994). One hundred and fifty-eight neurons responded to stimulation. Forty-five of these responded in more than one modality. Although unimodal cells were concentrated in three different areas, considerable overlap occurred, and bimodal and trimodal cells were interspersed throughout the three areas. Similarly, neurons that respond to different combinations of viscosity, taste, fat, astringency and temperature have also been described (Kadohisa *et al.* 2004; Rolls *et al.* 2003b; Verhagen *et al.* 2003b).

One striking feature of many of the multimodal neurons in the primate OFC is their selective responses for appropriately paired (i.e., congruent) stimuli. For example, a cell responding optimally to a glucose solution responded optimally to fruit-related odors rather than to salmon or other odors, that were less appropriate for a sweet tasting solution (Rolls and Baylis 1994; Rolls 2001). Similarly, a bimodal taste/visual cell responding to glucose also responded to the sight of sweet foods, such as a banana or a syringe, that had been used to deliver blackcurrant juice, but not to the sight of pliers, which had never been associated with a sweet taste. One taste/smell cell responded to neither the smell nor the taste of any of the palatable food items. Instead, this neuron fired maximally in response to the aversive stimuli of quinine and caprylic acid, suggesting a dissociation between neurons that respond to either hedonically positive or negative stimuli (or if considered in terms of ingestive processing, nutritive versus toxic stimuli).

## Humans

Similar mechanisms of chemosensory integration have been identified in the human OFC. Examination of Figs. 6.3, 6.9, and 6.11 clearly shows that stimulation with either tastes or smells can produce activation in the overlapping regions of the human OFC. While such findings do not confirm whether these areas specifically integrate information across different modalities, other fMRI studies have addressed this question more directly. By examining the neural interaction between smells and tastes (de Araujo et al. 2003; Small et al. 2004), or between smells and pictures (Gottfried and Dolan 2003), response supra-additivity has been demonstrated in the human OFC. Cross-modal response enhancement of this type may reflect the hemodynamic analog of multisensory integration at the cellular level, as originally demonstrated in cat superior collicular neurons exposed to visual, auditory, and tactile stimuli (Stein and Meredith 1993). Recently, the phenomenon has been extended to human imaging studies of audio-visual integration (Calvert et al. 2000). The present results suggest that supra-additive response patterns are applicable to chemosensory modalities, and may provide a powerful approach to investigating the assembly of flavor representations (and food identity) in the human brain.

As with sensory integration in the primate OFC, the response of the human OFC to multiple chemosensory inputs is affected by prior experience. In the study by Gottfried and Dolan (2003), greater activity was observed in the anterior medial OFC to semantically congruent compared to incongruent cross-modal stimuli (e.g., a picture of a double-decker bus and the odor of diesel, versus the picture of cheese and the smell of fish), and subjects detected the odor significantly more quickly when it occurred in conjunction with a congruent compared to an incongruent picture. Small and colleagues identified a response in the caudal OFC that was significantly greater to a sweet vanilla (congruent) versus a salty vanilla (incongruent) solution. Similarly, de Araujo and colleagues (2003) noted a significant correlation between subjects' ratings of taste-odor congruency and activation in the anterior medial OFC. The marked dependence on congruency that characterizes the OFC response indicates that chemosensory integration in the OFC is heavily influenced by prior experience with both food and non-food stimuli.

A factor unique to taste-odor integration is the route of odorant administration. Ordinarily when odors are perceived with taste, they reach the olfactory epithelium via the retronasal route (in which vapors in the mouth rise through the nasopharynx). In contrast, when subjects sniff, odorants reach the epithelium via the nostrils (orthonasally). Although in both cases odorants reach the epithelium, increasing evidence suggests this has a significant impact on how odorants and tastes are integrated (Burdach et al. 1984; Heilmann and Hummel 2001; Halpern and Sun 2002; Heilmann et al. 2002; Small et al. 2004) (but see Sakai et al. (2001); Pierce and Halpern (1996)). In the first neuroimaging study to directly examine the integration of olfactory and gustatory information, Small et al. (1997a) demonstrated decreased activity in the caudal OFC when an orthonasal smell was presented simultaneous with a taste, relative to when the smell and taste were presented in isolation. Perhaps in this situation, the orthonasal smell and the taste compete for OFC processing. While the specific mechanisms leading to such differences remain to be elucidated, the data strongly suggest that the neural integration of taste and smell into a unitary flavor percept is dependent either upon the two stimuli being perceived as coming from a common source (i.e., a drink) versus two separate sources (i.e., a cotton wand and filter paper), or upon the route of olfactory administration (orthonasal versus retronasal).

It is worth noting in this regard that ortho vs. retronasal olfaction differentially map to the anticipatory vs. consummatory phases of feeding. Berridge (1996) has described the affective states related to these two processes as 'wanting' and 'liking', respectively, and has proposed that separable neural substrates support each component. In the sensory realm, this dichotomy reflects distal vs. contact stimulation. Taste, temperature, texture and trigeminal irritation (as encountered in spicy food) are all contact sensations, signifying the receipt of a food reward and are associated with liking. In contrast, the sight or aroma of a food is a distal sensory experience that signifies the availability of a food and are associated with wanting. The olfactory system is unique in that it reflects both distal sensation (orthonasal) and contact sensation (retronasal). This observation led Rozin (1982) to propose that 'olfaction is the only dual sensory modality, in that it senses both objects in the external world and objects in the body (mouth)'.

To study this difference Small *et al.* (2005) placed two tubes in the nasal cavity using endoscopic guidance, allowing them to deliver a pulse of vaporized odorant via either route while avoiding simultaneous engagement of oral sensory experiences normally associated with retronasal olfaction. Comparison of brain activity evoked to the same food odor, sensed orthonasally versus retronasally, revealed two different patterns of OFC activity. Activity in the medial OFC and the subcallosal region responded preferentially during retronasal perception, whereas activity in the lateral OFC responded preferentially during orthonasal perception (see Fig. 6.13). A similar dissociation was not observed when comparing orthonasal versus retronasal presentation of three equally pleasant and intense non-food odors (Small *et al.* 2005), suggesting that the context in which a sensory stimulus is normally experienced can markedly alter the neural response it evokes. The emergence of this segregation for food odors is remarkable in that it occurs even in the absence of any current oral sensation.

If the lateral response to orthonasal olfaction, and the more medial response to retronasal olfaction are taken as indices of anticipatory vs. consummatory responses, then these data may be interpreted as suggesting a lateral vs. medial distinction between wanting and liking systems within the OFC. While some additional data support this distinction (Small *et al.* 2001c; Kringelbach *et al.* 2003b), not all data are consistent with such a model. Nevertheless, it is clear from work in rodents (Schoenbaum *et al.* 2003), primates (Schultz *et al.* 1998, 2000; Tremblay and Schultz 1999, 2000), and humans (Berns *et al.* 2001; O'Doherty *et al.* 2002, 2003) that the OFC plays a role in both the prediction and anticipation of food reward, making it a key structure for both wanting and liking and their interaction.

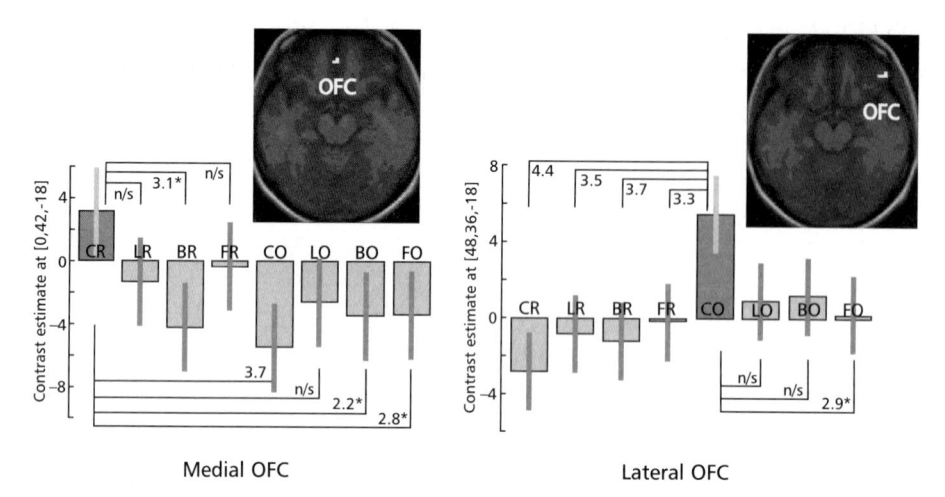

**Fig. 6.13** Dissociation of orbital regions responding preferentially to orthonasal versus retronasal perception of an appetitive food odor (chocolate). Left: axial section showing preferential response in the medial orbitofrontal cortex (OFC) to retronasally perceived chocolate odor (CR) compared to orthonasally perceived chocolate odor (CO) as well as to ortho- or retronasal perception of three similarly pleasant and intense nonfood odors (LR = lavender retro; BR = butanol retro; FR = farnesol retro; LO = lavender ortho; BO = butanol ortho; FO = farnesol ortho). Right: axial section showing preferential response in the lateral OFC to orthonasal perception of chocolate odor versus all other odorants. T-map thresholded at $p < 0.001$ and a cluster threshold of $k < 3$. Graphs show response (parameter estimates) from each of the eight odorant conditions minus each associated baseline condition at the co-ordinate defined on the y-axis. Response is in arbitrary units. Numbers below or above lines joining the bars indicate z-values associated with direct contrasts of CO with each of the 7 other odorant conditions * indicates that the z-value is only significant uncorrected across the small volume of interest. n/s = nonsignificant.

## 6.5. Regulation of feeding behavior

Given the importance of the OFC in representing reward and flavor, it is useful to consider the OFC representation of taste within the context of a more general model of food reward. For a more in-depth consideration of these topics the reader is referred to (Del Parigi et al. 2002b; Saper et al. 2002; Kringelbach and Rolls 2004).

### Relative reward and food selection

The OFC appears critical in coding the relative reward value of different foods (Tremblay and Schultz 1999; Arana et al. 2003; Kilgore et al. 2003; Hinton et al. 2004). For example, in monkeys Tremblay and Schultz have shown that the same OFC neuron will respond to a preferred apple over a less preferred fruit loop, but when an apple is pitted against a more preferred raisin, the neuron switches its responsiveness to the raisin (Tremblay and Schultz 1999). In humans, Arana and colleagues had subjects imagine that they were in a restaurant looking at different menus while measuring brain activity with PET (Arana *et al.* 2003). All of the menus were individually tailored according to the subject's preference. In some trials subjects simply looked at the menu while in others they were required to select items. They found that the rostral gyrus rectus showed a greater response to high-incentive menus and responded more when a choice was required. In contrast, the lateral OFC responded more when subjects had to suppress responses to alternative desirable items to select their most preferred. The implication of this result is that the OFC not only encodes the relative reward value of the menus but also contributes to food selection. Studies in both monkeys and humans with OFC impairment have been observed to show alterations in food selection (Baylis and Gaffan 1991; Ikeda *et al.* 2002). However, it is unclear if these changes reflect actual disruption of the evaluation of the relative value of foods, or is simply a consequence of perceptual deficits or impairments in associative learning. Recent data indicate that monkeys with OFC lesions show stable food preferences in selecting among palatable foods that they are familiar with (Izquierdo 2004), indicating that at a minimum, some form of preferential food selection is maintained, even though such animals may lose the ability to adaptively use information about food preferences to guide behavioral choices, especially when the value of one of the foods changes (Baxter *et al.* 2000).

## Internal state

Internal state markedly influences the affective value of chemosensory stimuli (Cabanac 1971; Rolls *et al.* 1983; Berridge 1991). Specifically, ongoing or recent consumption of a food leads to a temporary reduction in its reward value. Two different processes may come in to play in this context. First, a general process of alliesthesia may develop in which fullness leads to a general reduction in the pleasantness of food (Cabanac 1971). Second, a more narrow satiety for specific foods, but not to other foods, develops during the consumption of a given food. This latter phenomenon has been termed sensory-specific satiety (Rolls *et al.* 1983), because factors, such as similarity of color, shape, flavor and texture, appear more important than nutritive content in directing this reduction in affective value. There is compelling evidence that the OFC plays a critical role in this process. At the neurophysiological level, unimodal gustatory and olfactory responses to specific foods are greatly attenuated by satiety (Nakano *et al.* 1984; Rolls *et al.* 1989; Critchley and Rolls 1996), while leaving responses to other foods intact. Sensory-specific satiety-related declines have also been seen for visual responses to food items, confirming the multimodal nature of the phenomenon (Nakano *et al.* 1984; Rolls *et al.* 1989; Critchley and Rolls 1996). In contrast to the OFC, single cell recordings of the insular-operculum primary taste region do not show evidence of a sensory-specific satiety effect, which led Rolls and colleagues to conclude that this process first emerges at the level of the OFC. Data from humans suggest that insular responses to foods do actually decrease with repetition (Small *et al.* 2001c; Gautier 2001), but this may reflect the more general process of alliesthesia, rather than the sensory-specific satiety.

Substantial evidence indicates that the human OFC also changes its response during the satiation of specific foods. Two studies have reported on the brain response to flavored food or drink as subjects ate or drank that substance to satiety or beyond. Responses in the medial OFC to palatable food or drink decreased as that substance was eaten to satiety (and beyond), and these decreases correlated with declining pleasantness ratings (Kringelbach et al. 2003; Small et al. 2001c). In contrast, activity in the lateral OFC, extending into the ventrolateral prefrontal cortex, increased with satiety, and correlated inversely with pleasantness ratings (Small et al. 2001c). These data thus suggest that there are both decreases and increases in activity with satiety, and that a medial/lateral segregation distinguishes these responses. Although no published studies have examined satiety to pure gustatory stimuli in the olfactory domain in humans, O'Doherty et al. (2000) scanned participants as they smelled either banana or vanilla odor, both before and after eating bananas to satiety. Post-satiety odor-evoked activations to the banana smell decreased in the posterior OFC (right side, 4 subjects; left side, 1 subject), without any corresponding OFC signal decline in response to the vanilla smell. O'Doherty et al. did not observe specific areas that became more active with satiety. However, several additional studies examining brain activity in states of hunger and satiety confirm that satiety is accompanied by both areas of increased and decreased activity.

Tataranni and colleagues used PET to measure resting brain activity after a 36-hour fast (condition 1: hungry), followed by a 2ml sample of a liquid meal (condition 2: taste) and then a satiating amount of a liquid meal (conditioned 3: satiety) (Gautier et al. 1999; Tataranni et al. 1999; Gautier et al. 2001; Del Parigi, et al. 2002a). Activity in the caudal OFC was greater during hunger compared to satiety, while activity in the dorsal and ventral prefrontal cortex, extending into the anterior-lateral most part of the OFC, was greater in satiety compared to hunger. The authors have proposed that the prefrontal cortex signals satiation by inhibiting limbic/paralimbic areas. Interestingly, changes in both caudal and more lateral areas appear amplified in obese subjects (Gautier et al. 2001; Del Parigi et al. 2002b; Tataranni and Del Parigi 2003).

In order to follow the time course of changes following eating, Lui et al. (2000) used fMRI to assess the change in activity within a single medially placed sagittal slice following ingestion of 75 g glucose solution. Between 1- and 3-minute post ingestion, activity in the medial OFC extending onto the medial surface began to decrease, with further decreases occurring between 7- and 13-minute post ingestion. The decreasing activity was located in the same region as the decreasing activity Small and colleagues reported in response to eating chocolate to beyond satiety (Small et al. 2001c) and also overlapped with the region reported by Tataranni and colleagues as less active in full versus hungry states (Tataranni et al. 1999).

Additional support for the hypothesis that more anterior-lateral areas play an inhibitory role comes from the menu selection study by Arana and colleagues described above (Arana et al. 2003). Whereas caudo-medial OFC activity was observed when subjects viewed preferred vs nonpreferred menus, lateral OFC activity was observed when participants had to suppress responses to alternative desirable items in order to select their most preferred item (Arana et al. 2003). A follow-up study examined the interaction between what the authors termed intrinsic and extrinsic determinants of food reward (Hinton *et al.* 2004). Intrinsic factors were defined as contributions by homeostatic mechanisms (e.g., hunger, satiety), and extrinsic factors as contributions by variables related to the food, such as food preferences and the associated experience-based incentive properties of food. The follow-up study was identical to the earlier experiment, except that subjects performed the task while hungry and while full. Consequently, they were able to evaluate the effect of intrinsic factors (internal state) upon extrinsic factors evaluated by their task. In addition to replicating their initial findings, and in accordance with previous studies comparing hunger and satiety, they found that the lateral OFC was more responsive when viewing and choosing items from a menu when full compared to hungry. More importantly, they found that intrinsic factors (i.e., hunger vs. satiety) had a far greater effect upon neural response than extrinsic factors (imagining high versus low incentive foods). The only region of the brain in which an interaction was observed between intrinsic and extrinsic factors was in a region of the anterior orbital gyrus, which was preferentially activated by viewing and choosing food from high versus low incentive menus when hungry compared to when full (Fig. 6.14).

To summarize, the findings from studies of lean subjects eating food (Liu *et al.* 2000; Small *et al.* 2001c; Kringelbach *et al.* 2003), at rest in the state of hunger or satiety (Gautier *et al.* 1999; Tataranni *et al.* 1999; Del Parigi *et al.* 2002a; Tataranni and Del Parigi 2003), and imagining and selecting food items while hungry and full (Arana *et al.* 2003; Hinton *et al.* 2004), all point to a role for the medial OFC in guiding feeding behavior while hungry, which declines with fullness, while the anterior-lateral OFC becomes increasingly engaged when individuals are full. These data suggest that the OFC plays both a general regulatory role, that is sensitive to homeostatic factors (alliesthesia), as well as a sensory-specific role in regulating feeding behavior. The possibility that the lateral OFC extending into the ventrolateral prefrontal cortex (areas 47/12) plays an inhibitory role in feeding (Del Parigi *et al.* 2002b) is consistent with the more general models of lateral OFC/ventrolateral PFC functions that emphasize its engagement when a selected action requires the suppression of previously rewarded responses (Elliott *et al.* 2000). Additionally, the findings of Hinton and colleagues suggest that the OFC processing in the anterior orbital gyrus is important in integrating both intrinsic and extrinsic factors to food reward.

The general differentiation between anterior/lateral and medial OFC in response to satiety is interesting, in that it appears to at least partially resemble the general medial-pleasant/lateral-unpleasant trends observed for valence (Kringelbach and Rolls 2004). Although we have argued that this distinction does not provide a full explanation of response distributions in the chemosensory domain, the general trend has been observed in many paradigms and modalities, ranging from cross-modal paradigms of episodic memory (Gottfried *et al.* 2004) and associative learning (Gottfried *et al.* 2002b) to the processing of faces (O'Doherty *et al.* 2003) and abstract monetary reinforcers (O'Doherty *et al.* 2001a). That stimuli in the hungry state (when they have the highest appetitive value), engage more medial areas, while stimuli in the full or satiated state (when they have lower appetitive value) engage more lateral areas, appears consistent with this

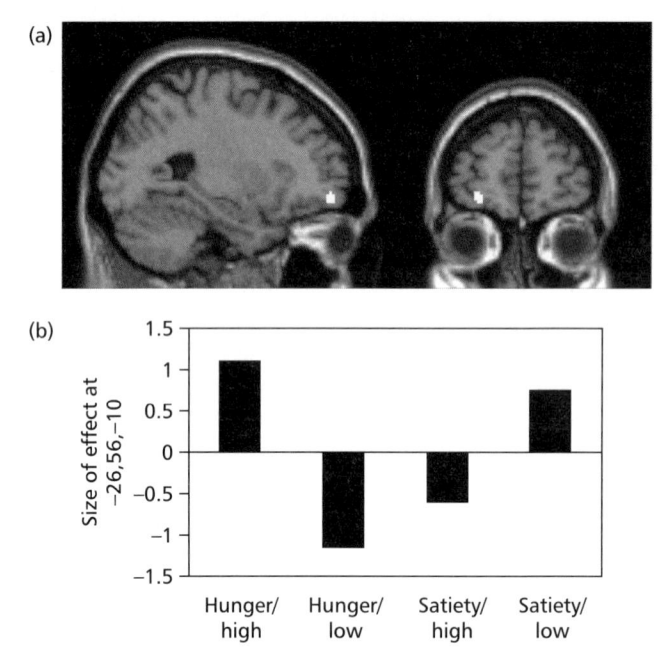

**Fig. 6.14** Interaction of incentive value and hunger state in the anterior OFC. The interaction between incentive value and hunger state. (A) Sagittal and coronal sections depicting a distinct area of the left OFC, which showed increased activation in the interaction between hunger state and incentive value. (B) The pattern of interaction between hunger state and incentive value is shown by this condition-specific plot: the conditions were hunger/high incentive, hunger/low incentive, satiety/high incentive and satiety/low incentive. Adapted with permission from Hinton et al. (2004).

pattern, though in the satiated state, the food stimuli are often less pleasant rather than specifically unpleasant.

Finally, it is worth considering how the aforementioned results may contribute to our understanding of eating disorders. Food reward is multifaceted with evidence for separable circuits representing food wanting, food liking, and signals of internal state such as hunger and fullness (Berridge 1996). A remarkable feature of the OFC is that it appears to encode all aspects of food reward. This contrasts with the representation of food reward in the amygdala, which appears highly specific to incentive salience encoding (O'Doherty et al. 2002; Gottfried et al. 2003; Small et al. 2005). One implication of this confluence of function in the OFC is that an integrated output computed here represents a final common pathway for food reward supporting food selection. Although dysfunction in any one aspect of food reward (e.g., enhanced palatability vs. increased hunger) could lead to aberrant eating behaviors, if the OFC represents a final common pathway, abnormal responses in the OFC are likely to be associated with multiple forms of eating disorders. Further, it is possible that the precise location of such responses may be associated with either overeating or undereating. There is indirect support in the literature for such a hypothesis. For example, while Uher *et al* (2004) reported that an abnormal ventral medial prefrontal response to symptom-provoking stimuli was a common feature of both anorexia and bulimia nervosa, the findings of Del Parigi and Tataranni point towards abnormal ventrolateral responses in obese subjects (Del Parigi *et al.* 2002a).

## 6.6. Conclusion

In their seminal paper on the connections of the OFC, Carmichael and Price wrote:

> At first consideration, there appears to be a discrepancy between the suggestion from the lesion results that the orbital cortex has a behavioral role in stimulus-reward association and the indication from the connectional data that there is a sensory hierarchy related to feeding. On further consideration, however, such apparent discrepancy may be an artifact of experimental methodology. In fact, the hierarchical processing of chemical and visceral sensations and the formation of reward may be the same thing. (Carmichael and Price 1996, p. 205).

Although we would not go so far as to say that the hierarchical processing of chemical visceral sensations is synonymous with the formation of reward, the data reviewed in this chapter make clear that the processing of chemosensory stimuli, the formation of reward, and the regulation of feeding are intimately intertwined in the OFC. Human feeding behavior is extremely complex, and includes both anticipatory (wanting) and consummatory (liking) factors. Numerous processes must be accomplished just to get to the point of eating a food. For instance, the acquisition and selection of items for ingestion requires an ability to identify and associate odors and visual objects with their taste or flavor properties, to learn how to properly acquire the items and/or make them palatable, to select one potential food over another, to learn when one food item is no longer good, to learn when it is no longer safe or advantageous to acquire a food, and to gauge the value of specific foods and foods in general, given recent experience and current internal state. While obviously not all of these processes are accomplished by a single brain region, the specificity of its olfactory and gustatory processing, its ability to integrate these representations with visual sensations and other oral sensations, its coding of changes in stimulus-food reward contingencies and the modification of responses by hunger and satiety, ideally position the OFC to implement executive-level control in the regulation of feeding.

Although much of this book focuses on the OFC's role in non-chemosensory behavior, the fundamental processing properties that allow the OFC to flexibly influence food acquisition and consumption may apply to the OFC's involvement in other cognitive and behavioral domains. We have characterized several of these properties, such as high stimulus specificity, separable responses to positive and negatively valenced stimuli, engagement during judgements about hedonics, multisensory integration, sensitivity to prior experience with stimulus combinations, and differential coding based on internal state. To the extent that these properties generalize to other behavioral domains, they may critically influence how the OFC differentially contributes to neurobehavioral functions relative to other brain regions, and may causally direct the phenomenological features of neuropsychiatric symptoms associated with OFC dysfunction.

## References

Allen, W. F. (1940) Effect of ablating the frontal lobes, hippocampi, and occipito-parieto-temporal (excepting piriform areas) lobes on positive and negative olfactory conditioned reflexes. *American Journal of Physiology* **128**:754–771.

Allen, W. F. (1943a) Distribution of cortical potentials resulting from insufflation of vapors into the nostrils and from stimulation of the olfactory bulbs and the piriform lobe. *American Journal of Physiology* **139**:553–555.

Allen, W. F. (1943b) Results of prefrontal lobectomy on acquired and on acquiring correct conditioned differential responses with auditory, general cutaneous and optic stimuli. *American Journal of Physiology* **139**:525–531.

Anderson, A. K., Christoff, K., Stappen, I., Panitz, D., Ghahremani, D. G., Glover, G., Gabrieli, J. D., and Sobel, N. (2003) Dissociated neural representations of intensity and valence in human olfaction. *Nature Neuroscience* **6**:196–202.

Arana, F. S., Parkinson, J. A., Hinton, E., Holland, A. J., Owen, A. M., and Roberts, A. C. (2003) Dissociable contributions of the human amygdala and orbitofrontal cortex to incentive motivation and goal selection. *Journal of Neuroscience* **23**:9632–9638.

Barbas, H. (1993) Organization of cortical afferent input to orbitofrontal areas in the rhesus monkey. *Neuroscience* **56**:841–864.

Barry, M. A., Gatenby, J. C., Zeiger, J. D., and Gore, J. C. (2001) Hemispheric dominance of cortical activity evoked by focal electrogustatory stimuli. *Chemical Senses* **26**:471–482.

Baylis, L. L., Rolls, E. T., and Baylis, G. C. (1995) Afferent connections of the caudolateral orbitofrontal cortex taste area of the primate. *Neuroscience* **64**:801–812.

Beckstead, R. M., Morse, J. R., and Norgren, R. (1980) The nucleus of the solitary tract in the monkey: projections to the thalamus and brain stem nuclei. *Journal of Comparative Neurology* **190**:259–282.

Benjamin, R. M. and Jackson, J. C. (1974) Unit discharges in the mediodorsal nucleus of the squirrel monkey evoked by electrical stimulation of the olfactory bulb. *Brain Research* **75**:181–191.

Berns, G. S., McClure, S. M., Pagnoni, G., and Montague, P. R. (2001) Predictability modulates human brain response to reward. *Journal of Neuroscience* **21**:2793–2798.

Berridge, K. C. (1991) Modulation of taste affect by hunger, caloric satiety, and sensory-specific satiety in the rat. *Appetite* **16**:103–120.

Berridge, K. C. (1996) Food reward: brain substrates of wanting and liking. *Neuroscience and Biobehavioral Reviews* **20**:1–25.

Buckner, R. L., Goodman, J., Burock, M., Rotte, M., Koutstaal, W., Schacter, D., Rosen, B., and Dale, A.M. (1998) Functional-anatomic correlates of object priming in humans revealed by rapid presentation event-related fMRI. *Neuron* **20**:285–296.

Burdach, K. J., Kroeze, J. H., and Koster, E. P. (1984) Nasal, retronasal, and gustatory perception: an experimental comparison. *Perception and Psychophysics* **36**:205–208.

Cabanac, M. (1971) Physiological role of pleasure. *Science* **173**:1103–1107.

Carmichael, S. T., Clugnet, M. C., and Price, J. L. (1994) Central olfactory connections in the macaque monkey. *Journal of Comparative Neurology* **346**:403–434.

Carmichael, S. T. and Price, J. L. (1994) Architectonic subdivision of the orbital and medial prefrontal cortex in the macaque monkey. *Journal of Comparative Neurology* **346**:366–402.

Carmichael, S. T. and Price, J. L. (1995) Sensory and premotor connections of the orbital and medial prefrontal cortex of macaque monkeys. *Journal of Comparative Neurology* **363**:642–664.

Carmichael, S. T. and Price, J. L. (1996) Connectional networks within the orbital and medial prefrontal cortex of macaque monkeys. *Journal of Comparative Neurology* **371**:179–207.

Carpenter, R. H. S. (2003) *Neurophysiology*, 4th edn. London: Arnold Publishers.

Cerf-Ducastel, B., Van de Moortele, P. F., MacLeod, P., Le Bihan, D., and Faurion, A. (2001) Interaction of gustatory and lingual somatosensory perceptions at the cortical level in the human: a functional magnetic resonance imaging study. *Chemical* Senses 26:371–383.

Chiavaras, M. M. and Petrides, M. (2000) Orbitofrontal sulci of the human and Macaque monkey brain. *Journal of Comparative Neurology* **422**:35–54.

Critchley, H. D. and Rolls, E. T. (1996a) Olfactory neuronal responses in the primate orbitofrontal cortex: analysis in an olfactory discrimination task. *Journal of Neurophysiology* **75**:1659–1672.

Critchley, H. D. and Rolls, E. T. (1996b) Hunger and satiety modify the responses of olfactory and visual neurons in the primate orbitofrontal cortex. *Journal of Neurophysiology* **75**:1673–1686.

Dade, L. A., Zatorre, R. J., Evans, A. C., and Jones-Gotman, M. (2001) Working memory in another dimension: functional imaging of human olfactory working memory. *Neuroimage* **14**:650–660.

Dade, L. A., Zatorre, R. J., and Jones-Gotman, M. (2002) Olfactory learning: convergent findings from lesion and brain imaging studies in humans. *Brain* **125**:86–101.

de Araujo, I. E. and Rolls, E. T. (2004) Representation in the human brain of food texture and oral fat. *Journal of Neuroscience* **24**:3086–3093.

de Araujo, I. E., Rolls, E. T., Kringelbach, M. L., McGlone, F., and Phillips, N. (2003) Taste-olfactory convergence, and the representation of the pleasantness of flavor, in the human brain. *European Journal of Neuroscience* **18**:2059–2068.

de Olmos, J., Hardy, H., and Heimer, L. (1978) The afferent connections of the main and the accessory olfactory bulb formations in the rat: an experimental HRP-study. *Journal of Comparative Neurology* **181**:213–244.

Del Parigi, A., Chen, K., Gautier, J. F., Salbe, A. D., Pratley, R. E., Ravussin, E., Reiman, E. M., and Tataranni, P. A. (2002a) Sex differences in the human brain's response to hunger and satiation. *American Journal of Clinical Nutrition* **75**:1017–1022.

Del Parigi, A., Gautier, J. F., Chen, K., Salbe, A. D., Ravussin, E., Reiman, E., and Tataranni, P. A. (2002b) Neuroimaging and obesity: mapping the brain responses to hunger and satiation in humans using positron emission tomography. Annals of the New York Academy of Sciences 967:389–397.

Delwiche, J. F. (2004) The impact of sensory and physical interactions on perceived flavor. *Food Quality and Preferences* **15**:137–146.

Distel, H., Ayabe-Kanamura, S., Martinez-Gomez, M., Schicker, I., Kobayakawa, T., Saito, S., and Hudson, R. (1999) Perception of everyday odors—correlation between intensity, familiarity and strength of hedonic judgement. *Chemical Senses* **24**:191–199.

Doty, R. L., Shaman, P., and Dann, M. (1984) Development of the University of Pennsylvania Smell Identification Test: a standardized microencapsulated test of olfactory function. *Physiology and Behavior* **32**: 489–502.

Dusser de Barenne, J.G. (1933) "Corticalization" of function and functional localization in the cerebral cortex. *Archives of Neurology and Psychiatry* **40**: 884–901.

Eichenbaum, H., Morton, T. H., Potter, H., and Corkin, S. (1983) Selective olfactory deficits in case H.M. *Brain* **106**:459–472.

Elliott, R., Dolan, R. J., and Frith, C. D. (2000) Dissociable functions in the medial and lateral orbitofrontal cortex: evidence from human neuroimaging studies. *Cerebral Cortex* **10**:308–317.

Engel, R. (1928) Experimental investigations of the dependence of pleasantness and unpleasantness upon the strength of the stimulus in the case of taste. *Archiv Fuer die Gesamte Psychologie* **64**:1–36.

Eslinger, P. J., Damasio, A. R., and Van Hoesen, G. W. (1982) Olfactory dysfunction in man: anatomical and behavioral aspects. *Brain and Cognition* **1**:259–285.

Faurion, A., Cerf, B., Le Bihan, D., and Pillias, A. M. (1998) fMRI study of taste cortical areas in humans. *Annals of the New York Academy of Sciences* **855**:535–545.

Faurion, A., Cerf, B., Van De Moortele, P. F., Lobel, E., Mac Leod, P., and Le Bihan, D. (1999) Human taste cortical areas studied with functional magnetic resonance imaging: evidence of functional lateralization related to handedness. *Neuroscience Letters* **277**:189–192.

Francis, S., Rolls, E. T., Bowtell, R., McGlone, F., O'Doherty, J., Browning, A., et al. (1999) The representation of pleasant touch in the brain and its relationship with taste and olfactory areas. *Neuroreport* **10**:435–459.

Frey, S. and Petrides, M. (1999) Re-examination of the human taste region: a positron emission tomography study. *European Journal of Neuroscience* **11**:2985–2988.

Gautier, J. F., Chen, K., Uecker, A., Bandy, D., Frost, J., Salbe, A. D., et al. (1999) Regions of the human brain affected during a liquid-meal taste perception in the fasting state: a positron emission tomography study. *American Journal of Clinical Nutrition* **70**:806–810.

Gautier, J. F., Del Parigi, A., Chen, K., Salbe, A. D., Bandy, D., Pratley, R. E., et al. (2001) Effect of satiation on brain activity in obese and lean women. *Obesity Research* **9**:676–684.

Glendinning, J. I. (1994) Is the bitter rejection response always adaptive. *Physiology and Behavior* **56**:1217–1227.

Gottfried, J. A., Deichmann, R., Winston, J. S., and Dolan, R. J. (2002) Functional heterogeneity in human olfactory cortex: an event-related functional magnetic resonance imaging study. *Journal of Neuroscience* **22**:10819–10828.

Gottfried, J. A. and Dolan, R. J. (2003) The nose smells what the eye sees: crossmodal visual facilitation of human olfactory perception. *Neuron* **39**:375–386.

Gottfried, J. A. and Dolan, R. J. (2004) Human orbitofrontal cortex mediates extinction learning while accessing conditioned representations of value. *Nature Neuroscience* **7**:1142–1152.

Gottfried, J. A. and Zald, D. H. (2005) On the scent of human olfactory orbitofrontal cortex: Meta-analysis and comparison to nonhuman primates, *Brain Research: Brain Research Reviews* **50**: 287–304.

Gottfried, J. A., O'Doherty, J., and Dolan, R. J. (2002) Appetitive and aversive olfactory learning in humans studied using event-related functional magnetic resonance imaging. *Journal of Neuroscience* **22**:10829–10837.

Gottfried, J. A., O'Doherty, J., and Dolan, R. J. (2003) Encoding predictive reward value in human amygdala and orbitofrontal cortex. *Science* **301**:1104–1107.

Gottfried, J. A., Smith, A. P., Rugg, M. D., and Dolan, R. J. (2004) Remembrance of odors past: human olfactory cortex in cross-modal recognition memory. *Neuron* **42**:687–695.

Gottfried, J. A., Winston, J.S., and Dolan, R. J. (2006) Dissociable codes of odor quality and odorant structure in human piriform cortex. *Neuron* **49**: 467–479.

Grill-Spector, K. and Malach, R. (2001) fMR-adaptation: a tool for studying the functional properties of human cortical neurons. *Acta Psychologia* **107**:293–321.

Haberly, L.B. (1998) Olfactory cortex. In; Shepherd, G. M. (ed.) *The Synaptic Organization of the Brain*, pp. 377–416. New York: Oxford University Press.

Halpern, B. P. and Sun, B. C. (2002) Asymmetric interactions between heterogenous retronasal and orthonasal odorant pairs. Paper presented at: Association for Chemoreception Sciences (Sarasota Florida, Chemical Senses)

Heilmann, S. and Hummel, T. (2001) Olfactory event-related potentials to ortho- and retronasal stimulation. Paper presented at: Achems (Sarasota, Fl)

Heilmann, S., Strehle, G., Rosenheim, K., Damm, M., and Hummel, T. (2002) Clinical assessment of retronasal olfactory function. *Archives of Otolaryngology—Head and Neck Surgery* **128**:414–418.

Hellekant, G., Danilov, Y., and Minomiva, Y. (1997) Primate sense of taste: Behavioral and single chorda tympani and glossopharyngeal nerve fiber recordings in the Rhesus monkey, *Macaca mulata*. *Journal of Neurophysiology* **77**:978–993.

Hendricks, S. J., Brunjes, P. C., and Hill, D. L. (2004) Taste bud cell dynamics during normal and sodium-restricted development. *Journal of Comparative Neurology* **472**:173–182.

Hill, D. L., Mistretta, C. M., and Bradley, R. M. (1986) Effects of dietary NaCl deprivation during early development on behavioral and neurophysiological taste responses. *Behavioral Neuroscience* **100**:390–398.

Hill, D. L. and Przekop, P. R., Jr. (1988) Influences of dietary sodium on functional taste receptor development: a sensitive period. *Science* **241**:1826–1828.

Hinton, E. C., Parkinson, J. A., Holland, A. J., Arana, F. S., Roberts, A. C., and Owen, A. M. (2004) Neural contributions to the motivational control of appetite in humans. *European Journal of Neuroscience* **20**:1411–1418.

Hughlings-Jackson, J. and Beevor, C. E. (1890) Case of tumor of the right temporo-sphenoidal lobe bearing on the localization of the sense of smell and on the interpretation of a particular variety of epilepsy. *Brain* **12**:346–357.

Hughlings-Jackson, J. and Stewart, P. (1899) Epileptic attacks with a warning of a crude sensation of smell and with the intellectual aura (dreamy state) in a patient who had symptoms pointing to gross organic disease of the right temporo-sphenoidal lobe. *Brain* **22**:534–549.

Izquierdo, A., Suda, R. K., and Murray, E. A. (2004) Bilateral orbital prefrontal cortex lesions in rhesus monkeys disrupt choices guided by both reward value and reward contingency. *Journal of Neuroscience* **24**: 7540–7548.

Jones-Gotman, M. and Zatorre, R. J. (1988) Olfactory identification deficits in patients with focal cerebral excision. *Neuropsychologia* **26**:387–400.

Jones-Gotman, M. and Zatorre, R. J. (1993) Odor recognition memory in humans: role of right temporal and orbitofrontal regions. *Brain and Cognition* **22**:182–198.

Kadohisa, M., Rolls, E. T., and Verhagen, J. V. (2004) Orbitofrontal cortex: neuronal representation of oral temperature and capsaicin in addition to taste and texture. *Neuroscience* **127**:207–221.

Katz, D. B., Simon, S. A., and Nicolelis, M. A. (2002) Taste-specific neuronal ensembles in the gustatory cortex of awake rats. *Journal of Neuroscience* **22**:1850–1857.

Kilgore, W. D. S., Young, A. D., Femia, L. A., Bogorodzki, P., Rogowsak, J., and Yurgelun-Todd, D. A. (2003) Cortical and limbic activation during viewing of high- versus low-calorie foods. *Neuroimage* **19**:1381–1394.

Kinomura, S., Kawashima, R., Yamada, K., Ono, S., Itoh, M., Yoshioka, S., et al. (1994) Functional anatomy of taste perception in the human brain studied with positron emission tomography. *Brain Research* **659**:263–266.

Kobayakawa, T., Endo, H., Ayabe-Kanamura, S., Kumagai, T., Yamaguchi, Y., Kikuchi, Y., et al. (1996) The primary gustatory area in human cerebral cortex studied by magnetoencephalography. *Neuroscience Letters* **212**:155–158.

Kobayakawa, T., Ogawa, H., Kaneda, H., Ayabe-Kanamura, S., Endo, H., and Saito, S. (1999) Spatio-temporal analysis of cortical activity evoked by gustatory stimulation in humans. *Chemical Senses* **24**:201–209.

Kobayashi, M., Sasabe, T., Takeda, M., Kondo, Y., Yoshikubo, S., Imamura, K., et al. (2002) Functional anatomy of chemical senses in the alert monkey revealed by positron emission tomography. *European Journal of Neuroscience* **16**:975–980.

Krimm, R. F. and Hill, D. L. (1999) Early dietary sodium restriction disrupts the peripheral anatomical development of the gustatory system. *Journal of Neurobiology* **39**:218–226.

Kringelbach, M. L. and Rolls, E. T. (2004) The functional neuroanatomy of the human orbitofrontal cortex: evidence from neuroimaging and neuropsychology. *Progress in Neurobiology* **72**:341–372.

Kringelbach, M. L., O'Doherty, J., Rolls, E. T., and Andrews, C. (2003) Activation of the human orbitofrontal cortex to a liquid food stimulus is correlated with its subjective pleasantness. *Cerebral Cortex* **13**: 1064–1071.

Liu, Y., Gao, J. H., Liu, H. L., and Fox, P. T. (2000) The temporal response of the brain after eating revealed by functional MRI. *Nature* **405**:1058–1062.

McClure, S. M., Berns, G. S., and Montague, P. R. (2003) Temporal prediction errors in a passive learning task activate human striatum. *Neuron* **38**:339–346.

McClure, S. M., Li, J., Montague, L. M., and Montague, P. R. (2004) Neural correlates of behavioral preference for culturally familiar drinks. *Neuron* **44**:379–387.

Kourtzi, Z. and Kanwisher, N. (2001) Representation of perceived object shape by the human lateral occipital complex. *Science* **293**:1506–1509.

Mesulam, M.M. and Mufson, E. J. (1982a) Insula of the old world monkey. I. Architectonics in the insulo-orbito-temporal component of the paralimbic brain. *Journal of Comparative Neurology* **212**:1–22.

Mesulam, M. M. and Mufson, E. J. (1982b) Insula of the old world monkey. III: Efferent cortical output and comments on function. *Journal of Comparative Neurology* **212**:38–52.

Montague, L. M. and Berns, G. S. (2002) Neural economics and the biological substrates of valuation. *Neuron* **36**:265–284.

Mora, F., Avrith, D. B., Phillips, A. G., and Rolls, E. T. (1979) Effects of satiety on self-stimulation of the orbitofrontal cortex in the rhesus monkey. *Neuroscience Letters* **13**:141–145.

Morecraft, R. J., Geula, C., and Mesulam, M. M. (1992) Cytoarchitecture and neural afferents of orbitofrontal cortex in the brain of the monkey. *Journal of Comparative Neurology* **323**:341–358.

Morrot, G., Brochet, F., and Dubourdieu, D. (2001) The color of odors. *Brain and Language* **79**:309–320.

Moskowitz, H. W., Kumaraiah, V., and Sharma, K. N. (1975) Cross-cultural differences in simple taste preferences. *Science* **190**:1217–1218.

Mufson, E. J. and Mesulam, M. M. (1982) Insula of the old world monkey. II: Afferent cortical input and comments on the claustrum. *Journal of Comparative Neurology* **212**:23–37.

Nakano, Y., Oomura, Y., Nishino, H., Aou, S., Yamamoto, T., and Nemoto, S. (1984) Neuronal activity in the medial orbitofrontal cortex of the behaving monkey: modulation by glucose and satiety. *Brain Research Bulletin* **12**:381–385.

Naito, J., Kawamura, K., and Takagi, S. F. (1984) An HRP study of neural pathways to neocortical olfactory areas in monkeys. *Neuroscience Research* **1**:19–33.

Nauta, W. J. (1971) The problem of the frontal lobe: a reinterpretation. *Journal of Psychiatric Research* **8**:167–187.

O'Doherty, J., Rolls, E. T., Francis, S., Bowtell, R., McGlone, F., Kobal, G., Renner, B., and Ahne, G. (2000) Sensory-specific satiety-related olfactory activation of the human orbitofrontal cortex. *Neuroreport* **11**:893–897.

O'Doherty, J., Kringelbach, M. L., Rolls, E. T., Hornak, J., and Andrews, C. (2001a) Abstract reward and punishment representations in the human orbitofrontal cortex. *Nature Neuroscience* **4**:95–102.

O'Doherty, J., Rolls, E. T., Francis, S., Bowtell, R., and McGlone, F. (2001b) Representation of pleasant and aversive taste in the human brain. *Journal of Neurophysiology* **85**:1315–1321.

O'Doherty, J. P., Deichmann, R., Critchley, H. D., and Dolan, R. J. (2002) Neural responses during anticipation of a primary taste reward. *Neuron* **33**:815–826.

O'Doherty, J. P., Dayan, P., Friston, K., Critchley, H., and Dolan, R. J. (2003a) Temporal difference models and reward-related learning in the human brain. *Neuron* **38**:329–337.

O'Doherty, J., Winston, J. S., Critchley, H., Perrett, D., Burt, D. M., and Dolan, R. J. (2003b) Beauty in a smile: the role of medial orbitofrontal cortex in facial attractiveness. *Neuropsychologia* **41**:147–155.

Ogawa, H., Ito, S., and Nomura, T. (1985) Two distinct projection areas from tongue nerves in the frontal operculum of macaque monkeys as revealed with evoked potential mapping. *Neuroscience Research* **2**:447–459.

Ongur, D. and Price, J. L. (2000) The organization of networks within the orbital and medial prefrontal cortex of rats, monkeys and humans. *Cerebral Cortex* **10**:206–219.

Ongur, D., Ferry, A. T., and Price, J. L. (2003) Architectonic subdivision of the human orbital and medial prefrontal cortex. *Journal of Comparative Neurology* **460**:425–449.

Penfield, W. and Jasper, H. (1954) *Epilepsy and the Functional Anatomy of the Human Brain*. Boston: Little Brown and Co.

Petrides, M. and Pandya, D. (1994) Comparitive architectonic analysis of the human and macaque frontal cortex. In: Boller F., and Grafman J. (eds) *Handbook of Neuropsychology*, pp. 17–58. Amsterdam: Elsevier.

Peirce, J. and Halpern, B. P. (1996) Orthonasal and retronasal odorant identification based upon vapor phase input from common substances. *Chemical Senses* **21**:529–43.

Pittman, D. W. and Contreras, R. J. (2002) Dietary NaCl influences the organization of chorda tympani neurons projecting to the nucleus of the solitary tract in rats. *Chemical Senses* **27**:333–341.

Poellinger, A., Thomas, R., Lio, P., Lee, A., Makris, N., Rosen, B.R., and Kwong, K. K. (2001) Activation and habituation in olfaction—an fMRI study. *Neuroimage* **13**:547–560.

Potter, H. and Nauta, W. J. (1979) A note on the problem of olfactory associations of the orbitofrontal cortex in the monkey. *Neuroscience* **4**:361–367.

Potter, H. and Butters, N. (1980) An assessment of olfactory deficits in patients with damage to prefrontal cortex. *Neuropsychologia* **18**:621–628.

Powell, T. P. S., Cowan, W. M., and Raisman, G. (1965) The central olfactory connexions. *Journal of Anatomy* **99**:791–813.

Pribram, K. H., Lennox, M. A., and Dunsmore, R. H. (1950) Some connections of the orbito-fronto-temporal, limbic and hippocampal areas of Macaca mulatta. *Journal of Neurophysiology* **13**:127–135.

Pribram, K. H. and MacLean, P. D. (1953) Neuronographic analysis of medial and basal cerebral cortex. II. Monkey. *Journal of Neurophysiology* **16**:324–340.

Price, J.L. (1973) An autoradiographic study of complementary laminar patterns of termination of afferent fiber to the olfactory cortex. *Journal of Comparative Neurology* **150**:87–108.

Price, J. L., Carmichael, S. T., Carnes, K., Clugnet, M. C., and Kuroda, M. (1991) Olfactory input to the prefrontal cortex. In: J. Davis and H. Eichenbaum (eds). *Olfaction: A model system for computational neuroscience*, pp. 101–120. MIT Press, Cambridge, MA.

Pritchard, T. C., Edwards, E. M., Smith, C. A., Hilgert, K. G., Gavlick, A. M., Maryniak, T. D., et al. (2005) Gustatory neural responses in the medial orbitofrontal cortex of the old world monkey. *Journal of Neuroscience* **25**:6047–56.

Pritchard, T. C., Hamilton, R. B., Morse, J. R., and Norgren, R. (1986) Projections of thalamic gustatory and lingual areas in the monkey, Macaca fascicularis. *Journal of Comparative Neurology* **244**:213–228.

Pritchard, T. C., Hamilton, R. B., and Norgren, R. (1989) Neural coding of gustatory information in the thalamus of Macaca mulatta. *Journal of Neurophysiology* **61**:1–14.

Ray, J. P. and Price, J. L. (1993) The organization of projections from the mediodorsal nucleus of the thalamus to orbital and medial prefrontal cortex in macaque monkeys. *Journal of Comparative Neurology* **337**:1–31.

Rempel-Clower, N. L. and Barbas, H. (1998) Topographic organization of connections between the hypothalamus and prefrontal cortex in the Rhesus monkey. *Journal of Comparative Neurology* **398**:393–419.

Rolls, E.T. and Baylis, L.L. (1994) Gustatory, olfactory, and visual convergence within the primate orbitofrontal cortex. *Journal of Neuroscience* **14**:5437–5452.

Rolls, B. J., Rolls, E. T., Rowe, E.A., and Sweeney, K. (1981a) Sensory specific satiety in man. *Physiology and Behavior* **27**:137–142.

Rolls, B.J., Rowe, E.A., Rolls, E.T., Kingston, B., Megson, A., and Gunary, R. (1981b) Variety in a meal enhances food intake in man. *Physiology and Behavior* **26**:215–221.

Rolls, E. T., Rolls, B. J., and Rowe, E. A. (1983) Sensory-specific and motivation-specific satiety for the sight and taste of food and water in man. *Physiology and Behavior* **30**:185–192.

Rolls, E. T., Sienkiewicz, Z. J., and Yaxley, S. (1989) Hunger modulates the responses to gustatory stimuli of single neurons in the caudolateral orbitofrontal cortex of the macaque monkey. *European Journal of Neuroscience* **1**:53–60.

Rolls, E. T., Yaxley, S., and Sienkiewicz, Z. J. (1990) Gustatory responses of single neurons in the caudolateral orbitofrontal cortex of the macaque monkey. *Journal of Neurophysiology* **64**:1055–1066.

Rolls, E. T., Critchley, H. D., Mason, R., and Wakeman, E. A. (1996a) Orbitofrontal cortex neurons: role in olfactory and visual association learning. *Journal of Neurophysiology* **75**:1970–1981.

Rolls, E. T., Critchley, H. D., and Treves, A. (1996b) Representation of olfactory information in the primate orbitofrontal cortex. *Journal of Neurophysiology* **75**:1982–1996.

Rolls, E. T., Critchley, H. D., Browning, A. S., Hernadi, I., and Lenard, L. (1999) Responses to the sensory properties of fat of neurons in the primate orbitofrontal cortex. *Journal of Neuroscience* **19**:1532–1540.

Rolls, E. T., O'Doherty, J., Kringelbach, M. L., Francis, S., Bowtell, R., and McGlone, F. (2003a) Representations of pleasant and painful touch in the human orbitofrontal and cingulate cortices. *Cerebral Cortex* **13**:308–317.

Rolls, E. T., Verhagen, J. V., and Kadohisa, M. (2003b) Representation of the texture of food in the primate orbitofrontal cortex: neurons responding to viscosity, grittiness, and capsaicin. *Journal of Neurophysiology* **90**:3711–3724.

Rolls, E. T., Kringelbach, M. L., and De Araujo, I. E. (2003c) Different representations of pleasant and unpleasant odors in the human brain. *European Journal of Neuroscience* **18**:695–703.

Royet, J. P. and Plailly, J. (2004) Lateralization of olfactory processes. *Chemical Senses* **29**:731–745.

Royet, J. P., Koenig, O., Gregoire, M. C., Cinotti, L., Lavenne, F., Le Bars, D., Costes, N., Vigouroux, M., Farget, V., Sicard, G., Holley, A., Mauguiere, F., Comar, D., and Froment, J. C. (1999) Functional anatomy of perceptual and semantic processing for odors. *Journal of Cognitive Neuroscience* **11**:94–109.

Royet, J. P., Zald, D., Versace, R., Costes, N., Lavenne, F., Koenig, O., and Gervais, R. (2000) Emotional responses to pleasant and unpleasant olfactory, visual, and auditory stimuli: a positron emission tomography study. *Journal of Neuroscience* **20**:7752–7759.

Royet, J. P., Hudry, J., Zald, D. H., Godinot, D., Gregoire, M.C., Lavenne, F., Costes, N., and Holley, A. (2001) Functional neuroanatomy of different olfactory judgements. *Neuroimage* **13**:506–519.

Royet, J. P., Plailly, J., Delon-Martin, C., Kareken, D. A., and Segebarth, C. (2003) fMRI of emotional responses to odors: influence of hedonic valence and judgement, handedness, and gender. *Neuroimage* **20**:713–728.

Rozin, P. (1982) "Taste-smell confusions" and the duality of the olfactory sense. *Perception and Psychophysics* **31**:397–401.

Russchen, F. T., Amaral, D. G., and Price, J. L. (1987) The afferent input to the magnocellular division of the mediodorsal thalamic nucleus in the monkey, Macaca fascicularis. *Journal of Comparative Neurology* **256**:175–210.

Sakai, N., Kobayakawa T. Gotow, N. Saito, S., and Imada, S. (2001) Enhancement of sweetness ratings of aspartame by a vanilla odor presented either by orthonasal or retronasal routes. *Perceptual Motor Skills* **92**:1002–1008.

Saper, C. B., Chou, T. C., and Elmquist, J. K. (2002) The need to feed: homeostatic and hedonic control of eating *Neuron* **36**:199–211.

Savic, I. (2002) Imaging of brain activation by odorants in humans. *Current Opinion in Neurobiology* **12**:455–461.

Savic, I., Gulyas, B., Larsson, M., and Roland, P. (2000) Olfactory functions are mediated by parallel and hierarchical processing. *Neuron* **26**:735–745.

Schoenbaum, G., Setlow, B., and Ramus, S. J. (2003) A systems approach to orbitofrontal cortex function: recordings in rat orbitofrontal cortex reveal interactions with different learning systems. *Behavioral Brain Research* **146**:9–29.

Schultz, W., Tremblay, L., and Hollerman, J. R. (1998) Reward prediction in primate basal ganglia and frontal cortex. *Neuropharmacology* **37**:421–429.

Schultz, W., Tremblay, L., and Hollerman, J. R. (2000) Reward processing in primate orbitofrontal cortex and basal ganglia. *Cerebral Cortex* **10**:272–284.

Scott, T. R. and Plata-Salaman, C. R. (1999) Taste in the monkey cortex. *Physiology and Behavior* **67**:489–511.

Scott, T. R., Yaxley, S., Sienkiewicz, Z. J., and Rolls, E. T. (1986a) Gustatory responses in the frontal opercular cortex of the alert cynomolgus monkey. *Journal of Neurophysiology* **56**:876–890.

Scott, T. R., Yaxley, S., Sienkiewicz, Z. J., and Rolls, E. T. (1986b) Gustatory responses in the nucleus tractus solitarius of the alert cynomolgus monkey. *Journal of Neurophysiology* **55**:182–200.

Shipley, M. T. and Adamek, G.D. (1984) The connections of the mouse olfactory bulb: a study using orthograde and retrograde transport of wheat germ agglutinin conjugated to horseradish peroxidase. *Brain Research Bulletin*, **12**, 669–688.

Shipley, M. T. and Ennis, M. (1996) Functional organization of olfactory system. *Journal of Neurobiology* **30**:123–176.

Shuler, M. G., Krimm, R. F., and Hill, D. L. (2004) Neuron/target plasticity in the peripheral gustatory system. *Journal of Comparative Neurology* **472**:183–192.

Slotnick, B. (2001) Animal cognition and the rat olfactory system. *Trends in Cognitive Neuroscience* **5**:216–222.

Small, D. M. and Prescott, J. (2005) Odor/taste integration and the perception of flavor. *Experimental Brain Research* **166**:345–357.

Small, D. M., Jones-Gotman, M., Zatorre, R. J., Petrides, M., and Evans, A. C. (1997a) Flavor processing: more than the sum of its parts. *Neuroreport* **8**:3913–3917.

Small, D. M., Jones-Gotman, M., Zatorre, R. J., Petrides, M., and Evans, A. C. (1997b) A role for the right anterior temporal lobe in taste quality recognition. *Journal of Neuroscience* **17**:5136–5142.

Small, D. M., Zald, D. H., Jones-Gotman, M., Zatorre, R. J., Pardo, J. V., Frey, S., et al. (1999) Human cortical gustatory areas: a review of functional neuroimaging data. *Neuroreport* **10**:7–14.

Small, D. M., Zatorre, R. J., and Jones-Gotman, M. (2001a) Changes in taste intensity perception following anterior temporal lobe removal in humans. *Chemical Senses* **26**:425–432.

Small, D. M., Zatorre, R. J., and Jones-Gotman, M. (2001b) Increased intensity perception of aversive taste following right anteromedial temporal lobe removal in humans. *Brain* **124**:1566–1575.

Small, D. M., Zatorre, R. J., Dagher, A., Evans, A. C., and Jones-Gotman, M. (2001c) Changes in brain activity related to eating chocolate: from pleasure to aversion. *Brain* **124**:1720–1733.

Small, D. M., Gregory, M. D., Mak, Y. E., Gitelman, D., Mesulam, M. M., and Parrish, T. (2003) Dissociation of neural representation of intensity and affective valuation in human gustation *Neuron* **39**:701–711.

Small, D. M., Voss, J., Mak, Y. E., Simmons, K. B., Parrish, T., and Gitelman, D. (2004) Experience-dependent neural integration of taste and smell in the human brain. *Journal of Neurophysiology* **92**:1892–1903.

Small, D. M., Gerber, J., Mak, Y. E., and Hummel, T. (2005) Differential neural responses evoked by orthonasal versus retronasal odorant perception in humans. *Neuron* **18**:593–605.

Smith, D. V. and Scott, T. R. (2003) Gustatory Neural Coding. In: Doty, R.L. (ed.). *Handbook of Olfaction and Gustation*, pp. 731–758. Philadelphia: Marcel Dekker.

Smith, D. V. and Travers, J. B. (1979) A metric for the breadth of tuning of gustatory neurons. *Chemical Senses* **4**:215–229.

Smith-Swintosky, V. L., Plata-Salaman, C. R., and Scott, T. R. (1991) Gustatory neural coding in the monkey cortex: stimulus quality. *Journal of Neurophysiology* **66**:1156–1165.

Stein, B. E. (1998) Neural mechanisms for synthesizing sensory information and producing adaptive behaviors. *Experimental Brain Research* **123**:124–135.

Steiner, J. E., Glaser, D., Hawilo, M. E., and Berridge, K. C. (2001) Comparative expression of hedonic impact: affective reactions to taste by human infants and other primates. *Neuroscience and Biobehavioral Reviews* **25**:53–74.

Tanabe, T., Iino, M., Ooshima, Y., and Takagi, S.F. (1974) An olfactory area in the prefrontal lobe. *Brain Research* **80**:127–130.

Tanabe, T., Iino, M., and Takagi, S.F. (1975a) Discrimination of odors in olfactory bulb, piriform-amygdaloid areas, and orbitofrontal cortex of the monkey. *Journal of Neurophysiology* **38**:1284–1296.

Tanabe, T., Yarita, H., Iino, M., Ooshima, Y., and Takagi, S. F. (1975b) An olfactory projection area in orbitofrontal cortex of the monkey. *Journal of Neurophysiology* **38**:1269–1283.

Tataranni, P. A. and Del Parigi, A. (2003) Functional neuroimaging: a new generation of human brain studies in obesity research. *Obesity Reviews* **4**:229–238.

Tataranni, P. A., Gautier, J. F., Chen, K., Uecker, A., Bandy, D., Salbe, A. D., *et al.* (1999) Neuroanatomical correlates of hunger and satiation in humans using positron emission tomography. *Proceedings of the National Academy of Sciences of the United States of America* **96**:4569–4574.

Thorpe, S. J., Rolls, E. T., and Maddison, S. (1983) The orbitofrontal cortex: neuronal activity in the behaving monkey. *Experimental Brain Research* **49**:93–115.

Tremblay, L. and Schultz, W. (1999) Relative reward preference in primate orbitofrontal cortex. *Nature* **398**:704–708.

Tremblay, L. and Schultz, W. (2000) Modifications of reward expectation-related neuronal activity during learning in primate orbitofrontal cortex. *Journal of Neurophysiology* **83**:1877–1885.

Turner, B. H., Gupta, K. C., and Mishkin, M. (1978) The locus and cytoarchitecture of the projection areas of the olfactory bulb in Macaca mulatta. *Journal of Comparative Neurology* **177**:381–396.

Uher, R., Murphy, T., Brammer, M. J., Dalgleish, T., Ng, V. W., Andrew, C. M., *et al.* (2004) Medial prefrontal cortex activity associated with symptom provocation in eating disorders. *American Journal of Psychiatry* **161**:1238–1246.

Verhagen, J. V., Rolls, E. T., and Kadohisa, M. (2003) Neurons in the primate orbitofrontal cortex respond to fat texture independently of viscosity. *Journal of Neurophysiology* **90**:1514–1525.

von Bonin, G. and Green, J. R. (1949) Connections between orbital cortex and diencephalon in the macaque. *Journal of Comparative Neurology* **90**:243–254.

Walker, A. E. (1940) A cytoarchitectural study of the prefrontal area of the macaque monkey. *Journal of Comparative Neurology* **73**:59–86.

Wang, G. J., Volkow, N. D., Telang, F., Jayne, M., Ma, J., Rao, M., *et al.* (2004) Exposure to appetitive food stimuli markedly activates the human brain. *Neuroimage*, **21**:1790–1797.

Wise, R. A. (2002) Brain reward circuitry: insights from unsensed incentives. *Neuron* **36**:229–240.

Yarita, H., Iino, M., Tanabe, T., Kogure, S., and Takagi, S. F. (1980) A transthalamic olfactory pathway to orbitofrontal cortex in the monkey. *Journal of Neurophysiology* **43**:69–85.

Yaxley, S., Rolls, E. T., and Sienkiewicz, Z. J. (1988) The responsiveness of neurons in the insular gustatory cortex of the macaque monkey is independent of hunger. *Physiology and Behavior* **42**:223–229.

Yaxley, S., Rolls, E. T., and Sienkiewicz, Z. J. (1990) Gustatory responses of single neurons in the insula of the macaque monkey. *Journal of Neurophysiology* **63**:689–700.

Zald, D. H. (2003) The human amygdala and the emotional evaluation of sensory stimuli. *Brain Research Reviews* **41**:88–123.

Zald, D. H. and Pardo, J. V. (1997) Emotion, olfaction, and the human amygdala: amygdala activation during aversive olfactory stimulation. *Proceedings of the National Academy of Sciences USA* **94**:4119–4124.

Zald, D., and Pardo, J. V. (2000) Cortical activation induced by intraoral stimulation with water in humans. *Chemical Senses* **25**:267–275.

Zald, D. H., Lee, J. T., Fluegel, K. W., and Pardo, J. V. (1998) Aversive gustatory stimulation activates limbic circuits in humans. *Brain* **121**:1143–1154.

Zald, D. H., Hagen, M. C., and Pardo, J. V. (2002) Neural correlates of tasting concentrated quinine and sugar solutions. *Journal of Neurophysiology* **87**:1068–1075.

Zatorre, R. J. and Jones-Gotman, M. (1991) Human olfactory discrimination after unilateral frontal or temporal lobectomy. *Brain* **114**:71–84.

Zatorre, R. J., Jones-Gotman, M., Evans, A.C., and Meyer, E. (1992) Functional localization and lateralization of human olfactory cortex. *Nature* **360**:339–340.

Zatorre, R. J., Jones-Gotman, M., and Rouby, C. (2000) Neural mechanisms involved in odor pleasantness and intensity judgements. *Neuroreport* **11**:2711–2716.

Zellner, D. A. and Durlach, P. (2002) What is refreshing? An investigation of the color and other sensory attributes of refreshing foods and beverages. *Appetite* **39**:185–186.

Chapter 7

# Involvement of primate orbitofrontal neurons in reward, uncertainty, and learning

Wolfram Schultz and Leon Tremblay

## 7.1. Introduction

The functions of the orbitofrontal cortex (OFC) have been classically defined by the effects of lesions in humans and animals and by the anatomical connections to other parts of the brain with better-known functions. Globally, orbitofrontal functions can be characterized as being related to emotional processes underlying the reaction of the individual to environmental events. These functions include such aspects as reward, punishment, control of risk and learning. Several chapters in this book describe the data obtained using these techniques in great detail. As these functions become better known, it will be important to see how the activity of individual neurons encode the component processes that contribute to these functions. We would like to understand how neurons respond to external events with motivational significance, and we would hope to find neurons whose activity reflects representations of predictable future events of motivational significance that adapt to the ongoing experience. We have very little knowledge about how single neurons might manage the uncertainty of motivational events and contribute to controlled decision-making in situations of risk and ambiguity. We need to create model situations of behavioral challenges and investigate the activity of orbitofrontal neurons in these situations to hopefully obtain information that would be relevant in real-life situations as well. This chapter will review studies on neurophysiological recordings on behaving monkeys performed in our laboratory that should elucidate some of these issues, and we will also describe studies in other laboratories to obtain a more comprehensive picture of the current state of knowledge. To provide comparative data on another, closely connected brain structure, we will briefly describe reward studies on single neurons in the striatum.

## 7.2. Behavioral tasks and types of neural relationships

Experimental psychologists have developed specific behavioral tasks to test particular behavioral processes in isolation and with a minimum of interfering variables, and neuroscientists have used these tasks to study the involvement of specific brain structures in individual behavioral processes. A good behavioral paradigm for assessing the function of the frontal cortex, and the closely associated basal ganglia, is the delayed response task, which tests spatial processing, working memory, movement preparation and execution, prediction of reward, and goal-directed behavior (Jacobsen and Nissen 1937; Divac *et al.* 1967). In order to relate our findings to the previous lesion data based on these tests, we used several versions of spatial and conditional delayed response tasks and investigated the motivational properties of single neurons in the OFC of macaque monkeys. In such tasks, an initial visual instruction cue indicates the spatial position of the target of an arm movement for a brief period. Following a delay of several seconds, during which the spatial information is absent, a neutral stimulus appears which tells the subject to make its response to the remembered target to receive a liquid or food reward. In our adaptations of the delayed response task, we used different initial instruction cues that, in addition to spatial information, also predicted which reward the animal would receive for responding correctly. We also introduced a delay of 2–3 s between the correct response and the delivery of reward to provide the animal with a specific reward expectation period after the required behavioral reactions had been performed. Thus, the delayed response task is an instrumental (operant) task in which the reward is contingent upon the correct response of the animal. The association of the initial instruction cue is Pavlovian, and the instruction is a predictor for a specific reward without the animal responding differentially for each reward.

Orbitofrontal neurons in area 11, rostral area 13, and lateral area 14 display mainly three types of activations during delayed response tasks. These consist of responses to instructions, activations preceding rewards, and responses to rewards (Fig. 7.1). Neurons in the dorsolateral prefrontal cortex and striatum show additional activations preceding the movement, following the movement-triggering stimulus, and during the movement itself (Fuster 1973; Hikosaka *et al.* 1989a–c; Funahashi *et al.* 1990; Apicella *et al.* 1992). These additional task relationships are found less frequently in the OFC (Tremblay and Schultz 1999; Hikosaka and Watanabe 2000). It is a consistent finding in our studies and those of others that activations in the OFC show pronounced relationships to rewards. This result is true for all forms of task relationships and will be described in detail in this article.

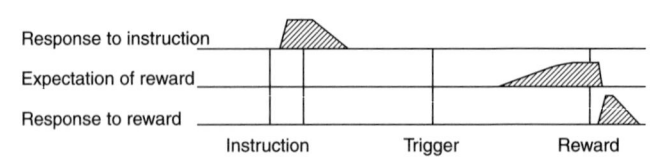

**Fig. 7.1** Schematic overview of the main forms of activity in primate OFC during the performance of a delayed response task. The instruction cue indicates which behavioral reaction to perform and which reward to expect, the trigger releases the behavioral reaction, and the reward consists of a drop of liquid or a piece of food. In unrewarded trials, the reward is replaced by an auditory stimulus indicating correct trial performance. Reproduced with permission from Tremblay and Schultz (2000b).

## 7.3. Reward discrimination

### Background

The earliest report about reward-related activity in orbitofrontal neurons comes from Rosenkilde *et al.* (1981) who report responses to liquid reward following performance of a spatial delayed response task used in Fuster's investigations of dorsolateral prefrontal function. About half of the tested orbitofrontal neurons showed increased activity after correct behavioral reactions and no responses after incorrect behavior. The relationship to reward, rather than more sensory properties of the objects, is demonstrated by the observation that satiation of the specific reward objects reduced the neuronal responses while responses to nondevalued rewards were maintained (Thorpe *et al.* 1983; Rolls *et al.* 1989). Nearly one third of orbitofrontal neurons with responses to the delivery of liquid reward discriminated between different volumes of the same reward (Wallis and Miller 2003).

Responses were also observed to conditioned stimuli that predicted liquid reward. These responses discriminated between reward and punishment, suggesting reinforcer specificity (Thorpe *et al.* 1983). The neural responses to reward-predicting stimuli reversed frequently in parallel with the reversal of stimulus-reward associations (Rolls *et al.* 1996). As with responses to the reward itself, satiation differentially reduced the responses to stimuli that predicted the sated reward but failed to influence responses to stimuli predicting nondevalued rewards, suggesting that the neurons were coding the reward value and not simply the prediction of the occurrence of an event irrespective of its motivational value or valence (Critchley and Rolls 1996).

Sustained activations preceding the expected outcome of liquid reward have been observed in several studies in rats (Schoenbaum *et al.* 1998) and monkeys (Hikosaka and Watanabe 2000; Wallis and Miller 2003), and about one half of these neurons discriminated between rewards and punishers, between different rewards, between reward volumes, and between reward and no-reward. Reward-discriminating activity may precede even much earlier events in a task, such as initial instruction cues, if such stimuli are known to lead to predictable outcomes (Hikosaka and Watanabe 2004).

### Reward vs. no-reward

#### Behavioral task

To obtain a more comprehensive overview of orbitofrontal activities, we used a modified delayed go/no-go task, and investigated the discrimination between rewards, the relationship to arm movements and the withholding of responding. The task included a rewarded movement trial, a rewarded nonmovement (no-go) trial as a movement control, and an unrewarded movement trial as a reward control (Tremblay and Schultz, 2000a). As instructions, we used three colored, fractal instruction pictures which appeared on a computer monitor for 1.0 s in a central position in front of the animal. Each picture was specific for one trial type, namely rewarded movement, rewarded nonmovement and unrewarded movement. A red, square trigger stimulus presented 2.5–3.5 s after instruction onset required the animal to execute or withhold a reaching movement, depending on trial type. The trigger stimulus was the same in all three trial types. In rewarded movement trials, the animal released the resting key and touched a small lever below the trigger stimulus to receive a small quantity of apple juice after a delay of 1.5 s. In rewarded nonmovement trials, the animal remained motionless on the resting key for 2.0 s and received the same liquid reward after a further 1.5 s. In unrewarded movement trials, the animal reacted as in rewarded movement trials, but correct performance was followed by a 1 kHz sound after 1.5 s rather than liquid reward. One of the two rewarded trial types followed each correctly performed unrewarded trial. Animals were required to perform the unrewarded movement trial, as incorrect performance resulted in trial repetition. The sound constituted a conditioned auditory reinforcer, as it ameliorated task performance and predicted a rewarded trial, but it was not an explicit reward, hence the simplifying term 'unrewarded' movement. Thus, each instruction was a unique stimulus in each trial, indicating the behavioral reaction to be performed following the trigger (execution or withholding of movement) and predicting the type of reinforcer (liquid or sound). Each trial contained two delay periods, namely the instruction-trigger delay during which the animal remembered the type of instruction and prepared the behavioral reaction, and the trigger-reward or trigger-sound delay during which the animal could expect the reward or auditory reinforcer, respectively.

### Neural activity

We tested 505 orbitofrontal neurons in this experiment and found that 188 of them (37%) showed changes in activity in relation to one or more events of the delayed response task used (Tremblay and Schultz 2000a).

Instruction responses occurred in 99 task-related neurons, and 60% of them discriminated between the types of reinforcer, irrespective of the execution or withholding of movement. Thus, 38 neurons responded preferentially in rewarded movement trials, or rewarded nonmovement trials, or both rewarded trial types. Twenty-two neurons responded preferentially in unrewarded movement trials but not in any rewarded trial type, indicating a relationship to the type of reinforcer or the type of behavioral reaction, which was not further distinguished. By contrast, only three neurons showed a strict movement relationship in responding preferentially in both types of movement trial, irrespective of the reward. The responses had median latencies of 90–110 ms and durations of 280–700 ms, which varied insignificantly between familiar and learning trials ($p > 0.05$; Mann-Whitney U-Test).

Activations preceding the reward or sound occurred in 51 of the 188 task-related neurons. These activations began occasionally before, but usually after the trigger stimulus, at 1188–1282 ms before the reward or sound, and subsided $< 500$ ms after the reward or sound. Delayed reward prolonged the activations, and earlier reward terminated the activations (Fig. 7.2). Thus the activations were time-locked to the reward and not to the instruction or trigger stimuli, suggesting that they were related to the upcoming reward rather than constituting delayed responses to the preceding stimuli. All

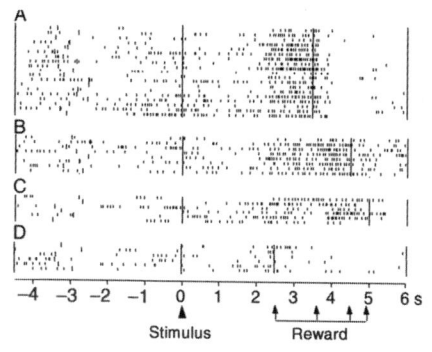

**Fig. 7.2** Activity during the expectation of reward in an orbitofrontal neuron. Delayed reward prolongs the activation, which subsides only after the reward has occurred (b, c), whereas earlier reward shortens the activity (d). (a): trials with the usual stimulus-reward interval. Chronological sequence of trials is shown from top to bottom in each of a–d. Dots in rasters denote the time of occurrence of neuronal impulses, referenced to trigger stimulus onset. Each line of dots represents one trial. Reproduced with permission from Tremblay and Schultz (2000a).

activations reflected the type of reinforcer, occurring in both liquid-rewarded trial types but not in sound-reinforced trials (41 neurons), or only in one of the rewarded trial types (6 neurons). Only four activations preceded preferentially the sound reinforcer in unrewarded movement trials.

Responses following the delivery of the reward or sound reinforcer occurred in 67 of the 188 task-related neurons. These responses were unlikely related to mouth movements, as licking movements occurred also at other task periods during which these neuronal responses were not observed. All of these responses reflected the type of reinforcer, occurring mostly in both liquid-rewarded trial types but not in unrewarded trials (62 neurons), or only in rewarded movement trials (2 neurons). Only three responses occurred preferentially after the sound reinforcer in unrewarded movement trials.

## Different rewards

### Behavioral task

We used a modified spatial delayed response task to compare neuronal activity between different liquid or food rewards (Tremblay and Schultz 1999). Each instruction picture contained two types of information; (i) the visual aspects indicated which one of two liquid or food rewards would be delivered for correct behavior at the trial end; (ii) the left or right position on the computer monitor indicated the target of arm movement to be performed in reaction to the subsequent trigger stimulus. The trigger stimulus consisted of two simultaneously appearing, identical red squares at the left and right positions of the instructions. Thus, the trigger indicated the time of responding without indicating the reward or movement target. Touching the lever on the side previously indicated by the instruction resulted after a brief delay in delivery of the corresponding liquid or food reward. To assess the influence of visual features on neuronal responses, we used several pairs of instruction pictures for indicating the same liquid or food rewards. Only two instruction pictures with their associated two liquid or two food rewards were presented in a given block of trials. Liquid rewards were grenadine and apple juice in one animal, and orange and grape juice in a second animal. Food rewards became available in a small box located to the right of the monitor following computer-controlled opening of its door. Foods were raisins (most preferred), small apple morsels (intermediately preferred) and sugar-honey cereal (least preferred). Rewards and target positions alternated randomly in each block of trials, with maximally three consecutive identical trials. Reward preferences were assessed in separate blocks of a choice version of the spatial delay task before or after recording from each neuron. In these trials two different instructions for two rewards appeared simultaneously and in random alternation between left and right target positions, allowing the animal to touch the lever corresponding to its reward of choice following the trigger stimulus.

### Neural activity

We tested 1424 orbitofrontal neurons in this experiment and found that 432 of them (33%) showed changes in activity in relation to one or more events of this delayed response task (Tremblay and Schultz 1999). All three principal types of task-related orbitofrontal activations discriminated between liquid rewards and between food rewards (Fig. 7.3). Activations occurred exclusively, or were significantly higher, for one liquid reward compared to the other liquid reward in 150 of 218 instruction responses (69%), 65

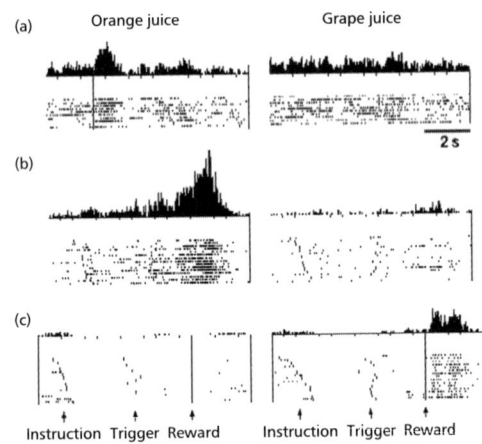

**Fig. 7.3** Reward-discriminating activity in the three principal types of activity of orbitofrontal neurons during performance in the spatial delayed response task. (a) Response to instruction. (b) Activity during the expectation of liquid reward. (c) Response to reward. a–c show three different neurons. Only one reward was delivered in any given trial, and trials alternated randomly between the two rewards. They are separated for analysis according to reward and rank-ordered according to instruction-reward intervals. Data from Tremblay and Schultz (2000a).

of 160 activations preceding reward (41%) and 76 of 146 reward responses (52%). These differences were unrelated to eye or licking movements before or after reward delivery which varied inconspicuously between trials using the two rewards. By contrast, only 7 of 218 instruction responses discriminated between left and right movement targets, and none of the 50 neurons tested with several instruction sets showed significantly different responses to different instructions indicating the same reward (although it should be noted that visual responses were indeed observed in other orbitofrontal neurons in later experiments). Thus the instruction responses reflected much better the predictable rewards than the spatial or visual features of the instructions.

Conclusions

The reviewed data demonstrate solid and reproducible activities in orbitofrontal neurons in response to rewards and reward-related stimuli and during the expectation of predictable rewards. About one third of orbitofrontal neurons in monkeys are activated during the performance of delayed response and visual discrimination tasks, indicating that these tasks can effectively challenge some of the functions of the OFC in this species. In these tasks, about one half of the neurons show different activations depending on which reward will be delivered for correct behavioral performance. Thus, the neurons distinguish between reward and no-reward, between reward and punishment, between different rewards, and between different volumes of the same reward. These discriminations occur not only after the reward has been delivered, but also in response to reward-predicting stimuli and during various periods during which rewards can be expected, based on previous experience with the task situation. Although it is a constant worry for the experimenter to design and choose a behavioral task that effectively challenges the functions of the investigated brain structure, it seems that a considerable number of orbitofrontal neurons are in fact tuned to rewards rather than spatial or object task components following behavioral training. This consideration suggests that the processing of reward information can be considered as a major function of the monkey OFC.

## 7.4. Reward preference and uncertainty

### Rationale

Survival in uncertain environments requires organisms to be sensitive to hundreds of different reward objects. The fine discrimination between large numbers of rewards, or between many small variations in reward size, would require enormous information-processing capacities. However, only a limited number of rewards, or quantity of a reward, is likely to occur at any given instance, and it would appear economical to devote the processing capacities only to those rewards that are actually likely to occur at a given moment. In the absence of further information, the uncertainty about which specific rewards are available at a given moment is high. Thus it would be advantageous to reduce the uncertainty by providing advance information about the probability distribution of currently available rewards. Predictive stimuli could provide such advance information, and associative learning theories have generated a framework for understanding the mechanisms of acquiring such predictions. A neural system could discriminate fine gradations among its inputs and have good stimulus-response gains if its elements could focus their discriminative capacity onto those rewards that are most likely to occur. The dynamic adaptation of information processing to the current needs would greatly enhance its efficacy, keep the energy demands of the brain within sustainable ranges and contribute to the survival of the individual in a competitive environment.

Individuals often attribute a value to a reward, or the quantity of a reward, which allows them to rank the available rewards according to their value. The ranking is expressed behaviorally by the preference for each reward relative to the others, which gave rise to the term of utility that is presently being used in microeconomics (Bernoulli 1738; von Neumann and Morgenstern 1944). It is often observed that preferences are not fixed to individual rewards, like physical properties, and do not take all possible rewards into account. Rather, preferences can be flexible, and are expressed relative to those rewards or reward quantities that are available at a given moment (Kahneman and Tversky, 1984). In order to set the preferences with optimal accuracy and obtain the most preferred rewards, individuals would profit from advance information that reduces the uncertainty by predicting what specific rewards, or quantities of rewards, are likely to occur within the next foreseeable future. For example, if apples and carrots were known to be available presently, and nothing else, an animal may prefer apple and forego carrot, whereas in the presence of bananas, the individual might prefer banana and forego apple. In this way, predictive stimuli reducing the uncertainty would lead to the explicit setting of preferences among the limited numbers of predictable rewards.

## Behavior

We assessed reward preferences behaviorally in a choice version of the spatial delayed response task in which two instruction pictures, instead of one, were shown simultaneously on a computer monitor above a left and a right lever, respectively (Tremblay and Schultz 1999). Each picture was associated with a different reward. Following the trigger stimulus, the animal chooses one of the rewards by touching the corresponding lever with its hand. We used three food rewards (A–C) but made only two of them available in a given trial block (A–B, B–C, or A–C). Thus each trial block employed a specific distribution of rewards, which was indicated to the animal during the first several trials by the specific instruction cues predicting the rewards. Thus in a trial block using rewards A and B, the trial instruction cue was either an A or a B reward-predicting stimulus, and the information was complete after both trial types had been initiated at least once within the block, which happened within the first four trials at most. Animals showed clear preferences in every comparison and shifted their preferences according to the currently available reward selection, choosing reward A over B, B over C, and A over C in 90–100% of trials, independent of the side of lever touch. Thus reward B was chosen less frequently when reward A was available but more frequently when reward C was available in a given trial block. Apparently, the preference for reward B was relative and depended on the reward with which it was being compared.

## Neural activity

The question arose whether there would be a neural correlate for the shift of reward preferences according to the currently available rewards. Would neurons in a major reward structure, such as the OFC show a graded, stable and specific signal for each reward irrespective of which other rewards were available within a given situation, or would neurons shift their processing according to the current reward distribution and thus improve their coding efficiency?

We tested 65 neurons in the OFC, all of which showed changes in activity during the performance of the task and discriminated between the different food rewards. The animals performed in the standard delay task with single instruction pictures indicating a single, preferred or nonpreferred food reward. Only two rewards were used in each trial block, which alternated semi-randomly between trials. (Reward preferences were checked in separate choice trials before or after each neuron was recorded.) The activity

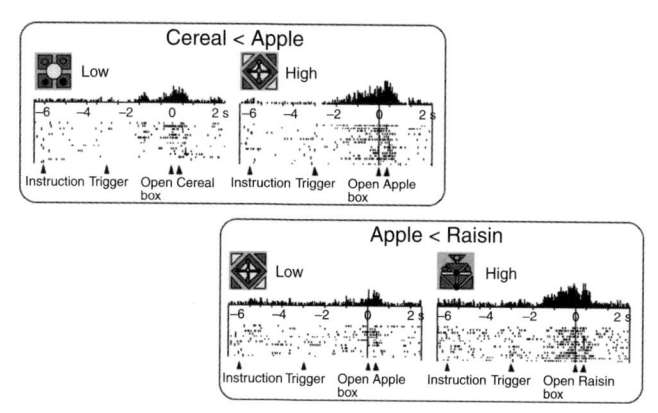

**Fig. 7.4** Shift of reward coding in orbitofrontal neuron with change in reward distribution. The activity increased during the expectation period preceding the relatively more preferred food reward (apple in top, raisin in bottom). Although the reward was physically identical in top and bottom panels (a small piece of apple), its relative motivational value differed dependent on the other reward available in the same trial block (high in top, low in bottom), as expressed by the animal's, choice behavior in separate trials. The different reward comparisons at top and bottom were run in separate trial blocks. Each reward was specifically predicted by an instruction picture, which is shown above the histograms. Thus the occurrence of the cereal-predicting stimulus predicted to the animal a cereal-apple block, whereas the raisin-predicting picture predicted an apple-raisin block (the apple-predicting stimulus was common for both blocks and thus could not contribute to the discrimination between the two trial blocks). Lever touch following the movement trigger stimulus induced 2 s later the opening of a food box (arrow) from which the animal collected the reward (arrow). Reproduced with permission from Tremblay and Schultz (1999).

of 40 of the 65 orbitofrontal neurons depended indeed on which other reward was available in a given trial block.

The neuron in Fig. 7.4 showed sustained activation during the expectation of a reward at the trial end. It showed significantly higher activation preceding apple (the behaviorally preferred reward) compared to cereal (nonpreferred) (top). When apple was compared with raisin in a different trial block (bottom), the same neuron was significantly more activated preceding raisin (now preferred) than apple (nonpreferred). Neuronal activations with apple occurred only when this reward was the preferred one (Fig. 7.4 center, top vs. bottom). Thus, this orbitofrontal neuron appeared to be activated with any reward that was relatively more preferred than the available alternative, no matter which of the different reward combinations was used. These shifts were also seen with the other types of task relationships, namely reward-predicting instruction responses (12 of 28 neurons), and responses following the reward (7 of 17 neurons). There were also neurons that showed higher activity for less preferred rewards, and some of them showed also the shift in response (6 of 28 instruction responses, 5 of 20 prereward activations and 2 of 17 reward responses).

## Conclusions

These results demonstrate how some orbitofrontal neurons shift their reward-discriminating activity into the range of the currently predicted rewards (Fig. 7.5), both losing and gaining responses depending on which other rewards are available. These orbitofrontal neurons do not appear to code the physical properties of rewards which are fixed and do not depend on other rewards. Their processing apparently does not take into account rewards that are not currently available. The obvious advantage of this mechanism is the focusing of processing onto those limited rewards that are available with a reasonable probability. As a result, the neural responses can fully use the available, limited range of neural activity, thus allowing the neurons to dynamically adapt and optimize their stimulus-response gain by adapting the probability distribution of the neural response to the probability distribution of the rewards.

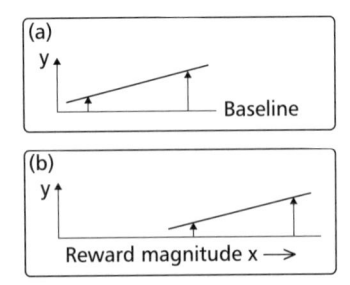

**Fig. 7.5** Schematics of the shift of stimulus-response function in OFC into the range of the currently used rewards. A different prediction *p* would shift an assumed linear part of the stimulus response function, according to $y = a + b(x+p)$.

## 7.5. Learning

### Background

Very few studies examined orbitofrontal neurons during learning. The studies used go/no-go tasks in which animals emitted a behavioral response to one conditioned stimulus (CS), and refrained from moving to a different CS, in order to receive a predicted reward or avoid a predicted punisher. Rolls *et al.* (1996) used differential odor and visual CSs for licking movements rewarded with sweet liquid (go) and for withholding licking to avoid aversive saline (no-go). The study tested the reversal of existing CS-outcome associations in monkeys and found that 35 of 45 orbitofrontal neurons reversed their differential, reinforcer-disciminating responses at about the same time as the animal reversed its go/no-go behavioral responses. Schoenbaum *et al.* (1998, 1999) tested odor CSs in rats during the differential acquisition of associations between novel odor CSs with rewarding sucrose and aversive quinine solutions involving the execution or withholding of nose-poking responses (go and no-go). The study found that the responses to the CSs in 96 of 328 orbitofrontal neurons came to discriminate between the two outcomes during learning, and most of them became differential only at the same time as the animal reached the behavioral learning criterion (90% correct performance in 20 trials). Furthermore, 66 of 328 orbitofrontal neurons developed differential activations during the expectation of the two types of reinforcers, and these changes occurred before the animals reached the learning criterion.

### Modifications of reward predictions

#### Rationale

Learning about the occurrence of a future reward or learning about an action leading to a reward involves the acquisition or change of predictions of outcome. Learning depends on the extent to which the outcomes occur differently than predicted, being governed by the discrepancy or 'error' between outcome and prediction (prediction error). Thus, as conceptualized in the Rescorla-Wagner learning model (Rescorla and Wagner 1972), learning proceeds as long as outcomes are not fully predicted and the prediction error is positive. Learning decreases as outcomes become increasingly predicted, and it ends when all outcomes are fully predicted. Thus learning can be graphed in an asymptotic learning curve. By contrast, behavior undergoes extinction when a predicted outcome is omitted and the prediction error is negative. Whereas dopamine neurons signal bidirectional reward prediction errors that can be used for learning (Hollerman and Schultz 1998), orbitofrontal neurons show only unidirectional activations to unpredicted rewards. However, the majority of reward-related activity in orbitofrontal neurons precedes the occurrence of reward, being related to the prediction and expectation of reward (Tremblay and Schultz 1999, 2000a). The issue is whether these forms of anticipatory activity would reflect the changes in reward prediction occurring during learning.

Our learning studies on orbitofrontal neurons were motivated by the finding of substantial processing of reward information by orbitofrontal neurons, by the notion that learning constitutes one of the major functions of rewards, and by the fact that very few studies had investigated orbitofrontal neurons during learning. We employed a learning set situation in which only a single task component changed at a time and animals learned the new reward contingencies within a few trials (Harlow 1949; Gaffan *et al.* 1988). The animals performed in a delayed go/no-go task in which the initial instruction picture predicted the reward and informed about the required behavioral reaction. Learning trials involved the presentation of a novel instruction picture in each type of trial, and animals learned the reward prediction and behavioral significance by trial and error. This procedure allowed us to investigate single neurons during complete learning episodes, and to compare their activity with performance in trials using familiar, well-learned instruction pictures. We were interested to see how orbitofrontal neurons would respond to novel stimuli and change their reward expectation-related activity during learning as predictions about future rewards are being set up and stimuli become known.

### Behavior

Our behavioral task employed three different trial types of rewarded arm movement, a rewarded nonmovement trial and an unrewarded movement trial with sound reinforcer (Tremblay and Schultz 2000b). The initial instruction picture was presented on a computer monitor which predicted whether the animal would receive a liquid reward or the auditory reinforcer, and which behavioral reaction (movement or non-movement) was required to obtain that outcome. During learning, three new instruction pictures were presented in the three trial types, respectively, whereas all other task events and the global task structure remained unchanged. Thus learning consisted in associating each new instruction picture (a) with liquid reward or conditioned auditory reinforcement, and (b) with executing or withholding the movement. Whereas each familiar instruction consisted of a single fractal picture, each learning instruction was composed of two partly superimposed and subtracted, simple geometric shapes that were drawn randomly from 256 images (64 shapes in red, yellow, green or blue), resulting in a total of 65536 possible stimuli in five colors.

Animals first learned all three trial types with familiar pictures at >90% correct. Then the instruction stimuli for the three trial types were replaced by novel pictures. When >90% correct performance was reached with a set of three learning stimuli, those stimuli were discarded and three novel stimuli were introduced. At the onset of neuronal recordings, the two animals tested had learned 625 and 77 problems, respectively. The repeated learning sessions resulted in a learning set, in which learning occurred largely within the first few trials, and approached asymptotic performance within 5–10 trials of each trial type. Medians of correct performance in the first 15 trials of each type were 73–93%, performance in familiar trials being >98%.

Rewarded movements in familiar trials showed consistently shorter reaction times and longer return times (from lever touch back to resting key), as compared to unrewarded movements (Fig. 7.6). After completing a rewarded movement, the animal kept its hand on the lever until the reward was delivered, without being required to do so, whereas after unrewarded movements the hand left the lever before the reinforcing sound. With new stimuli, return times were initially similar to typical rewarded movements in both rewarded and unrewarded movement trials. However, return times differentiated between rewarded and unrewarded movements after the first few correct trials of each type ($p < 0.001$) (Fig. 7.6). Erroneous movements in nonmovement trials were usually performed with return times typical for rewarded movements. During initial learning, forearm muscle activity between key release and return to resting key was similar in all movement trials and resembled the muscle activity seen in familiar rewarded movements. Later during each learning problem, muscle activity in unrewarded movement trials approached a pattern that was typical for familiar unrewarded movements. Thus, both the movement parameters and the muscle activity in initial learning trials were typical for rewarded movements and subsequently differentiated between rewarded and unrewarded movements. We concluded from these observations that the animals had a default expectation of reward in initial learning trials, which differentiated subsequently according to the type of reinforcer. The animal's preexisting reward expectation appeared to become adapted to the currently expected outcome according to the experience of the animal with the new stimuli.

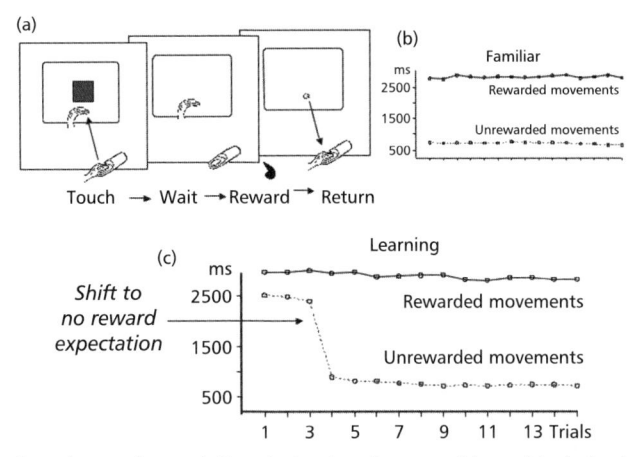

**Fig. 7.6** Adaptation of reward expectation during learning, as evidenced by behavioral reactions. (a) Movement phase of the behavioral task. Following release of the resting key, the animal's hand touched the target key (left), upon which the trigger stimulus disappeared (center) and reward or an auditory reinforcer was delivered, and the animal subsequently returned its hand to the resting key (right). (b) In familiar trials, the animal's return to the resting key differed depending on delivery of reward (top) or auditory reinforcement (bottom). Ordinate indicates median return times in ms (from release of resting key until return to resting key). (c) During learning, the animal's behavioral reaction after touching the target key initially resembled that in familiar rewarded trials (trials 1–3) and differentiated subsequently according to reward vs. auditory reinforcement. The shift after trial 3 is the result of averaging and varied with individual learning problems. The trial contained also an initial instruction and a subsequent trigger stimulus (not shown). Data from Tremblay et al. (1998).

### Neural activity

We studied 148 orbitofrontal neurons that showed statistically significant task-related activations in one or several of the three trial types (rewarded or unrewarded arm movements, rewarded non-movement trials) (Tremblay and Schultz 2000b). Every neuron was tested with familiar instruction stimuli and in at least one separate block of three novel instructions, thus allowing us to study each neuron during a full learning episode and compare the activity with that found during familiar performance. The learning-related changes consisted primarily in increases or decreases of task-related activity, although the analysis according to the type of task relationship revealed a number of more specific changes.

Instruction responses occurred in 93 of the 148 neurons. In 56 of the 93 neurons, response magnitudes increased significantly during learning compared to familiar trials (Fig. 7.7). Of these, 18 neurons showed responses only during learning and not in familiar trials. Twenty-four neurons showed responses in more trial types during learning compared to familiar performance, with a consequential reduction in trial selectivity. These changes were well reproduced with multiple sets of new learning pictures in 12 of 13 neurons tested. In 14 neurons that showed responses to all three instructions in

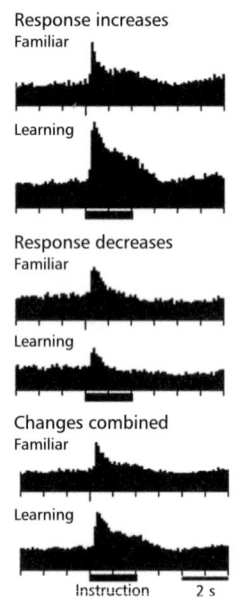

**Fig. 7.7** Changes of population responses to instructions during learning. Top: Increase of responses in 38 neurons which showed responses in famillar trials. Center: Decrease of responses in 26 neurons. Bottom: Averaged changes in neurons showing responses exclusively in learning trials, increased responses during learning, and decreased responses (82 neurons). Only data from rewarded movement trials are shown. In each display, histograms from the first 15 rewarded movement trials recorded with each neuron were added, and the resulting sum was divided by the number of neurons. Reproduced with permission from Tremblay and Schultz (2000b).

familiar trials, an increase of response occurred during learning in all three trial types which remained present during the > 100 learning trials tested. Averaging of the learning-related increases revealed that the changes amounted to substantially longer response durations (Fig. 7.7 top). The opposite change, response decrease during learning, was observed in 26 of the 93 neurons and amounted to slightly shortened population responses (Fig. 7.7 center). Not surprisingly, the combined averaging of increases and decreases together resulted only in a minor learning change (bottom).

Activations preceding reinforcers occurred in 39 of the 148 neurons. They were observed in one or both rewarded trial types and not with unrewarded movements, or exclusively in unrewarded movement trials. Most of these activations were maintained during learning (23 neurons). However, they occurred also in initial unrewarded movement trials and subsided at about the same time or a few trials before the animals made the behavioral change towards performing unrewarded movements with the appropriate parameters (Fig. 7.8). Such activations occurred also with erroneous movements in nonmovement trials. Analogous changes in the opposite direction were observed with activations occurring exclusively in unrewarded movement trials. During initial learning trials, the activations were absent in unrewarded movement trials and appeared after a few trials when the animal's behavior after target acquisition suggested acquisition of an appropriate expectation of no-reward. In addition to these changes, nine neurons showed significantly increased activations irrespective of adaptation of movement parameters, including five neurons that were activated exclusively during learning. Seven neurons showed decreased activations in learning compared to familiar trials.

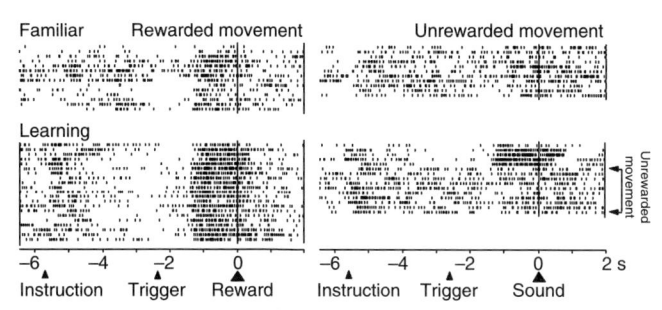

**Fig. 7.8** Adaptation of reward expectation-related neural activation. This orbitofrontal neuron was activated before the liquid reward in familiar rewarded movement and nonmovement trials (only movement trials shown). During initial learning trials, the neuron was also activated in unrewarded movement trials which were performed with parameters of rewarded movements (return times; top trials in bottom right). The activation disappeared when the animal performed the unrewarded movement appropriately, as judged from return of the hand to the resting key (bracket to the right). In addition, activations in correctly performed rewarded movement trials were increased during learning compared to familiar trials. Rewarded movement, nonmovement and unrewarded movement trials alternated semirandomly during the experiment and were separated for analysis. Familiar and learning trials were performed in separate blocks. Trial sequence is shown from top to bottom in each raster. Reproduced with permission from Tremblay and Schultz (2000b).

Responses to liquid reward occurred in 44 of the 148 neurons in one or both rewarded trials during familiar performance, whereas the auditory reinforcer in unrewarded movement trials failed to elicit any response in this sample of neurons. Of these, 10 neurons showed significantly increased responses, six neurons showed decreased responses, and the remaining 28 neurons had unaltered reward responses during learning. Four neurons responded exclusively during learning.

## Conclusions

The reviewed data suggest that orbitofrontal neurons respond in a flexible manner to stimuli signalling rewards and aversive outcomes. In the intuitively most understandable form, orbitofrontal neurons acquire differential responses to visual or olfactory stimuli associated with different types of reinforcer to be expected or avoided. These responses develop during the learning of novel stimuli (Schoenbaum *et al.* 1999; Tremblay *et al.* 2000b) and during the reversal of existing differential associations of stimuli with rewards and punishers (Rolls *et al.* 1996). In a similar way, activations develop during the expectation of reward when learning novel stimuli (Schoenbaum *et al.* 1998). In addition, some orbitofrontal neurons show increased responses during the learning period, and often during the subsequent consolidation period of several types of trials, compared to familiar stimuli (Tremblay *et al.* 2000b). It is possible that these differences are due to response reduction with overtraining in familiar trials, or else that attentional or consolidation mechanisms play a role in increasing neural responsiveness during learning episodes. By contrast, some orbitofrontal neurons show reduced responses during learning compared to familiar stimuli (Tremblay and Schultz 2000b), possibly reflecting stronger neural associations and better established predictions of outcomes that develop through longer experience. As an interesting consequence, the mixture of increases and decreases in neural responsiveness during learning may produce falsely minor or negative effects in neural magnetic resonance imaging studies, in which the blood oxygenation resulting from activity in groups of neurons is measured on a whole, and the decreased activities could cancel the effects of increased activities.

An interesting form of learning-related change consists of the adaptation of reward expectations at behavioral and neural levels. The changes of overt behavioral reactions, such as arm movements and muscle activity, reveal how the animal's internal reward expectations adapt to the current values of reward prediction while the new stimuli are being learned across successive trials. These behavioral changes are preceded or paralleled by changes in neural activations preceding the reward, which may constitute a neural correlate for the adaptation of expectations to the new task contingencies. These observations suggest a neural mechanism for adaptive learning in which existing expectation-related activity is matched to the new condition, rather than acquiring all relationships to task contingencies from scratch. The mechanism would involve an initial setup of predictions about an environmental event and subsequently only process the deviations from the predictions (prediction error), rather than continuously setting up complete representations of all sensory inputs (Rao and Ballard 1999). The behavioral consequence would be the well-known error-driven learning by which prediction errors are being reduced and the animal's internal expectations adapt to the current predictions.

## 7.6. Comparison with striatum

### Types of neural relationships

The striatum (caudate nucleus, putamen and ventral striatum including nucleus accumbens) receives strong axonal projections from the OFC that are primarily directed to the ventromedial striatum (Selemon and Goldman-Rakic 1985; Haber *et al.* 1995). The same part of the striatum also receives strong projections from the amygdala (Russchen *et al.* 1985; Fudge *et al.* 2002). It is therefore natural to assume that the slowly discharging neurons in the ventral striatum would process reward information in similar ways as the orbitofrontal neurons, including showing responses to reward-predicting instructions, activations during the expectation of reward, and responses following the reception of reward (Fig. 7.9). However, the striatum also receives inputs from a large number of cortical areas involved in many aspects of movements and behavioral outputs. Indeed, neural activities were found in great sophistication in relation to arm and eye movements in the more dorsal parts of caudate and putamen to which these projections are directed. Thus striatal neurons are activated during the instruction, preparation and execution of specific movements to specific spatial targets (Crutcher and DeLong 1984; Alexander 1987; Hikosaka *et al.* 1989a–c; Alexander and Crutcher 1990a,b; Crutcher and Alexander 1990; Kimura 1990; Romo *et al.* 1992; Schultz and Romo 1992) (Fig. 7.9). Thus the five main forms of task relationships comprised instruction responses, sustained activity during the movement preparatory delay, responses to the movement-eliciting trigger stimulus merging with activity preceding and accompanying the movement itself, sustained activations during the expectation of the reward, and responses following the delivery of the reward. However, what came entirely unexpectedly was a substantial conjunction of reward- and movement-related activity that was not seen in orbitofrontal neurons, the details of which will be described here.

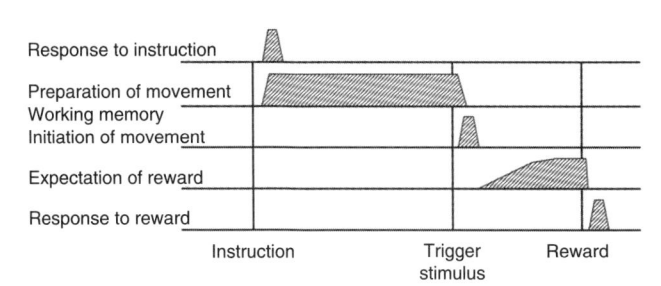

**Fig. 7.9** Schematic overview of the main forms of activity in primate striatum during the performance of a delayed response task.

## Reward discrimination

### Reward vs. no-reward

Monkeys performed the same go/no-go delayed response task that was used for studying orbitofrontal neurons (Hollerman *et al.* 1998). The task involved a rewarded arm reaching movement, the same arm movement but unrewarded, and a rewarded nonmovement trial. The kind of reinforcement (reward or auditory reinforcer) was indicated by the initial instruction cue which also signalled the type of behavioral reaction to be performed (execution or withholding of arm movement). The task allowed us to investigate the same neurons during the preparation and execution of the same arm movements in the presence and the absence of the expectation of liquid reward. We found that most of the 259 task-related striatal neurons recorded were sensitive to the type of reinforcer expected at trial end, most of them being preferentially or exclusively activated in rewarded as compared to unrewarded movements (58–80% of neurons showing any of the five forms of task relationship). Only few striatal neurons showed higher task-related activations in unrewarded compared to rewarded trials. Fig. 7.10 shows a typical neuron that was activated during the preparation of movements in rewarded trials but showed no activation during the preparation of an unrewarded movement nor during the same task period in rewarded non-movement trials. Thus the neuron showed a conjunction of a motor process (movement preparation) and reward expectation.

Even stronger reward dependency of movement-related activity was observed when reward was associated asymmetrically with movements to different spatial positions. Many striatal neurons showed spatial preferences when all directions were equally well

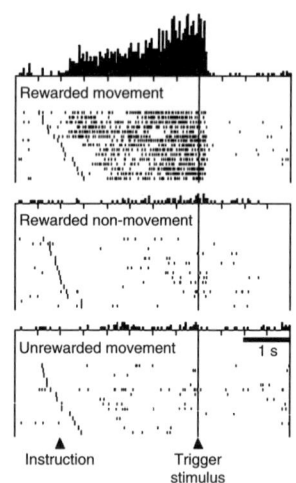

**Fig. 7.10** Neural activity reflecting the expectation of reward during the preparation of movement. This caudate neuron was activated preceding the trigger stimulus only in rewarded movement trials. Neuronal activity is referenced to trigger onset, which in movement trials elicited the reaching movement. Reward or sound reinforcement was delivered 1.5 s after lever touch in movement trials, and 3.0 s after the trigger in non-movement trials. Trials are rank-ordered according to instruction-trigger intervals. Reproduced with permission from Hollerman *et al.* (1998).

rewarded, but they shifted their spatial preference into the currently rewarded direction when only one of the several directions of eye movement was rewarded (Kawagoe et al. 1998), indicating a precedence of reward associations over spatial positions of stimuli and suggesting a particularly prominent role of the striatum in reward-related processes. In a task involving multiple movements, striatal neurons showed activity discriminating between rewards, no-rewards and cocaine drug reward during several trial epochs (Bowman et al. 1996, Shidara et al. 1998). The group of tonically discharging striatal neurons showed differential responses to reward-predicting instructions when tested against unrewarded stimuli (Kimura et al. 1984) or aversive events (air puff) (Ravel et al. 1999). Neurons discriminating between reward- and punisher-related CSs were also found in the rat striatum (Setlow et al. 2003; Roitman et al. 2005).

### Different rewards

Monkeys performed a spatial delayed response task in which the initial instruction cue indicated which kind of liquid reward would be delivered for correct performance of the arm reaching movement (Hassani et al. 2001). In separate choice trials we tested the animal's preference for the individual rewards to determine the relative motivational value of each reward. We found that 17–49% of 610 neurons showed significant differences in activity between different liquid rewards, consisting frequently but not exclusively of stronger activations with more preferred rewards compared to non-preferred rewards. Neurons that discriminated between different reward-predicting stimuli discriminated frequently also between spatial positions of stimuli, indicating that the two forms of discrimination can occur in the same neurons.

### Different reward magnitudes

In a spatial delayed response task the initial instruction cue indicated which quantity of the same liquid reward would be delivered for correct performance of the arm reaching movement (Cromwell and Schultz 2003). We found that 55% of 374 neurons showed significant differences in task-related activation between at least two different magnitudes of reward during one or several of the different trial epochs that were not related to the visual objects per se.

### Reward preference

The shift of reward-discriminating activity observed in the OFC might constitute a specific mode of processing in this part of the cortex, or it might represent a more widespread phenomenon occurring also in other reward centers of the brain. To investigate this possibility, we studied some neurons in the striatum while shifting the choices of available rewards according to the ranked preferences of the animals (Cromwell et al. 2005). As with orbitofrontal neurons, we compared neural activity for the same reward between two different reward combinations. We tested 20 striatal neurons that discriminated between different rewards or reward magnitudes and labelled the rewards according to the animal's preferences as $A > B > C$ as in the orbitofrontal study. Twelve of the twenty neurons shifted their activity according to the currently tested rewards. For example, a neuron would be less activated for the same intermediate reward B in A–B compared to B–C blocks. Within both blocks, activation was strongest with whatever was the more preferred or larger of the two rewards, irrespective of the actual physical reward parameters. Shifts in activations occurred irrespective of kind or amount of reward, and irrespective of activation increases or decreases with more preferred reward. Thus neural reward coding in some striatal neurons can shift depending on the currently available rewards and in a similar manner as in orbitofrontal neurons. We have found a similar adaptation in dopamine neurons coding errors in reward prediction (Tobler *et al.* 2005), which together would indicate a general mechanism by which reward neurons adapt their input–output relationships to the range of currently available rewards.

## Learning

We investigated 205 slowly discharging, task-related neurons in the anterior striatum (Tremblay *et al.* 1998), using the same learning set paradigm based on the delayed go/no-go task as for orbitofrontal neurons (Tremblay and Schultz 2000b). The learning-related changes in striatal neurons were in general similar to those observed in orbitofrontal neurons, although some differences were noted. One half to two thirds of the tested neurons showed adaptations of existing reward expectation activity during learning, and increases or decreases in task-related activity during learning compared to familiar performance. Some neurons failed to respond in early learning trials and acquired the responses over the course of subsequent trials, whereas other neurons responded to stimuli exclusively during the learning period. Striatal neurons showed more movement preparation-related activity than orbitofrontal neurons, and this

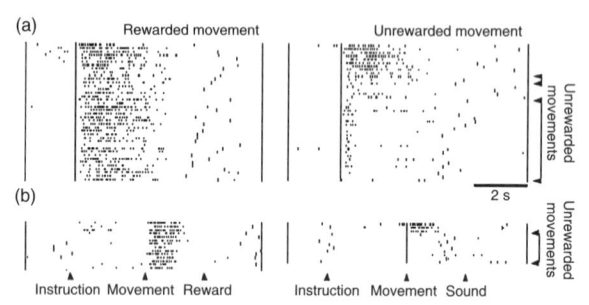

**Fig. 7.11** Adaptation of reward expectation-related activity during learning in two caudate neurons. (a) Activation during the preparation of movement between instruction and movement onset. (b) Activation during the execution of movement. During familiar performance (not shown), both neurons were activated in rewarded movement trials, but not in unrewarded movement trials. During initial learning trials, both neurons also showed activations in unrewarded movement trials, which subsided when the animal had acquired the meaning of the appropriate instruction cues and performed these trails with parameters typical for unrewarded movements (return times; brackets to the right). Trial sequence is shown from top to bottom in each raster. Reproduced with permission from Tremblay *et al.* (1998).

contrast led to two notable differences. First, the movement preparatory activity was influenced by the expectation of reward and adapted to new reward contingencies like other reward expectation activity. It occurred during all initial learning trials and became restricted to rewarded trials when the significance of the instruction pictures became known. These adaptations demonstrate that reward representations in striatal neurons during the preparation of movements can undergo dynamic changes. Second, striatal neurons showed changes in activity that predicted exactly whether the animal would perform a movement or a nonmovement reaction, irrespective of the chosen action being right or wrong according to the instruction. This result suggests that the movement preparatory activity constituted an intrinsic property of these neurons and was not simply a result of the overtraining.

In addition to the changes in activity expectation and movement preparation related to striatal neurons in monkeys and rats acquire discriminating responses to reward- or punisher-predicting stimuli during associative Pavlovian or operant learning (Aosaki *et al.* 1994; Setlow *et al.* 2003). During T-maze learning, neurons in the rat striatum showed progressively earlier activity in the maze as learning advanced (Jog *et al.* 1999). Behavioral reversal of stimulus-outcome associations resulted in reversed neural responses in neurons discriminating between reward- and punisher-predicting olfactory stimuli (Setlow *et al.* 2003). Associating an aversive stimulus, such as a loud or air puff-related noise, with a liquid reward changed the response to the stimulus (Ravel *et al.* 2003). Behavioral extinction of conditioned stimuli by withholding reward resulted in reduction of neural responses (Aosaki *et al.* 1994). Compatible with the motor and sensory functions of the striatum, neurons in this structure not only respond to motivationally significant stimuli but can also acquire responses to stimuli indicating the direction of behavioral reactions, as seen in a conditional oculomotor task (Pasupathy and Miller 2005). Neurons in the posterior striatum responded more strongly to novel visual stimuli compared to familiar stimuli, and these neurons generally failed to respond to rewards and reward-predicting stimuli (Brown *et al.* 1995). Taken together, striatal neurons readily acquire responses that are sensitive to the outcome and movement contingencies of external stimuli.

## Conclusions

Similar to the OFC, striatal neurons are substantially involved in the processing of reward information, to the extent that both structures can be considered as important components of the brain's reward system. Neurons in the dorsal and ventral striatum signal the reception of rewards and reward-predicting stimuli, and are activated during the expectation of rewards. Apart from the responses to reward itself, these activities are learned by stimulus-reward associations, probably in a Pavlovian manner. Reward-related activity in the striatum discriminates between rewards, motivationally neutral events and aversive events, and the behavioral reversal of associations between stimuli and reinforcers of opposing valence induces the reversal of discriminating neural responses. Although the potential reward inputs from the OFC and the amygdala are primarily directed towards the ventral striatum, reward-related activities are found throughout the entire anterior striatum, and the sources of these activities are not entirely clear.

The main differences between striatal and orbitofrontal reward-related activities occur in relation to arm and eye movements. There are no substantial neural correlates for motor processes in the OFC, whereas neural activity in the striatum shows clear relationships to movement parameters. Interestingly, the movement processes in striatal neurons are substantially modulated by expected rewards. More detailed testing suggests relationships to goal-directed mechanisms in which the rewarding goal and the contingency between movement and goal appear to be represented during the movement leading to the goal. Correspondingly, during learning, striatal neurons show adaptations of expectation-related activity that concern both reward and motor processes.

These comparisons might allow us to speculate on the potential origin of neural mechanisms underlying voluntary actions towards rewarding outcomes. If we assume that these processes are executed by the dorsolateral prefrontal cortex, there are at least two anatomically plausible routes by which the necessary information can arrive in this structure. Orbitofrontal inputs could provide substantial motivational information to the dorsolateral prefrontal cortex, concerning rewards and probably punishers. Inputs from the striatum via pallidum, substantia nigra reticulata and the ventroanterior thalamus might carry integrated information about movements and rewards to the dorsolateral cortex.

## 7.7. General conclusions

Although lesion studies in humans suggest that the OFC is involved in a large array of behavioral functions related to the emotional components of voluntary behavior and decision-making, neurophysiological work in behaving monkeys can only investigate a small selection of these functions. These limits are not only due to the much less elaborated functions of the monkey compared to the human cortex but also reflect the constraints imposed by controlled neurophysiological studies which basically restrict testing to events that occur within time spans of seconds to minutes and are observable in an objective, nonverbally communicable manner. Nevertheless, the neurophysiological studies have consistently revealed, in monkeys and rats, that orbitofrontal neurons show substantial covariations in activity with reward-related processes. It can be reasonably well conjectured that reward functions constitute considerable components of orbitofrontal contributions to behavioral output, and investigating these mechanisms should provide a good step towards understanding the role of the OFC in voluntary behavior.

These results from neurophysiological studies, together with previous lesioning and psychopharmacological studies in monkeys and rats, and neuroimaging studies in humans, suggest the OFC as one of the main reward centers of the brain. Other major reward centers include the dorsal and ventral striatum, which show additional movement-related activity not found in the OFC, the amygdala for which only limited knowledge on reward processes is available, and the dopamine neurons, which differ from the OFC in processing a reward prediction error in their large majority. In addition, neurons in a number of frontal and parietal cortical regions show modulations of movement-related activity by rewards and are activated during specific phases of decision-making for rewards, the details of which would go beyond this article.

Orbitofrontal neurons respond to rewards and reward-related stimuli, and these responses are differentially and dynamically attributed through learning to events of specific motivational valence (rewards and punishers). This conclusion stems from experiments on monkeys and rats involving novel learning, repeated learning after set formation, and reversal learning. On the other hand, orbitofrontal neurons do not seem to be involved in motor processes, neither in their execution nor their initiation, planning or preparation. These results suggest a parcellation of prefrontal mechanisms into motivational and movement processes which appears to follow the unfortunate distinction of brain functions between rational-cognitive and emotional-motivational processes. The separation is most likely exaggerated, and future work should take both cognitive and motivational mechanisms into account rather than considering them as distinct entities. The neural activities in the striatum are examples on how motivational and motor processes can converge to produce mechanisms related to the acquisition of rewarding goals.

As expected from a cortical structure, orbitofrontal neurons do not only respond to stimuli as they occur in the environment, they are also activated in anticipation of future events, in particular future rewards. These results suggest that orbitofrontal neurons have access to representations of the outside world and are not only driven by actually occurring events. This 'inner world' of representations changes with experience, and the expectation-related activity of orbitofrontal neurons shows appropriate adaptations during learning that reflect the transition from previous or default representations to expectations that are updated by the latest experience. This result, obtained in learning experiments, relates to a major way of brain operation in which not all inputs are processed continuously in all structures. Rather, sensory inputs give rise to predictions, and only deviations from these predictions are processed, thus reducing considerably the information processing and hence energy consumption necessary to obtain an accurate view of the world. Without much surprise, the OFC appears to participate in this mechanism.

Whether it is responses to past events or activations anticipating future events, orbitofrontal neurons appear to dynamically adjust their task-related activities according to the currently used distribution of rewards. Apart from the taste neurons in this structure, the observed activities do not appear to be specific for a particular reward object, but rather reflect the value of the currently used reward relative to the predicted distribution. The advance information provided by the reward-predicting stimuli helps to reduce the uncertainty about which rewards will be available in the immediate future and would allow orbitofrontal neurons to enlarge or restrict their processing according to the current level of uncertainty. By managing the uncertainty about rewards in this way, orbitofrontal neurons can optimize the coding of reward within the biological constraints of brain function.

By comparison, neurons in the striatum process reward information in a similar manner as orbitofrontal neurons, including the adaptation of expectations to novel stimuli during learning and the management of uncertainty. However, different from the OFC, the striatum is often considered as a part of the motor system and contains many neurons that appear to be primarily involved in the preparation and execution of movements. Even these neurons show considerable modulations by rewards, pointing to a way by which reward information can be linked to behavioral output mechanisms.

## Acknowledgement

This work was supported by the Wellcome Trust and the Swiss National Science Foundation.

## References

Alexander G. E. (1987) Selective neuronal discharge in monkey putamen reflects intended direction of planned limb movements. *Experimental Brain Research* **67**:623–34.

Alexander G. E. and Crutcher M. D. (1990a) Preparation for movement: Neural representations of intended direction in three motor areas of the monkey. *Journal of Neurophysiology,* **64**:133–50.

Alexander G. E. and Crutcher M. D. (1990b) Neural representations of the target (goal) of visually guided arm movements in three motor areas of the monkey. *Journal of Neurophysiology,* **64**:164–78.

Aosaki T., Tsubokawa H., Ishida A., Watanabe K., Graybiel A. M., and Kimura M. (1994) Responses of tonically active neurons in the primate's striatum undergo systematic changes during behavioral sensorimotor conditioning. *Journal of Neuroscience,* **14**:3969–84.

Apicella P., Scarnati E., Ljungberg T., and Schultz W. (1992) Neuronal activity in monkey striatum related to the expectation of predictable environmental events. *Journal of Neurophysiology,* **68**:945–60.

Bernoulli D. (1738) Specimen theoriae novae de mensura sortis. Comentarii Academiae Scientiarum Imperialis Petropolitanae (Papers Imperial Academy of Science St. Petersburg), **5**:175–92. Translated as: Exposition of a new theory on the measurement of risk (1954) *Econometrica,* **22**:23–36.

Bowman E. M., Aigner T. G. and Richmond B. J. (1996) Neural signals in the monkey ventral striatum related to motivation for juice and cocaine rewards. *Journal of Neurophysiology,* **75**:1061–73.

Brown V. J., Desimone R. and Mishkin M. (1995) Responses of cells in the caudate nucleus during visual discrimination learning. *Journal of Neurophysiology,* **74**:1083–94.

Critchley H. G. and Rolls E. T. (1996) Hunger and satiety modify the responses of olfactory and visual neurons in the primate orbitofrontal cortex. *Journal of Neurophysiology,* **75**:1673–86.

Cromwell H. C. and Schultz W. (2003) Effects of expectations for different reward magnitudes on neuronal activity in primate striatum. *Journal of Neurophysiology,* **89**:2823–38.

Cromwell H. C., Hassani O. K., and Schultz W. (2005) Relative reward processing in primate striatum. *Experimental Brain Research,* **162**:520–25.

Crutcher M. D. and Alexander G. E. (1990) Movement-related neuronal activity selectively coding either direction or muscle pattern in three motor areas of the monkey. *Journal of Neurophysiology,* **64**:151–63.

Crutcher M. D. and DeLong M. R. (1984) Single cell studies of the primate putamen. II. Relations to direction of movement and pattern of muscular activity. *Experimental Brain Research,* **53**:244–58.

Divac I., Rosvold H. E., and Szwarcbart M. K. (1967) Behavioral effects of selective ablation of the caudate nucleus. *Journal of Comparative and Physiological Psychology*, **63**:184–90.

Fudge J. L., Kunishio K., Walsh P., Richard C., and Haber S. N. (2002) Amygdaloid projections to ventromedial striatal subterritories in the primate. *Neuroscience*, **110**:257–75.

Funahashi S., Bruce C. J., Goldman-Rakic P. S. (1990) Visuospatial coding in primate prefrontal neurons revealed by oculomotor paradigms. *Journal of Neurophysiology* **63**:814–31.

Fuster J. M. (1973) Unit activity of prefrontal cortex during delayed-response performance: Neuronal correlates of transient memory. *Journal of Neurophysiology*, **36**:61–78.

Gaffan E. A., Gaffan D. and Harrison S. (1988) Disconnection of the amygdala from visual association cortex impairs visual reward association learning in monkeys. *Journal of Neuroscience*, **8**:3144–50.

Haber S., Kunishio K., Mizobuchi M., and Lynd-Balta E. (1995) The orbital and medial prefrontal circuit through the primate basal ganglia. *Journal of Neuroscience* **15**:4851–67.

Harlow H. F. (1949) The formation of learning sets. Psychological Review, **56**:51–65.

Hassani O. K., Cromwell H. C., and Schultz W. (2001) Influence of expectation of different rewards on behavior-related neuronal activity in the striatum. *Journal of Neurophysiology*, **85**:2477–89.

Hikosaka O., Sakamoto M., and Usui S. (1989a) Functional properties of monkey caudate neurons. I. Activities related to saccadic eye movements. *Journal of Neurophysiology*, **61**:780–98.

Hikosaka O., Sakamoto M., and Usui S. (1989b) Functional properties of monkey caudate neurons. II. Visual and auditory responses. *Journal of Neurophysiology*, **61**:799–13.

Hikosaka O., Sakamoto M., and Usui S. (1989c) Functional properties of monkey caudate neurons. III. Activities related to expectation of target and reward. *Journal of Neurophysiology*, **61**:814–32.

Hikosaka K. and Watanabe M. (2000) Delay activity of orbital and lateral prefrontal neurons of the monkey varying with different rewards. *Cerebral Cortex*, **10**:263–71.

Hikosaka K. and Watanabe M. (2004) Long-range and short-range reward expectancy in the primate orbitofrontal cortex. European *Journal of Neuroscience*, **19**:1046–54.

Hollerman J. R., Tremblay L., and Schultz W. (1998) Influence of reward expectation on behavior-related neuronal activity in primate striatum. *Journal of Neurophysiology*, **80**:947–63.

Jacobsen C. F. and Nissen H. W. (1937) Studies of cerebral function in primates: IV. The effects of frontal lobe lesions on the delayed alternation habit in monkeys. *Journal of Comparative and Physiological Psychology*, **23**:101–12.

Jog M. S., Kubota Y., Connolly C. I., Hillegaart V. and Graybiel A. M. (1999) Building neural representations of habits. *Science* **286**:1745–49.

Kahneman D., Tversky A. (1984) Choices, Values, and Frames. American Psychologist **4**:341–50.

Kimura M. (1990) Behaviorally contingent property of movement-related activity of the primate putamen. *Journal of Neurophysiology,* **63**:1277–96.

Kimura M., Rajkowski J., and Evarts E. (1984) Tonically discharging putamen neurons exhibit set-dependent responses. *Proceedings of the National Academy of Science (USA)*, **81**:4998–5001.

von Neumann J. and Morgenstern O. (1944) *The Theory of Games and Economic. Behavior*. Princeton University Press, Princeton.

Pasupathy A. and Miller E. K. (2005) Different time courses of learning-related activity in the prefrontal cortex and striaum. *Nature*, **433**:873–76.

Rao R. P. N. and Ballard D. H. (1999) Predictive coding in the visual cortex: a functional interpretation of some extra-classical receptive-field effects. *Nature Neuroscience*, **2**:79–87.

Ravel S., Legallet E., and Apicella P. (1999) Tonically active neurons in the monkey striatum do not preferentially respond to appetitive stimuli. *Experimental Brain Research*, **128**:531–34.

Ravel S., Legallet E., and Apicella P. (2003) Responses of tonically active neurons in the monkey striatum discriminate between motivationally opposing stimuli. *Journal of Neuroscience*, **23**:8489–97.

Rescorla R. A. and Wagner A. R. (1972) A theory of Pavlovian conditioning: Variations in the effectiveness of reinforcement and nonreinforcement. In: A. H. Black and W. F. Prokasy, eds. *Classical Conditioning II: Current Research and Theory*, pp. 64–99. New York: Appleton Century Crofts,.

Roitman M. F., Robert A., Wheeler R. A., and Carelli R. M. (2005) Nucleus accumbens neurons are innately tuned for rewarding and aversive taste stimuli, encode their predictors, and are linked to motor output. *Neuron*, **45**:587–97.

Rolls E. T., Sienkiewicz Z. J., and Yaxley S. (1989) Hunger modulates the responses to gustatory stimuli of single neurons in the caudolateral orbitofrontal cortex of the macaque monkey. *European Journal of Neuroscience*, **1**:53–60.

Rolls E. T., Critchley H. D., Mason R. and Wakeman E. A. (1996) Orbitofrontal cortex neurons: role in olfactory and visual association learning. *Journal of Neurophysiology*, **75**:1970–81.

Romo R., Scarnati E., and Schultz W. (1992) Role of primate basal ganglia and frontal cortex in the internal generation of movements: II. Movement-related activity in the anterior striatum. *Experimental Brain Research*, **91**:385–395.

Rosenkilde, C. E., Bauer, R. H., and Fuster, J. M. Single cell activity in ventral prefrontal cortex of behaving monkeys. *Brain Research*, **209**:375–94.

Russchen F. T, Bakst I., Amaral D. G., and Price J. L. (1985) The amygdalostriatal projections in the monkey. An anterograde tracing study. *Brain Research*, **329**:241–57.

Schoenbaum G., Chiba A. A., and Gallagher M. (1998) Orbitofrontal cortex and basolateral amygdala encode expected outcomes during learning. *Nature Neuroscience*, **1**:155–59.

Schoenbaum G., Chiba A. A., and Gallagher M. (1999) Neural encoding in orbitofrontal cortex and basolateral amygdala during olfactory discrimination learning. *Journal of Neuroscience*, **19**:1876–84.

Schultz W. and Romo R. (1992) Role of primate basal ganglia and frontal cortex in the internal generation of movements: I. Preparatory activity in the anterior striatum. *Experimental Brain Research*, **91**:363–84.

Selemon L. D. and Goldman-Rakic P. S. (1985) Longitudinal topography and interdigitation of corticostriatal projections in the rhesus monkey. *Journal of Neuroscience*, **5**:776–94.

Setlow B., Schoenbaum G., and Gallagher M. (2003) Neural encoding in ventral striatum during olfactory discrimination learning. *Neuron*, **38**:625–36.

Shidara M., Aigner T. G., and Richmond B. J. (1998) Neuronal signals in the monkey ventral striatum related to progress through a predictable series of trials. *Journal of Neuroscience*, **18**:2613–25.

Thorpe S. J., Rolls E. T., and Maddison S. (1983) The orbitofrontal cortex: neuronal activity in the behaving monkey. *Experimental Brain Research*, **49**:93–115.

Tremblay L. and Schultz W. (1999) Relative reward preference in primate orbitofrontal cortex. *Nature*, **398**:704–8.

Tremblay L. and Schultz W. (2000a) Reward-related neuronal activity during go-nogo task performance in primate orbitofrontal cortex. *Journal of Neurophysiology*, **83**:1864–76.

Tremblay L. and Schultz W. (2000b) Modifications of reward expectation-related neuronal activity during learning in primate orbitofrontal cortex. *Journal of Neurophysiology*, **83**:1877–85.

Tremblay L., Hollerman J. R., and Schultz W. (1998) Modifications of reward expectation-related neuronal activity during learning in primate striatum. *Journal of Neurophysiology*, **80**:964–77.

Wallis J. D. and Miller E. K. (2003) Neuronal activity in primate dorsolateral and orbital prefrontal cortex during performance of a reward preference task. European *Journal of Neuroscience*, **18**:2069–81.

Chapter 8

# From associations to expectancies: orbitofrontal cortex as gateway between the limbic system and representational memory

Matthew Roesch and Geoffrey Schoenbaum

> Although our picture of the anatomical connections of the orbital prefrontal cortex is less detailed and precise than that of the dorsolateral areas, it is already clear that one or more orbital subareas ... has a knowledge base different from that available to the dorsolateral region, e.g., interoceptive and olfactory stimuli, and likewise has connections with motor centers, including the autonomic musculature and endocrine mechanisms. .....
> A primary defect in accessing the central representations of reward and punishment ... could explain the non-specific profile of cognitive deficits following orbital lesions.
> *Patricia Goldman-Rakic (1987).*

## 8.1. Introduction

Since Dr. Goldman-Rakic wrote these words, there has been an explosion of interest in the orbitofrontal cortex (OFC). We now know a great deal not only about the anatomical connections of the OFC in rats and in primates, but also about the behavioral and neurophysiological characteristics of this prefrontal subdivision. These findings point increasingly to a role for the OFC in emotional learning within a circuit that includes the amygdala and other limbic structures. Less clear, however, is the unique contribution of this prefrontal region to processing inside (and outside) this circuit. Indeed work has focused on the functional similarities between the OFC and the amygdala. In this chapter, we will review these findings to establish the involvement of the OFC in the associative learning mediated by this circuit; however we will also draw important distinctions between the critical role of limbic areas in associative *learning* and the pivotal role of the OFC in the *control* of this information and of it's application to govern behavior. We will argue that the OFC allows associative information, particularly information about the value of likely outcomes, to be manipulated in representational memory and integrated with non-associative variables concerning subsequent behavior, current context and internal state. The resultant 'expectancies' then influence processing in downstream limbic areas as well as other prefrontal regions, thereby promoting voluntary, cognitive, and goal-directed (e.g., not stimulus-driven) behavior and facilitating new learning.

As much as any other brain region, the OFC has been a mystery. Even within the prefrontal cortex, the least understood cortical region, the OFC was the last to be explored. Indeed, when Patricia Goldman-Rakic elegantly described the circuitry of the primate prefrontal cortex and its role in regulating behavior representational (e.g. working or scratchpad) memory nearly 20 years ago, so little was known about the OFC that it warranted less than a page in her 44-page essay (Goldman-Rakic 1987). Yet as we will argue, her much-ignored description of orbitofrontal function remains a cogent summary.

In this chapter we will describe recent findings that support a critical role for the OFC in affective processing as part of a circuit that includes the amygdala and other limbic structures. However like Dr. Goldman-Rakic, we will argue that the pivotal contribution of the OFC to processing within this circuit, and without, goes beyond simple associative learning about likely outcomes and encompasses the integration of this information with non-associative variables such as subsequent plans, current context and internal state to generate what we will refer to as 'expectancies' for future events. Such expectancies have been hypothesized by learning theorists (Dickinson 1989) to provide an internalized or inner representation of the consequences that may follow a particular act, which may then be applied by other parts of the brain to govern behavior and facilitate new learning. In describing the findings that support this idea, we will draw distinctions from neurophysiological and behavioral data concerning the respective roles of the OFC and limbic areas in generating and using these expectancies and show that a deficit in this function can explain the specific profile of cognitive deficits following orbitofrontal versus amygdalar lesions.

## 8.2. The rat OFC

Before beginning, it is worth noting that this chapter will discuss findings from rats and primates. Thus it is important to define the particular prefrontal regions within the rat brain that we will refer to as the OFC and the likely homologies that these areas have to the primate orbital prefrontal subdivision. A diagram of the critical brain regions and important connections that are consistent across species is presented in Fig. 8.1. This conceptualization begins with the observations of Rose and Woolsey (1948), who proposed that the prefrontal cortex might be defined by the projections of mediodorsal thalamus (MD) rather than by 'stratiographic analogy' (Cajal 1988). This definition, based on the observation that the defining granular layer of the primate prefrontal cortex

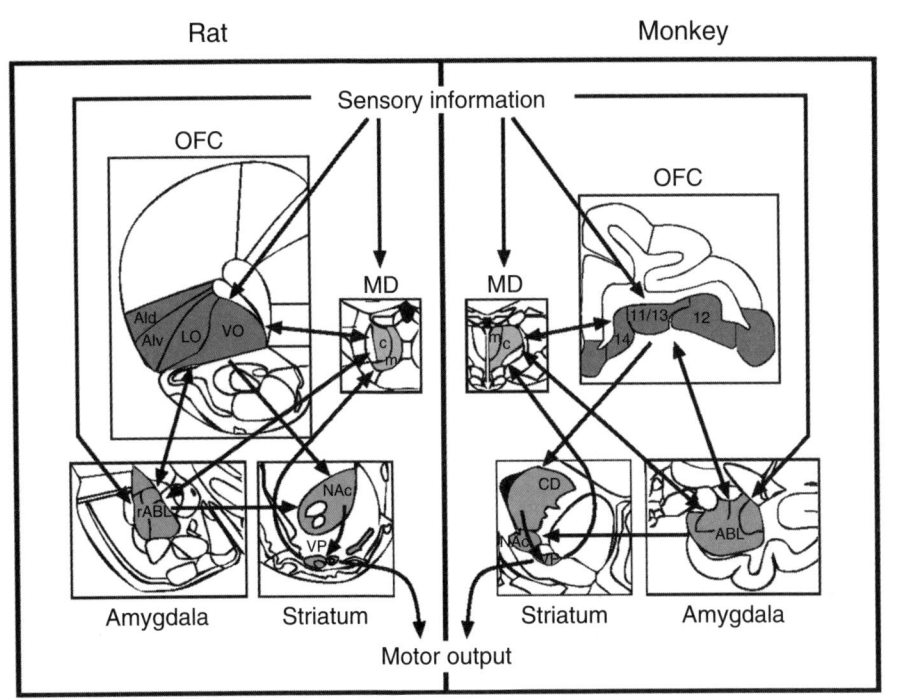

**Fig. 8.1** (Also see Color Plate 17.) Anatomical relationships of the OFC (OFC; ●) in rat and monkey. Based on their pattern of connectivity with mediodorsal thalamus (MD; ●), amygdala (●) and striatum, the orbital and agranular insular areas in rat prefrontal cortex are homologous to primate OFC. In both species, the OFC receives robust input from sensory cortices, associative information from the amygdala and outputs to the motor system through striatum. Each box illustrates a representational coronal section. AId, dorsal agranular insula; AIv, ventral agranular insula; c, central; m, medial; ABL, basolateral amygdala; rABL, rostral basolateral amygdala; CD, caudate; NAc, nucleus accumbens core; VP, ventral pallidum; LO, lateral orbital; VO, ventral orbital, including ventrolateral and ventromedial orbital regions.

reflects thalamic input, provides a foundation on which to define putative prefrontal homologues in non-primate species such as rats, which do not have a prominent granular cell layer.

In the rat, the MD can be divided into three segments (Krettek and Price 1977b; Groenewegen 1988). Projections from the medial and central segments of the MD define a region that includes the orbital areas and the ventral and dorsal agranular insular cortices (Leonard 1969; Krettek and Price 1977b; Kolb 1984; Groenewegen 1988). These regions of the MD in rat receive direct afferents from the amygdala, the medial temporal lobe, and the ventral pallidum/ventral tegmental area, and olfactory input from the piriform cortex (Krettek and Price 1977b; Groenewegen 1988; Ray and Price 1992). This pattern of connectivity is similar to that of the medially-located, magnocellular division of the primate MD, which defines the orbital prefrontal subdivision in primates (Kievit and Kuypers 1977; Goldman-Rakic and Porrino 1985; Russchen *et al.* 1987). Thus a defined region in the orbital area of the rat prefrontal cortex is likely to receive input from thalamus very similar to that reaching the primate orbital prefrontal cortex. Based in part on this pattern of input, the projection fields of medial and central MD in the orbital and agranular insular areas of the rat prefrontal cortex have been proposed as homologous to the primate orbitofrontal region (Leonard 1969; Groenewegen 1988; Preuss 1995; Ongur and Price 2000; Schoenbaum and Setlow 2001). These areas in rodents include the dorsal and the ventral agranular insular cortex, as well as the lateral and ventrolateral orbital regions. Notably this conception of the rat OFC does not include the medial or ventromedial orbital cortex, which lies along the medial wall of the hemisphere. This region has patterns of connectivity with the MD thalamus and other areas that are more similar to other regions on the medial wall.

Other important connections serve to highlight the similarity between the rat OFC, defined above, and the primate OFC. Perhaps most notable are reciprocal connections with the basolateral complex of the amygdala (ABL), a region thought to be involved in affective or motivational aspects of learning (Brown and Schafer 1888; Kluver and Bucy 1939; Weiskrantz 1956; Everitt and Robbins 1992; LeDoux 1996; Holland and Gallagher 1999; Davis 2000; Gallagher 2000; Baxter and Murray 2002). In primate, these connections have been invoked to explain certain similarities in behavioral abnormalities resulting from damage to either the OFC or the amygdala (Jones and Mishkin 1972; Gaffan and Murray 1990; Fuster 1997; Bechara *et al.* 1999; Baxter *et al.* 2000). Reciprocal connections between basolateral the amygdala and areas within the rat OFC, particularly the agranular insular cortex (Krettek and Price 1977a; Kolb 1984; Kita and Kitai,1990; Shi and Cassell 1998), suggest that interactions between these structures may have a similar importance for regulation of behavioral functions in rats as well. In addition, in both rats and primates, the OFC provides a strong efferent projection to ventral striatum, overlapping with innervation from limbic structures such as the amygdala and subiculum (Groenewegen *et al.* 1987; 1990; McDonald 1991; Haber *et al.* 1995). As outlined in Fig. 8.1, the specific circuitry connecting the OFC, limbic structures and ventral striatum presents a striking parallel across species that suggests possible similarities in functional interactions among these major components of the forebrain (Groenewegen *et al.* 1990; McDonald 1991; O'Donnell 1999). In addition, the OFC receives direct projections from the piriform cortex as well as olfactory-related regions of MD in both species (Yarita *et al.* 1980; Cinelli *et al.* 1985; Takagi 1986; Price *et al.* 1991; Barbas 1993; Carmichael *et al.* 1994).

## 8.3. The limbic-prefrontal gateway

Connectivity often offers important clues to the critical function of a brain region. Just as ventral striatum is critically positioned to serve as an interface between limbic and motor systems (Mogenson *et al.* 1980), the OFC is uniquely located to serve as the gateway between these limbic areas, which are concerned with passively encoding associations between cues and likely outcomes or consequences, and the active, representational memory systems of the prefrontal cortex. As we have just outlined, the OFC receives robust input from sensory cortices and associative information about predicted outcomes from the amygdala. These inputs allow the OFC to capture the significance of environmental stimuli. The OFC also has reciprocal connections to other areas within prefrontal cortex thought to be involved in the regulation of behavior through representational memory, and it sends output to ventral striatum. The OFC is therefore well positioned to integrate information regarding planned actions, context and internal state with associative input concerning likely outcomes or consequences to generate expectations regarding the value of future events.

This role is evident in the firing activity of OFC neurons, which appears to reflect the value of predicted outcomes, events or consequences. Encoding of predicted outcomes is most obviously present during sampling of cues that are fully predictive of reward or punishment. This was first appreciated in the primate OFC. Although activity of some neurons in the OFC was first noted to reflect the sensory properties of olfactory cues (Tanabe et al. 1975; Onoda et al. 1984), it was quickly apparent that these selective firing patterns could be heavily influenced by the associative significance of the cues. In one of the first reports on neural activity in this area in awake behaving animals, Thorpe et al. (1983) showed that activity was strongly dependent on the meaning of the stimulus, rather than its physical characteristics. For example, some neurons in the OFC responded to the sight of a syringe when it was used to deliver saline to the monkey, but failed to respond when glucose was made available in the same syringe. This was true even though the syringe was unchanged in appearance. Moreover, it was only after the monkey had discovered the new contents of the syringe that the firing patterns changed.

Subsequent findings in rats have confirmed that cue-selective firing in the OFC reflects predicted outcomes. For example, in rats trained to perform an 8-odor discrimination task, in which 4 odors were associated with reward and 4 odors were associated with non-reward, we found that OFC neurons were more strongly influenced by the associative significance of the odor cues than by the actual odor identities (Schoenbaum and Eichenbaum 1995a). This was evident in the firing properties of odor responsive neurons, of which 77% discriminated between differently-valenced odors but only 38% discriminated between similarly-valenced, rewarded odors (Fig. 8.2). The predominant influence of valence was also clear in the encoding properties of the ensembles, which often failed to identify the precise odor that had been presented on a trial but rarely ($<$1% with $>$100 cells) failed to identify the valence associated with that odor (Fig. 8.3b). Put another way, an analysis of the average bits of unique information provided by each neuron indicated that an ensemble of 5364 neurons was necessary to correctly encode odor identity on 90% of the trials, whereas only 40 neurons were required to encode odor valence at that level (Schoenbaum and Eichenbaum 1995b). Indeed in comparison with encoding in the piriform cortex in the same rats, ensembles in the OFC had slightly less information about the identity of the odor cues, but they were far better at decoding

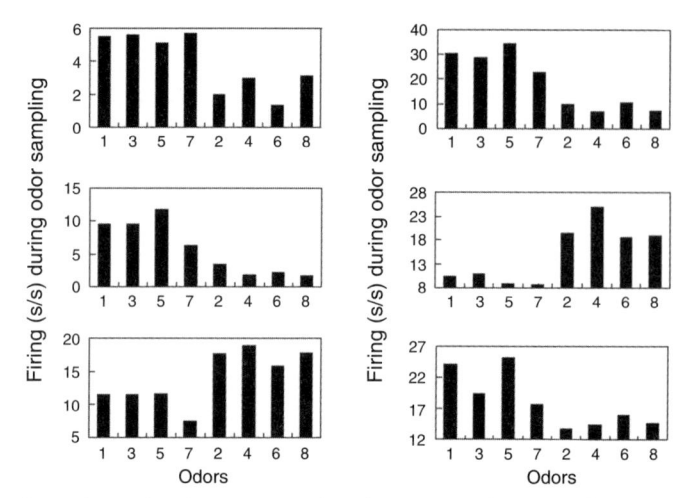

**Fig. 8.2** Encoding of associated outcome rather than sensory features of cues in OFC revealed by neural activity during odor sampling. Neurons in rat OFC were recorded during performance of an 8-odor go, no-go odor discrimination task. Activity is shown in spikes/second during odor sampling for 6 different neurons recorded in OFC. Each panel shows activity for a different neuron to each of the 8 different odors presented in each session. Rewarded 'positive' odors 1+, 3+, 5+, and 7+ are on the left of each panel, and non-rewarded 'negative' odors 2-, 4-, 6-, and 8- are on the right of each panel. Note that most neurons typically differentiated between the odors based on valence.

the valence of the odor that was presented on each trial. Rolls and colleagues have reported similar results from monkeys trained to discriminate 8 different reward-associated odors from two odors associated with delivery of an aversive fluid, showing that in monkeys as in rats, the identity of the odor cues is less important than their associative significance (Critchley and Rolls 1996b).[1]

The remarkably promiscuous nature of encoding in the OFC was confirmed by Ramus and Eichenbaum (2000), who recorded from OFC neurons in a paradigm in which the

---

[1] Some primate studies report substantial olfactory sensory encoding in the OFC. However these studies typically fail to manipulate the odor cues' associative significance. When this is done, the representations of odor cues in the primate OFC turn out to be substantially based on the associative meaning of the odor cue and are not directly driven by the odors' sensory qualities. For example Rolls and colleagues (Critchley and Rolls 1996a) have reported that 7/9 neurons responsive to food odors change their firing in response to satiety. Similarly when odor-selective neurons were recorded in a reversal task, these researchers found that 19/28 odor-selective neurons changed firing selectivity during reversal (Rolls et al. 1996). These results would seem to suggest that in monkeys, as in rats, the identity of the odor cues is less important than their associative significance in driving neural activity in the OFC. Of course primates differ from rats and thus it may be that the primate OFC maintains distinct sensory representations of cues—separate from associative significance—that we have not identified in rats (see Chapter 6).

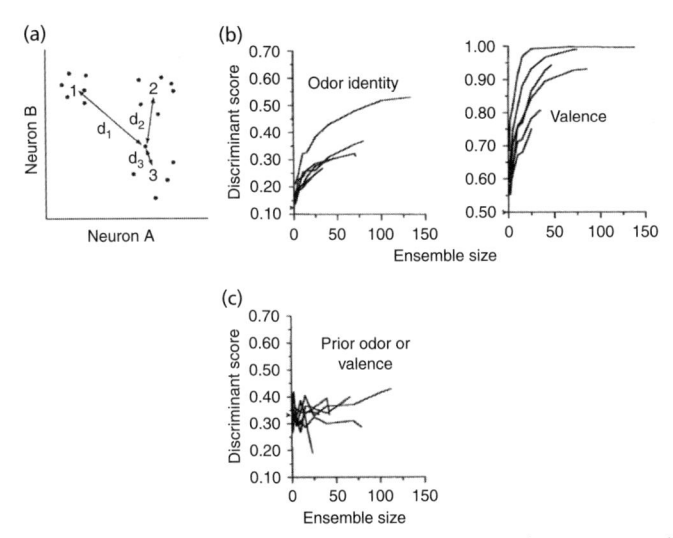

**Fig. 8.3** Ensemble encoding in OFC during odor sampling in an 8 odor go, no-go discrimination task. A modified linear discriminant analysis was performed to determine how well activity in populations of odor responsive orbitofrontal neurons in each rat could identify attributes of the odor cue on each trial and of the sequence of odor cues on preceding or subsequent trials. (a) Illustration of ensemble analysis. This analysis creates a space using the activity of each neuron as one of $n$ dimensions. The population response on each trial is plotted in this space, and then the average population vector (1, 2, 3 or positive vs. negative, etc.) for each item in a discrimination (odor, valence, predictedness, etc.) is determined. The population vector for the response on each trial is then classified as belonging to the nearest cluster in this space by calculating the distance to each average vector ($d_1$, $d_2$, $d_3$). The discriminant score is calculated by comparing the classification of each trial to the actual identity. (b) Discriminant score for ensembles of orbitofrontal neurons during discrimination performance. Populations composed of odor responsive neurons performed better at identifying the valence than the identity of the odor cues. (c) The same neurons performed poorly at identifying the odor presented on the preceding trial or on whether the prior trial had been rewarded or not.

identity was made irrelevant. They trained rats on an 8-odor continuous delayed-non-match-to-sample task. In this task, the rat is rewarded not according to the identity of the odor cue, as in a discrimination task, but rather based on whether the odor presented on the trial differs from the odor presented on the prior trial. As a result, the same odor will be rewarded on one trial, when it is different from the odor just presented ('non-match'), and non-rewarded on the next trial, when it is presented a second time ('match'). Under these conditions, the relevant construct associated with reward is not odor identity but rather the 'match' or 'non-match' attribute of the cue. They found that 64% of the odor responsive neurons discriminated the match/non-match attribute of the odor cues, while only 16% fired selectively to one of the particular odors. Further analysis of this

16% revealed that about half of these neurons (21 of 43) also fired differentially to their preferred odor, based on whether the odor was similar to (match) or different from (non-match) the previously sampled odor, while only 13 neurons were selective strictly based on the identity of the odor cue. Thus under these conditions, OFC neurons show essentially no cue selectivity at all and instead encode the retrospective comparison to the odor on the preceding trial. Interestingly such retrospective encoding, comparing the valence or identity of the current odor to that on the preceding trial, was not present in the OFC during discrimination performance when such a comparison is meaningless (Fig. 8.3c) (Schoenbaum and Eichenbaum 1995b).

Neural correlates such as these, and further reports of switches in cue-selectivity by neurons in the OFC during reversal learning (Thorpe *et al.* 1983; Rolls *et al.* 1996; Schoenbaum *et al.* 1999), led to the suggestion that networks in the OFC might function as a sort of association look-up table (Rolls 1996, 1997), implying that this prefrontal area is a repository in which the amygdala or other brain areas can store current associations between particular cues and outcomes. Although this proposition describes the cue-selective activity and the way that this activity appears to track outcomes, it fails to reflect the complex connectivity of this area or the neural activity observed in other trial periods not directly linked with the reward. Most importantly it does not capture the complex behavioral effects of damage to this area on behavior.

Instead, as we will outline, we would suggest that the neuronal activity observed in the OFC represents an expectancy of the value of the likely outcome. Thus the selective firing of these neurons does not simply reflect the fact that a particular cue has been reliably associated with a particular outcome *in the past*, but instead reflects the judgement of the animal given current circumstances that acting on that associative information will lead to that outcome *in the future*. Moreover, this judgement is represented as the value of that particular outcome relative to internal goals or desires, and these expectancies are updated constantly, so that the firing in the OFC reflects in essence the expected value of the subsequent state that will be generated given a particular response, whether that state is a primary reinforcer or simply a step towards that ultimate goal.

The resultant expectancies differ from simple associative encoding in two important ways. First they provide an internalized model of future reality that can be used to guide behavior, which does not require external cues for its maintenance, and second they provide an expectation of likely outcomes that can be compared to actual outcomes to facilitate learning in other brain regions. As we shall see, these attributes are reflected in the characteristics of neural activity in the OFC and in the critical role of the OFC in behavior.

## 8.4. Neuronal activity in OFC reflects expectations for likely outcomes

Evidence for the existence of outcome expectancies is readily apparent in animal behavior. For example, in 1928, Tinklepaugh illustrated that monkeys exhibit certain expectations when promised their favorite piece of fruit (Tinklepaugh 1928). In that study, monkeys performed a delayed-response task for different kinds of food reward. On some trials, the monkey was shown his favorite food, a banana, in one of two cups placed in front of him. During the delay period in which the monkey had to retain the spatial information regarding where the food was located, the preferred reward was secretly switched for a less preferred reward, such as a piece of lettuce. Upon discovering the lettuce in place of the banana, the monkey typically became angry, sometimes shrieking at the experimenter. Notably, the monkey did not become angry if it was given the lettuce so long as he had been shown the lettuce during the cue period. He became angry only when the reward was different than the one that was expected. This experiment shows that within the context of a given trial, the monkey was expecting the delivery of the specific reward, in this case a banana, besides remembering where it was presented and the behavioral response required to obtain it.

It is clear from this example that monkeys use cues (e.g., the place) and the current context (e.g., the task) to generate expectancies about what is likely to happen next and in particular about the value of the expected event. Other examples show that rats too generate such expectancies, and this is evident when expectations are violated in either direction. For example, rats that receive a large or more preferred reward unexpectedly show surprise and increased attention (Holland and Gallagher 1993). Simply encoding the associations between cues and rewards cannot provide the necessary information to generate these responses; rather they require the active maintenance in *representational* memory of the expected outcome—and of the value of that expected outcome relative to internal goals.

A growing number of studies suggest that the neural correlate of such an expectation is present and perhaps generated in the OFC. This is evident in human neuroimaging studies in which the OFC and associated brain regions show increased blood flow in anticipation of different outcomes (Breiter *et al.* 2001; O'Doherty *et al.* 2002). These studies demonstrate the involvement of the OFC in expectancy-related signaling and indicate that this signal is distributed to influence a network of brain regions. Moreover Mesulam and colleagues have reported that BOLD signal increases significantly in the OFC when expectancies for reward are violated as in the Tinklepaugh study described above (Nobre *et al.* 1999). Single unit recording studies in both rats and primates have extended these reports by showing that neural activity during a period of reward expectation reflects different outcome expectations and that this selective activity appears to encode the value of the expected outcome (Schoenbaum and Eichenbaum 1995a; Schoenbaum *et al.* 1998; Tremblay and Schultz 1999; Hikosaka and Watanabe 2000; Schoenbaum, Setlow, Saddoris *et al.* 2003; Wallis and Miller 2003; Hikosaka and Watanabe 2004; Roesch and Olson 2004b). For example, during odor discrimination performance, neurons in the rat OFC fire after responding but before reward delivery. In an early study, these activations differed between rewarded and non-rewarded responses; this differential firing was observed leading up to the response, in effect anticipating reward delivery (Schoenbaum and Eichenbaum 1995a).

Subsequent work, using a discrimination task that incorporated an explicit delay between responding and reward delivery (Fig. 8.4), demonstrated a number of critical

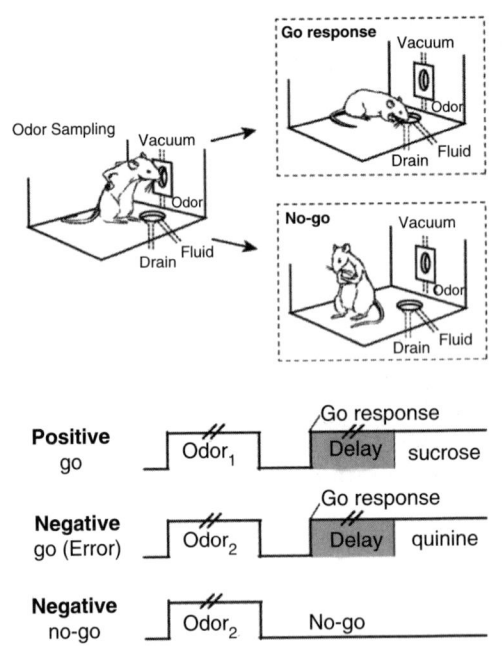

**Fig. 8.4** Schematic illustrating the odor discrimination task. The top panel illustrates the training apparatus and how it is used; the bottom panel illustrates the three types of trials that occur, depending on the rat's behavior and the relative timing of different events on those trials. On each trial the rat samples an odor cue from an odor port located in the wall of the chamber and then has 3 s to decide whether to respond to a nearby fluid well or make a 'no-go' response. A go response to a positive odor results in delivery of an appetitive sucrose solution; a go response to a negative odor results in delivery of an aversive quinine solution. Note the shaded regions prior to delivery of reinforcers. This area denotes a delay between a fluid well response and delivery of the reinforcer. During this delay, the rat must remain in the well, but there are no external cues to indicate the outcome of the response. As illustrated by the example in Fig. 8.5, this delay period allows us to isolate neural activity that maintains a representation of the expected outcome from responses evoked by the predictive odor cues.

properties of selective firing during this period of the trial (Schoenbaum *et al.* 1998, 2003a, 2004). First neural activity was maintained during this delay period. Second activity during the delay did not simply reflect prolonged firing to the odor cue, but rather, in rats that had learned to discrimination, this activity typically differed for different value outcomes. Third, the selective response was acquired, emerging early during learning at the same time that other subtle performance changes were observed related to the animal's speed of responding, and often reversing to track the outcome if the animals were presented with a reversal of the discrimination. In other words, this selective activity appeared to represent the expectation that the animal had, based on learning in the session, for the likely outcome of the response. Importantly this 'outcome-expectant' activity occurred in the absence of any external signaling cues and thus, by definition, reflected a representational memory trace. Many of these characteristics are evident in the example in Fig. 8.5.

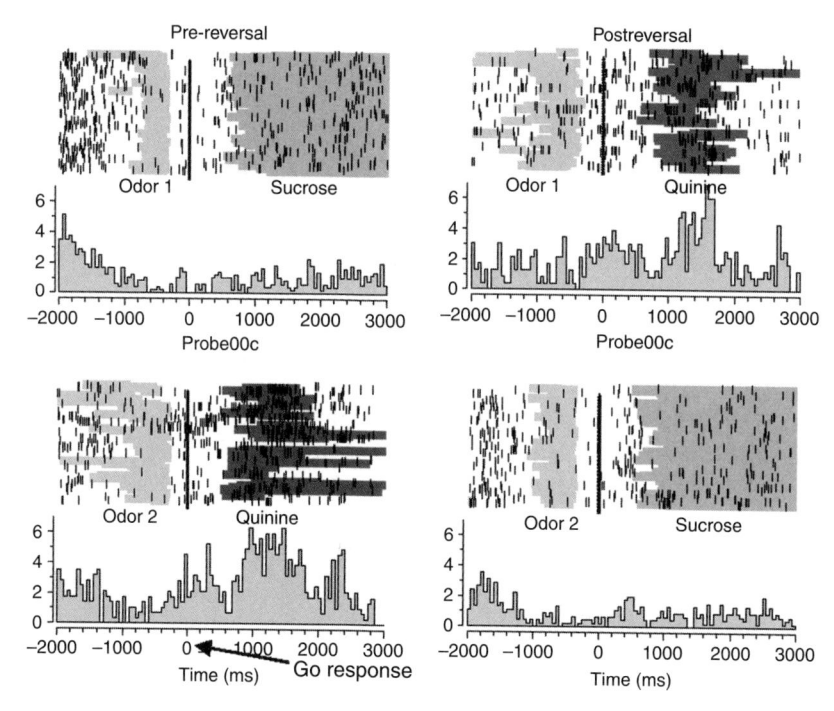

**Fig. 8.5** (Also see Color Plate 18.) Outcome-expectant activity in the OFC during acquisition (pre-reversal) and reversal (post-reversal) of a novel 2-odor discrimination problem. Activity from a single neuron is shown synchronized to the go-response at the fluid well on each trial. Activity is displayed in raster format at the top and as a peri-event time histogram at the bottom of each panel. No-go trials, which typically occurred after learning cannot be shown. Grey shading indicates odor sampling, green indicates sucrose presentation, and red quinine presentation. Note that this neuron develops a response after responding across a brief delay as the rat awaits the outcome in the well. This response is evident before quinine is delivered. Moreover it is acquired as the rat learns the discrimination and rapidly tracks the quinine after reversal.

Similar activations have been observed in primates performing instrumental arm or eye movements for reward, and in extremely simple experiments where free liquid reward is delivered at regular intervals outside any task parameters (Tremblay and Schultz 1999; Hikosaka and Watanabe 2000; Tremblay and Schultz 2000a,b; Wallis and Miller 2003; Roesch and Olson 2004b). As in rats, outcome-expectant activity in primates is directly related to the anticipated occurrence of reward and not other task factors. For example, this activity is un-correlated with spatial location or behavioral responses, and it is observed in the absence of other events that may signal the end of the trial. In addition, the time course of outcome-expectant activity is shortened or prolonged when reward is delivered earlier or later than expected, respectively. Finally, like behavioral expectations that only emerge through experience, outcome-expectant activity observed in the primate OFC develops during learning. In other words, selectivity only emerges after the monkey learns to anticipate which conditions result in reward delivery. In fact, when learning to discriminate rewarded from unrewarded conditions, monkeys initially expect to receive reward on all trials. This optimistic expectation is reflected in outcome-expectant OFC activity, which is often observed on every trial in OFC neurons early in learning (Tremblay and Schultz 2000b).

Such explicit outcome-expectant firing patterns in the OFC have also been linked closely to the value of the predicted reward and in particular to the value of a reward relative to the internal state of the animal. This was evident in a study done by Schultz and colleagues in monkeys (Tremblay and Schultz 1999); see Chapter 7, this volume. These authors demonstrated that monkeys had different relative preferences between different rewards (cabbage, raisins, grapes) and that these preferences were reflected in their choices between visual cues paired with the different rewards. They reported that OFC neurons fired selectively after responding in anticipation of the different rewards and that this selective activity was influenced by the *relative* preference that the monkey had shown previously for the two rewards in a given block of trials. In other words, activity in anticipation of a particular reward differed according to whether it was the preferred or non-preferred reward in the block. Hikosaka and Watanabe have reported similar results (Hikosaka and Watanabe 2000), as have Wallis and Miller (Wallis and Miller 2003). In each case, activity in the OFC reflected judgements regarding the *relative* value of the expected outcome. Such judgements require simple associative information to be integrated with information regarding internal state, an operation which presumably requires representational memory.

Of course the OFC is not the only brain region in which such correlates have been demonstrated. Many brain areas exhibit anticipatory firing activity (Watanabe 1996; Schoenbaum *et al.* 1998; Tremblay *et al.* 1998; Cromwell and Schultz 2003; Roesch and Olson 2003; Setlow *et al.* 2003; Wallis and Miller 2003; Roesch and Olson 2004b). However, it is likely that not all anticipatory activity reflects the value judgement we are assigning to this signal in the OFC. For example, Roesch and Olson (2003) have observed outcome-expectant activity in premotor areas, where it reflects the motivational modulation of motor signals. By contrast, outcome-expectant activity in the OFC is more closely related to value judgements in the same task (Roesch and Olson 2004b).

When value-related anticipatory activity has been identified elsewhere, such as in the ABL (Schoenbaum *et al.* 1998), the ventral striatum (Setlow *et al.* 2003) or other prefrontal regions (Wallis and Miller 2003), there is evidence that it appears in these areas only after it appears in the OFC. For example, in a re-analysis of data on outcome-expectant firing activity in the OFC and downstream regions of the amygdala and striatum published in several reports (Schoenbaum *et al.* 2003a; Setlow *et al.* 2003; Saddoris *et al.* 2004a), we examined when such firing appeared in OFC versus other areas (Schoenbaum *et al.* 2004). Comparing firing activity in anticipation of a neuron's preferred outcome with activity on trials of the oppositely valenced outcome, using a z-test to find the first trial on which firing differed, we found that outcome-expectant activity appeared first in the OFC and later in the ABL and ventral striatum, particularly for the aversive outcome that had to be avoided in the task (Fig. 8.6). Outcome-expectant neurons in the OFC became selective more quickly, exhibiting a significant increase in activity on trial 20 in the OFC compared to trial 25 for outcome-expectant neurons in the downstream limbic areas. Examination of the distribution of the first significant trial in the populations revealed that while the OFC neurons had a unimodal distribution, with a peak fewer than 10 trials, neurons in the limbic regions exhibited a bimodal distribution with a peak at 11–20 trials and another at 40–50 trials during learning. In addition outcome-expectant neurons in the OFC were significantly more likely to track the outcome across reversal than outcome-expectant neurons in the ABL and ventral striatum, which were more likely to become non-selective after reversal. This might occur if the outcome-expectant signal were generated first in the OFC and then transferred to these associated brain regions, where additional information might make the signal richer or more specific. Consistent with this suggestion, we have recently found that outcome-expectant firing in ABL is dependent on input from the OFC (Saddoris, et al. 2005). These findings will be discussed in more detail later in the chapter.

Interestingly, these results are precisely the opposite of what we have reported from these data during the cue-sampling period. During cue-sampling, we have found that selective activity appears first in ABL during learning and is *more* likely to reverse there than in the OFC (Schoenbaum *et al.* 1999, 2003a; Saddoris *et al.* 2005). Moreover cue-selective firing in the OFC is dependent on input from ABL, as we will discuss later in the chapter (Schoenbaum *et al.* 2003a). Thus when external cues are available to evoke associative representations, ABL takes the lead in driving activation in the network,

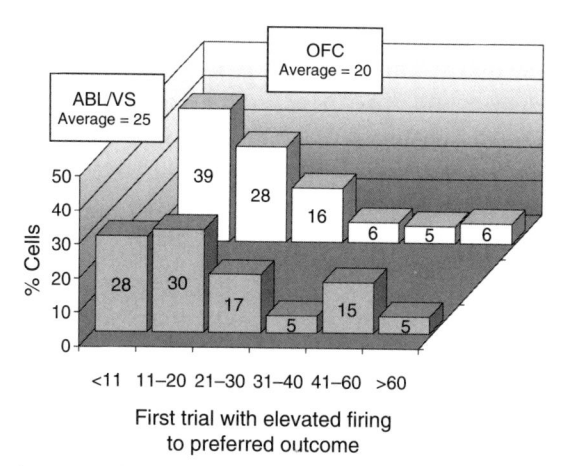

**Fig. 8.6** Rate of development of outcome-expectant activity in OFC compared to basolateral amygdala (ABL) and ventral striatum (VS). For this analysis, firing in anticipation of quinine on each negative trial was compared with the average firing in anticipation of sucrose in the surrounding 20 trial block. A z-test was applied to determine whether the value on the trial of interest fell outside of the normal distribution of values from the selection of oppositely valenced trials. The results indicated that outcome-expectant firing, at least on negative trials, emerged more rapidly in the OFC than in these downstream limbic areas.

whereas in the absence of external cues, as during the delay period prior to outcome delivery in our task, signals from the OFC predominate. This difference reflects an asymmetrical flow of information between the OFC and these limbic areas during these two trial periods, which are distinguished primarily by their representational memory demands (see Fig. 8.4).

Indeed if the OFC is generating outcome-expectancies during these explicit delay periods, then the significance of neural activity in other trial periods within the OFC must be re-evaluated. For example, activity that has been interpreted as reflecting cue significance or associated outcome may instead reflect the expectancy for reward that is generated by the associative history of a particular odor cue, given the current internal state, context and plan of action. Consistent with this proposal, cue-selective firing in the OFC has been shown to be strongly dependent on these non-associative variables.

Evidence comes from studies we and others have done looking at how neural activity in the OFC develops during discrimination learning and in subsequent reversals. These studies show clearly that selective firing to predictive cues in the OFC is strongly dependent on subsequent behavior. For example, Rolls et al. have reported that when primates acquire discrimination reversals, neurons will switch their odor preferences and that this activity reversal occurs in parallel with a shift in discrimination performance (Rolls 1996). Similarly in their continuous delayed-non-match-to-sample task, Ramus and Eichenbaum (2000) report that the proportion of the OFC neurons that were selective for the match/non-match comparison during cue sampling increased as performance increased. Importantly, this parallel is observed even when there is clear evidence of associative learning preceding accurate choice performance, confirming that encoding in the OFC can be dissociated from simple associative learning. For example, we have found that as rats acquire new 2-odor discrimination problems, in which one odor predicts sucrose and the other odor predicts quinine, the vast majority of cue-selective the OFC neurons develop selective responses to the odor cues only during accurate choice performance even though we observe significant changes in response latency very early in training (Schoenbaum et al. 1999; Schoenbaum et al. 2003a). Thus cue-selective firing in the OFC clearly does not simply reflect the cue-outcome associations but rather exhibits a strong dependence on the use of that information in controlling purposeful, directed behavior.

Moreover the same studies demonstrate that the particular context is often critical to cue-selective firing in OFC neurons. For example, we have found that although many cue-selective OFC neurons switch their cue preference during reversal learning, a much larger proportion becomes non-selective (Schoenbaum et al. 1999; Schoenbaum et al. 2003a). This is also the case in primate reversal studies (Thorpe et al. 1983; Rolls et al. 1996). Meanwhile during reversal learning, many previously non-selective OFC neurons become selective to the cues as the rats acquire the new cue-outcome associations (Schoenbaum et al. 1999; Schoenbaum et al. 2003a). This pattern of encoding is somewhat specific to the OFC, as we will discuss below, and suggests that the particular context—pre or post-reversal—plays a role in the selective response of these neurons as a population.

Analogous features were evident in well-trained rats performing the same discrimination day after day. In this study, Alvarez and Eichenbaum (2002) reported that there were ongoing changes in the population of neurons that were selective for the cues across days, even though the cue-outcome associations had not changed. These data show that networks in the OFC exhibit some specificity for the particular circumstances under which associative information about cues is recalled.

In addition, cue-selective firing in the OFC is particularly sensitive to internal state or the relative value that a particular outcome has according to the particular goals of the animal. This was initially suggested by reports that firing in the OFC to the odor or sight of food objects was affected by satiety (Critchley and Rolls 1996a). Subsequently Tremblay and Schultz (1999) have shown that, as was true for explicit outcome-expectant firing in the OFC, selective firing to cues that predict different rewards is also influenced by the relative preference of a monkey between the rewards that are available. Thus a particular neuron might fire more or less for a cue paired with a particular reward, depending upon whether it was preferred more or less than competing rewards. Similarly Roesch and Olson have recently demonstrated that cue-selective firing in the OFC tracks a number of metrics of outcome value. For example, neurons fired differently for a reward depending on its size, the time required to obtain it and the aversive consequences associated with alternative behavior (Roesch and Olson 2004a, 2005).

Finally, expectations not explicitly cued, but inherent within the experimental design, are also represented in the firing rates of OFC neurons. In both rats and primates, OFC neurons fire to multiple events within a behavioral trial in a variety of tasks (Schoenbaum and Eichenbaum 1995a; Lipton *et al.* 1999; Ramus and Eichenbaum 2000; Yonemori *et al.* 2000; Alvarez and Eichenbaum 2002). For example, we have described a variety of task-related activations defined by certain events during performance of an eight-odor discrimination task (Schoenbaum and Eichenbaum 1995a). Here we have focused thus far on activity to the odor cues and after responding. However the OFC neurons are also responsive in other trial periods, such as light onset, approach to the odor port, reward consumption or the start of the inter-trial interval. Although these task events are not designed to be discriminative cues, they do have predictive value because they reflect a sequence of behaviors that can be reliably expected to lead to reward. This relationship to reward may explain why the OFC neurons are activated by these events. Two further features of firing to these events are consistent with this hypothesis. First, these neurons typically respond not only to the particular events (light onset, odor poke) but also exhibit a ramping response in *anticipation* of these events (Fig. 8.7). Second, a post hoc analysis indicates that although the task design did not intentionally endow these events with any predictive value for reward, there were incidental relationships within the odor sequence that caused some of these events to be more or less predictive. For example, the probability for reward on the current trial changed depending on how many positive or negative trials had occurred in a row. This information was reflected in the animals' performance and in the firing activity of these neurons. For example, rats initiated trials more rapidly (or more slowly) when the trial sequence indicated that a positive (or negative) odor

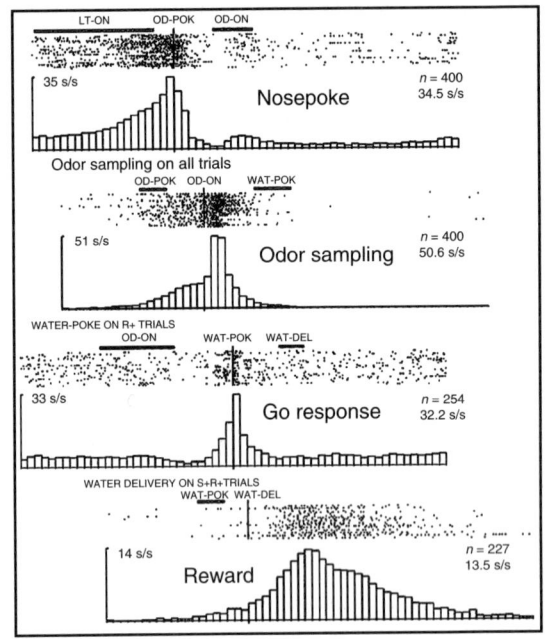

**Fig. 8.7** Neural activity in the OFC in anticipation of trial events. Neurons in rat OFC were recorded during performance of an 8-odor go, no-go odor discrimination task. Each panel shows activity is shown in four different orbitofrontal neurons, synchronized to four different task events. Activity is displayed in raster format at the top and as a peri-event time histogram at the bottom of each panel, and labels over each figure indicate the synchronizing event and any events that occurred before or after (light onset, odor poke, odor onset, water poke, water delivery). Note that these four neurons each fired in association with a different event, and the firing in each neuron increased in anticipation of that event.

would be predicted on the upcoming trial, and the neurons that fired in anticipation of light onset or nose-poke often fired differently (more or less) on such trials.

Such anticipatory or expectant firing has been observed in other studies of OFC neurons. For example Lipton *et al.* (1999) cued rats to the delivery of reward based on the presence or absence of a highly recognizable odor. These odors were always presented in one of four locations within a square testing apparatus. Thus, at the beginning of a trial, when rats were cued to respond at specific spatial location, they could reliably expect delivery of a particular odor, which in turn would predict reward. In this context, OFC neurons that responded selectively to odors showed increased firing not only when these odors were presented but also during a period immediately preceding odor delivery.

These neurophysiological data suggest that a general feature of neural activity in the OFC is to generate expectancies that provide an internalized or inner representation of impending events. These expectancies are strongly influenced but not synonymous with associative encoding about the value of those impending events. They differ in that they allow such associative information, presumably from limbic regions, access to representational memory, where that information can be integrated with the current context, the particular desires or goals at hand, and with alternate plans of action to meet those goals. These properties are reflected in the ability of OFC neurons to maintain information about expected outcomes in the absence of external cues. They are also evident in the influence of outcome preference and behavior—particularly purposeful behavior, directed at obtaining or avoiding particular outcomes—on the firing of OFC neurons. Importantly, this function is not restricted to cues that are explicitly paired with rewards but rather reflects an ongoing evaluation of the predictive value of the current state in light of current goals. As a result, we observe expectant activity in the OFC in response to other events that contain predictive value regarding the goal of the animal's actions. As we will see in the next section, the resultant neural signal plays a critical role in governing behavior and facilitating new learning.

## 8.5. Behavioral deficits after OFC lesions reflect the loss of outcome-expectant encoding

This notion that the OFC guides behavior through expectancies is consistent with the effects of OFC damage on behavior in humans, monkeys and rats. In humans, OFC damage results in inappropriate and maladaptive behavior (Harlow 1868). This deficit is often evident in social situations or settings in which it is necessary to inhibit otherwise appropriate, natural, 'prepotent', or otherwise stimulus-driven response tendencies. Although such deficits have been summarized as reflecting a critical role in response inhibition, the inability to inhibit responding in these particular situations may reflect a critical role for the OFC when the appropriate response cannot be selected using simple associations. This is often the case in social situations, where a myriad of contextual cues determine which information is relevant at any particular time, and in formal learning tasks in which the likely outcome has changed or is ambiguous. These settings require representational memory in which to manipulate associative information concerning predicted outcomes, so that it can be integrated with contextual cues and possible responses to determine the proper course of action. The critical contribution of the OFC would be to generate the expectancies that distill this information into a signal that can be utilized by other brain regions to guide responding and facilitate new learning.

This characterization is consistent with the findings in formal testing situations. For example, OFC damage has become closely associated with deficits in reversal learning across species. Humans, monkeys and rats with damage to the OFC are unable to rapidly learn reversals of previously acquired discriminations (Teitelbaum 1964; McEnaney and Butter 1969; Jones and Mishkin 1972; Rolls *et al.* 1994; Dias *et al.* 1997; Meunier *et al.* 1997; Ferry *et al.* 2000; Schoenbaum *et al.* 2002; Bohn *et al.* 2003; Brown and McAlonan 2003; Chudasama and Robbins 2003; Fellows and Farah 2003; Schoenbaum *et al.* 2003b; Izquierdo *et al.* 2004). This hallmark deficit has been cited to show that the OFC is critical for response inhibition. However in each of these reports it is notable that the same subjects showed little or no deficit in acquiring the original discrimination, often inhibiting the same response that they were later unable to inhibit in the reversal setting.

For example, in a go/no-go setting, OFC-lesioned rats learn to inhibit the prepotent go response normally during acquisition of the discriminations but then are impaired at inhibiting the same go response later after reversal (Schoenbaum *et al.* 2003b). In other words, when the relationships between the cues, responses and predicted outcomes were certain and unambiguous, so that they could be represented as simple associations, the OFC was not critical for appropriate behavior. The OFC was only required in order to appreciate and respond appropriately to new and conflicting stimulus-outcome associations. As we will discuss later, we believe this result reflects the important role that outcome-expectant signals from the OFC play in facilitating the acquisition of new associative information.

The OFC also appears to be important in learning and using information about outcomes that are distant, uncertain or probabilistic. For example, humans with damage to the OFC are unable to appropriately guide behavior based on the consequences of their actions in the Iowa gambling task (Bechara *et al.* 1997). In this task subjects must choose from decks of cards associated with rewards and penalties of different sizes. Bad decks have large rewards on each trial but occasionally also result in large penalties. Good decks have low rewards but no penalties. Normal subjects initially choose from the bad, high reward decks but subsequently bias their choices toward the good, low reward decks. Like normal subjects, patients with OFC damage initially respond more to the high-reward decks, indicating that they can learn to direct behavior to reflect different size rewards; however they fail to modify their responding to reflect the occasional large penalties that become apparent as the task proceeds. One explanation for this deficit is that they fail to appreciate or utilize the probabilistic penalties to guide their behavior. This deficit may reflect an inability to bring associative information concerning outcome value into representational memory to be integrated across time. The resulting behavior appears impulsive, stimulus-driven and inappropriate, but the underlying deficit is not one of response inhibition but rather is in generating expectancies in representational memory that reflect the value of expected outcomes.

Work in rats has confirmed the importance of the OFC in appreciating delayed or probabilistic rewards (Otto and Eichenbaum 1992; DeCoteau *et al.* 1997; Mobini *et al.* 2002; Kheramin *et al.* 2003; Winstanley *et al.* 2004). Some of these studies have given the animals an explicit choice between a small, immediate reward and a larger, more delayed or more probabilistic reward. Normal rats exhibit a 'breakpoint' at which they switch their preference from the larger reward to the smaller reward as the delay increases or the probability of getting the larger reward declines. Like patients in the aforementioned gambling task, rats with OFC lesions respond normally when given a simple choice between a large and a small reward but exhibit an abnormal responding when the delivery of the large reward is delayed or probabilistic. Thus, the OFC is not required to discriminate between different size rewards, which may reflect simple stimulus-driven responding based on associative information, but it is essential when it is necessary to integrate that same information across time, which requires the ability to hold associative information about predicted outcomes in representational memory where evaluations of the value of delayed or probabilistic outcomes can occur.

An even more direct demonstration of this role comes from reinforcer devaluation tasks. Unlike all of the above tasks, these tasks provide an assessment of the control of behavior by the value of an outcome that is not present during the critical probe test. As a result appropriate responding can only be based on an internalized representation of the value of the expected outcome, which requires the animal to generate precisely the type of expectancy that we have described in the OFC. For example, in a Pavlovian version of this procedure, illustrated in Fig. 8.8, rats are first trained to associate a light cue with food. After conditioned responses are established to the light, the food is then devalued by pairing it with illness induced by injection of lithium chloride to form a conditioned taste aversion. Normal animals subsequently decrease their conditioned responses to the light cue in an extinction probe session conducted after devaluation. The spontaneous decrease in conditioned responding after devaluation reflects the ability to learn about

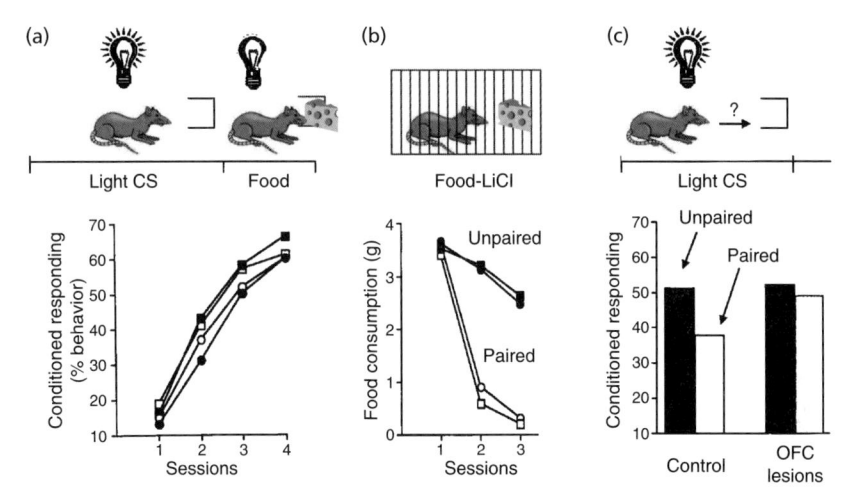

**Fig. 8.8** Effects of neurotoxic lesions of the OFC on performance in a reinforcer devaluation task. As illustrated (a, upper panel), control rats and rats with bilateral neurotoxic lesions of the OFC were trained to associate a conditioned stimulus (light CS) with an unconditioned stimulus (food US). Over several sessions (1–4), both lesioned (circles) and control (squares) rats developed a conditioned response at the food cup to the light (a, lower panel). This food cup response is represented as the percentage of total behavior. There was no effect of the lesion on the development of the food cup response. The rats then received presentations of the food item in their home cages followed by illness induced by lithium chloride (LiCl) injection (b, upper panel). Some rats in each group received paired presentations of food and illness (dark circles and squares), while others received unpaired presentations (light circles and squares). Rats that received paired presentations stopped consuming the food item (b, lower panel). Again no effect of lesion was observed. The following day the rats were returned to the training environment, and conditioned responses to the light cue were measured (c, upper panel). When exposed to the light CS (c, lower panel), control rats that had received paired presentations of food and illness reduced conditioned responses to the food cup relative to unpaired controls. Rats with orbitofrontal lesions did not show this decrease in conditioned responding as a result of reinforcer devaluation.

and use the light-food association, but it also requires an ability to link that original associative learning to newly acquired information about the current value of the outcome. This manipulation occurs without new learning, since the light cue is presented under extinction conditions, and thus requires representational memory (e.g., light-devalued food).

Rats with lesions to the OFC fail to show an effect of devaluation on conditioned responding in this paradigm (Gallagher *et al.* 1999) (Fig. 8.8). Unlike normal rats, these rats continue to respond at high rates in the extinction probe test despite the devaluation treatment. In other words, they respond to the light cue to obtain the food, even though they will not consume the food. This effect is observed despite normal acquisition of the conditioned response and of the conditioned taste aversion and normal extinction curves during the critical probe test. Thus the deficit does not reflect a general inability to inhibit conditioned responding. In addition, as we will discuss below, this effect is observed even when lesions are made after associative learning, indicating that the OFC is not solely involved in acquiring the cue-outcome association (Pickens *et al.* 2003). Rather the OFC has a critical role in the probe test, when it is necessary to control conditioned responding according to internal representations of the new value of the expected outcome.

Similar findings have been obtained in monkeys using an instrumental version of the reinforcer devaluation task (Izquierdo *et al.* 2004). In this task, monkeys were trained to discriminate between different visual objects to obtain different food rewards. Once the monkeys had learned the discriminations, they were then fed to satiety on one of the food rewards, thereby devaluing that food item. Monkeys were then given choices between visual objects associated with the devalued versus a non-devalued reward. Normal monkeys biased their choices away from objects associated with the devalued reward. The OFC-lesioned monkeys failed to show this biasing effect of devaluation on choice performance. Like the rat devaluation task, this task requires the monkeys to integrate the original learning with the newly acquired value of the outcomes to guide responding, and again the OFC was critical to this ability.

Of course, as Tinklepaugh's monkeys show (1928), the ability to generate representations of the value of expected outcomes contributes not only to performance but is also reflected in reactions to violations of that expectation. In addition, the ability to contrast expectations with actual outcomes is critical for associative learning (Rescorla and Wagner 1972). This role may account for the effect of OFC lesions on reversal learning, in which critical learning signals may depend upon the ability to recognise the difference between the expected reward and the actual outcome at the start of reversal learning. Another more explicit illustration comes from a recent paper that examined the role of the OFC in generating overt reactions when two outcomes must be compared (Camille *et al.* 2004). This study examined the impact of OFC damage on the emotional reactions of patients in a gambling task in which they were presented with a choice between two stimuli that predicted punishment or reward at varying levels of probability. In one situation, the subjects made the choice on each trial and received feedback only regarding the outcome that they had chosen. In the other situation, the subjects made their choice and then received feedback both about the outcome they chose and about the other un-selected outcome. After the feedback, the subjects rated their affect from sad to happy. The affective ratings from normal controls varied with the amount of reward that they received, but their rating was also affected when they were made aware of the amount of the unselected reward or punishment. For example, a reward made them happier and a loss made them less sad when they knew that they had avoided a large penalty. This effect of feedback about the un-selected outcome facilitated performance in the task by normal subjects. By contrast, although patients with OFC damage showed normal reactions to the outcomes they received, they failed to show any effect of knowledge about the un-selected outcome on their affective ratings or their performance. These deficits may reflect the patients' inability to hold associative information in representational memory to compare the relative value of the selected and un-selected outcomes.

## 8.6. Commonalities between orbitofrontal and amygdalar contributions

Thus far we have reviewed evidence supporting a critical role for the OFC in allowing associative information to access the representational memory systems of the prefrontal cortex. We have argued that the resultant expectancies provide a neural signal integrating associative information concerning simple cue-outcome associations with information about context, internal state, and future plans. This signal, which predicts the value of likely events and consequences, can be used by other brain regions to guide behavior and facilitate new learning. In support of this idea, we have discussed findings from recording studies, which show that neural activity in the OFC, in rats and primates, represents information about expected outcomes. This encoding is evident during explicit delays after responding but before outcome delivery, and the same signal is also present in other trial periods, including during the sampling of predictive cues and even during trial initiation periods. In each case, the neural activity typically precedes the event and reflects any information that event provides for the value of the subsequent state of the task.

We have also discussed behavioral findings consistent with our position. These reports show that the OFC is particularly critical when associative information concerning the value of outcomes must be used to guide behavior. In each case, this critical role is only evident when normal behavior requires associative information about outcomes to be maintained in representational memory. Thus, the OFC is not necessary for instrumental learning, Pavlovian conditioning, or even discriminative responding as long as the appropriate response can be determined from an unambiguous stimulus-response or even stimulus-outcome association. However the OFC is critical in each of these settings when that associative information, particularly regarding outcomes, must be manipulated or integrated with other information in order to generate the appropriate response. We believe that these operations require the representational memory systems of the prefrontal cortex to generate outcome-expectant signals observed in the OFC.

Yet much of the evidence we have described for the role of the OFC in anticipating future events and consequences can also be found in studies of amygdalar function, and in particular of the ABL. This is perhaps not surprising, given the strong reciprocal connections between the OFC and the ABL and the role that the ABL is proposed to play in associative learning. However it is worth considering how our proposition differentiates between the critical contributions of the OFC and ABL. To do this, we will first consider evidence that the two areas share a common role in encoding and using associative information about predicted outcomes. Then we will consider a number of recent findings that dissociate the roles of the two areas in how that information is applied to guide responding. We believe that the dissociations that have been demonstrated support a critical role for the ABL in acquiring and modifying cue-outcome associations, while the OFC is involved in allowing that information to be utilized in representational memory to either strengthen or inhibit more "stimulus-driven" responses mediated by the ABL and facilitate new learning when the actual outcome differs from expectations.

At first glance, there appears to be substantial apparent overlap in the functions of ABL and the OFC. For example, single units in the ABL, like those in the OFC, fire to cues based on their predictive value or association with outcome. This has been frequently noted in studies of fear conditioning in which a conditioned stimulus, such as a tone, is paired with delivery of an aversive outcome, such as footshock. A number of studies (Quirk *et al.* 1995; Quirk *et al.* 1997; Maren 2000) have reported that cells in the amygdala, particularly in the lateral nucleus within the basolateral complex, develop selective responses to the conditioned stimulus during fear conditioning. Similar correlates have also been reported to presentation of predictive cues in appetitive conditioning paradigms and also in instrumental situations (Fuster and Uyeda 1971; Cain and Bindra 1972; Sanghera *et al.* 1979; Nishijo *et al.* 1988; Muramoto *et al.* 1993; Schoenbaum *et al.* 1999; Toyomitsu *et al.* 2002; Belova *et al.* 2004; Saddoris, Gallagher *et al.* 2004). As in the OFC, cue-selective activity in the amygdala changes during reversal and extinction learning (Nishijo *et al.* 1988; Schoenbaum *et al.* 1999; Toyomitsu *et al.* 2002; Saddoris, Gallagher *et al.* 2004). For example, in two separate studies of encoding during odor discrimination learning, we have found that the majority of neurons in ABL switch their *acquired* odor preference as the rats are learning the reversed discrimination (Schoenbaum *et al.* 1999; Saddoris *et al.* 2005). Thus ABL neurons, like those in the OFC encode information about the predicted outcome during cue sampling.

Similarly a number of behavioral studies have found comparable deficits after OFC or amygdalar, and particularly ABL, lesions. As we have noted, OFC lesions cause deficits in using information about outcomes to guide behavior. This function is particularly evident in tasks that involve overt changes in the outcome value, such as reversal or devaluation, or in tasks that require the manipulation of the outcome information across delays. The amygdala, by virtue of its connections and long-standing involvement in associative learning, should be important for many of these operations. Consistent with this, amygdala lesions have been found to affect behavior after outcome devaluation, during second-order conditioning and conditioned reinforcement, and during discrimination tasks involving reversals, differing reward magnitudes or delayed/probabilistic rewards (Jones and Mishkin 1972; Kesner and Williams 1995; Hatfield *et al.* 1996; Malkova *et al.* 1997; Bechara *et al.* 1999; Parkinson *et al.* 2001; Cousens and Otto 2003; Winstanley *et al.* 2004). These parallels have been noted in rats and in primates. For example, Hatfield *et al.* (1996) found that ABL-lesioned rats were unable to spontaneously alter responding to a Pavlovian cue after devaluation of the predicted outcome. This effect is identical to that we have demonstrated following neurotoxic OFC lesions. Murray and colleagues have shown that the amygdala plays a similar role in monkeys, using the same selective-satiation task on which they tested the OFC-lesioned monkeys (Malkova *et al.* 1997). Again in this task, the monkeys learn a series of visual discriminations in which different visual cues are associated with different foods. The monkeys are then satiated on one food item and given choices between objects that lead to delivery of the satiated food and objects that lead to delivery of a non-satiated food. Like OFC-lesioned monkeys, monkeys with lesions of the amygdala fail to bias their choices toward objects that result in delivery of the non-satiated foods. Notably crossed lesions, in which the OFC is damaged in one hemisphere and the ABL in the other, produce a similar deficit, proving conclusively that serial processing in these areas is required (Baxter *et al.* 2000).

This parallel extends to a number of tasks in which the ability to use information about the predicted outcome is perhaps less explicit. For example, Hatfield *et al.* (1996) also found that the ABL was critical to the ability of first-order conditioned stimuli to support second-order conditioned responding. Thus rats with bilateral neurotoxic lesions of ABL developed conditioned responses normally to a light paired with food, but failed to transfer that conditioned responding to a tone that predicted presentation of the light cue. These data show that ABL lesions impair the ability to endow the first-order cue with motivational value. Although we are not aware of any attempts to assess a directly analogous function in primate amygdala or OFC, several studies by Roberts and colleagues have shown changes in the ability of a Pavlovian cue to serve as a conditioned reinforcer after lesions of these two regions in primates (Parkinson *et al.* 2001; Pears *et al.* 2003). In this conceptually analogous task, the monkey was first trained to associate an auditory tone with delivery of fluid reward. The tone stimulus then served as conditioned reinforcer to support subsequent instrumental responding. Primates with lesions to either the OFC or amygdala were impaired at maintaining normal responding to the tone, suggesting that these areas play a critical role in the control of behavior based on the motivational value of the conditioned reinforcer.

Furthermore both the ABL and OFC appear to be important for the facilitation of discriminative instrumental responding that is observed when each response leads consistently to different outcomes (Trapold 1970). This effect is evident when rats are trained to discriminate between two responses to receive reward based on predictive cues, which indicate the appropriate response. If the two responses, and thus the two cues, are paired with different rewards, rats will acquire the discrimination faster than if the same outcomes are delivered for each response. This effect is thought to reflect the use of outcome expectancies to promote differentiation of the two responses. Blundell *et al.* (2001)

showed that rats with lesions to the ABL failed to show enhanced learning in this situation. When each response led to the same rewards, lesioned rats learned the discrimination at the same rate as intact controls, but when each response led to a different reward, lesioned rats failed to exhibit the facilitated acquisition observed in controls. Holland and colleagues have recently duplicated this finding in rats with the OFC lesions (McDannald *et al.* 2005).

## 8.7. A model for differentiating between orbitofrontal and amygdalar contributions

These findings indicate that there is substantial overlap in the critical function of the ABL and OFC. Both areas appear to be necessary for normal task performance when that performance is critically dependent on acquiring and using associations between cues and outcomes. At the same time, however, an increasing number of recent studies have begun to note important differences between the functions of the OFC and ABL. These few reports—both behavioral and neurophysiological—suggest a distinction between these two areas in encoding such associative information and then applying it to direct behavior.

This distinction is particularly evident in single unit studies. These studies have demonstrated that, unlike activity in the OFC, cue-selective firing activity in the ABL occurs very quickly during learning, and can be separated from some behaviors. This is true in fear conditioning where selective firing to tone cues develops in the first few trials (Quirk *et al.* 1995, 1997). Similarly in our odor discrimination task, selective firing to the odor cues emerges after only 10–15 trials (Schoenbaum *et al.* 1999; Saddoris *et al.* 2004). This period of learning is characterized by very poor performance on the discrimination. Thus ABL neurons become cue-selective independent of the use of the cue significance to guide choice performance. This pattern of rapid encoding of cue-outcome associations is also evident when associations between cues and outcomes change. For example, we have found that the majority of odor-selective ABL neurons switch odor preference rapidly during reversal learning (Schoenbaum *et al.* 1999; Saddoris *et al.* 2004). Although there has been little examination of whether the amygdala neurons in primates exhibit rapid changes in cue-selectivity under these circumstances, the majority of the data available indicates that they in fact do. Notwithstanding one study published nearly a quarter century ago, concerning the responses of 9 cue-selective amygdala neurons that failed to reverse (Sanghera *et al.* 1979), there are now several reports showing that primate amygdala neurons do exhibit rapid changes in cue-selectivity in response to changes in cue-outcome associations. For example, Ono and colleagues have found that primate amygdala neurons that fire selectively to the sight of preferred food items will change firing immediately when those food items are made unappetizing by the addition of salt (Nishijo *et al.* 1988). More recently, a report using modern neurophysiological and behavioral techniques, has confirmed that primate amygdala neurons will reverse cue-preference during reversal learning in much the same proportions we have reported in rats (Patton et al. 2006).

This characteristic of rapid encoding of simple associative information stands in marked contrast to the neurophysiological characteristics we have outlined above that appear to define the OFC. In brief, although neurons in both regions acquire cue-selective firing during learning, cue-selectivity in the ABL emerges rapidly, whereas cue-selectivity in the OFC only emerges more slowly. Cue-selective firing in the ABL is independent of choice behavior, whereas selective firing in the OFC only appears during accurate choice performance. Finally while most neurons reverse cue-selectivity during reversal learning, only a quarter of cue-selective OFC neurons reverse; the vast majority instead become non-selective for the new odor-outcome contingencies. Thus OFC encoding appears to reflect conjunctions of items (particular odor-outcome pairings) or context (before or after reversal), and the rules for how this information is applied to direct discriminative responding, whereas activity in the ABL appears to more closely encode the actual cue-outcome associations.

Based on these differences we have recently compared the influence of the reciprocal connections between the OFC and ABL on encoding in each region in an odor discrimination and reversal learning task. These studies, summarized in Figure 8.9, revealed important asymmetries in how these two regions interact, asymmetries that are consistent with the differences we have proposed between these two areas. In one study, we lesioned ABL and recorded in the OFC as rats learned and reversed novel odor problems (Schoenbaum et al. 2003a). We found that input from the ABL was critical for the normal representation of outcome-related information in the OFC. In rats without ABL, the

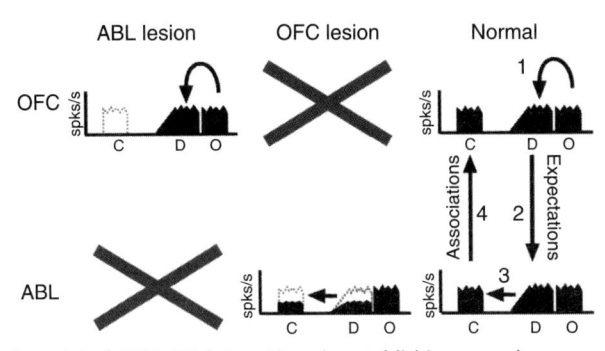

**Fig. 8.9** Simplified model of OFC-ABL interactions in establishing neural representations observed during learning in the odor discrimination task shown in Figure 8.4. Panels on left show correlates present in OFC or ABL after learning in rats with lesions the other region. ABL lesions have no effect on outcome-expectant activity in OFC but abolish the activation of these cells during cue-sampling. By contrast, OFC lesions disrupt outcome-expectant activity in ABL but neurons that do show this pattern still become cue-selective. Rightmost panel shows representations in an intact rat after learning; arrows/numbers indicate the directionality and order of formation during learning based on interpretation of the lesion/recording experiments. C, cue; D, delay before outcome; O, outcome.

proportion of cue-selective neurons in the OFC was significantly reduced, and the cue-selective neurons that did emerge were less likely to acquire a new cue preference during learning or across reversal as the rat learned new information about predicted outcomes. Instead the neurons tended to simply encode the identity of the odor cues. Furthermore, ABL lesions abolished the direct representation of the outcome that was observed in intact rats during cue-sampling. Specifically-lesioned rats failed to activate outcome-encoding neurons during cue-sampling even during accurate performance. This lack of outcome-related information in the OFC was associated with abnormal changes in response latency in the ABL-lesioned rats. Such changes in response latency reflect judgements regarding the value of expected outcomes (Crespi 1942; Holland and Straub 1979; Sage and Knowlton 2000). The effect of ABL lesions on response latencies in this task has since been confirmed in a separate behavioral report (Schoenbaum et al. 2003b). These behavioral data are shown in Fig. 8.10A; notably damage to either the OFC or ABL will produce these abnormal changes in response latencies, consistent with the proposal that they depend on information generated in ABL during learning and then utilized by the OFC.

However, much of the task-related firing activity normally present in the OFC remained unaffected. Neurons in the OFC in ABL-lesioned rats continued to fire to other task events, such as light onset, responding, and reward consumption in much the same manner and proportion as in intact rats. Thus input from the ABL was critical for the representation of information about the predicted outcome in the OFC, but even without this input, the OFC networks continued to generate expectancies for impending events although perhaps now devoid of information about the value or significance of these impending events.

The second study—which looked at the importance of reciprocal connections from the OFC to ABL—confirmed the importance of the OFC in generating these expectancies. In this study, we made lesions of the OFC and recorded in the ABL, again as rats learned and reversed novel odor discrimination problems (Saddoris et al. 2005). Since connections from the OFC to ABL are largely ipsilateral, we made unilateral lesions of the OFC to avoid changes in performance and reversal learning that would have occurred after a bilateral OFC lesion and recorded in the ABL ipsilateral to the OFC lesions. We found that without input from the OFC in that hemisphere, the outcome-expectant firing normally observed in ABL was significantly reduced. This is precisely the result one would expect if this signal reaches the ABL via input from the OFC.

In addition, in this same study we also found that the rapid cue-selective encoding during learning and reversal that characterizes the ABL in intact rats was significantly diminished; without OFC input, ABL neurons became cue-selective much more slowly. They did not default to a sensory representation of the cues, as we observed in the OFC when ABL input was removed, but rather they were delayed in the rate at which they developed cue-selective firing with learning. This effect was particularly evident during reversal learning. Slower associative encoding in the ABL as a result of OFC lesions, particularly during reversal, is consistent with the idea that learning in other structures should be facilitated by the outcome expectancies generated in the OFC. With a reduced or

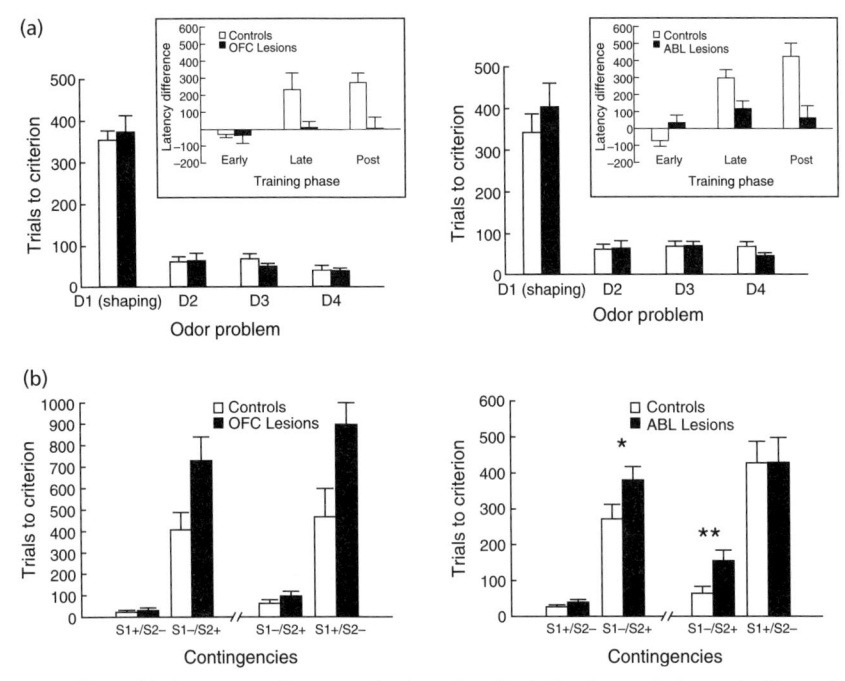

**Fig. 8.10** Effect of lesions on performance in the odor discrimination task shown in Figure 8.4. (a) Go, no-go choice performance and changes in response latency during acquisition of novel odor discrimination problems. Main figure shows average trials required to meet a 90% performance criterion for correct go, no-go performance across 4 problems. Inserts show the average difference in response latency on negative vs. positive trials (excluding no-gos) during acquisition of D2-D4 for an early trial block and a late trial block before the rats met the performance criterion and a post-criterion trial block after the rats acquired the problems. Both OFC (left panel) and ABL (right panel) lesioned rats showed normal acquisition of the discrimination problems. However lesioned rats failed to show the normal changes in response latency observed in controls during learning. (b) Go, no-go choice performance of OFC (left panel) and ABL (right panel) lesioned rats across serial reversals of an odor problem. Rats were required to demonstrate retention of the discrimination with the original contingencies (S1+/S2−) then acquire a reversal of the problem (S1−/S2+). Subsequently the rats were required to demonstrate retention of the problem with the altered or reversed contingencies (S1−/S2+) and then acquire a reversal back to the original contingencies (S1+/S2−). Rats with OFC lesions were impaired in acquiring both reversals. Rats with ABL lesions did not exhibit a reversal-specific impairment but rather were impaired in acquiring (*) and retaining the problem (**) when it was presented with the reversed contingencies (but not the original contingencies).

impoverished expectation for the outcome after responding, for example, it would be impossible to generate the error signals when expectations do not match reality. As a result, the neural network would be slower to encode the predictive significance of the cues. Interestingly, the resulting loss of rapidly flexible encoding in the ABL suggests that the reversal deficit that is the hallmark of OFC dysfunction may result from abnormally rigid encoding in the ABL after reversal. Consistent with this speculation, we have found that lesions of the OFC in this task produce robust reversal deficits (Fig. 8.10), but lesions of the ABL, which would abolish the putative neural substrate mediating this deficit, result in no impairment (Fig. 8.10b) (Schoenbaum *et al.* 2003b).

## 8.8. From neurophysiology to behavior

The model suggested by these neurophysiological studies (Fig. 8.9) provides potential explanations for deficits that have been reported after ABL and OFC lesions on a variety of tasks. As we have just described, our own data showing reversal impairments after OFC but not ABL lesions can be accounted for using this model for OFC-ABL interactions. A second very interesting example is the delayed discounting task (Mobini *et al.* 2002; Winstanley *et al.* 2004). As discussed earlier, several studies have reported abnormal 'breakpoints' in OFC-lesioned rats forced to choose between small, immediate rewards or larger, delayed rewards. As the delay increases, normal rats reach a breakpoint where they respond for the small reward rather than waiting for the larger reward. OFC-lesioned rats exhibit a shift in this breakpoint, in one study exhibiting 'impulsive' responding for the small reward at shorter delays than controls but in another study exhibiting 'perseverative' responding for the large reward at longer delays than controls. One key difference between these two reports is in when the lesions were made during training. Impulsive responding was observed when lesions were made before the rats learned the associations between the response levers and the large and small rewards. Perseverative responding was observed when lesions were made after substantial training. Our data suggests that the OFC would play two roles in this paradigm. One role would be to allow associative information from the ABL access to representational memory to be integrated across any delay. This role would be important in general performance as the delays for the large reward increased, and thus would be unaffected by when lesions were made. The second function would be to facilitate normal encoding by the ABL of the associations between the different manipulanda and the large and small rewards during the initial training, and thus would be most severely affected by lesions made before this training.

Viewed in these terms, the results of the two delayed discounting tasks reflect the differential effects of pre versus post-training OFC lesions on associative learning. Rats lesioned after initial training, as in the report by Winstanley, have normal associative representations but are simply unable to maintain them in representational memory so that the value of the large reward can be discounted. As a result, these rats would exhibit apparently 'perseverative' responding, waiting an abnormally long time for the large reward that they had learned was associated with a particular lever. By contrast, rats lesioned before any training, as in the report by Mobini, have an associative learning deficit that renders meaningless the loss of any delayed discounting function in the OFC. These rats fail to normally encode the lever-reward associations in the ABL. We would speculate that the facilitative effect of OFC input is probably particularly important for encoding information about outcomes that are separated from the predictive cues or events, as is the case for the large reward, which was often delayed in initial training. As a result, rats with pre-training lesions would exhibit apparently "impulsive" responding, selecting the small reward lever even at very short delays for the simple reason that the associative representations of value for this lever are better encoded. Notably rats that received OFC lesions before any training perform similarly to rats that receive ABL lesions after initial training; both exhibit impulsive responding for the small reward (Winstanley *et al.* 2004). This similarity is consistent with the argument that impulsive responding in both cases reflects the elimination of a basic associative learning function, supplied by the ABL but facilitated by the OFC.

This model is also consistent with results from the Iowa gambling task, mentioned earlier, which also emphasises the ability to distinguish between immediate and delayed outcomes. Patients with either OFC or ABL damage, which was of course sustained before any testing, respond impulsively in this task, maintaining an abnormal preference for the more likely outcome (Bechara *et al.* 1999). Moreover this deficit was accompanied by an inability on the part of patients in both groups to generate what the authors referred to as an 'anticipatory skin conductance response', which developed in normal subjects as they learned to expect of the large penalties when choosing from the bad deck. Interestingly although patients with OFC or ABL damage both lacked this skin conductance response in the gambling task, OFC-lesioned subjects retained the ability to exhibit conditioned changes in skin-conductance to a cue paired with a loud noise, indicating that simple associative learning was unimpaired in these patients. By contrast, patients with ABL damage failed to generate the conditioned changes in skin conductance, although they could respond to the loud noise itself. These data indicate that ABL damage disrupted associative learning mechanisms, whereas OFC damage left this simple capacity intact.

Our model is also consistent with other recent behavioral studies that have dissociated the effects of OFC and ABL lesions in several associative learning tasks. For example, recent reports show that the ABL is particularly important for learning information about outcomes and the OFC is important for applying that information to guide behavior, consistent with a role for the ABL in associative learning and for the OFC in generating expectancies to then guide behavior. Thus ABL lesions made before learning impair second-order conditioning (Hatfield *et al.* 1996), whereas lesions made after learning no longer cause impairment (Setlow *et al.* 2002). Similarly ABL lesions before (Burns *et al.* 1999) but not after conditioning (Malkova *et al.* 1997; Cousens and Otto 2003) causing deficits in the ability of the cue to serve as a conditioned reinforcer. One notable exception to this observation is a study by Parkinson *et al.* (2001), which found that post-training the amygdala lesions did disrupt conditioned reinforcement. Interestingly this study used a post-operative training schedule which included substantial ongoing exposure to the conditioned stimulus and the primary reinforcer, which may have introduced some contribution from ongoing learning. By contrast lesions of the OFC made after conditioning continue to affect the ability of cues to serve as a conditioned reinforcers irrespective of paradigm (Cousens and Otto 2003; Pears *et al.* 2003).

We have reported similar findings in the Pavlovian reinforcer devaluation paradigm illustrated in Figure 8.8. Although it was not emphasized earlier, this task is conducted in several stages. In the first stage, the rat learns the light-food association. In the second stage, the rat learns the food-illness association. In the third stage, the rat must access the light-food association, integrate it with new information regarding the value of the food outcome, and then use that information to reduce responding to the light cue during an extinction probe test. OFC or ABL lesions made prior to any learning cause marked impairments in the ability of the rats to reduce responding as a result of devaluation (Hatfield *et al.* 1996; Gallagher *et al.* 1999). However ABL lesions made after learning in the first stage no longer have any effect on behavior, whereas OFC lesions made at this stage continue to cause impairment (Pickens *et al.* 2003). Indeed, even when OFC lesions are made after the second stage, behavior continues to be impaired (Pickens *et al.* 2004). These data indicate that the ABL plays a critical role in encoding the original associations, while the OFC is critical for accessing those associations in representational memory to guide behavior in the probe test.

Of course, our model also predicts that many behaviors, which reflect associative information for outcome, may be affected by ABL but not OFC lesions. A review of the literature shows that this is indeed the case. For example, ABL lesions cause deficits in fear conditioning (LeDoux *et al.* 1990; Amorapanth *et al.* 2000; Davis 2000; Maren and Goosens 2001), but OFC lesions typically do not (Morgan and Ledoux 1999). ABL lesions also abolish CS-potentiated feeding (Petrovich *et al.* 2002; Holland and Gallagher 2003), OFC lesions do not (McDannald *et al.* 2005). In these cases, it appears other outflow tracts from the ABL (fearful behavior via central nucleus, CS-potentiated feeding via hypothalamus) mediate the use of associative information from ABL. These findings show that the OFC is not critical to associative learning per se, but only to the use of that information in particular situations. As we have tried to illustrate, situations in which the OFC makes a critical contribution typically require representational memory for that associative information in order to modulate or redirect, based on context-specific information, 'stimulus-driven' responding mediated by downstream limbic areas. As such, we would agree with Dr Goldman-Rakic in concluding that this prefrontal area, like other prefrontal regions, makes a critical contribution to behavior 'when internalized or inner models of reality are used to *govern* behavior' (Goldman-Rakic 1987).

## 8.9. Conclusions

Here we have reviewed findings from neurophysiological and behavioral studies, in rats and also in primates, in Pavlovian as well as instrumental settings regarding the critical role of the OFC and of the close connections between the OFC and limbic regions involved in associative learning. We believe these findings support the idea that the OFC functions as a gateway to allow associative information acquired by these downstream areas access to the representational memory. Such access is critical to enabling these simple associative structures to be supported in the absence of external cues, so that the information they encode about outcomes received *in the past* can be integrated with non-associative information regarding current context, internal state or goals, and behavioral strategies to determine the likelihood of receiving those outcomes *in the future*. The result of this operation is evident in neuroimaging and recording studies as outcome expectant activity that we have discussed. This 'expectancy' can then influence processing in downstream limbic areas as well as other prefrontal regions, thereby promoting voluntary, cognitive, and goal-directed (e.g., not stimulus-driven) behavior and facilitating new learning, which as we have outlined are critical functions of the the OFC.

## References

Alvarez, P. and Eichenbaum, H. (2002) Representations of odors in the rat OFC change during and after learning. *Behavioral Neuroscience* **116**:421–433.

Amorapanth, P., Ledoux, J. E., and Nader, M. A. (2000) Different lateral amygdala outputs mediate reactions and actions elicited by a fear-arousing stimulus. *Nature Neuroscience* **3**:74–79.

Barbas, H. (1993) Organisation of cortical afferent input to orbitofrontal areas in the rhesus monkey. *Neuroscience* **56**:841–864.

Baxter, M. G., and Murray, E. A. (2002) The amygdala and reward. *Nature Reviews Neuroscience* **3**:563–573.

Baxter, M. G., Parker, A., Lindner, C. C. C., Izquierdo, A. D., and Murray, E. A. (2000) Control of response selection by reinforcer value requires interaction of amygdala and OFC. *Journal of Neuroscience* **20**:4311–4319.

Bechara, A., Damasio, H., Tranel, D., Damasio, A. R. (1997) Deciding advantageously before knowing the advantageous strategy. *Science* **275**:1293–1294.

Bechara, A., Damasio, H., Damasio, A. R., Lee, G. P. (1999) Different contributions of the human amygdala and ventromedial prefrontal cortex to decision-making. *Journal of Neuroscience* **19**:5473–5481.

Blundell, P., Hall, G., and Killcross, S. (2001) Lesions of the basolateral amygdala disrupt selective aspects of reinforcer representation in rats. *Journal of Neuroscience* **21**:9018–9026.

Bohn, I., Giertler, C., and Hauber, W. (2003) NMDA receptors in the rat orbital prefrontal cortex are involved in guidance of instrumental behavior under reversal conditions. *Cerebral Cortex* **13**:968–976.

Breiter, H. C., Aharon, I., Kahneman, D., Dale, A., and Shizgal, P. (2001) Functional imaging of neural responses to expectancy and experience of monetary gains and losses. *Neuron* **30**:619–639.

Brown, S. and Schafer, E. A. (1888) An investigation into the functions of the occipital and temporal lobes of the monkey's brain. *Philosophical Transactions of the Royal Society of London B* **179**:303–327.

Brown, V. J. and McAlonan, K. (2003) Orbital prefrontal cortex mediates reversal learning and not attentional set shifting in the rat. *Behavioral Brain Research* **146**:97–130.

Burns, L. H. Everitt, B. J., and Robbins, T. W. (1999) Effects of excitotoxic lesions of the basolateral amygdala on conditioned discrimination learning with primary and secondary reinforcement. *Behavioral Brain Research* **100**:123–133.

Cain, D. P. and Bindra, D. (1972) Responses of amygdala single units to odors in the rat. *Experimental Neurology* **35**:98–110.

Cajal, R. Y. (1988) Studies on the fine structure of the regional cortex of rodents 1: Suboccipital cortex (retrosplenial cortex of Brodmann) [Trabajos del Laboratorio de Investigaciones Biologicas de la Universidad de Madrid, 20: 1–30, 1922]. In: *Cajal on the Cerebral Cortex: An Annotated Translation of the Complete Writings* (J DeFilipe, E. G. Jones, eds), pp. 524–546. New York: Oxford University Press.

Camille, N., Coricelli, G., Sallet, J., Pradat-Diehl, P., Duhamel, J.-R., and Sirigu, A. (2004) The involvement of the OFC in the experience of regret. *Science* **304**:1168–1170.

Carmichael, S. T., Clugnet, M.-C., and Price, J. L. (1994) Central olfactory connnections in the Macaque monkey. *Journal of Comparative Neurology* **346**:403–434.

Chudasama, Y. and Robbins, T. W. (2003) Dissociable contributions of the orbitofrontal and infralimbic cortex to Pavlovian autoshaping and discrimination reversal learning: further evidence for the functional heterogeneity of the rodent frontal cortex. *Journal of Neuroscience* **23**:8771–8780.

Cinelli, A. R., Moyano-Ferreyra, H., and Barragan, E. (1985) Reciprocal functional connections of the olfactory bulbs and other olfactory related areas with the prefrontal cortex. *Brain Research Bulletin* **19**:651–661.

Cousens, G. A. and Otto, T. (2003) Neural substrates of olfactory discrimination learning with auditory secondary reinforcement. I. Contributions of the basolateral amygdaloid complex and OFC. *Integrative Physiological and Behavioral Science* **38**:272–294.

Crespi, L. P. (1942) Quantitative variation of incentive and performance in the white rat. *American Journal of Psychology* **55**:467–517.

Critchley, H. D. and Rolls, E. T. (1996a) Hunger and satiety modify the responses of olfactory and visual neurons in the primate OFC. *Journal of Neurophysiology* **75**:1673–1686.

Critchley, H. D. and Rolls, E. T. (1996b) Olfactory neuronal responses in the primate orbitofrontal cortex: analysis in an olfactory discrimination task. *Journal of Neurophysiology* **75**:1659–1672.

Cromwell, H. C. and Schultz, W. (2003) Effects of expectations for different reward magnitudes on neuronal activity in primate striatum. *Journal of Neurophysiology* **89**:2823–2838.

Davis, M. (2000) The role of the amygdala in conditioned and unconditioned fear and anxiety. In: *The Amygdala: A Functional Analysis* (J. P. Aggleton, ed), pp. 213–287. Oxford: Oxford University Press.

DeCoteau, W. E. and Kesner, R. P., and Williams, J. M. (1997) Short-term memory for food reward magnitude: the role of the prefrontal cortex. *Behavioral Brain Research* **88**:239–249.

Dias, R., Robbins, T. W., and Roberts A. C. (1997) Dissociable forms of inhibitory control within prefrontal cortex with an analog of the Wisconsin card sort test: restriction to novel situations and independence from "on-line" processing. *Journal of Neuroscience* **17**:9285–9297.

Dickinson, A. (1989) Expectancy theory in animal conditioning. In: *Contemporary Learning Theories: Pavlovian Conditioning and the Status of Traditional Learning Theory* (S. B. Klein, R. R. Mowrer, eds), pp. 279–308. Hillsdale, NJ: Erlbaum.

Everitt, B. J. and Robbins, T. W. (1992) Amygdala-ventral striatal interactions and reward-related processes. In: *The Amygdala: Neurological Aspects of Emotion, Memory, and Mental Dysfunction* (J. P. Aggleton, ed), pp. 401–429. Oxford: John Wiley and Sons.

Fellows, L. K. and Farah, M. J. (2003) Ventromedial frontal cortex mediates affective shifting in humans: evidence from a reversal learning paradigm. *Brain* **126**:1830–1837.

Ferry, A. T., Lu, X. C., and Price, J. L. (2000) Effects of excitotoxic lesions in the ventral striatopallidal-thalmocortical pathway on odor reversal learning: inability to extinguish an incorrect response. *Experimental Brain Research* **131**:320–335.

Fuster, J. M. (1997) *The Prefrontal Cortex*, 3rd edition. New York: Lippin-Ravencott.

Fuster, J. M. and Uyeda, A. A. (1971) Reactivity of limbic neurons of the monkey to appetitive and aversive signals. *Electroencephalography and Clinical Neurophysiology* **30**:281–293.

Gaffan, D. and Murray, E. A. (1990) Amygdalar interaction with the mediodorsal nucleus of the thalamus and the ventromedial prefrontal cortex in stimulus-reward associative learning in the monkey. *Journal of Neuroscience* **10**:3479–3493.

Gallagher, M. (2000) The amygdala and associative learning. In: *The Amygdala: A Functional Analysis* (J. P. Aggleton, ed). Oxford: Oxford University Press.

Gallagher, M., McMahan, R. W., and Schoenbaum, G. (1999) OFC and representation of incentive value in associative learning. *Journal of Neuroscience* **19**:6610–6614.

Goldman-Rakic, P. S. (1987) Circuitry of primate prefrontal cortex and regulation of behavior by representational memory. In: *Handbook of Physiology: The Nervous System* (V. B. Mountcastle, F. Plum, S. R. Geiger, eds), pp. 373–417. Bethesda, MD: American Physiology Society.

Goldman-Rakic, P. S. and Porrino, L. J. (1985) The primate mediodorsal (MD) nucleus and its projection to the frontal lobe. *Journal of Comparative Neurology* **242**:535–560.

Groenewegen, H. J. (1988) Organization of the afferent connections of the mediodorsal thalamic nucleus in the rat, related to the mediodorsal-prefrontal topography. *Neuroscience* **24**:379–431.

Groenewegen, H. J., Vermeulen-van der Zee, E., te Kortschot, A., and Witter, M. P. (1987) Organisation of the projections from the subiculum to the ventral striatum in the rat. A study using anterograde transport of Phaseolus vulgaris leucoagglutinin. *Neuroscience* **23**:103–120.

Groenewegen, H. J., Berendse, H. W., Wolters, J. G., and Lohman, A. H. M. (1990) The anatomical relationship of the prefrontal cortex with the striatopallidal system, the thalamus and the amygdala: evidence for a parallel organisation. *Progress in Brain Research* **85**:95–118.

Haber, S. N., Kunishio, K., Mizobuchi, M., and Lynd-Balta, E. (1995) The orbital and medial prefrontal circuit through the primate basal ganglia. *Journal of Neuroscience* **15**:4851–4867.

Harlow, J. M. (1868) Recovery after passage of an iron bar through the head. *Publications of the Massachusetts Medical Society* **2**:329–346.

Hatfield, T., Han, J. S., Conley, M., Gallagher, M., and Holland, P. (1996) Neurotoxic lesions of basolateral, but not central, amygdala interfere with Pavlovian second-order conditioning and reinforcer devaluation effects. *Journal of Neuroscience* **16**:5256–5265.

Hikosaka, K. and Watanabe, M. (2000) Delay activity of orbital and lateral prefrontal neurons of the monkey varying with different rewards. *Cerebral Cortex* **10**:263–271.

Hikosaka, K. and Watanabe, M. (2004) Long- and short-range reward expectancy in the primate OFC. *European Journal of Neuroscience* **19**:1046–1054.

Holland, P. C. and Straub, J. J. (1979) Differential effects of two ways of devaluing the unconditioned stimulus after Pavlovian appetitive conditioning. *Journal of Experimental Psychology: Animal Behavior Processes* **5**:65–78.

Holland, P. C. and Gallagher, M. (1993) Amygdala central nucleus lesions disrupt increments, but not decrements, in conditioned stimulus processing. *Behavioral Neuroscience* **107**:246–253.

Holland, P. C. and Gallagher, M. (1999) Amygdala circuitry in attentional and representational processes. *Trends in Cognitive Sciences* **3**:65–73.

Holland, P. C. and Gallagher, M. (2003) Double dissociation of the effects of lesions of basolateral and central amygdala on conditioned stimulus-potentiated feeding and Pavlovian-instrumental transfer. *European Journal of Neuroscience* **17**:1680–1694.

Izquierdo, A. D., Suda, R. K., and Murray, E. A. (2004) Bilateral orbital prefrontal cortex lesions in rhesus monkeys disrupt choices guided by both reward value and reward contingency. *Journal of Neuroscience* **24**:7540–7548.

Jones, B. and Mishkin, M. (1972) Limbic lesions and the problem of stimulus-reinforcement associations. *Experimental Neurology* **36**:362–377.

Kesner, R. P. and Williams, J. M. (1995) Memory for magnitude of reinforcement: dissociation between amygdala and hippocampus. *Neurobiology of Learning and Memory* **64**:237–244.

Kheramin, S., Brody, S., Ho, M.-Y., Velazquez-Martinez, D. N., Bradshaw, C. M., Szabadi, E., Deakin, J. F. W., and Anderson, I. M. (2003) Role of the orbital prefrontal cortex in choice between delayed and uncertain reinforcers: a quantitative analysis. *Behavioral Processes* **64**:239–250.

Kievit, J. and Kuypers, H. G. J. M. (1977) Organisation of the thalamo-cortical connections to the frontal lobe in the Rhesus monkey. *Experimental Brain Research* **29**:299–322.

Kita, H. and Kitai, S. T. (1990) Amygdaloid projections to the frontal cortex and the striatum in the rat. *Journal of Comparative Neurology* **298**:40–49.

Kluver, H. and Bucy, P. C. (1939) Preliminary analysis of the temporal lobes in monkeys. *Archives of Neurology and Psychiatry* **42**:979–1000.

Kolb, B. (1984) Functions of the frontal cortex of the rat: a comparative review. *Brain Research Reviews* **8**:65–98.

Krettek, J. E. and Price, J. L. (1977a) Projections from the amygdaloid complex to the cerebral cortex and thalamus in the rat and cat. *Journal of Comparative Neurology* **172**:225–254.

Krettek, J. E. and Price, J. L. (1977b) The cortical projections of the mediodorsal nucleus and adjacent thalamic nuclei in the rat. *Journal of Comparative Neurology* **171**:157–192.

LeDoux, J. E. (1996) *The Emotional Brain*. New York: Simon and Schuster.

LeDoux, J. E., Cicchetti, P., Xagoraris, A., and Romanski, L. M. (1990) The lateral amygdaloid nucleus: sensory interface of the amygdala in fear conditioning. *Journal of Neuroscience* **10**:1062–1069.

Leonard, C. M. (1969) The prefrontal cortex of the rat. I. Cortical projections of the mediodorsal nucleus. II. Efferent connections. *Brain Research* **12**:321–343.

Lipton, P. A., Alvarez, P., Eichenbaum, H. (1999) Crossmodal associative memory representations in rodent OFC. *Neuron* **22**:349–359.

Malkova, L., Gaffan, D., and Murray, E. A. (1997) Excitotoxic lesions of the amygdala fail to produce impairment in visual learning for auditory secondary reinforcement but interfere with reinforcer devaluation effects in rhesus monkeys. *Journal of Neuroscience* **17**:6011–6020.

Maren, S. (2000) Auditory fear conditioning increases CS-elicited spike firing in lateral amygdala neurons even after extensive overtraining. *European Journal of Neuroscience* **12**:4047–4054.

Maren, S. and Goosens, K. A. (2001) Contextual and auditory fear conditioning are mediated by the lateral, basal, and central amygdaloid nuclei in rats. *Learning and Memory* **8**:148–155.

McDannald, M., Saddoris, M. P., Gallagher, M., and Holland, P. C. (2005) Lesions of OFC impair rats' differential-outcome expectancy learning but not conditioned stimulus-potentiated feeding. *Journal of Neuroscience* **25**:4626–4632.

McDonald, A. J. (1991) Organisation of the amygdaloid projections to the prefrontal cortex and associated striatum in the rat. *Neuroscience* **44**:1–14.

McEnaney, K. W. and Butter, C. M. (1969) Perseveration of responding and non-responding in monkeys with orbital frontal ablations. *Journal of Comparative Physiology and Psychology* **4**:558–561.

Meunier, M., Bachevalier, J., and Mishkin, M. (1997) Effects of orbital frontal and anterior cingulate lesions on object and spatial memory in rhesus monkeys. *Neuropsychologia* **35**:999–1015.

Mobini, S., Body, S., Ho, M.-Y., Bradshaw, C. M., Szabadi, E., Deakin, J. F. W., and Anderson, I. M. (2002) Effects of lesions of the OFC on sensitivity to delayed and probabilistic reinforcement. *Psychopharmacology* **160**:290–298.

Mogenson, G. J., Jones, D. L., and Yim, C. Y. (1980) From motivation to action: functional interface between the limbic system and the motor system. *Progress in Neurobiology* **14**:69–97.

Morgan, M. M. and Ledoux, J. E. (1999) Contribution of ventrolateral prefrontal cortex to the acquisition and extinction of conditioned fear in rats. *Neurobiology of Learning and Memory* **72**:244–251.

Muramoto, K., Ono, T., Nishijo, H., and Fukuda, M. (1993) Rat amygdaloid neuron responses during auditory discrimination. *Neuroscience* **52**:621–636.

Nishijo, H., Ono, T., and Nishino, H. (1988) Single neuron responses in alert monkey during complex sensory stimulation with affective significance. *Journal of Neuroscience* **8**:3570–3583.

Nobre, A. C., Coull, J. T., Frith, C. D., and Mesulam, M. M. (1999) OFC is activated during breaches of expectation in tasks of visual attention. *Nature Neuroscience* **2**:11–12.

O'Doherty, J., Deichmann, R., Critchley, H. D., Dolan, R. J. (2002) Neural responses during anticipation of a primary taste reward. *Neuron* **33**:815–826.

O'Donnell, P. (1999) Ensemble coding in the nucleus accumbens. *Psychobiology* **27**:187–197.

Ongur, D. and Price, J. L. (2000) The organization of networks within the orbital and medial prefrontal cortex of rats, monkeys and humans. *Cerebral Cortex* **10**:206–219.

Onoda, N., Imamura, K., Obota, F., and Iino, M. (1984) Response selectivity of neocortical neurons to specific odors in the rabbit. *Journal of Neurophysiology* **52**:638–652.

Otto, T. and Eichenbaum, H. (1992) Complementary roles of the orbital prefrontal cortex and the perirhinal-entorhinal cortices in an odor-guided delayed non-matching-to-sample task. *Behavioral Neuroscience* **106**:762–775.

Parkinson J. A., Crofts H. S., McGuigan M., Tomic D. L., Everitt B. J., Roberts A. C. (2001) The role of the primate amygdala in conditioned reinforcement. Journal of Neuroscience **21**:7770–7780.

Pears A, Parkinson J. A, Hopewell L, Everitt B. J, Roberts A. C. (2003) Lesions of the orbitofrontal but not medial prefrontal cortex disrupt conditioned reinforcement in primates. Journal of Neuroscience **23**:11189–11201.

Petrovich G. D., Setlow B, Holland P. C., and Gallagher M. (2002) Amygdalo-hypothalamic circuit allows learned cues to override satiety and promote eating. Journal of Neuroscience **22**:8748–8753.

Pickens C. L., Saddoris M. P., Gallagher M., and Holland P. C. (2004) Orbitofrontal lesions impair use of cue-outcome associations in a devaluation task. *Behavioral Neuroscience* **119**:317–322.

Pickens, C. L., Setlow, B., Saddoris, M. P., Gallagher, M., Holland, P. C., and Schoenbaum, G. (2003) Different roles for OFC and basolateral amygdala in a reinforcer devaluation task. *Journal of Neuroscience* **23**:11078–11084.

Preuss, T. M. (1995) Do rats have prefrontal cortex? The Rose-Woolsey-Akert programme reconsidered. *Journal of Comparative Neurology* **7**:1–24.

Price, J. L., Carmichael, S. T., Carnes, K. M., Clugnet, M-C., Kuroda, M., and Ray, J. P. (1991) Olfactory input to the prefrontal cortex. In: *Olfaction: A Model System for Computational Neuroscience* (J. Davis, H. Eichenbaum, eds), pp. 101–120. Cambridge MA: MIT Press.

Quirk, G. J., Repa, J. C., and LeDoux, J. E. (1995) Fear conditioning enhances short-latency auditory responses of lateral amygdala neurons: parallel recordings in the freely behaving rat. *Neuron* **15**:1029–1039.

Quirk, G. J., Armony, J. L., and LeDoux, J. E. (1997) Fear conditioning enhances different temporal components of tone-evoked spike trains in auditory cortex and lateral amygdala. *Neuron* **19**:613–624.

Ramus, S. J. and Eichenbaum, H. (2000) Neural correlates of olfactory recognition memory in the rat OFC. *Journal of Neuroscience* **20**:8199–8208.

Ray, J. P. and Price, J. L. (1992) The organisation of the thalamocortical connections of the mediodorsal thalamic nucleus in the rat, related to the ventral forebrain—prefrontal cortex topography. *Journal of Comparative Neurology* **323**:167–197.

Rescorla, R. A. and Wagner, A. R. (1972) A theory of Pavlovian conditioning: variations in the effectiveness of reinforcement and non-reinforcement. In: *Classical Conditioning II: Current Research and Theory* (Black, A. H. and Prokasy, W. F., eds), pp. 64–99. New York: Appleton-Century-Crofts.

Roesch, M. R. and Olson, C. R. (2003) Impact of expected reward on neuronal activity in prefrontal cortex, frontal and supplementary eye fields and premotor cortex. *Journal of Neurophysiology* **90**:1766–1789.

Roesch, M. R. and Olson, C. R. (2004a) Neuronal activity related to reward value and motivation in primate frontal cortex. *Science* **304**:307–310.

Roesch, M. R. and Olson, C. R. (2004b) Neuronal activity related to reward value and motivation in primate frontal cortex. *Science* **304**:307–310.

Roesch, M. R. and Olson, C. R. (2005) Neuronal activity in primate orbitofrontal cortex reflects the value of time. *Journal of Neurophysiology* **94**:2457–2471.

Rolls, E. T. (1996) The OFC. *Philosophical Transactions of the Royal Society of London B* **351**:1433–1443.

Rolls, E. T. (1997) Taste and olfactory processing in the brain. *Critical Reviews in Neurobiology* **11**:263–287.

Rolls, E. T., Hornak, J., Wade, D., and McGrath, J. (1994) Emotion-related learning in patients with social and emotional changes associated with frontal lobe damage. *Journal of Neurology, Neurosurgery, and Psychiatry* **57**:1518–1524.

Rolls, E. T., Critchley, H. D., Mason, R., and Wakeman, E. A. (1996) OFC neurons: role in olfactory and visual association learning. *Journal of Neurophysiology* **75**:1970–1981.

Rose, J. E. and Woolsey, C. N. (1948) The OFC and its connections with the mediodorsal nucleus in rabbit, sheep, and cat. *Res Pub Ass Nerv Ment Dis* **27**:210–232.

Russchen, F. T., Amaral, D. G., and Price, J. L. (1987) The afferent input to the magnocellular division of the mediodorsal thalamic nucleus in the monkey, Macaca fascicularis. *Journal of Comparative Neurology* **256**:175–210.

Saddoris, M. P., Gallagher, M., Schoenbaum, G. (2005) Rapid associative encoding in basolateral amygdala depends on connections with OFC. *Neuron* **46**:321–331.

Sage, J. R. and Knowlton, B. J. (2000) Effects of US devaluation on win-stay and win-shift radial maze performance in rats. *Behavioral Neuroscience* **114**:295–306.

Sanghera, G. Rolls, E. T., and Roper-Hall, A. (1979) Visual responses of neurons in the dorsolateral amygdala of the alert monkey. *Experimental Neurology* **63**:610–626.

Schoenbaum, G. and Eichenbaum, H. (1995a) Information coding in the rodent prefrontal cortex. I. Single-neuron activity in OFC compared with that in piriform cortex. *Journal of Neurophysiology* **74**:733–750.

Schoenbaum, G. and Eichenbaum, H. (1995b) Information coding in the rodent prefrontal cortex. II. Ensemble activity in OFC. *Journal of Neurophysiology* **74**:751–762.

Schoenbaum, G. and Setlow, B. (2001) Integrating OFC into prefrontal theory: common processing themes across species and subdivision. *Learning and Memory* **8**:134–147.

Schoenbaum, G., Chiba, A. A., and Gallagher, M. (1998) OFC and basolateral amygdala encode expected outcomes during learning. *Nature Neuroscience* **1**:155–159.

Schoenbaum, G., Chiba, A. A., and Gallagher, M. (1999) Neural encoding in OFC and basolateral amygdala during olfactory discrimination learning. *Journal of Neuroscience* **19**:1876–1884.

Schoenbaum, G., Nugent, S., Saddoris, M. P., and Setlow, B. (2002) Orbitofrontal lesions in rats impair reversal but not acquisition of go, no-go odor discriminations. *Neuroreport* **13**:885–890.

Schoenbaum, G., Setlow, B., Saddoris, M. P., and Gallagher, M. (2003a) Encoding predicted outcome and acquired value in OFC during cue sampling depends upon input from basolateral amygdala. *Neuron* **39**:855–867.

Schoenbaum, G., Setlow, B., Nugent, S. L., Saddoris, M. P., and Gallagher, M. (2003b) Lesions of OFC and basolateral amygdala complex disrupt acquisition of odor-guided discriminations and reversals. *Learning and Memory* **10**:129–140.

Schoenbaum, G., Saddoris, M. P., Setlow, B., and Gallagher, M. (2004) Outcome-expectant activity in rat OFC provides parallel representations of appetitive and aversive outcomes. *Society for Neuroscience Abstracts* **30**:716.716.

Setlow, B., Gallagher, M., and Holland, P. (2002) The basolateral complex of the amygdala is necessary for acquisition but not expression of CS motivational value in appetitive Pavlovian second-order conditioning. *European Journal of Neuroscience* **15**:1841–1853.

Setlow, B., Schoenbaum, G., and Gallagher, M. (2003) Neural encoding in ventral striatum during olfactory discrimination learning. *Neuron* **38**:625–636.

Shi, C. J. and Cassell, M. D. (1998) Cortical, thalamic, and amygdaloid connections of the anterior and posterior insular cortices. *Journal of Comparative Neurology* **399**:440–468.

Takagi, S. F. (1986) Studies on the olfactory nervous system of the old world monkey. *Progress in Neurobiology* **27**:195–250.

Tanabe, T., Iino, M., and Takagi, S. F. (1975) Discrimination of odours in olfactory bulb, piriform-amygdaloid areas, and OFC of the monkey. *Journal of Neurophysiology* **38**:1284–1296.

Teitelbaum, H. (1964) A comparison of effects of orbitofrontal and hippocampal lesions upon discrimination learning and reversal in the cat. *Experimental Neurology* **9**:452–462.

Thorpe, S. J., Rolls, E. T., and Maddison, S. (1983) The OFC: neuronal activity in the behaving monkey. *Experimental Brain Research* **49**:93–115.

Tinklepaugh, O. L. (1928) An experimental study of representation factors in monkeys. *Journal of Comparative Psychology* **8**:197–236.

Toyomitsu, Y., Nishijo, H., Uwano, T., Kuratsu, J., and Ono, T. (2002) Neuronal responses of the rat amygdala during extinction and reassociation learning in elementary and configural associative tasks. *European Journal of Neuroscience* **15**:753–768.

Trapold, M. A. (1970) Are expectancies based upon different positive reinforcing events discriminably different? *Learning and Motivation* **1**:129–140.

Tremblay, L. and Schultz, W. (1999) Relative reward preference in primate OFC. *Nature* **398**:704–708.

Tremblay, L. and Schultz, W. (2000a) Reward-related neuronal activity during go-no go task performance in primate OFC. *Journal of Neurophysiology* **83**:1864–1876.

Tremblay, L. and Schultz, W. (2000b) Modifications of reward expectation-related neuronal activity during learning in primate OFC. *Journal of Neurophysiology* **83**:1877–1885.

Tremblay, L., Hollerman, J. R., and Schultz, W. (1998) Modifications of reward expectation-related neuronal activity during learning in primate striatum. *Journal of Neurophysiology* **80**:964–977.

Wallis, J. D. and Miller, E. K. (2003) Neuronal activity in primate dorsolateral and orbital prefrontal cortex during performance of a reward preference task. *European Journal of Neuroscience* **18**:2069–2081.

Watanabe, M. (1996) Reward expectancy in primate prefrontal neurons. *Nature* **382**:629–632.

Weiskrantz, L. (1956) Behavioral changes associated with ablations of the amygdaloid complex in monkeys. *Journal of Comparative Physiology and Psychology* **9**:381–391.

Winstanley, C. A., Theobald, D. E. H., Cardinal, R. N., and Robbins, T. W. (2004) Contrasting roles of basolateral amygdala and OFC in impulsive choice. *Journal of Neuroscience* **24**:4718–4722.

Yarita, H., Iino, M., Tanabe, T., Kogure, S., and Takagi, S. F. (1980) A transthalamic olfactory pathway to OFC in the monkey. *Journal of Neurophysiology* **43**:69–85.

Yonemori, M., Nishijo, H., Uwano, T., Tamura, R., Furuta, I., Kawasaki, M., *et al.* (2000) Orbital cortex neuronal responses during and odor-based conditioned associative task in rats. *Neuroscience* **95**:691–703.

# Chapter 9

# A componential analysis of the functions of primate orbitofrontal cortex

Angela C. Roberts and John Parkinson

## 9.1. Introduction

Profound disturbances in emotional social and behavior and poor decision making are the hallmarks of ventromedial prefrontal dysfunction in humans (Fuster 1989; Damasio *et al.* 1998), a region of the frontal lobes that includes the medial aspects of the OFC and the ventral aspects of the medial prefrontal cortex (PFC), the area of cortex lying within the cingulate gyrus, anterior to the genu of the corpus callosum. While similar changes in social and emotional behavior have been observed in monkeys with damage to the OFC (Butter *et al.* 1970; Franzen and Myers 1973; Raleigh and Steklis 1981), the most well described behavioral characteristic of such monkeys is their inflexible responding across a variety of different contexts (Mishkin 1964; Fuster 1989). This chapter will focus on the component psychological processes shown to be dependent upon the primate OFC, which together provide the flexibility necessary for complex emotional and social interactions within sophisticated primate societies.

The orbital surface of the frontal lobes in nonhuman primates is composed of a number of cytoarchitectonically distinct regions that have differing patterns of connections both within and outside the frontal lobes. These regions include areas 10, 11, 12/47, 13, and 14, according to the maps of Petrides and Pandya (1995), and Carmichael and Price (1994) in rhesus monkeys. Based on an extensive analysis of the connections of the OFC and medial PFC, Carmichael and Price (1996) have divided the regions into two networks. The orbital network includes areas 10o, 11l and 11m, the majority of 13 (l, m, and b), 12 (m, r and l) and agranular insula, and has been proposed to function as a system for sensory integration since together these regions receive highly processed information from all the major senses, visual, auditory, somatosensory, gustatory, olfactory and visceral. The remaining regions on the orbital surface, 13a, 14r and 14c, 12o, and an intermediate section of agranular insula (Iai), have extensive connections with the medial wall of the frontal lobes, including areas 32, 25, and 24a,b,c, and together have been proposed to form a medial network functioning as a visceromotor system since its component regions, whilst having little sensory input, provide the major cortical output to visceromotor structures in the hypothalamus and brainstem. In humans, it is damage to this combined network that is associated with disturbed emotional and social behavior.

It should be noted that in subsequent discussions of OFC ablation studies in rhesus monkeys, reference is made to the medial and lateral surfaces of the OFC as topographical areas which, based on the Carmichael and Price model, involve elements of both the medial PFC visceromotor network and the OFC sensory integration network. Wherever possible, the cytoarchitectonic areas (according to the maps of Carmichael and Price 1996) that are included within an ablated area are described. With respect to excitotoxic lesion studies of the OFC in the new world monkey, the common marmoset, the lesion includes dysgranular and granular cortical regions on the orbital surface, not including the neighbouring granular cortex lying within the ventrolateral convexity, but including the cortex within the ventromedial convexity (see Dias *et al.* 1996a for a description of the cytoarchitecture). Excitotoxic lesions of the marmoset medial PFC include dysgranular and agranular regions within the medial wall, but exclude the cortex below in the ventromedial convexity and above in the overlying dorsal, granular region. Based on our own unpublished anterograde and retrograde tracing studies, the marmoset OFC, like that in the rhesus monkey, is heavily interconnected with the sensory cortical regions, while the medial PFC is poor in sensory connections, but like that in the rat and rhesus monkey is interconnected with the limbic regions including the amygdala, parahippocampus and hypothalamus.

## 9.2. Early studies on discrimination learning

The cortex lying on the orbital surface has been the most commonly studied region in monkeys to date, with far less attention given to the cortex lying ventrally within the medial wall. The importance of the primate OFC in adaptive behavior has been recognized since the publication of a series of influential experiments in the 1960's and early 1970s (Butter *et al.* 1963; Mishkin 1964; Butter 1969; Iversen and Mishkin 1970). Prior to this, many examples of 'behavioral disinhibition' following ablations of the frontal cortex had been reported, and these had been variously described as 'perseverative interference between antagonistic responses', 'a loss of act inhibition', 'a disinhibition of inhibitory reflexes' (Mishkin 1964). In all cases such descriptions attempted to characterize the nature of the repetitive, inappropriate responding that occurred following such ablations. Indeed, it had even been suggested that different forms of behavioral disinhibition existed, associated with ablations of the distinct regions of the PFC—that of the dorsolateral PFC, that included the inferior prefrontal convexity, resulting in a perseveration of response tendencies, and that of the OFC, resulting in motivational disinhibition. However, upon the use of more restricted ablations within the PFC, behavioral inhibition became the product of one particular region, area 12/47, while the medial aspects of the orbitofrontal surface were recognized as performing a different function.

Specifically, area 12/47 (including not only that part lying on the lateral orbital surface but also the dorsal sector lying on the ventrolateral surface, together known as the inferior prefrontal convexity) as well as the very lateral extreme of area 11 were shown to be involved in the suppression of a previously established cognitive or 'affective' set (Mishkin 1964). This was based on the effects of ablations of this region on a range of discrimination tests, including 'go, no-go', one-trial object discrimination and object discrimination reversal (Butter 1969; Iversen and Mishkin 1970) tests in which an animal had to inhibit a previously acquired response preference or stimulus preference. We now know, however, that the failure to suppress previously established cognitive sets can itself be fractionated; lesions of the ventrolateral PFC in marmosets, including the ventrolateral convexity, impair specifically the ability to shift attentional sets or higher-order rules, while lesions of the OFC cause a selective deficit in reversing stimulus-reward associations (Dias *et al.* 1996b). (Whether it turns out that shifting attentional sets and reversing stimulus-reward associations are dependent upon the dorsal and ventral aspects of area 12/47, respectively in both rhesus monkeys and humans, remains to be determined).

In contrast, selective ablation of the medial orbitofrontal surface in rhesus monkeys, including area 13, orbital 14, much of area 11 and the tip of orbital 10, did not disrupt 'go, no-go' performance, and induced a different pattern of impairment on a series of object reversals to that seen following ablation of the lateral orbitofrontal surface. Thus, following medial orbitofrontal ablation the deficit was present across all reversals rather than, in the case of lateral orbitofrontal ablation, being present on the first one or two reversals only. In addition, unlike the lateral orbitofrontal ablation, the medial orbitofrontal ablation did not cause persistent, repetitive responding to the previously rewarded object, but resulted instead in the majority of errors being made while learning to respond to the previously unrewarded object; a pattern of deficit recently replicated by Izquierdo *et al.* (2004). Thus it was hypothesized that the medial orbitofrontal region, consistent with its dense reciprocal connections with the amygdala, was involved in processing affective information, distinct from the suppression of inappropriate cognitive sets, a function performed by the inferior prefrontal convexity (Iversen and Mishkin 1970). However, not entirely consistent with such a functional dissociation was the finding that prolonged responding during extinction of an instrumental lever press response was associated with ablation of the posteromedial orbitofrontal surface (including the caudal sector of areas 13 and 14) but not the inferior prefrontal convexity (Butter 1969). This led Butter to propose that motivational processes may contribute to behavioral suppression, a deficit not dissimilar from the motivational disinhibition described a few years earlier.

While these findings highlight differential contribution of the regions within lateral and medial sectors of the OFC to object discrimination reversal and extinction, they leave the precise nature of their contributions unclear, not least because such tests involve the formation of many different types of association, both Pavlovian and instrumental, that may contribute to successful performance.

## 9.3. **Psychological considerations**

### Pavlovian associations between an object and the affective and sensory properties of the reward

In the tests for object discrimination reversal and instrumental extinction, that have been described so far, many types of associations may contribute to a successful performance.

Thus, in a typical object discrimination task for primates, two objects are presented in each trial, with the location of each object on the animal's right and left varying pseudo-randomly across trials. A response to one object leads to reward and a response to the other is unrewarded. Repeated pairing of an object with the reward will lead to the object becoming a conditioned stimulus (CS), such that it will elicit Pavlovian conditioned responses, when presented alone in the absence of the reward. These responses may include behavioral approach towards the CS itself (Hearst and Jenkins 1974), as well as approach responses towards the food. In an object discrimination task, the location of the CS and the food are spatially overlapping, but in contexts in which they are spatially segregated, it has been shown that the presentation of a CS can lead to approach responses to the CS, independently of the approach responses to the food source itself, even if the CS is in the opposite direction to the food reward (Bussey et al. 1997). Indeed, these behaviorally distinct responses have been shown to be under differential neural control, excitotoxic lesions of the central nucleus of the amygdala disrupting the former, but not the latter (Hatfield et al. 1996). Conditioned responses towards the CS appear to be elicited automatically as a result of the relationship of the CS with the reward, and are not under instrumental control (see Williams and Williams 1969; Mackintosh 1974). Such CS-elicited approach responses to both the food and the CS itself could conceivably underlie responding in the object discrimination task where only a simple displacement response of an object is required in order to gain access to reward. However, much of this work on Pavlovian approach responses has been performed in rats and pigeons, and whether these responses are as prominent in monkeys is less clear.

Another association that may be particularly prominent in an object discrimination task is the association between the object and the sensory properties of the reward, independent of the reward's affective value. Thus having displaced the rewarded object, the next sensory experience is the sight and touch of the food rather than its affective taste properties. Under these circumstances it may be the predictive and discriminative properties of the object with respect to the sensory and informational properties of the reward that guides responding (Medin 1977; Gaffan 1979; Williams 1994).

## Instrumental associations

In addition to the development of Pavlovian approach responses, the act of displacing the object and retrieving the food reward will also lead to instrumental conditioning. At least two different forms of learning may underlie instrumental conditioning, one of which has been proposed to be declarative in nature and to underlie goal directed actions, the other reflects a form of stimulus-response, procedural or habit learning (Dickinson and Balleine 1994 ). With respect to the latter, the retrieval of the food reward following the displacement of the object will act to strengthen the association between the object and the displacement response, but the reward per se doesn't enter into the association. This type of stimulus-induced responding is insensitive to changes in the incentive value of the reward unless new learning is allowed to take place, and is insensitive to changes in the contingency between the response and the reward. In contrast, those forms of learning that fulfill the criterion of goal directed actions include knowledge of the contingent relationship between object displacement (the response) and the outcome (reward) and knowledge about the incentive value of that outcome. Consequently, such actions are sensitive to the degradation of the contingency or the devaluation of the outcome.

Under some circumstances, the object itself, by virtue of its association with the reward may become a conditioned reinforcer, and thus the animal may select that object as a consequence of its reinforcing value. This effect may be due to the CS acting to retrieve the reinforcing value of the primary reward with which it has been paired (Holland and Rescorla 1975), and/or may be a consequence of the CS eliciting a general affective response. That the conditioned stimuli may possess such general affective properties has been demonstrated by the phenomenon of transreinforcer blocking (Dickinson and Dearing 1979). Blocking is a feature of Pavlovian conditioning in which conditioning to one CS is blocked by the presence of another CS that already predicts the *same* unconditioned stimulus (US). The finding that conditioning of a CS to foot shock is blocked by the presence of another CS associated with the absence of an appetitive event is an example of transreinforcer blocking. Since these two aversive events share no common properties other than their aversiveness, the blocking effect must depend upon an association between the CS and general affect, suggesting that affective states can be independent of a particular reinforcer. Recently we have shown that a conditioned reinforcer can indeed support new learning in rats despite the devaluation of the primary reinforcer with which it was originally paired, suggesting that at least, under certain circumstances, the underlying representation supporting responding may reflect the activation of a central positive motivational state (Parkinson *et al.* 2005).

These different Pavlovian and instrumental associations that have been described so far, and that may support object discrimination learning, may equally support learning of the lever press response used by Butter (1969)—with the lever itself acting as the object—although the relative effectiveness of these different associations in controlling responding may well have differed between the two tasks. In the instrumental lever response task, an animal did not have to select one of two responses, as in the object discrimination task. Instead, responding on a single lever led to food reward. In addition, unlike the object discrimination task in which each correct response leads to reward, by the end of lever press training, responding was maintained on a variable interval schedule in which the first response after a variable delay resulted in reward, a schedule in which the association between responding and reward is weakened, making it more likely that responding is under a greater control stimulus-response habits as opposed to goal directed actions (Dickinson and Balleine 1994).

Since any one of these processes, or a combination of them, may contribute to a successful acquisition of an object discrimination task or an instrumental lever pressing task, it is perhaps not surprising that object discrimination learning and instrumental response learning, per se, are relatively robust and not easily impaired by discrete lesions of specific neural structures. Moreover, the relative strength of the different associations and thus their relative control over responding in any one study will depend upon a number of different task parameters, such as the salience of the objects, the incentive value of the reward and the precise contingencies between objects, responses and the reward.

## Reversal and extinction

Upon reversal of the reward contingencies or upon extinction, a particular stimulus or signal acquires a second meaning (i.e. no reward) that becomes available along with the first meaning (i.e. reward) (see Rescorla 2001 and Delamater 2004 for a review of evidence that the original association remains intact). As discussed by Bouton (2004) such stimuli are ambiguous, and as such their current meaning is determined by the current context. Thus, while the associations formed during the initial learning appear relatively independent of the context in which they were presented, the new learning that takes place during the extinction (or reversal) is relatively context specific, with the type of new associations formed again dependent upon the particular task employed. There are a number of important factors that will contribute to the extinction or reversal of a response, whether it be Pavlovian or instrumental in nature (Rescorla 2001; Delamater 2004). First, a negative affective response is likely to develop due to the loss of reward, and will function to produce various associative changes, opposite to those induced by the original positive affective event, with the intensity of this negative event depending in part upon the intensity of the previously obtained reward (contrast effect). The evidence that the loss of reward acts as an aversive event is supported by the finding that animals will learn a response to escape from it, and that the neutral stimuli presented at the time of the reward loss become themselves capable of promoting escape learning (Rescorla 2001). Second, since there is no explicit sensory event associated with the loss of reward, any associations formed will be dominated by the responses generated by the animal itself in response to the reward loss, such as the negative affect described above. A particularly prominent hypothesis is that extinction may result in an inhibitory association forming specifically between the original CS and the previously rewarded response, that is, the response most likely to be emitted early in extinction or reversal, and which becomes associated with the aversive event. In support of this, it has been shown that, following extinction, although the response associated with the original stimulus-reward association is no longer expressed, the original stimulus-reward association can continue to motivate behaviors (and thus other responses) distinct from those expressed at the time of the contingency change (Rescorla 2001). In a recent fear conditioning study in rats such an inhibitory process, mediated by the medial PFC and acting at the level of the amygdala, has been proposed by Quirk and Gehlert (2003). Finally, it is conceivable that the ability to determine the contingent relationship between the reinforcer and either the object (Rescorla and Wagner 1972) or one's actions (Dickinson and Balleine 1994) provides the animal with the ability to detect changes in contingency rapidly and thus act accordingly. A rapid detection of such a change may then lead to the engagement of active inhibitory control processes that suppress the previously correct, but now, incorrect associations between stimuli and responses.

Thus, in order to understand the functions of the primate OFC, it is important that their role in these different types of processing are examined. A considerable knowledge has already been obtained with respect to the role of different nuclei within the amygdala of rats with respect to some of these processes, and more recently, the contribution of the primate OFC to these same processes has begun to be explored.

## 9.4. OFC and the affective properties of stimuli

The process by which the stimuli in the environment acquire a conditioned reinforcement value by virtue of their association with primary reinforcement, and subsequently themselves act as goals for instrumental actions has been shown to be dependent upon the amygdala in both rats and monkeys (Cador et al. 1989; Burns et al. 1993; Parkinson et al. 2001). In the laboratory, conditioned reinforcers, such as light paired with food, sex or drugs, can produce a high rate of instrumental responding over a protracted period of time in both monkeys and rats under lean schedules of primary reward, i.e., second-order schedules (Goldberg 1973; Katz 1979; Everitt et al. 1989; Arroyo et al. 1998). In addition, it can support new instrumental learning in the absence of primary reinforcement (Mackintosh 1974). Thus, conditioned reinforcers may underlie a great deal of human behavior (Williams 1994) acting as sub-goals to bridge temporal gaps between rewards and contributing to complex social phenomena, including drug dependence (Altman et al. 1996) and decision making (Damasio 1998). Recently, in a new world monkey, the common marmoset, the contribution of the OFC to this process has been examined and compared to that of the medial PFC that, like the OFC, also has dense connections with the amygdala. Thus, marmosets were trained to respond on a fixed ratio schedule of five responses in order to gain access to a fluid reward, each response being accompanied by an auditory CS, that also accompanied reward presentation. Following excitotoxic lesions of either the OFC or medial PFC, animals were required to make more and more responses before receiving a CS presentation, and only after five CS presentations did animals gain access to the primary reward (see Fig. 9.1a for experimental protocol). The extent to which the CS maintains responding on this lean schedule of primary reward can be determined by omitting the CS, while still presenting the reward itself after the appropriate number of responses. Marmosets with excitotoxic lesions of the OFC (Pears et al. 2003), similar to excitotoxic lesions of the amygdala (Parkinson et al. 2001), were insensitive to the removal of the CS in contrast to the marked decline in responding apparent in both the medial PFC-lesioned animals and sham-operated controls (see Fig. 9.1b). Moreover, these same OFC lesioned marmosets were also impaired in acquiring a new visual pattern discrimination to gain access to the CS in the absence of primary reward (Fig. 9.1c). The spared performance of marmosets with lesions of the medial PFC confirms that it is the OFC, along with the amygdala, that is important in guiding goal-directed actions based on the affective value of conditioned stimuli.

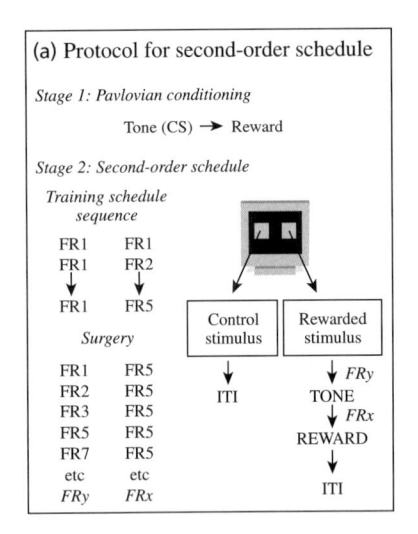

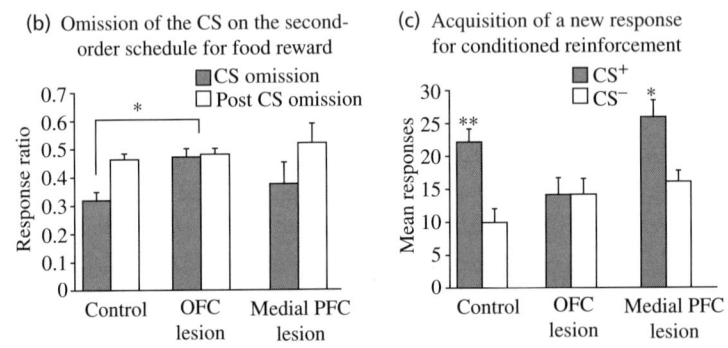

**Fig. 9.1** (a) For the second-order schedule study animals first learned that a tone was associated with banana juice reward (Pavlovian stage 1). They then learned that a response to a stimulus presented on either the left or right of a touch sensitive computer screen resulted in the presentation of the tone (CS) and access to the reward. Subsequently they had to make more and more responses to gain access to the reward (FRx;1–5) but each response resulted in the tone CS (FRy; 1, Stage 2). After surgery, 5 tone CS presentations always resulted in reward but the number of responses before receiving a tone presentation was increased every fourth session from 1 to 2, 3, 5 etc. (FRy). Animals continued to progress up the schedule until they failed to gain a single reward on two consecutive sessions. (b) Performance on the CS omission test is presented as the ratio of responses following omission of the CS (CS omission phase) relative to the two days immediately prior to CS omission (pre-CS omission phase) that is, CS omission/(CS omission + pre-CS omission). A score of 0.5 indicates that responding was equivalent to the pre-CS omission phase, while a score below 0.5 indicates a decline in responding. For comparison purposes performance on the two days following the CS-omission phase in which the CS was present (post-CS omission phase) is also shown. * indicates that the response ratio for the CS omission phase in the OFC-lesioned group was significantly greater than the control group at $p < 0.05$. (c) Mean number of responses made to a visual stimulus associated with a $CS^+$ compared to a stimulus associated with a $CS^-$ in control, OFC and medial PFC-lesioned groups. While animals from the control and medial PFC lesioned groups showed differential responding to the stimulus associated with the $CS^+$ compared to that associated with the $CS^-$, OFC lesioned animals responded equally to the two stimuli. **and * indicate that the responding for the $CS^+$ was significantly greater than that for the $CS^-$ in control and medial PFC lesioned groups at $p < 0.01$ and 0.05 respectively.

As described above, recent evidence from our laboratory supports the hypothesis that acquisition of a new response for conditioned reinforcement may depend upon the CS evoking a general positive affective state, independent of the specific reinforcer with which it was paired (Parkinson *et al*. 2005) and hence the impaired responding for conditioned reinforcers seen in marmosets with lesions of the amygdala and OFC may reflect the contribution of these structures to the process by which general affective value can guide behavior. However, the importance of these structures in the ability of the CS to evoke a representation of the incentive value of the specific primary reinforcer with which it is associated and to guide accordingly has also been shown. Thus rhesus monkeys, having learned to displace particular objects in order to gain access to reward,

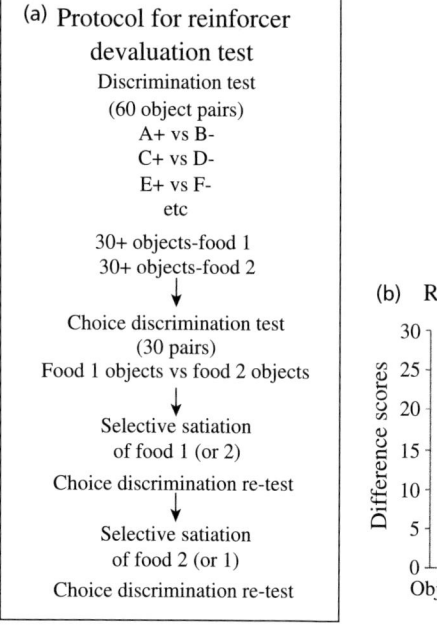

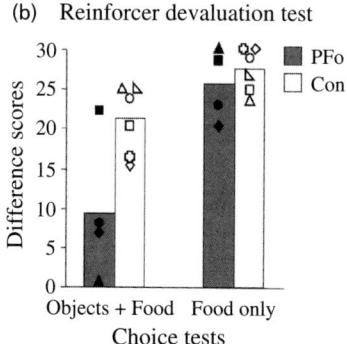

**Fig. 9.2** (a) For the reinforcer devaluation test animals were first presented with the same 60 pairs of objects on repeated sessions until they had learned which object from each pair of objects was associated with reward. Half the rewarded objects were associated with one food type and the other half with another food type. They then received a series of choice tests in which rewarded objects associated with the different food types were presented in pairs and animals could select which object and thus which food type they preferred. If animals had been sated on one or other of the food types immediately, prior to the choice test, then their subsequent choice of object revealed a preference for the food type that had not been sated, compared to baseline when animals had not been sated on either food. For comparison, animals also received a choice test between the two different foods following satiation on one or other of the foods. (b) Mean difference scores of control and OFC (PFo) lesioned groups on the two types of choice test, that is, objects overlying the food reward (Objects + Food) and food reward presented alone (Food only). The OFC lesioned group were impaired at avoiding the objects associated with the devalued food reward but were unimpaired at avoiding the devalued food itself. Redrawn with permission from Izquierdo *et al*. (2004). Copyright 2004 by the Society for Neuroscience).

subsequently are able to select between these different objects based upon the memory of the current incentive value of the particular rewards with which they are associated (see Fig. 9.2a for experimental protocol). For example, if the monkeys are satiated on one of the rewards, e.g. peanuts, then subsequently in an extinction test they will select primarily those objects associated in the past with the other reward, e.g. raisins, and avoid objects associated in the past with peanuts. In contrast, excitotoxic lesions of the amygdala (Malkova *et al.* 1997) as well as ablation of the medial region of the OFC (Fig. 9.2b; Izquierdo *et al.* 2004) impairs this ability and following satiation on one of the rewards, monkeys no longer select the object associated with the other reward. Instead they choose randomly. Indeed, Baxter *et al.* (2000) have shown that this process is dependent upon an interaction between the amygdala and OFC since unilateral lesions of each structure in opposite hemispheres, resulting in the disconnection of the two structures, also disrupts this process. This ability of the CS to gain access to the current incentive value of the reinforcer and guide behavior is not specific to instrumental responding, it also impacts on Pavlovian responding, an effect also shown to be dependent upon the amygdala (Hatfield *et al.* 1996) and OFC in rats (Gallagher *et al.* 1999). However, while the findings discussed so far highlight the importance of the OFC in the process by which stimuli in the environment take on reinforcer-specific, as well as general, motivational significance and guide responding they do not elucidate the unique function of this structure since these processes also appear dependent upon the amygdala.

## 9.5. OFC and its proposed role in goal selection

A clue as to the differential contribution of the OFC and amygdala to the motivational control of behavior came from a study in humans comparing the response of patients with ventromedial PFC damage to those with amygdala damage on the Iowa gambling task (Bechara *et al.* 1999). Both groups of patients performed poorly on the task and both groups failed to show autonomic responses in anticipation of a negative outcome. However, while PFC damaged patients did show an autonomic response upon receipt of the negative outcome, in the form of a loss of a large sum of money, amygdala damaged subjects did not. One explanation for this difference was that while the amygdala was involved generally in assessing the motivational significance of stimuli, the ventromedial PFC was involved specifically in more complex situations; for example, when stimuli were associated with both reward and punishment. However, excitotoxic lesions of the OFC in rats have been shown to disrupt the acquisition of conditioned approach to an appetitive CS using a simple Pavlovian procedure (Chudasama and Robbins 2004), an effect similar to that seen following excitotoxic lesions of the central nucleus of the amygdala, ruling out the selective involvement of the OFC solely in more complex learning situations with multiple competing outcomes.

An alternative proposal is suggested by more recent studies in rats comparing the effects of lesions of the OFC and amygdala *before* and *after* an animal has acquired knowledge about the motivational significance of environmental cues. These studies reveal that a lesion of the basolateral nucleus of the amygdala only disrupts the ability of a CS from gaining access to the current incentive value of a reinforcer and using this information to guide behavior if it is present prior to the initial learning of the stimulus-reinforcer association. No effects are seen if the amygdala is lesioned after learning has taken place but before changes are made to the incentive value of the primary reinforcer (Pickens *et al.* 2003). In contrast, lesions of the OFC disrupt performance whether lesioned before or after the initial acquisition of the stimulus-reinforcer association. These results suggest that the basolateral nucleus of the amygdala may be critical for forming representations that link cues to the incentive properties of outcome but not for maintaining such representations, updating them with new incentive information or for using them to guide behavior. In contrast, the OFC would certainly appear important for one or both of the latter processes as well as perhaps for the initial acquisition. However, such a difference between the OFC and the amygdala does not appear to be the case in all appetitive contexts. For example, excitotoxic lesions of the amygdala as well as the OFC impair the acquisition of a new response for a conditioned reinforcer and disrupt responding on a second-order schedule maintained by CS's even if the initial acquisition of the stimulus-reinforcer association has been learned prior to the lesion (Cador *et al.* 1989; Parkinson *et al.* 2001; Pears *et al.* 2003). It has been proposed that additional learning about the CS-primary reinforcer association post lesion (see task design in Fig. 9.1a) may have contributed to the performance of the controls on the second-order schedule in (Parkinson *et al.* 2001), thus disadvantaging the amygdala-lesioned animals that would not have been able to learn from this post lesion experience, thereby explaining their poorer performance. This is far less likely though in the case of (Cador *et al.* 1989) in which post lesion exposure to the CS-primary reinforcer relationship was minimal. Instead, these contrasting findings raise the possibility that the ability of a CS to reinforce a new response (Cador *et al.* 1989), as opposed to modifying existing responses (Pickens *et al.* 2003), requires the continued contribution of the amygdala.

One hypothesis that is supported by current experimental findings proposes that affective information from the amygdala may be used by the OFC to generate expectancies and select goals. Thus, a number of studies have shown that neurons in the OFC of monkeys and rats signal the value of expected outcomes based on past experience (Rolls 1998; Schoenbaum *et al.* 1998; Tremblay and Schultz 1999; Hikosaka and Watanabe 2000; Roesch and Olson 2004). In addition, a series of elegant studies by Schoenbaum and colleagues comparing neuronal activity in both the amygdala and OFC, have shown that while neurons in the rat amygdala encode the motivational significance of an olfactory cue prior to the knowledge being reflected in the behavior of the animal, the encoding of such information in the rat OFC only develops at the same time as its apparent effects on behavior (Schoenbaum *et al.* 1999); the activity in the latter being dependent upon an intact amygdala (Schoenbaum *et al.* 2003; see Roesch and Schoenbaum chapter for detailed discussion). Moreover, a proportion of these cue selective neurons are active not only during the presentation of the cue but also following the response in expectation of a particular motivationally significant outcome and, as a consequence, have been hypothesized to encode the specific incentive value of the expected outcome predicted by the cue. In contrast, other cue-selective neurons do not show such anticipatory firing with respect to the outcome and thus have been proposed to encode instead the acquired affective value of the conditioned cue itself, that is, the cues conditioned reinforcing properties (Schoenbaum *et al.* 2003). Finally neurons in the OFC can process the relative value of currently anticipated outcomes (Tremblay and Schultz 1999). Indeed, based on these findings Montague and Berns (2002) have developed a predictor-valuation model of decision making in which the OFC integrates information relating to various rewards and punishments and their predictors, and despite the variable nature of these, which may include food, money, sex, produces a common neural currency on which decisions can be made as to desirable future outcomes.

This hypothesis may help to explain the differing effects that excitotoxic lesions of the amygdala and OFC have on second-order schedules of responding for food reinforcement in marmosets (Pears *et al.* 2003). Thus, while both the OFC and the amygdala contributed to the process of conditioned reinforcement and lesions of these structures resulted in marmosets being insensitive to omission of the CS, only lesions of the amygdala disrupted overall responding on the second order schedule per se. In contrast, the responding of OFC-lesioned marmosets was equivalent to controls at each level of the schedule (compare Figs. 9.3a and b). Since responding on such schedules is under the control of both the CS and the primary reinforcer and the CS can take on general positive reinforcing properties independent of the current value of the reward, then there are at least two possible explanations for the data. The first makes the assumption that the reinforcing efficacy of the conditioned and primary reinforcers is summative, while the second appeals to a mechanism of competition and regulatory control. Thus according to the first, the OFC will enable control animals to anticipate both primary and conditioned reinforcement and the resulting level of motivation will be able to sustain quite high levels of responding. Following an amygdala lesion, however, information about the value of the conditioned reinforcer will be lost, but expectations about the value of the primary reinforcer may still be generated by the OFC, based, in part, on the conditioning of the general environment to reward via hippocampal circuitry. However, such contextual cues will be more weakly associated with reward and as the response demands increase for an amygdale-lesioned marmoset, their responding will collapse at a much earlier time point than controls. In contrast, with a lesion of the OFC, the animal cannot generate expectations about either the primary or the conditioned reinforcer and therefore this information cannot be matched against the overall response demands of the task to determine when responding should cease. Consequently, responding is more likely to be controlled by habit-like stimulus-response associations, which may well be able to sustain responding on quite high schedules particularly if the response demands are increased relatively gradually, as in the study by Pears *et al.* (2003). In contrast, according to the second explanation there may be competition between the stimuli that control responding on the schedule, including the conditioned and primary reinforcer, competition that is regulated by the

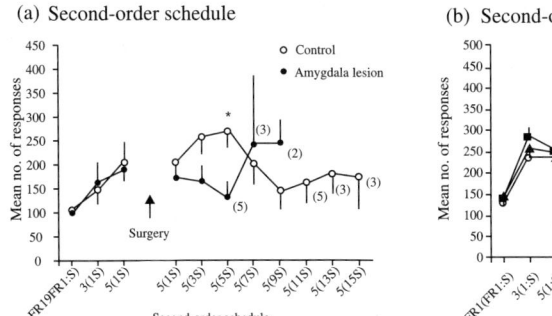

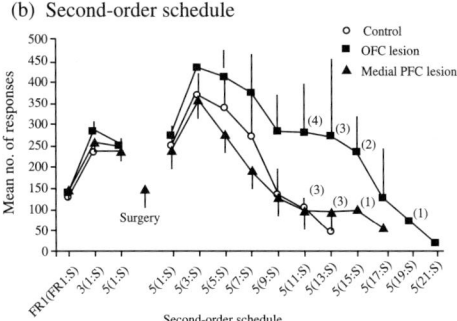

**Fig. 9.3** Mean number of responses (± sem) on each level of the second-order schedule both pre- and post-surgery for (a) the control (n = 6) and amygdala lesioned (n = 6) groups and (b) the control (n = 5), OFC (n = 5), and medial PFC (n = 4) lesioned groups. Numbers in parenthesis indicate the numbers of animals remaining in each group at each level of the schedule. FRx(FRy:S) denotes 'y' number of responses before presentation of the CS (S) and 'x' number of CS's before receiving primary reinforcement. Thus, 5(7:S) reflects the level at which 7 responses are required before presentation of a CS and 5 such CS's must be gained, and thus a total of 35 responses made, before obtaining primary reinforcement. While amygdala lesioned animals dropped out of the schedule at far lower response demands than controls, this was not the case for the OFC or medial PFC lesioned animals. This was despite the fact that the responding of the OFC lesioned group, like the amygdala-lesioned group was insensitive to omission of the CS.

OFC. Indeed evidence for competition between exteroceptive and interoceptive cues within the OFC has been reported in a recent functional neuroimaging study (Hurliman et al. 2005). A loss of this regulation following OFC lesions may result in general disinhibition and greater than normal control over responding by stimuli that would normally have far less control. Such an effect however may not be seen following amygdala lesions since the OFC is intact and thus still able to regulate control but in the absence of the conditioned reinforcer.

The hypothesis that the OFC anticipates outcomes and can thereby determine ones goal for action predicts that neural activations within the amygdala and OFC should be differentiated according, respectively, to whether a subject is just presented with multiple incentive items to consider but upon which no decision has to be made, compared to the situation in which an explicit expectancy-based preference judgement is required. To test this prediction we showed human subjects a series of restaurant menus that varied in incentive value, specifically tailored for each individual, and in half the trials subjects were asked to make a selection from the menu (Arana et al. 2003). Consistent with the prediction, the amygdala was differentially activated in subjects when viewing high-incentive compared to low incentive menus regardless of whether a choice was required (Fig. 9.4a). The activity in this region also varied as a function of individual subjective ratings of incentive value (Fig. 9.4b), consistent with other imaging studies that have activated the amygdala with motivational stimuli (Hamann *et al.* 1999; LaBar *et al.* 2001; Morris and Dolan 2001; O'Doherty *et al.* 2002; Zald *et al.* 2002; Anderson *et al.* 2003). In contrast, neighbouring and possibly overlapping regions of the OFC were recruited both during incentive judgements and goal selection. Activity in the medial region of the OFC showed a greater response to high incentive menus compared to low (Fig. 9.4c), and also a greater response when making a choice, than when not (Fig. 9.4d), with the latter activity also correlating with subjective ratings of difficulty (Fig. 9.4e). This is therefore consistent with the hypothesis that the medial region of OFC represents an area of convergence of sensory, value and other mnemonic information relating to prospective outcomes to allow for goal selection. In addition to activity

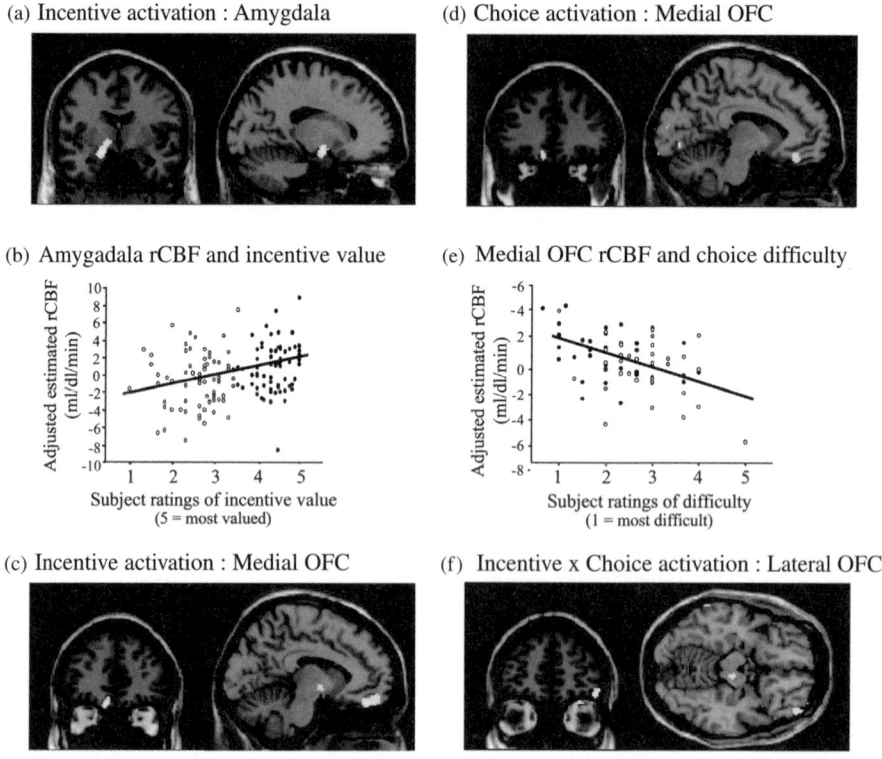

**Fig. 9.4** A region of the left amygdala ($x = -16, y = -4, z = -14$) (a) and a region of the left medial OFC ($x = -8, y = 44, z = -20$) (c) showed significantly increased rCBF in the high incentive condition compared to the low incentive condition in the restaurant menu task, with the rCBF in the amygdala being positively correlated with subjective ratings of incentive value (b). In addition, an area of left medial OFC ($x = -8, y = 36, z = 16$) (d) also showed significantly greater activity in the choice trials compared to the no choice trials with activity in this region positively correlated with subjective ratings of difficulty (e). Finally, a region of the lateral OFC ($x = 48, y = 52, z = -14$) (f) showed significantly increased activity for the incentive x choice interaction, specifically when subjects selected between high-incentive alternatives.

in this medial orbitofrontal region, the lateral orbitofrontal cortex was also activated but specifically when making choices between high incentive menus (Fig. 9.4f), choices that subjects rated as being the most difficult to make. Under these circumstances, when there is not a single, most preferred item, then perhaps goal selection requires some form of active inhibitory process to suppress the alternative, highly attractive items (see below for discussion on inhibition).

## 9.6. OFC and the discriminative properties of stimuli

Besides conditioned motivational information being used by the OFC to generate expectancies and contribute to goal selection, non-affective information about the relationship between stimuli may also contribute to these processes, dependent upon the connections of the OFC, not with the amygdala but with higher-order sensory cortices. Thus, as described earlier, a CS in the environment can be associated not only with the affective properties of food reward but also with their sensory properties and thus, in some circumstances it is the non-affective, discriminative properties of a CS that may guide responding. Indeed, we have proposed that in some studies of conditioned reinforcement in which the effects of amygdala lesions have failed to disrupt responding (Malkova *et al.* 1997; Gaffan *et al.* 1989), performance may have been more under the control of the discriminative, sensory properties of the CS rather than its reinforcing properties (Parkinson *et al.* 2001). Moreover, as described above, and as proposed by Gaffan and Harrison (1987), these discriminative properties of a CS can contribute to performance on object discrimination tasks in which animals learn about the relationship between the stimulus and the perceptual, as well as the motivational properties of the reward. This is particularly likely in traditional tests of object discrimination learning using the Wisconsin General Test Apparatus (WGTA), when the food reward is seen and touched by the animal before the intrinsic incentive properties of the food, that is, its taste, are experienced. Evidence to support the proposal that the OFC can use non-affective as well as affective information to generate expectancies was provided by Schoenbaum *et al.* (2003) in which it was demonstrated that rat OFC neuronal activity that was sensitive to a reward related cue appeared more bound to the sensory, rather than the affective properties of the cue following basolateral amygdala lesions. Since associations between objects and the sensory properties of food reward may be classified as a form of stimulus-stimulus association and such associations have been shown to be dependent upon the perirhinal cortex (Murray and Richmond 2001) then projections from the perirhinal cortex to the OFC may provide the latter with such non-affective information which may be used to produce expectancies about food reward. This information may provide the animal with knowledge about the contingent relationship between the object and reward, and, as proposed recently by Izquierdo *et al.* (2004), interactions between the perirhinal cortex and OFC may be important in detecting changes in reward contingencies upon their reversal in an object discrimination task.

## 9.7. Discrimination reversal and extinction revisited

The role of the OFC in generating expectancies, comparing current anticipated reinforcers and selecting goals may be a product of integrated processing within the entire OFC. Nevertheless, the finding of distinct impairments following restricted ablations of selected regions of the primate OFC in tests of discrimination reversal and extinction (Butter 1969; Iversen and Mishkin 1970) argues for the existence of functional differentiation. Two, related proposals to have arisen from the neuroimaging literature regarding the role of the lateral OFC in discrimination reversal learning are (i) the evaluation of punishers and, in particular, those punishers that lead to (ii) a change in behavior, including inhibition of a prepotent response tendency (Elliott et al. 2001; Kringelback and Rolls 2004). A third proposal highlights the involvement of the lateral OFC, in particular the caudolateral OFC, in detecting a change in reward contingencies (O'Doherty et al. 2003). This latter study ruled out a role for the OFC specifically in response inhibition by virtue of the fact that the only OFC region to be activated immediately prior to a switch showed this activity in both, an instrumental and Pavlovian version of the reversal task; it being argued that in the Pavlovian task no response need be inhibited. However, a variety of responses including autonomic and somatic can accompany Pavlovian conditioning, responses that were not measured in the study. Thus there may well have been a need to inhibit such responses at the time of the reversal in the Pavlovian version and thus a role for this region in response inhibition cannot yet be ruled out.

Of relevance to this issue of inhibitory control are findings from both human functional neuroimaging and lesion studies that implicate the posterior regions of the inferior PFC, including areas 44, 45 and 47/12 (Petrides and Pandya 1995), in inhibition in a variety of contexts including cognitive set-shifting, memory retrieval and cancellation of an intended movement; the latter having been proposed to be required in stop signal and go/no-go tasks (reviewed by Aron et al. 2004) and indeed, discrimination reversal (Cools et al. 2001). The area of activation accompanying a reversal in Cools et al. (2001) is immediately adjacent to the lateral orbitofrontal activations reported in other discrimination reversal studies (Kringelbach and Rolls 2003; O'Doherty et al. 2003), lying just dorsal to, but at a similar antero-posterior level as these lateral orbitofrontal activations. The finding that various switching tasks appear to activate loci within the inferior prefrontal convexity either argues for a commonality of function within this region or multiple, related functions that all contribute to switching, including inhibitory control, detecting contingency change and evaluating punishment. Since the lesions in the studies by Iversen and Mishkin (1970) and Butter (1969) targeted the entire prefrontal convexity including the lateral OFC, an impairment in one or all of these functions may have contributed to the marked perseverative responding following a discrimination reversal.

In contrast to the impairment in reversal learning following ablation of the inferior prefrontal convexity, that following ablation of the medial orbital surface does not appear to be perseverative in nature and thus is likely to be of distinct origin. Since a particular object acquires a second meaning following a reversal of the reward contingencies, then perhaps the deficit seen following medial OFC lesions reflects the loss of the ability to represent both meanings concurrently, a function normally performed by the medial OFC. Certainly, it is the medial region that is consistently activated by rewarding stimuli and consistent with this dual meaning hypothesis is the finding that following reversal of the reward contingencies a distinct population of neurons in rat OFC develops activity to the reward-related cue, rather than the same cue-related neurons reversing their activity (Schoenbaum *et al.* 1999).

Deficits in reversal learning and extinction are not just restricted to damage to OFC, they have also been reported following lesions of the ventral sector of the medial wall in rats. Specifically, an excitotoxic lesion of the infralimbic region disrupts extinction of both a Pavlovian appetitive (Rhodes and Killcross 2004) and aversive (Morgan and LeDoux 1995; Quirk *et al.* 2000) CS, as well as disrupting an appetitive discrimination reversal task (Chudasama and Robbins 2003). While the precise homologues in monkeys and humans of regions within the medial wall of rats is still debated, it has been proposed that the infralimbic and prelimbic cortices are homologous to areas 25 and 32 respectively, in primates (Ongur and Price 2000; Preuss 1995; but see Vertes 2004). Thus, it would be predicted that the deficits underlying impaired reversal and extinction performance following infralimbic lesions in rats would differ from those underlying the reversal and extinction deficits associated with ablations of the OFC in rhesus monkeys and available data supports this conclusion. Following extinction of an appetitive response, the prolonged responding in monkeys with orbitofrontal ablation (Butter 1969) was present on the very first day of extinction, consistent with a failure to detect or act upon changes in reward contingencies. In contrast, rats with infralimbic lesions displayed equivalent reductions in conditioned responding to an appetitive or aversive CS on the first day of extinction compared to controls, suggesting that their ability to detect or act upon a change in reinforcement contingencies was intact. However, they showed far greater spontaneous recovery on the second day of extinction, 24hrs later (Quirk *et al.* 2000; Rhodes and Killcross 2004). Since contextual cues play an important role in extinction learning (Bouton and Peck 1989) it has been proposed that this spontaneous recovery may be due to the lesion interfering with the manner in which contextual cues modulate the inhibitory associations in extinction (Rhodes and Killcross 2004). A similar proposal may account for the impairment in discrimination reversal learning that also accompanies infralimbic lesions in rats (Chudasama and Robbins 2003). The deficit appears to be due to a difficulty in learning to respond to the previously unrewarded stimulus rather than inhibiting responding to the previously rewarded stimulus, a pattern similar to that reported following ablation of the medial orbitofrontal region in rhesus monkeys (Izquierdo *et al.* 2004). However, since contextual cues will contribute to reversal learning, as they do in extinction learning, then the possibility remains that spontaneous recovery of the previously learned discrimination, as occurs in extinction, may underlie the reversal deficit in Chudasama and Robbins (2003). This could be determined by comparing performance on the first and second days of reversal learning.

## 9.8. OFC and autonomic arousal

The discussion so far has focused on the role of the OFC in processes of reward expectancy, goal selection and adaptation and how these processes may contribute to positive affective behavior. However, when stimuli in the environment predict reward, the positive affective state that is induced is reflected not only in behavior itself but also within central and peripheral arousal systems. Thus, changes in peripheral somatic and autonomic activity are important for preparing the body to support ongoing affective behavior and feedback from these peripheral changes has been hypothesized to contribute to the perception of arousal (Schacter and Singer 1962), to engage appraisal processes including the self-perception of emotional state (Ekman *et al.* 1983), to enhance emotional memories (McGaugh 2004) and to influence decision making (Nauta 1971; Damasio 1998). To this end, visceral feedback has been shown to act upon the nucleus of the solitary tract, the primary visceral relay station in the brainstem and from there to influence activity in the ascending noradrenergic projections to the forebrain, where it has an impact on memory consolidation via the amygdala (for a review, see McGaugh 2004) and on attentional processing via the cholinergic neurons of the basal forebrain (for review, see Berntson *et al.* 2003). The close relationship between visceral feedback and the ascending noradrenergic and cholinergic pathways suggests, that at least at one level, there may be coordination between peripheral and central arousal mechanisms upon induction of an affective state, important for the integration of adaptive bodily responses with ongoing emotional and attentional processing.

What then is the relationship between the OFC and these central and peripheral arousal systems? First, the OFC is intimately linked to the internal milieu, receiving information about the current internal state of the organism as well as having influence over it (Price *et al.* 1996; Barbas *et al.* 2003). Second, the OFC, along with the medial PFC, has a relatively privileged position among the cortical areas in its regulation of the ascending central arousal pathways including noradrenaline (Arnsten and Goldman-Rakic 1984; Jodo *et al.* 1998), acetylcholine (Mesulum and Mufson 1984; Zaborsky *et al.* 1997), dopamine (Sesack *et al.* 1989) and serotonin (Hajos *et al.* 1998). Thus the OFC is in a position to coordinate peripheral and central arousal mechanisms. The interaction of the OFC and these central arousal pathways is discussed elsewhere in this book by Robbins *et al.* while the role of the human ventromedial PFC with respect to peripheral arousal mechanisms is discussed by Naqvi, Tranel and Bechara. Below, in the context of positive affective behavior, peripheral arousal mechanisms and their neural control will be described using a novel experimental procedure that measures behavioral and peripheral arousal simultaneously in a freely moving marmoset.

Previously, autonomic arousal had been studied most commonly with respect to fear conditioning in rats, and in this context, conditioned autonomic arousal was disrupted by excitotoxic lesions of the central nucleus of the amygdala (Iwata *et al.* 1986) and following ablation of the medial PFC (Frysztak and Neafsey 1994). In other studies the effects of lesions to these structures on conditioned autonomic activity have been replicated even though conditioned bradycardia, as opposed to conditioned accelerative responses have been studied (Powell *et al.* 1997). In contrast, autonomic arousal in appetitive conditioning paradigms has seldom been reported. Instead a commonly observed conditioned autonomic response is heart rate deceleration (Hunt and Campbell 1997). Such decelerations probably reflect the contribution of the parasympathetic division of the autonomic system and, when seen in response is heart rate deceleration to a novel stimulus, have been described as cardiac orienting responses. They have been proposed to reflect the engagement of attentional processes that promote the detection and processing of the sensory stimulus (Kapp *et al.* 1992; Middleton *et al.* 1999). In contrast, increases in blood pressure, under sympathetic control, reflecting the motivational, arousing state induced by the CS (Cannon 1927) tend to be reported only in experimental appetitive procedures using high-incentive rewards (Randall *et al.* 1975) and in aversive procedures they are more likely seen in freely moving animals in which the sympathetic system isn't already operating at a higher level due to the stress of the restraint procedure (Iwata *et al.* 1986). In the only studies to have examined the neural mechanisms underlying appetitively conditioned autonomic changes, electrolytic lesions of the medial PFC were shown to produce a mild attenuation of a HR deceleration but only when made before (McLaughlin and Powell 1999), but not after, conditioning (McLaughlin and Powell 2001).

Using a radiotelemetry receiver, that can measure blood pressure and heart rate remotely by receiving signals from a probe inserted into the descending aorta, we have shown that anticipation of highly preferred food leads to a rise in both blood pressure and heart rate in marmosets and that a further rise, over and above that seen during the anticipatory phase, accompanies the actual consumption of the food (Braesicke *et al.* 2005; Fig. 9.5). In the anticipatory phase the marmosets are given the opportunity to view, but not gain access to, these foods and thus, the physical characteristics of the foods, e.g. shape and color, most likely act as appetitive CS's to induce positive arousal in the animals in the expectation of gaining access to these foods. No such responses are seen when marmosets view their daily, staple diet of pellets. This rise in blood pressure to the sight of food is not activity dependent and is a sympathetically driven, positive arousing response, as we have shown that it is blocked by an injection of the B-adrenergic receptor blocker, sotolol (Reekie *et al.* unpublished findings). It is dependent upon an intact amygdala as combined excitotoxic lesions of the basolateral nucleus and central nucleus disrupt the anticipatory rise (Fig. 9.5c) even though the same lesion leaves behavioral arousal intact, as measured by the marmosets approach and responding at the window through which the food can be viewed (Fig. 9.5d). Also left intact after the lesion is the autonomic arousal associated with the consumption of the food (Fig. 9.5e) highlighting the relative selectivity of the effects of the amygdala lesion on anticipatory peripheral arousal. The neural substrates underlying consummatory arousal are as yet unknown but may include the parabrachial nucleus of the brainstem and the shell region of the nucleus accumbens (see Berridge 2003 for a review of the neural mechanisms underlying the hedonic aspects of taste).

In contrast to the effects of amygdala lesions, preliminary findings, examining the effects of excitotoxic lesions of the OFC and medial PFC (Reekie *et al.* 2004), suggest that neither of these prefrontal regions are involved in the control of autonomic arousal in either the anticipatory or consummatory periods (Fig. 9.5f–h). Thus, while the expression

# 256 | FUNCTIONS OF PRIMATE ORBITOFRONTAL CORTEX

**(a) Apparatus**

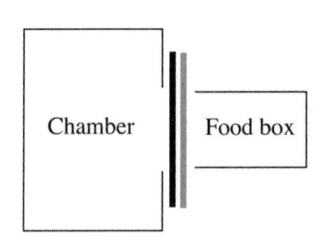

**(b) Task protocol**

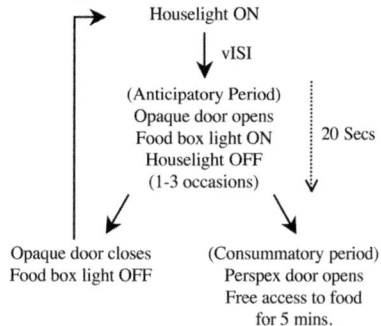

Houselight ON
↓
vISI
↓
(Anticipatory Period)
Opaque door opens
Food box light ON
Houselight OFF
(1–3 occasions)

20 Secs

↙ ↘

Opaque door closes
Food box light OFF

(Consummatory period)
Perspex door opens
Free access to food
for 5 mins.

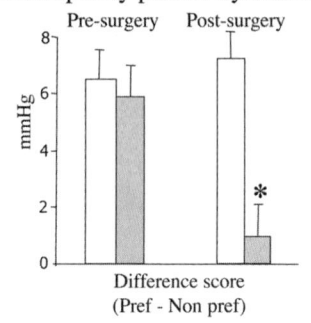

**(c) Anticipatory period: Systolic BP**

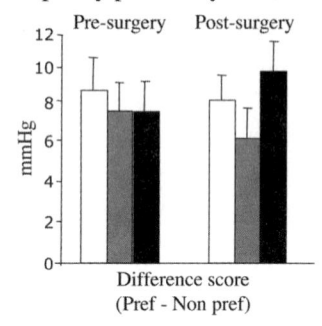

Pre-surgery   Post-surgery

mmHg

*

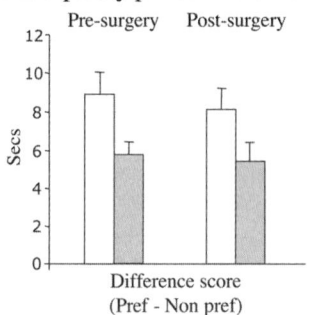

Difference score
(Pref – Non pref)

**(d) Anticipatory period: Behavior**

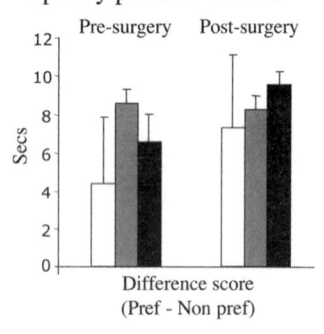

Pre-surgery   Post-surgery

Secs

Difference score
(Pref – Non pref)

**(e) Consummatory period: Systolic BP**

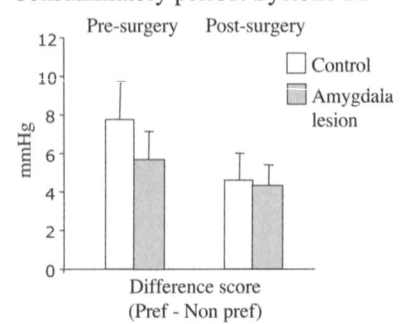

Pre-surgery   Post-surgery

☐ Control
▨ Amygdala lesion

mmHg

Difference score
(Pref – Non pref)

**(f) Anticipatory period: Systolic BP**

Pre-surgery   Post-surgery

mmHg

Difference score
(Pref – Non pref)

**(g) Anticipatory period: Behavior**

Pre-surgery   Post-surgery

Secs

Difference score
(Pref – Non pref)

**(h) Consummatory period: Systolic BP**

Pre-surgery   Post-surgery

☐ Control
▨ OFC Lesion
■ MPFC Lesion

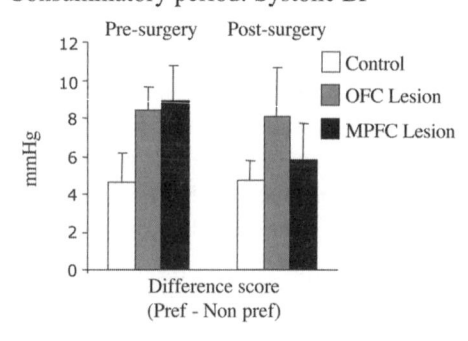

mmHg

Difference score
(Pref – Non pref)

**Fig. 9.5** Legend shown opposite

of conditioned autonomic arousal in a simple Pavlovian setting is dependent upon the amygdala it is not dependent upon either the OFC or medial PFC, consistent with the findings of Bechara *et al* in patients (1999). Whether these prefrontal regions are involved in the acquisition of this conditioned response remains to be determined. However, as mentioned earlier, the finding that the acquisition of Pavlovian conditioned approach to an appetitive CS is disrupted by orbitofrontal lesions in rats (Chudasama and Robbins 2003) raises the possibility that the OFC may contribute to specific components of the Pavlovian conditioned response, perhaps those components that are attentional, but not motivational in nature? In this respect it would be of interest to determine whether OFC lesions also disrupt the conditioned orienting response to an appetitive CS, proposed to be part of an attentional circuit that includes the central nucleus of the amygdala (Holland and Gallagher 1999).

Further studies are now required to determine the relationship between the PFC, autonomic activity and behavior in more complex reward settings, such as in the anticipation of the outcome of responses and changing reward contingencies. For example, what is the relationship between autonomic activity and object discrimination reversal or extinction and what is the neural basis to this relationship? Our own preliminary findings, using an appetitive Pavlovian task, reveal that the anticipatory autonomic arousal to a CS associated with high-incentive food shows a sharp reduction if marmosets receive, instead of the high-incentive food, low incentive food pellets. This decline is far more rapid than that seen at the end of the CS period on a partial reinforcement schedule in which not all CS's are followed by food reward (Reekie, Braesicke, Man and Roberts, unpublished findings). The neural basis to this effect is, though, unknown. Another important question to address is the nature of the involvement of the OFC in any peripheral feedback effect. Thus, considerable knowledge has already been gained with respect to the role of the amygdala in mediating the effects of peripheral arousal on long term memory (McGaugh 2004) but what of the effects of peripheral arousal on attentional mechanisms and motivational mechanisms and what part, if any, does the OFC play in mediating these effects?

**Fig. 9.5** A schematic diagram of the apparatus used in the autonomic study illustrates the separation of the food box from the rest of the chamber by a transparent door (grey line) and an opaque door (black line) (a). The task procedure for each session is summarized in (b). During the anticipatory period the opaque door opens and the animal can view the food inside the box. In the consummatory period, at the end of the session, both the opaque and transparent doors open and the animal gains access to the food. Mean changes in systolic BP (± SEM) during the anticipatory period (20 s), compared to the preceding baseline period (20 s) are shown for both controls ($n = 4$) and amygdala lesioned ($n = 4$) groups in (c). * indicates a significant difference of $p = 0.005$ between the control and amygdala-lesioned groups post-surgery. Mean duration of food box directed behavior (± SEM) including looking, approaching and responding at the box during the anticipatory period is shown in (d) and mean changes in systolic BP (± SEM) during the first minute of the consummatory period, compared to the immediately preceding anticipatory period, is shown in (e). Preliminary findings from three animals each in control, the OFC- and medial PFC-lesioned groups for systolic BP (f) and food box-directed behavior (g) during the anticipatory period and for systolic BP during the consummatory period (h) reveal no significant lesion effects.

## 9.9. Conclusions

Evidence is accruing consistent with an involvement of the primate OFC in integrating information about rewards and punishments and their predictors and using this information to select goals for action. This function is dependent upon interactions between the OFC and related brain structures, including the amygdala for processing affective information about explicit conditioned reinforcers and other structures (e.g., the hippocampus), for contextually evoked representations of reinforcement and the perirhinal cortex for non-affective information about stimuli-stimulus associations, the latter possibly contributing to contingency learning. Integration of this information within the OFC may subsequently lead to goal selection with the affective and attentional processing of the goal, and its predictors, being enhanced by the central and peripheral arousal mechanisms that are modulated by the OFC.

## References

Altman, J., Everitt, B. J., Glautier, S., Markou, A., Nutt, D., Oretti, R., Phillips, G. D., and Robbins, T. W. (1996) The biological, social and clinical bases of drug-addiction—commentary and debate. *Psychopharmacology* **125**:285–345.

Anderson, A. K., Christoff, K., Stappen, I., Panitz, D., Ghahremani, D. G., Glover, G., Gabrieli, J. D., and Sobel, N. (2003) Dissociated neural representations of intensity and valence in human olfaction. *Nature Neuroscience* **6**:196–202.

Arana, F. S., Parkinson, J. A. Hinton, E., Holland, A., Owen, A., and Roberts, A. C. Dissociable Contributions of the Human Amygdala and Orbitofrontal Cortex to Incentive Motivation and Goal Selection (2003) *Journal of Neuroscience* **23**:9632–38.

Arnsten, A. F. T. and Goldman-Rakic, P. S. (1984) Selective prefrontal cortical projections to the region of the locus coeruleus and raphé nuclei in the rhesus monkey. *Brain Research* **306**:9–18.

Aron, A. R., Robbins, T. W., and Poldrack, R. A. (2004) Inhibition and the right inferior frontal cortex. Trends in Cognitive Sciences **8**:170–1766.

Arroyo, M., Markou, A., Robbins, T. W., and Everitt, B. J. (1998) Acquisition, maintenance and reinstatement of intravenous cocaine self- administration under a second-order schedule of reinforcement in rats: effects of conditioned cues and continuous access to cocaine. *Psychopharmacology* **140**:331–44.

Barbas, H., Saha, S., Rempel-Clower, N., and Ghashghaei, T. (2003) Serial pathways from primate prefrontal cortex to autonomic areas may influence emotional expression. *BMC Neuroscience* **4**:25.

Baxter, M. G., Parker, A., Lindner, C. C., Izquierdo, A. D., and Murray, E. A. (2000) Control of response selection by reinforcer value requires interaction of amygdala and orbital prefrontal cortex. *Journal of Neuroscience* **20**:4311–19.

Bechara, A., Damasio, H., Damasio, A. R., and Lee, G. P. (1999) Different contributions of the human amygdala and ventromedial prefrontal cortex to decision-making. *Journal of Neuroscience* **19**:5473–5481.

Berntson, G. G., Sarter, M., and Cacioppo, J. T. (2003) Ascending visceral regulation of cortical affective information processing. *European Journal Neuroscience* **18**:2103–2109.

Berridge, K. C. (2003) Pleasures of the brain. *Brain and Cognition* **52**:106–128.

Bouton, M. (2004) Context and behavioral processes in extinction. *Learning and Memory* **11**:485–494.

Bouton M., and Peck C. (1989) Context effects on conditioning, extinction and re-instatement in an appetitive conditioning preparation. *Animal Learning and behavior* **17**:188–198.

Braesicke, K., Parkinson, J. A., Reekie, Y., Mann, M. S., Hopewell, L., Pears, A., *et al.* (2005) Autonomic arousal in an appetitive context in primates: a behavioral and neural analysis. *European Journal of Neuroscience* 21:1733–1740.

Burns, L. H., Robbins, T. W., and Everitt, B. J. (1993) Differential effects of excitotoxic lesions of the basolateral amygdala, ventral subiculum and medial prefrontal cortex on responding with conditioned reinforcement and locomotor activity potentiated by intra-accumbens infusions of d-amphetamine. *Behavioral Brain Research* 55:167–183.

Bussey, T. J., Everitt, B. J., and Robbins, T. W. (1997) Dissociable effects of cingulate and medial frontal cortex lesions on stimulus-reward learning using a novel Pavlovian autoshaping procedure for the rat:implications for the neurobiology of emotion. *Behavioral Neuroscience* 111:908–919.

Butter, C. M. (1969) Perseveration in extinction and in discrimination reversal tasks following selective frontal ablations in macaca mulatta. *Physiology and Behavior* 4:163–171.

Butter, C. M., Mishkin, M., and Rosvold, H. E. (1963) Conditioning and extinction of a food rewarded response after selective ablations of frontal cortex in rhesus monkeys. *Experimental Neurology* 7:65–75.

Butter, C. M., Snyder, D. R., and McDonald, J. A. (1970) Effects of orbital frontal lesions on aversive and aggressive behaviors in rhesus monkeys. *Journal of Comparative Physiological Psychology* 72:132–144.

Cador, M., Robbins, T. W., and Everitt, B. J. (1989) Involvement of the amygdala in stimulus reward associations—interaction with the ventral striatum. *Neuroscience* 30:77–86.

Cannon, W. B. (1927) The James-Lange theory of emotions: A critical examination and an alternative theory. *American Journal of Psychology* 39:106–124.

Carmichael, S. T. and Price, J. L. (1994) Architectonic subdivision of the orbital and medial prefrontal cortex in the macaque monkey. *Journal of Comparative Neurology* 346:366–402.

Carmichael, S. T. and Price, J. L. (1996) Connectional networks within the orbital and medial prefrontal cortex of macaque monkeys. *Journal of Comparative Neurology* 371:179–207.

Chudasama, Y. and Robbins, T. W. (2003) Dissociable contributions of the orbitofrontal and infralimbic cortex to Pavlovian autoshaping and discrimination reversal learning:further evidence for the functional heterogeneity of the rodent frontal cortex. *Journal of Neuroscience* 23:8771–8780.

Cools, R., Clark, L., Owen, A. M., and Robbins, T. W. (2002) Defining the neural mechanisms of probabilistic reversal learning using event-related functional magnetic resonance imaging. *Journal of Neuroscience* 22:4563–4567.

Damasio, A. R. (1998) The somatic marker hypothesis and the possible functions of the prefrontal cortex. In: Prefrontal Cortex: Cognitive and Executive Functions (Roberts A. C., Robbins T. W., Weiskrantz L., eds), pp. 27–50. Oxford: Oxford University Press.

Delamater, A. R. (2004) Experimental extinction in Pavlovian conditioning:behavioral and neuroscience perspectives. *Quarterly Journal Experimental Psychology* 57B:97–132.

Dias, R. Robbins, T. W. and Roberts, A. C. (1996a) Primate analogue of the Wisconsin Card Sort Test:effects of excitotoxic lesions of the prefrontal cortex in the marmoset. *Behavioral Neuroscience* 110:872–886.

Dias, R. Robbins, T. W. and Roberts, A. C. (1996b) Dissociation in the prefrontal cortex of affective and attentional shifting. *Nature* 380:69–72

Dickinson, A. and Balleine, B. (1994) Motivational control of goal-directed action. *Animal Learning and Behavior* 22:1–18.

Dickinson, A. and Dearing, M. F. (1979) Appetitive-aversive interactions and inhibitory processes. In: Mechanisms of Learning and Motivation. (Dickinson A., Boakes R. A., eds), pp. 203–231. Erlbaum, Hillsdale, New Jersey.

Ekman, P., Levenson, R. W., and Friesen, W. V. (1983) Autonomic nervous system activity distinguishes among emotions. *Science* 221:1208–1210.

Elliott, R. D., Dolan, R. J., and Frith, C. D. (2000) Dissociable functions in the medial and lateral orbitofrontal cortex:evidence from human neuroimaging studies. *Cerebral Cortex* **10**:308–317.

Everitt, B.J., Cador, M., and Robbins, T. W. (1989) Interactions between the amygdala and ventral striatum in stimulus reward associations—studies using a 2nd-order schedule of sexual reinforcement. *Neuroscience* **30**:63–75.

Franzen, E. A. and Myers, R. E. (1973) Neural control of social behavior: prefrontal and anterior temporal cortex. *Neuropsychologia* **11**:141–157.

Frysztak, R. J. and Neafsey, E. J. (1994) The effect of medial frontal cortex lesions on cardiovascular conditioned emotional responses in the rat. *Brain Research* **643**:181–193.

Fuster, J. M. (1989) The prefrontal cortex:Anatomy, Physiology and Neuropsychology of the Frontal Lobe. Second edition. New York: Raven Press.

Gaffan, D. (1979) Acquisition and forgetting in monkeys' memory of informational object-reward associations. *Learning and Motivation* **10**:419–444.

Gaffan, D., Gaffan, E. A., and Harrison, S. (1989) Visual-visual associative learning and reward-association learning in monkeys: the role of the amygdala. *Journal of Neuroscience* **9**:558–64.

Gaffan, D. and Harrison, S. (1987) Amygdalectomy and disconnection in visual learning for auditory secondary reinforcement by monkeys. *Journal of Neuroscience* **7**:2285–2292.

Gallagher, M. (2000) The amygdala and associative learning. In: The Amygdala: A Functional Analysis (Aggleton J.P., ed), pp. 311–329. Oxford: Oxford University Press.

Gallagher, M., McMahan, R. W., and Schoenbaum, G. (1999) Orbitofrontal cortex and representational memory. *Journal of Neuroscience* **19**:6610–6614.

Goldberg, S. R. (1973) Comparable behavior maintained under fixed-ratio and second-order schedules of food presentation, cocaine injection or d-amphetamine injection in the squirrel monkey. *Journal Pharmacol Exp Ther* **186**:18–30.

Hajos, M., Gartside, S. E., Varga, V., and Sharp, T. (1998) An electrophysiological and neuroanatomical study of the medial prefrontal cortical projection to the midbrain raphé nuclei in the rat. *Neuroscience* **87**:95–108.

Hall, R. E., Livingston, R. B., and Bloor, C. M. (1977) Orbital cortical influences on cardiovascular dynamics and myocardial structures in conscious monkeys. *Journal of Neurosurgery* **46**:638–647.

Hamann, S. B., Ely, T. D., Grafton, S. T., and Kilts, C. D. (1999) Amygdala activity related to enhanced memory for pleasant and aversive stimuli. *Nature Neuroscience* **2**:289–293.

Hatfield, T., Han, J.S., Conley, M., Gallagher, M., and Holland, P. (1996) Neurotoxic lesions of basolateral, but not central, amygdala interfere with Pavlovian 2nd-order conditioning and reinforcer devaluation effects. *Journal of Neuroscience* **16**:5256–5265.

Hearst, E. and Jenkins, H. M. (1974) Sign tracking: the stimulus-reinforcer relation and directed action. Monograph of the Psychonomic Society.

Hikosak, K. and Watanabe, M. (2000) Delay activity of orbital and lateral prefrontal neurons of the monkey varying with different rewards. *Cerebral Cortex* **10**:263–271.

Holland, P. C. and Rescorla, R. A. (1975) Second-order conditioning with food unconditioned stimulus. Journal of Comparative and Physiological Psychology **88**:459–467.

Holland, P. C. and Gallagher, M. (1999) Amygdala circuitry in attentional and representational processes. *Trends in Cognitive Sciences* **3**:65–73.

Hunt, P. S. and Campbell, B. A. (1997) Autonomic and behavioral correlates of appetitive conditioning in rats. *Behavioral Neuroscience* **111**:494–502.

Hurliman, E., Nagode, J. C., and Pardo, J. V. (2005) Double dissociation of exteroceptive and interoceptive feedback systems in the orbital and ventromedial prefrontal cortex of humans. *Journal of Neuroscience* **25**:4641–4648.

Iversen, S. D. and Mishkin, M. (1970) Perseverative interference in monkeys following selective lesions of the inferior prefrontal convexity. *Experimental Brain Research* **11**:376–386.

Iwata, J., LeDoux, J. E., Meeley, M. P., Arneric, S., and Reis, D. J. (1986) Intrinsic neurons in the amygdaloid field projected to by the medial geniculate body mediate emotional responses conditioned to acoustic stimuli. *Brain Research* **383**:195–214.

Izquierdo, A., Suda, R. K., and Murray, E. A. (2004) Bilateral orbital prefrontal cortex lesions in rhesus monkeys disrupt choices guided by both reward value and reward contingency *Journal of Neuroscience* **24**:7540–7548.

Jodo, E., Chiang, C., and Aston-Jones (1998) Potent excitatory influence of prefrontal cortex activity on noradrenergic locus coeruleus neurons. *Neuroscience* **83**:63–79

Kapp, B. S., Whalen, P. J., Supple, W. F., and Pascoe, J. P. (1992) Amygdaloid contributions to conditioned arousal and sensory information processing. In: The Amygdala: neurobiological aspects of emotion, memory and mental dysfunction (Aggleton, J.P. ed) pp. 229–254. New York: Wiley-Liss.

Katz, J. L. (1979) A comparison of responding maintained under second-order schedules of intramuscular cocaine injection or food presentation in squirrel monkeys. *Journal of Experimental Animal Behavior* **32**:419–31.

Kringelbach, M. L. and Rolls, E. T. (2003) Neural correlates of rapid reversal learning in a simple model of human social interaction. *Neuroimage* **20**:1371–83.

LaBar, K. S., Gitelman, D. R., Parrish, T. B., Kim, Y. H., Nobre, A. C., and Mesulam, M. M. (2001) Hunger selectively modulates corticolimbic activation to food stimuli in humans. *Behavioral Neuroscience* **115**:493–500.

Mackintosh, N. J. (1974) The Psychology of Animal Learning. Academic Press. London.

Malkova, L., Gaffan, D., and Murray, E. A. (1997) Excitotoxic lesions of the amygdala fail to produce impairment in visual learning for auditory secondary reinforcement but interfere with reinforcer devaluation effects in rhesus monkeys. *Journal of Neuroscience* **17**:6011–20.

McGaugh, J. L. (2004) The amygdala modulates the consolidation of memories of emotionally arousing experiences. *Annual Review Neuroscience* **27**:1–28.

McLaughlin, J. and Powell, D. A. (1999) Pavlovian heart rate and jaw movement conditionin in the rabbit:effects of medial prefrontal lesions. *Neurobiology of Learning and Memory* **71**:150–166.

McLaughlin, J. and Powell, D. A. (2001) Posttraining prefrontal lesions impair jaw movement conditioning performance, but have no effect on accompanying heart rate changes. *Neurobiology of Learning and Memory* **78**:279–293.

Medin, D. L. (1977) Information processing and discrimination learning set. In: *Behavioral Primatology* (Schrier A. M., ed.), pp. 33–69. Mahwah, NJ: Erlbaum.

Mesulam, M. M. and Mufson, E. J. (1984) Neural inputs into the nucleus basalis of the substantia innominata (Ch. 4) of the rhesus monkey. *Journal of Comparative Neurology* **107**:253–274.

Middleton, H. C., Sharma, A., Agouzoul, D., Sahakian, B. J., and Robbins, T. W. (1999) Contrasts between the cardiovascular concomitants of tests of planning and attention. *Psychophysiology* **36**:610–618.

Mishkin, M. (1964) Perseveration of central sets after frontal lesions in man. In: The Frontal granular cortex and behavior (Warren J. M., Akert K., eds), pp. 219–294. New York: McGraw-Hill.

Montague, P. R. and Berns, G. S. (2002) Neural economics and the biological substrates of valuation. *Neuron* **36**:265–284.

Morgan, M. and LeDoux, J. (1995) Differential contribution of dorsal and ventral medial prefrontal cortex to acquisition and extinction of conditioned fear in rats. *Behavioral Neuroscience* **109**:681–688.

Morris, J. S. and Dolan, R. J. (2001) Involvement of human amygdala and orbitofrontal cortex in hunger- enhanced memory for food stimuli. *Journal Neuroscience* **21**:5304–5310.

Murray, E. A. and Richmond, B. J. (2001) Role of perirhinal cortex in object perception, memory and associations. Current Opinion in *Neurobiology* **11**:188–193.

Nauta, W. J. (1971) The problem of the frontal lobe: a reinterpretation. *Journal Psychiatric Research* **8**:167–187.

O'Doherty, J. P., Deichmann, R., Critchley, H. D., and Dolan, R. J. (2002) Neural responses during anticipation of a primary taste reward. *Neuron* **33**:815–826.

O'Doherty, J. P., Critchley, H. D., Deichmann, R., and Dolan, R. J. (2003) Dissociating valence of outcome from behavioral control in orbital and ventral prefrontal cortices. *Journal of Neuroscience* **27**:7931–7939.

Ongur, D. and Price, J. L. (2000) The organization of networks within orbital and medial prefrontal cortex of rats, monkeys and humans. *Cerebral Cortex* **10**:206–219.

Parkinson, J. A., Crofts, H. S., McGuigan, M., Tomic, D. L., Everitt, B. J., and Roberts, A. C. (2001) The role of the primate amygdala in conditioned reinforcement. *Journal of Neuroscience* **21**:7770–7780.

Parkinson, J. A., Roberts, A. C., Everitt, B. J., and Di Ciano, P. (2005) Acquisition of instrumental conditioned reinforcement is resistant to the devaluation of the unconditioned stimulus. *Quarterly Journal of Experimental Psychology* **58B**:19–30.

Pears, A., Parkinson, A., Hopewell, L. J., Everitt, B. J., and Roberts, A. C. (2003) Lesions of the orbitofrontal, but not medial prefrontal cortex, disrupt conditioned reinforcement in primates. *Journal of Neuroscience* **23**:11189–11201.

Petrides, M. and Pandya, D. N. (1994) Comparative cytoarchitectonic analysis of the human and the macaque frontal cortex. In Handbook of *Neuropsychology*, Vol 9 (Boller F., Grafman J., eds), pp. 17–58. Amsterdam:Elsevier.

Pickens, C. L., Setlow, B., Saddoris, M. P., Gallagher, M., Holland, P. C., and Schoenbaum, G. (2003) Different roles for orbitofrontal cortex and basolateral amygdala in a reinforcer devaluation task. *Journal of Neuroscience* **23**:11078–11084.

Powell, D. A., Chachich, M., Murphy, V., McLaughlin, J., Tebbutt, D., and Buchanan, S. L. (1997) Amygdala-prefrontal interactions and conditioned bradycardia in the rabbit. *Behavioral Neuroscience* **111**:1056–1074.

Preuss, T. M. (1995) Do rats have prefrontal cortex? The Rose-Woolsey-Akert program reconsidered. *Journal of Cognitive Neuroscience* **7**:1–24.

Price, J. L., Carmichael, S. T., and Drevets, W. C. (1996) Networks related to the orbital and medial prefrontal cortex; a substrate for emotional behavior? *Progress in Brain Research* **107**:523–36.

Quirk, G. J. and Gehlert, D. R. (2003) Inhibition of the amygdala: key to pathological states. *Annals of the New York Academy of Sciences* **985**:263–272.

Quirk, G., Russo, G., Barron, J., and Lebron, K. (2000) The role of ventromedial prefrontal cortex in the recovery of extinguished fear. *Journal of Neuroscience* **20**:6225–6231.

Raleigh, M. J. and Steklis, D. (1981) Effects of orbitofrontal and temporal neocortical lesions on the affiliative behavior of vervet monkeys. *Experimental Neurology* **73**:378–389.

Randall, D. C., Brady, J. V., and Martin, K. H. (1975) Cardiovascular dynamics during classical appetitive and aversive conditioning in laboratory primates. Pavlovian *Journal of Biological Science* **10**:66–75.

Reekie, Y., Braesicke, K., Parkinson, J. A., Mann, M., Schnell, C. R., and Roberts, A. C. (2004) Autonomic and behavioral arousal in appetitive Pavlovian conditioning in monkeys: effects of amygdala and orbitofrontal lesions. *Society for Neuroscience abstract* 84.7.

Rescorla, R. A. and Wagner, A. R. (1972) A theory of Pavlovian conditioning:variations in the effectiveness of reinforcement and non reinforcement. In Classical Conditioning II:Current research and theory (Black A.H., Prokasy W.F., eds), pp. 64–99. New York:Appleton-Century-Crofts.

Rescorla, R. A. (2001) Experimental extinction. In Handbook of contemporary learning theories (Mowrer, R. R. and Klein S. B. eds), pp. 119–154, Lawrence Erlbaum Associates, Mahwah.

Rhodes, S. E. and Killcross, S. (2004) Lesions of rat infralimbic cortex enhance recovery and reinstatement of an appetitive Pavlovian response. *Learning and Memory* **11**:611–616.

Roesch, M. R. and Olson, C. R. (2004) Neuronal activity related to reward value and motivation in primate frontal cortex. *Science* **304**:307–310.

Rolls, E. (1998) Neurophysiology and functions of the primate amygdala and the neural basis of emotion. In Roberts A. C., Robbins T. W., Weiskrantz L. (ed.) Prefrontal cortex: cognitive and executive functions. Oxford University Press, Oxford.

Rolls, E. T., Hornak, J., Wade, D., and McGrath, J. (1994) Emotion-related learning in patients with social and emotional changes associated with frontal lobe damage. *Journal of Neurology, Neurosurgery and Psychiatry* **57**:1518–1524.

Schachter, S. and Singer, J. E. (1962) Cognitive, social, and physiological determinants of emotional state. *Psychological Review* **69**:379–399.

Schoenbaum, G., Chiba, A. A., and Gallagher, M. (1998) Orbitofrontal cortex and basolateral amygdala encode expected outcomes during learning. *Nature Neuroscience* **1**:155–159.

Schoenbaum, G., Chiba, A. A., and Gallagher, M. (1999) Neural encoding in orbitofrontal cortex and basolateral amygdala during olfactory discrimination learning. *Journal of Neuroscience* **19**:1876–1884.

Schoenbaum, G., Setlow, B., Saddoris, M. P., and Gallagher, M. (2003) Encoding predicted outcome and acquired value in orbitofrontal cortex during cue sampling depends upon input from basolateral amygdala. *Neuron* **39**:855–867.

Sesack, S. R., Deutch, A. Y., Roth, R. H., and Bunney, B. S. (1989) Topographical organization of the efferent promections of the medial prefrontal cortex in the rat: an anterograde tract-tracing study with *Phaseolus vulgaris* leucoagglutinin. *Journal of Comparative Neurology* **290**:213–242.

Tremblay, L. and Schultz, W. (1999) Relative reward preference in primate orbitofrontal cortex. *Nature* **398**:704–708.

Vertes (2004) Differential projections of the infralimbic and prelimbic cortex in the rat. *Synapse* **51**:32–58.

Williams, B. A. (1994) Conditioned reinforcement: Neglect or outmoded explanatory construct? *Psychonomic Bulletin and Review* **1**:457–475.

Williams, D. R. and Williams, H. (1969) Auto-maintenance in the pigeon:sustained pecking despite contingent non-reinforcement. *Journal of Experimental Analysis of Behavior*, **12**:511–520.

Zaborszky, L., Gaykema, R. P., Swanson, D. J., and Cullinan, W. E. (1997) Cortical input to the basal forebrain. *Neuroscience* **79**:1051–1078.

Zald, D. H., Hagen, M. C., and Pardo, J. V. (2002) Neural correlates of tasting concentrated quinine and sugar solutions. *Journal Neurophysiology* **87**:1068–1075.

Chapter 10

# The role of human orbitofrontal cortex in reward prediction and behavioral choice: insights from neuroimaging

John P. O'Doherty and Raymond J. Dolan

## 10.1. Introduction

The orbitofrontal cortex (OFC) is arguably one of the least understood regions of the human brain. Interest in this region has increased dramatically in recent decades as evidenced in an increasing number of laboratories conducting lesion and neurophysiology studies on this region, in both rats and monkeys. The functions of the OFC have also begun to be probed in humans, through the application of laboratory-based neuropsychological tasks designed to tap behavioral deficits following orbitofrontal lesions that had previously been reported only anecdotally (Bechara *et al.* 2000). By far the biggest impetus to the study of this region has been the emergence of powerful functional neuroimaging techniques such as fMRI and PET. These have enabled the most detailed account yet of the response properties of intact human OFC. In this review we set out some of the main conclusions that can be drawn concerning the functions of the OFC derived from human functional neuroimaging research carried out over the past ten years. In many cases, neuroimaging studies have confirmed previous findings from neurophysiology in animals or human lesion studies, but in others they have led to novel insights into the functions of this erstwhile mysterious brain region.

Clues as to the functions of the OFC can be derived from its anatomical connectivity. The OFC is highly inter-connected with sub-cortical structures involved in affective processing such as the amygdala and ventral striatum (Carmichael and Price 1995a), consistent with a role for the OFC in reward- and affect-related processing. On the other hand, the OFC as part of the prefrontal cortex is also highly interconnected with other sectors within the prefrontal cortex (Carmichael and Price 1996). Given that prefrontal cortex has a generic role in the flexible control of behavior (Miller and Cohen 2001) it is likely that the OFC, in addition to its role in reward and affect, shares functional commonalities with other parts of the prefrontal cortex. In the course of this review we will argue that as a key component of the reward system, the OFC is involved in representing stimulus-reward value as well as maintaining a representation of expected reward value, functions which it may share in common with the amygdala and ventral striatum. Furthermore, the OFC responds following errors in reward prediction, i.e., signals that may underlie learning of reward predictions. The OFC's contribution to the flexible control of behavior we suggest is likely to reflect its role in facilitating rapid changes in behavior in the face of changing reinforcement contingencies and the associated changes in the value of expected outcomes.

## 10.2. **Reward value representations in human OFC**

One of the best-established findings regarding the OFC is its involvement in coding for the reward-value of stimuli (see chapter by Rolls in this volume). Neurons in this region respond to a particular taste or odor when an animal is hungry but decrease their firing rate once the animal is satiated and the corresponding food is no longer rewarding (Rolls *et al.* 1989; Critchley and Rolls 1996). Imaging studies in humans have not only confirmed these findings in both the olfactory and gustatory domains (O'Doherty *et al.* 2000, 2001b; Gottfried *et al.* 2002a; Anderson *et al.* 2003; Small *et al.* 2003), but also substantially extended our understanding of the role of the OFC in coding stimulus reward value in four distinct ways. Firstly, neuroimaging data have shown that the OFC codes the reward value of stimuli across the range of sensory modalities, including somatosensory, auditory and visual domains (Blood *et al.* 1999; O'Doherty *et al.* 2003b; Rolls *et al.* 2003b; see Fig. 10.1). Secondly, unlike in animals, it is possible to ask humans

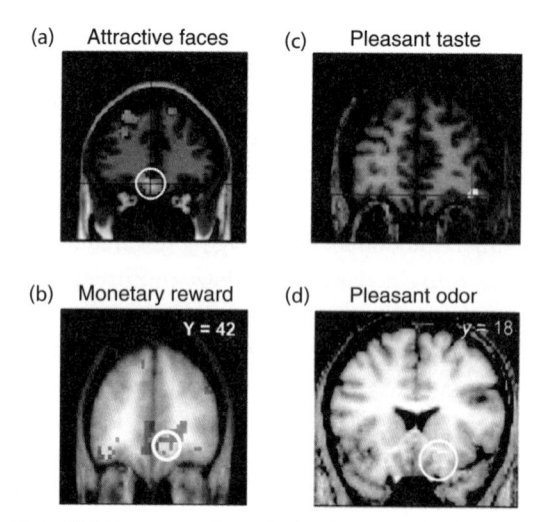

**Fig. 10.1** (See Color Plate 23.) Representation of stimulus reward-value in human OFC. This figure illustrates OFC activation found in four different fMRI studies each of which involved a different type of reward. (a) Activation map showing an area of medial OFC that was found to have enhanced responses to presentation of attractive compared to unattractive faces. Adapted from O'Doherty *et al.* 2003b. (b) Activity in medial OFC following receipt of abstract monetary reward (compared to monetary loss). Data derived from a study by O'Doherty *et al.* (2003a). (c) Shows a region of anterior lateral OFC responding to receipt of a pleasant glucose taste. Reproduced with permission from O'Doherty, Deichmann *et al.* 2002. (d) Shows a region of posterior medial OFC responding following presentation of a pleasant odor. Reproduced with permission from Gottfried, Decinmann *et al.* 2002a.

what stimuli they find rewarding, and indeed assay reward value through subjective measures such as preference rankings or pleasantness ratings. Such stimulus evaluations have been found to have a close correspondence with evoked activity in OFC, indicating that this region not only discriminates between rewards and non-rewards, but also that activity in this region scales with subjective experience of sensory pleasure (Kringelbach *et al.* 2003; O'Doherty *et al.* 2003b). Thirdly, it has been shown that the OFC not only responds to rewards in each sensory modality, but also to abstract rewards not tied to a particular sensory modality, such as money or social praise (Elliott *et al.* 1997; Breiter *et al.* 2001; Knutson *et al.* 2001). Finally, the spatiotemporal resolution of imaging techniques while crude compared to single unit neurophysiology, nevertheless enables responses across the entire OFC to be measured at one time, especially once technical obstacles in measuring signals in this area with fMRI are overcome (see chapter by Stenger in this volume). This makes it possible to address the question of whether different subregions of the OFC have distinct functional contributions in a manner that cannot be accomplished with any other technique. It is the question of regional specialisation within the OFC that we will now address.

## 10.3. **Functional segregation within the OFC modality of reward**

There are good arguments for a hypothesis that separate representations of reward value exist for different types of reinforcer. This is because the value of a reinforcer is determined, in large part, through its association with a specific underlying motivational or drive state. For instance, water has a very different value depending on whether one is thirsty or satiated. Similarly, salted peanuts have much greater value when one is hungry than when one is full. The value of these two different reinforcers need to be represented separately, lest one ends up in the absurd situation of finding salted peanuts highly rewarding when thirsty, and water highly rewarding when hungry. It is only by having separate representations for each reward that is possible to modulate reward value independently as a function of different underlying motivational states.

However, there is an equally good argument for the existence of a unitary representation of reward value that codes for the value of any type of stimulus irrespective of the modality in which it is encoded. In order to make behavioral decisions between different types of available reward (which clearly occurs all the time in the course of everyday life), it is necessary to somehow encode these different rewards in the same representational space or 'currency' (Montague and Berns 2002). For instance, imagine on an evening out you are faced with the dilemma of choosing between the competing goals of going to a restaurant when hungry or foregoing dinner and going to a concert instead. Clearly such a decision requires representation of the value of both goal states. However, in order to make a comparison between them their relative value have to be expressed in a common representational space.

It seems parsimonious to have an arrangement whereby there is a common representation of reward value to facilitate behavioral decision-making, as well as a distinct representations of value for different types of reward, to facilitate goals pertaining to specific

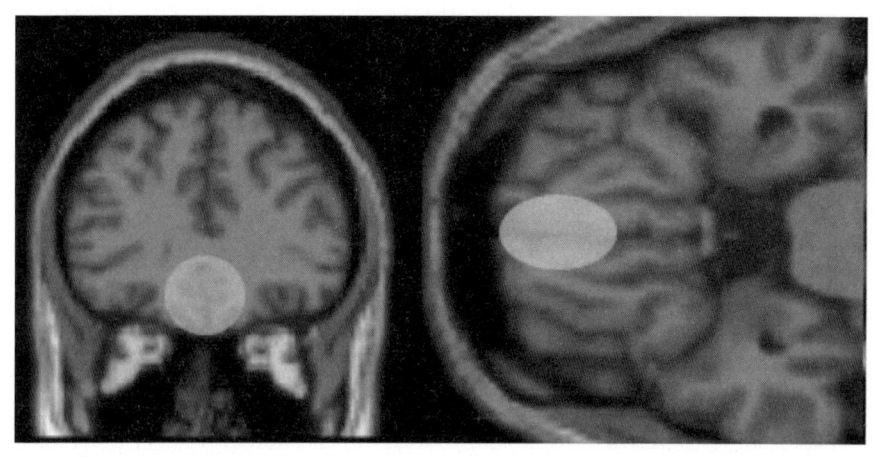

**Fig. 10.2** Illustration of the location of a region proposed to be involved in encoding a common representation of stimulus reward value. The area is located in the anteromedial OFC, and is circled in this figure (superimposed on coronal and axial anatomical slices through the OFC).

motivational states. Although the OFC is a multi-modal area, different sensory pathways project to different sub-regions of the OFC (Carmichael and Price 1995b). Thus, one possible basis for functional segregation within OFC is through sensory modalities engaging distinct sub-regions. Surprisingly, in spite of a substantial number of imaging studies of reward, a clear evidence for functional segregation as a function of reward modality is lacking. This is in large part due to the fact that very few studies have compared responses to different reward modalities within the same paradigm and in the same subjects (see Royet *et al.* 2000 for an exception). Without such direct comparisons, it is difficult to determine whether differences in the spatial localisation of activations reported between studies is a function of the reward modality, of arbitrary differences in task or subjects, or results from random variation between studies. By the same token, evidence for the existence of a supramodal unitary representation of reward value is similarly lacking. However, one region that does appear to be activated more consistently than any other to different types of reward is anterior medial OFC (O'Doherty *et al.* 2001a, 2003a, b; Rolls *et al.* 2003a) Consequently, this sector of the OFC is at present a potential candidate region for mediating a representation of unitary stimulus reward value (the locus of the proposed area is illustrated in Fig. 10.2).

## 10.4. Valence: rewards vs punishments

Another potential basis for functional specialization is that pertaining to the valence of the stimulus: whether it is rewarding (pleasant) or punishing (aversive). The suggestion that there are two distinct motivational systems, an appetitive (or reward-based) and an aversive (or punishment-based) system was first proposed by Konorski (1948), and by others since (Davidson 1992; Rolls 2000). These systems are usually considered to be mutually inhibitory, and are characterized by two different classes of behavior: approach in the case of appetitive stimuli, and avoidance in the case of aversive stimuli. The proposal of distinct appetitive and aversive systems has led to neuroanatomical hypotheses as to the locus of such distinct systems in the brain. These include, a hypothesis of hemispheric specialization for reward and punishment (Davidson 1992), differential involvement of the amygdala in aversive processing, and of most relevance to the present chapter, the suggestion of a differential involvement of the medial OFC in reward, and lateral OFC in punishment (O'Doherty et al. 2001a).

O'Doherty et al. (2001), reported a medial versus lateral dissociation in the OFC responses to reward and punishment during performance of a probabilistic reversal learning task in which subjects could win or lose abstract monetary reward. Medial sectors of the OFC were found to respond to abstract monetary reward, and a part of the lateral OFC was found to respond to monetary loss. Moreover activity in these regions correlated with the magnitude of the money won or lost: the anterior medial OFC correlated positively with the magnitude of money won, while the lateral OFC correlated positively with the magnitude of money lost. A similar dissociation was found in a fMRI study of facial attractiveness in which both high and low attractiveness faces were presented to subjects while they performed an unrelated gender judgment task. Faces high in attractiveness recruited medial OFC whereas low attractive faces recruited the lateral OFC (O'Doherty et al. 2003b). Small et al. (2001) reported a similar dissociation between the medial and the lateral OFC during the consumption of a chocolate meal to satiety. The medial OFC responded during early stages of feeding, when the chocolate had high reward value, whereas enhanced lateral OFC activity was only evident when subjects were reaching satiety and the chocolate went from being pleasant to aversive. A number of imaging studies of olfaction have reported a similar medial versus lateral dissociation, with the medial OFC responding to pleasant odors and lateral to aversive odors (Gottfried et al. 2002b; Anderson et al. 2003; Rolls et al. 2003a). A meta-analysis of neuroimaging studies by Kringelbach and Rolls (2004) also found evidence across diverse studies of a medial-lateral trend in the OFC whereby the medial OFC tended to respond more to rewards, and lateral to punishment.

While these studies appear to support a medial—lateral dissociation, a number of other studies have failed to report such a dissociation (see also Gottfried et al. Chapter 6 this volume). For instance, Elliott et al. (2003) used a block fMRI design to measure neural responses to parametrically varied quantities of monetary gain and loss. Significant activity was reported in both the medial and the lateral OFC to both monetary gain and loss. In a further study of probabilistic reversal learning by O'Doherty et al. (2003a), monetary gains were associated with activity in both the medial and the lateral OFC, whereas monetary loss recruited a part of the posterior lateral OFC only if this was followed by a switch in behavioral strategy on the subsequent trial. These contradictory findings pose a challenge to the simple valence hypothesis involving medial and lateral

OFC function. In order to understand the possible bases of such discrepant findings it may be important to consider two factors. Firstly, many studies that have failed to report OFC functional dissociations are complex gambling or decision-making tasks in which a number of distinct processes are likely to be engaged beyond a mere coding for outcome value, such as anticipation or expectation of reward, selection of appropriate behavioral responses, detecting change in contingencies, and/or implementing changes in behavioral strategy. As we will see in later sections, the OFC seems to play a role in a number of these different processes. Thus, the degree to which these different processes are engaged in a given task could contribute to discrepancies in reported results between studies.

A rewarding or punishing outcome can in some instances provide a clear signal as to an appropriate behavioral strategy, but in other instances this may not be the case. Consider a simple reversal learning task in which one stimulus is rewarded and the other stimulus is punished. The subjects' task is to choose the stimulus that is rewarded and avoid the stimulus that is punished. Subsequently, contingencies are reversed so that the previously rewarded stimulus is punished and the previously punished stimulus is now rewarded. Once subjects obtain a punishment after choosing a particular stimulus then this is a clear signal to switch their choice of stimulus. Thus, there is a clear mapping between the rewarding and punishing outcomes and an appropriate behavioral strategy: following a reward—keep doing what you are doing; following a punishment—stop what you were doing and do something else. This is a natural (and axiomatic) mapping where a reward serves to increase the probability of a given behavior, and a punishment serves to decrease the probability of a given behavior, a strategy known as 'win-stay/lose-shift'. Indeed this is a computationally trivial strategy that does not require calculation of expected values and, in many naturalistic decision-making scenarios, may be a perfectly reasonable basis for choice behavior. For example, if a foraging animal samples a plant with a nutritious and pleasant tasting fruit it is sensible to visit that plant again, whereas if it samples from a plant with fruit that has a bitter aversive taste, it makes sense to not go back to that plant again (at least for a while). The literature on avoidance learning suggests that animals learn very quickly (within one or a few trials) to avoid an aversive stimulus, but that once established, avoidance behavior is strongly resistant to extinction (Solomon and Wynne 1953).

However, in some gambling or decision-making tasks, a monetary loss does not automatically signal that behavior should be switched. With reference to the previous reversal example, if one introduces a probabilistic component to the task whereby sometimes one is rewarded and punished following choice of either stimulus, but that the 'correct' stimulus is rewarded more than punished and the 'incorrect' stimulus is punished more than rewarded, then this introduces an element of ambiguity or uncertainty into the meaning of a rewarding or punishing outcome. It is no longer the case that the reward or punishment itself is a straightforward cue as to what behavioral strategy should be adopted because a punishing outcome can occur following choice of the correct stimulus as well as following choice of the incorrect stimulus. This is the case in many other types of decision-making task besides reversal, such as in gambling tasks where it is advantageous to sustain monetary loss in the short term in order to gain monetarily in the long run. However, such complexities introduce a decoupling between the reward and punishment obtained on a given trial and the competing behavioral strategies of persisting in a given response following a reward versus avoiding that response following a punishment. In future studies it will be interesting to examine neural activity in the medial and the lateral OFC under circumstances when rewarding or punishing outcomes are unambiguous signals, involving a simple mapping between reward and punishments and appropriate behavioral strategies (justifying a win-stay/lose shift strategy), compared to a situation where rewards and punishments are ambiguous signals.

## 10.5. Expected reward value

While in some circumstances it is possible to select behavior purely on the basis of the immediate outcome, we have also seen that in other circumstances such a strategy is non-adaptive. Many real-life decision-making scenarios require choices be made between actions which yield rewards with varying probabilities and/or varying magnitudes. In order to solve such decision problems it is necessary to compute the expected future reward that will follow from choosing a particular action and use these estimated values to guide action selection (for a discussion of neurophysiological evidence on expected reward representations in the OFC, see Chapter 7, by Schultz and Tremblay, this volume).

Reward prediction in its simplest form can be studied through the phenomenon of classical conditioning, in which an arbitrary cue (or conditioned stimulus) takes on predictive value by being associated with subsequent delivery of an affectively significant or unconditioned stimulus (which can be a reward or punishment, or strictly speaking, an appetitive or aversive stimulus). Neuroimaging studies have implicated the OFC alongside other structures such as the amygdala and ventral striatum in reward prediction. An example is a study by O'Doherty *et al.* (2002), where arbitrary fractal stimuli were presented and followed, after a variable interval, by either a pleasant taste (glucose), an affectively neutral taste (control tasteless solution) or by an aversive taste (saline). This design enabled responses during both prediction and receipt of reward to be evaluated separately. Significant effects were found in the antero-lateral OFC during anticipation, as well as receipt, of reward (Fig. 10.3a). These results have subsequently been confirmed in other paradigms, using different types of reward. For instance, Gottfried *et al.* (2002b) studied neural responses to cues associated with subsequent delivery of either a pleasant or aversive odor where each cue was followed on 50% of occasions by a specific odor. Significant orbitofrontal responses (in the anterior central OFC) were found to the predictive cues associated with the pleasant and aversive odors. These findings suggest that the OFC may be involved in maintaining predictions for negatively valenced as well as positively valenced stimuli.

It is also important to consider the content of predictive value representations in the OFC. A conditioned stimulus (CS) can access different aspects of an associated unconditioned stimulus (UCS), such as its sensory properties, its general affective properties (pleasant or aversive), or its specific reward value. To delineate predictive representations that access the specific value of the associated reward, Gottfried *et al.* (2003) performed a study in which predictive cues were associated with one of two food-related odors, and subjects were scanned whilst being presented with such cues before and after feeding to satiety on one of the corresponding foods, thereby selectively devaluing the odor of the food eaten. Predictive responses in the anterior central OFC were found to track the specific value of the corresponding odors, indicating that the reward value and not the sensory properties of a stimulus is coded in this region (Fig. 10.3b).

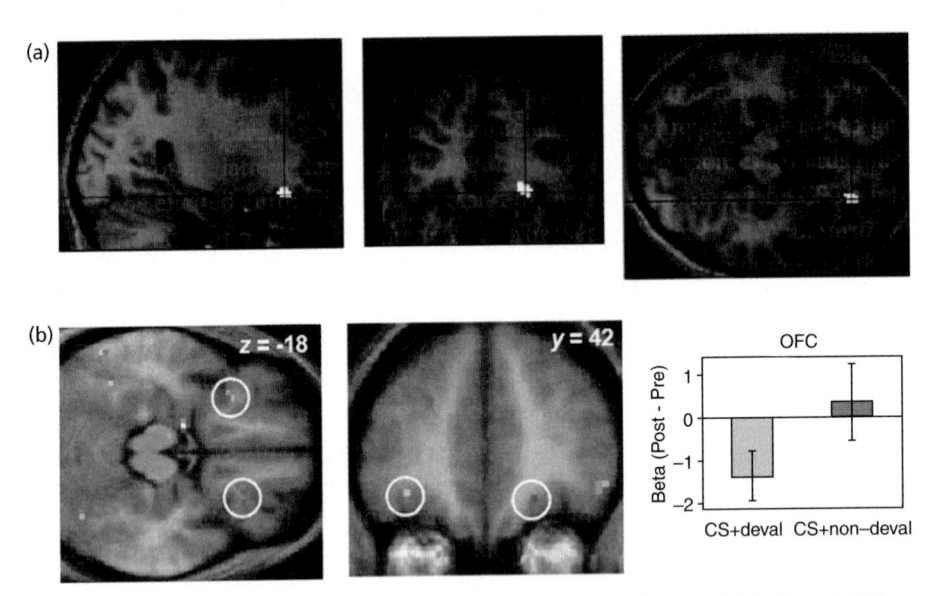

**Fig. 10.3** (Also see Color Plate 20.) Predictive reward value coding in OFC. (a) Region of OFC responding during anticipation of a pleasant taste reward (1M Glucose). The contrast shown is for anticipation of pleasant taste compared to anticipation of a mildly aversive taste (saline). Adapted from O'Doherty *et al.* (2002). (b) Results are shown from a classical conditioning paradigm in which arbitrary visual cues were paired with two food-related odors. Following devaluation of one of the odors (by feeding to satiety on the associated food), neural responses to the predictive cue associated with the devalued odor decreased selectively from pre to post-satiety in OFC (and other brain regions such as amygdala and ventral striatum). Coronal and axial sections are shown through OFC (left two panels). A plot of parameter estimates is also shown (right most panel) illustrating the change in activity in this region from pre-post satiety for the cue associated with the devalued odor (CS+deval) and the cue associated with the non-devalued odor (CS+nondeval). This indicates that predictive value responses in these regions are linked to the specific value of the associated reward. Adapted from Gottfried *et al.* (2003).

## 10.6. Distinct representations for expectation and receipt of reward

Do predictive stimuli access the same or distinct neural representations as that elicited by the reward itself? In order to appreciate the importance of this question it is useful to consider the ultimate function of predictive value representations. As alluded to earlier, predictions enable behavior to be organized prospectively so that an organism is prepared in advance for the occurrence of an affectively significant event. Many such responses can be considered to be reflexive, in that they are automatically elicited by a conditioned stimulus. The paradigmatic example of this is the conditioned salivatory response that Pavlov observed in his food conditioned dogs (Pavlov 1927). In this example as in others the conditioned responses are identical to those elicited by the unconditioned stimulus. For example Pavlov's dogs salivate to the food itself and then also come to salivate to the conditioned stimulus after learning. A central question in learning theory concerns the nature of CS encoding. Pavlov (1927) proposed that a CS constitutes a 'stimulus substitute' for the UCS in that it elicits the same response that occurs following presentation of the UCS (see also Jenkins and Moore 1973). Stimulus substitution could be a very useful mechanism for enabling the animal to know *what* is predicted.

However, not all conditioned responses are identical to those elicited by the unconditioned stimulus. For example, a CS for food reward can involve approach and orientation responses distinct from those produced by the UCS itself (Zener 1937; Holland 1977; Mackintosh 1983). Consequently stimulus substitution may not be the only mechanism by which a CS acquires predictive value. Indeed it would be extremely useful for the animal to have a predictive mechanism that signals an impending behaviorally significant event without eliciting a representation of the event itself. In effect this would enable an animal to distinguish cues that predict a stimulus from the actual UCS itself. In many instances different behavioral responses are appropriate when anticipating a rewarding or punishing event than when experiencing it. If stimulus substitution were to be the only mechanism in place then a CS would be indistinguishable from the UCS from the point of view of the animal. Thus, a light cue predicting food would be treated as if it were the food itself and the animal would attempt to consume it. Intriguingly, this type of behavior has been observed in some instances (Jenkins and Moore 1973). However, given that in many cases, animals (including humans) can distinguish a predictive cue from the UCS itself, as indicated by distinct behavioral responses in these two cases, it seems likely from a behavioral standpoint that there are at least two distinct associative mechanisms in the brain, one based on stimulus substitution and the other uniquely signalling prediction.

At the neural level this issue speaks directly to the content of predictive value representations in regions such as the OFC. According to stimulus substitution, a predictive cue should elicit exactly the same neural activity as that elicited by the reward or punishment itself. Thus, in regions mediating stimulus substitution there should be no difference between activity elicited by the UCS and that elicited by the CS. On the contrary, a brain region mediating a CS-unique representation, should respond exclusively to the predictive cue and never respond to the reward or punishment itself. A recent study by O'Doherty et al. (2003d) directly addressed this question. In this study a total of 6 different cues were repeatedly presented to subjects during an fMRI scan. One cue was reliably associated with the subsequent delivery of a pleasant juice reward (blackcurrant juice), another cue was always followed by an affectively neutral taste (tasteless control solution), and yet another cue was followed by a strongly aversive flavor (cold black tea with 0.5M salt added). In addition to these 'reliable' predictors, three other cues were non-predictive in that they were followed randomly by any one of these three outcomes. Stimulus substitution related responses were distinguished from CS-specific responses by testing for regions that responded to the predictive cue in the 'predictive trials' and determining whether that region responded also to actual delivery of reward or punishment in the 'non-predictive' trials. If so, then this region mediated stimulus substitution and if not, then the region responded exclusively during prediction i.e. it was CS-specific.

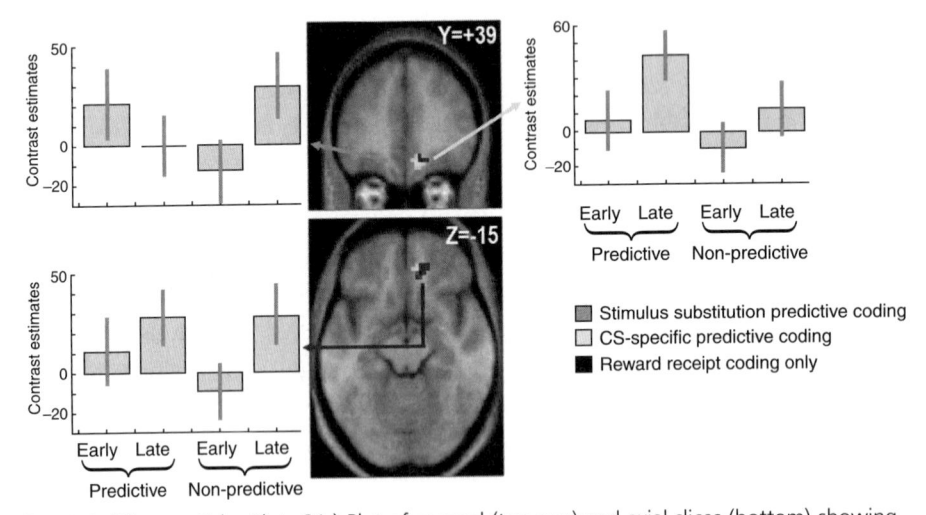

**Fig. 10.4** (Also see Color Plate 21.) Plot of coronal (top row) and axial slices (bottom) showing areas of OFC demonstrating predictive or outcome specific responses during trials in which a rewarding juice stimulus is presented compared to an affectively neutral one. The region shown in green shows a change in the timing of activity as a function of predictability—responding at the time of the outcome (late) in non-predictive trials, and at the time of the cue presentation (early) in predictive trials. This region therefore demonstrates stimulus-substitution. The area shown in yellow shows late responses in the predictive trials but not in the non-predictive trial. This is likely to reflect either tonic anticipatory neural activity, or else a late phasic response before the reward is delivered. As there is no response in this area in the non-predictive trials, this is a form of CS-specific representation. Areas in blue show a conjunction of late component effects in non-predictive and late component effects in predictive trials, indicating areas likely to be involved in responding to the unconditioned stimulus itself irrespective of predictability. Contrast plots of parameter estimates are also shown for each area depicting early and late responses in the predictive and non-predictive trials.

Using this approach, OFC was found to show CS-specific and stimulus-substitution responses, with distinct regions (of the medial and the mid-central OFC) showing each type of response profile. Furthermore, additional regions of the medial OFC were found to code exclusively for receipt of the reward itself and did not show prediction related activity (see Fig. 10.4). These results are compatible with a role for the OFC in both types of reward prediction, indicating that this region signals not only what is predicted, but also distinguishes between predictions and receipt of reward. Other evidence consistent with a role for the OFC in stimulus substitution is reported by Cox et al. (2005).

## 10.7. Prediction error responses in OFC

How does the OFC, and indeed other brain regions, acquire predictive value representations? Contemporary models of animal learning consider that learning occurs via a prediction error which signals a discrepancy between expected and actual reward (or punishment) (Rescorla and Wagner 1972). In one extension of this theory—temporal difference learning, predictions are formed about the expected future reward in a trial, and a prediction error reports differences in successive predictions of future reward (Sutton and Barto 1990). Single unit studies in non-human primates implicate phasic

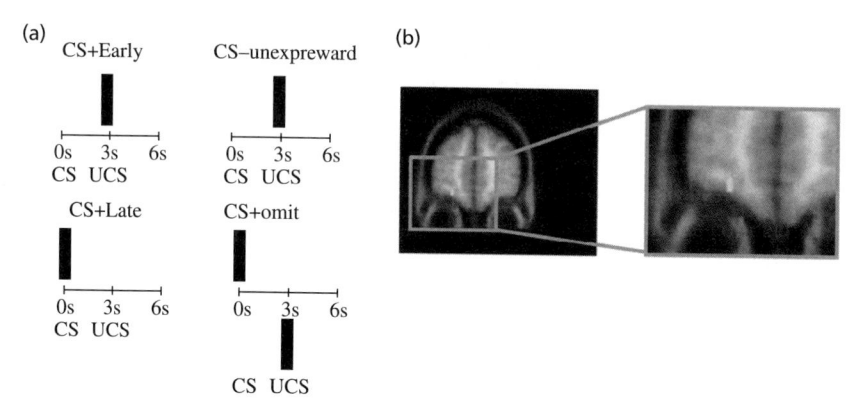

**Fig. 10.5** Prediction error signals in human OFC during a classical conditioning paradigm in which in one trial type (CS+) an arbitrary visual cue is associated 3 secs later with delivery of a taste reward (1M glucose), and in another trial type (CS−) a different cue is followed by no taste. In addition, occasional 'surprise' trials occur in which the CS+ is presented but the reward is omitted (CS+omit), and the CS− is presented but a reward is unexpectedly delivered (CS−unexpreward). (a) Schematic of putative temporal difference prediction error (PE) signals during the experiment. During early CS+ trials (before learning is established) the PE signal should occur at the time of delivery of the reward, whereas by late CS+ trials (post-learning) the signal should have switched to the time of presentation of the CS. On CS+omit trials a positive PE signal should occur at the time of presentation of the CS, but a negative PE signal should occur at the time the reward was expected (CS+omit). CS−unexpreward trials should be associated with a positive signal at the time the reward is presented. (b) Region of OFC showing a significant correlation with the temporal difference prediction error signal. Adapted with permission from O'Doherty et al. 2003c.

activity within dopaminergic neurons as a possible neural substrate of this signal (Schultz *et al.* 1997). Over the course of learning the signal shifts its responses from the reward to the CS, unexpected omission of reward results in a decrease in activity from baseline (a negative prediction error), whereas unexpected presentation of reward results in an increase in activity (positive prediction error).

Human neuroimaging studies of classical conditioning for reward report prediction error signals in prominent target areas of dopamine neurons, including the OFC (Berns *et al.* 2001; O'Doherty *et al.* 2003c) (Fig. 10.5). Dopamine neurons could facilitate learning of value predictions in these areas by gating plastic changes between sensory and reward representations. The value signal in temporal difference learning could either be stimulus-substitution based or CS-specific, depending on whether the model considers the actual reward received on a given trial to be represented in the same way as the learned value signal or distinct from it. Temporal difference learning is flexible, in that it can dynamically cope with changes in reward contingencies on a trial by trial basis. The finding, that neural prediction error signals are present in the OFC and throughout the reward network, is consistent with the possibility that this mechanism is used to mediate flexible learning and updating of stimulus-reward associations, a function often ascribed to the OFC (Rolls 2000). One caveat is that dopaminergic input into cortex may not exhibit temporal characteristics consistent with a teaching signal as found in the striatum (Seamans and Yang 2004). Consequently, it is an open question as to whether such associative learning takes place in the OFC itself, or occurs elsewhere and is then propagated to the OFC via recurrent connections.

## 10.8. Behavioral choice and the OFC

In order to choose between different actions it is necessary to maintain a representation of the likely predicted future reward associated with each action. Such predictions need to be compared and evaluated to select the action with the highest overall predicted reward value. Lesions of the orbital and the ventral medial prefrontal cortex in humans produce impairments on tasks designed to assess an ability to choose optimal actions associated with different overall levels of reward and modify behavior flexibly following a change in reward associations (Bechara *et al.* 1994; Rolls *et al.* 1994; Hornak *et al.* 2004). These studies suggest a role for the OFC in guiding action selection for reward. However, the precise functional contribution of the OFC to the decision-making process is not immediately clear on the basis of those lesion studies. For example, impairments in such patients could be due simply to an inability to maintain representations of the predicted reward necessary to guide decision-making processes instantiated elsewhere. Alternatively, the OFC could play a role in the actual decision-making process itself, engaging in an actual comparison process between the expected values of different actions and in the selection of a specific action on the basis of this evaluation.

As we have seen, there is now considerable evidence to implicate the OFC in maintaining predictions of future reward, consistent with the first possibility. However, a study by

O'Doherty et al. (2003a), also provides evidence to support the second possibility—a role for the OFC in the actual decision-making process itself. This study involved reversal learning, a task described previously, in which subjects must choose between two different actions that yield rewards and punishments with different probabilities. One action is advantageous in that choice of that action has a high probability of reward (70%) and a low probability of punishment (30%), whereas the other action is disadvantageous in that choice of that action yields reward with a low probability (30%) and punishment with high probability (70%). Occasionally, the contingencies reverse such that subjects must evaluate on a trial by trial basis if contingencies have changed and switch their choice action in order to perform adaptively. The design of the study enabled responses to reward and punishment to be dissociated from signals related to behavioral choice. Thus, in this task subjects can make one of two decisions: maintain responding to the current stimulus or switch their choice of stimulus.

In order to separate out the decision process from rewarding and punishing feedback, trials in which a punishment is obtained and followed by a switch in stimulus choice (punish_switch) were evaluated separately from trials in which a punishment is obtained and subjects maintain responding to the current stimulus (punish_no switch). Regions involved in behavioral choice (stay vs. switch) were identified by comparing punish_switch trials to reward (no_switch) and/or punish (no_switch) trials. A region of the medial and the anterior lateral OFC was found to respond on trials in which the subject maintained responding on the subsequent trial (irrespective of whether the outcome was a reward or a punishment) (see Fig. 10.6.) A different region of the posterior lateral OFC (contiguous with anterior insula) was found to respond following a punishment on trials in which subjects switched their choice of stimulus on the subsequent trial, but not otherwise. These findings suggest that the OFC is involved in behavioral choice with different sectors of the OFC signalling the different behavioral strategies—some regions signal that on-going behavior should be maintained, whereas other regions signal that behavior should be changed. Additional evidence consistent with a role for the OFC in behavioral choice has also reported by others. This includes a study by Arana et al. (2003), in which the OFC was recruited while subjects were choosing items from a restaurant menu, as described in more detail below. Furthermore, Ernst et al. (2004) reported a specific role for the OFC in responding during selection (choice) of gambles in a simple two choice decision-making task, again compatible with a role for this region in behavioral choice.

## 10.9. Distinguishing the functional contributions of OFC from other components of the reward network

The OFC is part of a highly interconnected network of brain regions involved in reward-related processing that includes the amygdala, ventral and dorsal striatum, and dopaminergic midbrain among others. One important question over the next ten years will be to determine the distinct contributions of these different components of the

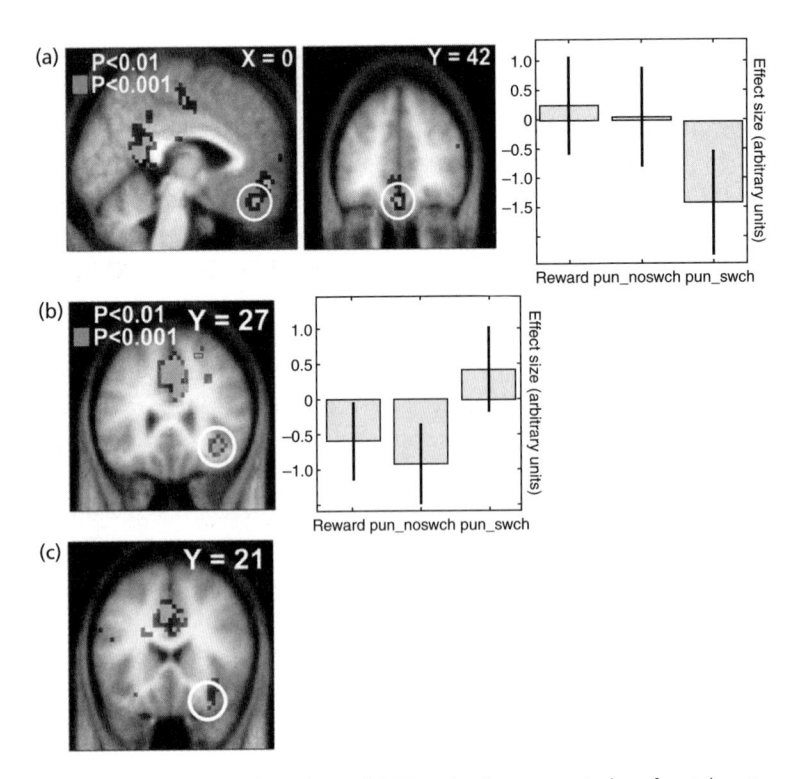

**Fig. 10.6** (Also see Color Plate 22.) Regions of OFC and adjacent ventral prefrontal cortex involved in behavioral choice. Adapted with permission from O'Doherty et al. 2003a).
(a) Sagittal and axial slices through a region of medial OFC that is involved in signaling that a behavioral response should be maintained. The plot of parameter estimates (right of figure) indicates that this region does not respond to rewards or punishments per se, but shows greater responses on rewarding and punishing trials if the subject does not switch their behavior compared to punishing trials followed by a switch in behavior (pun_swch). (b) Region of anterior insula extending into postserior lateral OFC that shows enhanced responses following a punishment if on the subsequent trial the subject switches their choice of stimulus (pun_swch) compared to rewarding or punishing trials were no such switch of behavior occurs. (c) Region of posterior lateral OFC that shows enhanced responses on punished trials following a switch in behavior compared to punished trials followed by no switch in behavior.

reward network. Currently a number of different hypotheses exist in the literature, especially as regards separating the functions of the OFC from that of the amygdala. One suggestion is that these two structures differ in the degree to which they are involved in behavioral choice (Arana et al. 2003). In a PET study, Arana et al. (2003) scanned subjects while they perused fictional restaurant menus containing either high or low incentive items (based on individual preferences). In one condition subjects had to choose specific items, whereas in another condition they merely had to read the menu. The OFC activity was related to choice of menu items (in some instances modulated by incentive value), whereas amygdala responses were related to incentive value, but responses in the amygdala were independent of behavioral choice. Another perspective focuses on the role of the amygdala and the OFC in processsing different components of sensory input. According to this view, the amygdala is involved in coding for the intensity of a stimulus whereas the OFC encodes its valence (Anderson *et al.* 2003; Small *et al.* 2003). Yet another view emphasizes a role for amygdala in the initial acquisition of predictions, in contrast with the OFC which is proposed instead to maintain more flexible predictive representations that can be updated rapidly following changes in contingencies (Rolls 2000; Morris and Dolan 2004).

This latter idea relates to a proposed distinction between the amygdala and ventromedial prefrontal cortex in acquisition and extinction of conditioned fear responses. Conditioned fear associations in the amgydala may not be erased during extinction (as would be predicted by a Rescorla-Wagner or temporal difference learning rule), but appear to remain permanently encoded in that region (Milad and Quirk 2002). Extinction at the behavioral level may occur by preventing expression of conditioned responses via active inhibition of the amygdala (Milad *et al.* 2004). The medial prefrontal cortex in the rat is implicated in this active inhibitory process and recent functional neuroimaging studies propose a role for the ventromedial prefrontal cortex (bordering the orbital surface) in extinction of fear conditioning in humans (Gottfried and Dolan 2004; Phelps *et al.* 2004).

We note that extinction itself represents a form of new learning. Whatever the precise contribution of the OFC (or adjacent medial prefrontal cortices) to extinction learning, the evidence we have discussed for a temporal difference prediction error signal in this region suggests that the OFC is also involved in actively learning reward predictions rather than merely learning to inhibit previously acquired associations. This raises the possibility that there are at least two distinct associative learning systems in the brain—an inflexible possibly amygdala based system, and a flexible system which uses prediction errors to actively update value predictions in the ventral striatum and the OFC. One may speculate that somehow a control mechanism must be in place to enable only one of these two systems to control behavior in the event that they provide opposing predictions (as would happen for example during extinction or reversal learning). This control mechanism could take the form of an active inhibition of one or other system. Further work is needed to discriminate between these different viewpoints. A useful direction in future imaging studies will be to make use of new developments in imaging methodology such as dynamic causal modeling that enable changes in connectivity between different regions as a function of task or stimulus input to be explored (Friston *et al.* 2003). Such a technique could not only aid in determining the distinct contributions of the amygdala and OFC, but also help to characterize the nature of the interactions between these regions during reward-related processing.

## 10.10. Combining functional neuroimaging with lesion studies: the problem of causality

A significant limitation of functional neuroimaging is that notwithstanding its power in establishing correlations between neural activity and behavior, it is extremely difficult using neuroimaging alone to determine causality. For this, it is necessary to resort to other means including lesion studies or other experimental techniques, such as transcranial magnetic stimulation (TMS) or neuropharmacology. With respect to the OFC, technical and physiological limitations preclude the use of TMS and no one neurotransmitter selectively targets OFC. Consequently, human lesion studies (in spite of the many limitations of this approach) are likely to remain the predominant approach in probing causal links between human OFC function and behavior. One fruitful direction for future research lies in combining neuroimaging with lesion studies, as this offers the means to determine the effects of a lesion in one part of the reward network (e.g ventral striatum) on the neural representations in another part of the network (e.g. the OFC). Such studies could provide insight into the causal nature of the interactions between different components of the reward network, thereby complementing the connectivity approaches described above, as well as complementing recent approaches along similar lines involving the combination of single-unit recording and lesion studies in animals (Schoenbaum et al. 2003; see Chapter 8, this volume).

## 10.11. Conclusions

In order to survive, most animals including humans need to be able to learn and flexibly adapt their behavior so that optimal choices can be made in an uncertain environment. In this review we describe evidence from human neuroimaging which suggests that the OFC contributes to this process in at least four distinct ways: by encoding the reward value of a stimulus, maintaining flexible representations of predicted reward value (using both stimulus substitution and CS-specific coding mechanisms), encoding errors in reward prediction and signaling future behavioral choice. Moreover, the OFC shows heterogenous response profiles with distinct regions mediating each of these functions. While much remains unknown about the OFC, particularly as regards differentiating its functions from those of other brain areas, human functional imaging studies have made a considerable contribution toward understanding this most enigmatic of brain areas.

## References

Anderson, A. K., Christoff, K., Stappen, I., Panitz, D., Ghahremani, D. G., Glover, G., et al. (2003) Dissociated neural representations of intensity and valence in human olfaction. *Nat Neurosci* **6**:196–202.

Arana, F. S., Parkinson, J. A., Hinton, E., Holland, A. J., Owen, A. M., and Roberts, A.C. (2003) Dissociable contributions of the human amygdala and orbitofrontal cortex to incentive motivation and goal selection. *J Neurosci* **23**:9632–9638.

Bechara, A., Damasio, A. R., Damasio, H., and Anderson, S. W. (1994) Insensitivity to future consequences following damage to human prefrontal cortex. *Cognition* **50**:7–15.

Bechara, A., Tranel, D., and Damasio, H. (2000) Characterization of the decision-making deficit of patients with ventromedial prefrontal cortex lesions. *Brain* **123**(Pt 11): 2189–2202.

Berns, G. S., McClure, S. M., Pagnoni, G., and Montague, P. R. (2001) Predictability modulates human brain response to reward. *J Neurosci* **21**:2793–2798.

Blood, A. J., Zatorre, R. J., Bermudez, P., and Evans, A. C. (1999) Emotional responses to pleasant and unpleasant music correlate with activity in paralimbic brain regions. *Nat Neurosci* **2**:382–387.

Breiter, H. C., Aharon, I., Kahneman, D., Dale, A., and Shizgal, P. (2001) Functional imaging of neural responses to expectancy and experience of monetary gains and losses. *Neuron* **30**:619–639.

Carmichael, S. T. and Price, J. L. (1995a) Limbic connections of the orbital and medial prefrontal cortex in macaque monkeys. *J Comp Neurol* **363**: 615–641.

Carmichael, S. T. and Price, J. L. (1995b) Sensory and premotor connections of the orbital and medial prefrontal cortex of macaque monkeys. *J Comp Neurol* **363**:642–664.

Carmichael, S.T. and Price, J.L. (1996) Connectional networks within the orbital and medial prefrontal cortex of macaque monkeys. *J Comp Neurol* **371**:79–207.

Cox, S. M., Andrade, A., and Johnsrude, I. S. (2005) Learning to like: a role for human orbitofrontal cortex in conditioned reward. *J Neurosci* **25**(10):2733–40.

Critchley, H. D. and Rolls, E. T. (1996) Hunger and satiety modify the responses of olfactory and visual neurons in the primate orbitofrontal cortex *J Neurophysiol* **75**:1673–1686.

Davidson, R. J. (1992) Anterior cerebral asymmetry and the nature of emotion. *Brain Cogn* **20**:125–151.

Elliott, R., Frith, C. D., and Dolan, R. J. (1997) Differential neural response to positive and negative feedback in planning and guessing tasks. *Neuropsychologia* **35**:1395–1404.

Elliott, R., Newman, J. L., Longe, O. A., and Deakin, J. F. (2003) Differential response patterns in the striatum and orbitofrontal cortex to financial reward in humans: a parametric functional magnetic resonance imaging study. *J Neurosci* **23**:303–307.

Ernst, M., Nelson, E. E., McClure, E. B., Monk,C. S., Munson, S., Eshel, N., *et al.* (2004) Choice selection and reward anticipation: an fMRI study. *Neuropsychologia* **42**:1585–97.

Friston, K. J., Harrison, L., and Penny, W. (2003) Dynamic causal modelling. *Neuroimage* **19**:1273–1302.

Gottfried, J. A. and Dolan, R. J. (2004) Human orbitofrontal cortex mediates extinction learning while accessing conditioned representations of value. *Nat Neurosci* **7**:1144–1152.

Gottfried, J. A., Deichmann, R., Winston, J. S., and Dolan, R. J. (2002a) Functional heterogeneity in human olfactory cortex: an event-related functional magnetic resonance imaging study. *J Neurosci* **22**: 10819–10828.

Gottfried, J. A., O'Doherty, J., and Dolan, R. J. (2002b) Appetitive and aversive olfactory learning in humans studied using event-related functional magnetic resonance imaging. *J Neurosci* **22**:10829–10837.

Gottfried, J. A., O'Doherty, J., and Dolan, R. J. (2003) Encoding predictive reward value in human amygdala and orbitofrontal cortex. *Science* **301**:1104–1107.

Holland, P. C. (1977) Conditioned stimulus as a determinant of the form of the Pavlovian conditioned response. *J Exp Psychol Anim Behav Process* **3**:77–104.

Hornak, J., O'Doherty, J., Bramham, J., Rolls, E. T., Morris, R. G., Bullock, P. R., *et al.* (2004) Reward-related reversal learning after surgical excisions in orbito-frontal or dorsolateral prefrontal cortex in humans. *J Cogn Neurosci* **16**:463–478.

Jenkins, H. M. and Moore, B. R. (1973) The form of the autoshaped response with food or water reinforcers. *J Exp Anal Behav* **20**:163–181.

Knutson, B., Fong, G. W., Adams, C. M., Varner, J. L., and Hommer, D. (2001) Dissociation of reward anticipation and outcome with event-related fMRI. *Neuroreport* **12**:3683–3687.

Konorski, J. (1948) *Conditioned reflexes and neuron organization*. Cambridge: Cambridge University Press.

Kringelbach, M. L., O'Doherty, J., Rolls, E. T., and Andrews, C. (2003) Activation of the human orbitofrontal cortex to a liquid food stimulus is correlated with its subjective pleasantness. *Cereb Cortex* **13**:1064–1071.

Mackintosh, N. J. (1983) *Conditioning and Associative Learning*. Oxford: Clarendon Press.

Milad, M. R. and Quirk, G. J. (2002) Neurons in medial prefrontal cortex signal memory for fear extinction. *Nature* **420**:70–74.

Milad, M. R., Vidal-Gonzalez, I., and Quirk, G. J. (2004) Electrical stimulation of medial prefrontal cortex reduces conditioned fear in a temporally specific manner. *Behav Neurosci* **118**:389–394.

Miller, E. K. and Cohen, J. D. (2001) An integrative theory of prefrontal cortex function. *Annu Rev Neurosci* **24**:167–202.

Montague, P. R. and Berns, G. S. (2002) Neural economics and the biological substrates of valuation. *Neuron* **36**:265–284.

Morris, J. S. and Dolan, R. J. (2004) Dissociable amygdala and orbitofrontal responses during reversal fear conditioning. *Neuroimage* **22**:372–380.

O'Doherty, J., Rolls, E. T., Francis, S., Bowtell, R., McGlone, F., Kobal, G., et al. (2000) Sensory-specific satiety-related olfactory activation of the human orbitofrontal cortex. *Neuroreport* **11**:893–897.

O'Doherty, J., Kringelbach, M. L., Rolls, E. T., Hornak, J., and Andrews, C. (2001a) Abstract reward and punishment representations in the human orbitofrontal cortex. *Nat Neurosci* **4**:95–102.

O'Doherty, J., Rolls, E. T., Francis, S., Bowtell, R., and McGlone, F. (2001b) Representation of pleasant and aversive taste in the human brain. *J Neurophysiol* **85**:1315–1321.

O'Doherty, J. P., Deichmann, R., Critchley, H. D., and Dolan, R. J. (2002) Neural responses during anticipation of a primary taste reward. *Neuron* **33**:815–826.

O'Doherty, J., Critchley, H., Deichmann, R., and Dolan, R. J. (2003a) Dissociating valence of outcome from behavioral control in human orbital and ventral prefrontal cortices. *J Neurosci* **23**:7931–7939.

O'Doherty, J., Winston, J., Critchley, H., Perrett, D., Burt, D. M., and Dolan, R. J. (2003b) Beauty in a smile: the role of medial orbitofrontal cortex in facial attractiveness. *Neuropsychologia* **41**:147–155.

O'Doherty, J. P., Dayan, P., Friston, K., Critchley, H., and Dolan, R. J. (2003c) Temporal difference models and reward-related learning in the human brain. *Neuron* **38**:329–337.

O'Doherty, J. P., Winston, J., Seymour, B., and Dolan, R. J. (2003d) Common and distinct neural responses to predicted and unpredicted pleasant and aversive stimuli in the human brain. Society For Neuroscience Annual Meeting. Abstract

Pavlov, I. P. (1927) *Conditioned reflexes*. Oxford: Oxford University Press.

Phelps, E. A., Delgado, M. R., Nearing, K. I., and LeDoux, J. E. (2004) Extinction learning in humans: role of the amygdala and vmPFC. *Neuron* **43**:897–905.

Rescorla, R. A. and Wagner, A. R. (1972) A theory of Pavlovian conditioning: variations in the effectiveness of reinforcement and nonreinforcement. In: *Classical Conditioning II: Current Research and Theory* (Black, A.H. and Prokasy, W. F., eds.), pp. 64–99. New York: Appleton Crofts.

Rolls, E. T. (2000) The orbitofrontal cortex and reward. *Cereb Cortex* **10**:284–294.

Rolls, E. T., Hornak, J., Wade, D., and McGrath, J. (1994) Emotion-related learning in patients with social and emotional changes associated with frontal lobe damage. *J Neurol Neurosurg Psychiatry* **57**:1518–1524.

Rolls, E. T., Kringelbach, M. L., and de Araujo, I. E. (2003a) Different representations of pleasant and unpleasant odors in the human brain. *Eur J Neurosci* **18**:695–703.

Rolls, E. T., O'Doherty, J., Kringelbach, M. L., Francis, S., Bowtell, R., and McGlone, F. (2003b) Representations of pleasant and painful touch in the human orbitofrontal and cingulate cortices. *Cereb Cortex* **13**:308–317.

Rolls, E. T., Sienkiewicz, Z. J., and Yaxley, S. (1989) Hunger modulates the responses to gustatory stimuli of single neurons in the caudolateral orbitofrontal cortex of the macaque monkey. *Eur J Neurosci* **1**:53–60.

Royet, J. P., Zald, D., Versace, R., Costes, N., Lavenne, F., Koenig, O., et al. (2000) Emotional responses to pleasant and unpleasant olfactory, visual, and auditory stimuli: a positron emission tomography study. *J Neurosci* **20**:7752–7759.

Schoenbaum, G., Setlow, B., Saddoris, M. P., and Gallagher, M. (2003) Encoding predicted outcome and acquired value in orbitofrontal cortex during cue sampling depends upon input from basolateral amygdala. *Neuron* **39**:855–867.

Schultz, W., Dayan, P., and Montague, P. R. (1997) A neural substrate of prediction and reward. *Science* **275**:1593–1599.

Small, D. M., Gregory, M. D., Mak, Y. E., Gitelman, D., Mesulam, M. M., and Parrish, T. (2003) Dissociation of neural representation of intensity and affective valuation in human gustation. *Neuron* **39**:701–711.

Small, D. M., Zatorre, R. J., Dagher, A., Evans, A. C., and Jones-Gotman, M. (2001) Changes in brain activity related to eating chocolate: from pleasure to aversion. *Brain* **124**:1720–1733.

Solomon, R. L. and Wynne, L. C. (1953) Traumatic avoidance learning: Acquisition in normal dogs. *Psychological Review* **61**:353–385.

Sutton, R. S. and Barto, A. G. (1990) Time derivative models of Pavlovian reinforcement. In: *Learning and Computational Neuroscience: Foundations of Adaptive Networks* (Gabriel, M. and Moore, J. eds.), pp. 497–537. Cambridge, MA: MIT Press.

Zener, K. (1937) The significance of behavior accompanying conditioned salivary secretion for theories of the conditioned response. *Am J Psychol* **50**:384–403.

# Chapter 11

# Memory processes and the orbitofrontal cortex

Matthias Brand and Hans J. Markowitsch

## 11.1. Introduction

Memory constitutes one of the most important brain functions involved in nearly all human activities of daily life. Various brain lesions can affect memory processes either partly or entirely as can be seen in amnesic patients suffering from brain damage of different locations and etiologies. Even though widespread brain structures and fiber connections are involved in memory processes, some brain regions are particularly important for specific memory functions: the so-called bottleneck-structures; (e.g. Brand and Markowitsch 2003b; Markowitsch 1995). To evaluate the involvement of those brain regions, it is of crucial importance to distinguish different memory types. Those differentiations can be made regarding time, content or process. Limbic and paralimbic regions as well as parts of the prefrontal cortex are necessarily engaged in some of the memory functions. Whenever information to be encoded or retrieved has an emotional or personal connotation, the orbitofrontal cortex (OFC) is of crucial interest. Specifically, the OFC is involved in autobiographical memory that is mostly emotional and of personal significance by its very nature.

In this chapter, we first briefly describe a content-based distinction of memories. Thereafter, we give an overview of memory-brain relations with a focus on prefrontal and limbic regions. We will present the specific role of the OFC in emotional and autobiographical memory processes and its disturbance or malfunction in patients suffering from neurological or psychiatric diseases. A short description of confabulations and their associations with the OFC is given at the end of the chapter.

## 11.2. Content-based classification of memory

Long-term memory information can be subdivided into different memory systems based on content (e.g. declarative vs. non-declarative, see Squire 1987). One common classification is that of Tulving and Markowitsch (see Markowitsch 2003b). They make a distinction between five sub-systems of long-term memory that concern different types of memory content: (1) Procedural memory comprises motor, sensory, and cognitive skills (e.g. riding a bike); (2) priming means an improved reproduction or recognition of information unconsciously experienced before; (3) perceptual memory allows the recognition of an object or other stimuli on the basis of familiarity without the need to refer to its name or function (= presemantic); (4) semantic memory consists of facts (world, school knowledge); and (5) episodic memory which will be defined and explained later in this chapter. These memory systems are hierarchically organized and involve more or less separate neural networks for their processes which are encoding, consolidating, storing, and retrieving (overview in Markowitsch 2000a, 2000b, 2003b). This chapter particularly addresses the involvement of the OFC in episodic and—to a lesser degree—semantic memory.

The highest system in the hierarchy is episodic memory. It consists of event information stored with contextual information (time, place, and situation), and therefore comprises large parts of our own autobiography. More precisely, episodic memory constitutes the conjunction of subjective time, autonoëtic consciousness and the experiencing self (Tulving 2005). Episodic memories usually have an emotional tone. Though the terms 'episodic memory' and 'autobiographical memory' are often used as synonyms, autobiographical memory also consists of facts of our biography, such as our own name or the name of the street we lived in when we went to school, which, however, constitute information stored in the semantic memory system (see Fig. 11.1). The distinction between autobiographic-episodic and autobiographic-semantic information becomes clear when referring to Tulving's (2005) definition which requires for episodic memory the ability of mental time traveling, the feeling of personal involvement and engagement and a reflection with regard to oneself.

Episodic memory nowadays is seen as being autonoëtic (while noëtic means the conscious retrieval of information without involvement of self-related cognitions, autonoëtic, in this context, stands for conscious remembering that requires an awareness of one's own self). Tulving defines episodic memory as 'a recently evolved, late-developing, and early-deteriorating past-oriented memory system [...], and probably unique to humans. It makes possible mental time travel through subjective time, from the present to the past, thus allowing one to re-experience, through autonoëtic awareness, one's own previous

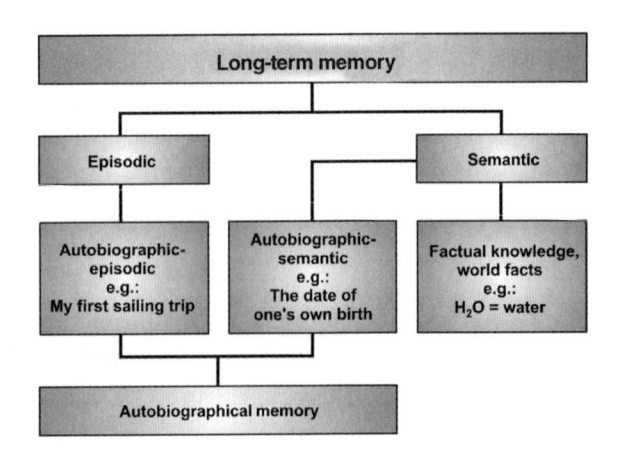

**Fig. 11.1** Autobiographic-episodic, autobiographic-semantic, and semantic long-term memory (see text for details).

experiences. [. . .]. The essence of episodic memory lies in the conjunction of three concepts—the self, autonoëtic awareness, and subjectively sensed time.' (Tulving 2002, p. 5, see also Tulving 2005). A recent description of brain correlates of autonoëtic consciousness and its relation to autobiographical memory is given in Fujiwara and Markowitsch (2005).

## 11.3. Memory processes and the brain

During memory processes the following steps of information processing can be differentiated: information uptake, encoding, consolidation, storage and retrieval. These memory processes involve both overlapping neural networks as well as distinct brain structures, depending on the content of information. Beyond some basal networks necessary for all kinds of memory processes, several brain regions act as so-called bottleneck-structures (Brand and Markowitsch 2003b). This means that these structures are obligatory for reliable memory formation, which will be described in the following paragraphs. We will focus on different parts of the prefrontal cortex, particularly those of the OFC, and their engagement in encoding and retrieval of episodic information.

### Encoding and the orbitofrontal cortex

For encoding (and consolidation) of episodic and semantic memory parts of the medial temporal lobe, the medial diencephalon, and—at least partially—the basal forebrain and parts of other telencephal limbic structures, such as the cingulate gyrus, are assumed to be the primary neural correlates. These structures and interconnecting tracts are organized in two separable but interconnected circuits: the Papez circuit and the basolateral amygdaloid circuit (see Fig. 11.2). The Papez circuit (Fig. 11.2a) consists of the hippocampal formation, the mammillary bodies, the anterior thalamic nuclei, and fiber tracts such as the fornix, the superior thalamic pedunculi, and the cingulum. The basolateral amygdaloid circuit (Fig. 11.2b) comprises the amygdala, the mediodorsal nucleus of the thalamus, parts of the basal forebrain (subcallosal area) and several interconnecting fibers such as the ventral amygdaloid tract. The OFC is interconnected with multiple structures of the Papez and the amygdaloid circuit as well as further limbic and paralimbic regions, like the hippocampal complex and surrounding structures, the amygdala, the medial thalamus, specific nuclei of the hypothalamus, the ventromedial part of the caudate nucleus and the ventral striatum/nucleus accumbens. (For a summary of the most important connections between the OFC and memory-related structures, see Fig. 11.3). Further interconnections assumed to be of particular importance for processing specific memory functions are mentioned in the respective sections of this chapter.

The classical view of the Papez circuit's functions—claimed by Papez himself (Papez 1937)—was that these structures were engaged in emotional processing and therefore emotional memory. However, current research emphasizes the role of the amygdaloid circuit, rather than the Papez circuit, in processing the emotional connotation of memory information (e.g. Zald 2003). Patients with circumscribed damage of the amygdaloid structures often demonstrate particular deficiencies in forming emotional memory

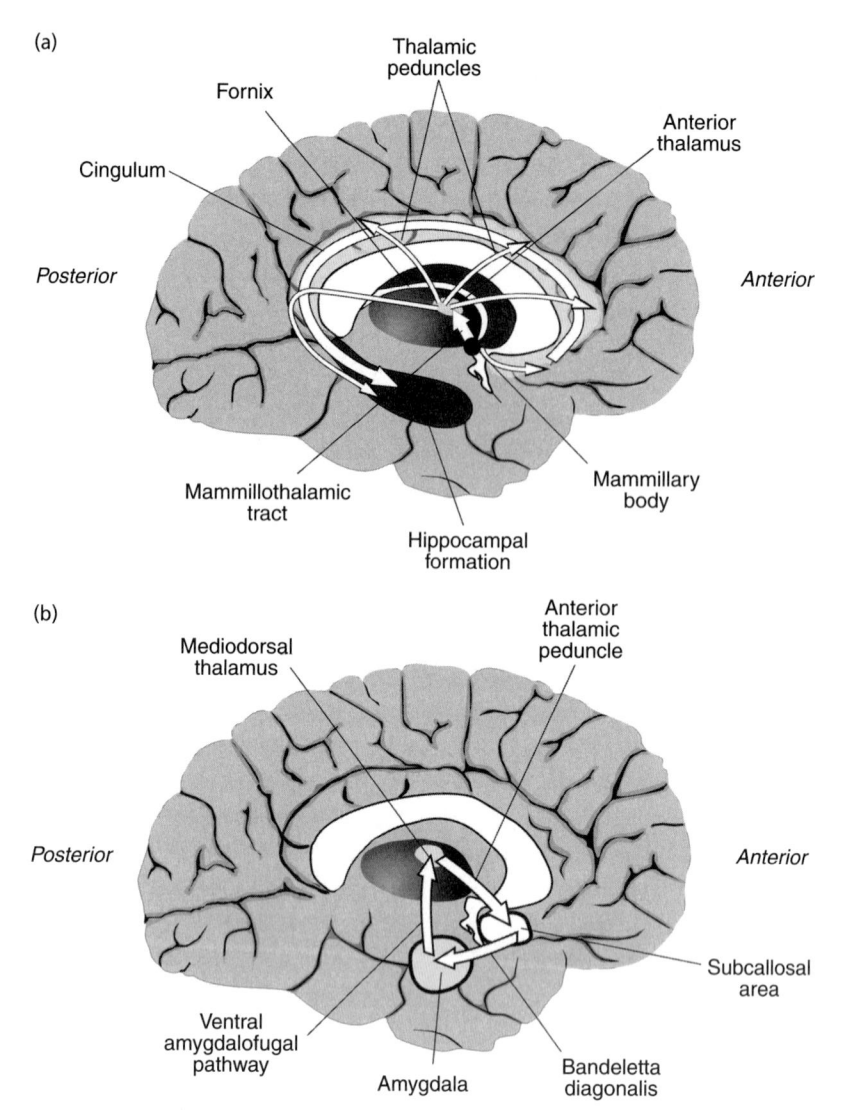

**Fig. 11.2** Two separate but interconnected circuits engaged in the transfer of episodic (and partly semantic) information from short-term to long-term memory. In (a) the Papez circuit is presented; (b) represents the amygdaloid circuit (see text for further information). Modified with permission from Fig. 5 of Brand and Markowitsch (2003b).

(Siebert *et al.* 2003). In contrast, the Papez circuit is now regarded to be broadly involved in information transfer from short-term to long-term stores (encoding and consolidation). Even very small lesions restricted to structures of the Papez circuit can result in severe anterograde memory impairments, involving an inability to learn new information (Calabrese *et al.* 1995).

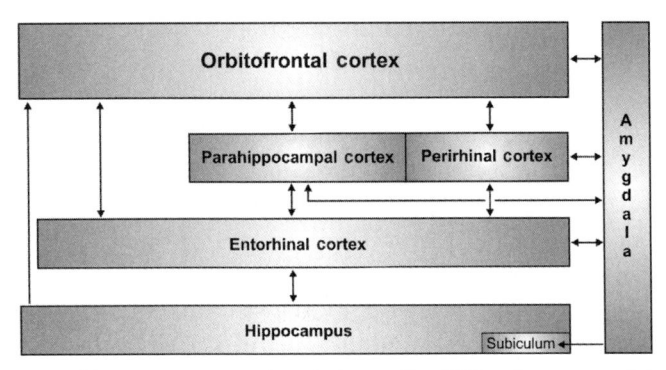

**Fig. 11.3** Summary of important connections between the OFC and memory-related structures. Figure modified from Fig. 4.9 of Zald and Kim (2001). Also see this reference for a detailed distinction of interconnections of specific OFC parts.

Fletcher and Henson (2001) reviewed functional neuroimaging studies on memory encoding and retrieval and suggested an important role of the prefrontal cortex for both episodic and semantic memory acquisition. They specifically addressed the role of the ventrolateral prefrontal cortex, which they defined as including parts of areas 44, 45 as well as the most lateral portions of the OFC (area 47). For the ventrolateral region, left-sided encoding-related activations were found when subjects had to make semantic decisions (Demb *et al.* 1995; Kapur *et al.* 1994; Vandenberghe *et al.* 1996), paired associations and in general verbal encoding (Shallice *et al.* 1994; Wagner *et al.* 1998), distinction between novel versus repeated words (Kopelman *et al.* 1998), proactive interference (Fletcher *et al.* 2000), and intentional learning of words (Grady *et al.* 1998). Instead, a selective activation of the right ventrolateral prefrontal cortex was found during encoding of non-verbal material (Wagner *et al.* 1998). Elliott *et al.* (2000) suggested an important role of the lateral OFC in memory processes that is its engagement in selection of stimuli based on a feeling of 'rightness' (that is most likely related to a feeling of familiarity with a stimulus and to guessing).

More specific engagement of the OFC proper occurs with specific types of material. Studies with animals as well as those with humans highlighted the impact of the OFC on processing olfactory stimuli and on encoding olfactory information (Brodmann area 13; see Markowitsch 1995; Sobel *et al.* 1998; see also Gottfried *et al.*, Chapter 6, this volume).

Beyond OFC's relevance for processing olfactory information, some recent neuroimaging studies point to its engagement in encoding face associations (for the involvement of the OFC in performing a face working memory task, see the study of Courtney *et al.* 1996). In an $^{15}$O-positron emission tomographic (PET) study of Herholz *et al.* (2001) subjects had to learn face-name associations during scanning. The results indicate a left lateralized network (beyond bilateral occipital associations cortices) comprising inferior temporal gyrus, the inferior part of pre- and post-central gyrus, and the OFC (Brodmann area 11). The authors suggest that orbitofrontal activation during face-name encoding most likely reflects identification and encoding of facial emotional expression. This argument is in line with animal studies that observed face sensitive neurons in the orbitofrontal region in monkeys (cf. Rolls 2004) and reports of patients with frontal lobe damage involving the OFC, who are impaired in facial expression identification (e.g., Hornak et al. 1996). Additionally, functional neuroimaging investigations found the OFC activated during processing facial expressions (Dolan et al. 1996) and during processing of famous faces (Gorno-Tempini et al. 1998).

A more general engagement of the OFC during encoding of visual material was demonstrated in a PET-study by Frey and Petrides (2000). They presented abstract visual material to their participants during the scanning procedure and asked the subjects to encode the information for a later retrieval after the scan. Compared to the control condition (presentation of three familiar abstract pictures with the instruction to not encode them) the right OFC (Brodmann area 11) was activated (in addition to bilateral fusiform and entorhinal/perirhinal cortices) during the encoding condition. The authors argue that during encoding—when requirements for executive processes, such as organizing and monitoring of the material to be encoded, are eliminated or less important—the OFC is more engaged than are other frontal lobe regions (such as the dorsolateral prefrontal cortex). Furthermore, they argue that the right OFC has a relative dominance for visual material compared to the left OFC, which is more involved in encoding of verbal information (see above). The results of the study of Frey and Petrides are in line with other neuroimaging studies that emphasize the engagement of the OFC in encoding processes most likely due to its strong connections with the medial temporal lobes and other limbic areas (see also the PET-study of Frey et al. 2004 in which the involvement of the OFC in auditory encoding was examined). Those connections are described in more detail in the following section.

## Retrieval and the OFC

As mentioned above, the orbitofrontal part of the prefrontal cortex influences emotional processing and therefore is presumed to be particularly involved in emotional memory. One reason for this lies in the connections between orbitofrontal and limbic structures (see Figure 11.4), such as the amygdala (cf. Barbas 2000; Zald and Kim 2001)—the structure *par excellence* engaged in perception-emotion processing (cf. LeDoux 2000). Therefore, some authors refer to the OFC as 'expanded limbic system' (Nauta 1979, Nieuwenhuys 1996). In the context of memory retrieval the connections between the anterior temporal pole and the OFC via the ventral branch of the uncinate fascicle (see Figure 11.5) are of crucial interest. This regional combination in the right hemisphere triggers the retrieval of episodic (autobiographical) events, while the one of the left hemisphere primarily triggers retrieving facts (Fink 2003, Fink et al. 1996; Habib et al. 2003, Lepage et al. 1998; see Brand and Markowitsch 2003a). The uncinate fascicle seems to be (gender-independent) asymmetrically organized; it is larger and contains more fibers in the right compared to the left hemisphere (Highley et al. 2002). Therefore, we can conclude that connections between frontal and anterior temporal regions are more extensive within the right hemisphere. This could reflect a correlate of the stronger engagement of right-sided limbic structures of the medial temporal lobe and the right

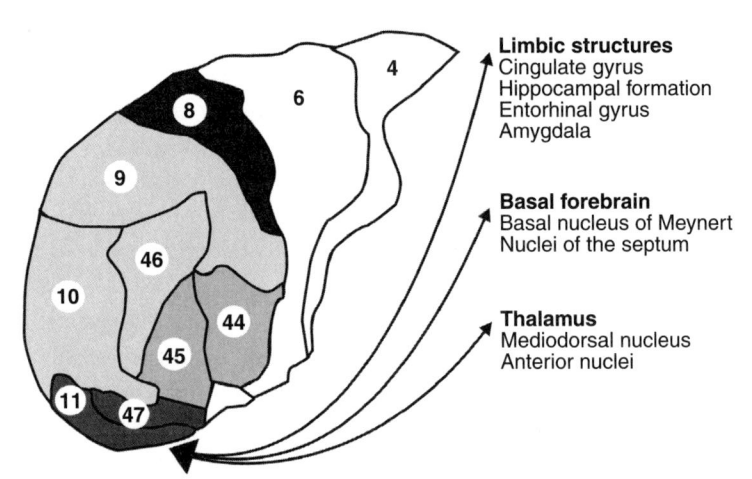

**Fig. 11.4** The main interactive connections between the orbitofrontal cortex and structures of the limbic system, the basal forebrain, and the thalamus. These associations are seen as primary neural correlates for integrating emotional and self-related information necessary for the formation and retrieval of autobiographical-episodic memories (numbers indicate Brodmann areas according to Brodmann 1914).

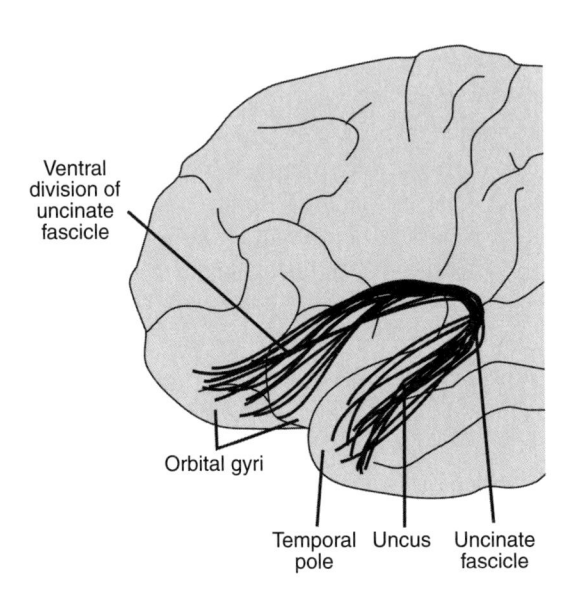

**Fig.11.5** The ventral part of the uncinate fascicle connects the temporal pole with the orbitofrontal cortex. This fiber tract is more extensive on the right side, probably reflecting the dominant role of right-sided limbic structures for autobiographical-episodic memory retrieval (see text). Modified with permission from Fig. 1 of Markowitsch (1995).

prefrontal cortex regions in retrieval of episodic memories—a speculation which makes sense when considering that both emotional and fact-based information has to be combined and synchronized for episodic, but not for semantic memory retrieval.

A recent functional magnetic resonance imaging (fMRI) study (Smith *et al.* 2004) emphasized the role of limbic circuits involving the medial part of the OFC for the retrieval of emotional but not neutral material. During scanning, the authors presented to healthy subjects neutral pictures either in an emotional or in a neutral context for initial encoding and subsequent retrieval. Findings showed an emotion-specific but valence-independent activation pattern including the left amygdala, the insular cortex, the medial OFC, and the middle frontal gyrus for correct recognition of pictures learned in an emotional background. Retrieval of neutral words learned previously in an emotional context also activated OFC but only for words embedded in a positive learning context and not for words embedded in a negative learning context (Maratos *et al.* 2001).

In addition to the relevance of the OFC for retrieval of emotional material, the OFC is also engaged in retrieving only marginally emotional stimuli as evidenced by studies that investigated neural responses for familiar compared to novel stimuli. In this context, Xiang and Brown (2004) discovered in a single cell study in primates neural responses in the ventromedial prefrontal cortex and the OFC (in addition to anterior cingulate gyrus) but not in the dorsolateral prefrontal cortex. Results emphasize the specific aspect of 'familiarity processing' for memory retrieval associated with the ventral prefrontal cortex (indicated by the response latencies of orbitofrontal neurons that started firing after the neurons of the inferior temporal cortex). As outlined, the role of the OFC is stronger (but not exclusive) for visual compared to verbal material in encoding processes. A highly similar distinction of preferences for retrieval of verbal and non-verbal material associated with the OFC is reported by Bunge *et al.* (2004). They revealed that the right pars orbitalis of the inferior frontal gyrus (in addition to anterior hippocampus with a dominance for the right side) was activated during associative retrieval of visual material (paired abstract pictures). However, this pattern was primarily seen for retrieval of 'strong pairs', meaning that these pairs had been shown 11 times during pre-scanning session, but not for the 'weak pairs' which were presented only four times during the learning phase prior to the scanning. Accordingly, the results do not exclusively support the idea that the OFC is just a key structure for retrieval of visual material. Rather, they indicate that the OFC is primarily engaged in successful retrieval of strong compared to weak associations (strong associations require less retrieval effort, that is less of a controlled organization of the to-be-recalled information). This argument converges with the study by Frey and Petrides (2000) in which right OFC activation was found during encoding of visual information when executive processes of encoding had been eliminated.

As defined by Tulving (2002), one of the main characteristics of episodic memory is that contents are stored with respect to time and place during information acquisition (see section 11.2). The time and place aspect of episodic memory retrieval was studied by Fujii *et al.* (2004) using PET. Participants experienced various events before the scanning procedure with an experimental manipulation of the encoding variables 'time' (the events were experienced at different times, that means some of them were experienced before a 15-minute recess, the others thereafter), and 'place' (different rooms for experiencing the events). During scanning, subjects had to retrieve the events from the different conditions (time, place and simple recognition). Results showed that retrieval of the time aspect of the events was linked to activations of the right posterior orbitofrontal and the left inferior parietal cortex. In contrast, place retrieval activated the right parietal cortex, right posterior cingulate gyrus, left precentral region and right cerebellum. The authors suggest that retrieval of time and place information of episodic content is supported by distinct neural networks (in addition to a general retrieval network comprising, for instance, the middle frontal gyrus). Following this argument, it seems plausible that beyond the described role of the OFC for emotional memory retrieval, the (right) posterior OFC is also crucially involved in the processing of time aspects necessary to embed the specific to-be-retrieved episodic information in a coherent time-place relation. Accordingly, the OFC, due to its involvement in processing of emotional and other specific episodic information, plays a major role for autobiographical memory. This topic is discussed in the following section.

## 11.4. Autobiographical memory and the OFC

### Autobiographical memory and the OFC in healthy subjects

Autobiographical memories are inherently personal and usually have an emotional tone (see Markowitsch 2000a,b, 2003a,b). One can argue that the sum of our autobiographical memories builds our personality and our self (Conway and Pleydell-Pearce 2000). Remembering events of one's own autobiography requires retrieval of information stored in episodic long-term memory, the associated emotional tone and a sense of self and past life history (Tulving 2001). The quality of re-experiencing former personal events depends on the retrieval of the emotional connotation of these episodes (e.g. Markowitsch 2003b). To become autobiographical, episodic information has to be (a) evaluated regarding its emotional valence and (b) related to the self across the personal time axis (Larsen *et al.* 1996).

Given that retrieval of autobiographical memories is associated with emotional and self-related processes, the OFC should be crucially involved. However, due to the complex nature of autobiographical episodic memory, involving aspects of self, emotion, and general memory functions, a number of interacting brain regions is engaged. Consequently, a bilateral network of brain regions was found to be activated during autobiographical memory processing in neuroimaging studies with healthy volunteers, including the ventrolateral, dorsolateral and orbitofrontal/ ventromedial prefrontal cortex, temporal pole, lateral and medial temporal cortex (with hippocampal formation and parahippocampal gyrus, temporo-parietal junction area) and posterior cingulate/ retrosplenial cortex as well as parts of the cerebellum (cf. Fink *et al.* 1996, Maguire 2001, Markowitsch *et al.* 1997, 2000). However, there is also evidence for a sensitivity of the ventrolateral prefrontal cortex for the remoteness of memories showing increased activity for recent compared to remote autobiographical events (Maguire *et al.* 2001). Similarly, a differential engagement of right and left limbic structures in autobiographical memory retrieval was found which likely depends on the age of memory. Maguire and Frith (2003b), for example, revealed a temporal gradient only for the right hippocampus with decreasing activity accompanying increasing remoteness of the memories. An age-related engagement of the right compared to the left hippocampus in autobiographical memory retrieval was also demonstrated in the study of Maguire and Frith (2003a) showing that the right hippocampus is more activated in older relative to younger adults.

Some recent neuroimaging studies with normal subjects investigated brain correlates of autobiographical memory retrieval with respect to the different emotional connotations of the information to be retrieved. For instance, Piefke *et al.* (2003) studied retrieval of recent and remote autobiographical memories for negative and positive events using fMRI in healthy subjects. They administered a semi-structured interview before scanning and presented sentences for each participant during fMRI describing the previously reported events. Each participant was asked to narrate ten positive and ten negative events from their recent past as well as the same number of positive and negative events which had happened during their childhood. From each single event, six sentences describing the episode as clearly as possible were generated and used during the fMRI investigation. It was found that remembering positive relative to negative events and irrespective of the time-axis (recent versus remote) activated bilaterally the medial part of the OFC (in addition to the temporal pole and the entorhinal region). On the other hand, negative relative to positive memories led to activations of the right middle temporal gyrus. The role of the medial part of the OFC for autobiographical memory retrieval of positive events is also confirmed by the study of Markowitsch *et al.* (2003). They examined healthy subjects using fMRI and showed them a fixed set of 36 key words to induce remembering of specific autobiographical events (18 positive, 18 negative). The stimuli and the answers given in a pre-scanning interview were grouped along the time-axis: first 12 years of life, age between 12 and 18 years, and age 18 years to the present. Results showed that, across all time periods, retrieval of positive autobiographical events contrasted with negative events activated the left hippocampal region and ventromedial prefrontal cortex (or—more specifically—the subgenual Brodmann area 25) extending into parts of the medial OFC. In contrast, remembering negative episodes was related to a symmetrical activation of the lateral OFC, possibly extending into ventrolateral parts of the prefrontal cortex. Additional activations were reported for the lateral temporal cortex (right) and the cerebellum (left), even though these activations were less prominent compared to the lateral orbitofrontal activation. Figure 11.6 presents the distinct activation patterns within the OFC region for positive and negative autobiographical memory retrieval.

The results of Markowitsch *et al.* (2003) are in accordance with neuroanatomical studies emphasizing (at least partially) separate neural networks including parts of the OFC for positive and negative autobiographical memory retrieval. In particular, the medial part (close to the ventromedial prefrontal cortex) is connected with the hippocampal formation, the parahippocampal gyrus, and the posterior part of the cingulate and retrosplenial region. In contrast, the lateral orbitofrontal area is linked to the amygdala as well as to sensory and premotor areas (Cavada *et al.* 2000). Of these structures the amygdala is associated with experiencing fear and other negative emotions (LeDoux 2000; Morris and Dolan 2004; Siebert *et al.* 2003). Accordingly, the lateral part of the OFC, which is strongly interconnected with the amygdala, receives input from this limbic structure and likely

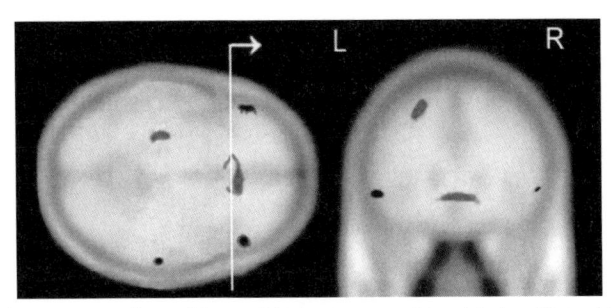

**Fig. 11.6** Activation pattern of autobiographical memory retrieval in healthy subjects. Activations of the lateral orbitofrontal cortex (dark grey) were linked to retrieval of negatively toned episodes while medial orbitofrontal activations (bright grey) were demonstrated for retrieval of positive events. Modified with permission from Fig. 1 of Markowitsch *et al.* (2003).

integrates this information in autobiographical retrieval (but see also Maguire and Frith 2003b, who did not obtain a direct link between activity of the dorsal amygdala and emotional valence or intensity of autobiographical memories). However, it is unclear which specific process of autobiographical memory retrieval is mediated by the OFC. As mentioned above, autobiographical-episodic memory comprises emotional and self-related processes beyond general retrieval functions. The aforementioned studies (Markowitsch *et al.* 2003; Piefke *et al.* 2003) emphasize the importance of the OFC for emotional aspects of autobiographical memories with the medial part being sensitive for positive events and the more lateral OFC being linked to negatively toned episodes. Additionally, there is varying evidence for the involvement of the orbitofrontal region (and ventral medial frontal cortex) in self-referential cognitive processes (Johnson *et al.* 2002; Zysset *et al.* 2002, 2003, but see Schmitz *et al.* 2004), though the main region of self-reflection seems to be the medial prefrontal (both ventral and more dorsal parts) and the posterior cingulate (retrosplenial) gyrus (e.g. Johnson *et al.* 2002). The right ventral prefrontal region is also linked to an 'episodic-memory-retrieval-mode' (Lepage *et al.* 2000).

In summary, a number of studies suggest that the OFC is a key region involved in autobiographical memory, representing self-related and emotional aspects of autobiographic-episodic memories.

## Autobiographical memory in patients with orbitofrontal damage

Due to the close connections between emotion, self, and autobiographical memory, we also expect that patients with lesions restricted to the OFC show impaired autobiographical memory, at least with regard to vivid and detailed re-experiencing of episodes from their personal past. A case study of Levine *et al.* (1998) gives support for this hypothesis. The described male patient suffered from OFC damage, including a disruption of parts of the uncinate fascicle caused by traumatic brain injury. The patient developed retrograde amnesia (i.e. forgetting of remote memories) which initially covered both semantic and episodic remote memories. However, the retrieval deficit of semantic information improved, resulting in isolated autobiographical-episodic retrograde amnesia at the time of examination. Both, structural and functional neuroimaging methods support the assumption of a link between these kinds of retrograde amnesias with damage of the OFC and disruption of the frontotemporal connection. Based on results from an Autobiographical Interview test (Levine et al. 2002), Levine (2004) suggests that, relative to controls, patients with ventral prefrontal lesions (including the OFC but not the dorsolateral prefrontal cortex) are impaired in recall of so-called internal details of their own autobiography (reflecting information about specific events, perceptual information and mental states associated with episodic recall). In contrast, recall of external details (the term external is used to describe factual and nonspecific information that is not embedded in a specific time and space context) seems unaffected in those patients. In the same vein, patients with lesions restricted to the dorsolateral prefrontal cortex are predicted to be unimpaired in retrieving internal details and seem to produce more external details compared to control subjects. Levine (2004) argues that selective ventral prefrontal cortex lesions disrupt connections with medial temporal regions leading to specific episodic memory deficits due to self-regulation impairments. Those lesions might also lead to a reduced ability to perform 'mental time travel' resulting in deficits in consciously re-experiencing episodes from the past. This conclusion is in accordance with the aforementioned interpretation of neuroimaging results obtained from healthy subjects. These results indicate that the OFC is (in addition to other brain regions, primarily limbic structures) involved in re-experiencing emotional and self-related aspects of personal memories.

The aspect of time in retrieving autobiographical-episodic memories (see Fujii et al. 2004), also deteriorated in patients with Alzheimer's disease (Joray et al. 2004) who have orbitofrontal volume decrease (beyond global cortical reductions). Associations between orbitofrontal damage, autobiographical memory deficits and alterations of autonoëtic consciousness are also reported for patients with Alzheimer's disease and frontotemporal dementia (Piolino et al. 2003). These studies emphasize the important role of the OFC in integrating autobiographical-episodic memory and self-related processes.

## Autobiographical memory and the OFC in psychiatric patients

Patients with psychiatric diseases of different etiologies typically have functional brain abnormalities rather than structural alterations. However, there is evidence for both structural and functional brain changes in many psychiatric conditions, including patients with schizophrenia (Staal et al. 2000), depression (Bremner et al. 2002), and post-traumatic stress disorder (PTSD) (Bremner 1999; Elzinga and Bremner 2002).

To investigate functional brain correlates of specific autobiographical memories in PTSD, Driessen et al. (2004) used fMRI to examine retrieval of traumatic and negative but non-traumatic autobiographical events in two groups of patients, one with borderline Personality Disorder (BPD) and another one with BPD and additional PTSD. During the fMRI investigation they displayed individual keywords that induced remembering of specific autobiographical episodes (the keywords were obtained from a semi-structured interview). Two of four episodes were rated as traumatic by the participants and fulfilled DSM IV criteria for traumatic events (American Psychiatric Association 1994; see the criteria for diagnosis of PTSD). The other two episodes were described by the participants as negative but non-traumatic, and these events did not meet the above mentioned DSM IV criteria. The keywords to induce re-experiencing the individual episodes were defined by the participants themselves. In all subjects (patients with BPD with and without PTSD) the OFC, in addition to Broca's region, as well as medial occipital and anterior temporal regions, were significantly activated when traumatic relative to non-traumatic events were compared. However, in the differential analysis of activation patterns in the sub-group of patients with PTSD compared to patients without PTSD, bilateral lateral orbitofrontal activation was only found in the sub-group without current PTSD. In contrast, in the patients with PTSD the OFC was only slightly activated (with a right-sided dominance). In these patients, activations of sensory-motor and temporal regions as well as of the amygdala (bilateral) preponderated. The results support the assumption that the OFC is involved in emotional autobiographical memories and show that in patients with a non-organic causation of brain abnormalities, (e.g. in PTSD, see Nutt and Malizia 2004), specific functional alterations of lateral orbitofrontal activation during retrieval of traumatic compared to negative but non-traumatic events can occur (for further trauma-related activations of limbic structures see Liberzon et al. 1999; hippocampal alterations in PTSD patients are described, e.g. in Elzinga and Bremner 2002 and Bremner et al. 2003a). The finding of decreased activation in the orbitofrontal region, as described by Driessen et al. (2004), is congruent with a PET-study of Bremner et al. (2003b) investigating neural correlates of retrieval of neutral versus emotional word pairs in patients with PTSD. The authors demonstrated decreased activations in PTSD patients compared to healthy participants in a large network of frontal regions including the orbitofrontal and medial prefrontal regions as well as anterior cingulate gyrus, left hippocampal formation and inferior temporal gyrus. Instead, in the patient group compared to control subjects, increased activations were observed in left posterior cingulate gyrus and left middle frontal gyrus, visual and motor cortices as well as left inferior parietal region. However, during retrieval of the neutral word pairs, no differences in neural activity between patients and control subjects were seen. As suggested by the authors of this study, the results give further evidence for PTSD-related dysfunctional neural networks necessarily involved in emotional and memory processes. Beyond this interpretation one can expand the arguments in the direction of deficient (or blocked) self-related episodic memory functions as indicated by a decrease of activations in regions involved in autobiographical memory retrieval (see above). According to this idea, one can speculate that processing of even simple but emotionally valenced word-pairs is suppressed by a blockade of self-related cognitions reflected by a decrease of orbitofrontal and other frontal regions (such as the anterior cingulate gyrus).

Orbitofrontal abnormalities as well as other, primarily prefrontal, brain changes were also obtained in patients with schizophrenia (e.g. Liu et al. 2002; Staal et al. 2000; Weinberger and Faith Berman 1996). Neuropsychologically, executive dysfunctions (Bersani *et al.* 2004; Bustini *et al.* 1999; Stratta *et al.* 2004) and memory deficits (Aleman *et al.* 1999, Cirillo and Seidman 2003; Hill *et al.* 2004; Holthausen *et al.* 2003) are two of the most prominent symptoms of schizophrenia. Fahim *et al.* (2004) examined a monozygotic twin pair with one suffering from schizophrenia and the remaining twin being healthy. They investigated neural correlates of encoding of emotional pictures and found that the OFC was activated to a higher degree in the healthy twin. This result is in accordance with previous studies that showed structural and functional changes in the orbitofrontal region possibly linked to memory dysfunctions. Orbitofrontal changes in schizophrenia are also likely linked to autobiographical memory and emotional impairments. Corcoran and Frith (2003) reported impaired autobiographical episodic memory in patients with schizophrenia that correlated with their deficits in 'theory of mind (ToM)' functions. ToM, also known as 'mindreading' (Baron-Cohen 1995; Baron-Cohen *et al.* 2001), refers to the ability to infer other people's (or one's own) mental states, such as thoughts, intentions, and emotions, and is therefore considered to be an essential skill of social cognition or social intelligence (cf. Adolphs 2001, 2003). The results of Corcoran and Frith (2003) indicate that patients with schizophrenia are deficient in the application of contextual information to retrieve specific social information. This deficit can lead to both autobiographical disturbances (when information about personal events and situations has to be retrieved) and deficits in less person-related judgments, as required in ToM tasks. Further evidence for impaired autobiographical memory in schizophrenia comes from studies comparing autonoëtic remembering and noëtic knowing (e.g. Sonntag *et al.* 2003). Sonntag *et al.* suggest that patients with schizophrenia suffer from impaired strategies to regulate information of autonoëtic awareness with respect to the information's relevance. Impaired autobiographical memory in schizophrenia is also reported by Riutort *et al.* (2003). They used autobiographical fluency and autobiographical enquiry tasks and revealed that both personal-episodic as well as personal-semantic memory were deteriorated in patients with schizophrenia compared to control subjects. Additionally, the patients retrieved less specific details of autobiographical memories (e.g. time and context information being less accurate), a pattern that is also seen in depression (review and meta-analysis in Van Vreeswijk and De Wilde 2004). In depression an over-generalization of remote autobiographical memory often occurs. Instead of narrating temporally and contextually distinctive episodes, patients with depression tend to report categorical descriptions of summarized repeated occasions (Barnhofer *et al.* 2002; de Decker *et al.* 2003; Williams 1996). Using $^{15}$O-PET, Liotti *et al.* (2002) revealed decreased activations in the medial OFC when inducing a sad mood in depressive patients by stimuli obtained from the patients' autobiographical memory scripts. Results indicate that in depression specific pathways involved in transient mood changes and emotional autobiographical memory are dysfunctional. Those disturbances can be seen as neural correlates for autobiographical memory deficits, such as over-generalization of negative information and negatively toned re-experiencing of remote events.

In summary, we conclude that in patients with psychiatric diseases, in the absence of circumscribed brain lesions, functional prefrontal abnormalities can result in memory disorders. Behavioral deficits are frequently seen in the most complex memory system, autobiographical memory, which necessarily includes emotional information and self-related processes. Therefore, it can be affected by emotional disturbances and deficits in autonoëtic consciousness. The OFC thus appears to be one of the key structures for understanding alterations of self-related cognitions and autobiographical memory in psychiatric patients.

## 11.5. Confabulations and the orbitofrontal cortex

As outlined above, the OFC is involved in several memory functions and dysfunctions, due to numerous (largely interactive) connections between limbic and paralimbic regions. A recent neuroimaging study by Treyer *et al.* (2003) highlighted the importance of the OFC for adjusting thoughts and behavior to current stimuli presentation. The healthy participants performed different recognition tasks while brain activation was measured with $H_2^{15}O$-PET. The task comprised four blocks, each consisting of three runs. The three runs of one block were composed of the same stimuli (e.g. line drawings) that were arranged in a different order in each run. Subjects had to decide whether a picture was new or already seen but only within the currently ongoing run (e.g. current relevance in the third run although all pictures were already seen in the first two runs). Results indicated that the left OFC, together with the ventral striatum (nucleus accumbens), parts of the caudate nucleus, substantia nigra, and medial thalamus, is crucially engaged in selecting memories of current relevance (see also the study of Schnider *et al.* 2002, analyzing event-related potentials). The finding of Treyer *et al.* is in accordance with reports of patients with lesions extending into the (posterior) OFC showing problems in memory selection (Schnider 2001). Those patients often exhibit deficits in traditional memory tests and produce confabulations (i.e. plausible but imagined memories that likely fill in gaps in what is remembered) comprising both parts of true past events as well as fictitious elements. An explanation for such deficits is that these patients fail to suppress activated memory aspects that are irrelevant for the ongoing reality. The phenomenon of confabulation is seen in patients with distinct orbitofrontal lesions as well as in patients suffering from basal forebrain damage extending into the posterior OFC. One of the most common etiologies of basal forebrain lesions is an aneurysm rupture of the anterior communicating artery (ACoA) (Borsutzky *et al.* 2000). The so-called ACoA-syndrome consists of the cardinal symptoms of anterograde amnesia, confabulation and personality changes (cf. Böttger *et al.* 1998; De Luca and Diamond 1995). However, some case studies point to the fact that in patients suffering from all three symptoms, the brain lesion is not restricted to the basal forebrain but rather comprises the medial ventral frontal cortex (mainly the posterior part of the OFC) e.g. Beeckmans *et al.* 1998; Eslinger and Damasio 1984. Conversely, in the rehabilitation of confabulations patients with lesions restricted to the orbitofrontal region and not the basal forebrain, and patients with only minimal basal forebrain damage show better neuropsychological outcome and recover earlier from confabulating than patients who suffer more extensive lesions to the basal forebrain (Schnider *et al.* 2000). Accordingly, we suggest that OFC damage is more involved in personality changes and in anterograde amnesia in ACoA-patients while basal forebrain lesions play a major role for the production of confabulations.

The Capgras syndrome (Capgras and Reboul-Lachaux 1923) and the phenomenon of reduplicative paramnesia (Signer 1994) are special syndromes of distorted memory (cf. Markowitsch 2000b). Patients with Capgras syndrome confabulate extensively. They feel that a close relative is replaced by an impostor but as an exact double. Although they recognize obvious familiarity in appearance and behavior, they strongly believe that relatives, friends or pets have been substituted. In those patients, confabulations most likely represent the attempt of rationalisation of their delusional symptoms. The Capgras syndrome can occur in the course of paranoid-hallucinatory schizophrenia, affective disorders or dementia as well as due to specific brain damages or dysfunctions. Limbic lesions spreading into frontal brain regions are the most frequent neurological etiology of Capgras syndrome (cf. Mesulam 1995) which is in accordance with the mentioned role of basal forebrain and orbitofrontal damage in the causation of confabulations in ACoA-syndrome.

## 11.6. Conclusion

A wide network of brain structures and fiber tracts is involved in processing memory, depending on the aspects of time, content and emotionality. Principal structures associated with episodic and autobiographical memory (in principle for both encoding and retrieval) are parts of the limbic system, paralimbic regions and prefrontal areas. Within the frontal lobe, the OFC plays an important role in encoding and retrieving episodic and in particular autobiographic-episodic memories, due to their emotional and self-related nature. In processing such memories, the lateral and medial parts of this prefrontal region are most likely differentially engaged, with the medial part being linked to positive and the lateral part to negative memories. The relevance of the OFC for emotional and autobiographical memory can be seen in neuroimaging studies with healthy participants as well as in those investigating neural correlates of autobiographical memory disturbances in psychiatric patients. From our point of view, we think that the OFC's contribution to memory is more than just an executive control or working memory function. Instead, on the basis of the summarized neuroimaging and patient studies, we propose that the main role of the OFC in (autobiographical) memory lies in the mediation between specific memories, memory-related emotions and a feeling of self-awareness or autonoëtic consciousness.

## References

**Adolphs, R.** (2001) The neurobiology of social cognition. *Current Opinion in Neurobiology* **11**:231–239.
**Adolphs, R.** (2003) Cognitive neuroscience of human social behaviour. *Nature Review Neuroscience* **4**:165–178.

Aleman, A., Hijman, R. de Haan, E. H., and Kahn, R. S. (1999) Memory impairment in schizophrenia: a meta-analysis. *American Journal of Psychiatry* **156**:1358–1366.

American Psychiatric Association (1994) *Diagnostic and Statistical Manual of Mental Disorders*. Washington DC: APA.

Barbas, H. (2000) Connections underlying the synthesis of cognition, memory, and emotion in primate prefrontal cortices. *Brain Research Bulletin* **52**:319–330.

Barnhofer, T. de Jong-Meyer, R., Kleinpass, A., and Nikesch, S. (2002) Specificity of autobiographical memories in depression: an analysis of retrieval processes in a think-aloud task. *British Journal of Clinical Psychology* **41**:411–416.

Baron-Cohen, S. (1995) *Mindblindness*. Cambridge, MA: MIT Press.

Baron-Cohen, S., Wheelwright, S., Hill, J., Raste, Y. and Plumb, I. (2001) The "Reading the Mind in the Eyes" Test revised version: a study with normal adults, and adults with Asperger syndrome or high-functioning autism. *Journal of Child Psychology and Psychiatry* **42**:241–251.

Beeckmans, K., Vancoillie, P., and Michiels, K. (1998) Neuropsychological deficits in patients with an anterior communicating artery syndrome: a multiple case study. *Acta Neurologica Belgica* **98**:266–278.

Bersani, G., Clemente, R., Gherardelli, S., and Pancheri P. (2004) Deficit of executive functions in schizophrenia: relationship to neurological soft signs and psychopathology. *Psychopathology* **37**:118–123.

Borsutzky, S., Brand, M., and Fujiwara, E. (2000) Basal forebrain amnesia. *Neurocase* **6**:377–391.

Böttger, S., Prosiegel, M., Steiger, H., and Yassouridis, A. (1998) Neurobehavioural disturbances, rehabilitation outcome, and lesion site in patients after rupture and repair of anterior communicating artery aneurysm. *Journal of Neurology, Neurosurgery, and Psychiatry* **65**:93–102.

Brand, M. and Markowitsch, H. J. (2003a) Amnesia: neuroanatomic and clinical issues. In: T. Feinberg, T. and Farah, T. (eds.) *Behavioral neurology and neuropsychology*, pp. 431–443. New York: McGraw-Hill.

Brand, M. and Markowitsch H. J. (2003b) The principle of bottleneck structures. In: Kluwe, R. H., Lüer, G., and Rösler, F. (eds.) *Principles of learning and memory*, pp. 171–184. Basel: Birkhäuser.

Bremner, J. D. (1999). Does stress damage the brain? *Biological Psychiatry* **45**:797–805.

Bremner, J. D., Vythilingam, M., Vermetten, E. *et al.* (2002) Reduced volume of orbitofrontal cortex in major depression. *Biological Psychiatry* **51**:273–279.

Bremner, J. D., Vythilingam, M., Vermetten, E. *et al.* (2003a) MRI and PET study of deficits in hippocampal structure and function in women with childhood sexual abuse and posttraumatic stress disorder. *American Journal of Psychiatry* **160**:924–932.

Bremner, J. D., Vythilingam, M., Vermetten, E. *et al.* (2003b) Neural correlates of declarative memory for emotionally valenced words in women with posttraumatic stress disorder related to early childhood sexual abuse. *Biological Psychiatry* **53**:879–889.

Brodmann, K. (1914) Physiologie des Gehirns. In: von Bruns, P. (ed.) *Neue deutsche Chirurgie (Bd. 11, Tl.1)*, pp. 85–426. Stuttgart: Enke.

Bunge, S. A., Burrows, B., and Wagner, A. D. (2004) Prefrontal and hippocampal contributions to visual associative recognition: Interactions between cognitive control and episodic retrieval. *Brain and Cognition* **56**:141–152.

Bustini, M., Stratta, P., Daneluzzo, E., Pollice, R., Prosperini, P., and Rossi, A. (1999) Tower of Hanoi and WCST performance in schizophrenia: problem-solving capacity and clinical correlates. *Journal of Psychiatric Research*, **33**:285–290.

Calabrese, P., Markowitsch, H. J., Harders, A. G., Scholz, A., and Gehlen, W. (1995) Fornix damage and memory: a case report. *Cortex* **31**:555–564.

Capgras, J., and Reboul-Lachaux, J. (1923) L'illusion des 'sosies' dans un délire systématisé chronique. *Bulletin de la Société clinique de médecine mentale* **11**:6–16.

Cavada, C., Company, T., Tejedor, J., Cruz-Rizzolo, R. J., and Reinoso-Suarez, F. (2000) The anatomical connections of the macaque monkey orbitofrontal cortex. A review. *Cerebral Cortex* **10**:220–242.

Cirillo, M. A., and Seidman, L. J. (2003) Verbal declarative memory dysfunction in schizophrenia: from clinical assessment to genetics and brain mechanisms. *Neuropsychology Review* **13**:43–77.

Conway, M. A., and Pleydell-Pearce, C. W. (2000) The construction of autobiographical memories in the self-memory system. *Psychological Review* **107**:261–288.

Corcoran, R., and Frith, C. D. (2003) Autobiographical memory and theory of mind: evidence of a relationship in schizophrenia. *Psychological Medicine* **33**:897–905.

Courtney, S. M., Ungerleider, L. G., Keil, K., and Haxby, J. V. (1996) Object and spatial visual working memory activate separate neural systems in human cortex. *Cerebral Cortex* **6**:39–49.

de Decker, A., Hermans, D., Raes, F., and Eelen, P. (2003) Autobiographical memory specificity and trauma in inpatient adolescents. *Journal of Clinical Child and Adolescent Psychology* **32**:22–31.

De Luca, J., and Diamond, B. J. (1995) Aneurysm of the anterior communicating artery: A review of neuroanatomical and neuropsychological sequelae. *Journal of Clinical and Experimental Psychology* **17**:100–121.

Demb, J. B., Desmond, J. E., Wagner, A. D., Vaidya, C. J., Glover, G. H., and Gabrieli, J. D. (1995) Semantic encoding and retrieval in the left inferior prefrontal cortex: a functional MRI study of task difficulty and process specificity. *Journal of Neuroscience* **15**:5870–5878.

Dolan, R. J., Fletcher, P., Morris, J., Kapur, N., Deakin, J. F., and Frith, C. D. (1996) Neural activation during covert processing of positive emotional facial expressions. *Neuroimage* **4**:194–200.

Driessen, M., Beblo, T., Mertens, M. *et al.* (2004) Post-traumatic stress disorder and fMRI activation patterns of traumatic memory in patients with borderline personality disorder. *Biological Psychiatry* **55**:603–611.

Elliott, R., Dolan, R. J., and Frith, C. D. (2000) Dissociable functions in the medial and lateral orbitofrontal cortex: evidence from human neuroimaging studies. *Cerebral Cortex* **10**:308–317.

Elzinga, B. M., and Bremner, J. D. (2002) Are the neural substrates of memory the final common pathway in posttraumatic stress disorder (PTSD)? *Journal of Affective Disorders* **70**:1–17.

Eslinger, P., and Damasio, A. (1984) Behavioral disturbances associated with rupture of anterior communicating artery. *Seminars in Neurology* **3**:385–389.

Fahim, C., Stip, E., Mancini-Marie, A., and Beauregard, M. (2004) Genes and memory: the neuroanatomical correlates of emotional memory in monozygotic twin discordant for schizophrenia. *Brain and Cognition* **55**:250–253.

Fink, G. R. (2003) In search of one's own past: the neural bases of autobiographical memories. *Brain* **126**:1509–1510.

Fink, G. R., Markowitsch, H. J., Reinkemeier, M., Bruckbauer, T., Kessler, J., and Heiss, W.-D. (1996) Cerebral representation of one's own past: neural networks involved in autobiographical memory. *Journal of Neuroscience* **16**:4275–4282.

Fletcher, P. C., Shallice, T., and Dolan, R. J. (2000) 'Sculpting the response space'—an account of left prefrontal activation at encoding. *NeuroImage* **12**:404–417.

Fletcher, P. C., and Henson, R. N. A. (2001) Frontal lobes and human memory: Insights from functional neuroimaging. *Brain* **124**:849–881.

Frey, S., Kostopoulos, P., and Petrides, M. (2004) Orbitofrontal contribution to auditory encoding. *NeuroImage* **22**:1384–1389.

Frey, S., and Petrides, M. (2000) Orbitofrontal cortex: A key prefrontal region for encoding information. *Proceedings of the National Academy of Sciences of the United States of America* **97**:8723–8727.

Fujii, T., Suzuki, M., Okuda, J. et al. (2004) Neural correlates of context memory with real-world events. *NeuroImage* **21**:1596–1603.

Fujiwara, E., and Markowitsch, H. J. (2005) Autobiographical disorders. In: Feinberg, T. and Keenan, J. (eds.) *The lost self: pathologies of the brain and identity*. New York: Oxford University Press.

Gorno-Tempini, M. L., Price, C. J., Josephs, O. et al. (1998) The neural systems sustaining face and proper-name processing. *Brain* **121**:2103–2118.

Grady, C. L., McIntosh, A. R., Rajah, M. N., and Craik, F. I. (1998) Neural correlates of the episodic encoding of pictures and words. *Proceedings of the National Academy of Sciences of the United States of America* **95**:2703–2708.

Habib, R., Nyberg, L., and Tulving, E. (2003) Hemispheric asymmetries of memory: the HERA model revisited. *Trends in Cognitive Sciences* **7**:241–245.

Herholz, K., Ehlen, P., Kessler, J., Strotmann, T., Kalbe, E., and Markowitsch, H. J. (2001) Learning face-name associations and the effect of age and performance: a PET activation study. *Neuropsychologia* **39**:643–650.

Highley, J. R., Walker, M. A., Esiri, M. M., Crow, T. J., and Harrison, P. J. (2002) Asymmetry of the uncinate fasciculus: a postmortem study of normal subjects and patients with schizophrenia. *Cerebral Cortex* **12**:1218–1224.

Hill, S. K., Beers, S. R., Kmiec, J. A., Keshavan, M. S., and Sweeney, J. A. (2004) Impairment of verbal memory and learning in antipsychotic-naive patients with first-episode schizophrenia. *Schizophrenia Research* **68**:127–136.

Holthausen, E. A., Wiersma, D., Sitskoorn, M. M., Dingemans, P. M., Schene, A. H., and van den Bosch, R. J. (2003) Long-term memory deficits in schizophrenia: primary or secondary dysfunction? *Neuropsychology* **17**:539–547.

Hornak, J., Rolls, E. T., and Wade, D. (1996) Face and voice expression identification in patients with emotional and behavioural changes following ventral frontal lobe damage. *Neuropsychologia* **34**:247–261.

Johnson, S. C., Baxter, L. C., Wilder, L. S., Pipe, J. G., Heiserman, J. E., and Prigatano, G. P. (2002) Neural correlates of self-reflection. *Brain* **125**:1808–1814.

Joray, S., Herrmann, F., Mulligan, R., and Schnider, A. (2004) Mechanism of disorientation in Alzheimer's disease. *European Neurology* **52**:193–197.

Kapur, S., Craik, F. I., Tulving, E., Wilson, A. A., Houle, S., and Brown, G. M. (1994) Neuroanatomical correlates of encoding in episodic memory: levels of processing effect. *Proceedings of the National Academy of Sciences of the United States of America* **91**:2008–2011.

Kopelman, M. D., Stevens, T. G., Foli, S., and Grasby, P. (1998) PET activation of the medial temporal lobe in learning. *Brain* **121**:875–887.

Larsen, S. F., Thompson, C. P., and Hansen, T. (1996) Time in autobiographical memory. In: Rubin, D. C. (ed.) *Remembering or past*, pp. 129–156. New York: Cambridge University Press.

LeDoux, J. E. (2000) The amygdala and emotion: A view through fear. In: Aggleton, J. P. (ed.) *The Amygdala*, pp. 289–310. Oxford: Oxford University Press.

Lepage, M., Ghaffar, O., Nyberg, L., and Tulving, E. (2000) Prefrontal cortex and episodic memory retrieval mode. *Proceedings of the National Academy of Sciences of the United States of America* **97**:506–511.

Lepage, M., Habib, R., and Tulving, E. (1998) Hippocampal PET activations of memory encoding and retrieval: the HIPER model. *Hippocampus* **8**:313–322.

Levine, B. (2004) Autobiographical memory and the self in time: brain lesion effects, functional neuroanatomy, and lifespan development. *Brain and Cognition* **55**:54–68.

Levine, B., Black, S. E., Cabeza, R. et al. (1998) Episodic memory and the self in a case of isolated retrograde amnesia. *Brain* **121**:1951–1973.

Levine, B., Svoboda, E., Hay, J. F., Winocur, G., and Moscovitch, M. (2002) Aging and autobiographical memory: dissociating episodic from semantic retrieval. *Psychology and Aging* **17**:677–689.

Liberzon, I., Taylor, S. F., Amdur, R. *et al.* (1999) Brain activation in PTSD in response to trauma-related stimuli. *Biological Psychiatry* **45**:817–826.

Liotti, M., Mayberg, H. S., McGinnis, S., Brannan, S. L., and Jerabek, P. (2002) Unmasking disease-specific cerebral blood flow abnormalities: mood challenge in patients with remitted unipolar depression. *American Journal of Psychiatry* **159**:1830–1840.

Liu, Z., Tam, W. C., Xie, Y., and Zhao, J. (2002) The relationship between regional cerebral blood flow and the Wisconsin Card Sorting Test in negative schizophrenia. *Psychiatry and Clinical Neurosciences* **56**:3–7.

Maguire, E. A. (2001) Neuroimaging studies of autobiographical event memory. *Philosophical Transactions of the Royal Society of London—Series B* **356**:1441–1451.

Maguire, E. A., and Frith, C. D. (2003a) Aging affects the engagement of the hippocampus during autobiographical memory retrieval. *Brain* **126**:1511–1523.

Maguire, E. A., and Frith, C. D. (2003b) Lateral asymmetry in the hippocampal response to the remoteness of autobiographical memories. *Journal of Neuroscience* **23**:5302–5307.

Maguire, E. A., Henson, R. N., Mummery, C. J., and Frith, C. D. (2001) Activity in prefrontal cortex, not hippocampus, varies parametrically with the increasing remoteness of memories. *NeuroReport* **12**:441–444.

Maratos, E. J., Dolan, R. J., Morris, J. S., Henson, R. N., and Rugg, M. D. (2001) Neural activity associated with episodic memory for emotional context. *Neuropsychologia* **39**:910–920.

Markowitsch, H. J. (1995) Which brain regions are critically involved in the retrieval of old episodic memory? *Brain Research Reviews* **21**:117–127.

Markowitsch, H. J. (2000a) The anatomical bases of memory. In: Gazzaniga, M. S. (ed.) *The new cognitive neurosciences*, pp. 781–795. Cambridge: MIT Press.

Markowitsch, H. J. (2000b) Memory and amnesia. In: Mesulam, M.-M. (ed.) *Principles of behavioral and cognitive Neurology*, pp. 257–293. New York: Oxford University Press.

Markowitsch, H. J. (2003a). Autonoëtic consciousness. In: David, A. S., and Kircher, T. (eds.) *The self in neuroscience and psychiatry*, pp. 180–196. Cambridge: Cambridge University Press.

Markowitsch, H. J. (2003b) Psychogenic amnesia. *NeuroImage* **20**:S132-S138.

Markowitsch, H. J., Thiel, A., Kessler, J., von Stockhausen, H.-M., and Heiss, W.-D. (1997) Ecphorizing semi-conscious episodic information via the right temporopolar cortex—a PET study. *Neurocase* **3**:445–449.

Markowitsch, H. J., Thiel, A., Reinkemeier, M., Kessler, J., Koyuncu, A., and Heiss, W-D. (2000) Right amygdalar and temporofrontal activation during autobiographic, but not during fictitious memory retrieval. *Behavioural Neurology* **12**:181–190.

Markowitsch, H. J., Vandekerckhove, M. M., Lanfermann, H., and Russ, M. O. (2003) Engagement of lateral and medial prefrontal areas in the ecphory of sad and happy autobiographical memories. *Cortex* **39**:643–665.

Mesulam, M.-M. (1995) Notes on the cerebral topography of memory and memory distortion: a neurologist's perspective. In: Schacter, D. L. (ed.) *Memory distortion*, pp. 379–385. Cambridge, MA:Harvard University Press.

Morris, J. S., and Dolan, R. J. (2004) Dissociable amygdala and orbitofrontal responses during reversal fear conditioning. *NeuroImage* **22**:372–380.

Nauta, W. J. H. (1979) Expanding borders of the limbic system concept. In: Rasmussen, T. and Marino, R (eds.) *Functional neurosurgery*, pp. 7–23. New York: Raven Press.

Nieuwenhuys, R. (1996) The greater limbic system, the emotional motor system and the brain. In: Holstege, G., Bandler, R. and Saper, C. B. (eds.) *The emotional motor system (Progress in Brain Research, Vol. 107)*, pp. 551–580. Amsterdam:Elsevier.

Nutt, D. J., and Malizia, A. L. (2004) Structural and functional brain changes in posttraumatic stress disorder. *Journal of Clinical Psychiatry* **65**:11–17.

Papez, J. W. (1937) A proposed mechanism of emotion. *Archives of Neurology and Psychiatry* **38**:725–743.

Piefke, M., Weiss, P. H., Zilles, K., Markowitsch, H. J., and Fink, G. R. (2003) Differential remoteness and emotional tone modulate the neural correlates of autobiographical memory. *Brain* **126**:650–668.

Piolino, P., Desgranges, B., Belliard, S. et al. (2003) Autobiographical memory and autonoetic consciousness: triple dissociation in neurodegenerative diseases. *Brain* **126**:2203–2219.

Riutort, M., Cuervo, C., Danion, J. M., Peretti, C. S., and Salame, P. (2003) Reduced levels of specific autobiographical memories in schizophrenia. *Psychiatry Research* **117**:35–45.

Rolls, E. T. (2004) The functions of the orbitofrontal cortex. *Brain and Cognition* **55**:11–29.

Schmitz, T. W., Kawahara-Baccus, T. N., and Johnson, S. C. (2004) Metacognitive evaluation, self-relevance, and the right prefrontal cortex. *NeuroImage* **22**:941–947.

Schnider, A. (2001) Spontaneous confabulation, reality monitoring, and the limbic system—a review. *Brain Research Reviews* **36**:150–160.

Schnider, A., Ptak, R., von Daniken, C., and Remonda, L. (2000) Recovery from spontaneous confabulations parallels recovery of temporal confusion in memory. *Neurology* **55**:74–83.

Schnider, A., Valenza, N., Morand, S., and Michel, C. M. (2002) Early cortical distinction between memories that pertain to ongoing reality and memories that don't. *Cerebral Cortex* **12**:54–61.

Shallice, T., Fletcher, P., Frith, C. D., Grasby, P., Frackowiak, R. S., and Dolan, R. J. (1994) Brain regions associated with acquisition and retrieval of verbal episodic memory. *Nature* **368**:633–635.

Siebert, M., Markowitsch, H. J., and Bartel, P. (2003) Amygdala, affect, and cognition: Evidence from ten patients with Urbach-Wiethe disease. *Brain* **126**:2627–2637.

Signer, S. F. (1994) Localization and lateralization in the delusion of substitution. Capgras symptom and its variants. *Psychopathology* **27**:168–176.

Smith, A. P., Henson, R. N., Dolan, R. J., and Rugg, M. D. (2004) fMRI correlates of the episodic retrieval of emotional contexts. *NeuroImage* **22**:868–878.

Sobel, N., Prabhakaran, V., Desmond, J. E. et al. (1998) Sniffing and smelling: separate subsystems in the human olfactory cortex. *Nature* **392**:282–286.

Sonntag, P., Gokalsing, E., Olivier, C. et al. (2003) Impaired strategic regulation of contents of conscious awareness in schizophrenia. *Consciousness and Cognition* **12**:190–200.

Squire, L. R. (1987). *Memory and Brain*. New York: Oxford University Press.

Staal, W. G., Hulshoff, Pol, H. E., Schnack, H. G., Hoogendoorn, M. L., Jellema, K., and Kahn, R. S. (2000) Structural brain abnormalities in patients with schizophrenia and their healthy siblings. *American Journal of Psychiatry* **157**:416–421.

Stratta, P., Arduini, L., Daneluzzo, E., Rinaldi, O. di, Genova, A., and Rossi, A. (2004) Relationship of good and poor Wisconsin Card Sorting Test performance to illness duration in schizophrenia: a cross-sectional analysis. *Psychiatry Research* **121**:219–227.

Treyer, V., Buck, A., and Schnider, A. (2003) Subcortical loop activation during selection of currently relevant memories. *Journal of Cognitive Neuroscience* **15**:610–618.

Tulving, E. (2001) Episodic memory and common sense: how far apart? *Philosophical Transactions of the Royal Society of London—Series B* **356**:1505–1515.

Tulving, E. (2002) Episodic memory: from mind to brain. *Annual Reviews of Psychology* **53**:1–25.

Tulving, E. (2005) Episodic memory and autonoesis: uniquely human? In: Terrace, H. and Metcalfe, J. (eds.) *The Missing Link in Cognition: Evolution of Self-knowing Consciousness*, pp. 3–56. New York: Oxford University Press.

van Vreeswijk, M. F. and de Wilde, E. J. (2004). Autobiographical memory specificity, psychopathology, depressed mood and the use of the Autobiographical Memory Test: a meta-analysis. *Behavioural Research and Therapy* **42**:731–743.

Vandenberghe, R., Price, C., Wise, R., Josephs, O., and Frackowiak, R. S. (1996) Functional anatomy of a common semantic system for words and pictures. *Nature* **383**:254–256.

Wagner, A. D., Poldrack, R. A., Eldridge, L. L., Desmond, J. E., Glover, G. H., and Gabrieli, J. D. (1998) Material-specific lateralization of prefrontal activation during episodic encoding and retrieval. *NeuroReport* **9**:3711–3717.

Weinberger, D. R., and Faith Berman, K. (1996) Prefrontal function in schizophrenia: confounds and controversies. *Philosophical Transactions of the Royal Society of London—Series B* **351**:1495–1503.

Williams, J. M. G. (1996) Depression and the specificity of autobiographcial memory. In: Rubin, D. C. (ed.) *Remembering our past*, pp. 244–267. New York: Cambridge University Press.

Xiang, J. Z., and Brown, M. W. (2004) Neuronal responses related to long-term recognition memory processes in prefrontal cortex. *Neuron* **42**:817–829.

Zald, D. H. (2003) The human amygdala and the emotional evaluation of sensory stimuli. *Brain Research Brain Research Reviews* **41**:88–123.

Zald, D. H., and Kim, S. W. (2001) The orbitofrontal cortex. In: Salloway, S. P., Malloy, P. F. and Duffy, J. D. (eds.) *The frontal lobes and neuropsychiatric illness*, pp. 33–69. Washington DC:American Psychiatric Press.

Zysset, S., Huber, O., Ferstl, E., and von Cramon, D. Y. (2002) The anterior frontomedian cortex and evaluative judgment: an fMRI study. *NeuroImage* **15**:983–991.

Zysset, S., Huber, O., Samson, A., Ferstl, E. C., and von Cramon, D. Y. (2003) Functional specialization within the anterior medial prefrontal cortex: a functional magnetic resonance imaging study with human subjects. *Neuroscience Letters* **335**:183–186.

# Chapter 12

# The role of lateral orbitofrontal cortex in the inhibitory control of emotion

Christine I. Hooker and Robert T. Knight

## 12.1. Introduction

The prefrontal cortex (PFC) governs the executive control of information processing and behavioral expression, including the ability to selectively attend to and maintain information, inhibit irrelevant stimuli and impulses, and evaluate and select the appropriate response (Knight *et al.* 1995; Miller and Cohen 2001). This cognitive and behavioral control facilitates successful achievement of complex goal-directed behaviors. Some evidence suggests that multiple regions of the PFC have the capacity to perform multiple types of executive control functions (i.e. evaluate, maintain, inhibit, or select) suggesting a lack of regional specificity (Duncan and Owen 2000). However, other evidence indicates that particular regions of the PFC are biased toward specific functions and information domains (Shimamura 2000; Muller *et al.* 2002).

In particular, evidence indicates that the orbitofrontal cortex (OFC) participates in the executive control of information processing and behavioral expression by inhibiting neural activity associated with irrelevant, unwanted, or uncomfortable (e.g. painful) information, sensations, or actions (Shimamura 2000). The role of the OFC in inhibition has gained increasing prominence in the literature due to the dramatic rise in research investigating the neural correlates of social and emotional processing. Most investigations of social or emotional processing reveal that the OFC is involved; however, the exact role that it plays is still debated. Here, we review evidence that the lateral OFC, extending to the ventrolateral PFC, facilitates successful goal-oriented behavior by inhibiting the influence of emotional information in the context of physical sensation, selective attention, emotion regulation, judgement and decision-making and social relationships.

## 12.2. The mechanism of inhibition

Inhibition is the process of suppressing or restraining an action, sensation, feeling, thought, or desire. It is a general mechanism that is employed over many different types of stimuli and behavior. Inhibition is usually volitional, such as consciously suppressing the impulse to yell at a bad driver on the road or the desire to eat a piece of chocolate cake when you are on a diet. However, inhibition can also occur automatically, such as the suppression of neural activity that happens without full conscious awareness. For example, this automatic suppression can be observed in the decrease in neural response to the second sensory stimulus in a rapid sequence of stimuli. This automatic suppression or 'gating' of sensory information regulates the amount of information that the organism receives at one time, thereby facilitating the ability to process and organize that information efficiently (Adler et al 1982; Waldo and Freedman 1986; Knight et al. 1989; Freedman et al. 1996). Data suggests that both volitional and automatic inhibition involve the interaction of 'top-down' inhibitory control mechanisms originating from orbitofrontal, and other prefrontal areas, and 'bottom-up' sensory and stimuli based properties represented in primary sensory (e.g. V1, V2, somatosensory, etc.) and association (e.g. inferior temporal lobe for visual objects) cortex (Knight et al. 1989).

Several proposed theories provide a framework for understanding the role of the OFC in inhibition. The Dynamic Filtering Theory proposes that the PFC, as a whole, operates as a dynamic filtering mechanism for the multitude of incoming sensory information by maintaining selected neural activations and inhibiting (or 'filtering') others (Shimamura 2000). According to this view, different regions of the PFC govern different domains of information, with the OFC providing a dynamic filter for affective information (as opposed to cognitive information which is governed by lateral PFC (Shimamura 2000)).

The Disruption Theory, on the other hand, proposes that negative affect is received as an 'alarm signal,' which automatically instigates a conscious, evaluative process of that negative situation or stimulus. This evaluation inhibits (or 'disrupts') negative affect through orbital- and ventral PFC-mediated 'top-down' neural projections that suppress neural activity associated with negative affect (Lieberman 2003; Eisenberger and Lieberman 2004). Other theoretical frameworks emphasize a broader role for the OFC in the inhibition (Iverson and Mishkin 1970; Roberts and Wallis 2000) and integration (Bechara et al. 2000) of emotional information on behavior. Here, we consider recent research using multiple experimental paradigms in light of these theoretical frameworks.

## 12.3. OFC anatomy

In order to exert inhibitory influence, the OFC must receive information about sensory stimuli in the internal (e.g. physical sensations) and external environment. After monitoring this incoming information, the OFC uses its neural projections to suppress the neural activity associated with those stimuli in order to regulate their impact on the organism's behavior. The anatomy of the OFC is particularly suited for this function. The OFC receives neural inputs from every sensory modality—olfactory, gustatory, vision, auditory, and somatosensory—and is ideally suited to monitor information from multiple sources. In addition, the OFC has direct projections to the primary and secondary sensory cortices, and can modulate the strength of the neural signal coming from that sensory cortex, regulating the influence of that sensory signal on the rest of the brain, and, ultimately, on behavior (Barbas 2002; Kringelbach and Rolls 2004).

In addition to its reciprocal connections to primary and secondary sensory cortices, the OFC has immense reciprocal connections with subcortical structures, such as the amygdala, thalamus, and periaquaductal gray area, that are central in emotion processing, and thus has the perfect architecture for modulating neural activity associated with affective information and affectively motivated behavior (Barbas 2002; Kringelbach and Rolls 2004).

The best evidence for the functional use of this neural architecture in inhibitory control is a dynamic relationship between neural structures in which increased activity in the OFC is specifically related to decreased activity in the neural region where that stimulus is represented. In this chapter, we review evidence for such a dynamic relationship.

For ease of discussion in the chapter, we will refer to the region shown in red as the dorsolateral PFC (DLPFC) [BA 9, 46], green as the medial PFC (MPFC) [BA 8], purple as the ventral medial PFC (VMPFC) [BA 25, 32], turquoise as the ventrolateral PFC (VLPFC) [BA 44, 45, 47], light blue as the medial OFC (BA 11, 12) and navy blue as the lateral OFC (BA 47, 11). See Color plate 19 to view in color.

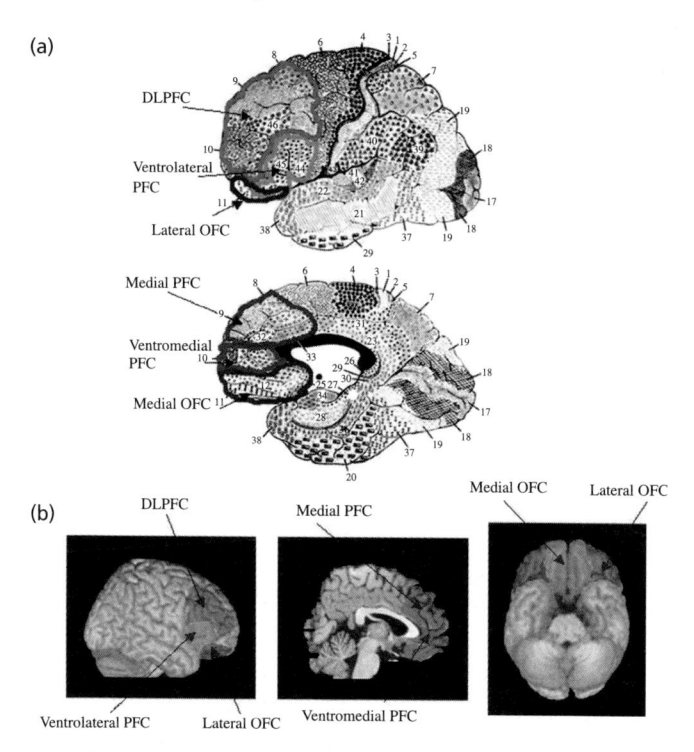

**Fig.12.1** (Also see Color Plate 19.) (a). Original Brodmann map colorized to highlight different Brodmann areas as well as basic prefrontal regions. Adapted with permission from Mark Dubin (http://spot.colorado.edu/~dubin/talks/brodmann). (b) Basic prefrontal regions shown on a canonical brain. Area shown in red is DLPFC, in pink is ventrolateral PFC, in purple is ventromedial PFC, in green is medial PFC, in light blue is medial OFC, and in dark blue is lateral OFC.

## 12.4. Regulation of sensation

Data from multiple sources illustrate that the OFC suppresses neural activity related to aversive or painful sensations. Rule *et al.* (2002) investigated neurophysiological response to unpredictable, mildly aversive sensory events in healthy adults, DLPFC lesion patients and OFC lesion patients (Shimamura 2000; Rule *et al.* 2002). Scalp event-related potentials were recorded while subjects received mild shocks to the wrist (i.e. somatosensory stimuli) or heard loud bursts of noise while watching a silent movie. For both the somatosensory and the auditory conditions the OFC patients showed enhanced neural activity, observed as larger P300 amplitude, in response to the sensory stimuli, when compared to the DLPFC group as well as healthy controls. Interestingly, the DLPFC lesion patients had reduced ERP responses to the sensory stimuli, a finding that has been demonstrated before and suggests that the DLPFC is important for the maintenance, or 'up-regulation,' of sensory stimuli.

Additionally, both the DLPFC and healthy control participants showed habituation of neural responding, reflected in decreasing P300 amplitudes, at later presentations of the aversive somatosensory stimuli. In contrast, for the OFC lesion patients, neural response remained exaggerated over successive trials and did not habituate (Rule *et al.* 2002). The process of habituation facilitates goal-oriented behavior by enhancing attention to novel stimuli in the environment, subsequently allowing attentional resources to be re-allocated once initial processing of the stimuli has been accomplished. If habituation does not occur, the person may become overwhelmed with sensory input and have a difficult time allocating attentional and neural resources to other tasks. This increased neural noise in the system can lead to impaired attention as well as greater interference of non-relevant information, ultimately leading to reduced memory capacity and a disruption in goal-oriented behavior.

This enhanced neural responding to both auditory and somatosensory stimuli occurred 300 ms after the onset of the stimulus and was observed most robustly at the Pz electrode. This component (i.e. the P300) most likely reflects neural activity from the temporo-parietal junction (TPJ) a region of multimodal association cortex (Soltani and

**Fig 12.2.** The graph shows mean P3 amplitude to somatosensory and auditory stimuli for orbitofrontal cortex (OFC) patients and controls and for dorsolateral prefrontal (DLPFC) patients and controls. As compared with other groups, the OFC patients exhibit grossly disinhibited P3 responses.
Reproduced with permission from Rule *et al.* (2002) with permission from Psychonomic Society Publications.

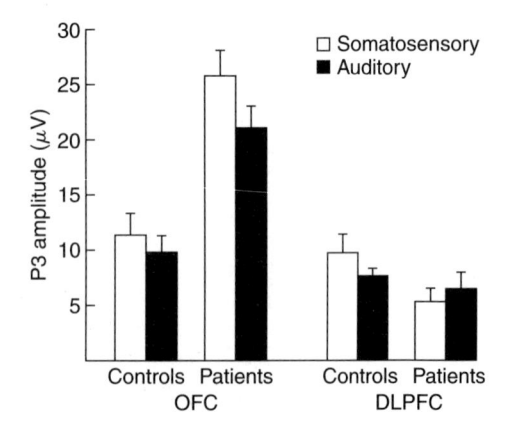

Knight 2000). However, enhanced neural response in the OFC patients was also observed at posterior frontal electrodes (Fz) 150 ms after the somatosensory stimulus (i.e. the N150). These data suggest that the OFC normally inhibits aversive sensations very early in sensory processing by regulating neural activity in posterior sensory association areas. Interestingly, OFC lesion patients have normal posterior P300 amplitudes in nonemotional tasks (Hartikainen *et al.* 2000b), suggesting that, consistent with the Dynamic Filtering Theory, the OFC specifically regulates automatic processing of emotional and/or aversive information.

A subsequent study measured multiple types of emotional and physiological response, including heart rate, skin conductance, and facial expression, when OFC patients received a surprise loud burst of noise in an unanticipated startle task (Roberts *et al.* 2004). OFC lesion patients displayed more surprise behavior and reported more fear than the normal controls in response to the noise. This disinhibition of behavior is consistent with the idea that the OFC may act as a filtering system to aversive sensory inputs such as an acoustic startle.

## 12.5. Suppression and modulation of pain

The ability to suppress neural activity in response to physical sensation is most important if that physical sensation is painful. Acute and chronic pain can be highly disruptive to people's lives and an immense amount of research effort is devoted to understanding components of pain treatment, especially the effects of placebo, since placebo treatment can lead to pain relief without damaging side effects (Harden 2005; Wager 2005). Multiple studies investigating the effect of placebo treatment on pain show evidence of a dynamic interaction such that increased lateral OFC/VLPFC activity is related to a decrease in neural activity in neural regions associated with the origin of the painful sensation, which is then related to a decrease in self-reported pain symptoms (Petrovic and Ingvar 2002; Petrovic *et al.* 2002; Lieberman *et al.* 2004; Wager *et al.* 2004). For example, Lieberman *et al.* (2004) tested chronic pain patients pre- and post-placebo treatment. They found that activity in the dorsal ACC, a region associated with the affective component of pain symptoms, decreased between pre- and post-testing, whereas right VLPFC activity increased between pre- and post-treatment. Importantly, at post-treatment, right VLPFC activity was negatively correlated with dorsal ACC activity and positively correlated with symptom improvement. The authors interpret these results as consistent with the Disruption Theory and suggest that the right VLPFC mediates a conscious, cognitive evaluation and expectation associated with placebo (i.e. 'I am getting treatment, so my pain symptoms will not be as bad') and these expectations suppress activity in pain associated brain regions. Wager *et al.* (2004), provides evidence consistent with this idea. They show that increased neural activity in the lateral OFC (bilaterally) and DLPFC (bilaterally) during anticipation of pain predict less pain related activity in the thalamus, insula and dorsal ACC. In addition, greater activity in both the OFC and DLPFC was related to less pain. These findings add a temporal component and suggest that the OFC can inhibit pain sensation by increasing activity in the anticipation of a painful stimulus.

Interestingly, this same pattern of brain activity was also observed during social pain. Subjects played a computer 'CyberBall' game in which they believed they were playing with two other people on computers outside of the scanner (Eisenberger *et al.* 2003). Over time, the subject was excluded from the game and watched while the two (supposed) outside participants played with each other. The dorsal ACC was active during the exclusion period, and activity in this region was positively correlated with the amount of distress that the subject felt, suggesting that this dorsal ACC region also registers social pain. In addition, right VLPFC activity was negatively correlated with dorsal ACC activity, suggesting that the VLPFC was regulating neural activity related to the social distress of being excluded. These findings converge with the idea that right VLPFC mediates cognitive evaluation that mitigates neural activity associated with the affective component of pain.

However, it is not clear that a conscious evaluation of pain expectancy, or a reappraisal of the pain situation is necessary for the recruitment of lateral OFC/VLPFC activity to dampen pain related activity, since engaging in an unrelated cognitive task, such as a spatial maze, shows the same interaction of increased lateral OFC activity and decreased pain related activity in the somatosensory cortex as well as diminished discomfort (Petrovic *et al.* 2000). In addition, electrical stimulation of the lateral OFC leads to a decrease in pain-related behaviors in rats (Zhang *et al.* 1997), suggesting that the lateral OFC can suppress pain-related activity through neural projections, and conscious evaluation is not necessary.

## 12.6. Regulation of affective information in selective attention

In order to successfully achieve stated goals, it is crucial to selectively attend to information relevant to the goal, and inhibit information that is irrelevant. Evidence suggests that the lateral OFC facilitates selective attention by controlling the interference of irrelevant emotional information in the environment on spatial attention. For example, Hartikainen *et al.* (2001) showed that OFC lesion patients could not regulate the influence of emotional primes on target detection. During the task, participants had to identify whether a briefly presented triangle in the right or left visual field was upright or inverted. Just prior to this target triangle, a positive, negative, or neutral picture prime (from the IAPS collection) was centrally presented. Subjects were told to ignore the picture and do their best on the task. Due to an inability to inhibit the influence of the emotional primes, OFC patients' reaction times were particularly sensitive to the emotional priming conditions (Hartikainen and Knight 2001; Hartikainen *et al.* 2001). Healthy participants' reaction times were speeded up in positive emotion priming conditions and slowed down in negative emotion priming conditions (Hartikainen *et al.* 2000a). This effect of emotion prime on reaction time was significantly enhanced for patients with OFC damage (Hartikainen *et al.* 2001). The patients showed quicker reaction times to detect the target immediately after a positive picture and slower reaction times to detect the target immediately after a negative picture.

In addition, OFC lesion patients showed a different pattern of electrophysiological response, especially during negative emotion trials. Patients had enhanced neural activity in posterior regions (i.e. P200 amplitude) for the negative emotional stimuli, and, an abnormal pattern of slow wave neural activity during target detection, including increased slow wave positivity at posterior scalp sites, and reduced slow wave negativity at frontal sites. Together these data suggest that the OFC is involved in inhibitory modulation of posterior cortical brain regions when emotion influences attention. Without this top-down modulation, neural regions that process emotional information are neurally disinhibited, and, as a result, the emotional information has an exaggerated effect on attention.

This role for the lateral OFC in facilitating selective attention by inhibiting interference is also demonstrated in fMRI research. Vuilleumier *et al.* (2001) showed a display of two faces and two houses arranged horizontally and vertically around a fixation cross. Subjects were given instructions either to attend to the houses and ignore the faces, or attend to the faces and ignore the houses, and then identify whether the two attended stimuli were matching or not. The faces had either a neutral or fearful expression. The results showed that the lateral OFC was engaged when fearful faces had to be ignored in order to make a judgement about the houses. However, the lateral OFC was not active when ignoring neutral faces. In addition, posterior visual association cortex showed modulation based on spatial attention: activity in the fusiform gyrus face processing area increased when subjects attended to faces and decreased when faces were ignored. The findings suggest that the lateral OFC is involved in inhibiting the irrelevant emotional information to assist the subject in completing the task, and it further indicates that lateral OFC activity may facilitate enhanced attention by suppressing the neural activity associated with the representation of the interfering stimulus (Vuilleumier *et al.* 2001).

Bishop *et al.* (2004) looked at the influence of state anxiety on this task with the idea that people high in anxiety will have a harder time ignoring irrelevant threatening stimuli in the environment. They found that the left lateral OFC/VLPFC as well as left DLPFC were both involved in ignoring the fearful face. Specifically, they found an inverse relationship between state anxiety and VLPFC/DLPFC activity during blocks of trials that necessitated ignoring the fearful face. Healthy, non-anxious people show increasing activity in these regions in anticipation of a 'to be ignored' fearful stimulus, suggesting that these subjects are employing VLPFC activity to help them suppress the interfering influence of the threat related stimulus when the emotional information is not relevant to the task. However, people high in anxiety did not recruit these regions to help them ignore the fearful stimuli, suggesting a failure in top down regulatory control in anxiety (Bishop *et al.* 2004).

## 12.7. Emotion regulation

The relationship between OFC mediated top-down inhibitory control and bottom-up representational activity in posterior and subcortical areas has been illustrated several times in studies of emotion regulation.

Emotion regulation refers to a variety of behavioral processes that individuals use to influence which emotions they have, when they have them, and how they experience and express them. It includes decreasing, maintaining or increasing both negative and positive emotions through rationalization, reappraisal, and suppression (Gross 2002). Investigations of the neural basis of emotion regulation suggest that multiple PFC regions, including the lateral OFC/VLPFC, control emotional experience and expression.

The neural basis of emotion regulation through suppression has been explored with feelings of anger, sadness, sexual arousal, and general negativity (Beauregard *et al.* 2001; Blair 2001; Levesque *et al.* 2003). For example, the OFC (BA 11) and VLPFC (BA 47) are active when subjects try to decrease their erotic feelings to erotic films as well as when they decrease their feelings of sadness to sad films (Levesque *et al.* 2003; Beauregard *et al.* 2001). Additionally, greater activity in the right OFC and right DLPFC was associated with more intense feelings of sadness during the suppression condition, indicating that more intense feelings of sadness required more OFC mediated inhibitory strength to suppress.

Reappraisal is a cognitive evaluation in which a person consciously re-interprets the meaning of a situation in order to change their emotional response to it. For example, the experience of getting fired from a job is usually distressing. A person's appraisal of the situation can increase or decrease negative affect associated with such an event. Engaging in a reappraisal strategy to decrease negative emotions associated with the event would include evaluations highlighting positive outcomes and emotions: 'This is a blessing in disguise; the job was not right for me anyway. Now I am free to find a job that is better suited to me, and I can be happier.' Reappraising the situation in such a way that increases distress would focus on negative outcomes and emotions: 'Getting fired is the worst thing that could happen. I am worthless. I may never work again. I will not be able to support myself.' Reappraisal as a means of decreasing distress is a relatively effective, healthy strategy for coping with emotionally difficult or aversive events (Gross 2002).

Decreasing distress by reappraisal employs VLPFC- and lateral OFC-mediated top-down regulation strategies which have their effect by decreasing neural activity associated with the representation of emotional feeling (Ochsner *et al.* 2002; Ochsner *et al.* 2004; Phan *et al.* 2005). For example, subjects were shown negative or neutral scenes and instructed to either allow themselves to feel whatever emotions came up for them or to reappraise the situation depicted in the scene in order to decrease their emotional response (Ochsner *et al.* 2002; Phan *et al.* 2005). When people allowed themselves to feel the full negative impact of negative scenes, regions associated with emotional feeling, such as the medial OFC, insula, and amygdala, were active; however, when subjects reappraised negative scenes so that their interpretation decreased their negative feelings, the VLPFC and the DLPFC were active (Ochsner *et al.* 2002, Phan *et al.* 2005). In addition, lateral OFC (Phan *et al.* 2005) and VLPFC (BA 44, 46) (Ochsner *et al.* 2002) activity during reappraisal was inversely correlated with activation in the amygdala. This is specific neural evidence that activity in the VLPFC has inhibitory effects on the neural representation of those feelings in the amygdala. Furthermore, an increase in bilateral VLPFC and bilateral DLPFC activity was related to a decrease in self-reported negative affect

(Phan *et al.* 2005), illustrating that this neural dynamic ultimately has its effect on emotional experience.

Though both VLPFC and DLPFC activity have been observed in emotion regulation, there is evidence that lateral OFC is particularly involved in strategies employed to decrease negative affect. In a direct comparison of using reappraisal to increase negative affect (i.e. up-regulation) as compared to decrease negative affect (i.e. down-regulation), Ochsner *et al.* (2004) found that the lateral OFC was more active when using reappraisal to decrease as opposed to increase negative affect, and the DLPFC was more active when using reappraisal to increase as opposed to decrease negative affect. Consistent with the idea that amygdala activity is related to emotional experience, the amygdala was more active when subjects used reappraisal to increase their distress during picture viewing and less active when using reappraisal to decrease their distress. This modulation of amygdala activity occurred within 2 seconds of employing the reappraisal strategy (Ochsner *et al.* 2004). In addition, Phan *et al.*(2005) found that as activity level in the DLPFC and amygdala increased, so did the experience of negative affect. Together these studies suggest that the DLPFC is involved in broad regulatory control of experience, especially in bringing emotional representations on line, maintaining them, and strengthening them, whereas the lateral OFC is involved in the inhibition of these cortically represented feelings and images.

## 12.8. Controlling the influence of mood congruent bias

In the absence of regulation, feelings and moods can influence and distort perception, attention and judgement. Specifically, people have a bias towards attending to factors in their environment that are congruent with their mood, a phenomenon known as mood-congruent bias (Bradley and Mogg 1994; Bradley *et al.* 1995). For example, anxious people are quicker than non-anxious people to detect threatening stimuli in the environment; spider phobics are quicker than snake phobics to detect photographs of spiders among snakes and snake phobics are quicker than spider phobics to detect snakes among spiders (Ohman *et al.* 2001a, b). Depressed people are more likely than non-depressed people to interpret a neutral situation as negative (Gotlib and Olson 1983; Gotlib and McCann 1984; Gotlib and McCabe 1992). This has also been shown in the context of social interactions; after an experimental mood manipulation, subjects in a sad mood are more likely to judge a neutral face as sad and subjects in a happy mood are more likely to judge a neutral face as happy (Halberstadt and Niedenthal 2001; Innes-Ker and Niedenthal 2002). Without inhibiting this distorting influence, mood congruent bias can distort perception and perpetuate emotional disorders such as depression, and anxiety.

Researchers are starting to investigate the neural influences of mood congruent bias. For example, in a go/no-go task, depressed, non-depressed and healthy subjects were presented with sad, happy and neutral words (Elliott *et al.* 2002). In certain blocks of trials subjects responded by button press to happy words and withheld response to sad words or they responded to sad words and withheld response to happy words.

Prior research indicates that depressed individuals are more attentive to mood-congruent sad words, and therefore have more difficulty ignoring these words when they have to be ignored. This additional effort needed to overcome mood-congruent bias is reflected in neural response; depressed individuals had more right lateral OFC/VLPFC activity when ignoring sad words (and attending to happy words) as compared to ignoring happy (and attending to sad words).

Mood-congruent bias has also been shown in risk-taking behavior; failure to regulate the influence of mood can lead to bad choices. Behaviorally, when people are in a good mood, they are more likely to underestimate the risk in a situation and engage in risky behavior, and, conversely, when people are angry or upset, they are more likely to overestimate risk and avoid risky behavior (Lerner and Keltner 2001). Therefore the mood you are feeling at a particular point in time will influence your judgement and behavior and this influence needs to be controlled in order to make good decisions. Beer *et al.* (in press) investigated the influence of negative mood priming on risk-taking behavior in a hypothetical roulette/gambling task. Behavioral results showed that negative mood priming before making a bet led to a more conservative choice than neutral priming. Importantly, they found an inverse correlation with the left lateral OFC and the influence of mood priming on betting. The more activity there was in the left lateral OFC, the less influence mood had on their choice of bet. This shows that the lateral OFC regulated the influence of emotion on decision-making. Interestingly, the left lateral OFC was also involved when subjects were instructed to use the emotional information as part of their betting strategy. This suggests that the lateral OFC may be involved in inhibiting emotional infor-mation when it is irrelevant and integrating it when it is relevant to the decision (Beer *et al.* in press). Additional research is needed to illustrate the relative role of the lateral OFC in inhibiting as compared to integrating emotional information during decision-making.

## 12.9. **Inhibitory control in attitude regulation and memory**

Lateral OFC- and VLPFC-mediated inhibition also regulates the influence of *a priori* belief and initial emotional response on reasoning and judgement. Goel and Dolan (2003) investigated the influence of *a priori* belief bias on logical reasoning. Deductive reasoning is the process of drawing valid conclusions from a given set of premises. Although it should be a 'closed system', drawing only from the given facts and information (thus safe from the influence of *a priori* beliefs, feelings, and intuitions), one's belief about the world can influence validity judgements about presented arguments. People are more likely to believe a logical argument when it fits with their *a priori* beliefs than the same type of argument when it conflicts with their beliefs. Consequently, if the logical conclusion is consistent with beliefs about the world, the beliefs are facilitatory with the logical task (and people do better), whereas a similarly logical argument in which the logical conclusion is inconsistent with the person's belief, they are more likely to disagree with the logical argument (and falsely reject it as invalid). Thus, rational and appropriate reasoning about a given set of facts or information often requires inhibiting a personal belief system to make a valid judgement. Without inhibiting one's personal belief, he is susceptible to belief bias in reasoning resulting in mistakes in logic and decision-making (Evans *et al.* 1983; Newstead *et al.* 1992; Evans *et al.* 2001; Morley *et al.* 2004).

Goel and Dolan (2003) had people identify the validity of a set of logical arguments. Some of the arguments were congruent with common beliefs about the world and some were equally valid arguments but contradictory to common world beliefs. Using fMRI they found that right OFC/VLPFC (BA 47, 45) was more active for correctly inhibiting belief bias as compared to error trials in which the belief bias led to faulty reasoning and the wrong conclusion. The VMPFC showed the opposite pattern. It was less active during the correct inhibition of belief bias but more active when the *a priori* beliefs led to incorrect reasoning (Goel and Dolan 2003). These findings illustrate that right OFC/VLPFC activity was related to resistance to belief bias and suggests that this region may have facilitated this resistance by inhibiting the influence of the *a priori* belief.

Initial emotional reactions to 'hot' political or social topics can also cloud logical reasoning, and, in many cases, these initial reactions need to be suppressed. Cunningham *et al.* (2004) investigated the neural circuits of making explicit (good vs. bad) and implicit (abstract vs. concrete) judgements of socially relevant topics such as abortion, gun control, and sex education. They found that activity in the right lateral OFC was significantly correlated with the amount that people felt that they had to control or suppress their initial response to the concept. Interestingly, post-scanning questionnaires revealed that topics which required the suppression of automatic responses were often those topics that had both positive and negative qualities. Therefore these topics required more evaluation before a judgement could be made (Cunningham *et al.* 2003, 2004), suggesting that, perhaps, the process of evaluation suppresses the influence of emotional response on judgement.

These studies are consistent with memory research showing that left VLPFC activity facilitates working memory by inhibiting proactive interference (D'Esposito *et al.* 1999). Proactive interference occurs when a recent but irrelevant memory interferes with current recognition or recall. Looking for a parked car in an often used parking lot is a common situation that can create proactive interference, since the memory of parking in the past can interfere with the recall of the car's current location (Badre and Wagner 2005). Experimenters manipulate susceptibility to proactive interference in item recognition tasks by 'luring' subjects with items that were recently presented in the task, but are not the target on that specific trial. Greater left VLPFC activity is associated with a correct response on trials containing a familiar ('lure') item. In other words, greater VLPFC activity is associated with resistance to proactive interference (D'Esposito *et al.* 1999, Badre and Wagner 2005). In addition, damage to the left VLPFC is associated with enhanced susceptibility to proactive interference (Thompson-Schill *et al.* 2002).

Together, these studies indicate that the inhibitory influence of the lateral OFC/VLPFC is not necessarily restricted to affective information, as is proposed by the Dynamic Filtering Theory (Shimamura 2000), nor is it confined to negative (as opposed to positive) affect, as is proposed by the Distraction Theory (Lieberman 2003). However, these studies do show that the lateral OFC and VLPFC help maintain goal oriented focus by suppressing the influence of interfering information. Furthermore, data on both belief bias and evaluative judgements are consistent with the idea that the VLPFC mediates a conscious, evaluative process, as is proposed by the Distraction Theory, which may suppress the influence of irrelevant information.

## 12.10. Self-regulation in social contexts

Appropriate and successful social behavior requires on-line monitoring of the impact and appropriateness of one's comments and actions. Individuals must manage the conflict between inner impulses and external social conventions. Cues from the environment (e.g. the formality of a black-tie dinner) or social cues from another person (e.g. an angry response to a behavior) provide information that is important for regulating one's own social behavior in that context. If, in the course of a social interaction, excessively friendly comments cause embarrassment or excessively aggressive comments cause anger, then one must inhibit those comments and develop a different way of communicating in order to build social relationships.

Numerous clinical anecdotes chronicle impulsive, selfish, disinhibited, socially inappropriate behavior of patients with lesions to the OFC. In one of the few empirical studies on inappropriate behavior in this patient group, Beer *et al.* (2003) investigated the regulation of social behavior in OFC lesion patients. The focus of these studies was to investigate whether self-conscious emotions, such as embarrassment, shame, guilt and pride, would help regulate social behavior in OFC patients as it does in healthy, normally functioning individuals. The results show that OFC patients are able to generate and feel self-conscious emotions but they do not use these emotions appropriately to regulate behavior (Beer *et al.* 2003).

In a self-disclosure task, subjects were presented with a set of emotional terms and asked to define each one, and give an example of when they felt that way. OFC patients inappropriately disclosed more intimate details than was necessary for the completion of the task. Similarly, in a teasing task, subjects were told to make up a nickname for two experimenters based on their initials. OFC patients exhibited inappropriately intimate and hostile teasing behavior such as sustained eye contact, intrusive body posture, playful gestures and prosody. In addition, they less frequently exhibited appropriate apologetic teasing behavior such as verbal apologies, submissive body posture, and blushing. Overall, they tended to tease strangers in an overly familiar way more suited to an intimate friendship or romantic relationship. In addition, they showed more pride in their teasing behavior, even though their behavior was inappropriate. These controlled experiments illustrate that the OFC patients are more likely to act inappropriately, and do not regulate their behavior based on on-line cues from the social situation.

Anger is another emotion that regulates behavior in social situations. When someone looks at you with an angry expression, it communicates disapproval and suggests that you should modify your current behavior. This has led to a suggestion that we have a social response reversal (SRR) system that is modulated by the disapproving negative cues of others (Blair and Cipolotti 2000). Lateral OFC may be involved in social reversals in both identifying that the behavior is wrong and in inhibiting that behavior and reversing course (Blair 2003, 2004). For example, OFC lesion patients have difficulty identifying social reversal cues such as anger and embarrassment (Blair and Cipolotti 2000; Blair 2003; Beer *et al.* 2003), and in identifying actions that would make other people angry, such as identifying a social *faux pas* (Stone *et al.* 1998) or feelings of anger and embarrassment in story protagonists (Blair and Cipolotti 2000). In addition, neruroimaging studies show more right lateral OFC activity in response to angry expressions as compared to other facial expressions (Blair *et al.* 1999; Blair 2003). These data suggest that in addition to providing the neural mechanism of inhibition, the lateral OFC may also register and implement signals indicating the need to inhibit, such as an angry or embarrassed facial expression.

Furthermore, an inability to regulate an angry emotional response to others can manifest in aggressive tendencies, and these aggressive tendencies in high-risk individuals may be controlled by lateral OFC/ VLPFC function (Blair 2001, 2004). Aggression research divides aggression into two types: reactive aggression in which a frustrating or threatening event triggers an impulsive aggressive act and instrumental (or proactive) aggression which is purposeful and goal directed (Barratt *et al.* 1999). Damage to OFC and medial PFC is associated with increased risk for the display of reactive aggression especially when the lesion occurs in childhood (Grafman *et al.* 1996; Anderson *et al.* 1999; Eslinger *et al.* 2004), whereas this damage is not associated with instrumental aggression. In addition, patients with a history of reactive aggression have reduced resting metabolism in lateral OFC (BA 47) (Raine *et al.* 1998a, b; Goyer *et al.* 1994), suggesting that lower levels of neural activity in this region are associated with problems controlling aggressive tendencies.

## 12.11. Conclusions

The lateral OFC extending to the VLPFC shows regional and functional specificity in controlling information-processing and behavioral expression through inhibition. In particular, this neural region modulates the influence of emotional information on sensation, attention, judgement, emotional experience and expression by suppressing neural activity associated with the interfering or unwanted stimulus. The research reviewed here is consistent with well-documented evidence showing that the lateral OFC/VLPFC is the primary neural system governing the inhibition of habitual motor responses (Casey *et al.* 1997, 2001; Braver *et al.* 2001; Aron *et al.* 2003, 2004), and reversing or extinguishing reinforcement related associations (Iversen and Mishkin 1970; Roberts and Wallis 2000; Rolls 2000). Current research reveals that the lateral OFC regulates behavior by inhibiting the influence of a broad scope of sensations, feelings, thoughts, and actions.

## 12.12. Future directions

There are several issues regarding the inhibitory mechanism that need to be resolved in future research. First, it is not clear if there is a distinction between the inhibitory function of the lateral OFC (BA 11, 12, 47) and the VLPFC, particularly the more superior portions of the inferior frontal gyrus (BA 44, 45). Many neuroimaging studies show inhibitory activity that extends across both regions (Levesque et al. 2003). Other studies of inhibitory control reveal two separate peaks of activity in the OFC and VLPFC (e.g. Phan et al. 2005), and, in other cases, the same inhibitory task shows peak activity in the lateral OFC (Ochsner et al. 2004) or the VLPFC (Ochsner et al. 2002) depending on the baseline condition.

Also, it is not clear if there are distinct functions of the right versus left lateral OFC/VLPFC during inhibition. Many inhibitory tasks show bilateral activations during inhibition (Wager et al. 2004). Other tasks show right but not left activity (Blair et al. 1999; Eisenberger et al. 2003; Goel and Dolan 2003; Levesque et al. 2003; Lieberman et al. 2004), whereas similar tasks show left but not right (Bishop et al. 2004; Beer et al. in press). Though it is tempting to claim that more emotional tasks, such as emotion regulation, may employ right hemisphere systems (Levesque et al. 2003), and more cognitive tasks, such as the inhibition of proactive interference, use left hemisphere systems, this does not seem to be the case, since some cognitive tasks, such as modulation of belief reasoning, show only right hemisphere involvement (Goel and Dolan 2003), and some emotional tasks, such as inhibiting a fearful face, show only left hemisphere involvement (Bishop et al. 2004).

Understanding the distinction between the OFC and the VLPFC, as well as understanding the laterality effects during inhibitory tasks, will be enhanced by neuropsychological studies of frontal lobe lesion patients. Historically, most studies do not have enough well-defined lesion patients to compare right and left lesion groups. However, a recent study with lesion patients reveals that the right VLPFC (BA 44, 45) is the region most associated with the ability to inhibit a motor response (Aron et al. 2003, 2004). Additional work needs to be done in order to see whether the right VLPFC is also the most primary region for the inhibition of emotion across a wide array of tasks.

Most of the research reviewed here investigates the regulation of negative information or negative affect. Few studies have investigated the neural substrates of inhibiting positive information or positive affect. Future work should help illuminate any differences in the neural substrates of regulating the effects of positive and negative affect on behavior.

Finally, evidence is clear that the lateral OFC/VLPFC governs inhibitory control. However, several studies suggest that it may play a dual role in both the inhibition and integration of emotional information on behavior (e.g. Beer et al. in press). Future work may reveal whether the lateral OFC/VLPFC is necessary for integration in addition to inhibition, and if so, what exact neural mechanisms or neuronal sub-populations facilitate the management of both of these processes.

## Acknowledgments

This work was supported by NIH grants MH71746 (C.I.H), 066737 (R.T.K), NS21135 (R.T.K), as well as the James S. McDonnell Foundation.

## References

Adler, L. E., Pachtman, E., Franks, R. D., Pecevich, M., Waldo, M. C. and Freedman, R. (1982) Neurophysiological evidence for a defect in neuronal mechanisms involved in sensory gating in schizophrenia. *Biological Psychiatry* **17**:639–54.

Anderson, S. W., Bechara, A., Damasio, H., Tranel, D. and Damasio, A. R. (1999) Impairment of social and moral behavior related to early damage in human prefrontal cortex. *Nature Neuroscience* **2**:1032–7.

Aron, A. R., Fletcher, P. C., Bullmore, E. T., Sahakian, B. J. and Robbins, T. W. (2003) Stop-signal inhibition disrupted by damage to right inferior frontal gyrus in humans. *Nature Neuroscience* **6**:115–6.

Aron, A. R., Robbins, T. W. and Poldrack, R. A. (2004) Inhibition and the right inferior frontal cortex. *Trends in Cognitive Science* **8**:170–7.

Badre, D. and Wagner, A. D. (2005) Frontal Lobe Mechanisms that Resolve Proactive Interference. *Cerebral Cortex*.

Barbas, H. (2002) Anatomic basis of functional specialization in prefrontal cortices in primates. In: Grafman, J.( ed.) *Handbook of neuropsychology* Vol. 7:1–27. Amsterdam: Elsevier Science.

Barratt, E. S., Stanford, M. S., Dowdy, L., Liebman, M. J. and Kent, T. A. (1999) Impulsive and premeditated aggression: a factor analysis of self-reported acts. *Psychiatry Research* **86**:163–73.

Beauregard, M., Levesque, J. and Bourgouin, P. (2001) Neural correlates of conscious self-regulation of emotion. *Journal of Neuroscience* **21**:RC165.

Bechara, A., Damasio, H. and Damasio, A. R. (2000) Emotion, decision-making and the orbitofrontal cortex. *Cerebral Cortex* **10**:295–307.

Beer, J. S. (in press) The importance of emotion-cognition interactions for social adjustment: insights from the oribitofrontal cortex. In: Harmon-Jones, E. and Winkielman, P.(eds.) *The Handbook of Social Neuroscience*. New York: Guildford.

Beer, J. S., Heerey, E. A., Keltner, D., Scabini, D. and Knight, R. T. (2003) The regulatory function of self-conscious emotion: insights from patients with orbitofrontal damage. *Journal of Personality and Social Psychology* **85**:594–604.

Beer, J. S., Knight, R. T. and D'Esposito, M. (in press) Controlling the integration of emotion and cognition: the role of frontal cortex in distinguishing helpful from hurtful emotional information. *Psychological Science*.

Bishop, S., Duncan, J., Brett, M. and Lawrence, A. D. (2004) Prefrontal cortical function and anxiety: controlling attention to threat-related stimuli. *Nature Neuroscience* **7**: 184–8.

Blair, R. J. (2001) Neurocognitive models of aggression, the antisocial personality disorders, and psychopathy. *Journal of Neurology, Neurosurgery, and Psychiatry* **71**:727–31.

Blair, R. J. (2003) Facial expressions, their communicatory functions and neuro-cognitive substrates. *Philosophical Transactions of the Royal Society of London B Biological Sciences* **358**:561–72.

Blair, R. J. (2004) The roles of orbital frontal cortex in the modulation of antisocial behavior. *Brain and Cognition* **55**:198–208.

Blair, R. J. and Cipolotti, L. (2000) Impaired social response reversal. A case of "acquired sociopathy". *Brain* **123** (**Pt** 6): 1122–41.

Blair, R. J., Morris, J. S., Frith, C. D., Perrett, D. I. and Dolan, R. J. (1999) Dissociable neural responses to facial expressions of sadness and anger. *Brain* **122** (**Pt** 5): 883–93.

Bradley, B. P. and Mogg, K. (1994) Mood and personality in recall of positive and negative information. *Behavioral Research and Therapy* **32**:137–41.

Bradley, B. P., Mogg, K. and Williams, R. (1995) Implicit and explicit memory for emotion-congruent information in clinical depression and anxiety. *Behavioral Research and Therapy* **33**:755–70.

Braver, T. S., Barch, D. M., Gray, J. R., Molfese, D. L. and Snyder, A. (2001) Anterior cingulate cortex and response conflict: effects of frequency, inhibition and errors. *Cerebral Cortex* **11**:825–36.

Casey, B. J., Castellanos, F. X., Giedd, J. N., Marsh, W. L., Hamburger, S. D., Schubert, A. B., Vauss, Y. C., Vaituzis, A. C., Dickstein, D. P., Sarfatti, S. E. and Rapoport, J. L. (1997) Implication of right frontostriatal circuitry in response inhibition and attention-deficit/hyperactivity disorder. *Journal of American Acadamy of Child and Adolescent Psychiatry* **36**:374–83.

Casey, B. J., Forman, S. D., Franzen, P., Berkowitz, A., Braver, T. S., Nystrom, L. E., Thomas, K. M. and Noll, D. C. (2001) Sensitivity of prefrontal cortex to changes in target probability: a functional MRI study. *Human Brain Mapping* **13**:26–33.

Cunningham, W. A., Johnson, M. K., Gatenby, J. C., Gore, J. C. and Banaji, M. R. (2003) Neural components of social evaluation. *Journal of Personality and Social Psychology* **85**:639–49.

Cunningham, W. A., Raye, C. L. and Johnson, M. K. (2004) Implicit and explicit evaluation: FMRI correlates of valence, emotional intensity, and control in the processing of attitudes. *Journal of Cognitive Neuroscience* **16**:1717–29.

D'Esposito, M., Postle, B. R., Jonides, J. and Smith, E. E. (1999) The neural substrate and temporal dynamics of interference effects in working memory as revealed by event-related functional MRI. *Proceedings of the National Academy of Sciences U S A* **96**:7514–9.

Duncan, J. and Owen, A. M. (2000) Common regions of the human frontal lobe recruited by diverse cognitive demands. *Trends in Neuroscience* **23**:475–83.

Eisenberger, N. I. and Lieberman, M. D. (2004) Why rejection hurts: a common neural alarm system for physical and social pain. *Trends in Cognitive Science* **8**:294–300.

Eisenberger, N. I., Lieberman, M. D. and Williams, K. D. (2003) Does rejection hurt? An FMRI study of social exclusion. *Science* **302**:290–2.

Elliott, R., Rubinsztein, J. S., Sahakian, B. J. and Dolan, R. J. (2002) The neural basis of mood-congruent processing biases in depression. *Archives of General Psychiatry* **59**:597–604.

Eslinger, P. J., Flaherty-Craig, C. V. and Benton, A. L. (2004) Developmental outcomes after early prefrontal cortex damage. *Brain and Cognition* **55**: 84–103.

Evans, J. S., Barston, J. L. and Pollard, P. (1983) On the conflict between logic and belief in syllogistic reasoning. *Memory and Cognition* **11**:295–306.

Evans, J. S., Handley, S. J. and Harper, C. N. (2001) Necessity, possibility and belief: a study of syllogistic reasoning. *Quarterly Journal of Experimental Psychology A* **54**:935–58.

Freedman, R., Adler, L. E., Myles-Worsley, M., Nagamoto, H. T., Miller, C., Kisley, M., McRae, K., Cawthra, E. and Waldo, M. (1996) Inhibitory gating of an evoked response to repeated auditory stimuli in schizophrenic and normal subjects. Human recordings, computer simulation, and an animal model. *Archives of General Psychiatry* **53**:1114–21.

Goel, V. and Dolan, R. J. (2003) Explaining modulation of reasoning by belief. *Cognition* **87**:B11–22.

Gotlib, I. H. and McCabe, S. B. (1992) An information-processing approach to the study of cognitive functioning in depression. *Progress in Experimental Personality and Psychopathology Research* **15**:131–61.

Gotlib, I. H. and McCann, C. D. (1984) Construct accessibility and depression: an examination of cognitive and affective factors. *Journal of Personality and Social Psychology* **47**:427–39.

Gotlib, I. H. and Olson, J. M. (1983) Depression, psychopathology, and self-serving attributions. *British Journal of Clinical Psychology* **22** (Pt 4):309–10.

Goyer, P. F., Andreason, P. J., Semple, W. E., Clayton, A. H., King, A. C., Compton-Toth, B. A., Schulz, S. C. and Cohen, R. M. (1994) Positron-emission tomography and personality disorders. *Neuropsychopharmacology* **10**:21–8.

Grafman, J., Schwab, K., Warden, D., Pridgen, A., Brown, H. R. and Salazar, A. M. (1996) Frontal lobe injuries, violence, and aggression: a report of the Vietnam Head Injury Study. *Neurology* **46**:1231–8.

Gross, J. J. (2002) Emotion regulation: affective, cognitive, and social consequences. *Psychophysiology* **39**:281–91.

Halberstadt, J. B. and Niedenthal, P. M. (2001) Effects of emotion concepts on perceptual memory for emotional expressions. *Journal of Personality and Social Psychology* **81**:587–98.

Harden, R. N. (2005) Chronic neuropathic pain. Mechanisms, diagnosis, and treatment. *Neurologist* **11**:111–22.

Hartikainen, K. M. and Knight, R. T. (2001) Lateral and orbital prefrontal cortex contributions to attention. In: Polich, J.(ed.) *Detection of change: Event-related potential and fMRI findings*. Kluwer Academic.

Hartikainen, K. M., Ogawa, K. H. and Knight, R. T. (2000a) Transient interference of right hemispheric function due to automatic emotional processing. *Neuropsychologia* **38**:1576–80.

Hartikainen, K. M., Ogawa, K. H., Soltani, M. and Knight, R. T. (2001) Effects of emotional stimuli on event-related potentials and reaction times in orbitofrontal patients. *Brain and Cognition* **47**:339–341.

Hartikainen, K. M., Ogawa, K. H., Soltani, M., Pepitone, M. and Knight, R. T. (2000b) Altered emotional influence on visual attention subsequent to orbital frontal damage in humans. *Society for Neuroscience Abstracts* **26**:2023.

Innes-Ker, A. and Niedenthal, P. M. (2002) Emotion concepts and emotional states in social judgement and categorization. *Journal of Personality and Social Psychology* **83**:804–16.

Iversen, S. D. and Mishkin, M. (1970) Perseverative interference in monkeys following selective lesions of the inferior prefrontal convexity. *Experimental Brain Research* **11**:376–86.

Knight, R. T., Grabowecky, M. F. and Scabini, D. (1995) Role of human prefrontal cortex in attention control. *Advances in Neurology* **66**:21–34; discussion 34–6.

Knight, R. T., Scabini, D. and Woods, D. L. (1989) Prefrontal cortex gating of auditory transmission in humans. *Brain Research* **504**:338–42.

Kringelbach, M. L. and Rolls, E. T. (2004) The functional neuroanatomy of the human orbitofrontal cortex: evidence from neuroimaging and neuropsychology. *Progress in Neurobiology* **72**:341–72.

Lerner, J. S. and Keltner, D. (2001) Fear, anger, and risk. *Journal of Personality and Social Psychology* **81**:146–59.

Levesque, J., Eugene, F., Joanette, Y., Paquette, V., Mensour, B., Beaudoin, G., Leroux, J. M., Bourgouin, P. and Beauregard, M. (2003) Neural circuitry underlying voluntary suppression of sadness. *Biological Psychiatry* **53**:502–10.

Lieberman, M. D. (2003) Reflective and reflexive judgment processes: a social cognitive neuroscience approach. In Forgas, J.P., Williams, K.R., von Hippel, W. (eds.) *Social Judgments: Explicit and Implicit Processes*, pp. 44–67. New York: Cambridge University Press.

Lieberman, M. D., Jarcho, J. M., Berman, S., Naliboff, B. D., Suyenobu, B. Y., Mandelkern, M. and Mayer, E. A. (2004) The neural correlates of placebo effects: a disruption account. *Neuroimage* **22**:447–55.

Miller, E. K. and Cohen, J. D. (2001) An integrative theory of prefrontal cortex function. *Annual Review of Neuroscience* **24**:167–202.

Morley, N. J., Evans, J. S. and Handley, S. J. (2004) Belief bias and figural bias in syllogistic reasoning. *Quarterly Journal of Experimental Psychology A* **57**:666–92.

Muller, N. G., Machado, L. and Knight, R. T. (2002) Contributions of subregions of the prefrontal cortex to working memory: evidence from brain lesions in humans. *Journal of Cognitive Neuroscience* **14**:673–86.

Newstead, S. E., Pollard, P., Evans, J. S. and Allen, J. L. (1992) The source of belief bias effects in syllogistic reasoning. *Cognition* **45**:257–84.

Ochsner, K. N., Bunge, S. A., Gross, J. J. and Gabrieli, J. D. (2002) Rethinking feelings: an FMRI study of the cognitive regulation of emotion. *Journal of Cognitive Neuroscience* **14**:1215–29.

Ochsner, K. N., Ray, R. D., Cooper, J. C., Robertson, E. R., Chopra, S., Gabrieli, J. D. and Gross, J. J. (2004) For better or for worse: neural systems supporting the cognitive down- and up-regulation of negative emotion. *Neuroimage* **23**:483–99.

Ohman, A., Flykt, A. and Esteves, F. (2001) Emotion drives attention: detecting the snake in the grass. *Journal of Experimental Psychology General* **130**:466–78.

Ohman, A., Lundqvist, D. and Esteves, F. (2001) The face in the crowd revisited: a threat advantage with schematic stimuli. *Journal of Personality and Social Psychology* **80**:381–96.

Petrovic, P. and Ingvar, M. (2002) Imaging cognitive modulation of pain processing. *Pain* **95**:1–5.

Petrovic, P., Kalso, E., Petersson, K. M. and Ingvar, M. (2002) Placebo and opioid analgesia—imaging a shared neuronal network. *Science* **295**:1737–40.

Petrovic, P., Petersson, K. M., Ghatan, P. H., Stone-Elander, S. and Ingvar, M. (2000) Pain-related cerebral activation is altered by a distracting cognitive task. *Pain* **85**: 19–30.

Phan, K. L., Fitzgerald, D. A., Nathan, P. J., Moore, G. J., Uhde, T. W. and Tancer, M. E. (2005) Neural substrates for voluntary suppression of negative affect: a functional magnetic resonance imaging study. *Biological Psychiatry* **57**:210–9.

Raine, A., Meloy, J. R., Bihrle, S., Stoddard, J., LaCasse, L. and Buchsbaum, M. S. (1998a) Reduced prefrontal and increased subcortical brain functioning assessed using positron emission to mography in predatory and affective murderers. *Behavioral Science and Law* **16**:319–32.

Raine, A., Phil, D., Stoddard, J., Bihrle, S. and Buchsbaum, M. (1998b) Prefrontal glucose deficits in murderers lacking psychosocial deprivation. *Neuropsychiatry, Neuropsychology, and Behavioral Neurology* **11**:1–7.

Roberts, A. C. and Wallis, J. D. (2000) Inhibitory control and affective processing in the prefrontal cortex: neuropsychological studies in the common marmoset. *Cerebral Cortex* **10**:252–62.

Roberts, N. A., Beer, J. S., Werner, K. H., Scabini, D., Levens, S. M., Knight, R. T. and Levenson, R. W. (2004) The impact of orbital prefrontal cortex damage on emotional activation to unanticipated and anticipated acoustic startle stimuli. *Cognitive, Affective, and Behavioral Neuroscience* **4**:307–16.

Rolls, E. T. (2000) The orbitofrontal cortex and reward. *Cerebral Cortex* **10**:284–94.

Rule, R. R., Shimamura, A. P. and Knight, R. T. (2002) Orbitofrontal cortex and dynamic filtering of emotional stimuli. *Cognitive, Affective, and Behavioral Neuroscience* **2**:264–70.

Shimamura, A. P. (2000) The role of prefrontal cortex in dynamic filtering. *Psychobiology* **28**:207–218.

Soltani, M. and Knight, R. T. (2000) Neural origins of the P300. *Critical Reviews of Neurobiology* **14**:199–224.

Stone, V. E., Baron-Cohen, S. and Knight, R. T. (1998) Frontal lobe contributions to theory of mind. *Journal of Cognitive Neuroscience* **10**:640–56.

Thompson-Schill, S. L., Jonides, J., Marshuetz, C., Smith, E. E., D'Esposito, M., Kan, I. P., Knight, R. T. and Swick, D. (2002) Effects of frontal lobe damage on interference effects in working memory. *Cognitive, Affective, and Behavioral Neuroscience* **2**:109–20.

Vuilleumier, P., Armony, J. L., Driver, J. and Dolan, R. J. (2001) Effects of attention and emotion on face processing in the human brain: an event-related fMRI study. *Neuron* **30**:829–41.

Wager, T. D. (2005) Expectations and anxiety as mediators of placebo effects in pain. *Pain* **115**:225–6.

Wager, T. D., Rilling, J. K., Smith, E. E., Sokolik, A., Casey, K. L., Davidson, R. J., Kosslyn, S. M., Rose, R. M. and Cohen, J. D. (2004) Placebo-induced changes in FMRI in the anticipation and experience of pain. *Science* **303**:1162–7.

Waldo, M. C. and Freedman, R. (1986) Gating of auditory evoked responses in normal college students. *Psychiatry Research* **19**:233–9.

Zhang, Y., Tang, J., Yuan, B. and Jia, H. (1997) Inhibitory effects of electrically evoked activation of ventrolateral orbital cortex on the tail-flick reflex are mediated by periaqueductal gray in rats. *Pain* **72**:127–135.

# Chapter 13

# Visceral and decision-making functions of the ventromedial prefrontal cortex

Nasir Naqvi, Daniel Tranel, and Antoine Bechara

## 13.1. Introduction

The orbital and mesial prefrontal cortices have been implicated in a range of affective processes, including hedonic and anticipatory responses to rewards and punishments (Kringelbach and Rolls 2004), subjective states of desire (London *et al.* 2000; Egan *et al.* 2003; Wang *et al.* 2004), basic emotions (Dougherty *et al.* 1999; Damasio *et al.* 2000; Phan *et al.* 2002; Hornak *et al.* 2003) and social emotions (Berthoz *et al.* 2002; Camille *et al.* 2004; Shamay-Tsoory *et al.* 2005). In neuropsychological studies, it has been shown that acquired damage to the ventromedial prefrontal cortex (VMPFC), a region at the junction of the orbitofrontal and mesial prefrontal cortices, often leads to profound alterations in the ability to make advantageous decisions in the personal, social and financial domains (Damasio 1994; Eslinger and Damasio 1985), suggesting that the orbital and mesial prefrontal cortices play an important role in executive functions, as well. What links all of these processes together is a need for the organism to evaluate the biological significance of a stimulus and to use this information to guide behavior. In addition, all of these processes are associated with alterations in the visceral state, such as changes in heart rate, blood pressure, gut motility and glandular secretion.

Changes in the visceral state may be considered as a form of anticipation of the bodily impact of objects and events in the world. Visceral responses to biologically relevant stimuli allow an organism to maximize the survival value of situations that may impact the state of the internal milieu. These include events that promote homeostasis, such as an opportunity to feed or engage in social interaction, as well as events that disrupt homeostasis, such as a physical threat or a signal of social rejection. Visceral responses are just one component of a broader emotional response system that also includes changes in the endocrine and skeletomotor systems, as well as changes within the brain that alter the perceptual processing of biologically relevant stimuli (Damasio 1994).

William James (1884) initially proposed that visceral responses to biologically relevant stimuli are a necessary component of the subjective experience of emotion. Specifically, James proposed that subjective emotional feeling—the hedonic meaning that is attributed to objects and events in the world—arises from the sensory feedback of the visceral responses that are elicited by those objects and events. According to the somatic marker hypothesis

(Damasio 1994), the sensory mapping of visceral responses not only contributes to feelings, but is also important for the execution of highly complex, goal-oriented behavior. In this view, visceral responses function to 'mark' potential choices as being advantageous or disadvantageous. This process aids in decision-making in which there is a need to weigh positive and negative outcomes that may not be predicted decisively through 'cold' rationality alone. Both the Jamesian view and the somatic marker hypothesis hold that the brain contains a system that translates the sensory properties of external stimuli into changes in the visceral state that reflect their biological relevance. We propose that this is the essential function of the VMPFC—a function that ties control of the visceral state to affect and decision-making.

In this chapter we review evidence that the VMPFC plays a role in eliciting visceral responses that are related to the value of objects and events in the world. We start by discussing anatomical and physiological evidence that the VMPFC can both influence the state of the viscera and also register changes in the viscera elicited by biologically relevant stimuli. We then review the results of lesion studies in humans showing that the VMPFC is necessary for eliciting visceral responses to certain forms of emotional stimuli. Finally, we review evidence supporting the somatic marker hypothesis, showing that the visceral responses that are controlled by the VMPFC play a role in guiding decision-making in the face of uncertain reward and punishment.

## 13.2. Visceral functions of the VMPFC

### Ventromedial prefrontal cortex vs. orbitofrontal cortex

The terms ventromedial prefrontal cortex (VMPFC) and orbitofrontal cortex (OFC) are often used interchangeably in the literature, even though these do not refer to identical regions. For this reason, it is necessary to clarify exactly what we mean when we use these terms. The OFC is the entire cortex occupying the ventral surface of the frontal lobe, dorsal to the orbital plate of the frontal bone. We have used the term VMPFC to designate a region that encompasses medial portions of the OFC along with ventral portions of the medial prefrontal cortex. The VMPFC is an anatomical designation that has arisen because lesions that occur in the basal portions of the anterior fossa, which include meningiomas of the cribiriform plate and falx cerebri and aneurysms of the anterior communicating and anterior cerebral arteries, frequently lead to damage in this area. Often, this damage is bilateral. With respect to the cytoarchitectonic fields identified in the human orbitofrontal and medial prefrontal cortices by Price and colleagues (Ongur et al. 2003), the VMPFC comprises area 14 and medial portions of areas 11 and 13 on the orbital surface and areas 25, 32 and caudal portions of area 10 on the mesial surface. The VMPFC excludes lateral portions of the OFC, namely area 47/12, as well as more dorsal and posterior regions of areas 24 and 32 of the medial prefrontal cortex. The VMPFC is thus a relatively large and heterogeneous area. The location of the human VMPFC is illustrated in Fig. 13.1.

Though it is problematic to compare rodent and human cytoarchitectonic maps of the frontal lobe, the prelimbic and infralimbic cortices located on the medial wall of rodent

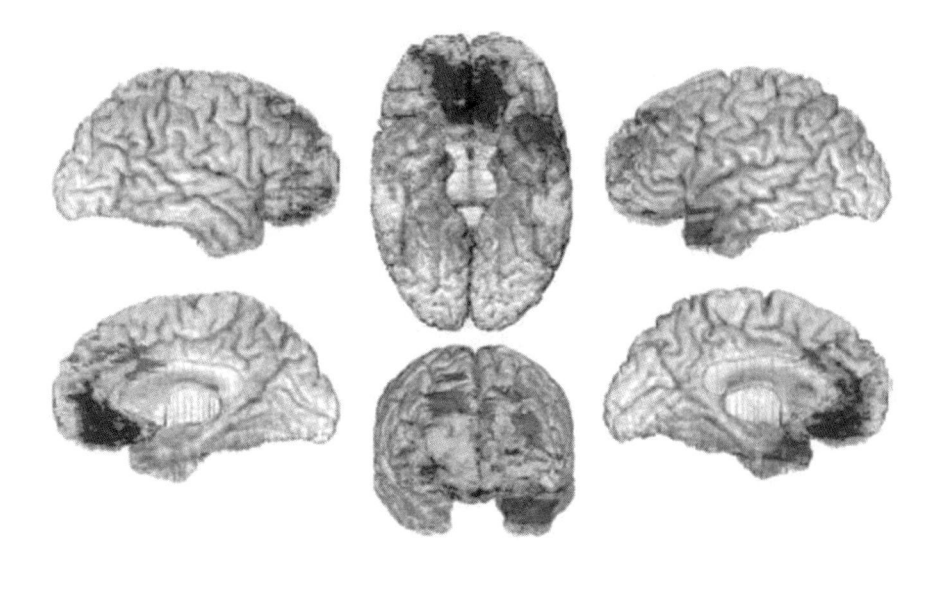

**Fig. 13.1** (Also see Color Plate 24.) The location of the VMPFC. A map showing areas of the brain that are damaged in patients who show impairments in visceral response and decision-making. The colors reflect the number of subjects with damage in a given voxel. The region of greatest overlap is the VMPFC. Note the involvement of medial wall and medial orbitofrontal areas and the relative absence of involvement of the lateral orbitofrontal areas. Also note the relative sparing of the frontal pole. Reproduced with permission from Bechara *et al.* (1998).

frontal cortex are considered by some authors to be homologous to Brodmann's areas 32 and 25, respectively (Vogt and Peters 1981; Krettek and Price 1977; Conde *et al.* 1995; Gabbott *et al.* 1997). These regions together may therefore be considered as the rodent equivalent of VMPFC, although it is important to note that this designation does not include any part of the OFC, which in rodents is situated on the dorsal bank of the rhinal sulcus.

Viewing the VMPFC as a single region may blur the distinction between functions subserved by the OFC on the one hand, and the ventral portion of the medial prefrontal cortex on the other. Recent evidence in both rodents (Chudasama and Robbins 2003) and nonhuman primates (Pears *et al.* 2003) suggests that these regions subserve distinct motivational and learning functions. Thus, lesions of the VMPFC in humans may disrupt more than one process. This is important to keep in mind when inconsistencies arise between animal studies and human studies. These differences are also important when comparing human lesion studies, which tend to examine the functions of relatively large regions, and functional imaging studies, which reveal more focused patterns of activity.

The VMPFC encompasses different regions that have been identified in functional imaging studies. The VMPFC includes the medial prefrontal area identified by Raichle and colleagues (Raichle *et al.* 2001) to be deactivated by a broad range cognitive tasks that require focused attention, reflecting a high level of resting activity that is suspended during goal-directed behavior. The VMPFC encompasses the medial orbitofrontal area identified by Rolls and colleagues in which activity is consistently related to the 'reward value' of hedonically positive stimuli (reviewed in Kringelbach and Rolls 2004) (This area is distinct from more lateral orbitofrontal areas identified by these authors to be activated by punishers that signal a need to change behavior (see the Chapter 5 on reward functions of the OFC by Rolls in this volume for a thorough discussion of this topic).) In addition, the VMPFC includes the subgenual cingulate cortex, an area that has been implicated through functional imaging studies in the pathogenesis of mood disorders (Drevets *et al.* 1997) (see Chapter 21, this volume that examines the role of subgenual area 25 in mood disorders). It remains to be seen whether these seemingly distinct functions reflect the operation of a single brain area that can be called the VMPFC, or instead are due to separate mental processes mediated by three functionally distinct areas encompassed by the VMPFC.

There are two points that need to be considered when interpreting functional imaging studies of the VMPFC. Firstly, as with all cognitive functions, activation or deactivation may show that this region is engaged by a particular function, but does not mean that it is necessary for the function to be performed. So, for example, even though activity in the VMPFC can be shown to be correlated with the reward value of hedonically positive stimuli, it still remains to be seen whether lesions in the human VMPFC disrupt the subjective experience of pleasure. Furthermore, in fMRI studies, the VMPFC undergoes significant BOLD signal dropout due to its location near an air-tissue interface. For this reason, the failure to detect activation or deactivation of the VMPFC using fMRI should not be taken as evidence that the VMPFC is not involved in the function under investigation, unless special procedures have been implemented to overcome signal dropout (see the chapter by Stenger *et al.* in this volume for a discussion of this problem and its potential solutions). For example, an fMRI study of decision-making by Fukui *et al.* (2005) using a task on which subjects with VMPFC damage are impaired (discussed below) did not show activation of the VMPFC. An earlier PET study by Ernst *et al.* (2002) using the same task did find activation in the VMPFC. While one may cite differences in the experimental conditions or data analysis in order to explain this discrepancy, perhaps the most parsimonious explanation is that the Fukui *et al.* study did not use procedures to counteract BOLD signal dropout in the VMPFC, which was not an issue in the Ernst *et al.* study. This points to the larger possibility that the VMPFC is involved in a broader set of functions than would be indicated by fMRI studies alone.

## Visceral connectivity of the VMPFC

The first anatomical studies on the subcortical connectivity of the prefrontal cortex were performed in macaques (Devito and Smith 1964; Johnson *et al.* 1968; Nauta and

Haymaker 1969). Though somewhat limited in their detail, these studies were able to show projections from the prefrontal cortex, in particular from the posterior orbitofrontal region, to the lateral hypothalamus and the periaqueductal gray region (PAG), regions that play a role in visceral, endocrine and somatomotor control. Later, more detailed studies in rodents found that the prelimbic and infralimbic cortices project to the lateral hypothalamus (Sesack et al. 1989) and the PAG (Neafsey et al, 1986). Direct lateral hypothalamic projections from the medial prefrontal region of rodents have also been demonstrated through electrical stimulation studies (Kita and Oomura 1981). Projections have also been shown from the infralimbic and prelimbic cortices to posterior portions of the brainstem, such as the parabrachial nucleus (PBN) and the nucleus of the solitary tract (NTS) (van der Kooy et al. 1982, 1984; Terreberry and Neafsey 1983, 1987). These are regions that receive visceral sensory input from the vagus and glossopharyngeal nerves and send output to other brainstem regions that control autonomic reflexes. Thus, the VMPFC is in a position to govern both simple visceral responses that are organized within the caudal brainstem, as well as relatively complex somatovisceral response patterns that are organized within the PAG and hypothalamus.

More recent tract tracing studies in macaques (An et al. 1998; Ongur et al. 1998) have revealed that projections from the VMPFC to the hypothalamus and PAG are topographically organized. These studies have shown that areas located on the medial wall project to the dorsolateral column of the PAG and the anterior and ventromedial hypothalamic areas, while areas located on the medial orbitofrontal surface project to the ventrolateral column of the PAG and lateral hypothalamic areas. The columns of the PAG have been shown to mediate distinct patterns of somatovisceral response (reviewed in Bandler and Shipley 1994). For example, chemical stimulation of the lateral column of the PAG elicits active avoidance and aggressive behaviors and their visceral concomitants, such as threat display and fight or flight, along with tachycardia and hypertension. In contrast, stimulation of the ventrolateral column elicits more passive strategies, such as freezing, accompanied by bradycardia and hypotension. Thus, different subregions of the VMPFC may have control over complex, pre-organized somatovisceral response patterns that differentiate between various classes of biological stimuli.

Nauta (1971) and then Neafsey (1990) proposed that the ventral prefrontal cortex represents a distinct visceromotor output region. Price and colleagues (Carmichael and Price 1996; Ongur and Price 2000) later refined this concept, synthesizing the previous anatomic literature on the prefrontal cortex with their own anatomical studies in macaques (see the chapter by Price in this volume for a review of this work). In their conception, the ventral prefrontal cortex (a region they term the orbitomedial prefrontal cortex) is composed of functionally distinct orbital and medial networks. The orbital network is essentially a sensory input area that receives afferents from late sensory cortices for vision, audition, olfaction, taste and visceral sensation and has reciprocal connections with the dorsolateral prefrontal cortex. The medial network is essentially a visceromotor output area that sends projections to subcortical structures that are involved in emotional and motivational processes, such as the amygdala and the nucleus accumbens, as well as regions of the brainstem and hypothalamus that directly govern the state of the viscera. In between the orbital and medial networks lies a transitional zone, which is interconnected with both the orbital and medial networks that may function to transfer information between the orbital and medial networks. According to this scheme, the VMPFC, as we define it, corresponds largely to the medial and intermediate networks—areas that are largely concerned with translating highly processed sensory inputs into visceromotor output.

## Visceral motor functions of the VMPFC

Much of the early evidence for the role of the VMPFC in the control of visceral functions came from studies examining the cardiorespiratory effects of stimulation in this area in cats and macaques (Spencer 1894; Smith 1938; Bailey and Sweet 1940; Delgado and Livingston 1948; Anand and Dua 1956). These studies showed that stimulation of the posterior and medial regions of the OFC leads to arrest of respiration, accompanied by changes in heart rate and blood pressure. Similar effects were also found after stimulation of the human OFC (Livingston *et al.* 1948; Chapman *et al.* 1949, 1950). In addition to effects upon cardiovascular function, stimulation of the macaque OFC has been shown to cause an increase in salivation (Kaada 1951; Showers and Crosby 1958), peripheral vasodilation (Delgado and Livingston 1948) and penile erection (Dua and MacLean 1964), indicating broad effects of the VMPFC upon visceral functions. Studies in rodents and cats have shown that stimulation of prelimbic and infralimbic cortices leads to changes in the visceral state. These include both increases and decreases in blood pressure (Lofving 1961), as well as decreases in heart rate (Burns and Wyss 1985; Hardy and Holmes 1988), and decreases in gastric motility (Hurley-Gius and Neafsey 1986).

Further support for the role of the VMPFC in visceral motor functions comes from lesion studies. Early studies in humans (Luria 1973; Luria and Homskaya 1970; Luria *et al.* 1964) and nonhuman primates (Kimble *et al.* 1965) examined the effects of relatively large lesions of the frontal lobes on visceral functions. More detailed lesion studies (Frysztak and Neafsey 1991, 1994) were subsequently performed in rodents, and showed that damage to the medial prefrontal cortex leads to alterations of conditioned heart rate and respiratory responses. Our own laboratory has performed studies examining the effects of lesions in an array of cortical areas on visceral responses (Tranel 2000; Tranel and Damasio 1994). These studies demonstrated that a number of cortical regions, including the VMPFC, but also the anterior cingulate cortex and the right inferior parietal cortex, are necessary for the generation of visceral responses to sensory stimuli. These studies also showed that the role of the VMPFC in governing visceral responses is specific to stimuli with emotional or social content (discussed in greater detail below).

More recent evidence for the visceral functions of the VMPFC comes from functional imaging studies. In a study using positron emission tomography (PET) Critchley *et al.* (2000a) showed that neural activity in the VMPFC co-varies with mean arterial pressure during isometric exercise and mental arithmetic tasks. Using functional magnetic resonance imaging (fMRI), Critchley *et al.* (2000b) again showed that activity in the VMPFC co-varies with visceral responses, this time with skin conductance response during anticipation and receipt of monetary rewards. In this study, skin conductance responses were modelled as discrete events, which allowed for the correlation with brain activity both preceding and following the responses. Using this approach, it was possible to show that activity in the VMPFC was related to both the generation of skin conductance responses as well as the afferent mapping of skin conductance responses, indicating both visceral motor and visceral sensory functions for the VMPFC. Using fMRI, it has also been shown that activity in the VMPFC is correlated with skin conductance response across a variety of cognitive states, including the resting state (Nagai *et al.* 2004; Patterson *et al.* 2002). These studies suggest that the visceral functions of the VMPFC are not specific to emotional stimuli, contrary to the results of the lesion studies described above. However, just because activity in the VMPFC is related to visceral responses in a given context does not mean that it is necessary for the generation of visceral responses in that context.

## Visceral sensory functions of the VMPFC

Both the somatic marker hypothesis and the Jamesian view of emotion place importance upon the sensory representations of the visceral responses that are elicited by biologically relevant stimuli. An important question, therefore, regards how visceral responses, once generated by the VMPFC and deployed in the body, are mapped in the brain. Visceral sensations are represented at multiple levels of the neuroaxis, including the spinal cord, brainstem, hypothalamus, thalamus and cortex (reviewed in Craig 2002). Each of these stages of visceral representation may have a specific role to play in affective and executive processes.

The sensory representation of the viscera within the insular cortex, in particular the right anterior insular cortex, has been proposed by Damasio (2000) and by Craig (2002) to play a special role in conscious emotional feelings. The right insular cortex, along with right somatosensory cortices, have also been proposed by Damasio (1994) to be a component of the somatic marker network for decision-making. The anterior (agranular) insular cortex projects to a number of areas involved in emotion and motivation, including the amygdala (Stefanacci and Amaral 2002) and the nucleus accumbens (Chikama *et al.* 1997). The anterior insular cortex also projects to the VMPFC, both via the orbital network (Carmichael and Price 1996) and through direct projections to medial network areas (Gabbott *et al.* 2003). In addition, recent anatomical evidence suggests that the right anterior insular cortex has evolved special functions in higher primates (Allman 2003; Craig 2002), consistent with a role in conscious feelings. Indeed, the right anterior insular cortex has been shown to be activated by a number of subjective feeling states (Lane *et al.* 1997; Damasio *et al.* 2000; Brody *et al.* 2002; Phan *et al.* 2002; Critchley *et al.* 2004; Pelchat *et al.* 2004). Thus, the visceral sensory representation within the right anterior insular cortex may play a role in the feelings that accompany decision-making, such as 'hunches' and 'gut feelings', that may guide decision-making in the face of uncertainty.

According to the somatic marker hypothesis (Damasio 1994; Bechara and Damasio 2005), visceral sensory signals can also influence decision-making by acting upon brainstem nuclei for ascending neurotransmitter systems, including dopaminergic, serotonergic and noradrenergic systems. These neurotransmitter systems exert widespread influence on the function of the prefrontal cortex, including the dorsolateral, medial and orbital prefrontal cortices, and on subcortical structures including the amygdala and the ventral and dorsal striatum. Through these projections, ascending neurotransmitter systems play a role in multiple attentional, executive and motivational processes (reviewed in Berridge 1998; reviewed in Robbins 2000; reviewed in Rahman *et al.* 2001). There is evidence for direct visceral sensory inputs to these nuclei (Cedarbaum and Aghajanian 1978; Phillipson 1979; Bianchi *et al.* 1998) and that visceral states can influence neurotransmitter release from these nuclei (reviewed in Berntson *et al.* 2003). The somatic marker hypothesis holds that visceral states, via their influence on ascending neurotransmitter systems, can influence decision-making via both by promoting the maintenance of specific goals in working memory and by biasing of behavior toward these goals. In this framework, brainstem neurotransmitter nuclei may also be engaged by 'as-if' loops. Here, areas such as the VMPFC, instead of triggering visceral responses in the body that feed back to brainstem neurotransmitter nuclei, facilitate neurotransmitter release via direct brainstem projections that bypass the body. This triggers neurotransmitter release 'as-if' a visceral response had been expressed in the body.

As discussed above, functional imaging evidence (Critchley *et al.* 2000b) indicates that the VMPFC, in addition to its role in the generation of visceral responses, plays a role in the sensory representation of the visceral responses. The VMPFC may receive information about visceral responses from the insular cortex, or via ascending neurotransmitter systems. The visceral responses mapped within the VMPFC may be those that are themselves induced within the VMPFC, or those that are induced in other areas that trigger visceral response to biologically relevant stimuli, such as the amygdala. This function may allow the VMPFC to compare the sensory representations of the visceral state with an efferent copy of the visceral response induced evoked by a biologically relevant stimulus. Differences between these two inputs may signal that a goal has been achieved—i.e., a behavioral consequence has occurred that has directly impacted the state of the viscera.

The VMPFC also has available to it information regarding the sensory consequences of innately pleasurable consummatory behaviors that impinge directly upon the body, such as feeding, as well as innately aversive bodily states that result from actual or potential tissue damage (nociception). For example, it has been shown that activity in the VMPFC is correlated with the subjective pleasantness of stimuli such as taste (Kringelbach *et al.* 2003) the oral sensations elicited by water (de Araujo *et al.* 2003) and pleasant touch (Rolls *et al.* 2003). In addition, activity in the VMPFC is correlated with subjective ratings of the intensity of thermal pain (Craig *et al.* 2000). All of these stimuli have direct homeostatic relevance and are signaled through a distinct sensory channel that includes the insular cortex (Craig 2002; Norgren 1990). This suggests that the VMPC represents the hedonic value of the visceral sensory signals generated by innately rewarding or punishing consummatory behavior (so-called primary reinforcers).

Functional imaging evidence (reviewed in Kringelbach 2004) supports a role for the hedonic representations within the VMPFC in the feelings of pleasure and pain that accompany good or bad outcomes of goal-directed behavior. However, human lesion evidence indicating that the VMPFC mediates the subjective hedonic impact of interoceptive stimuli is lacking. A further possibility is that interoceptive/visceral signals within the VMPFC are important for learning to associate the hedonic consequences of behavior with the particular courses of action that precede them (reward learning). This is supported by the findings from functional imaging studies that the VMPFC is activated when feedback indicating a correct choice is signaled specifically via an interoceptive route (Hurliman *et al*. 2005). In this way, interoceptive signals generated by primary reinforcers may form the basis for representations of more abstract rewards and punishments that are elicited within the VMPFC.

## 13.3. Decision-making functions of the VMPFC

### Lesions in the VMPFC lead to changes in real-life decision-making

Our laboratory's interest in the functions of the VMPFC was fuelled by observations in neurological patients that lesions in this area led to profound impairments in personality and real-life decision-making capabilities. The effects of damage in the VMPFC are typified by patient EVR, who was one of the earliest patients with VMPFC damage to be comprehensively studied in our laboratory. The history and initial neuropsychological evaluation of EVR were published previously (Eslinger and Damasio 1985). Briefly stated, after undergoing damage to the VMPFC, EVR acquired a profound impairment in his ability to make advantageous choices in his personal and social life. Despite this impairment, EVR demonstrated an intact intelligence, with normal performance on tests of language, memory, reasoning and fund of knowledge. Furthermore, EVR was not impaired in his ability to reason verbally about the consequences of his behavior; when presented with hypothetical social and ethical problems, he was capable of describing the proper course of action (Saver and Damasio 1991). Patients with VMPFC damage who were studied subsequently showed a similar pattern of deficit, with impaired decision-making in the personal and social domains despite intact intelligence (Tranel 2002). The findings from such patients indicates that damage to the VMPFC does not alter the fund of knowledge regarding appropriate social conduct, nor does it cause an inability to retrieve this knowledge, but rather leads to an inability to mark the consequences of one's own actions as being good or bad, as being in one's best interests or not (for a more detailed discussion of the real-life impairments in personal and social function caused by damage in the VMPFC, see the chapter by Koenigs and Tranel in this volume).

### The genesis of the somatic marker hypothesis

At the time when we first observed the real-life decision-making deficits of patients with VMPFC damage, a good deal of evidence for the visceral motor functions of the VMPFC, described above, had already accumulated. The question then arose as to whether the decision-making deficits caused by VMPFC damage were related to its visceromotor functions. Nauta (Nauta 1971) had by then proposed that the guidance of behavior by the frontal lobes was linked to the interoceptive and visceromotor functions of this area. Specifically, he hypothesized that the prefrontal cortex, broadly defined, functioned to compare the affective responses evoked by the various choices for behavior and to select the option that 'passed censure by an interoceptive sensorium.' According to Nauta, the 'interoceptive agnosia' suffered by patients with frontal lobe damage could explain their impairments in real-life, as well as their poor performance on various tests of executive function, including the Wisconsin card sort task. This model was meant to explain the function of the prefrontal cortex as a whole. Furthermore, it was meant as a broad explanation of executive function deficits, not of a specific deficit in decision-making within the social and personal domains. However, the deficits of patients with damage in the VMPFC were limited to the personal and social domains; patients like EVR showed marked impairments in their real-life personal and social functioning, but had intact intelligence. Indeed, these patients performed normally on standard laboratory tests of executive function such as the Wisconsin card sort task.

This background helped to shape a more specific formulation, deemed the 'somatic marker hypothesis' (Damasio 1994). According to this hypothesis, patients with damage in the VMPFC make poor decisions in part because they are unable to elicit somatic (visceral) responses that "mark" the consequences of their actions as positive or negative. In this framework, the VMPFC functions to elicit visceral responses that reflect the anticipated value of the choices. Though this function is specific to the VMPFC, it draws upon information about the external world that is represented in multiple higher-order sensory cortices. Furthermore, this function is limited to specific types of decision making, in particular those situations where the meaning of events is implied and the consequences of behavior are uncertain. These are situations, such as social interactions and decisions about one's personal and financial life, where the consequences of behavior have emotional value, i.e. they can be experienced as subjective feelings and can also increase or decrease the likelihood of similar behavior in the future (they are rewarding or punishing). Furthermore, these are situations where the rules of behavior are not explicit, but yet require some form of mental deliberation in real time in order to navigate them successfully. This form of reasoning is distinct from reasoning that does not require the weighing of positive and negative consequences, or in which the outcomes of decisions are known with a high degree of certainty. The somatic marker framework, in addition to explaining the specificity of the impairments in patients with VMPFC damage, leads to testable hypotheses about the kinds of information represented within the VMPFC and the relationship of this information to the state of the viscera.

## Lesions of the VMPFC impair visceral responses to emotional stimuli

One of the first empirical tests of the somatic marker hypothesis came from studies examining the effects of VMPFC lesions on the visceral response to complex visual stimuli (Damasio et al. 1990). Though it was known that the VMPFC played a role in the elicitation of visceral responses, the precise behavioral context in which visceral functions

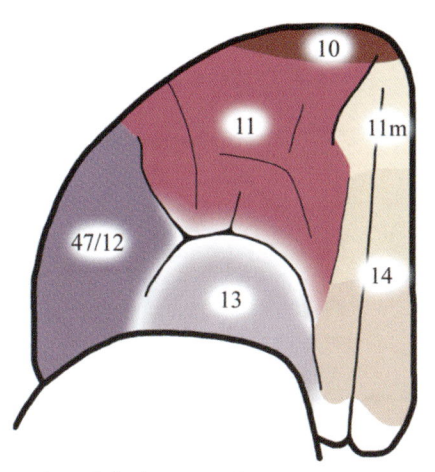

**Plate 1** Schematic reconstruction of the human orbital surface with architectonic areas shown in different colors. See p.31 for details.

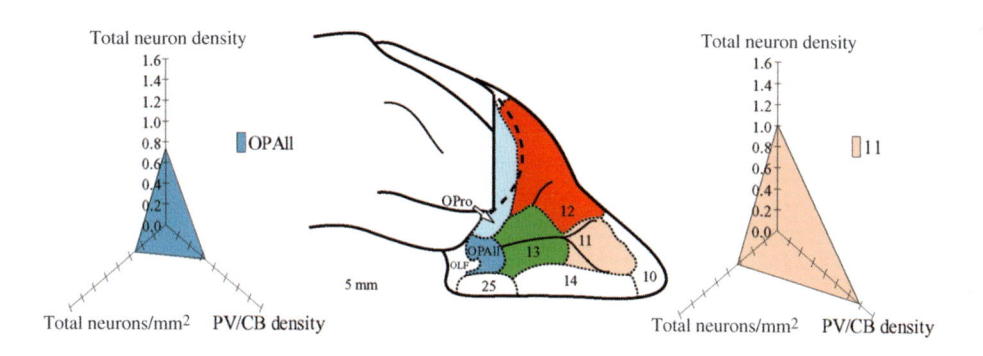

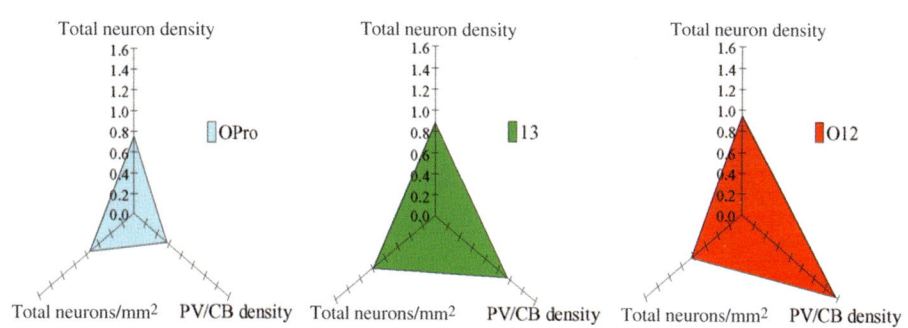

**Plate 2** Quantitative architectonic profiles of orbitofrontal cortices of the rhesus monkey. Areas are shown on the basal surface of the right hemisphere (top center). Fingerprints of the distinct orbitofrontal areas were constructed quantitatively on the basis of features that discriminate among areas, including density of neurons, total neurons/mm² and the ratio of parvalbumin (PV) to calbindin (CB) neurons. The temporal pole at top center was rendered transparent (thick dotted lines) to show the underlying orbitofrontal cortex. The map is according to Barbas and Pandya (1989). Abbreviations: O12, orbital area 12; OPro, orbital proisocortex (dysgranular cortex); OPAll, orbital periallocortex (agranular cortex). See p.58 for details.

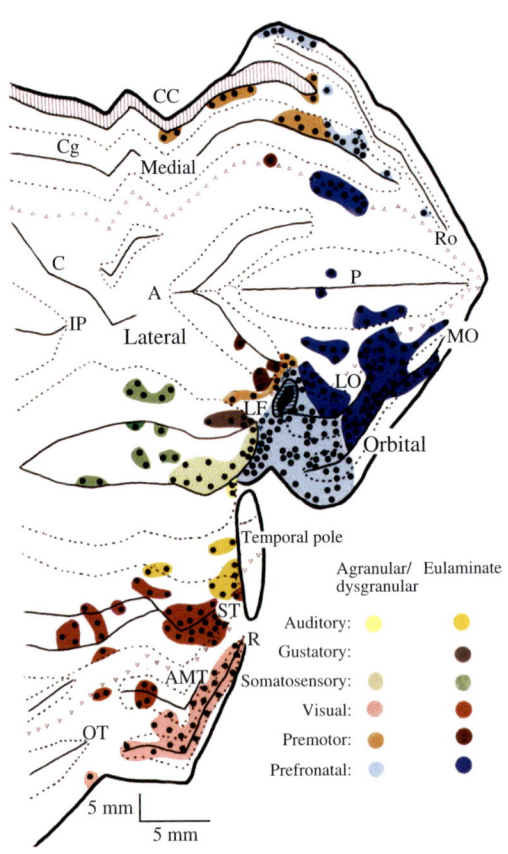

**Plate 3** Diverse cortical projections in orbitofrontal cortex. Unfolded two-dimensional map of the prefrontal cortex showing the origin of projection neurons (black dots) directed to orbitofrontal area OPro on the basal surface (black area, injection site; surround horizontal lines, halo of injection site). Projection neurons originate in visual association (red overlay), gustatory (purple), somatosensory (green), auditory (yellow), premotor (brown) and prefrontal cortices (blue). See p.61 for abbreviations.

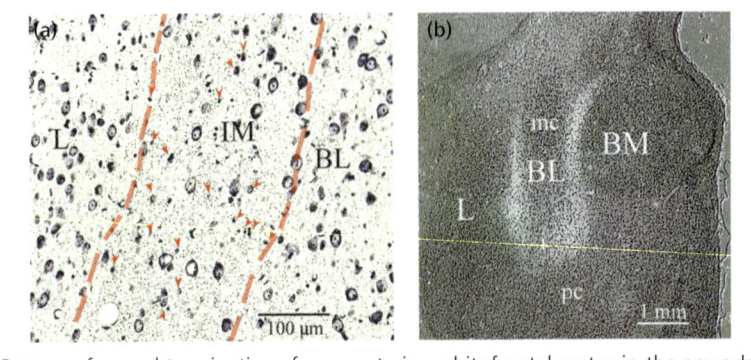

**Plate 4** Pattern of axonal terminations from posterior orbitofrontal cortex in the amygdala.
(a) Brightfield photomicrograph showing the intercalated masses of the amygdala (IM)
(b) Darkfield photomicrograph showing axons from area OPro targeting heavily IM (white grain),
See p.65 for details.

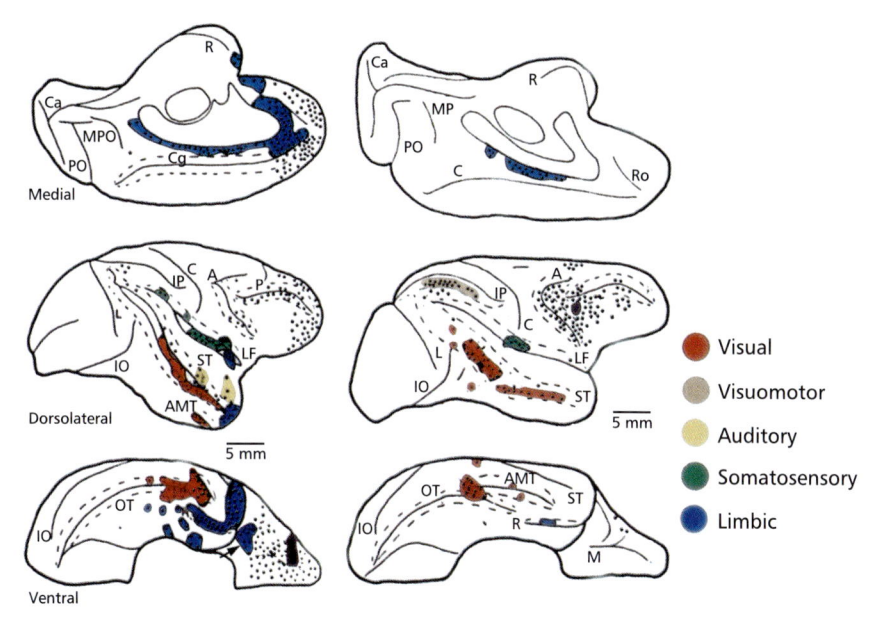

**Plate 5** Sequential pathways for the flow of information in prefrontal cortices. Comparison of the connections of an orbitofrontal area (left) and a lateral prefrontal area (right). (Left) Projection neurons (dots) directed to orbitofrontal area 11 (black area, bottom) originate from several sensory association cortices, as well as from posterior OFC (arrow), a bidirectional pathway. In turn, area 11 is connected with lateral prefrontal cortices, such as area 46 in the principalis region (P). (Right) Projection neurons directed to the ventral part of lateral area 46 (black area, center) originate predominantly from visual (red overlay) and visuomotor (brown) cortices. See p.77 for details.

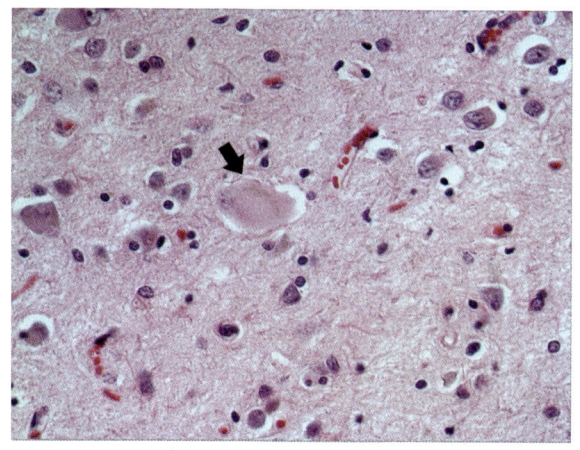

**Plate 6** Section of cingulate cortex from a patient with the Pick disease variant of FTD. Note a single 'balloon cell' (arrow) among smaller phenotypically normal neurons (hematoxylin and eosin-stained section). See p.624 for details.

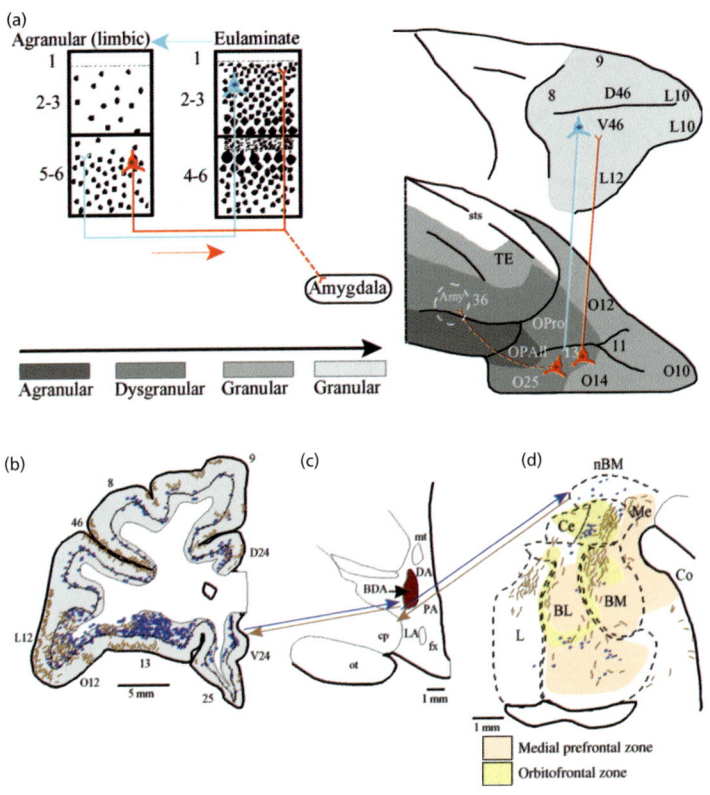

**Plate 7** Laminar-specific pathways underlying the perception and expression of emotions. (a) Eulaminate prefrontal areas (such as areas 46 and 8 on the lateral surface) receive detailed sensory input from early-processing sensory areas (not shown) and target the middle-deep layers of the agranular/dysgranular (limbic) prefrontal cortices (blue), such as posterior OFC. Neurons from the same layers, particularly layer 5, issue robust projections to the amygdala, and 'feedback' projections to lateral eulaminate areas (red). (b) Medial and orbitofrontal cortices are bidirectionally connected with hypothalamic autonomic centers (c), which are also connected with the amygdala (d). See p.78 for details.

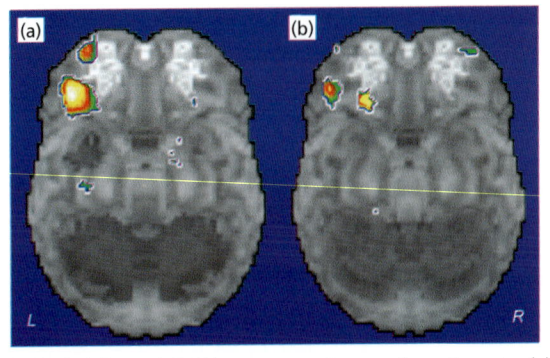

**Plate 8** Activations in the left lateral OFC in response to aversive odorants. (a) Left lateral OFC activation during exposure to highly aversive sulfides. (b) Left putative olfactory region and weaker left lateral OFC activations during exposure to mildly unpleasant (and less intense) odorants. See p.137 for details.

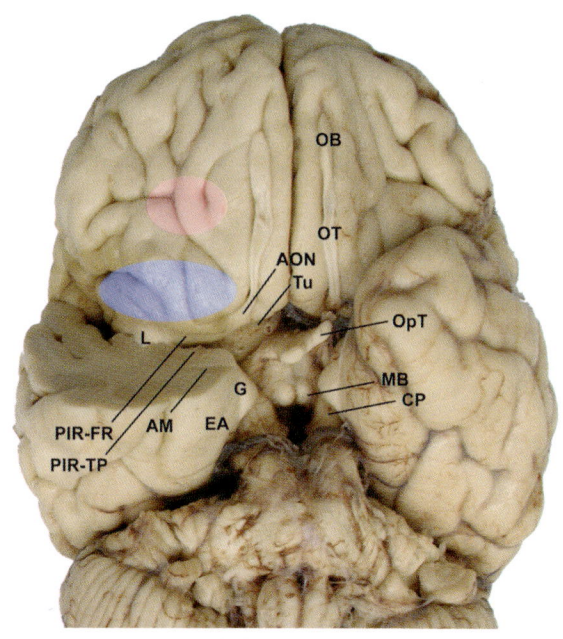

**Plate 9** Macroscopic view of the human ventral forebrain and medial temporal lobes. The blue oval signifies the cytoarchitectural homologues to the monkey olfactory region, whereas the pink oval denotes the putative human olfactory region, as defined by neuroimaging findings. See p.127 for details.

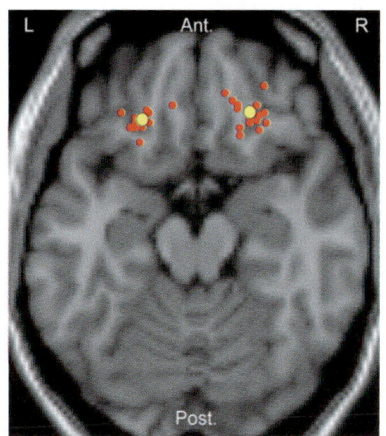

**Plate 10** Localization of putative olfactory OFC in humans. The peak coordinates (red dots) from 12 neuroimaging studies of basis olfactory processing are plotted on an MRI scan in Talairach space. The yellow dots indicate the mean voxel coordinates for right and left hemisphere. See p.132 for details.

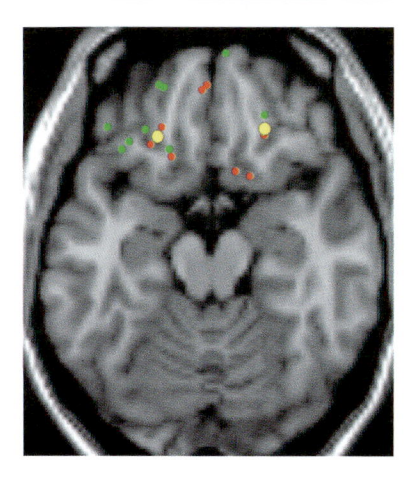

**Plate 11** Olfactory hedonics in human OFC. The peak neural responses to pleasant (8 voxels; red dots) and unpleasant (9 voxels; green dots) odors were identified from six imaging studies of olfactory hedonic processing. The yellow circles indicate the putative olfactory OFC areas. See p.138 for details.

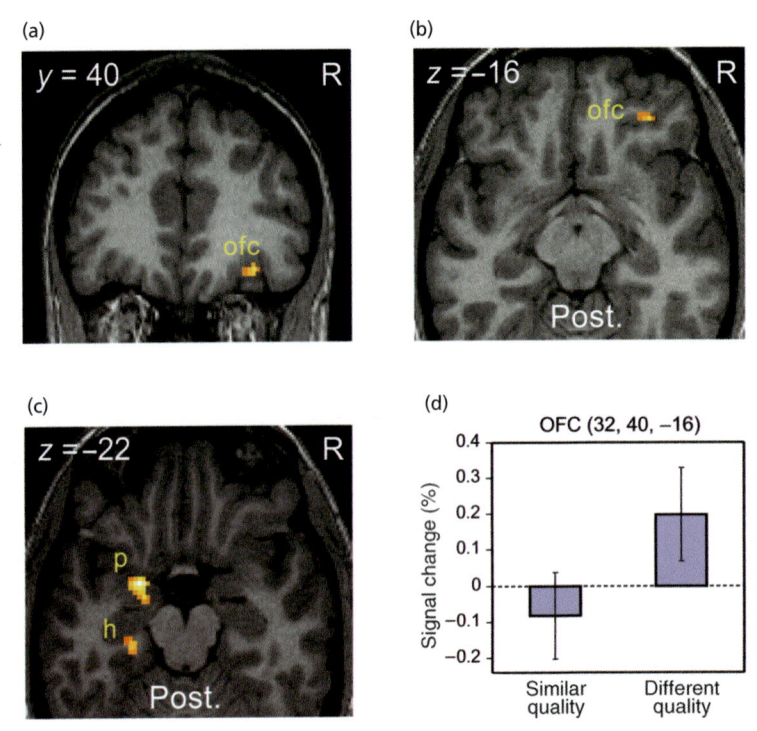

**Plate 12** Neural representations of odor quality in human OFC. (A–C) Comparison of qualitatively different vs. qualitatively similar odorants revealed significant activity in right OFC, piriform cortex (p) and hippocampus/parahippocampal cortex (h) (D) Group-averaged plots of the peak fMRI signal from right OFC indicate decreased responses to qualitatively similar vs. qualitatively dissimilar odorants. See p.140 for details.

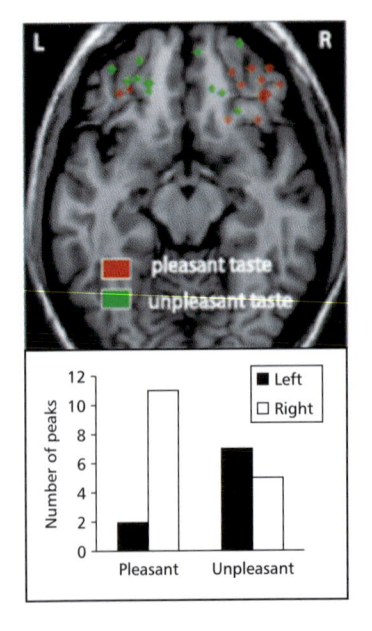

**Plate 13** Lateralization of responses to pleasant and unpleasant tastes. Top: 25 activation foci from 8 analyses of pleasant taste (red: either pleasant taste > neutral taste or pleasant taste > unpleasant taste) and 8 analyses of unpleasant taste (green: either unpleasant taste > neutral taste or unpleasant taste > pleasant taste) collected from 5 published studies and plotted onto an anatomical image. ($z = -14$ in MNI coordinates). Bottom: A bar graph showing the number of peaks ($y$ axis) that fell in the left or right OFC for the pleasant and unpleasant analyses. See p.150 for details.

(a) i  ii

(b) i  ii

■ Odor-evoked activations   ■ Learning-evoked activations

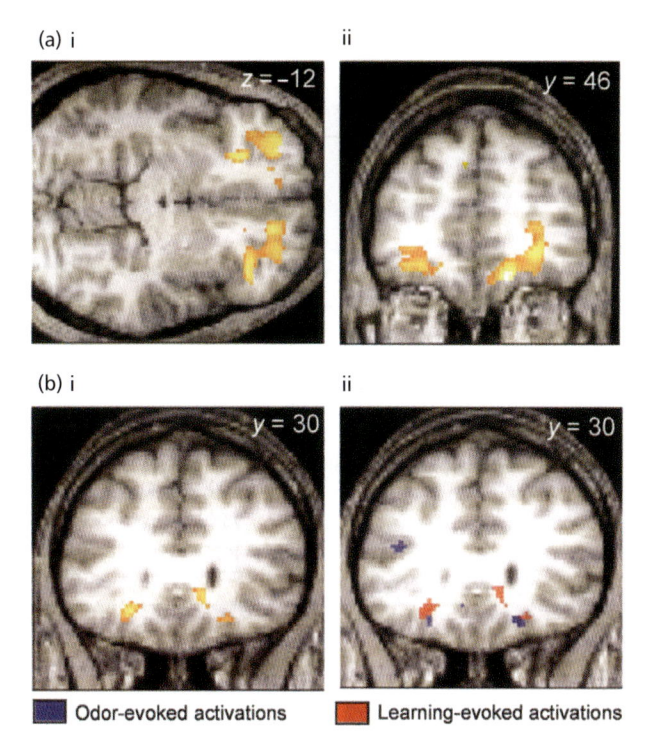

**Plate 14** Caudal–rostral pattern of functional heterogeneity in human OFC. In an fMRI paradigm of visual-olfactory associative learning, learning-evoked activity was detected across extensive areas of rostral OFC ((a)i, (a)ii), but was also observed in caudal OFC (b)i. In panel (b)ii the learning-related responses in caudal OFC (illustrated in red) overlap odor-evoked responses in *blue*. See p.144 for details.

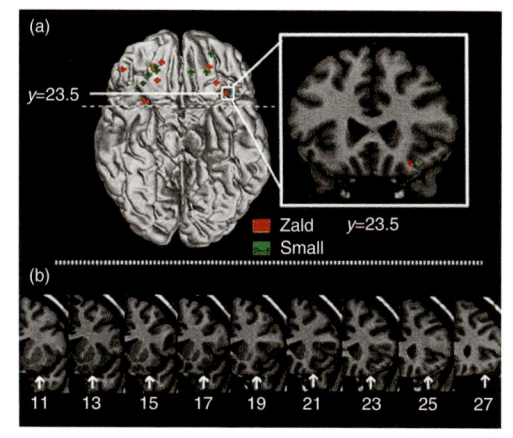

(a)

y=23.5

■ Zald   y=23.5
■ Small

(b)

11  13  15  17  19  21  23  25  27

**Plate 15** The orbitofrontal gustatory region in the humans (a) Crosses represent peaks from three gustatory studies projected onto a surface rendering with anterior temporal lobes removed. Based on anatomical descriptions in non-human primates the proposed gustatory region is at the transition between the ventral agranular insular cortex and the caudal OFC at its lateral point. The two peaks in the small white box fall within a homologous region of the human brain. (b) Coronal sections illustrating the transition from agranular insular cortex (sections 11–17) to the cortex in the caudolateral OFC (19–25). The arrows in 19–25 represents an analogous region to the 'secondary gustatory area' described in monkeys. See p.147 for details.

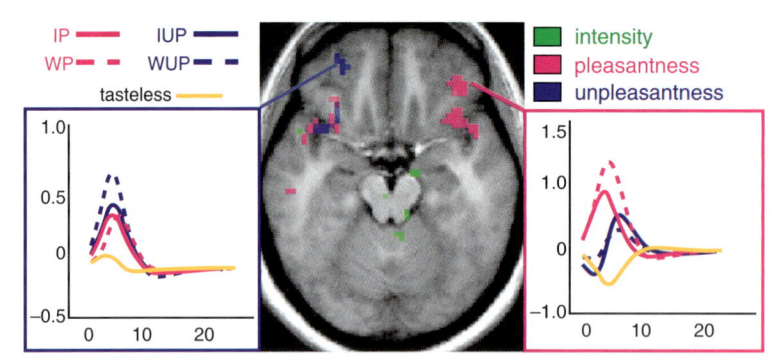

**Plate 16** Dissociation of taste intensity and affective coding. Green: activity related to taste intensity, pink: activity related to taste pleasantness and blue: activity related to taste unpleasantness. Note the valence-specific responses and lack of intensity response in the OFC. In contrast, activity related to intensity but not valence is seen in the amygdala. The graphs depict the fitted hemodynamic response function from the unpleasant (left—blue) and pleasant (right—pink) responses in the OFC. The beige line indicates tasteless. Dotted lines = weak solutions, solid lines = strong solutions. See p.151 for details.

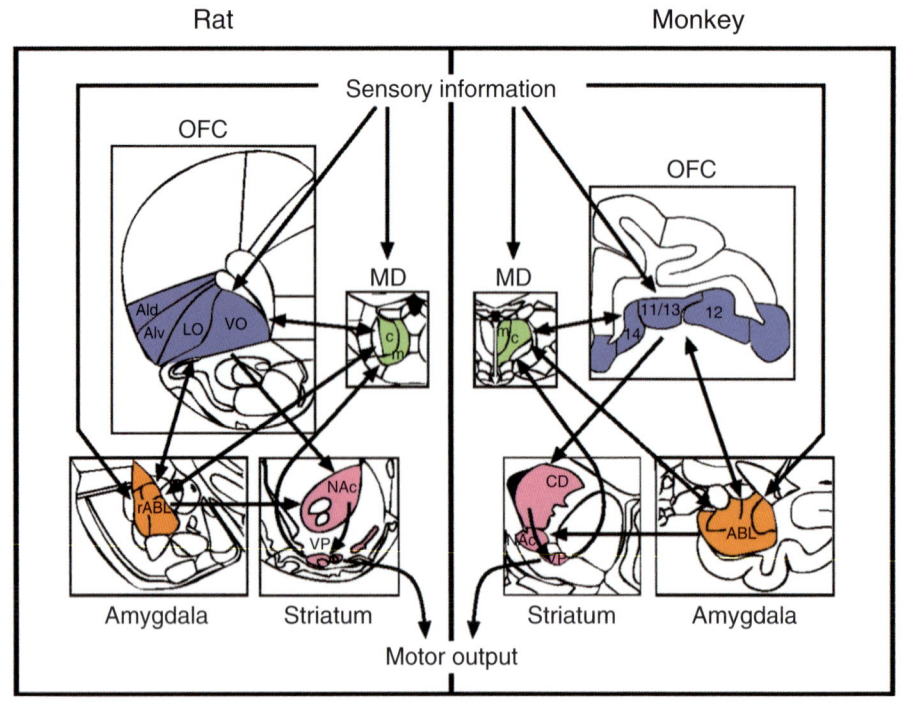

**Plate 17** Anatomical relationships of the OFC (blue) in rat and monkey. Based on their pattern of connectivity with mediodorsal thalamus (MD; green), amygdala (red) and striatum (red), the orbital and agranular insular areas in rat prefrontal cortex are homologous to primate OFC. See p. 201 for details.

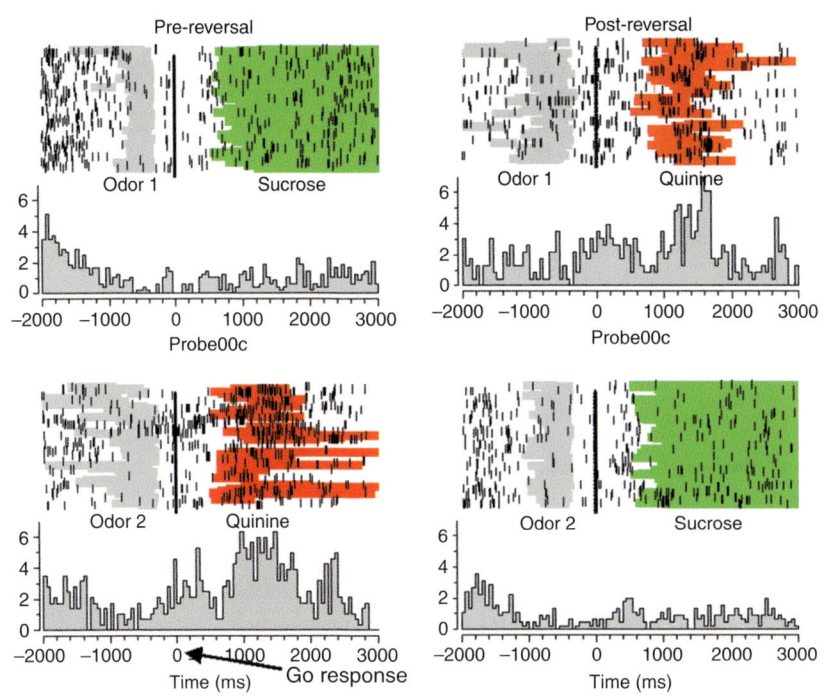

**Plate 18** Outcome-expectant activity in orbitofrontal cortex during acquisition (pre-reversal) and reversal (post-reversal) of a novel 2-odor discrimination problem. Grey shading indicates odor sampling, green indicates sucrose presentation, and red quinine presentation. See p.209 for details.

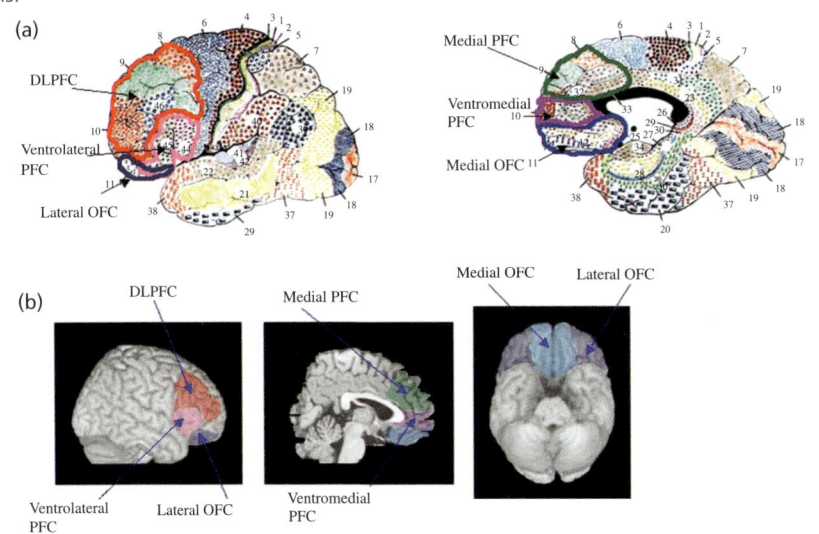

**Plate 19** (a) Original Brodmann map colorized to highlight different Brodmann areas as well as basic prefrontal regions. (b) Basic prefrontal regions shown on a canonical brain. Area shown in red is DLPFC, pink is ventrolateral PFC, purple is ventromedial PFC, green is medial PFC, light blue is medial OFC, and dark blue is lateral OFC. See p.309 for details.

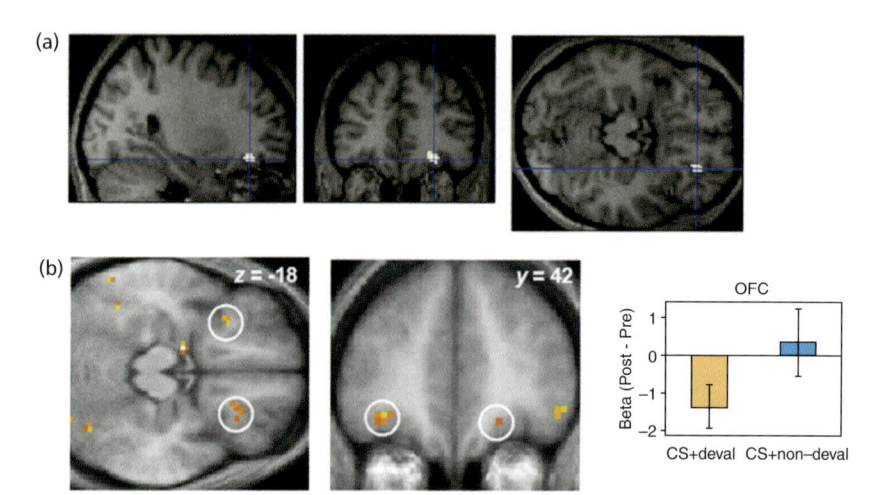

**Plate 20** Predictive reward value coding in OFC. (a) Region of OFC responding during anticipation of a pleasant taste reward (1M Glucose) compared to anticipation of a mildly aversive taste (saline). (b) Results from a classical conditioning paradigm in which arbitrary visual cues were paired with two food-related odours. Following devaluation of one of the odors (by feeding to satiety on the associated food), responses to the predictive cue associated with the devalued odor decreased selectively from pre to post-satiety in OFC. The right most panel shows the change in activity from pre-post satiety for the cue associated with the devalued odor (CS+deval) and the non-devalued odour (CS+nondeval) respectively. See p.272 for details.

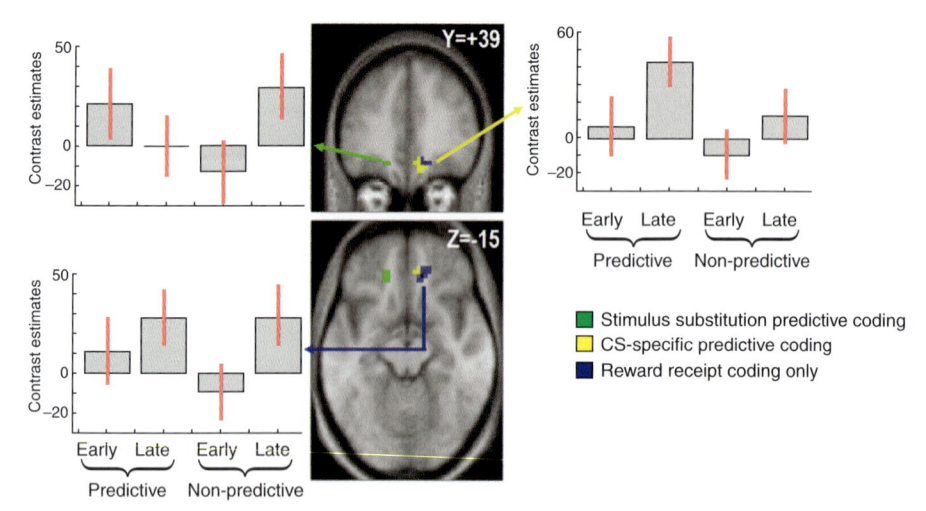

**Plate 21** Plot of coronal (top row) and axial slices (bottom) showing areas of OFC demonstrating predictive or outcome specific responses during trials in which a rewarding juice stimulus is presented compared to an affectively neutral one. The region shown in green shows a change in the timing of activity as a function of predictability. The area shown in yellow shows late responses in the predictive trials but not in the non-predictive trial. Areas in blue show a conjunction of late component effects in non-predictive and late component effects in predictive trials, indicating areas likely to be involved in responding to the unconditioned stimulus itself irrespective of predictability. See p.274 for details.

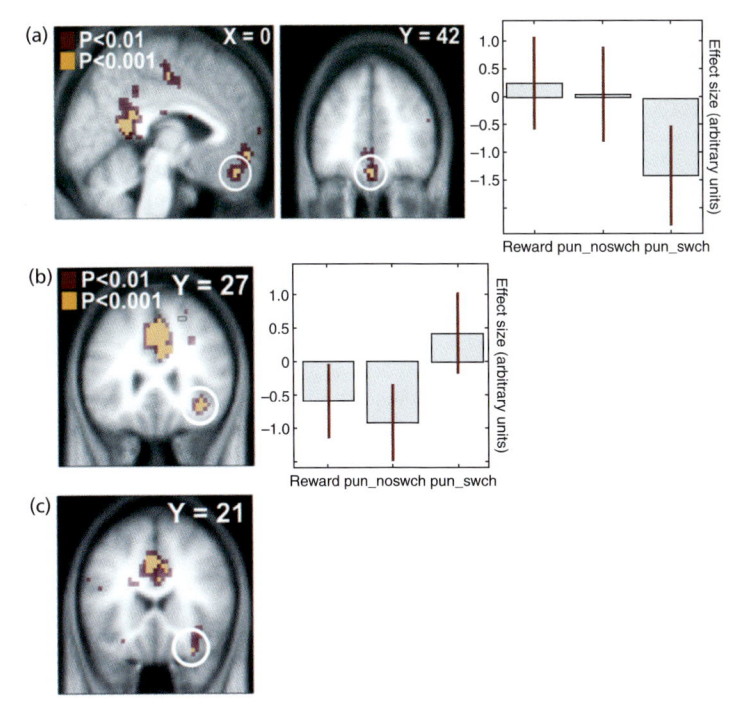

**Plate 22** Regions of OFC and adjacent ventral prefrontal cortex involved in behavioral choice. (a) region of medial OFC that is involved in signaling that a behavioral response should be maintained. This region does not respond to rewards or punishments per se, but shows greater responses on rewarding and punishing trials if the subject does not switch their behavior compared to punishing trials followed by a switch in behavior (pun_swch). (b) Region of anterior insula extending into postserior lateral OFC that shows enhanced responses following a punishment if on the subsequent trial the subject switches their choice of stimulus (pun_swch) compared to rewarding or punishing trials were no such switch of behavior occurs. (c) Region of posterior lateral OFC that shows enhanced responses on punished trials following a switch in behavior compared to punished trials followed by no switch in behavior. See p.278 for details.

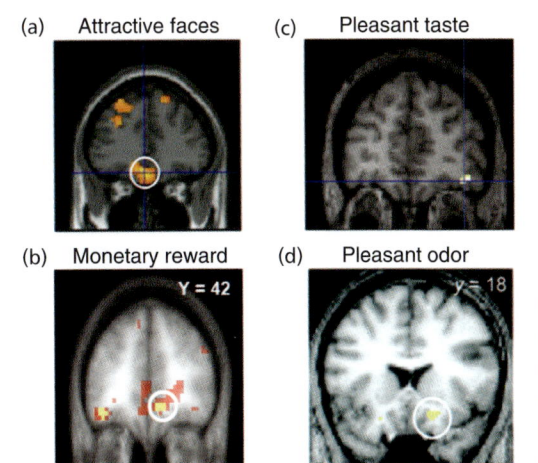

**Plate 23** Representation of stimulus reward-value in human OFC. This figure illustrates OFC activation found in four different fMRI studies each of which involved a different type of reward. See p.266 for details.

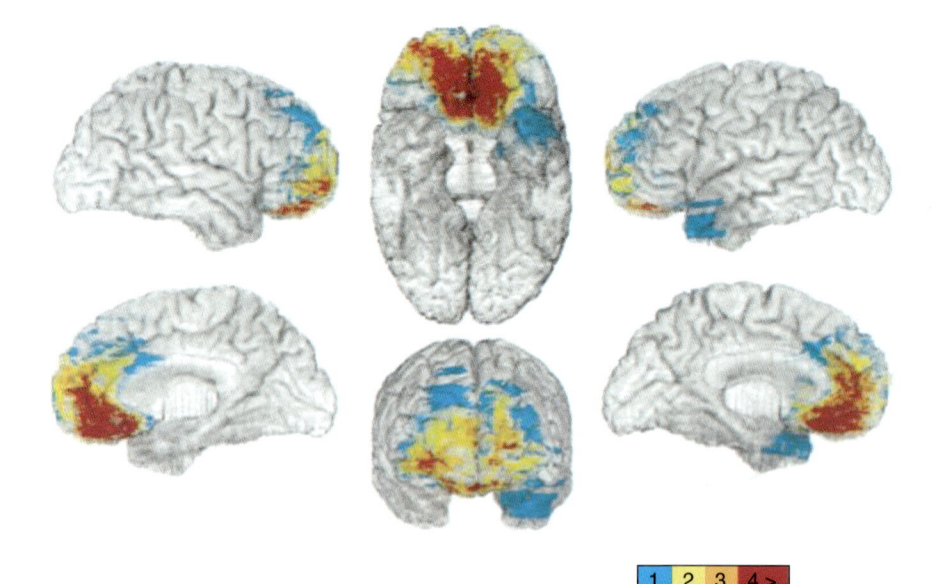

**Plate 24** The location of the VMPFC. A map showing areas of the brain that are damaged in patients who show impairments in visceral response and decision-making. The colors reflect the number of subjects with damage in a given voxel. See p.327 for details.

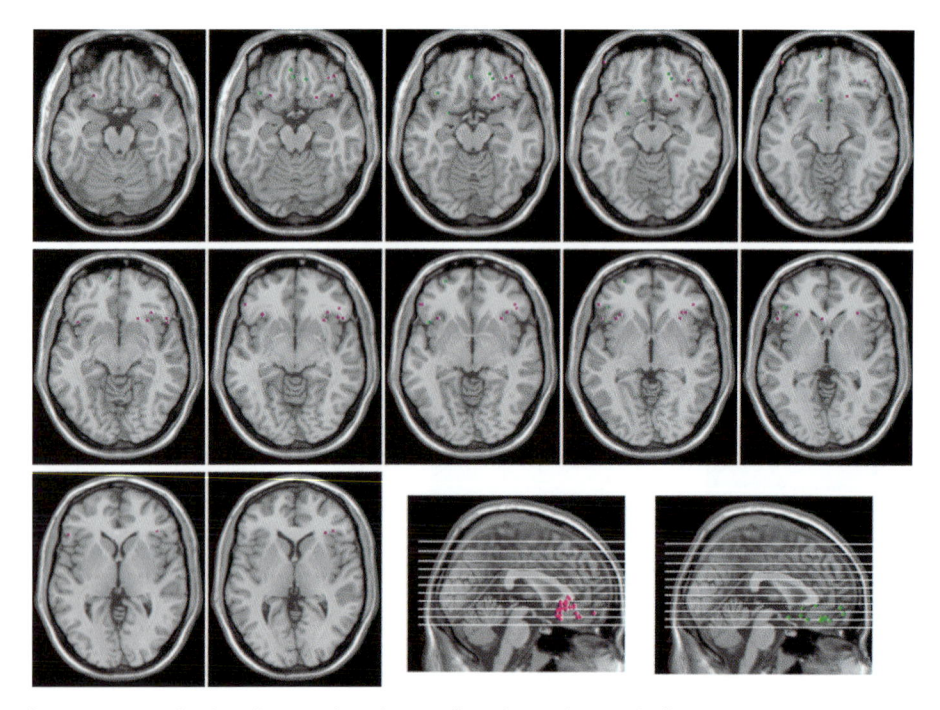

**Plate 25** OFC activations in neuroimaging studies of emotion. Loci of OFC activations in studies utilizing script-driven imagery are shown in green. Loci of OFC activations utilizing face viewing paradigms are shown in purple. See p.383 for details.

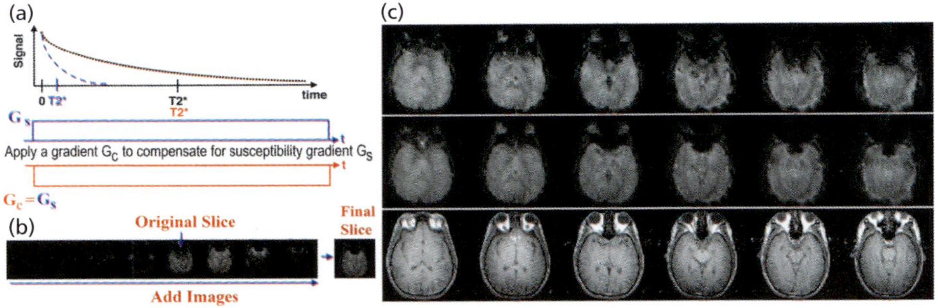

**Plate 26** (a) Diagram of T2* decay from a susceptibility gradient $G_S$. If an opposing compensation gradient $G_C$ is applied during the data acquisition, it will cancel out the effects of the susceptibility induced gradient $G_S$, so that the signal decay curve from region with susceptibility (in red) now overlaps with that from nonsusceptible regions (black) instead of showing its original rapid decay (blue dashed). (b) Series of eight images of one slice each acquired with a different $G_C$. Each slice contains images with different amounts of correction and the final image is obtained by summation. (c) Slices acquired reversed spiral, gradient compensated reversed spiral, and spin echo acquisitions at 3T from top to bottom, respectively. See p.438 for details.

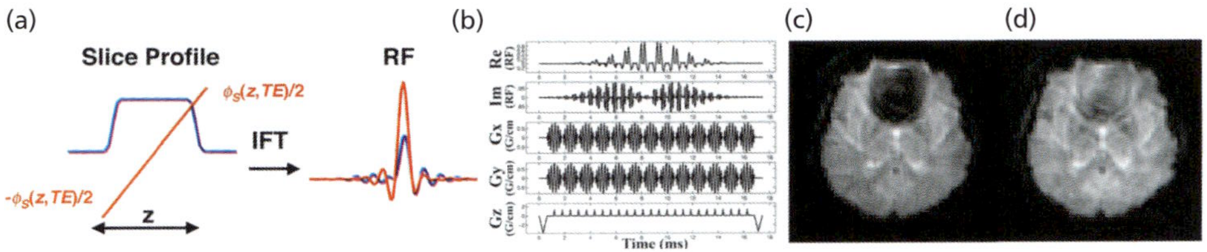

**Plate 27** (a) Phase correction can be put into a 1D tailored RF pulse by using the Fourier transform. (b) A 3D tailored RF pulse that applies the correction only in the OFC. (c) Image acquired without the 3D tailored RF pulse. (d) Same image acquired using the 3D tailored RF pulse. See p.440 for details.

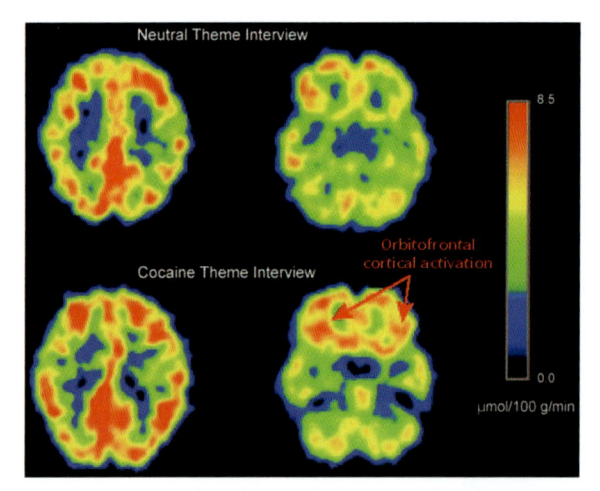

**Plate 28** OFC activation in active cocaine abusers during a cocaine theme interview and a neutral theme interview, as measured by FDG PET. See p.490 for details.

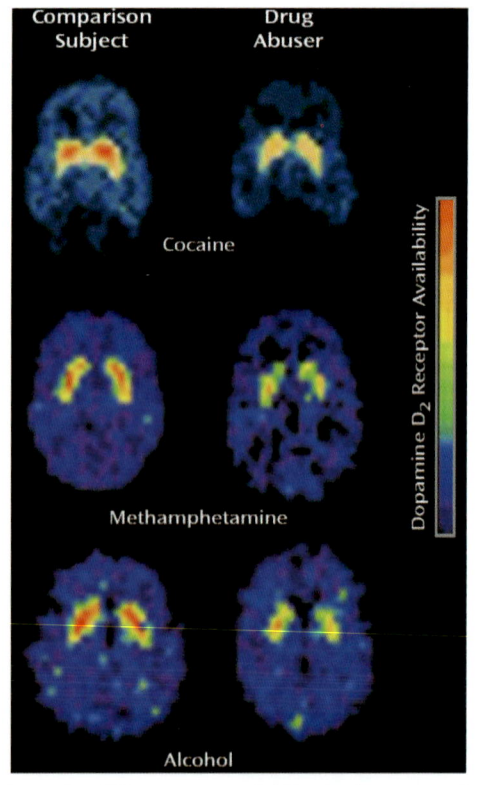

**Plate 29** Lower striatal dopamine D2 receptor availability in drug users during withdrawal from cocaine, methamphetamine, and alcohol than in normal comparison subjects. See p.495 for details.

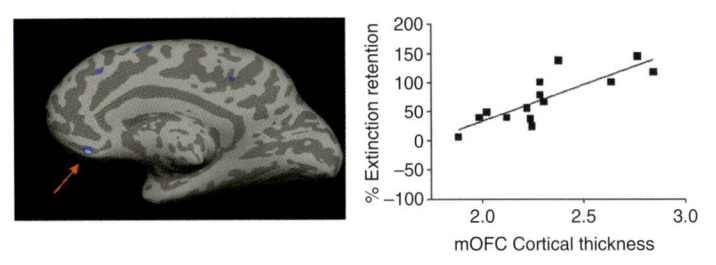

**Plate 30** Regions with positive correlations between cortical thickness and extinction retention (indicated by arrow) and regression plots for the correlations between percent extinction retention and cortical thickness in the mOFC. Threshold is set at $p < 0.01$ (dark blue) to $p < 0.001$ (cyan blue). See p.527 for details.

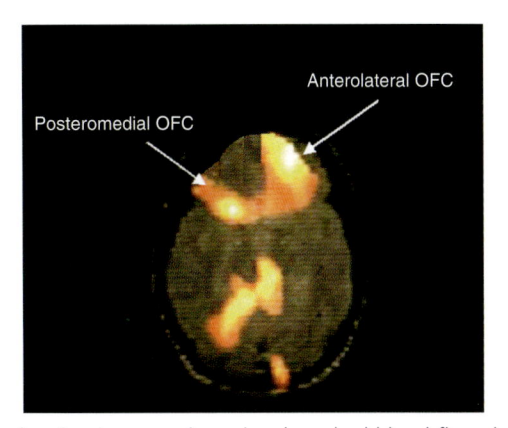

**Plate 31** A PET image showing increases in regional cerebral blood flow during symptom provocation (versus a control state) in subjects with OCD. Note that extracted values showed that OCD symptom severity was directly correlated with anterolateral OFC activation, where as severity of anxiety in OCD was inversely correlated with posteromedial OFC activation. Adapted with permission from Rauch et al. (1994). See p.533 for details.

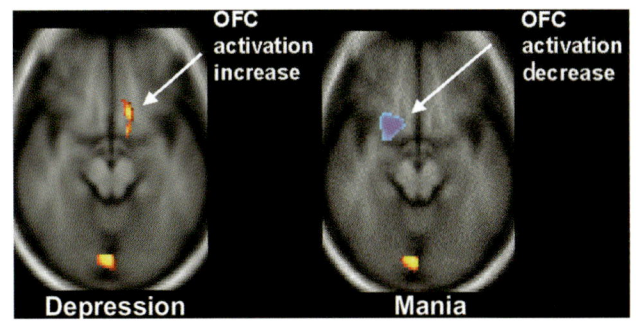

**Plate 32** The left image demonstrates the relative increases in left caudal OFC activation in a group of BD participants who were depressed at the time of scanning as compared to BD participants who were euthymic at scanning. The right image demonstrates relative decreases in right caudal OFC in BD mania as compared to euthymia. Reproduced with permission from Blumberg et al. (2003c). Images are presented in radiological convention, i.e. the right side of the image displays the left side of the brain. See p.560 for details.

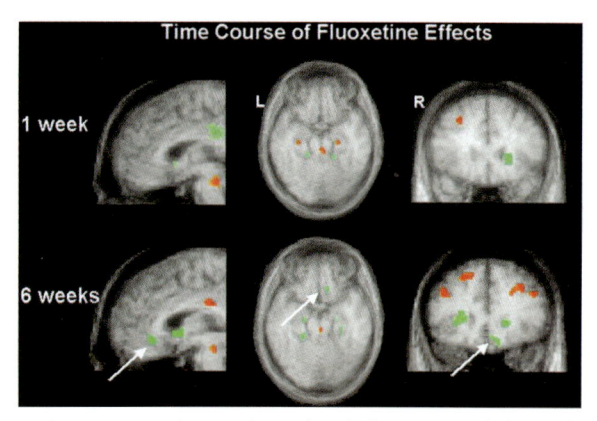

**Plate 33** The images demonstrate changes in regional glucose metabolism after 1 week and 6 weeks of fluoxetine treatment for MDD. Areas depicting decreases in metabolism are shown in green and areas of increased metabolism are shown in red. Decreases in subgenual prefrontal cortex metabolism (white arrows) are detected after 6 weeks of treatment. See p.565 for details.

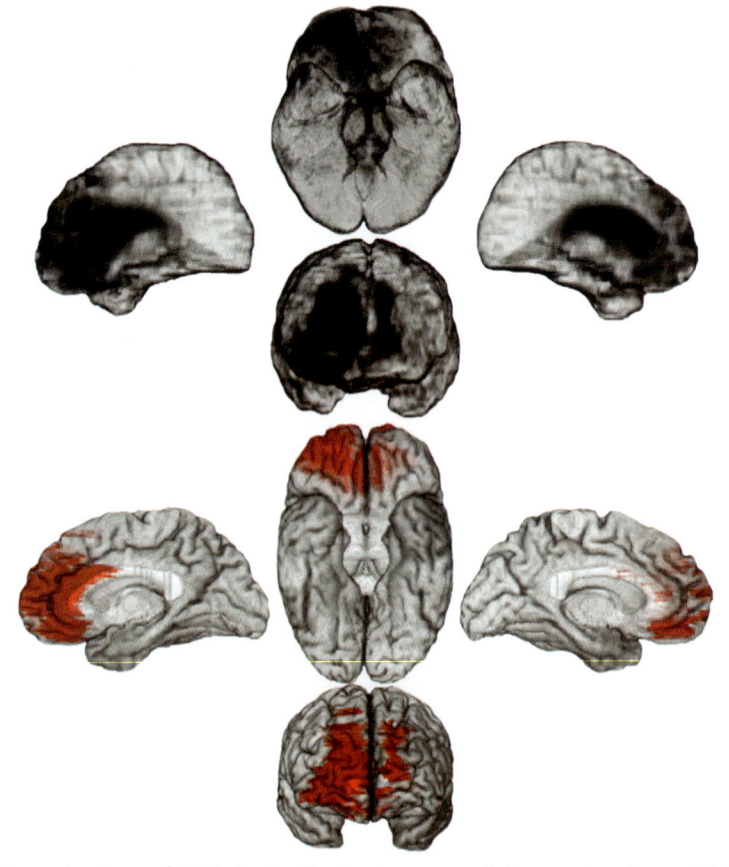

**Plate 34** Reconstructions of EVR's brain. The top images are 3-D reconstructions of EVR's brain from a computerized tomography CT scan. The bottom images depict EVR's lesion (in red) on a standard reference brain. See p.599 for details.

were engaged by the VMPFC was not known. In other words, it was possible that the VMPFC functioned to elicit visceral responses to all forms of stimuli, or only certain types of stimuli, such those that, like social signals, possessed emotional value that was often largely implicit. According to the somatic marker hypothesis, the visceral responses that were mediated by the VMPFC were especially related to the implied meaning of social stimuli.

To address this hypothesis, our laboratory performed a study that examined the visceral responses of a group of patients with damage in the VMPFC, all of whom, like patient EVR, showed a pattern of behavior that reflected an inability to make advantageous decisions in the personal and social realms despite intact intellectual functioning. The subjects were shown a series of affectively charged pictures, including pictures of mutilations, disasters and sexual images, along with a series of neutral pictures. The patients were tested in 2 conditions: one in which they watched the stimuli passively and another in which they were asked to describe the pictures in terms of their emotional content. After each stimulus, the skin conductance response (SCR), an index of sympathetic arousal, was measured. In addition, SCRs to 'orienting' stimuli, including loud noises and deep breaths, were measured. The responses of patients with VMPFC damage were compared to the responses of patients with damage to regions outside the VMPFC, as well as the responses of neurologically intact comparison subjects.

It was found that patients with VMPFC damage were significantly impaired in their visceral responses to emotional pictures when required to view them passively, compared to both neurologically intact and brain-damaged comparison subjects. However, when required to comment on the content of the pictures, the visceral responses of the VMPFC patients to the emotional pictures were largely intact. In addition, the VMPFC patients showed intact SCRs to orienting stimuli. These results indicate that the VMPFC plays a role in the elicitation of visceral responses to biologically relevant stimuli. Moreover, this impairment is specific to stimuli for which emotional meaning must be decoded through cognitive processes, and does not extend to stimuli that elicit visceral responses because they are innately aversive or arousing (e.g., a loud noise), or to stimuli that are physiological elicitors of visceral responses (e.g., a deep breath). The results also imply that the VMPFC mediates the visceral response to emotional stimuli when the evaluation of these stimuli does not require verbal mediation, i.e., when it is implicit.

## Lesions in the VMPFC lead to impairments in the Iowa Gambling Task

Once it was known that patients with VMPFC damage were abnormal both in their capacity to make decisions and in their ability to respond viscerally to the emotional meaning of certain stimuli, it still remained to be shown that these two abnormalities were linked. Up to this point, the behavioral abnormalities of these patients, which were striking in real life, had largely eluded conventional neuropsychological and laboratory tests. Thus, it was important to develop a laboratory test that simulated the real-life decisions in which patients with VMPFC damage failed. This test factored in reward and punishment, as well as the uncertainty and risk that accompany many real-life decisions. In addition, this test required participants to reason and deliberate the outcome of choices 'on-line,' in real time.

This task was developed in our laboratory. The test, referred to as the Iowa Gambling Task (IGT), is described in detail elsewhere (Bechara *et al.* 2000). Briefly, in the IGT, subjects are required to choose between 4 decks of cards: A, B, C and D. Choosing from any of the decks always results in the payout of money. Occasionally, choosing from any of the decks can also result in the loss of money. Decks A and B offer a large payout on each choice but also an occasional large loss. Decks C and D offer a small payout on each choice, but also an occasional small loss. These reward and punishment contingencies are designed such that choosing consistently from decks C and D (the advantageous decks) results in an eventual gain of money, whereas choosing consistently from decks A and B (the disadvantageous decks) results in an eventual loss of money. Subjects are instructed to attempt to win as much money as possible. Subjects are not instructed about the reward and punishment contingencies of each deck, but are instructed that some of the decks are 'better' than others. In this task, there is a requirement to weigh the consequences of immediate gain against the consequences of loss in the future. To arrive at this knowledge, subjects cannot simply reason about the value of each deck; they ultimately succeed by developing 'hunches' about each deck over time that may help guide them to the correct strategy. In this way, the IGT simulates the demands of real-life decisions.

We initially applied this task to a group of patients with VMPFC lesions, along with neurologically intact subjects and patients with damage in regions outside the VMPFC (Bechara *et al.* 1994). Subjects in all groups initially sampled from both the advantageous and disadvantageous decks, and, as would be expected based on the higher payout, tended to prefer the disadvantageous decks. Over time, neurologically intact and brain-damaged comparison subjects developed a strategy in which they avoided the disadvantageous decks and chose more often from the advantageous decks, which led to a higher overall gain of money over the course of the task. In contrast, subjects with VMPFC damage continued to pick preferentially from the disadvantageous decks, and as a result, lost money over the course of the task. These data are illustrated in Fig. 13.2. The results demonstrate that the VMPFC is critical for guiding decisions in the face of uncertain reward and punishment.

### Lesions in the VMPFC disrupt visceral responses during the Iowa Gambling Task

To test the role of somatic markers in the decision-making process, we performed an experiment in which we coupled the IGT with the measurement of the SCR (Bechara *et al.* 1996). In this experiment, we measured the SCR during two distinct epochs with respect to the moment of choice: an anticipatory epoch, during which the subject deliberated the consequences of their choice, and a reward/punishment epoch, which followed the payoff that was delivered with each choice as well as the loss of money that occasionally followed choices. It was found that in normal subjects, SCRs were elicited during both the

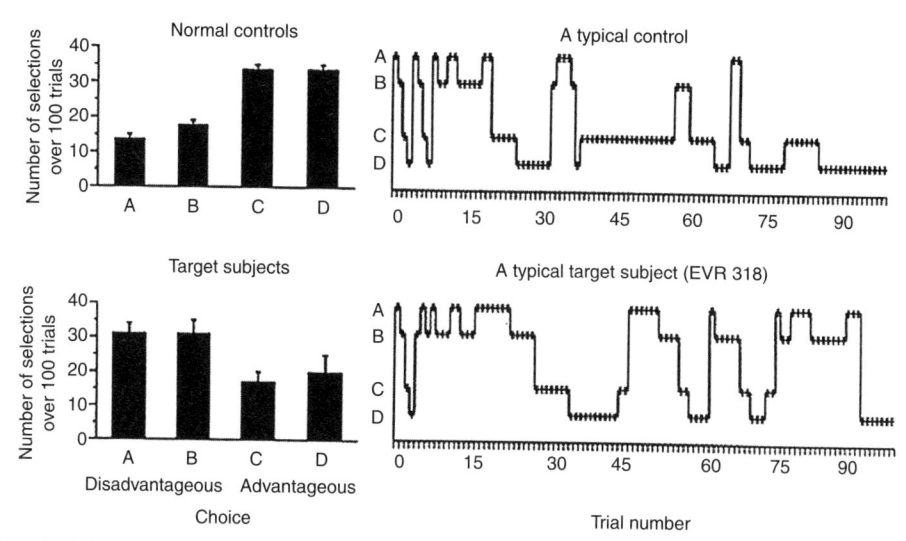

**Fig. 13.2** Damage to the VMPFC impairs performance on the IGT. Comparison of normal controls and vmPFC lesion subjects. On the left is the total number of choices made from each deck. On the right are profiles of card selections from representative subjects. Subjects with VMPFC damage fail to shift away from choosing from the disadvantageous decks, leading to an overall poorer performance (greater loss of money). Error bars represent the standard error of the mean. Reproduced with permission from Bechara et al. (1994).

anticipatory and reward/punishment epochs. Furthermore, it was found that anticipatory responses were larger before choosing from the disadvantageous decks than they were before choosing from the disadvantageous decks. In patients with VMPFC damage, these anticipatory responses were significantly impaired. Not only did they not differentiate between choices from advantageous and disadvantageous decks, they were also markedly reduced in amplitude. At the same time, SCRs, which followed winning, and losing money were largely intact. These data are illustrated in Figure 13.3. The results indicate that the impairment caused by VMPFC damage is specific to the visceral responses that index the anticipated value of choices, and do not impact the visceral responses that reflect the appreciation of outcomes of choices. If anticipatory visceral responses are necessary for guiding choice behavior, then a failure to deploy anticipatory visceral responses could explain why patients with VMPFC lesions perform poorly on the IGT.

An important question regards the information content of visceral responses that are elicited by the VMPFC. If somatic markers are to be useful in guiding decision-making processes involving uncertain reward and punishment, then they should provide information about both the valence of an anticipated outcome (e.g. whether a choice will result in winning or losing money) as well as information about the magnitude of the anticipated outcome (e.g. how much money will be won or lost). Our results using

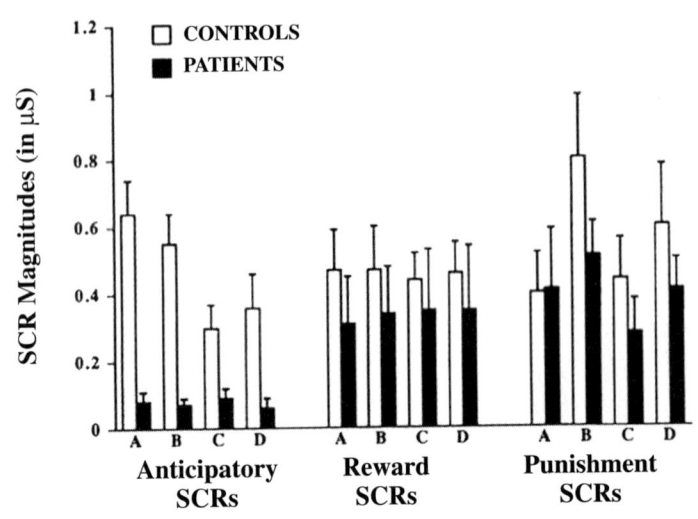

**Fig. 13.3** Damage to the VMPFC impairs anticipatory visceral responses, but not visceral responses to reward or punishment. Anticipatory SCRs are measured during the interval preceding each choice. Reward SCRs are measured after choices for which money is gained, but not lost. Punishment SCRs are measured after choices for which there is both gain and loss of money. Error bars represent the standard error of the mean. Anticipatory SCRs, which discriminate between advantageous decks (C and D) and disadvantageous decks (A and B) in normal comparison participants, are markedly impaired in subjects with VMPFC damage. Responses to reward and punishment are relatively spared by VMPFC damage. Reproduced with permission from Bechara et al. (1996).

the IGT show that the VMPFC triggers anticipatory visceral responses to both advantageous and disadvantageous decks. These responses are larger for disadvantageous decks than for advantageous decks, though they are still deployed for the advantageous decks. Further experiments from our laboratory (Bechara et al. 2002) and others (Tomb et al. 2002) have shown that when the reward-punishment contingencies are reversed, with the disadvantageous decks paying out a lower quantity of reward, rather than doling out a higher punishment, SCRs are now greater to the advantageous decks than to the disadvantageous decks. This suggests that the SCR is not merely an index of the potential 'badness' of choices. Rather, the SCR can index the magnitude of both the anticipated negative outcome of a choice as well as the magnitude of the anticipated positive outcome of a choice. It seems, however, that the SCR does not differentiate the anticipated valence of the outcomes. This is consistent with work by others (Lang et al. 1993) showing that the SCR does not differentiate the hedonic valence of emotional stimuli, but does index the magnitude of the arousal that they elicit. This would mean that some other signal is required in order assess the valence of the anticipated outcome. Our laboratory (Rainville et al. in press) and others (Fowles et al. 1982; Bradley and Lang 2000) have provided evidence that cardiovascular responses, such as changes in heart rate, can provide information that distinguishes between positive and negative emotional states.

Such signals can combine with those reflected in the SCR to provide information about both the perceived valence of the future outcome of a choice, as well as its perceived magnitude. Future experiments may examine how visceral signals reflecting distinct channels of autonomic outflow combine to influence decision-making.

## Visceral responses that signal the correct strategy that, precede explicit knowledge of the correct strategy

According to the somatic marker hypothesis, the VMPFC mediates an implicit representation of the anticipated value of choices that is distinct from an explicit awareness of the correct strategy. To test this idea, we performed a study (Bechara et al. 1997) in which we examined the development of SCRs over time in relation to subjects' knowledge of the advantageous strategy in the IGT. In this study, the IGT was administered as before, but this time, the task was interrupted at regular intervals and the subjects were asked to describe their knowledge about 'what was going on' in the task and about their 'feelings' about the task. Here, it was shown that normal subjects began to choose preferentially from the advantageous decks before they were able to report why these decks were preferred over the disadvantageous decks. They then began to form 'hunches' about the correct strategy, which corresponded to their choosing more from the advantageous decks than from the disadvantageous decks. Finally, some subjects reached a 'conceptual' stage, where they possessed explicit knowledge about the correct strategy (i.e., to choose from decks C and D because they pay less, but also result in less punishment). As before, normal subjects developed SCRs preceding their choices that were larger for the disadvantageous decks than for the advantageous decks. This time, it was also found that the SCR discrimination between advantageous and disadvantageous decks preceded the development of conceptual knowledge of the correct strategy. In fact, the SCR discrimination between advantageous and disadvantageous decks even preceded the development of hunches about the correct strategy. In contrast to the normal subjects, subjects with damage in the VMPFC failed to switch from the disadvantageous decks to the advantageous decks, as was previously shown. In addition, as previously shown, subjects this group failed to develop anticipatory responses that discriminated between the disadvantageous and advantageous decks. Furthermore, patients with VMPFC damage never developed 'hunches' about the correct strategy. These results are shown in Fig. 13.4. Together, they suggest that anticipatory visceral responses that are governed by the VMPFC precede emergence of advantageous choice behavior, which itself precedes explicit knowledge of the advantageous strategy. This suggests that signals generated by the VMPFC, reflected in visceral states, may function as a non-conscious bias toward the advantageous strategy.

Recently, other investigators have questioned whether it is necessary to invoke visceral responses as constituting non-conscious biasing signals (Maia and McClelland 2004). By using more detailed questions to probe subjects' awareness of the attributes of each of the decks in the IGT, this study showed that subjects possess explicit knowledge of the advantageous strategy at an earlier stage in the task than was shown in the Bechara et al. (1997)

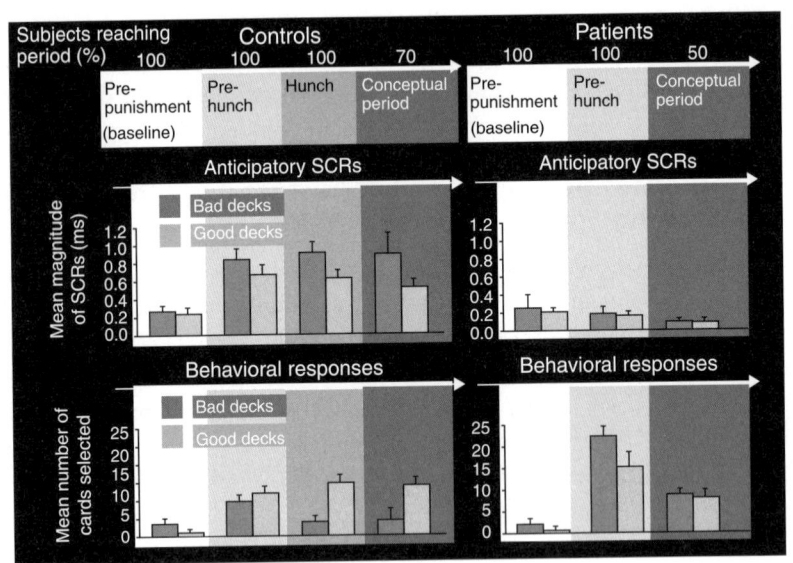

**Fig. 13.4** Anticipatory visceral responses that discriminate between advantageous and disadvantageous choices precede knowledge of the correct strategy. The pre-punishment period covered the start of the game when subjects sampled the decks and before they encountered the first loss. The pre-hunch period consisted of the next series of cards when subjects continued to choose cards from various decks, but professed no notion of what was happening in the game. The hunch period (never reached in patients) corresponded to the period when subjects reported 'liking' or 'disliking' certain decks, and 'guessed' which decks were risky or safe, but were not sure of their answers. The conceptual period corresponded to the period when subjects were able to articulate accurately the nature of the task and tell for certain which were the good and bad decks, and why they were good or bad. Error bars represent the standard error of the mean. Reproduced with permission from Bechara *et al.* (1997).

study. Furthermore, the Maia and McClelland study found that subjects began to make advantageous choices at around the same time that they reported knowledge of the correct strategy. These findings suggest that non-conscious somatic marker processes are not required in order to explain how decision-making occurs. A response to this study has been published elsewhere (Bechara *et al.* 2005), along with a rebuttal by Maia and McClelland (Maia and McClelland 2005). Two points bear discussion here. Firstly, since this study did not measure visceral responses nor examine the effects of brain damage, it does not disprove the hypothesis that somatic markers mediated by the VMPFC play a role in decision-making; it only shows that conscious awareness of the correct strategy occurs at around the same time as advantageous decision-making. Secondly, both the Bechara *et al.* (1997) study and the Maia and McClelland (2005) study found that some subjects continue to make disadvantageous choices despite being able to report the correct strategy. This pattern bears an uncanny resemblance to the way in which subjects with lesions in the VMPFC are able to report the correct strategies for personal and social decision-making, despite their severe deficits in the actual execution of personal and social behavior in real life (Saver and Damasio, 1991). Indeed, this clinical observation provided the initial impetus to hypothesize a role for covert biasing processes in decision-making in the first place. This indicates that, in both the IGT and in real life, conscious knowledge of the correct strategy may not be enough to guide advantageous decision-making.

Thus, some process that operates independently of conscious knowledge of the correct strategy (i.e. somatic markers) must be invoked in order to explain fully how individuals make advantageous decisions. Indeed, it seems likely that this process can sometimes bias behavior that goes against what a person consciously thinks to be the correct strategy. That non-conscious biasing processes may not precede conscious knowledge in time is potentially an important finding, but it does not provide a basis for rejection of the fundamental role of somatic markers as non-conscious biases of behavior.

## The decision-making functions of the VMPFC are different from the decision-making functions of the amygdala

Like patients with damage to the VMPFC, patients with bilateral damage to the amygdala also demonstrate impairments in their ability to make advantageous choices in their personal and social lives (Tranel and Hyman 1990; Adolphs *et al.* 1995). The amygdala, like the VMPFC, has been strongly implicated in emotional and motivational processes (reviewed in Cardinal 2002; reviewed in LeDoux 2000). There is much in common between the amygdala and the VMPFC in terms of cortical and subcortical connectivity. In particular, the amygdala receives information from higher-order sensory cortices for vision, olfaction, audition and visceral sensation (reviewed in Amaral *et al.* 1992), and sends output to subcortical sites that regulate the state of the viscera, including nuclei of the brainstem and hypothalamus (Price and Amaral 1981). Thus, like the VMPFC, the amygdala is positioned to receive multiple sensory inputs pertaining to biologically relevant stimuli and to trigger changes in the visceral state. This suggests that the amygdala plays a role similar to that of the VMPFC in decision-making. However, there are important differences between the amygdala and VMPFC in terms of their visceral and decision-making functions.

One source of evidence regarding the distinct roles of the amygdala and the VMPFC in decision-making comes from an experiment in which we administered the IGT to a group of subjects with bilateral amygdala damage (Bechara *et al.* 1999). Their performance on this task, along with their SCRs, were compared to a group of subjects with VMPFC damage, as well as a group of neurologically intact subjects. Similar to subjects with VMPFC damage, subjects with damage to the amygdala performed poorly on the IGT, failing to shift towards choosing more frequently from the advantageous decks. When examining the SCRs, however, there were different patterns of deficit in amygdala-lesioned subjects versus VMPFC-lesioned subjects. As discussed above, patients with VMPFC damage fail to deploy anticipatory visceral responses before their choices, responses which, in normal subjects, are larger before choosing from the disadvantageous decks than before choosing from the advantageous decks. As also noted previously, VMPFC-lesioned subjects continue to have normal SCRs to receiving reward and punishment. In contrast, patients with amygdala damage fail to deploy SCRs during both the anticipatory period and in response to receiving rewards and punishments. These data are shown in Figure 13.5. This suggests that the decision-making deficit in patients with amygdala damage is due to an inability to respond viscerally to rewards and punishments. This is different from the deficit in patients with VMPFC damage, who possess an inability to viscerally anticipate uncertain rewards and punishments, but who are normal in their ability to respond viscerally to rewards and punishments once they are received.

To examine further the distinction between the affective-visceral functions of the amygdala and VMPFC, the same subjects underwent a Pavlovian conditioning paradigm (Bechara et al. 1999). Here, it was found that patients with bilateral amygdala damage failed to acquire conditioned SCRs. In contrast, patients with VMPFC damage were not different from neurologically intact subjects in their ability to acquire conditioned SCRs.

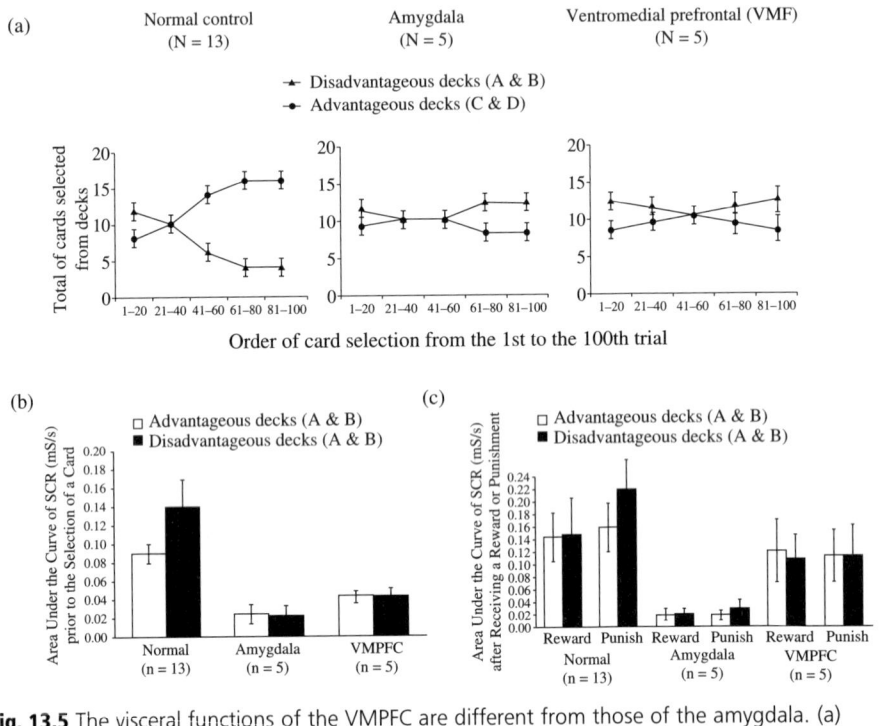

**Fig. 13.5** The visceral functions of the VMPFC are different from those of the amygdala. (a) Behavioral performance on the IGT. Both VMPFC- and amygdala-lesioned subjects fail to switch to choosing preferentially from the advantageous decks. (b) Both VMPFC and amygdala-lesioned subjects fail to produce anticipatory SCRs. (c) VMPFC-lesioned subjects produce normal SCRs to reward and punishment, but amygdala-lesioned subjects fail to produce SCRs to reward and punishment. Error bars represent the standard error of the mean. Not depicted are results showing that VMPFC-lesioned subjects produce normal SCRs to a classically conditioned stimulus, whereas amygdala-lesioned subjects fail to produce classically conditioned SCRs. Reproduced with permission from Bechara et al. (1999).

Both the amygdala and VMPFC patients were normal in their SCRs to the unconditioned stimulus. This indicates that the amygdala, but not the VMPFC, is required for the acquisition of Pavlovian conditioning (however, see the chapter by Dolan *et al.* in this volume for a discussion of the role of the VMPFC in Pavlovian conditioning processes). In other words, the visceral responses to stimuli that acquire hedonic value through simple associative learning processes are not mediated by the VMPFC, but are mediated by the amygdala. This parallels the dissociation between the amygdala and VMPFC with respect to the visceral response to reward and punishment in the IGT.

The distinction between the affective-visceral functions of the amygdala and VMPFC may be conceptualized in terms of the demands of processing biologically relevant stimuli on attention and working memory. In this view, the amygdala triggers visceral responses to stimuli whose biological significance can be decoded in relatively automatic fashion, such as winning or losing money, conditioned stimuli that reliably predict aversive and pleasurable events in the future, and stimuli with innate biological significance, such as the sight of spiders and snakes and facial expressions of fear. We refer to this class of stimuli as 'primary inducers'. The visceral responses to primary inducers can be elicited quickly and without 'thought' or complex attention. The VMPFC, in contrast, triggers visceral responses to stimuli whose biological significance must be decoded through a deliberative process. We refer to this class of stimuli as 'secondary inducers'. This includes thoughts of future loss or gain, particularly when loss or gain is uncertain, as well as the recollection of pleasant or unpleasant events from the past. Secondary inducers, which may not be present within the sensory field, must be brought 'in mind' in order to elicit a visceral response. Thus, VMPFC functions are related to attention and working memory and also to the recall of episodic memories.

## 13.4. Conclusion: the somatic marker hypothesis

### Somatic markers as executive processes

According to the somatic marker hypothesis (Damasio 1994; Bechara and Damasio 2005), the visceral responses elicited during decision-making, both during the contemplation of the future outcome of a choice and after the outcome of a choice has been signaled, aid in guiding decisions towards advantageous choices and away from disadvantageous choices. The process that is assessed by the IGT is ultimately a learning process, one in which knowledge of the correct strategy evolves over time. In this view, visceral responses to receiving reward and punishment, which are mediated by the amygdala, contribute to the encoding of the predictive value of the sensory cues and actions that preceded reward and punishment. Over time, through this encoding, subjects learn the association between a given choice and its outcome. This learning may precede explicit awareness of the contingencies between specific choices and their outcome. This learning is expressed by the VMPFC, which evokes learned representations of the predictive value of a choice in the period before a choice is made, when the outcomes of various choices are weighed against each other as they are held 'in mind'. The representation of predictive value is based upon the visceral response that is triggered within the VMPFC, an emotional response that 'marks' the value of options for behavior based upon past experience.

Within the somatic marker framework, then, the VMPFC functions as a system that holds the affective-visceral properties of objects in mind during the planning and organization of behavior that is directed towards courses of action that are in the overall best interests of the organism. This function falls into the broader executive role of the prefrontal cortex, of which the VMPFC is a part. This role is supported by connections between the VMPFC and higher-order sensory cortices, as well as connections between the VMPFC and the dorsolateral prefrontal cortex (both of which are mediated by the orbital network). The connections with higher-order sensory cortices provide a route for finely processed information about the sensory properties of biologically relevant stimuli to reach the VMPFC. The connections with the dorsolateral prefrontal cortex link the functions of the VMPFC to executive processes that guide attention and prioritize action, allowing the VMPFC to serve as a buffer for the maintenance of information pertaining to the homeostatic value of goal objects (i.e. predictive value). Thus, the VMPFC is not involved in regulating global working memory processes, as indicated by the finding that damage to the VMPFC does not disrupt performance on broad tasks of working memory(Bechara *et al.* 1998). However, VMPFC function does require intact working memory processes, as indicated by the finding that damage in regions of the prefrontal cortex that play a global role in working memory impairs performance on the IGT (Manes *et al.* 2002; Clark *et al.* 2003).

Some tasks that call upon representations of predictive reward value but do not require this information to be held in working memory may also engage the VMPFC. For example, Fellows and Farah (2005) have shown that damage to the VMPFC disrupts both reversal learning and IGT performance. In contrast, damage to the dorsolateral prefrontal cortex impairs performance upon the IGT, but does not impair reversal learning. An important caveat in the comparison of this study with studies from our laboratory is that the Farah and Fellows study examined damage in posterior regions of VMPFC that also impinged upon basal forebrain structures, such as the nucleus accumbens. Our studies, in contrast, have found that lesions restricted to more anterior regions of the VMPFC that do not include the basal forebrain can alter performance on the IGT (Bechara *et al.* 1998). Thus, it is possible that the reversal learning deficits found in the Fellows and Farah study are attributable to damage in the basal forebrain, rather than to damage in the VMPFC. Notwithstanding this, it is possible that both reversal learning and the IGT require an ability to register that the predictive reward value of a stimulus has changed, as well as an ability to inhibit a previously rewarded response. However, unlike the IGT, reversal learning does not require that information about the predictive reward value of a stimulus be held in working memory. Thus, the VMPFC may be engaged by processes that invoke representations of predictive reward value, as well as by processes that require inhibition of a previously rewarded response. Such process, which may themselves rely upon somatic markers, could operate independently of working memory under certain experimental situations such as reversal learning. In real life, however, where decision-making usually requires holding representations of predictive reward value in mind over a delay, they are likely to work in concert with working memory processes.

## The role of feedback of somatic markers in decision-making

According to the somatic marker hypothesis, the afferent feedback of visceral responses is an important component of the decision-making process. In other words, the visceral responses elicited during the contemplation of choices are necessary for biasing behavior in the advantageous direction, as well as for 'gut feelings' and 'hunches' related to choices. The question arises, then, as to whether the visceral responses induced by the VMPFC are actually necessary for decision-making, or whether they are merely an epiphenomenal bodily reflection of the operation of certain mental processes.

One way to address this question is to directly manipulate the sensory feedback of the visceral state during performance of the IGT. A number of studies have attempted to do this. For example, North and O'Carroll (2001) have examined how cervical transection of the spinal cord affects performance on the IGT. This study found no effect of the manipulation on performance on the IGT. Since the spinal cord carries somatosensory and interoceptive information from the body to the brain (Craig 2002), this may be taken as evidence that the sensory feed back of interoceptive states does not contribute to decision-making. However, a great deal of the information about visceral states is conveyed to the central nervous system via the vagus nerve, which is spared by spinal transection. If visceral states play a special role in signaling homeostatic processes, which we believe that they do, then it is not surprising that spinal transection has a limited effect on decision-making. A study by Heims *et al.* (2004) examined more specifically the role of visceral states in decision-making. They showed that patients with pure autonomic failure, a peripheral nervous disorder that broadly disrupts the ability to deploy visceral responses, do not demonstrate impaired performance on the IGT. This can also be taken as evidence that visceral responses are not necessary for decision-making. However, this study did not actually measure visceral responses during the IGT, so it is possible that subjects still produced some form of visceral response during the task. Also, it is possible that, because pure autonomic failure develops slowly and manifests later in life, significant neural reorganization may take place in subjects with this disease, altering the normal mechanism of decision-making. A study by Martin *et al.* (2004) showed that electrical stimulation of the vagus nerve during the IGT, which largely affects visceral afferent signaling, can actually improve performance. This can be taken as evidence that visceral states do play a role in decision-making. However, this study was limited by the fact that most of the subjects study suffered long-standing epilepsy and many of them had lower than normal decision-making to begin with.

Functional imaging studies have provided circumstantial evidence of the role of visceral states in decision-making. As discussed above, the insular cortex is a visceral sensory region that has been hypothesized to play a role in decision-making by mapping the visceral responses that are induced by the VMPFC and the amygdala. A number of studies (reviewed in Craig 2002) have shown that activity in the insular cortex is correlated with changes in the visceral state. The insular cortex is also activated by decision-making tasks that involve uncertain reward and punishment and an evaluation of emotional information. For example, a PET study by Ernst et al. (2002) has shown that performance of the IGT, in addition to activating the VMPFC, also activates the insular cortex. Moreover, this study found that activity in the insular cortex was correlated with the performance on the IGT. The insular cortex has also been shown using fMRI to be activated during other decision-making tasks that involve uncertain reward and punishment (Critchley et al. 2001; Huettel et al. 2005). In addition, a study by Sanfey et al. (2003) found that the insular cortex is activated by the evaluation of the fairness of offers of money. Here, activity in the insular cortex was shown to be correlated with the tendency to reject unfair offers. While these studies did not examine visceral responses directly, they show that the insular cortex, an area that has previously established as a visceral sensory representation area, is engaged during decision-making, particularly when the decisions require an evaluation of emotional consequences that are uncertain.

Thus, on the balance, the evidence seems to favor the role of visceral states in decision-making, however, more definitive evidence is required to establish exactly how and under what circumstances visceral states contribute to decision-making. Though certain forms of decision-making may engage somatic marker processes, it may be that not all forms of decision-making require the elicitation and sensory mapping of visceral states. Indeed, the somatic marker hypothesis maintains that, under some conditions, "as if" representations of the visceral state, mediated by direct connections between the VMPFC and brainstem neurotransmitter nuclei, may be sufficient to guide decision-making in the advantageous direction. Also, decisions that do not require the weighing of rewarding or punishing consequence or in which the outcome is relatively certain may not engage somatic marker processes at all.

## Why somatic markers?

A strictly computational approach to decision-making may not require that the brain represent signals that are expressed within the body in order to compute the anticipated value of options for behavior (Rolls 1999; Maia and McClelland 2004, 2005). It is important to keep in mind, however, that human brains differ from computers in many ways, not the least of which is a concern for the promotion of survival through regulation of the internal milieu—which is the regulation of bodily process. All nervous systems contain representations of basic bodily processes, such as those that regulate energy demands, reproduction, fluid balance, temperature and the response to sickness and injury. Survival requires precise control over the state of these processes in order to maintain them within the narrow range that is compatible with life (i.e. homeostasis). The autonomic nervous system functions to make relatively rapid adjustments in the visceral state that maximize the survival value of events in the world that have the potential to impact homeostasis.

It can be argued that visceral responses operate merely as reflexes, acting independently of higher-order cognitive processes. Certainly, visceral reflexes that are implemented at the level of the spinal cord and brainstem do provide some benefit after the fact for reacting to events that challenge homeostasis. However, a greater survival advantage is conferred by the ability to predict the impact of events upon the internal milieu before they occur. To do this would require the brain to connect sensory and motor representations of the viscera with processes that govern perception, learning, memory and goal-directed behavior. It is clear, based upon a multitude of studies, many of which are reported in this volume, that the VMPFC plays a role in a number of such 'cognitive' processes. It is also clear that the VMPFC plays a role in the control and mapping of visceral states. The most parsimonious explanation would seem to be that the cognitive and visceral functions of the VMPFC are somehow linked.

According to the somatic marker hypothesis, the integration of visceral states into higher cognitive functions, such as decision-making, is the function of the VMPFC. This function has expanded in evolution, allowing for the planning of behaviors that are executed further into the future, and for which the outcomes of behavior in terms of rewards and punishments are more abstract. In rodents and nonhuman primates, as well as in humans, the VMPFC is involved in the planning of behaviors related to the most immediate and basic needs, such as food, water and sex. In humans, the VMPFC also plays a role in guiding behaviors for which choosing advantageously requires a deliberate concern for one's long-term well-being as well as knowledge of cultural norms and expectations. In this way, the VMPFC may function to link highly evolved human faculties, such as moral behavior, altruism, financial reasoning, creativity and a sense of purpose in one's work life and social relationships, to the basic mechanisms that govern survival and the maintenance of homeostasis.

## Acknowledgment

This research was supported by NINDS Program Project Grant P01 NS19632

## References

Adolphs, R., Tranel, D., Damasio, H., and Damasio, A. R. (1995) Fear and the human amygdala. *J Neurosci* **15(9)**:5879–5891.

Allman, J., Hakeem, A., Tetreault, N., and Semendeferi, K. (2003) The spindle neurons of frontoinsular cortex (area f1) are unique to humans and African apes. *Society for Neuroscience Abstracts* 725, 4.

Amaral, D. G., Price, J. L., Pitkanen, A., and Carmichael, S. T. (1992) Anatomical organization of the primate amygdaloid complex. In: J. Aggleton (ed.), *The Amygdala: Neurobiological Aspects of Emotion, Memory, and Mental Dysfunction*, pp. 1–66. New York: Wiley-Liss.

An, X., Bandler, R., Ongur, D., and Price, J. L. (1998) Prefrontal cortical projections to longitudinal columns in the midbrain periaqueductal gray in macaque monkeys. *J Comp Neurol* **401(4)**:455–479.

Anand, B. K. and Dua, S. (1956) Circulatory and respiratory changes induced by electrical stimulation of limbic system (visceral brain) *J Neurophysiol* **19(5)**:393–400.

Bailey, P. and Sweet, W. H. (1940) Effect on respirations, blood pressure and gastric motility of stimulation of the orbital surface of the frontal lobe. *J Neurophysiol* **3**:276–281.

Bandler, R. and Shipley, M. T. (1994) Columnar organization in the midbrain periaqueductal gray: Modules for emotional expression? *Trends Neurosci* **17(9)**:379–389.

Bechara, A. and Damasio, A. R. (2005) The somatic marker hypothesis: A neural theory of economic decision-making. *Games and Economic Behavior* **52**:336–372.

Bechara, A., Damasio, A. R., Damasio, H., and Anderson, S. W. (1994) Insensitivity to future consequences following damage to human prefrontal cortex. *Cognition* **50**:7–15.

Bechara, A., Tranel, D., Damasio, H., and Damasio, A. R. (1996) Failure to respond autonomically to anticipated future outcomes following damage to prefrontal cortex. *Cerebral Cortex* **6**:215–225.

Bechara, A., Damasio, H., Tranel, D., and Damasio, A. R. (1997) Deciding advantageously before knowing the advantageous strategy. *Science* **275(5304)**:1293–1295.

Bechara, A., Damasio, H., Tranel, D., and Anderson, S. W. (1998) Dissociation of working memory from decision making within the human prefrontal cortex. *J Neurosci* **18(1)**:428–437.

Bechara, A., Damasio, H., Damasio, A. R., and Lee, G. P. (1999) Different contributions of the human amygdala and ventromedial prefrontal cortex to decision-making. *J Neurosci* **19(13)**:5473–5481.

Bechara, A., Tranel, D., and Damasio, H. (2000) Characterization of the decision-making deficit of patients with ventromedial prefrontal cortex lesions. *Brain* **123(Pt 11)**:1189–2202.

Bechara, A., Dolan, S., and Hindes, A. (2002) Decision-making and addiction (part ii): Myopia for the future or hypersensitivity to reward? *Neuropsychologia* **40(10)**:1690–1705.

Bechara, A., Damasio, H., Tranel, D., and Damasio, A. R. (2005) The iowa gambling task and the somatic marker hypothesis: Some questions and answers. *Trends Cogn Sci* **9(4)**:159–162; discussion 162–154.

Berntson, G. G., Sarter, M., and Cacioppo, J. T. (2003) Ascending visceral regulation of cortical affective information processing. *Eur J Neurosci* **18(8)**:2103–2109.

Berridge, K. C. and Robinson, T. E. (1998) What is the role of dopamine in reward: Hedonic impact, reward learning, or incentive salience? *Brain Research and Brain Research Reviews* **28(3)**:309–369.

Berthoz, S., Armony, J. L., Blair, R. J., and Dolan, R. J. (2002) An fMRI study of intentional and unintentional (embarrassing) violations of social norms. *Brain* **125(Pt 8)**:1696–1708.

Bianchi, R., Corsetti, G., Rodella, L., Tredici, G., and Gioia, M. (1998) Supraspinal connections and termination patterns of the parabrachial complex determined by the biocytin anterograde tract-tracing technique in the rat. *J Anat* **193(Pt 3)**:417–430.

Bradley, M. M. and Lang, P. J. (2000) Measuring emotion: Behavior, feeling and physiology. In: R. Lane and L. Nadel (eds.), *Cognitive Neuroscience of Emotion*, pp. 242–276. New York: Oxford University Press.

Brody, A. L., Mandelkern, M. A., London, E. D., Childress, A. R., Lee, G. S., Bota, R. G., et al. (2002) Brain metabolic changes during cigarette craving. *Arch Gen Psychiatry* **59(12)**:1162–1172.

Burns, S. M. and Wyss, J. M. (1985) The involvement of the anterior cingulate region in blood pressure control. *Brain Res* **340**:71–77.

Camille, N., Coricelli, G., Sallet, J., Pradat-Diehl, P., Duhamel, J. R., and Sirigu, A. (2004) The involvement of the orbitofrontal cortex in the experience of regret. *Science* **304(5674)**:1167–1170.

Cardinal, R. N., Parkinson, J.A., Hall, J., and Everitt, B.J. (2002) Emotion and motivation: The role of the amygdala, ventral striatum and prefrontal cortex. *Neuroscience and Biobehavioral Reviews* **26**:321–352.

Carmichael, S. T. and Price, J. L. (1996) Connectional networks within the orbital and medial prefrontal cortex of macaque monkeys. *J Comp Neurol* **371(2)**:179–207.

Cedarbaum, J. M. and Aghajanian, G. K. (1978) Afferent projections to the rat locus coeruleus as determined by a retrograde tracing technique. *J Comp Neurol* **178(1)**:1–16.

Chapman, W. P., Livingston, R. B., and Livingston, K. E. (1949) Frontal lobotomy and electrical stimulation of orbital surface of frontal lobes; effect on respiration and on blood pressure in man. *Arch Neurol Psychiatry* **62(6)**:701–716.

Chapman, W. P., Livingston, R. B., and Livingston, K. E. (1950) Effect of frontal lobotomy and of electrical stimulation of the orbital surface of the frontal lobes and tip of the temporal lobes upon respirations and blood pressure in man. In M. Greenblatt, R. Arnot and H. C. Solomon (eds.), *Studies in Lobotomy*. New York: Grune and Stratton.

Chudasama, Y. and Robbins, T. W. (2003) Dissociable contributions of the orbitofrontal and infralimbic cortex to Pavlovian autoshaping and discrimination reversal learning: Further evidence for the functional heterogeneity of the rodent frontal cortex. *J Neurosci* **23(25)**:8771–8780.

Clark, L., Manes, F., Antoun, N., Sahakian, B. J., and Robbins, T. W. (2003) The contributions of lesion laterality and lesion volume to decision-making impairment following frontal lobe damage. *Neuropsychologia* **41(11)**:1474–1483.

Conde, F., Maire-Lepoivre, E., Audinat, E., and Crepel, F. (1995) Afferent connections of the medial frontal cortex of the rat. Ii. Cortical and subcortical afferents. *J Comp Neurol* **352(4)**:567–593.

Craig, A. D. (2002) How do you feel? Interoception: The sense of the physiological condition of the body. *Nat Rev Neurosci* **3(8)**:655–666.

Craig, A. D., Chen, K., Bandy, D., and Reiman, E. M. (2000) Thermosensory activation of insular cortex. *Nat Neurosci* **3(2)**:184–190.

Critchley, H. D., Corfield, D. R., Chandler, M. P., Mathias, C. J., and Dolan, R. J. (2000a) Cerebral correlates of autonomic cardiovascular arousal: A functional neuroimaging investigation in humans. *J Physiol* **523(Pt 1)**:259–270.

Critchley, H. D., Elliott, R., Mathias, C. J., and Dolan, R. J. (2000b) Neural activity relating to generation and representation of galvanic skin conductance responses: A functional magnetic resonance imaging study. *Journal of Neuroscience* **20(8)**:3033–3040.

Critchley, H. D., Mathias, C. J., and Dolan, R. J. (2001) Neural activity in the human brain relating to uncertainty and arousal during anticipation. *Neuron* **29(2)**:537–545.

Critchley, H. D., Wiens, S., Rotshtein, P., Ohman, A., and Dolan, R. J. (2004) Neural systems supporting interoceptive awareness. *Nat Neurosci* **7(2)**:189–195.

Damasio, A. R. (1994) *Descartes' error: Emotion, reason and the human brain*. New York: Putnam and Sons.

Damasio, A. R. (2000) *The feeling of what happens: Body and emotion in the making of consciousness*. Chicago: Harcourt.

Damasio, A. R., Grabowski, T. J., Bechara, A., Damasio, H., Ponto, L. L., Parvizi, J., et al. (2000) Subcortical and cortical brain activity during the feeling of selfgenerated emotions. *Nat Neurosci* **3(10)**:1049–1056.

Damasio, A. R., Tranel, D., and Damasio, H. (1990) Individuals with sociopathic behavior caused by frontal damage fail to respond autonomically to social stimuli. *Behav Brain Res* **41(2)**:81–94.

de Araujo, I. E., Kringelbach, M. L., Rolls, E. T., and McGlone, F. (2003) Human cortical responses to water in the mouth, and the effects of thirst. *Journal of Neurophysiology* **90(3)**:1865–1876.

Delgado, J. M. and Livingston, K. E. (1948) Some respiratory, vascular and thermal responses to stimulation of the orbital surface of the frontal lobe. *J Neurophysiol* **11**:39–55.

Devito, J. L., and Smith, O. A., Jr. (1964) Subcortical projections of the prefrontal lobe of the monkey. *J Comp Neurol* **123**:413–423.

Dougherty, D. D., Shin, L. M., Alpert, N. M., Pitman, R. K., Orr, S. P., Lasko, M., et al. (1999) Anger in healthy men: A pet study using script-driven imagery. *Biol Psychiatry* **46(4)**:466–472.

Drevets, W. C., Price, J. L., Simpson, J. R., Jr., Todd, R. D., Reich, T., Vannier, M., et al. (1997) Subgenual prefrontal cortex abnormalities in mood disorders. *Nature* **386(6627)**:824–827.

Dua, S. and MacLean, P. D. (1964) Localization for penile erection in medial frontal lobe. *Am J Physiol* **207**:1425–1434.

Egan, G., Silk, T., Zamarripa, F., Williams, J., Federico, P., Cunnington, R., et al. (2003) Neural correlates of the emergence of consciousness of thirst. *Proceedings of theNational Academy of Sciences of the United States of America*, **100(25)**:15241–15246.

Ernst, M., Bolla, K., Mouratidis, M., Contoreggi, C., Matochik, J. A., Kurian, V., et al. (2002) Decision-making in a risk-taking task: A PET study. *Neuropsychopharmacology* **26(5)**:682–691.

Eslinger, P. J. and Damasio, A. R. (1985) Severe disturbance of higher cognition after bilateral frontal lobe ablation: Patient evr. *Neurology* **35(12)**:1731–1741.

Fellows, L. K. and Farah, M. J. (2005) Different underlying impairments in decision-making following ventromedial and dorsolateral frontal lobe damage in humans. *Cereb Cortex* **15(1)**:58–63.

Fowles, D. C., Fisher, A. E., and Tranel, D. T. (1982) The heart beats to reward: The effect of monetary incentive on heart rate. *Psychophysiology* **19(5)**:506–513.

Frysztak, R. J. and Neafsey, E. J. (1991) The effect of medial frontal cortex lesions on respiration, "freezing," and ultrasonic vocalizations during conditioned emotional responses in rats. *Cereb Cortex* **1(5)**:418–425.

Frysztak, R. J. and Neafsey, E. J. (1994) The effect of medial frontal cortex lesions on cardiovascular conditioned emotional responses in the rat. *Brain Res* **643(1–2)**:181–193.

Fukui, H., Murai, T., Fukuyama, H., Hayashi, T., and Hanakawa, T. (2005) Functional activity related to risk anticipation during performance of the iowa gambling task. *Neuroimage* **24(1)**:253–259.

Gabbott, P. L., Dickie, B. G., Vaid, R. R., Headlam, A. J., and Bacon, S. J. (1997) Localcircuit neurones in the medial prefrontal cortex (areas 25, 32 and 24b) in the rat: Morphology and quantitative distribution. *J Comp Neurol* **377(4)**:465–499.

Hardy, S. G. and Holmes, D. E. (1988) Prefrontal stimulus-produced hypotension in rat. *Exp Brain Res* **73(2)**:249–255.

Heims, H. C., Critchley, H. D., Dolan, R., Mathias, C. J., and Cipolotti, L. (2004) Social and motivational functioning is not critically dependent on feedback of autonomic responses: Neuropsychological evidence from patients with pure autonomic failure. *Neuropsychologia* **42(14)**:1979–1988.

Hornak, J., Bramham, J., Rolls, E. T., Morris, R. G., O'Doherty, J., Bullock, P. R., et al. (2003) Changes in emotion after circumscribed surgical lesions of the orbitofrontal and cingulate cortices. *Brain* **126(Pt 7)**:1691–1712.

Huettel, S. A., Song, A. W., and McCarthy, G. (2005) Decisions under uncertainty: Probabilistic context influences activation of prefrontal and parietal cortices. *J Neurosci* **25(13)**:3304–3311.

Hurley-Gius, K. M. and Neafsey, E. J. (1986) The medial frontal cortex and gastric motility: Microstimulation results and their possible significance for the overall pattern of organization of rat frontal and parietal cortex. *Brain Res* **365(2)**:241–248.

Hurliman, E., Nagode, J. C., and Pardo, J. V. (2005) Double dissociation of exteroceptive and interoceptive feedback systems in the orbital and ventromedial prefrontal cortex of humans. *J Neurosci* **25(18)**:4641–4648.

James, W. (1884) What is an emotion? *Mind* **9**:188–205.

Johnson, T. N., Rosvold, H. E., and Mishkin, M. (1968) Projections from behaviorally-defined sectors of the prefrontal cortex to the basal ganglia, septum, and diencephalon of the monkey. *Exp Neurol* **21(1)**:20–34.

Kaada, B. R. (1951) Somato-motor, autonomic and electrocortical responses to electrical stimulation of the "rhinencephalon" and other structures in primates, cat and dog. *Acta Physiol Scand* **24(Suppl. 83)**:285.

Kimble, D. P., Bagshaw, M. H., and Pribram, K. H. (1965) The gsr of monkeys during orienting and habituation following selective and partial ablation of the cingulate and frontal cortex. *Neuropsychologia* **3**:121–128.

Kita, H. and Oomura, Y. (1981) Reciprocal connections between the lateral hypothalamus and the frontal complex in the rat: Electrophysiological and anatomical observations. *Brain Res* **213(1)**:1–16.

Krettek, J. E. and Price, J. L. (1977) The cortical projections of the mediodorsal nucleus and adjacent thalamic nuclei in the rat. *J Comp Neurol* **171(2)**:157–191.

Kringelbach, M. L. (2004) Food for thought: Hedonic experience beyond homeostasis in the human brain. *Neuroscience* **126(4)**:807–819.

Kringelbach, M. L. and Rolls, E. T. (2004) The functional neuroanatomy of the human orbitofrontal cortex: Evidence from neuroimaging and neuropsychology. *Prog Neurobiol* **72(5)**:341–372.

Kringelbach, M. L., O'Doherty, J., Rolls, E. T., and Andrews, C. (2003) Activation of the human orbitofrontal cortex to a liquid food stimulus is correlated with its subjective pleasantness. *Cereb Cortex* **3(10)**:1064–1071.

Lane, R. D., Reiman, E. M., Ahern, G. L., Schwartz, G. E., and Davidson, R. J. (1997) Neuroanatomical correlates of happiness, sadness, and disgust. *American Journal of Psychiatry* **154(7)**:926–933.

Lang, P. J., Greenwald, M. K., Bradley, M. M., and Hamm, A. O. (1993) Looking at pictures: Affective, facial, visceral, and behavioral reactions. *Psychophysiology* **30(3)**:261–273.

LeDoux, J. E. (2000) Emotion circuits in the brain. *Annu Rev Neurosci* **23**:155–184.

Livingston, R. B., Chapman, W. P., and Livingston, K. E. (1948) *Stimulation of Orbital Surface in Man Prior to Frontal Lobotomy*. Baltimore, Maryland: Williams and Wilkins.

Lofving, B. (1961) Cardiovascular adjustments induced from the rostral cingulate gyrus. *Acta Physiol Scand* **184**(Suppl.):1–82.

London, E. D., Ernst, M., Grant, S., Bonson, K., and Weinstein, A. (2000) Orbitofrontal cortex and human drug abuse: Functional imaging. *Cereb Cortex* **10(3)**:334–342.

Luria, A. R. (1973) The frontal lobes and the regulation of behavior. In: K. H. Pribram and A. R. Luria (eds.), *Psychophysiology of the Frontal Lobes*, pp. 3–26. New York: Academic Press.

Luria, A. R. and Homskaya, E. D. (1970) Frontal lobes and the regulation of arousal processes. In D. I. Mostofsky (ed.), *Attention: Contemporary theory and analysis*, pp. 303–330. New York: Appleton-Century-Crofts.

Luria, A. R., Pribram, K. H., and Homskaya, E. D. (1964) An experimental analysis of the behavioral disturbance produced by a left frontal arachnoidal endothelioma (meningioma). *Neuropsychologia* **2**:257–280.

Maia, T. V. and McClelland, J. L. (2004) A reexamination of the evidence for the somatic marker hypothesis: What participants really know in the Iowa gambling task. *Proc Natl Acad Sci U S A* **101(45)**:16075–16080.

Maia, T. V. and McClelland, J. L. (2005) The somatic marker hypothesis: Still many questions but no answers: Response to Bechara *et al*. *Trends Cogn Sci* **9(4)**:162–164.

Manes, F., Sahakian, B., Clark, L., Rogers, R., Antoun, N., Aitken, M., et al. (2002) Decision-making processes following damage to the prefrontal cortex. *Brain* **125(Pt 3)**:624–639.

Martin, C. O., Denburg, N. L., Tranel, D., Granner, M. A., and Bechara, A. (2004) The effects of vagus nerve stimulation on decision-making. *Cortex* **40(4–5)**:605–612.

Nagai, Y., Critchley, H. D., Featherstone, E., Trimble, M. R., and Dolan, R. J. (2004) Activity in ventromedial prefrontal cortex covaries with sympathetic skin conductance level: A physiological account of a "default mode" of brain function. *Neuroimage* **22(1)**:243–251.

Nauta, W. J. (1971) The problem of the frontal lobe: A reinterpretation. *J Psychiatr Res* **8(3)**:167–187.

Nauta, W. J. H. and Haymaker, W. (1969) Hypothalamic nuclei and fiber connections. In: W. Haymaker, E. Anderson and W. J. H. Nauta (eds.), *The hypothalamus*, pp. 136–209. Springfield, Illinois: Thomas.

Neafsey, E. J. (1990) Prefrontal cortical control of the autonomic nervous system: Anatomical and physiological observations. *Prog Brain Res* **85**:147–165; discussion 165–146.

Neafsey, E. J., Hurley-Gius, K. M., and Arvanitis, D. (1986) The topographical organization of neurons in the rat medial frontal, insular and olfactory cortex projecting to the solitary nucleus, olfactory bulb, periaqueductal gray and superior colliculus. *Brain Res* **377**(2):561–570.

Norgren, R. (1990) Gustatory system. In: G. Paxinos (ed.), *The human nervous system* (pp. 845–861). San Diego: Academic Press. North, N. T., and O'Carroll, R. E. (2001). Decision making in patients with spinal cord damage: Afferent feedback and the somatic marker hypothesis. *Neuropsychologia* **39**(5):521–524.

Ongur, D., An, X., and Price, J. L. (1998) Prefrontal cortical projections to the hypothalamus in macaque monkeys. *J Comp Neurol* **401**(4):480–505.

Ongur, D., Ferry, A. T., and Price, J. L. (2003) Architectonic subdivision of the human orbital and medial prefrontal cortex. *J Comp Neurol* **460**(3):425–449.

Ongur, D. and Price, J. L. (2000) The organization of networks within the orbital and medial prefrontal cortex of rats, monkeys and humans. *Cerebral Cortex* **10**(3):206–219.

Patterson, J. C., 2nd, Ungerleider, L. G., and Bandettini, P. A. (2002) Task-independent functional brain activity correlation with skin conductance changes: An fmri study. *Neuroimage* **17**(4):1797–1806.

Pears, A., Parkinson, J. A., Hopewell, L., Everitt, B. J., and Roberts, A. C. (2003) Lesions of the orbitofrontal but not medial prefrontal cortex disrupt conditioned reinforcement in primates. *J Neurosci* **23**(35):11189–11201.

Pelchat, M. L., Johnson, A., Chan, R., Valdez, J., and Ragland, J. D. (2004) Images of desire: Food-craving activation during fmri. *Neuroimage* **23**(4):1486–1493.

Phan, K. L., Wager, T., Taylor, S. F., and Liberzon, I. (2002) Functional neuroanatomy of emotion: A meta-analysis of emotion activation studies in PET and fMRI. *Neuroimage* **16**(2):331–348.

Phillipson, O. T. (1979) Afferent projections to the ventral tegmental area of tsai and interfascicular nucleus: A horseradish peroxidase study in the rat. *J Comp Neurol* **187**(1):117–143.

Price, J. L. and Amaral, D. G. (1981) An autoradiographic study of the projections of the central nucleus of the monkey amygdala. *J Neurosci* **1**(11):1242–1259.

Rahman, S. J., Sahakian, B. N., R., C., Rogers, R., and Robbins, T. (2001) Decision making and neuropsychiatry. *Trends Cogn Sci* **5**(6):271–277.

Raichle, M. E., MacLeod, A. M., Snyder, A. Z., Powers, W. J., Gusnard, D. A., and Shulman, G. L. (2001) A default mode of brain function. *Proc Natl Acad Sci U S A* **98**(2):676–682.

Rainville, P., Bechara, A., Naqvi, N., and Damasio, A. R. (in press) Basic emotions are associated with distinct patterns of cardiorespiratory activity. *Int J Psychophysiol*.

Robbins, T. W. (2000) Chemical neuromodulation of frontal-executive functions in humans and other animals. *Exp Brain Res* **133**(1):130–138.

Rolls, E. T. (1999) *The brain and emotion*. Oxford, UK: Oxford University Press.

Rolls, E. T., O'Doherty, J., Kringelbach, M. L., Francis, S., Bowtell, R., and McGlone, F. (2003) Representations of pleasant and painful touch in the human orbitofrontal and cingulate cortices. *Cerebral Cortex* **13**(3):308–317.

Sanfey, A. G., Rilling, J. K., Aronson, J. A., Nystrom, L. E., and Cohen, J. D. (2003) The neural basis of economic decision-making in the ultimatum game. *Science* **300**(5626):1755–1758.

Saver, J. L. and Damasio, A. R. (1991) Preserved access and processing of social knowledge in a patient with acquired sociopathy due to ventromedial frontal damage. *Neuropsychologia* **29**(12):1241–1249.

Sesack, S. R., Deutch, A. Y., Roth, R. H., and Bunney, B. S. (1989) Topographical organization of the efferent projections of the medial prefrontal cortex in the rat: An anterograde tract-tracing study with phaseolus vulgaris leucoagglutinin. *J Comp Neurol* **290**(2):213–242.

Shamay-Tsoory, S. G., Tomer, R., and Aharon-Peretz, J. (2005) The neuroanatomical basis of understanding sarcasm and its relationship to social cognition. *Neuropsychology* **19**(3):288–300.

Showers, M. J. and Crosby, E. C. (1958) Somatic and visceral responses from the cingulate gyrus. *Neurology* **8(7)**:561–565.

Smith, W. K. (1938) The representation of respiratory movements in the cerbral cortex. *J Neurophysiol* **1**:55–68.

Spencer, W. G. (1894) The effect produced upon respiration by faradic excitation of the cerebrum in monkey, dog, cat and rabbit. *T Phil Trans R Soc Lond Biol B* **185**:609–660.

Terreberry, R. R. and Neafsey, E. J. (1983) Rat medial frontal cortex: A visceral motor region with a direct projection to the solitary nucleus. *Brain Res* **278(1–2)**:245–249.

Terreberry, R. R. and Neafsey, E. J. (1987) The rat medial frontal cortex projects directly to autonomic regions of the brainstem. *Brain Res Bull* **19(6)**:639–649.

Tomb, I., Hauser, M., Deldin, P., and Caramazza, A. (2002) Do somatic markers mediate decisions on the gambling task? *Nat Neurosci* **5(11)**:1103–1104; author reply 1104.

Tranel, D. (2000) Electrodermal activity in cognitive neuroscience: Neuroanatomical and neuropsychological correlates. In: R. D. Lane and L. Nadel, (eds.), *Cognitive Neuroscience of Emotion. Series in Affective Science*. New York: Oxford University Press.

Tranel, D. (2002) Emotion, decision making, and the prefrontal cortex. In D. T. Stuss and R. T. Knight (eds.), *Principles of Frontal Lobe Function*, pp. 338–353. New York: Oxford University Press.

Tranel, D. and Damasio, H. (1994) Neuroanatomical correlates of electrodermal skin conductance responses. *Psychophysiology* **31(5)**:427–438.

Tranel, D. and Hyman, B. T. (1990) Neuropsychological correlates of bilateral amygdala damage. *Arch Neurol* **47(3)**:349–355.

van der Kooy, D., Koda, L. Y., McGinty, J. F., Gerfen, C. R., and Bloom, F. E. (1984) The organization of projections from the cortex, amygdala, and hypothalamus to the nucleus of the solitary tract in rat. *J Comp Neurol* **224(1)**:1–24.

van der Kooy, D., McGinty, J. F., Koda, L. Y., Gerfen, C. R., and Bloom, F. E. (1982) Visceral cortex: A direct connection from prefrontal cortex to the solitary nucleus in rat. *Neurosci Lett* **33(2)**:123–127.

Vogt, B. A. and Peters, A. (1981) Form and distribution of neurons in rat cingulate cortex: Areas 32, 24, and 29. *J Comp Neurol* **195(4)**:603–625.

Wang, G. J., Volkow, N. D., Telang, F., Jayne, M., Ma, J., Rao, M., *et al.* (2004) Exposure to appetitive food stimuli markedly activates the human brain. *Neuroimage* **21(4)**:1790–1797.

Chapter 14

# Intracranial electrophysiology of the human orbitofrontal cortex

Ralph Adolphs, Hiroto Kawasaki, Hiroyuki Oya, and Matthew A. Howard

## 14.1. Introduction

Ever since the behavioral observations of patients with damage encompassing the orbitofrontal cortex (OFC) were first undertaken (Brickner 1932; Ackerly and Benton 1948; Damasio 1994; Damasio *et al.* 1994), the functions of this region of the brain have been an enigma. That situation is changing rapidly, as this volume attests, with the rapid accrual of data from sensory and cognitive neuroscience (Cavada and Schultz 2000). Studies in animals have documented responses of neurons in the OFC to the reinforcing properties of stimuli in several sensory modalities (Rolls, Chapter 5, this volume; Schultz and Tremblay, Chapter 7, this volume). Taste and smell, in particular, are associated with their reward properties through networks that prominently include the OFC (Tanabe *et al.* 1975; Rolls 1999), a conclusion that is also supported by recent functional imaging studies in human participants (Zald *et al.* 1998; O'Doherty *et al.* 2001, 2002; Gottfried *et al.* 2003). But auditory (Hornak *et al.* 1996; Frey *et al.* 2000) and visual stimuli (Hornak *et al.* 1996) are also processed by the OFC, and again in relation to their rewarding or punishing contingencies.

While conceptual issues certainly have contributed to the mysteriousness of the OFC, there are also technical issues that raise difficulties. Specifically, a variety of different techniques have been used to study the functions of the OFC, and different techniques permit different kinds of inferences. Neurophysiological recordings give an unparalleled temporal resolution at the millisecond scale, and permit precise localization of the recording sites; however, these are constrained to yield only a very narrow view of this vast brain region. Functional imaging, by contrast, provides a much coarser resolution in both space and time, but also provides a much broader view that can reveal global functions of the OFC as well as macroscopic subdivisions and topographical relations within it. Another important difference between electrophysiology and functional imaging is that the electrophysiological signal is a direct measure of neuronal electrical activity, whereas functional imaging methods utilize quite indirect indices of such activity, whose relationship to neuronal activity still remains poorly understood (see Logothetis *et al.* 2001).

In this chapter we will discuss a series of studies from our laboratory that has investigated the functions of the OFC in humans using invasive neurophysiological methods. Such studies are exceedingly rare, because they cannot be performed solely in a research setting. The only circumstance in which they can be performed is a clinical setting: neurosurgical implantation of depth electrodes (Ojemann et al. 1998; Engel et al. 2005). There are several reasons why such depth electrodes are implanted, ranging from exploratory recordings to guide neurosurgery for intractable pain, to monitoring of seizures to guide epilepsy surgery; we here focus only on the latter.

We hope that our discussion of this rare technique for investigating the functions of the OFC in humans will complement the other chapters in this volume. Lesion studies, electrophysiological studies in monkeys, and imaging studies in both healthy and pathological human populations are discussed in some detail in the other chapters. Intracranial neurophysiology in humans provides an important link to the neurophysiological studies in monkeys, and an important adjunct to the interpretation both of fMRI studies of the human OFC as well as of ERP studies of its function obtained from electrophysiology recorded at the scalp. We also hope that our presentation of this method will highlight its advantages and encourage other investigators to pursue it in future studies.

## 14.2. The technique

We recorded responses to visually presented stimuli in patients who had chronically implanted depth electrodes for monitoring epilepsy. The patients had been diagnosed with disabling epilepsy that was not responsive to medication, and had consequently elected to undergo neurosurgical treatment for their epilepsy. The neurosurgical treatment consisted of several phases, prior to possible removal of pathological brain tissue, in which extra and intracerebral EEG monitoring provided information about the focus of the patient's seizures. Our research participants were selected from a small sample in whom seizures could not be sufficiently well-localized with scalp electrodes, and were therefore subsequently localized with depth electrodes, from which we obtained our research recordings. The implantation of hybrid clinical-research electrodes does not pose any additional risks to the patient above and beyond the risks associated with standard clinical depth electrode monitoring. Typically, the implanted electrodes remain in place 1–2 weeks until sufficient seizures have been recorded in order to provide the neurosurgeon with the clinical information necessary to guide the surgery.

Needless to say, there are several issues that require careful attention in such experiments. For instance, cognitive and sensory function should be assessed in all subjects through neuropsychological testing. In our patients, we carried out extensive neuropsychological testing both pre- and post-operatively to ensure that IQ, visual function, and other factors would not confound our data. No less important is careful documentation of the medication status of the patients. Typically, such patients are administered opiate analgesics immediately after the implantation surgery. But this medication is then tapered over the course of the next several days, and often stopped entirely (or reduced to the level of taking a few Tylenol a day). Anti-epileptic medication levels are usually high in these patients prior to the surgery (but not entirely efficacious, which is the reason that they are considering neurosurgical intervention for their epilepsy). However, the anti-epileptic medications are also tapered substantially after the electrode implantation. The reason for this is that the clinician actually wishes to increase the probability of seizures, since the whole point of implanting the electrodes is to record the abnormal electrical activity that constitutes a seizure, such that its source can be accurately localized within the brain and this information can then guide the neurosurgical decision regarding what tissue to resect. Finally, the patients will be administered antibiotics to prevent infection, but this is unlikely to be a source of concern with respect to the quality of the electrophysiological recordings. Towards the end of the monitoring period, if the patient has not yet had enough seizures to provide the clinical data needed, sleep deprivation is sometimes introduced as well in an effort to increase the probability of seizures.

The upshot of these considerations is now described. First, we select for research recordings only those patients with a neuropsychological profile that ensures the stimuli and task we plan to use can be processed validly at least at the behavioral level. Second, we wait several days after electrode implantation (ca. 4–5 days) for medications to be tapered, and the cognitive state of the patient to stabilize. There is then a window of a few days during which the best quality of recordings can be obtained: medications have been tapered, the patient is awake and alert, but the frequency of seizures is not so high as to interfere with the research and the patient is not sleep-deprived.

Finally, one must consider the possible pathology of the tissue from which the recordings are obtained. Since the reason for the electrode implantations is the clinical monitoring of epilepsy, it is possible that some of the recording sites are in fact in epileptogenic tissue. This can be ascertained by an examination of the electrical activity obtained from cells in that region: if seizure-like artifacts are observed on the recordings, then this would raise a red flag. Similarly, structural MR examination can reveal a structural abnormality that may be the cause of the seizures, located close to the recording sites—for instance, a tumor may be nearby or within the tissue from which the recordings are obtained. All of these considerations result in a further selection of recordings from only that tissue which appears otherwise healthy. In our studies, we include for further analysis only recordings obtained from regions that were not subsequently resected as the likely source of seizure activity (typically, the patients have multiple electrodes in multiple regions, since their seizure foci are unknown, thus increasing the probability that at least some sites will turn out to be distal to the subsequently determined seizure foci).

### Neuroanatomical localization

In addition to the superior temporal resolution that intracranial electrophysiology affords (a factor of about 1000 compared to functional MRI), it also yields very precise anatomical localization. It should be pointed out that the anatomy associated with human neurophysiology differs in two important respects from the analogous method in monkeys. While MRI is typically used these days in live monkeys to provide an on-line guide to localization, it is later possible to obtain histological verification. Examination of electrode tracks, or the locations of injected tracers, can be performed under a microscope on sections of the postmortem fixed brain of the animal. While in principle one could also conduct such histological examination on the brain tissue of humans, this would depend on en bloc resections of the tissue, and would of course require that the tissue from which the recordings were obtained coincided with the tissue that was subsequently resected as a seizure focus—exactly the situation that we noted above one wishes to avoid. So, in general, one is limited to obtaining structural MR scans of the patient's brain.

These MR scans are obtained both pre-implantation and post-implantation in our laboratory. The post-implantation scans provide the location of the electrodes and their recording sites. However, the metal in the electrodes, while providing a strongly contrasted signal to visualize their location, also introduces a substantial artifact, since it distorts the local magnetic field. As a consequence, the electrodes appear much wider and 'fuzzier' on the MR scan than they in fact are. The center of the electrode will correspond to the center of its image on the MR scan, but the artifact on the scan obscures the immediately adjacent brain tissue proximal to the recording contacts. This is where the pre-implantation scans help: since they were obtained in the absence of the electrode, but from the identical brain, they can be co-registered with the post-implantation scans to visualize precisely that brain tissue that is adjacent to the recording sites. Examples of such mappings are provided in Fig.14.1a, and illustrate some of the regions in the prefrontal cortex from which we have obtained recordings.

A second difference between intracranial electrophysiology in humans as compared to monkeys concerns the sampling of the neurons from which recordings are obtained. Once implanted, the electrodes in humans cannot be moved. It is thus not possible to search for neurons with particular response properties (on the other hand, this may provide a less biased sampling of response properties). To hedge one's bets of finding a region from which good recordings can be obtained, the electrodes typically have multiple recording sites: some designs have small bundles of wires protruding from their tip (e.g., those used by Itzhak Fried and his team at UCLA; e.g., Fried *et al.* 1997; Kreiman *et al.* 2000; Quiroga *et al.* 2005), others have multiple recording contacts dispersed along the shaft of the electrode (our design). The design of our hybrid clinical-research electrodes is illustrated in Fig. 14.1b (Howard *et al.* 1996).

The sparse sampling deserves further comment. Compared to functional imaging, where one is able to measure responses in the whole brain, in electrophysiology one's spatial window into the brain is narrow but of very high resolution. This is even more so the case in our studies in humans, since the electrodes cannot be moved to sample different regions. Given that we know that the OFC consists of a complex assembly of sectors that differ cytoarchitectonically, hodologically, and functionally (e.g., Öngür and Price 2000), the method thus provides us with a sparse sampling of the functional responses

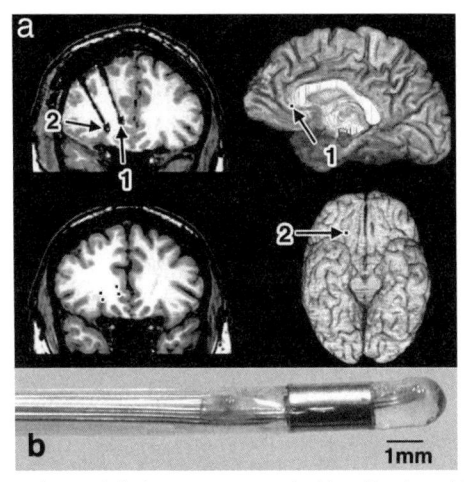

**Fig. 14.1** Recording electrodes and their neuroanatomical localization. (a) Neuroanatomical mapping of recording sites using MRI. Shown are two recording sites, indicated by the numbered arrows, in a patient who had electrodes implanted into the prefrontal cortex. Post-implantation scans (top left) show the actual electrodes in situ; pre-implantation scans (bottom left) show as small dots the mapped recording sites; three-dimensional reconstructions of the patient's brain (on the right) map the projections of these recording sites onto the surface of the brain.

(b) The hybrid research-clinical depth electrode (Howard et al. 1996). Electrodes (Radionics, Inc., Burlington, MA) had 2 recording sites at the tip, each consisting of a cluster of 4 high impedance contacts separated by 5 mm and a standard low impedance contact used solely for clinical monitoring. The high impedance contact locations can just be made out as the small set of 4 clustered metallic contacts to either side of the large clinical low-impedance contact. The electrode design permitted recording of single-unit activity from the high impedance contacts without introducing any additional risks to the patient. Electrodes consisted of a flexible tecoflex-polyurethane shaft (1.25 mm outer diameter), through which ran tefloncoated platinum-iridium high impedance electrode wires (50 micrometer cross-sectional diameter). The Pt-Ir wires were cut flush with the surface of the shaft to form the research recording contacts. Actual impedance of high impedance contacts during recordings ranged between 40 kOhm and 220 kOhm at 1 kHz.

within this region. Thus, Type-II errors may be quite common and result from simply not having sampled neurons with the requisite response properties. Statements about the 'absence' of a certain kind of response are thus not warranted.

## Analysis of responses

Analyses of neural activity need to take into account the fact that neural responses to external stimuli arise from the interaction between activity induced by stimuli and ongoing neural activity. This issue is likely to be especially prominent in the frontal cortex, where neuronal activity is often not tightly stimulus-locked, and where it can be driven not by the perceptual properties of a stimulus but rather by the motivational value with which that stimulus has come to be associated. The issue at the level of neuronal response is a reflection of the phenomenon at the psychological level: responses to the same external stimulus are not invariant over time, but instead depend on the continuously changing internal state of the subject, specified by a complex collection of factors such as emotion, attentiveness, memory, and so on. Since control over internal state is more difficult than control over the stimulus, the variability of responses is higher in those regions of the brain, like the prefrontal cortex, in which activity is influenced substantially by the former factor.

Recordings in the human prefrontal cortex are made more difficult also by two other factors. One is that the spontaneous rates of neurons in the OFC are typically rather low—a few spikes per second, on average. This would make it difficult to have sufficient statistical power to detect any decrease in the firing rate. Increase in the firing rate can be observed more readily, but many of the evoked responses we have found are also rather weak. To compound this difficulty still further, it is often too tedious for the patients to go through a large number of trials, and one must make do with a relatively small number. All of these issues result in a relatively low signal-to-noise ratio, much lower than what can be achieved in monkeys, where large trial repetitions can be achieved.

Several statistical approaches can be taken to the data. While parametric statistical methods, such as ANOVA, continue to be useful, the distributions of the data often violate some of the basic assumptions behind these standard statistical techniques. Thus data need to be transformed, or additional statistical methods need to be used to validate the findings. The most robust method is to use data-driven approaches, in which the actual distribution of the data is used to derive the statistics. Re-sampling methods provide one such approach, and one scheme is outlined in Fig. 14.2. Such re-sampling methods can be used as an initial step in the statistical analysis, and epochs within which significant responses were found can then be subjected to more conventional statistical treatments.

## 14.3. Responses to emotional visual stimuli

In one series of studies, we have investigated the response of neurons in the medial and the ventral human prefrontal cortex to pictures of facial expressions, or to pictures of complex emotional scenes. We have concentrated on these visual emotional stimuli for several reasons. First, they are a class of stimuli that are relatively easy to control, well-standardized, and we have used them extensively in other studies (e.g., lesion studies). Well-known stimulus sets are available both for emotional facial expressions (Ekman and Friesen 1976) and for emotional visual scenes (Lang et al. 1988). Second, responses to visual stimuli have been recorded in the OFC in monkeys, and it is known that the OFC receives highly processed visual information from visual cortices in the anterior temporal lobe.

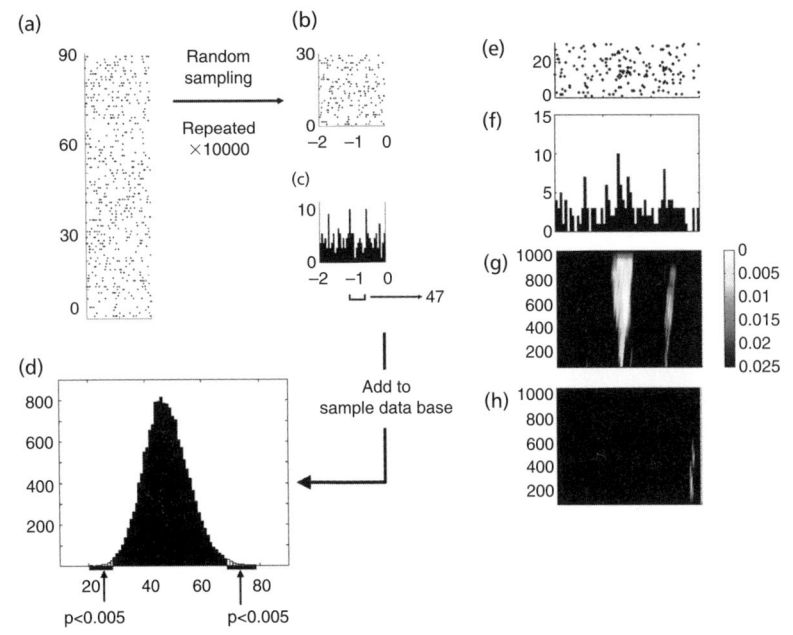

**Fig. 14.2** Data-driven calculation of the statistical significance of neuronal responses using resampling. As the first step of our statistical evaluation of the responses, we assessed the deviation of post-stimulus responses from the permutation distribution of the pre-stimulus response, for each neuron from which we recorded. The example shown used a window-width of 500 ms and step sizes of 50 ms. The prestimulus epoch consisted of −2000 to 0 ms. In the example shown, 30 stimuli were shown to the subject; consequently we drew 30 random samples (a,b) from the prestimulus epoch of a neuron and generated a histogram of their binned firing rates, as shown in (c). The number of spikes were counted in each 500 ms window, e.g., 47, as indicated on the bottom right in (c), and then added to the cumulative histogram shown in (d), which plots the number of occurrences versus spike counts in the 500 ms windows. The entire process was repeated 10000 times, yielding the large numbers of observations on the y axis of d and a relatively smooth distribution. The tails of the distribution corresponding to $P<0.005$ were then used to assign statistical significance to peristimulus changes in firing rate observed for that same neuron with that same bin-width (500 ms in the example). Using this method, initial data (rasters and histogram shown in (e) and (f)) could be converted to a probability matrix of significant excitatory (g) and inhibitory (h) responses (p-value legend to the right of panel (g). The time epochs defined by these significant responses to a particular emotion category were then used in the standard non-parametric and parametric statistical tests we used in subsequent stages.

It should be noted that, either for very small window widths or for neurons with very low spontaneous activity, the permutation distribution becomes skewed towards zero and truncated there (since negative firing rates are impossible), and we consequently lose statistical power in detecting significant decreases in firing rate. Thus, our findings may be biased towards finding significant rate increases rather than decreases, especially given the generally low spontaneous firing rates we observed.

## Responses to faces

Faces are well-suited to begin an investigation of the responses to emotional factors, for several reasons. They are ecologically valid stimuli that can signal specific, so-called 'basic' emotions, and there is evidence that the recognition of these emotions is rather reliable across different subjects. A well-known set of such stimuli that has been validated in many studies is the Pictures of Facial Affect, developed by Paul Ekman and colleagues (Ekman and Friesen 1976). In addition to their ecological validity, facial expressions of emotion provide a major advantage in studies of the emotional value of stimuli: many other classes of visual stimuli confound the emotional value of the stimulus with its perceptual (sensory visual) properties. If one is showing complex stimuli that differ in terms of their emotional value, one therefore needs to check that any differential neural responses could not be driven by correlated differences in their sensory properties, an issue to which we will return in the next section. Faces circumvent this difficulty, since it is possible to change the emotional expression of the face with minimal changes to other properties of the image: happy, fearful, or neutral expressions of the same person taken under the same lighting, for instance, vary minimally in terms of their mean luminance, spatial frequency, or color composition. They also do not vary in terms of other semantic information: the identity of the person, their age, and their identity can remain invariant despite large changes in their facial expression that signifies an emotion. Finally, because of these properties of faces, it is possible to create stimulus continua that are difficult to achieve otherwise. One can morph different emotional expressions to create subtle shades of emotions. This parametric variation in emotional expressions can then be used as a particularly powerful probe of the neural responses elicited by them.

Both lesion and functional imaging studies have examined responses to emotional faces in the human OFC. Lesions to this region impair the ability to trigger emotional responses to faces, as measured by the skin conductance response, for instance, but leave intact cognitive abilities to recognize the identity of the face (Tranel et al. 1995). However, recognition of the emotional expression shown on the face appears to be compromised by such lesions (Hornak et al. 1996), in line with increased activation in this region to conditioned autonomic responses when faces are used as the CS+ in Pavlovian conditioning studies (Morris et al. 1997). In other functional imaging studies, expressions of fear (Vuilleumier et al. 2001) or anger (Blair et al. 1999) have been reported to activate the OFC, especially on the right side, although there is also evidence that the OFC is activated by happy faces (Dolan et al. 1996) as well as attractive faces (O'Doherty et al. 2003). It is interesting to note that lesions of the right OFC appear to cause not only impairments in the recognition of facial emotion (Hornak et al. 1996), but also impairments in the ability to express such emotions on the patient's own face (Angrilli et al. 1999).

Fig. 14.3 shows responses we recorded in medial sectors of the human prefrontal cortex, on the right side, when facial expressions of different emotions were shown. Two aspects of these responses are noteworthy: they are rapid (occurring within about 120–150 ms after onset of the face), and they are differentiated by the emotion shown in

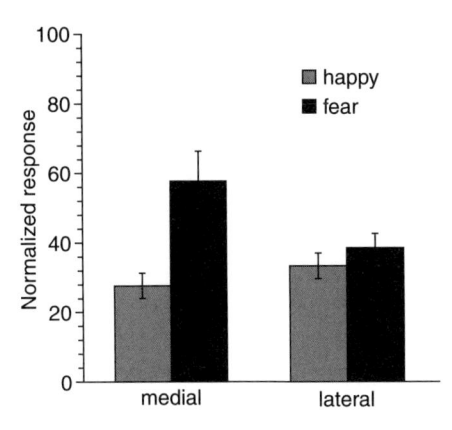

**Fig. 14.3** Responses to emotional facial expressions recorded in the human OFC. Shown are normalized responses recorded in the medial and lateral OFC of the patient whose brain is shown in Fig. 14.1a. Mean (and SEM) evoked responses are shown to facial expressions of happiness and of fear. Data from Kawasaki *et al.* (2001), copyright MacMillan Press.

the face. The recording sites were situated in medial OFC and in subgenual cingulate cortex. The first of these corresponds to area 11l in the medial OFC, whereas the second correspond to area10r, granular paracingulate cortex corresponding to the rostral part of monkey area 10 m (Öngür *et al.* 2003). Not only did we find responses modulated differentially by the emotion shown in the face, but the pattern of these responses varied with anatomical location (Kawasaki *et al.* 2001). Other intracranial recordings to emotional faces in the OFC have been carried out by Krolak-Salmon *et al.* (2004) and by Marinkovic *et al.* (2000). It is interesting to note that the first of these recorded responses with a latency much longer than ours, about 500 ms, likely reflecting different aspects of emotion processing that were recorded here, as also indicated by the fact that their ERPs depended on attention to the face stimuli.

## Responses to scenes

While faces provide the advantages of more easily dissociating emotional value from perceptual properties, as noted above, they do not provide particularly potent emotional stimuli. More potent stimuli can be found in scenes showing, for instance, pleasant, neutral, or aversive stimuli. One database of such scenes that has been the most widely used is the one developed by Peter Lang: the International Affective Picture System (IAPS) (Lang *et al.* 1988). These stimuli are complex images that show tremendous visual heterogeneity. The aversive pictures, for instance, include pictures of wartime, of mutilated bodies, of illness, of spiders, of snakes, of guns, of explosions, and so forth. If judiciously chosen, they do not share any simple features in common, vary widely in terms of low-level visual properties, and overlap in these domains with stimuli that signal pleasant or neutral emotions. The only factor that classifies them as belonging to the same, aversive, emotion category is the way that human observers judge them. We further verified that, indeed, our 3 emotion categories did not show systematic differences in low-level visual properties that might be driving the differential responses we observed. We calculated their luminance in different color channels, and their spatial frequencies, and checked that these did not differ among the stimulus categories. The very heterogeneity of the stimuli helped to ensure that no low-level properties or simple visual features would account for the differences. Some functional imaging studies have reported that the OFC can be activated by complex visual scenes regardless of emotional meaning, when compared to baseline activity (Taylor *et al.* 2000), perhaps related to more general orienting responses that also drive the skin conductance response (Williams *et al.* 2000).

It should be noted that such stimuli, while standardized in terms of their valence and arousal, to some extent confound these two dimensions. In particular, it is much easier to obtain negatively-valenced (aversive) images of high arousal, than it is to obtain positively-valenced (pleasant) images of the same arousal. Pictures of guns and mutilations, for example, are much more arousing than pictures of babies, burgers, or puppies. So findings regarding the valence-specific responses of neurons could be due, at least in part, to responses driven by the co-varying arousal instead. When highly arousing pleasant stimuli are used, responses similar to those obtained with strongly aversive stimuli may be obtained. The problem is that the only category of visual stimuli that is pleasant and features such strong arousal are sexually explicit stimuli (pictures of nudes or of couples engaged in sex). A functional imaging study of the amygdala showed that such stimuli do indeed elicit large responses, whereas pleasant stimuli with lower arousal do not (Hamann *et al.* 2004); similar findings have been obtained also in the OFC (Karama *et al.* 2002). However, because of their possibly embarrassing nature, sexually explicit stimuli can also introduce confounds of their own, and are not always suitable as stimuli. For these and other reasons, we did not use them in our experiments.

The findings, which we obtained in the OFC in response to emotionally salient scenes from the IAPS, were both at the single-unit level as well as at the field potential level. At the single-unit level, we found responses of neurons in the OFC that showed rapid modulation specific to highly aversive images. The latency of these responses, as in the case of faces, was around 120–140 ms, a latency that is so rapid that essentially only feed-forward processing in the visual system could account for them. It is likely that the extensive visual connections between anterior temporal visual cortices and the OFC mediate these responses at the anatomical level (Seltzer and Pandya 1989). As Fig. 14.4 shows, the peristimulus-time histograms and rasters of single neurons show clear modulations by aversive visual stimuli. Most interesting is the apparently biphasic nature of the responses we recorded at this site: they feature an initial decrease in firing rate, followed by a prolonged increase. The rapid modulation of firing rate by the stimuli, with a latency around 120 ms, is illustrated further in Fig. 14.5, which plots the strength of the p-value describing the post-stimulus to the pre-stimulus rates as a function of the width of the window of analysis used, for the three categories of emotional scenes we used as stimuli (pleasant, neutral, and aversive) (Kawasaki *et al.* 2001).

Though striking, these findings at the level of the single neuron provide only a narrow window onto the functional organization of the OFC. One would like to record from

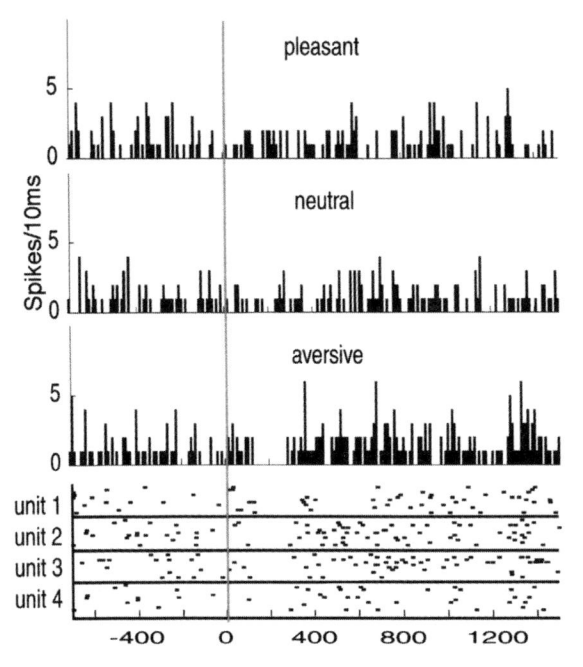

**Fig. 14.4** Single neuron responses to emotional scenes. Shown are peristimulus-time histograms (top) and single neuron rasters (bottom) in response to stimuli in three emotion categories. The rasters correspond to the PSTH for aversive stimuli only. Data from Kawasaki et al. 2001, copyright MacMillan Press.

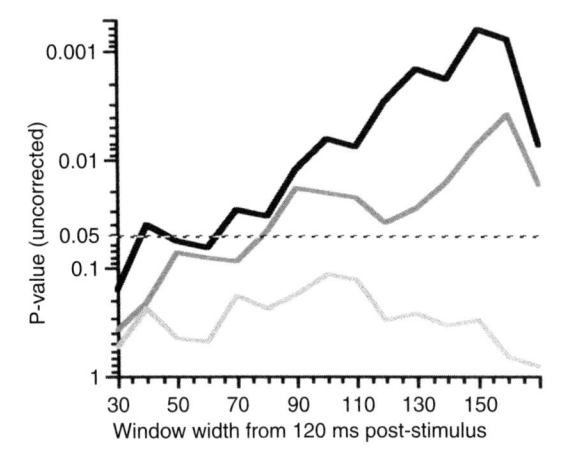

**Fig. 14.5** Selectivity to aversive scenes develops rapidly in the OFC. The three different curves plot the p-values obtained over time in comparing windows of increasing width (from 120 ms post-stimulus onset) to the pre-stimulus firing rate. Black line: aversive stimuli; dark grey line: pleasant stimuli; light grey line: neutral stimuli. Data from Kawasaki et al. 2001, copyright MacMillan Press.

small populations of neurons, in different regions, in multiple patients. In one recent study (Kawasaki et al. 2005), we recorded from 3 patients with electrodes implanted in the prefrontal cortex, showing them the same IAPS stimuli described above. A composite analysis of the responses that we found is given in Fig. 14.6. As the figure shows, there is no clear topography to the responses that we found; the proportion of responsive and selective neurons is quite low; and the predominant selectivity is for

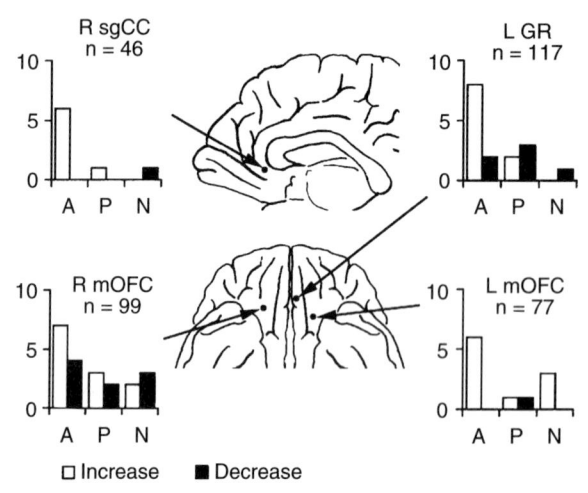

**Fig. 14.6** A composite summarizing the responses to emotional scenes recorded in the OFC of three patients. While each patient had multiple recording sites, only the four locations at which significant selective response were obtained are shown. For each location, the corresponding bar graphs depict the number of neurons (y-axis) that showed significant increases (light bars) or decreases (black bars) to aversive (A), pleasant (P), or neutral (N) stimuli. The total number of neurons recorded at each location is indicated at the top of each graph, together with the anatomical location; abbreviations—sgCC: subgenual cingulate cortex, GR: gyrus rectus, mOFC: medial OFC. Data from Kawasaki et al. (2005), copyright MIT Press.

aversive pictures. The lack of topography may result from the very sparse sampling of different regions, together with the small number of patients in whom recordings were obtained. It thus remains possible that a larger sampling would reveal topography; it also remains possible that other methods that measure more global neuronal activity in the OFC can better reveal such topography, as indeed has been reported in some other studies (Northoff et al. 2000). The apparent selectivity for aversive pictures needs to be interpreted in the light of the correlation between valence and arousal that we mentioned above. It may thus be that the responses we found were driven largely by arousal alone. A preliminary analysis of this question, however, has failed to confirm that the story is this simple, and we believe that neurons in the OFC code for more complex, and fine-grained, emotional attributes than a simple arousal or valence dimension.

The findings reviewed above are consistent with the large literature from animal studies. Neurons within the OFC have been shown to respond selectively to a variety of sensory stimuli based on their emotional significance, independently of their basic perceptual properties (Hikosaka and Watanabe 2000; Rolls 1999; Thorpe et al. 1983) and more related to whether they are rewarding or punishing (Rosenkilde et al. 1981). For instance, the responses of neurons in the OFC are influenced by altering the value of a sensory stimulus by changing the motivational state of the animal (Rolls et al. 1989) (e.g., by satiating it or depriving it of food when presenting a food-related stimulus), or by placing it in a comparison with other available rewards (Tremblay and Schultz 1999). Similar results have been obtained using functional imaging of the OFC in human studies: activation changes when the value of a food stimulus is changed (Small et al. 2001), and activation for faces correlates with the perceived attractiveness of the face (O'Doherty et al. 2003).

While the OFC is a large neural territory, there have been some suggestions of topography of the value representations of neurons within it. In monkeys, some topography of neuronal responses to specific rewards have been reported; but these were in dorsolateral rather than orbital prefrontal cortex (Watanabe et al. 2002). In humans, studies have indicated that medial and ventral sectors of the prefrontal cortex may be especially important for processing aversive emotions (Marinkovic et al. 2000; Kawasaki et al. 2001): facial expressions of fear (Vuilleumier et al. 2001) and anger (Blair et al. 1999) activate the right OFC in functional imaging studies, and recognition of facial expressions of anger is disrupted by TMS applied to medial prefrontal regions (Harmer et al. 2001).

Other studies have pointed to differential processing of positive or negative emotions by the left or the right frontal lobes, respectively (Borod et al. 1998; Canli 1999; Davidson 1992; Davidson and Irwin 1999; Royet et al. 2000), or by lateral versus medial sectors of prefrontal cortex (Northoff et al. 2000), raising the possibility that different emotions may be processed by neuroanatomically distinct regions at least at this relatively macroscopic level. Of particular interest is a functional imaging study of olfaction that showed that the OFC was more responsive to the valence than to the arousal of odors, with medial OFC encoding pleasantness, and lateral OFC encoding unpleasantness (Anderson et al. 2003). However, exactly the opposite valence encoding in the OFC was described by Northoff et al. (2000) using a combined fMRI/MEG technique and the IAPS images, perhaps reflecting differences between sensory modalities, or differential involvement of these sectors at different time scales. Nonetheless, both these studies (Anderson et al. 2003; Northoff et al. 2000) as well as others (Small et al. 2003) have suggested that the OFC is more important for processing valence information than it is for processing arousal information.

It is important to elaborate that not only do different spatial regions of the OFC and surrounding cortices appear to process different aspects of emotion, but that different processing will be engaged at different points in time, perhaps even within the same region. The shortest latencies we recorded approach 120 ms, just fast enough for feedforward processing in the human visual system, given that earliest responses to faces in occipital cortex may occur around 60–80 ms, and earliest responses in temporal visual cortex specialized for faces with latencies around 100 ms (Liu et al. 2002). The latter finding also nicely illustrates the different processing that can occur at different temporal scales: rapid coarse categorization of faces in the fusiform gyrus can take place with latencies as fast as 100 ms, whereas more subordinate categorization that discriminates between different faces appears to take around 170 ms (Liu et al. 2002). Our earliest responses recorded in the OFC may thus reflect a rather coarse initial detection of superordinate stimulus categories, such as arousal, with more detailed and context-dependent processing requiring substantially longer processing times (e.g., as in the study by Krolak-Salmon *et al.* (2004) mentioned earlier).

## 14.4. Responses to the expectation of reward or punishment

All of the above studies have the shortcoming that the perceptual properties of the stimuli and their emotional value are not experimentally dissociated. Studies in animals typically do dissociate these properties (Rolls, Chapter 5, this volume; Schultz, Chapter 7, this volume): an initially appetitive stimulus, for instance, can be made aversive by satiating the animal, or by associating it with an aversive unconditioned stimulus (Hikosaka and Watanabe 2000). Some similar experiments have also been done in humans using functional imaging: having people eat chocolate until they get sick of it will change the reward value of the chocolate, and change the corresponding response of neurons in the brain that encode this reward value (Small *et al.* 2001). Similarly, associating initially neutral visual stimuli with reward and punishment will modulate the conditioned value of those visual stimuli (Morris *et al.* 1997). These reward-value encodings correlate with activation of the OFC in humans, as described by O'Doherty in Chapter 10, this volume (Gottfried *et al.* 2003), and extend to quite abstract representations of reward or punishment (O'Doherty *et al.* 2001).

We recently carried out an intracranial recording study that used the same approach (Oya *et al.* 2005). The study was undertaken in the same patient whose brain was shown in Fig. 14.1a, but this time we analyzed field potentials rather than single-unit recordings in order to assess somewhat more broadly the electrophysiology related to ensembles of neurons. While the task we used was chosen because it has been demonstrated to provide a sensitive measure of damage to the OFC, it also provided the opportunity to examine neuronal responses to the rewarding values of stimuli dissociated from their sensory properties.

We used the Iowa Gambling Task (Bechara *et al.* 2000), as described by Naqvi *et al.* Chapter 13, this volume, and recorded from the medial OFC. In this task, subjects choose cards from decks in order to win or lose money. On the basis of feedback (experience with these decks over time), they develop an expectation of what is likely to happen when they choose a card from a given deck: on some decks, it is likely that one can incur a large monetary loss, whereas on others, one gradually makes more money than one loses. Over time, subjects will thus come to expect a certain outcome when they choose from a given deck. One parameter than can be used to measure this, a parameter that models of the learning algorithm often use, is the so-called 'reward-prediction error'. Simply put, this is the discrepancy between what one expects and what one actually gets. So, if you expect to win a lot of money from a given deck, because you've won a lot on it in the past, and you now suddenly lose money, there will be a large reward-prediction error. Similarly, if you expect to lose money on a given deck, and now suddenly win big, there will be a large reward-prediction error (of opposite sign to that in the previous example).

We modelled the subjects' card choices with a commonly used reinforcement learning algorithm (Sutton and Barto 1998; Dayan and Abbott 2001; Seymour et al. 2004) that predicts the best choice to be made on the basis of the statistical distribution of the winning and losing contingencies experienced on the four card decks. Our modeling of the subject's choices incorporated a trial-by-trial update of the action selection probability, ($p$, the probability of selecting from a particular card deck), from estimations of parameters that specify an update value of action values for each possible action ($Q$, the value associated with choosing from a particular card deck) and the degree of exploration or exploitation (Sutton and Barto 1998; Dayan and Abbott 2001). This assumes that the subject maintains expectations for each of the action values obtained after choosing from one of the four decks, and updates the probability of choosing from a particular deck on the next trial by comparing action values of all possible actions at the trial. An indicator of the subject's learning in the task is the average value of all of the possible actions (expected value, $V$; the mean action values of all four decks weighted by the probability of choosing from them). We note that this does not have quite the same meaning as 'state value' in actor-critic or temporal difference learning schemes, since we modeled the task as a static action choice. Rather, it reflects the overall utility of the subject's prospect, as formulated in prospect theory. As with the well-known Delta and Rescorla-Wagner rules, it is the error in reward (or punishment) prediction that plays a crucial role in updating the old estimate (the weights in the model). This yielded the subject's reward-prediction error, PE, corresponding to the difference between expected and obtained reward (see Oya et al. 2005 for details on the computational model and the data recorded in this study, of which we provide only a sampling here).

The findings were as follows: first, behavioral performance on the Iowa Gambling Task was normal. The patient learned to change his choice behavior so as to switch from initially preferring the risky decks, to later preferring the safe decks. This is reassuring further evidence that the overall function of the OFC in the patient was intact. We plotted the reward-prediction errors obtained from the above computational model during this task while recording from neurons in the medial prefrontal cortex. What we found was both expected and surprising. Consistent with our predictions, and consistent with data from functional imaging studies, we found that the electrophysiological signal recorded in the prefrontal cortex correlated with the magnitude of the reward-prediction error. Surprisingly, this correlation held only for those choices made from decks in which large losses were expected but not obtained (that is, in risky choices that were unexpectedly rewarded).

Specifically, we found that: (A) reward-prediction error correlated with the ERP's alpha band component recorded from medial prefrontal cortex; (B) the association between reward-prediction error and alpha-band ERP was strongly driven by choices that were anticipated to be risky but violated the expectation of punishment; (C) emotional response (anticipatory SCR) was negatively correlated with action values. It is important to note that the 'risky' trials in which punishment was omitted were perceptually indistinguishable from 'safe' trials with no punishment. In both cases, the point in time at which a punishment could have occurred showed the text 'please wait' instructing the subject to wait until the next trial. Yet we found a significant association between PE and alpha-band ERP in the risky but not in the safe case (Fig. 14.7). Similarly, while triggered by the expectation of possible punishment (i.e., the temporal onset of the punishment feedback epoch), the ERPs we recorded could not have encoded actual punishment magnitude, since none was administered on those trials we chose for analysis. We thus believe that the differences in electrophysiological activity we found must have been driven by the patient's expectations rather than the sensory stimuli presented: namely, the prediction error arising from the discrepancy between the two. This can be seen dramatically in the difference between the gray and black traces (the averaged intracranial ERPs) shown in Fig. 14.8.

These findings are consistent with a large emerging literature implicating the OFC in reinforcement learning, and in updating the expectations of punishment or of reward that are contingent on a particular behavior choice. Findings in animals ranging from rats (Schoenbaum et al. 1998) to monkeys (Roesch and Olson 2004) as well as humans corroborate this picture (O'Doherty et al. 2003). Another recent single-unit study in human anterior cingulate cortex also provided evidence that the reward value of stimuli, as well as the motor response, modulated neuronal responses (Williams et al. 2004).

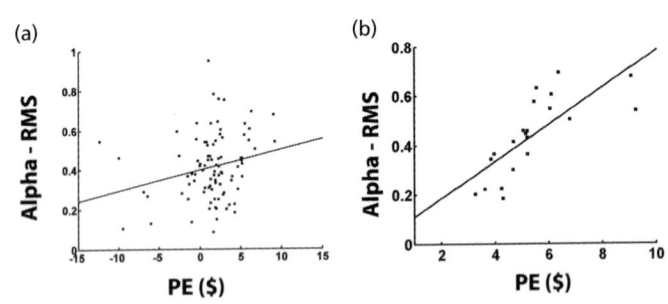

**Fig. 14.7** Correlations of response in the medial prefrontal cortex with reward prediction error. The recording site is location 1 in Figure 14.1 a, which showed large modulations of ERPs in response to winning or losing money on the Iowa Gambling Task; consequently, we analyzed responses from that location in more detail. Note that responses are field potentials, i.e., activity generated by a local assembly of neurons, and that we describe power only in the frequency band showing the largest modulation on the task, in the alpha-frequency range (around 10 Hz).

(a) Scatter plot showing correlation between reward-prediction error (PE, x-axis) and alpha band amplitude of the response recorded in the medial prefrontal cortex (y-axis). (N=91 trials from the total 100 trials of the Iowa Gambling Task, after trial rejection due to signal artifacts).

(b) Scatter plot showing the same correlation as in (a), but restricted to only those trials in which the patient chose from 'risky' decks (thus expecting punishment, decks A and B in the Iowa Gambling Task) on which no punishment in fact occurred (N=21). As can be seen, the correlation is strong only on those trials where a punishment was expected, but did not occur, resulting in a large reward-prediction error. Data from Oya et al. (2005), copyright The National Academy of Sciences.

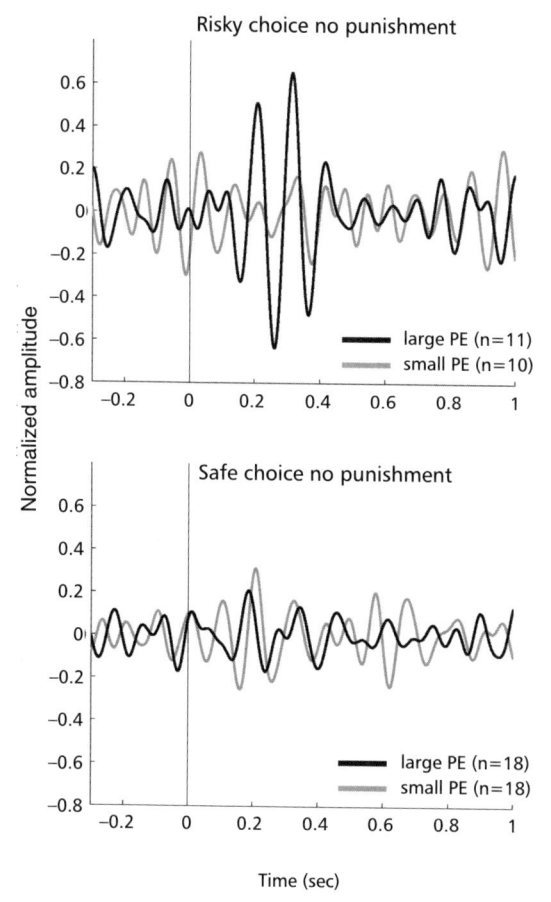

**Fig. 14.8** Responses in medial prefrontal cortex encode reward-prediction error conditional on the risk of punishment. The same data as in Fig. 14.7 are shown in terms of their averaged ERP amplitudes. The data are broken down for the responses obtained on those trials in which the patient chose from one of the risky decks, but received no punishment (top), and those trials on which the patient chose from one of the safe decks and received no punishment. Traces of responses are aligned in time to the patient's choice (decision to pick a card from one of the four decks on the Iowa gambling task), the vertical gray line at time=0.

Despite the fact that the stimulus on the screen and the behavioral outcome of the choice were identical in these two situations (top and bottom panels), responses differed as a function of the magnitude of the reward-prediction error (PE, broken down in terms of large and small, the black and gray traces). Once the patient had learned to expect punishment when choosing from a risky deck, but did not receive it when he actually chose from a risky deck, a large response was recorded (black trace in top graph). Data from Oya et al. (2005), copyright The National Academy of Sciences.

## 14.5. Conclusions

The last experiment described above indicates a specific role for the OFC in reward, and one which fits into a large emerging literature on reinforcement learning in this brain region (both in animal studies as well as from functional imaging in humans). In particular, the findings implicate the OFC in the continuous updating of expectations of reward and punishment based on the reward-prediction error signal that guides the acquisition of choice bias.

It is less clear what to say about the set of data that we discussed first: the responses driven by facial expressions and emotional scenes. Arguably, these could also be construed as providing expectations of reward and punishment, although such an inter-pretation would be complex and likely not rely on associations between behavioral choice and outcome, but rather on associations between the stimuli and an outcome (unlike in the Iowa Gambling Task, where it is really the patient's behavioral choice that is linked to the rewarding or punishing outcome, not the card decks as such). It is debatable to what extent the emotional value of faces or scenes would be acquired; certainly they are not learned in the context of the experiment. Thus, because we are not manipulating the rewarding value of the faces or scenes in the experiment, and because we are not linking their perception to a choice behavior of the patient, much less is known about how exactly they are linked to emotional value. They could be linked to an automatic emotional response to them (e.g., as indexed by skin conductance responses); they could be linked to a plan for a more complex behavior towards them (choosing to approach or avoid them, for example); or they could be linked to the anticipation of an expected outcome, regardless of any contingency on the behavior of the subject. Given the design of the experiments, we just don't know.

On the other hand, the faces and scenes provide us with ecologically valid stimuli that have strong and natural emotional value. It would be interesting to attempt in future studies to tease these different factors apart. For instance, if one chose highly intrinsically aversive stimuli, such as pictures of spiders or mutilations, and paired them with a pleasant outcome during the learning phase of an association task, would the very same neurons that initially responded selectively to aversive stimuli now no longer respond to those same stimuli? Or would a different set of neurons come into play that specifically associate the stimuli with their outcome, but that do not respond initially to their intrinsic emotional value? If it is the case that closely adjacent cells might be responsible for different aspects of reward processing, then intracranial recordings may well afford the necessary resolution to investigate this issue.

Overall, there is evidence that the OFC contains neurons with a rather complex range of response properties: some are driven more by the sensory properties of the stimuli, some more by their reinforcing properties, some by positive and some by negative value (Rosenkilde *et al.* 1981; Thorpe *et al.* 1983). Given the complex network of other brain structures with which the OFC is interconnected (Öngür and Price 2000), this should not come as a huge surprise. It appears that the OFC functions to associate the representations of a large class of sensory stimuli, in multiple sensory modalities, with both their unconditioned and their acquired value, likely positive as well as negative. Given this broad and flexible function, it is to be expected that one finds neurons whose responses can be modulated by a large heterogeneity of sensory stimuli associated with a large range of values. Ultimately, what we will need to have a comprehensive understanding of the functions of the OFC is a detailed account of the responses of neurons within it, of the change of their response over time as experience with a stimulus and its behavioral outcome is gained, and of both the inputs to and outputs from this region of the brain to other components of the network that processes stimulus value.

Given the complexity of the associations between stimuli in multiple sensory modalities, and their unconditioned and acquired value, we can also see how the OFC can play a prominent role in regulating social behavior, as noted in the introduction to this chapter. Social stimuli (other people) are complex, both perceptually and in terms of their value: we need to discriminate and remember the way a large number of different people's faces look, and we need to link their appearance to the value that each specific person has for us. The latter is largely acquired through experience with that person (although we also have stereotypes and biases for or against people just on their initial appearance), and can indeed change: our previous friends can become our foes and conversely, and all of this must be stored in some fashion in the brain. There is thus a clear connection between a function for the OFC in representing the reinforcing properties of stimuli (Rolls, Chapter 5, this volume), and its regulation of social behavior (Koenigs and Tranel, Chapter 23, this volume): the association between the perceptual properties of stimuli and their contingent associations with value (either through stimulus-stimulus association, or through associations between the behavior triggered by the stimulus and its outcome) plays a critical role also in social behavior. Neurons in the OFC should thus respond to socially relevant stimuli (social scenes, emotional faces, as we have used), and damage to this region of the brain should thus compromise plans and actions guided by the value of social stimuli.

## References

Ackerly, S. S. and Benton, A. L. (1948) Report of a case of bilateral frontal lobe defect. *Res Publ Assoc Res Nerv Ment Dis* **27**:479–504.

Anderson, A. K., Christoff, K., Stappen, I., Panitz, D., Ghahremani, D. G., Glover, G., et al. (2003) Dissociated neural representations of intensity and valence in human olfaction. *Nat Neurosci* **6**:196–202.

Angrilli, A., Palomba, D., Cantagallo, A., Maietti, A., and Stegagno, L. (1999) Emotional impairment after right orbitofrontal lesion in a patient without cognitive deficits. *Neuroreport* **10**:1741–1746.

Bechara, A., Tranel, D., Damasio, H., Adolphs, R., Rockland, C., and Damasio, A. R. (1995) Double dissociation of conditioning and declarative knowledge relative to the amygdala and hippocampus in humans. *Science* **269**:1115–1118.

Bechara, A., Damasio, H., and Damasio, A. R. (2000) Emotion, decision-making and the orbitofrontal cortex. *Cereb Cortex* **10**:295–307.

Blair, R. J., Morris, J. S., Frith, C. D., Perrett, D. I., and Dolan, R. J. (1999) Dissociable neural responses to facial expressions of sadness and anger. *Brain* **122** (Pt 5):883–893.

Brickner, R. M. (1932) An interpretation of frontal lobe function based upon the study of a case of partial bilateral frontal lobectomy. Localization of function in the cerebral cortex. *Proceedings of the Association for Research in Nervous and Mental Disease (Baltimore)* **13**:259.

Cavada, C. and Schultz, W. (2000) The mysterious orbitofrontal cortex. foreword. *Cereb Cortex* **10**: 205.

Damasio, A. R. (1994) Descartes' error and the future of human life. *Sci Am* **271**:144.

Damasio, H., Grabowski, T., Frank, R., Galaburda, A. M., and Damasio, A. R. (1994). The return of Phineas Gage: clues about the brain from the skull of a famous patient. *Science* **264**:1102–1105.

Dayan, P. and Abbott, L. F. (2001) *Theoretical neuroscience: computational and mathematical modeling of neural systems*. Cambridge, MA:MIT Press.

Dolan, R. J., Fletcher, P., Morris, J., Kapur, N., Deakin, J. F., and Frith, C. D. (1996) Neural activation during covert processing of positive emotional facial expressions. *Neuroimage* **4**:194–200.

Ekman, P. and Friesen, W. (1976) *Pictures of facial affect*. Palo Alto, CA:Consulting Psychologists Press.

Engel, A. K., Moll, C. K., Fried, I., and Ojemann, G. A. (2005) Invasive recordings from the human brain: clinical insights and beyond. *Nat Rev Neurosci* **6**:35–47.

Frey, S., Kostopoulos, P., and Petrides, M. (2000) Orbitofrontal involvement in the processing of unpleasant auditory information. *Eur J Neurosci* **12**:3709–3712.

Fried, I., MacDonald, K. A., and Wilson, C. L. (1997) Single neuron activity in human hippocampus and amygdala during recognition of faces and objects. *Neuron* **18**: 753–765.

Gottfried, J. A., O'Doherty, J., and Dolan, R. J. (2003) Encoding predictive reward value in human amygdala and orbitofrontal cortex. *Science* **301**:1104–1107.

Hamann, S., Herman, R. A., Nolan, C. L., and Wallen, K. (2004) Men and women differ in amygdala response to visual sexual stimuli. *Nat Neurosci* **7**:411–416.

Harmer, C. J., Thilo, K. V., Rothwell, J. C., and Goodwin, G. M. (2001) Transcranial magnetic stimulation of medial-frontal cortex impairs the processing of angry facial expressions. *Nat Neurosci* **4**:17–18.

Hikosaka, K. and Watanabe, M. (2000) Delay activity of orbital and lateral prefrontal neurons of the monkey varying with different rewards. *Cereb Cortex* **10**:263–271.

Hornak, J., Rolls, E. T., and Wade, D. (1996) Face and voice expression identification in patients with emotional and behavioral changes following ventral frontal lobe damage. *Neuropsychologia* **34**:247–261.

Howard, M. A., III, Volkov, I. O., Granner, M. A., Damasio, H. M., Ollendieck, M. C., and Bakken, H. E. (1996) A hybrid clinical-research depth electrode for acute and chronic in vivo microelectrode recording of human brain neurons. Technical note. *J Neurosurg* **84**:129–132.

Karama, S., Lecours, A. R., Leroux, J. M., Bourgouin, P., Beaudoin, G., Joubert, S., et al. (2002) Areas of brain activation in males and females during viewing of erotic film excerpts. *Hum Brain Mapp* **16**:1–13.

Kawasaki, H., Kaufman, O., Damasio, H., Damasio, A. R., Granner, M., Bakken, H., et al. (2001) Single-neuron responses to emotional visual stimuli recorded in human ventral prefrontal cortex. *Nat Neurosci* **4**:15–16.

Kawaski, H., Adolphs, R., Oya, H., Kouach. C., Damasio, H., Kaufman, O., et al. (2005) Analysis of single-unit responses to emotional scenes in human ventromedial prefrontal cortex. *J Cog Neurosci* **17**:1509–1518.

Kreiman, G., Koch, C., and Fried, I. (2000) Category-specific visual responses of single neurons in the human medial temporal lobe. *Nat Neurosci* **3**:946–953.

Krolak-Salmon, P., Henaff, M. A., Vighetto, A., Bertrand, O., and Mauguiere, F. (2004) Early amygdala reaction to fear spreading in occipital, temporal, and frontal cortex: a depth electrode ERP study in human. *Neuron* **42**:665–676.

Lang, P. J. and Oehman, A. (1988) *The International Affective Picture System*. Gainesville, FL:University of Florida.

Liu, J., Harris, A., and Kanwisher, N. (2002) Stages of processing in face perception: an MEG study. *Nat Neurosci* **5**:910–916.

Logothetis, N. K., Pauls, J., Augath, M., Trinath, T., and Oeltermann, A. (2001) Neurophysiological investigation of the basis of the fMRI signal. *Nature* **412**:150–157.

Marinkovic, K., Trebon, P., Chauvel, P. and Halgren, E. (2000) Localized face processing by the human prefrontal cortex: face-selective intracerebral potentials and post-lesion deficits. *Cognitive Neuropsychology* **17**:187–199.

Morris, J. S., Friston, K. J., and Dolan, R. J. (1997) Neural responses to salient visual stimuli. *Proc Biol Sci* **264**:769–775.

Northoff, G., Richter, A., Gessner, M., Schlagenhauf, F., Fell, J., Baumgart, F., *et al.* (2000) Functional dissociation between medial and lateral prefrontal cortical spatiotemporal activation in negative and positive emotions: a combined fMRI/MEG study. *Cereb Cortex* **10**:93–107.

O'Doherty, J., Kringelbach, M. L., Rolls, E. T., Hornak, J., and Andrews, C. (2001a) Abstract reward and punishment representations in the human orbitofrontal cortex. *Nat Neurosci* **4**:95–102.

O'Doherty, J., Rolls, E. T., Francis, S., Bowtell, R., and McGlone, F. (2001b) Representation of pleasant and aversive taste in the human brain. *J Neurophysiol* **85**:1315–1321.

O'Doherty, J. P., Deichmann, R., Critchley, H. D., and Dolan, R. J. (2002) Neural responses during anticipation of a primary taste reward. *Neuron* **33**:815–826.

O'Doherty, J. P., Dayan, P., Friston, K., Critchley, H., and Dolan, R. J. (2003 a) Temporal difference models and reward-related learning in the human brain. *Neuron* **38**:329–337.

O'Doherty, J., Winston, J., Critchley, H., Perrett, D., Burt, D. M., and Dolan, R. J. (2003) Beauty in a smile: the role of medial orbitofrontal cortex in facial attractiveness. *Neuropsychologia* **41**:147–155.

Ojemann, G. A., Ojemann, S. G., and Fried, I.. (1998) Lessons from the human brain: neuronal activity related to cognition. *The Neuroscientist* **4**:285–300.

Ongur, D., Ferry, A. T., and Price, J. L. (2003) Architectonic subdivision of the human orbital and medial prefrontal cortex. *J Comp Neurol.* **460**:425–449.

Ongur, D. and Price, J. L. (2000) The organization of networks within the orbital and medial prefrontal cortex of rats, monkeys and humans. *Cereb Cortex* **10**:206–219.

Oya, H., Adolphs, R., Kawasaki, H., Bechara, A., Damasio, A., and Howard, M. A., III (2005) Electrophysiological correlates of reward prediction error recorded in the human prefrontal cortex. *Proc Natl Acad Sci USA* **102**:8351–8356.

Quiroga, R. Q., Reddy, L., Kreiman, G., Koch, C., and Fried, I. (2005) Invariant visual representation by single neurons in the human brain. *Nature* **435**:1102–1107.

Roesch, M. R. and Olson, C. R. (2004) Neuronal activity related to reward value and motivation in primate frontal cortex. *Science* **304**:307–310.

Rolls, E. T. (1999) *The Brain and Emotion*. New York:Oxford University Press.

Rolls, E. T., Sienkiewicz, Z. J., Yaxley, S. (1989) Hunger modulates the responses to gustatory stimuli of single neurons in the caudolateral orbitofrontal cortex of the macaque monkey. *Europ J Neurosci* **1**:53–60.

Rosenkilde, C. E., Bauer, R. H., and Fuster, J. M. (1981) Single cell activity in ventral prefrontal cortex of behaving monkeys. *Brain Res* **209**:375–394.

Schoenbaum, G., Chiba, A. A., and Gallagher, M. (1998) Orbitofrontal cortex and basolateral amygdala encode expected outcomes during learning. *Nat Neurosci* **1**:155–159.

Seltzer, B. and Pandya, D. N. (1989) Frontal lobe connections of the superior temporal sulcus in the rhesus monkey. *Journal of Comparative Neurology* **281**:97–113.

Seymour, B., O'Doherty, J. P., Dayan, P., Koltzenburg, M., Jones, A. K., Dolan, R. J., *et al*. (2004) Temporal difference models describe higher-order learning in humans. *Nature* **429**:664–667.

Small, D. M., Zatorre, R. J., Dagher, A., Evans, A. C., and Jones-Gotman, M. (2001) Changes in brain activity related to eating chocolate: from pleasure to aversion. *Brain* **124**:1720–1733.

Small, D. M., Gregory, M. D., Mak, Y. E., Gitelman, D., Mesulam, M. M., and Parrish, T. (2003) Dissociation of neural representation of intensity and affective valuation in human gustation. *Neuron* **39**:701–711.

Sutton, R. S. and Barto, A. G. (1998) *Reinforcement learning: an introduction*. Cambridge, MA: MIT Press.

Tanabe, T., Yarita, H., Tanabe, T., Yarita, H., Iino, M., Ooshima, Y., *et al*. (1975) Olfactory projection area in orbitofrontal cortex of the monkey. *J Neurophysiol* **38**:1269–1283.

Taylor, S. F., Liberzon, I., and Koeppe, R. A. (2000) The effect of graded aversive stimuli on limbic and visual activation. *Neuropsychologia* **38**:1415–1425.

Thorpe, S. J., Rolls, E. T., and Maddison, S. (1983) The orbitofrontal cortex: neuronal activity in the behaving monkey. *Exp Brain Res* **49**:93–115.

Tremblay, L. and Schultz, W. (1999) Relative reward preference in primate orbitofrontal cortex. *Nature* **398**:704–708.

Vuilleumier, P., Armony, J. L., Driver, J., and Dolan, R. J. (2001) Effects of attention and emotion on face processing in the human brain: an event-related fMRI study. *Neuron* **30**:829–841.

Watanabe, M., Hikosaka, K., Sakagami, M., and Shirakawa, S. (2002) Coding and monitoring of motivational context in the primate prefrontal cortex. *J Neurosci* **22**:2391–2400.

Williams, L. M., Brammer, M. J., Skerrett, D., Lagopolous, J., Rennie, C., Kozek, K., *et al*. (2000) The neural correlates of orienting: an integration of fMRI and skin conductance orienting. *Neuroreport* **11**:3011–3015.

Williams, Z. M., Bush, G., Rauch, S. L., Cosgrove, G. R., and Eskandar, E. N. (2004) Human anterior cingulate neurons and the integration of monetary reward with motor responses. *Nat Neurosci* **7**:1370–1375.

Zald, D. H., Lee, J. T., Fluegel, K. W., and Pardo, J. V. (1998) Aversive gustatory stimulation activates limbic circuits in humans. *Brain* **121** (Pt 6):1143–1154.

Chapter 15

# Orbitofrontal cortex activation during functional neuroimaging studies of emotion induction in humans

Darin D. Dougherty, Lisa M. Shin, and Scott L. Rauch

## 15.1. Introduction

In this chapter we will review orbitofrontal cortex (OFC) activation during functional neuroimaging studies of emotion induction in humans. While a comprehensive review of the field of emotion theories is beyond the scope of this chapter, we will start by briefly outlining some theories of emotion and follow with a review of the brain regions believed to be involved in mediating the different components of emotional processing. Functional neuroimaging studies utilizing affect induction paradigms in healthy volunteers that have yielded OFC findings will be reviewed in this context. Finally, we will discuss the means by which the OFC plays its role in emotional processing.

## 15.2. Theories of emotion

Scientists have been interested in human emotional experience for centuries. In ancient times, it was believed that amounts of specific fluids (e.g., bile) in the body were responsible for emotional states. Clinical observation has always played a central role in furthering our understanding of emotional experience. For example, in the context of the modern mental status examination, mood is regarded as the subjective account of the individual regarding their prevailing internal emotion state, while affect refers to the observable, objective, outward signs of emotional state. While mood and affect are generally congruent in healthy individuals, patients with brain lesions may exhibit an uncoupling of mood and affect. In the nineteenth century, William James and Carl Lange conceived of emotions as a conscious experience of an arousing stimulus with a focus on the role of emotions in the survival of the organism (James 1884). A few decades later, Walter Cannon conducted research on the physical manifestations of emotional experience and criticized the idea that autonomic activity triggers emotional experience (Cannon 1929). In the modern era, the 'circumplex model' of emotion was developed, whereby specific emotions (e.g., sadness, happiness, etc.) can be uniquely characterized based on two orthogonal axes indexing valence (positive/negative) and arousal level (high/low) (Lang 1995).

Most recently, appraisalist theorists have suggested three components that are essential for emotion perception (for review see Phillips *et al.* 2003): 1) the identification of the emotional significance of a stimulus; 2) the production of an affective state in response to said stimulus; and 3) the regulation of this affective state. Modern functional neuroimaging techniques have allowed investigators to begin to elucidate specific brain regions associated with these separate components of emotion perception. A large body of literature now suggests that as an individual progresses through these stages of information processing, the limbic brain regions mediate the identification of the emotional significance of a stimulus, while production and regulation of the affective state are subsequently mediated by the paralimbic, and then more neocortical brain regions. Specifically, the amygdala is principally involved in the identification of the emotional significance of a stimulus. Next, primarily ventral anterior paralimbic regions such as the ventral prefrontal cortex (PFC), the ventral striatum, and the rostroventral anterior cingulate cortex (ACC) are involved in the production of an affective state. This chapter focuses on the orbitofrontal cortex (OFC), a sub-territory of the ventral PFC, which is argued to be crucial for the production of affective states. Finally, neocortical regions, such as the dorsal PFC and the dorsal ACC, are implicated in the regulation of affective states.

## 15.3. Overview of the functional neuroanatomy of emotional processing

### Identification of the emotional stimulus

Activation of the amygdala has been reported in functional neuroimaging studies that have utilized a variety of means to expose the study participants to both positively and negatively valenced emotional stimuli (for reviews see Davis and Whalen 2001; Rauch *et al.* 2003; Zald 2003). A commonly replicated functional neuroimaging study result is the activation of the amygdala in humans during exposure to fearful human faces (Brieter *et al.* 1996; Morris et al. 1996; Phillips *et al.* 1997, 2001; Wright *et al.* 2001). It has been postulated (Davis and Whalen 2001; LeDoux 1996) that the amygdala is the first-line processor of emotional information and thus serves a crucial vigilance function for the survival of the organism (e.g., seeing a fearful face suggests imminent danger from an unknown but proximal source). The amygdala is interconnected with, among other regions, the insula and ventral PFC (including the OFC) (see Price Chapter 3, this volume; Barbas and Zikopoulos Chapter 4, this volume; Amaral and Price 1984; Amaral and Insuasti 1992; Carmichael and Price 1995; Chiba *et al.* 2001; Stefanacci and Amaral 2002; Price 2003). The amygdala, insula, and ventral PFC, along with other structures, are intimately involved in initiating autonomic responses to emotional stimuli (Asahina *et al.* 2003; Cechetto and Saper 1990; Williams *et al.* 2001). Therefore, in addition to the amygdala's direct effects on autonomic function, the connections between the amygdala and insula and ventral PFC help to generate the autonomic responses associated with emotional stimuli.

## Production of the affective state

The ACC can be divided into two divisions (for review see Bush et al. 2000): a dorsal division (dACC) and a rostroventral division (rACC). Production of the affective state has been hypothesized to be partially mediated by the rACC. Importantly, the rACC has extensive connections to the amygdala and OFC as well as other brain regions also involved in the mediation of emotional processing and autonomic function (Devinsky et al. 1995). The rACC has consistently been activated during functional neuroimaging studies involving emotional stimuli and is believed to be an important brain region for processing emotional information and producing affective states (for review, see Bush et al. 2000).

The role of the OFC itself in the production of affective states has also been examined using a variety of functional neuroimaging paradigms. Activation of the OFC has been reported during functional neuroimaging studies that have used a variety of emotional stimuli (see Table 15.1), including script-driven imagery, face viewing paradigms, combinations of script-driven imagery and face viewing, use of visual stimuli other than viewing faces, use of olfactory/taste paradigms (see Gottfried et al. Chapter 6, this volume), paradigms designed to assess moral and social function, and reward paradigms (see O'Doherty and Dolan Chapter 10, this volume).

Script-driven imagery paradigms are typically explicitly intended to produce (or reproduce) specific emotional states. Scripts can be standardized or individually tailored. In the case of individually tailored scripts, autobiographical events are often used to optimize salience and reliability of response. Before the imaging session, the subject (with the help of the investigators) composes autobiographical scripts associated with the most vivid experiences the subject can remember. For example, an autobiographical script of a sad experience may involve the recall of the death of a family member, while an autobiographical script of a happy experience may involve the recall of the birth of a child. The scripts are then typically recorded with audiotape and played back to subject during the imaging session. The subject is instructed to imagine himself/herself in the situation, as if it were actually happening and to maintain this state during the imaging session. Although autobiographical script-driven imagery paradigms produce particularly robust emotional responses due to the salience of the stimuli, the variance in the stimulus conditions due to incomplete/imperfect standardization represents a limitation. All script-driven imagery paradigms entail the production/recall of sensory images, and hence the stimuli are in some sense externally prompted and internally generated. Thus, whereas the brain activation patterns associated with script-driven imagery reflect production of the affective state and modulation of that state, the degree to which they reflect perception of an emotional stimulus must be qualified (i.e., the distinction between responding to internal versus external stimuli; see Reiman et al. 1997). Given that script-driven imagery

**Table 15.1** OFC findings in functional neuroimaging studies of emotion induction

| Authors | Method | Comparison | Directionality | Brain region | Spatial coordinates (x, y, z) |
|---|---|---|---|---|---|
| *Using script-driven imagery paradigms* | | | | | |
| Damasio et al. 2000 | PET-O15 | Sadness–neutral | Increase | Left OFC | −7, 17, −12; −1, 41, −16 |
| | | Happiness–neutral | Decrease | Right OFC | 21, 39, −14 |
| | | Anger–neutral | Decrease | Right OFC | 12, 38, −17 |
| | | Fear–neutral | Decrease | Right OFC | 18, 44, −14 |
| | | | Decrease | Left OFC | −4, 48, −17 |
| Dougherty et al. 1999 | PET-O15 | Anger–neutral | Increase | Left OFC | −36, 24, −16; −44, 18, −4 |
| Dougherty et al. 2004 | PET-O15 | Anger–neutral (controls) | Increase | Left OFC | −8, 62, −10 |
| Liotti et al. 2000 | PET-O15 | Anxiety–neutral | Increase | Left OFC | −24, 4, −14 |
| Markowitsch et al. 2003 | fMRI | Sad–nappy | Increase | Left OFC | |
| | | | Increase | Right OFC | |
| | | Happy–sad | Increase | Left OFC | |
| | | | Increase | Right OFC | |
| Pardo et al. 1993 | PET-O15 | Sad–neutral | Increase | Left OFC | −43, 33, −2; −27, 59, −4 |
| Shin et al. 2000 | PET-O15 | Guilt–neutral | Increase | Left OFC | −44, 16, −4 |
| *Using face viewing paradigms* | | | | | |
| Blair et al. 1999 | PET-O15 + sex discrimination | Anger–neutral | Increase | Right OFC | 40, 38, −12 |
| Gur et al. 2002 | fMRI + emotion and age discrimination | Anger–sad | Increase | Right OFC | 42, 42, −16 |
| | | Emotion discrimination–baseline | Increase | Left OFC | −48, 56, −12 |
| Lange et al. 2003 | fMRI + passive viewing, sex discrimination, and emotion discrimination | Age discrimination–emotion discrimination | Increase | Right OFC | 36, 40, −16 |
| | | Fear–neutral during gender decision task | Increase | Right OFC | 47, 23, −7 |
| | | Fear–neutral during emotion judgment task | Increase | Left OFC | −32, 26, −7 |

| | | | | | |
|---|---|---|---|---|---|
| McClure et al. 2004 | fMRI + threat rating | Anger–neutral (Adult F>Adult M) | Increase | Right OFC | 48, 32, −6 |
| | | Anger–fearful (Adult F>Adult M) | Increase | Right OFC | 46, 36, −4; 42, 32, 2; 52, 18, −8 |
| | | Anger–fixation (Adult F) | Increase | Right OFC | 32, 22, −4; 52, 18, −8; 36, 28, −8 |
| | | | Increase | Left OFC | −44, 18, −2; −50, 22, −2; −50, 28, 0 |
| | | Anger–fixation (Adult M) | Increase | Right OFC | 52, 22, −8; 32, 22, −4; 34, 28, −2 |
| | | | Increase | Left OFC | −50, 36, −6; −3, 22, 0 |
| | | Fearful–fixation (Adult M) | Increase | Right OFC | 52, 22, −8 |
| | | | Increase | Left OFC | −52, 36, −4 |
| | | Neutral–fixation (Adult M) | Increase | Right OFC | 38, 28, −4 |
| | | Angry–fixation (adolescent F) | Increase | Right OFC | 32, 20, −8; 32, 22, −4 |
| | | Angry–fixation (adolescent M) | Increase | Right OFC | 28, 22, −14; 34, 30, 4 |
| | | | Increase | Left OFC | −40, 18, −10 |
| | | Fear–fixation (adolescent F) | Increase | Left OFC | −34, 22, −4 |
| | | Neutral–fixation (adolescent F) | Increase | Right OFC | 24, 20, −16 |
| | | Subjective fear-nose width during fearful faces (adult>adolescents) | Increase | Right OFC | 40, 22, −18 |
| Monk et al. 2003 | fMRI + selective attention | Passive viewing fearful faces–passive viewing neutral faces (adolescents>adults) | Increase | Left OFC | −30, 26, −2; −24, 22, −18 |
| | | | Increase | Right OFC | 22, 22, −10; 22, 18, −14 |

*Using script-driven imagery and face viewing paradigms*

| | | | | | |
|---|---|---|---|---|---|
| George et al. 1995 | PET-O15 | Sad–neutral | Increase | Left OFC | −14, 18, −8 |
| George et al. 1996 | PET-O15 | Sad–happy (women) | Increase | Right OFC | 20, 44, −16 |
| Kimbrell et al. 1999 | PET O15 | Anger–neutral | Increase | Left OFC | −20, 34, −8 |
| | | Anxiety–neutral | Increase | Left OFC? | −33, 13, −12 |

*Using emotional visual stimuli other than faces*

| | | | | | |
|---|---|---|---|---|---|
| Nitschke et al. 2004 | fMRI + photograph | Own infant–unfamiliar infant | Increase | Left OFC | −40, 32, −13 |
| | | | Increase | Right OFC | 34, 31, −10 |
| | | | | | 44, 32, −15 |
| Northoff et al. 2000 | ROI-based fMRI/MEG + IAPS | Negative–neutral | Increase | Medial OFC | |
| | | Positive–neutral | Increase | Lateral OFC | |
| Ranote et al. 2004 | fMRI + video | Unknown infant–own infant | Increase | Bilateral OFC | |

Note that spatial coordinates are either in Talaraich space or Montreal Neurological Space, depending upon the conventions of the particular study cited.

paradigms fundamentally entail the production of affective states, it is not surprising that a high percentage of functional neuroimaging studies employing such paradigms (see Table 15.1) report OFC findings (Phan et al. 2002).

In contrast, the majority of functional neuroimaging studies that use a passive face viewing paradigm (Brieter et al. 1996; Isenberg et al. 1999; Phillips et al. 1998; Reiman et al. 1997; Taylor et al. 2000) report amygdala findings, but do not report OFC activation. This suggests that while rapidly repeating passive viewing of faces engages the amygdala in identifying the emotional significance of stimuli, it may not lead to the production of an emotional state. For the purposes of this chapter, we reviewed functional neuroimaging studies that utilized face-viewing paradigms that reported OFC findings (see Table 15.1). We excluded studies that utilized face viewing paradigms involving conditioning (Armony and Dolan 2002; Morris and Dolan 2004), pharmacologic challenge (Bentley et al. 2003), and more complex cognitive tasks (Kringlebach et al. 2003) as these issues are covered in detail in other chapters in this volume. Interestingly, when subjects were instructed to perform simple cognitive tasks, such as sex discrimination (Blair et al. 1999; Lange et al. 2003), age discrimination (Gur et al. 2002), emotion discrimination (Gur et al. 2002; Lange et al. 2003), rating threat (McClure et al. 2004), or assessing nose width (Monk et al. 2003) while viewing faces, OFC activation was more likely to be reported as was attenuation of the amygdala response. Indeed, some of these studies included a passive viewing component where amygdala activation was reported, but OFC activation was not reported in adult subjects (Lange et al. 2003; Monk et al. 2003). These data suggest an attenuation of the limbic (i.e., amygdala) response to these emotional stimuli by attentional and cognitive demands.

It is interesting to note that when one compares the loci of OFC activation seen in the studies listed in Table 15.1, with some exceptions, the OFC activations associated with script-driven imagery paradigms are more medial whereas the OFC activations associated with face viewing paradigms are more lateral (see Fig. 15.1). Lesion studies have suggested that lateral territories of the OFC may be central to social cognitive processes (Blair and Cipolotti 2000; Hornak et al. 1996; Stone et al. 1998), while neuroimaging studies have shown activation of the lateral OFC in response to violation of social norms (Berthoz et al. 2002). In contrast, a recent meta-analysis (Phan et al. 2002) suggests that emotion activation studies that involve a cognitive component tend to differentially activate the ACC and medial PFC. In addition, there is a preponderance of left-sided OFC activations in Table 15.1. This is somewhat inconsistent with emotion theories of laterality that suggest right hemisphere dominance for emotional processing (Borod 1993; Heilman et al. 1993) or theories that suggest that the left hemisphere is associated with positive emotions and approach while the right hemisphere is associated with negative emotions and withdrawal (Davidson et al. 1990; Silberman and Weingartner, 1986).

Two studies, which included both healthy adults and healthy adolescents, suggest developmental differences in the relationship between amygdala and OFC functions. Monk et al. (2003) conducted an fMRI study with seventeen mixed-gender healthy adult volunteers and seventeen mixed-gender healthy adolescent volunteers. The paradigm

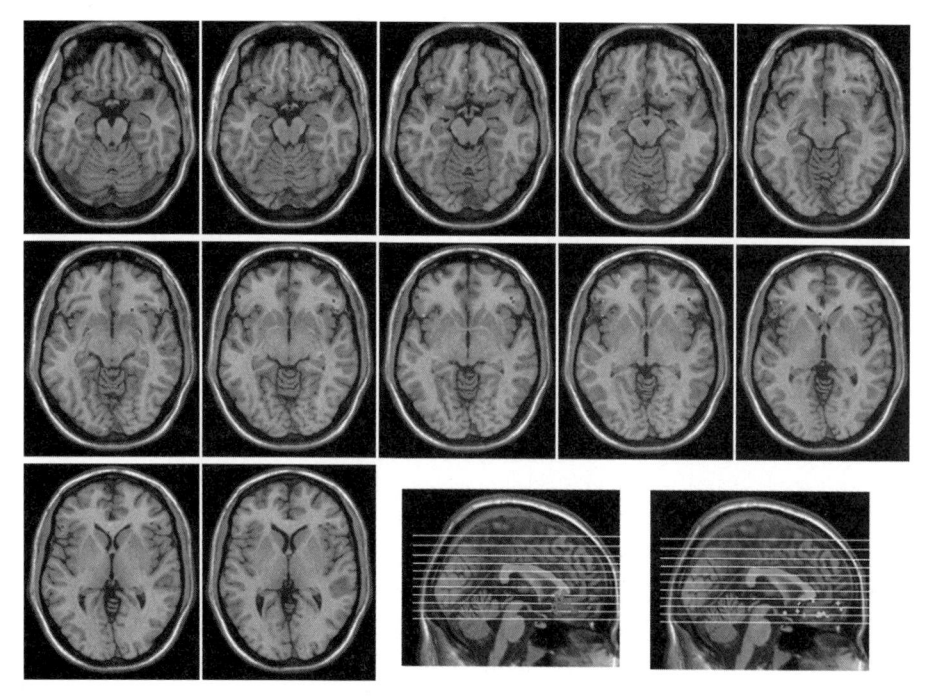

**Fig. 15.1** (Also see Color Plate 25.) Orbitofrontal cortex activations in neuroimaging studies of emotion. Loci of OFC activations in neuroimaging studies utilizing script-driven imagery are shown in green (white in above figure). Loci of OFC activations utilizing face viewing paradigms are shown in purple (black in above figure). Note that each colored circle represents a locus of activation from the studies listed in Table 15.1. The circles are overlaid onto a standardized magnetic resonance image.

involved passive viewing or selective attention during the viewing of emotional (angry, fearful, happy, neutral) faces. While viewing the faces, the subjects were instructed to: (A) focus their attention on their subjective response to the face; or (B) focus their attention on a physical feature of the face; or (C) they were allowed to passively view the face (unconstrained). During the viewing of fearful faces, the contrast of condition A minus condition B revealed significantly greater activation of the right OFC in the adults when compared to the adolescents. However, the contrast of condition C minus condition B revealed significantly greater activation of the bilateral OFC in adolescents when compared to adults. Thus, compared to adolescents, adults demonstrate greater activation of the OFC during attentional tasks while adolescents, compared to adults, demonstrate greater activation of the OFC during exposure to emotional stimuli. McClure *et al.* (2004) used fMRI to study seventeen healthy mixed-gender adolescent volunteers and seventeen healthy mixed-gender adult volunteers while they viewed emotional faces and rated the degree of threat conveyed by the faces. The OFC was activated across a variety of conditions in all groups (see Table 15.1), but the overarching finding was that the OFC was activated to a greater degree when processing unambiguous versus ambiguous threat stimuli. Furthermore, in adults, the women showed greater activation of the OFC than men while evaluating the degree of threat conveyed by emotional faces, whereas in the adolescents this gender difference was not present. In sum, initial neuroimaging studies on emotion processing indicate that age and sex are potentially important variables. However, further research is indicated to better establish the nature of these age and sex effects.

The vast majority of functional neuroimaging studies of emotion that utilize visual stimuli have entailed viewing faces with differing emotional expressions. However, a small number of studies have used other visual stimuli that are thought to produce emotional responses. Nitschke *et al.* (2004) conducted an fMRI study in which six female healthy volunteers viewed pictures of their recently born infant, an unfamiliar infant, or an anonymous adult. In the comparison of viewing a picture of their own infant minus viewing a picture of an unfamiliar infant, the mothers exhibited activation of bilateral OFC, and the OFC activation was directly correlated with positive mood ratings while viewing the picture of their own infant. Ranoate *et al.* (2004) conducted a similar fMRI study, except video clips were used instead of still pictures. In contrast to the Nitschke *et al.* study, OFC activation in the mothers was seen in the comparison of viewing a film clip of an unknown infant minus viewing a film clip of their own infant.

Lastly, the International Affective Picture System (IAPS) is a commonly used method of inducing emotion. This series of pictures includes positively valenced, negatively valenced, and emotionally neutral stimuli; valence and arousal ratings have been well-established. Most studies using the IAPS in conjunction with passive viewing paradigms have not shown engagement of the OFC. As is the case with face viewing paradigms, it is likely that this is because these pictures are typically presented within a passive viewing paradigm, and more frequent OFC activation would emerge if subjects were actually asked to make online ratings of the stimuli. Nonetheless, Northoff *et al.* (2000) performed a study that included ten mixed-gender healthy volunteers and combined fMRI with magnetoencephalography (MEG) measurements while the subjects viewed IAPS pictures that did result in OFC activation. MEG allows for much greater temporal resolution (in the millisecond range) than fMRI. The authors found that exposure to the negatively valenced pictures was associated with strong and early medial OFC activation while exposure to the positively valenced pictures was associated with later and weaker lateral OFC activation.

## Regulation of the affective state

Once the significance of an emotional stimulus has been identified and an affective state has been produced, the regulation of this affective state involves neocortical brain regions. The dorsolateral PFC (DLPFC) is activated during cognitive tasks and this activation is generally associated with reduced activity in ventral PFC regions (for a review see Drevets and Raichle 1998). This reciprocal relationship between dorsal and ventral PFC regions allows for dorsally mediated cognitive control of ventrally mediated affective states. Therefore, one role of the OFC in the regulation of affective states may involve the OFC responding to dorsally mediated cognitive control by dampening the autonomic responses that were generated soon after the identification of the emotional significance of the initial stimulus.

## 15.4 Factors in the role of the OFC in emotional experience

### Amygdala–OFC relationship

The amygdala and OFC share extensive connections (see Price Chapter 3, this volume, Barbas and Zikopoulos Chapter 4, this volume). Shortly after the amygdala responds to emotional stimuli, OFC activation follows (Kawasaki *et al.* 2001; Krolak-Salmon, *et al.* 2004; Streit *et al.* 1999 and 2001). In this context, both the amygdala and OFC are central to the generation of autonomic responses to these stimuli (Bechara *et al.* 1999 and 2000; Cechetto and Saper 1990; Nagai *et al.* 2004; Tranel and Damasio 1994). According to the somatic marker hypothesis, this autonomic state alerts the body to the response to emotional stimuli and results in the conscious production of the affective state (Damasio 1996).

Because the amygdala and OFC share extensive connections, and because the relationship between these structures is so crucial for emotional processing, it follows that feedback between these structures must occur. There is, in fact, evidence that these two structures are mutually inhibitory in that activity in one structure inhibits activity in the other structure (Garcia *et al.* 1999; Hariri *et al.* 2000; Milad and Quirk 2002). Thus, following activation of the amygdala during the identification of the emotional significance of a stimulus, the OFC is activated. This OFC activation then inhibits further amygdala activation. This likely serves at least two functions. First, it returns the amygdala to a baseline 'ready' state so that it is available to serve its vigilance function. Second, this shutdown of the amygdala allows the affective processing stream to continue as the OFC participates in generating an autonomic response to the stimulus.

An amygdalocentric model of depression posits that the amygdala is chronically hyperactive (Drevets *et al.* 1992) or hyperresponsive (Sheline *et al.* 2001) resulting in deactivation of PFC regions, including the OFC (Dougherty and Rauch 1997; Whalen *et al.* 2002). Recent work in healthy volunteers (Dougherty *et al.* 2004) suggests that, during emotion induction, an inverse relationship between the amygdala and OFC is observed. However, in subjects with depression this relationship is not present, and in subjects with depression plus anger attacks the relationship is reversed (Dougherty *et al.* 2004).

### OFC–dorsal PFC relationship and its role in regulation of the affective state

As described above, dorsal PFC and ACC regions are central to regulating affective states. These regions are activated during cognitive tasks, and a reciprocal relationship exists between ventral and dorsal PFC and ACC regions, which allows dorsal PFC and ACC regions to regulate the affective state, that is mediated by the ventral PFC and ACC (Drevets and Raichle 1998). A growing number of functional neuroimaging studies have assessed mood regulation and the interaction between mood and cognition. These studies are informative regarding the relationship between the OFC and dorsal PFC.

The majority of functional neuroimaging studies of mood regulation demonstrated activation of dorsal PFC regions during mood regulation (Beauregard et al. 2001; Levesque et al. 2003 and 2004; Ochsner et al. 2002; Pietrini et al. 2000), and one demonstrated corresponding decreases in activity in the amygdala and OFC (Ochsner et al. 2002). However, some of the studies also resulted in activation of the OFC (Levesque 2003 and 2004; Pietrini et al. 2000). One interesting study by Pietrini and colleagues (2000) demonstrated a decrease in regional cerebral blood flow (rCBF) in the OFC when the study participants imagined an unrestrained aggressive response to a slight, but an increase in OFC rCBF when they imagined restraining themselves following exposure to the slight. In this case, the authors postulated that the OFC was activated in order to inhibit an aggressive behavioral response following anger induction.

Other functional neuroimaging studies have examined the relationship between mood and cognition (Baker et al. 1997; Keightley et al. 2003; Maratos et al. 2001; Smith et al. 2004). The study by Baker et al. (1997) revealed OFC activation during both elated and depressed mood induction and dorsal PFC activation during a verbal fluency task. When mood induction and the verbal fluency task were studied simultaneously, the OFC activation was no longer present and the activation in the dorsal PFC was attenuated. Keightley et al. (2003) also reported decreased activity in the amygdala and the OFC during exposure to emotional stimuli while simultaneously performing cognitive tasks. In contrast to having study participants perform a cognitive task during emotion induction, other studies have focused on cognitive tasks involving emotionally valenced material. For example, two studies (Maratos et al. 2001; Smith et al. 2004) assessed episodic memory for emotionally valenced words and pictures, respectively. Maratos et al. (2001) describe an increase in OFC activity only during positive word retrieval, while Smith et al. (2004) reported an increase in OFC activity during retrieval of any emotionally valenced pictures. These elegantly designed studies demonstrate the inhibitory effects of cognition on brain regions associated with emotional processing.

## OFC and autonomic regulation

A number of brain regions, including the amygdala, OFC, ACC, and insula, are involved in the generation of autonomic responses (Cechetto and Saper 1990; Williams et al. 2001 Asahina et al. 2003). Lesions encompassing the OFC lead to decreased skin conductance responses (SCRs) (Bechara et al. 1999; Tranel and Damasio 1994) while functional neuroimaging studies have repeatedly implicated the OFC in the generation of SCRs (Frederickson et al. 1998; Critchley et al. 2000; Williams et al. 2001; Patterson et al. 2002; Nagai et al. 2004). According to the 'somatic marker' hypothesis, feedback to the brain from the body via autonomic responses allows the brain to respond to stimuli in an adaptive manner (Damasio 1994). In this way, the OFC plays a central role in emotional processing. The OFC, along with other brain regions, serves to transmit the emotional significance of a stimulus (initially detected by the amygdala) to the organism via autonomic responses. These autonomic responses are central to the production of an affective state. In turn, the brain responds to the autonomic state in a regulatory manner that involves more dorsal neocortical brain regions.

## 15.5. Conclusion

In this chapter, we have reviewed the processes underlying emotional perception. These include: 1) the identification of the emotional significance of a stimulus; 2) the production of an affective state; and 3) the regulation of the affective state. A large body of literature, including functional neuroimaging studies in healthy volunteers, suggests that the OFC is crucial for the second step outlined above, the production of an affective state. Lastly, we reviewed the mechanisms by which the OFC likely contributes to the production of an affective state. These mechanisms include the close relationship between the amygdala and the OFC, the reciprocal relationship between ventral limbic and paralimbic structures (including the OFC) and dorsal paralimbic and neocortical brain regions, and the role of the OFC in the generation of autonomic states.

## References

Amaral, D. G. and Price, J. L. (1984) Amygdalo-cortical projections in the monkey (Macaca fascicularis). *J Comp Neurol* **230**:465–496.

Amaral, D. G. and Insuasti, R. (1992) Retrograde transport of D-[H]-aspartate injected into the monkey amygdaloid complex. *Exp Brain Res* **88**:375–388.

Arashina, M. Suzuki, A., Mori, M., Kanesaka, T., and Hattori, T. (2003) Emotional sweating response in a patient with bilateral amygdala damage. *Int J Psychophysiol* **47**:87–93.

Armony, J. L. and Dolan, R. J. (2002) Modulation of spatial attention by fear-conditioned stimuli: an event-related fMRI study. *Neuropsychologia* **40**:817–826.

Baker, S. C. Frith, C. D., and Dolan, R. J. (1997) The interaction between mood and cognitive function studied with PET. *Psychol Med* **27**:565–578.

Beauregard, M., Levesque, J., and Bourgouin, P. (2001) Neural correlates of conscious self-regulation of emotion. *J Neurosci* **21**:RC165.

Bechara, A., Damasio, H., Damasio, A. R., and Lee, G. P. (1999) Different contributions of the human amygdala and ventromedial prefrontal cortex to decision-making. *J Neurosci* **19**:5473–5481.

Bechara, A., Damasio, H., and Damasio, A. R. (2000) Emotion, decision-making and the orbitofrontal cortex. *Cereb Cortex* **10**:295–307.

Bentley, P., Vuilleumier, P., Thiel, C. M., Driver, J., and Dolan, R. J. (2003) Cholinergic enhancement modulates neural correlates of selective attention and emotional processing. *Neuroimage* **20**(1):58–70.

Berthoz, S., Armony, J., Blair, R. J. R., and Dolan R. (2002) Neural correlates of violation of social norms and embarrassment. *Brain* **125**:1696–1708.

Blair, R. J. R., Morris, J. S., Frith, C. D., Perrett, D. I., and Dolan, R. J. (1999) Dissociable neural responses to facial expressions of sadness and anger. *Brain* **122**:883–893.

Blair, R. J. R., and Cipolotti, L. (2000) Impaired social response reversal: a case of "acquired sociopathy". *Brain* **123**:1122–1141.

Borod, J. C. (1993) Cerebral mechanisms underlying facial, prosodic and lexical emotional expression: a review of neuropsychological studies and methodological issues. *Neuropsychology* **7**:445–463.

Breiter, H. C., Etcoff, N. L., Whalen, P. J., Kennedy, W. A., Rauch, S. L., Buckner, R. L. *et al*. (1996) Response and habituation of the human amygdala during visual processing of facial expression. *Neuron* **17**:875–887.

Bush, G., Luu, P., and Posner, M. I. (2000) Cognitive and emotional influences in anterior cingulate cortex. *Trends Cogn Sci* **4**:215–222.

Cannon, W. B. (1929) *Bodily changes in pain, hunger, fear, and rage. Volume 2*. New York: Appleton.

Carmichael, S. T. and Price J. L. (1995) Limbic connections of the orbital and medial prefrontal cortex in macaque monkeys. *J Comp Neurol* **363**:615–641.

Cechetto, D. R., and Saper, C. B. (1990) Role of the cerebral cortex in autonomic function. In: Loewy, A. D. and Spyer, K. M. (eds), *Central Regulation of Autonomic Functions*. Oxford: Oxford University Press, pp. 208–223.

Chiba, T., Kayahara, T., and Nakano, K. (2001) Efferent projections of infralimbic and prelimbic areas of the medial prefrontal cortex in the Japanese monkey, Macaca fuscata. *Brain Res* **888**:83–101.

Critchley, H. D., Elliott, R., Mathias, C. J., and Dolan, R. J. (2000) Neural activity relating to generation and representation of galvanic skin conductance responses: a functional magnetic resonance imaging study. *J Neurosci* **20**:3033–3040.

Damasio, A. R. (1996) The somatic marker hypothesis and possible functions of the prefrontal cortex. *Philos Trans R Soc Lond B Biol Sci* **351**:1413–1420.

Damasio, A. R., Grabowski, T. J., Bechara, A., Damasio, H., Ponto, L. L. B., Parvizi, J., *et al.* (2000) Subcortical and cortical brain activity during the feeling of self-generated emotions. *Nat Neurosci* **3**:1049–1056.

Davidson, R. J., Ekman, P., Saron, C. D., and Senulis J. A. (1990) Approach-withdrawal and cerebral asymmetry: emotional expression and brain physiology. *I J Pers Soc Psychol* **58**:330–341.

Davis, M. and Whalen, P. J. (2001) The amygdala: vigilance and emotion. *Mol Psychiatry* **6**:3–34.

Devinsky, O., Morrell, M.J., and Vogt, B.A. *et al.* (1995) Contributions of the anterior cingulate cortex to behavior. *Brain* **118**:279–306.

Dougherty, D., and Rauch S. L. (1997) Neuroimaging and neurobiological models of depression. *Harvard Rev Psychiatry* **5**:138–159.

Dougherty, D. D., Shin, L. M., Alpert, N. M., Pitman, R. K., Orr, S. P., Lasko, M., *et al.* (1999) Anger in healthy men: a PET study using script-driven imagery. *Biol Psychiatry* **46**:466–472.

Dougherty, D. D., Rauch, S. L., Deckersbach, T., Marci, C., Loh, R., Shin, L. M., *et al.* (2004) Ventromedial prefrontal cortex and amygdala dysfunction during an anger induction positron emission tomography study in patients wit major depressive disorder with anger attacks. *Arch Gen Psychiatry* **61**:795–804.

Drevets, W. C. Videen, T. O., Price, J.L., Preskorn, S. H., Carmichael, S. T., and Raichle, M. E. (1992) A functional anatomical study of unipolar depression. *J Neurosci* **12**:3628–3641.

Drevets, W. C., and Raichle M. E. (1998) Reciprocal suppression of regional cerebral blood flow during emotional versus higher cognitive processes: implications for interactions between emotion and cognition. *Cognition and Emotion* **12**:353–385.

Frederickson, M., Furmark, T., Olsson, M. T., Fischer, H., Andersson, J., and Langstrom B. (1998) Functional neuroanatomical correlates of electrodermal activity: a positron emission tomographic study. *Psychophysiology* **35**:179–185.

Garcia, R., Vouimba, R.M., Baudry, M., and Thompson, R. F. (1999) The amygdala modulates prefrontal cortex activity relative to conditioned fear. *Nature* **402**:294–296.

George, M. S., Ketter, T. A., Parekh, P. I., Horwitz, B., Herscovitch, P., and Post, R. M. (1995) Brain activity during transient sadness and happiness in healthy women. *Am J Psychiatry* **152**:341–351.

George, M. S., Ketter, T. A., Parekh, P. I., Herscovitch, P., and Post R. M. (1996) Gender differences in regional cerebral blood flow during transient self-induced sadness or happiness. *Biol Psychiatry* **40**:859–871.

Gur, R. C., Schroeder, L., Turner, T., McGrath, C., Chan, R. M., Turetsky, B. I., *et al.* (2002) Brain activation during facial emotional processing. *Neuroimage* **16**:651–662.

Hariri, A. R., Bookheimer, S. Y., and Mazziotta, J. C. (2000) Modulating emotional responses: effects of a neocortical network on the limbic system. *Neuroreport* **11**:43–48.

Heilman, K. M., Bowers, D., and Valenstein, E. (1993). Emotional disorders associated with neurological diseases. In: Heilman, K. M. and Valenstein, E. (eds), *Clinical Neuropsychology*, pp. 461–498. New York: Oxford University Press.

Hornak, J., Rolls, E. T., and Wade D. (1996) Face and voice expression identification in patients with emotional and behavioral changes following ventral frontal damage. *Neuropsychologica* **34**:247–261.

Isenberg, N., Silbersweig, D., Engelien, A., Emmerich, S., Malavade, K., Beattie, B. *et al*. (1999) Linguistic threat activates the human amygdala. *Proc Natl Acad Sci* USA **96**:10456–10459.

James, W. (1884) What is an emotion? *Mind* **9**:188–205.

Kawasaki, H., Kaufman, O., Damasio, H., Damasio, A. R., Granner, M., Bakken, H., et al. (2001) Single-neuron responses to emotional visual stimuli recorded in human ventral prefrontal cortex. *Nat Neurosci* **4**:15–15.

Keightly, M. L. Winocur, G., Graham, S. J., Mayberg, H. S., Hevenor, S. J., and Grady, C. L. (2003) An fMRI study investigating cognitive modulation of brain regions associated with emotional processing of visual stimuli. *Neuropsychologia* **41**:585–596.

Kimbrell, T. A., George, M. S., Parekh, P. I., Ketter, T. A., Podell, D. M., Danielson, A. L., *et al*. (1999) Regional brain activity during transient self-induced anxiety and anger in healthy adults. *Biol Psychiatry* **46**:454–465.

Kringelbach, M. L., and Rolls, E. T. (2003) Neural correlates of rapid reversal learning in a simple model of human social interaction. *Neuroimage* **20(2)**:1371–1383.

Krolak-Salmon, P., Henaff, M.A., Vigfhetto, A., Bertrand, O., and Mauguiere, F. (2004) Early amygdala reaction to fear spreading in occipital, temporal, and frontal cortex: a depth electrode ERP study in human. *Neuron* **42**:665–676.

Lang, P. J. (1995) The emotion probe. Studies of motivation and attention. *American Psychologist* **50**:372–385.

Lange, K., Williams, L. M., Young, A. W., Bullmore, E. T., Brammer, M. J., Williams, S. C. R., *et al*. (2003) Task instructions modulate neural responses to fearful facial expressions. *Biol Psychiatry* **53**:226–232.

LeDoux J. E. (1996) *The Emotional Brain*. New York: Touchstone.

Levesque, J., Eugene, F., Joanette, Y., Paquette, V., Mensour, B., Beaudoin, G., *et al*. (2003) Neural circuitry underlying voluntarily self-regulation of sadness. *Biol Psychiatry* **53**:502–510.

Levesque, J., Joanette, Y., Mensour, B., Beaudoin, G., Leroux, J. M., Bourgouin P., *et al*. (2004) Neural basis of emotional self-regulation in childhood. *Neuroscience* **129(2)**:361–369.

Liotti, M., Mayberg, H. S., Brannan, S. K., McGinnis, S., Jerabek, P., and Fox, P. T. (2000) Differential limbic-cortical correlates of sadness and anxiety in healthy subjects: implications for affective disorders. *Biol Psychiatry* **48**:30–42.

Markowitsch, H. J., Vandekerckhovel, M. M., Lanfermann, H., and Russ, M. O. (2003) Engagement of lateral and medial prefrontal areas in the ecphory of sad and happy autobiographical memories. *Cortex* **39(4–5)**:643–665.

Maratos, E. J., Dolan, R. J., Morris, J. S., Henson, R. N. A., and Rugg, M. D. (2001) Neural activity associated with episodic memory for emotional context. *Neuropsychologia* **39**:910–920.

McClure, E. B. Monk, C. S., Nelson, E. E., Zarahn E., Leibenluft E., Bilder R. M., *et al*. (2004) A developmental examination of gender differences in brain engagement during evaluation of threat. *Biol Psychiatry* **55(11)**:1047–1055.

Milad, M. R. and Quirk, G. J. (2002). Neurons in medial prefrontal cortex signal memory for fear extinction. *Nature* **420**:70–74.

Monk, C. S., McClure, E. B., Nelson, E. E., Zarahn, E., Bilder, R. M., Leibenluft, E., *et al*. (2003) Adolescent immaturity in attention-related brain engagement to emotional facial expressions. *Neuroimage* **20(1)**:420–428.

Morris, J. S., Frith, C. D., Perrett, D. I., Rowland, D., Young, A. W., Calder, A. J. *et al*. (1996) A differential neural response in the human amygdala to fearful and happy facial expressions. *Nature* **383**:812–815.

Nagai, Y., Critchley, H. D., Featherstone, E., Trimble, M. R., and Dolan, R. J. (2004) Activity in ventromedial prefrontal cortex covaries with sympathetic skin conductance level: a physiological account of a 'default mode' of brain function. *Neuroimage* **22**:243–251.

Nitschke, J. B., Nelson, E. E., Rusch, B. D., Fox, A. S., Oakes, T. R., and Davidson, R. J. (2004) Orbitofrontal cortex tracks positive mood in mothers viewing pictures of their newborn infants. *Neuroimage* **21**(2):583–592.

Northoff, G., Richter, A., Gessner, M., Schlagen hauf, F., Fell, J., Baumgart, F., *et al*. (2000) Functional dissociation between medial and lateral prefrontal cortical spatiotemporal activation in negative and positive emotions: a combined fMRI/MEG study. *Cereb Cortex* **10**:93–107.

Oschner, K. N., Bunge, S. A., Gross, J. J., and Gabrieli, J. D. E. (2002) Rethinking feelings: an fMRI study of the cognitive regulation of emotion. *J Cogn Neurosci* **14**:1215–1229.

Pardo, J. V., Pardo, P. J., and Raichle, M. E. (1993) Neural correlates of self-induced dysphoria. *Am J Psychiatry* **150**:713–719.

Patterson, J. C., Ungerleider, L. G., and Bandettini, P. A. (2002) Task-independent functional brain activity correlation with skin conductance changes: an fMRI study. *Neuroimage* **17**:1797–1806.

Phan, K. L., Wager, T., Taylor, S. F., and Liberzon, I. (2002) Functional neuroanatomy of emotion: a meta-analysis of emotion activation studies in PET and fMRI. *Neuroimage* **16**:331–348.

Phillips, M. L., Young, A. W., Senior, C., Calder, A. J., Perrett, D., Brammer, M. *et al*. (1997) A specific neural substrate for perception of facial expressions of disgust. *Nature* **389**:495–498.

Phillips, M. L., Young, A. W., Scott, S. K., Calder, A. J., Andrew, C., and Giampietro, V. *et al*. (1998) Neural responses to facial and vocal expressions of fear and disgust. Proc R Soc Lond B *Biol Sci* **265**:1809–1817.

Phillips, M. L., Medford, N., Young, A. W., Williams, L., Williams, S. C. R., Bullmore E. T. *et al*. (2001) Time course of left and right amygdalar responses to fearful facial expressions. *Hum Brain Mapp* **12**:193–202.

Phillips, M. L., Drevets, W. C., Rauch, S. L., and Lane, R. (2003) Neurobiology of emotion perception I: the neural basis of normal emotion perception. *Biol Psychiatry* **54**:504–514.

Pietrini, P., Guazzelli, M., Basso, G., Jaffe, K., and Grafman, J. (2000) Neural correlates of imaginal aggressive behavior assessed by positron emission tomography in healthy subjects. *Am J Psychiatry* **157**:1772–1781.

Price, J. L. (2003) Comparative aspects of amygdala connectivity. *Ann NY Acad Sci* **985**:50–58.

Ranote, S., Elliott, R., Abel, K. M., Mitchell, R., Deakin, J. F., and Appleby, L. (2004) The neural basis of maternal responsiveness to infants: an fMRI study. *Neuroreport* **15**(11):1825–1829.

Rauch, S. L., Shin, L. M., and Wright, C. I. (2003) Neuroimaging studies of amygdala function in anxiety disorders. *Ann N Y Acad Sci* **985**:389–410.

Reiman, E. M., Lane, R. D., Ahern, G. L., and Schwartz, G. E. (1997) Neuroanatomical correlates of externally and internally generated human emotion. *Am J Psychiatry* **154**:918–925.

Schienle, A., Schafer, A., Stark, R., Walter, B., and Vaitl, D. (2005) Gender differences in the processing of disgust- and fear-inducing pictures: an fMRI study. *Neuroreport* **16**:277–280.

Sheline, Y. I., Brach, D. M., Donnelly, J. M., Ollinger, J. M., Snyder, A. Z., and Mintun, M. A. (2001) Increased amygdala response to masked emotional faces in depressed subjects resolves with antidepressant treatment: an fMRI study. *Biol Psychiatry* **50**:651–658.

Shin, L. M., Dougherty, D. D., Orr, S. P., Pitman, R. K., Lasko, M., Macklin, M. L., *et al*. (2000) Activation of anterior paralimbic structures during guilt-related script-driven imagery. *Biol Psychiatry* **48**:43–50.

Silberman, E. K., and Weingartner, H. (1986) Hemispheric lateralization of functions related to emotion. *Brain Cogn* **5**:322–353.

Smith, A. P., Henson, R. N., Dolan, R. J., and Rugg, M. D. (2004) fMRI correlates of the episodic retrieval of emotional contexts. *Neuroimage* **22**(2):868–878.

Stefanacci, L., and Amaral, D. G. (2002) Some observations on cortical inputs to the macaque monkey amygdala: an anterograde tracing study. *J Comp Neurol* **451**:301–323.

Stone, V. E., Baron-Cohen, S., and Knight, R. T. (1998) Frontal lobe contributions to the theory of mind. *J Cogn Neurosci* **10**:640–656.

Streit, M., Ioannides, A. A., Liu, L., Wolwer, W., Dammers, J., Gross, J., et al. (1999) Neurophysiological correlates of the recognition of facial expressions of emotion as revealed by magnetoencephalography. *Brain Res Cogn Brain Res* **7**:481–491.

Streit, M., Ioannides, A., Sinnerman, T., Wolwer, W., Dammers, J., Ailles, K., et al. (2001) Disturbed facial affect recognition in patients with schizophrenia associated with hypoactivity in distributed brain regions: a magnetoencephalography study. *Am J Psychiatry* **158**:1429–1436.

Taylor, S. F., Liberzon, I., and Koeppe, R. A. (2000) The effect of graded aversive stimuli on limbic and visual activation. *Neuropsychology* **38**:1415–1425.

Tranel, D. and Damasio, H. (1994) Neuroanatomical correlates of electrodermal skin conductance responses. *Psychophysiology* **31**:427–438.

Whalen, P. J., Shin, L. M., Somerville, L. H., McLean, A. A., and Kim, H. (2002) Functional neuroimaging studies of the amygdala in depression. *Sem Clin Neuropsychiatry* **7**:234–242.

Williams, L. M., Phillips, M. L., Brammer, M. J., Skerrett, D., Lagopoulos, J., Rennie, C., et al. (2001) Arousal dissociates amygdala and hippocampal fear responses: evidence from simultaneous fMRI and skin conductance recording. *Neuroimage* **14**:1070–1079.

Wright, C. I., Fischer, H., Whalen, P. J., McInerney, S. C., Shin, L. M., and Rauch, S. L. (2001) Differential prefrontal cortex and amygdala habituation to repeatedly presented emotional stimuli. *Neuroreport* **12**:379–383.

Zald, D. H. (2003) The human amygdala and the emotional evaluation of sensory stimuli. *Brain* Res Rev **41**:88–123.

# Chapter 16

# Neurochemical modulation of orbitofrontal cortex function

Trevor W. Robbins, Luke Clark, Hannah Clarke, and Angela C. Roberts

## 16.1. Introduction

Evidence from neuroanatomy, electrophysiological recording, neuropsychology (in both animals and humans) and functional neuroimaging encourages the view that the orbital and medial prefrontal cortex (PFC) is an anatomical complex with radically different functions from other portions of the frontal lobe, such as the dorsolateral (DL)-PFC, the frontal pole and the anterior cingulate cortex, whether in the primate or the rodent brain. In particular, the association of this complex with the regulation of emotional behavior has been emphasized, deriving in part from its connectivity with 'limbic' structures, such as the amygdala and hippocampus, as well as the ventral striatum, and also the descending innervation of emotional 'executive' structures within the hypothalamus and brain stem (Ongur and Price 2000). In this chapter, following Ongur and Price (2000), we will sometimes refer to the orbitofrontal cortex (OFC) and medial (m-) PFC together as the orbitomedial cortex (OM-PFC).

Damage to the OM-PFC complex (mainly to the ventromedial prefrontal cortex and OFC) produces impairments in decision-making cognition that have been related to failures to integrate visceral and cognitive processing (Damasio *et al.* 1990). These decision-making deficits have been hypothesized to depend on 'somatic markers', as indexed by autonomic variables such as the galvanic skin response (Bechara *et al.* 1994; Damasio 1994). The source of such signals has been debated, and may depend on inputs to the OM-PFC from the somatosensory cortex (Damasio 1994), However, another, less specific source of such somatic signals may in fact be the ascending monoaminergic systems, which have been linked to the regulation of arousal, mood, stress and reward-related processing.

The source of the projections of the monoamine and cholinergic systems to the PFC has been known for some time. The systems arise at different levels in the neuraxis, including from the hind-brain, mid-brain and basal forebrain. The main noradrenergic innervation of the OM-PFC in the rat originates in the locus coeruleus of the dorsal pons (Holets 1990). The innervation of the rat and the primate PFC by ascending serotoninergic neurons derives mainly from cells in the mesencephalic dorsal raphé nuclei (Azmitia 1995; Porrino and Goldman-Rakic 1982). The main DA projection to the OM-PFC arises from the mid-brain ventral tegmental area (Porrino and Goldman-Rakic 1982).

Bentivoglio and Morelli (2005) and Hurd and Hall (2005) provide up-to-date and complementary reviews of the detailed anatomical findings concerning cortical dopamine projections and receptor distributions, taking species differences into account. Finally, the nucleus of Meynert (nucleus magnocellularis in rats) of the basal forebrain is the main source of the ascending cholinergic projections to the OM-PFC (rats: Bigl *et al.* 1982; Mesulam *et al.* 1983; humans: Mesulam and Geula 1992).

The neurochemical influences of these systems have been relatively ignored in the context of OM-PFC function. However, recent evidence from neuropsychiatric disorders has resulted in them gaining more prominence. For instance, depression and obsessive-compulsive disorder are two psychiatric conditions associated with structural and functional abnormalities in the OM-PFC region, and these conditions are successfully treated with serotoninergic drugs such as the selective serotonin reuptake inhibitors (SSRIs) (Mayberg 1992, 2003; Griest and Jefferson 1998; Brody *et al.* 1999; Saxena and Rauch 2000; Lucey 2001; Drevets 2001). In addition, neuroimaging and postmortem studies of human drug abusers have indicated changes in functioning of the OM-PFC (Volkow and Fowler 2000). For example, methamphetamine abusers have been shown to have losses of striatal dopamine (DA) transporters but also reductions in serotonin (5-hydroxytryptamine, 5-HT) metabolites in the OFC (Wilson *et al.* 1996). In chronic cocaine abusers, changes in striatal raclopride binding (an index of dopamine D2 receptor availability) are correlated with changes in regional cerebral metabolism of the OFC (Volkow and Fowler 2000). In nonhuman primates, it has also been shown that mesocortical DA neurons projecting to structures within the OFC signal reward-related information including the omission of expected reward (the error-prediction signal), which affects the processing of information within this structure, such as preference among food objects (Hollerman *et al.* 2000).

The OFC has also been linked with the regulation of the cholinergic system, partly via its extensive back projections to the source of origin of these systems in the basal forebrain (Mesulam and Mufson 1984; Gaykema *et al.* 1991; Zaborszky *et al.* 1997). The monoamine systems similarly receive inputs from the OM-PFC, although most of the relevant evidence centres on m-PFC structures in the rat, such as the prelimbic (PL) and the infralimbic (IL) prefrontal cortex which may be homologous to the OM-PFC in primates (Vertes 2004). Using a combination of anatomical and electrophysiological techniques in rats, descending projections have been shown for noradrenaline (NA) (Jodo *et al.* 1998), 5-HT (Hajos *et al.* 1998; Celada *et al.* 2001) and DA (Sesack *et al.* 1989). Arnsten and Goldman-Rakic (1984) also identified descending projections from the PFC to the vicinity of the locus coeruleus and raphé nuclei in monkeys. This chapter will review evidence indicating this close association of the OM-PFC with the ascending neuromodulatory systems, in the context of evolving theories about its functions.

## 16.2. Decision-making cognition and reversal learning

It is first necessary to describe some of the behavioral functions of the OFC that have proven to be sensitive to neurochemical modulation. To date, most studies of the neurochemical modulation of the OM-PFC have focussed on the measurement of changes in behavioral responses on tasks that have been found to be sensitive to OM-PFC damage in humans or animals. These include tasks involving decision making and reversal learning.

Decision-making requires choice among several response alternatives with variable feedback or pay-offs in terms of gains and losses, for example, in money or in points earned in a gambling situation. The pay-offs may vary in terms of the magnitudes of reward or punishment, the probability of reward and punishment occurring, and also the delay between the choice and the pay-off. The most relevant sources of evidence for the role of the OFC have been provided by studies of brain-damaged patients and functional neuroimaging in normal volunteers.

## Human lesion studies

Two major paradigms have been used to assess decision-making performance, the Iowa Gambling Task (Bechara et al. 1994) and the Cambridge Gamble Task (Rogers et al. 1999a) which, while exhibiting many similarities, also have some important differences.

Impaired performance on the Iowa Gambling Task has provided some striking evidence for a role of the OM-PFC in decision-making. The task emphasizes the learning of reward and punishment contingencies associated with each of four card decks. Healthy subjects typically learn to avoid those decks associated with large, immediate rewards but also occasional, disastrously large penalties, in favour of 'safe' card decks associated with relatively smaller gains but only very minor penalties. Patients with bilateral damage to ventral regions of the OM-PFC do not acquire this preference, but continue to choose from the 'risky' decks. Thus, restricted regions of the OM-PFC have been implicated in the acquisition and retrieval of reinforcing feedback, whether positive or aversive, that guides response choice (Damasio 1994).

By contrast, the Cambridge Gamble Task requires no acquisition of such information, or its manipulation via working memory processes; all the information required to make a decision is presented on every trial in a visually explicit format on a touch sensitive screen. The subject is first required to guess the location of a yellow token hidden behind one of 10 boxes coloured either red or blue. The ratio of red to blue boxes varies across trials (9:1, 8:2, 7:3, or 6:4). Then, following each choice the subject is required to bet a proportion of their points on the decision, thus providing a direct measure of risk-taking behavior. The bets are offered to the subject in a fixed sequence, either ascending or descending, in counter-balanced series. Thus, it is also possible to assess impulsive or delay-averse responding, according to the pattern of choices on the ascending and descending series. Preference for risky behavior would be manifest as high bets on both the ascending and descending series, whereas delay-averse responding would be indicated by high bets in the descending series contrasting with low bets in the ascending series.

Several studies have shown that increased betting behavior in the Cambridge Gamble Task arises from frontal pathology. Specifically, increased betting was shown in three neurological conditions associated with OM-PFC disruption: i) patients with subarachnoid aneurysmal haemorrhage of the anterior communicating artery (Mavaddat et al. 2000), which is the major blood vessel supplying the medial and orbital sectors of the frontal lobes in humans; ii) patients with frontal variant, fronto-temporal dementia (Rahman *et al.* 1999), which targets the OFC from its earliest stages (Salmon *et al.* 2003); and iii) patients with large OM-PFC lesions (Manes *et al.* 2002) due to either stroke or tumour resection. In each of these studies the quality of decision-making in terms of the actual response choices was unimpaired in the patients, although there were some changes in latency. In the original study (Rogers *et al.* 1999a) patients with damage to the OM-PFC did show impaired probabilistic choice (and were slow to do so), but they were actually more conservative in their betting, possibly as a compensatory reaction to their impaired judgment.

The neural specificity of these alterations in decision-making, including pathological changes, is still being determined. Patients with amygdala lesions are also impaired on the Iowa Gambling Task, but they do not develop autonomic responses (such as galvanic skin response) to punishing feedback, unlike patients with OFC lesions (Bechara *et al.* 1999). Performance of the Iowa Gambling task was initially reported to be unimpaired by lesions of the dorsolateral (DL-) and dorsomedial (DM-) PFC that often impaired spatial working memory (Bechara *et al.* 1998). By contrast, Manes *et al.* (2002) did report deficits in performance of the Iowa Gambling Task in patients with DL- or DM-PFC lesions, which also affected aspects of spatial planning and working memory (Fig. 16.1). Further studies have found that the Iowa Gambling task was particularly susceptible to right-sided rather than left-sided PFC lesions (Tranel *et al.* 2002; Clark *et al.* 2003). Deficits were also particularly associated with right-sided PFC lesion groups on the Cambridge Gamble task (and a related variant which emphasizes conflict, the Cambridge Risk task) but these were smaller than the deficits on the Iowa Gambling Task. However, the study by Manes *et al.* (2002) indicated that performance on these tests, unlike that on the Iowa task, was not affected by DL- or DM-PFC damage, consistent with the hypothesis that these novel decision-making tasks did not recruit as many ancillary psychological processes as the Iowa task.

## Functional imaging studies in human volunteers

Converging evidence for a possible role of the OFC in decision-making cognition is provided by neuroimaging studies of normal healthy volunteers. Investigation of OFC function using functional magnetic resonance imaging (fMRI) has been hindered by susceptibility artefacts, where there is a loss of signal from tissue adjacent to air cavities due to inhomogeneity in the magnetic field. Recent developments have offered several techniques to improve OFC resolution (Cusack *et al.* 2005). Earlier studies employed positron emission tomography (PET) with radioactively labelled water ($H_2^{15}O$) or glucose, which avoids such artefacts. The Cambridge Gamble Task was adapted for PET neuroimaging to investigate the neural substrates of decision-making in healthy volunteers (Rogers *et al.* 1999b). The resultant Cambridge Risk differs from the Gamble Task in offering a fixed bet to either alternative, so that on some trials there was a clear conflict between choosing a more certain alternative for fewer points and a less certain alternative for more points. The main findings were that there were three sites of activation in the OFC, and that these were restricted to the right hemisphere- thus being consistent with the more

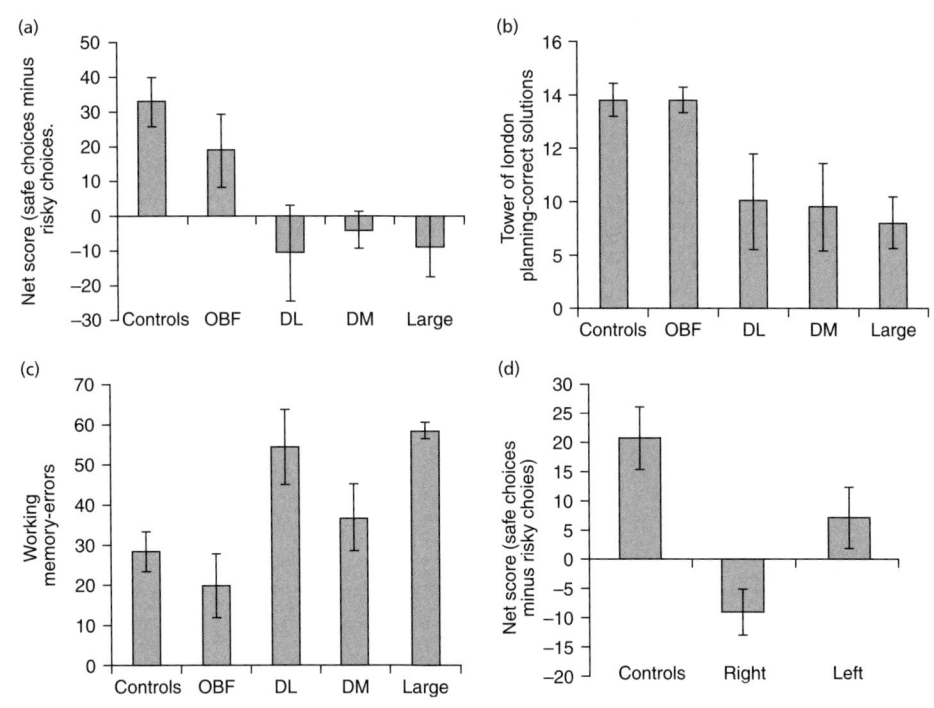

**Fig. 16.1** Neuropsychological functioning in lesion groups with discrete frontal lobe pathology. (a) In contrast to previous findings, patients with dorsolateral and dorsomedial PFC lesions were impaired on the Iowa Gambling Task. The figure shows total net score (the number of safe choices minus the number of risky choices) across the 100 trials of the task. A group with discrete OFC lesions (mainly left-sided) were unimpaired at the task. The dorsolateral, dorsomedial and large lesion cases were also impaired on traditional executive measures of spatial planning (the Tower of London test) (b) and self-ordered spatial working memory (c). In a follow-up study in a larger group of unilateral frontal lobe cases, decision-making impairments on the Iowa Gambling Task were found to be predominantly associated with right frontal lobe lesions (d). Reproduced with permission from Manes *et al.* (2002) and Clark *et al.* (2003).

deleterious effects of right-sided lesions (e.g. see Clark *et al.* 2003, above). Another PET investigation used a similar subtractive method to analyse the activations produced by the Iowa Gambling task (Ernst *et al.* 2002). As might have been predicted by the more widespread effects in patients with frontal lobe lesions on this task compared with the Risk task (c.f. Manes *et al.* 2002), Ernst *et al.* also found a widespread frontal response, probably reflecting such additional requirements of the Iowa task as learning and working memory.

## Functions of the OFC in humans and other animals

Decision-making has not been a major area of study in investigations of the functions of the OFC in other animals, although there are some indications that this may change radically in the near future. Several lines of evidence have shown that single units within the OFC process certain aspects of reinforcement, probably including the representation of goal objects and errors during reversal learning (Thorpe et al. 1983; Rolls 1999; Hollerman et al. 2000). Choice mechanisms can be examined in animals (and humans) by titrating two options that pit a higher magnitude of reward against either an uncertain or delayed outcome (Mazur 1987). In a delay-discounting (or 'delayed gratification') procedure, for example, the subject is asked to choose between a small reward available after a short delay, versus a larger reward available at a longer delay. By manipulating the long delay over successive trials, it is possible to infer an 'indifference point' where the two options are valued equally. A similar procedure may be employed to investigate probability discounting, where a small but likely reward is paired with a larger but less certain reward, as in the Cambridge Risk Task. So far, experimental studies in rats have focused mainly on delay discounting. Choosing the small immediate reward has often been interpreted as an example of impulsive responding, which clearly resembles the 'myopia for future outcomes' seen in brain-damaged patients with ventromedial PFC lesions (e.g. Bechara et al. 1994).

Interconnected functional circuitry that includes the nucleus accumbens core subregion, the basolateral amygdala and lateral OFC in rats has been defined by lesion studies (Mobini et al. 2000; Cardinal et al. 2001; Winstanley et al. 2004). However, there are some puzzles to resolve in these findings. Whereas lesions of the nucleus accumbens core and basolateral amygdala produce a striking preference for small, immediate reinforcement, selective damage to the OFC can have the opposite consequence (Winstanley et al. 2004, but see also Mobini et al. 2000). This latter effect may possibly result from a tendency towards perseveration, as the test procedure requires the rat to switch away from the currently preferred lever providing large magnitude reinforcement, to express a preference instead for the immediate, but lower magnitude, reinforcer.

Perseveration is a common concomitant of responding in other, apparently less complex, situations, for example, in object or visual reversal learning, where the contingencies governing reinforcement for choosing one of the two stimuli are reversed in animals previously trained to discriminate them on the basis of reinforcement (macaques: Butter 1969; Iversen and Mishkin 1970; Jones and Mishkin 1972; marmosets: Dias et al. 1996, 1997; rats: Chudasama and Robbins 2003: see Zald and Kim (2001) for a review). Thus, at least part of the deficit in reversal learning following OFC lesions depends on an impaired capacity to suppress the tendency to respond to the previously reinforced stimulus. This is sometimes observed during serial reversal learning, where there are successive reversals of reinforcement contingencies between the two stimuli. The other main contribution to reversal learning deficits comes from: (i) the failure to associate the alternative stimulus with reinforcement, which is often impaired by damage to distinct parts of the prefrontal cortex, e.g., IL-PFC in rats (Chudasama and Robbins 2003); and ii) more speculatively, an impaired overtraining reversal effect, where reversal is enhanced by the attentional processes hypothetically recruited by extended practice on the original discrimination (Bussey et al. 1996).

Do such considerations apply to human decision-making in tests such as the Iowa Gambling Task? Clearly, the reinforcement contingencies involved in the Iowa task are much more complex, as a particular card deck is associated with both reinforcement and punishment, usually delivered according to different schedules. However, choice behavior on this task can be characterized as initially exhibiting a preference for the seemingly attractive, 'risky' decks, to be replaced in normal subjects by the more sober choice of the small pay-off, but 'safe' decks. This shift in behavior can thus be construed as a global switch in choice behavior, possibly analogous to reversal learning, which may be impaired in patients with OM-PFC lesions (e.g. see Clark et al. 2003). Evidence for this view has recently been adduced by Fellows and Farah (2005) who eliminated the element of reversal in the Iowa gambling task, by shuffling the cards so that patients would not initially become biased to the risky deck by high reward pay-offs. This manoeuvre had the effect of greatly reducing the task deficit in a group of OM-PFC-lesioned patients, thereby adding support to the hypothesis that some of the earlier impairments could be explained in terms of a reversal learning deficit. The same manipulation did not remediate the deficit in a second group with DL-PFC lesions, perhaps consistent with the notion that these patients display Iowa Gambling Task deficits related to impairments in broader executive function (attention, working memory and planning) (Fig. 16.1). This hypothesis has clear merit in explaining why reversal learning impairments in animals and decision-making deficits in humans may perhaps be related to homologous neural circuitry including the OFC, across species. It has led to a number of studies that have focused on the role of the OFC and related structures in reversal learning in humans.

## Reversal learning in humans

The evidence relating the frontal cortex, and more specifically the OFC, to reversal learning in humans is now overwhelming. Daum et al. (1991) showed that patients with diverse frontal lesions failed a visual discrimination reversal test, whilst acquiring the initial discrimination normally: patients with temporal lobe damage, by contrast, exhibited the opposite pattern. Evidence for a specific involvement of the OM-PFC came from a study by Rolls et al. (1994) who showed reversal impairments in patients with large OM-PFC lesions compared to lesion patients without damage in this region. Patients with OM-PFC damage showed the perseverative tendency often exhibited by monkeys with OFC lesions, this tendency also correlating with questionnaire ratings of impulsive and socially inappropriate symptoms. It is significant that the majority of the unilateral lesioned patients studied by Rolls et al. in fact, had damage to the right hemisphere, consistent with the possibility that reversal learning is governed by the same hemisphere as performance on the more complex decision-making tasks (see Tranel et al. 2002; Clark et al. 2003). Two further studies have substantiated the likely involvement of the OM-PFC in reversal learning. Fellows and Farah (2003) showed that patients with damage limited to this area were impaired specifically in reversal: they also included a group of patients with DL-PFC lesions with no such deficit. The study by Rahman et al. (1999) that investigated patients with frontal variant, fronto-temporal dementia, also showed the patients to be impaired in visual reversal learning, as well as exhibiting the impairments in decision-making described above. These deficits were particularly striking as the impairments in reversal were greater than in shifting attentional set, which is more sensitive in general to frontal lobe damage (Owen et al. 1991).

## Functional imaging studies of reversal learning in human volunteers

Neuroimaging data have shown convergence with the lesion data, although a number of the studies have been hampered by methodological concerns. Studies directly examining reversal learning have included both PET (with $H_2^{15}O$) and event-related fMRI methodologies. An early study by Rogers *et al.* (2000) was unsuccessful in demonstrating changes in regional cerebral blood flow (rCBF) during visual reversal learning, probably because of the necessity to employ a blocked imaging design likely to show reduced sensitivity to transient shifts in behavioral performance (although changes in attentional shifting in the dorsolateral and anterior PFC were nevertheless apparent).

Event-related fMRI studies have had greater success. Thus, Nagahama *et al.* (2001) reported changes in the BOLD response during reversal shifts in the posteroventral prefrontal cortex, distinct from those areas implicated in attentional set-shifting. However, these investigators limited their imaging field of view to more dorsal brain regions above the AC-PC axis, and thus did not evaluate the role of the OFC *per se*. An elegant event-related fMRI study by O'Doherty *et al.* (2001) used a reversal task to assess reinforcement and punishment during fMRI, finding dissociable effects in the medial and lateral parts of the OFC, respectively. However, their effects of punishment were confounded with the requirement to make a reversal shift. A probabilistic reversal task which thus employed spurious negative feedback on some trials has since shown significant signal changes in the ventrolateral PFC, m-PFC and ventral striatum (Fig. 16.3), although it was not possible in this study to investigate the OFC directly because of the susceptibility artefact (Cools *et al.* 2002). By subtracting the spurious probabilistic feedback from reversal errors accompanied by a switch, this study demonstrated that the ventrolateral prefrontal cortex response was related to switching *per se*. In this study, subjects initially acquired a 80:20 reward: punishment visual discrimination and then were subjected to repeated reversals when criterion discrimination learning was attained. These data appear consistent with human lesion studies demonstrating that ventral PFC lesions impair performance on probabilistic reversal paradigms (Hornak *et al.* 2004, Berlin *et al.* 2004). In summary, there is ample evidence to suggest that the OM-PFC is implicated in tasks such as reversal learning, where affective choices have to be switched, and this capacity may reflect part of the more sophisticated processing that occurs in more complex decision-making situations.

## 16.3. Neurochemical modulation of PFC function with particular reference to OM-PFC

Considerable evidence collected using rats and monkeys suggests that the cognitive functions of the PFC are subject to differential neuromodulation by the ascending catecholaminergic (dopaminergic and noradrenergic), serotoninergic or cholinergic

---

(Data from Crofts *et al.* 2001). (f). Examples of the various discriminations used in the attentional set-shifting paradigm in which the dimension that was relevant in the previous compound discrimination either remained relevant (ID shift) or became irrelevant (ED shift). In the reversal, the compound stimuli remained the same but the previously rewarded exemplar became unrewarded and *vice versa*.

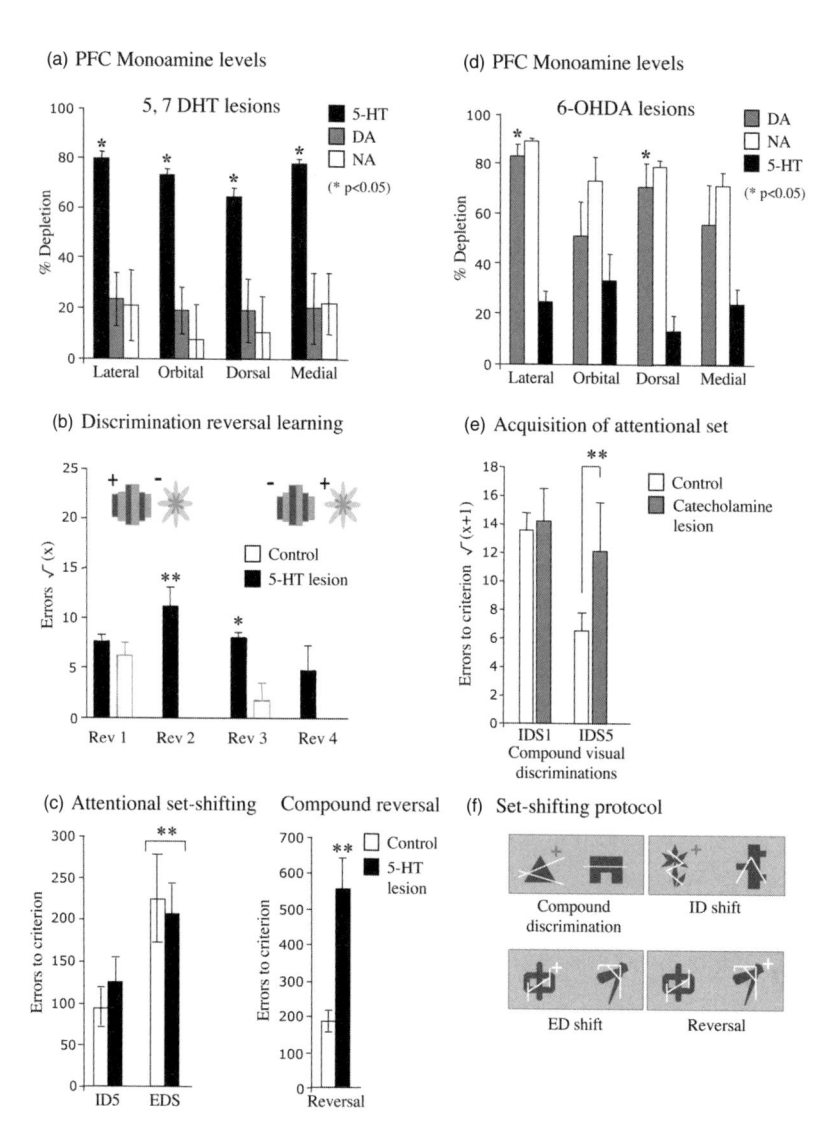

**Fig. 16.2** (a). Mean percentage depletions (±SEM) of serotonin (5-HT), dopamine (DA) and noradrenaline (NA) in the lateral PFC (Brodmann's area 9), orbitofrontal cortex (OFC), dorsal granular PFC (Brodmann's area 8) and pre-genual medial PFC of marmosets with 5,7 DHT lesions of the PFC. *Source:* Data from Clarke *et al.* 2005. (b). 5-HT lesioned animals made significantly more perseverative errors to the previously rewarded stimulus on a series of 4 discrimination reversals compared to the sham operated control animals. For each reversal the previously rewarded stimulus (+) became unrewarded and the previously unrewarded stimulus became rewarded, * p<0.005. Values shown are mean (±SEM). Data from Clarke *et al.* (2004). (c). 5-HT lesioned animals were not impaired in shifting attentional set from one dimension to another in a compound discrimination task. 5-HT lesioned animals were impaired on the subsequent compound reversal consistent with findings in (B). Values shown are mean (±SEM). *Source:* Data from Clarke *et al.* 2005). (d). Mean percentage depletions (±SEM) of dopamine (DA), noradrenaline (NA) and serotonin (5-HT) in the lateral PFC, orbitofrontal cortex, dorsal granular PFC and pre-genual medial PFC of marmosets with 6-OHDA lesions of the prefrontal cortex. Data from Crofts *et al.* (2001). (e). 6-OHDA lesioned animals did not improve performance across a series of compound discriminations (ID1-ID5) in which the same dimension (shapes or lines) remained relevant throughout.

systems. This modulation is likely to depend on the precise region of the PFC engaged by the task at hand and the degree of innervation of that region by these systems (see review by Lewis 2001). Although the monoamines innervate all the main sectors of the prefrontal cortex, including the OM-PFC, there is some apparent variability in the strength of this innervation and the distribution of receptors at the regional as well as laminar level (e.g. Goldman-Rakic et al. 1990; Gebhard et al. 1995). Thus for example, tyrosine hydroxylase immunoreactivity (enyzme for amino-acid precursor of both DA and NA) and dopamine $\beta$-hydroxylase is highly expressed in Brodmann area 9, as compared to other PFC regions in the rhesus monkey, especially area 11 (Lewis 2001). There are often considerable laminar differences in the density of different receptors associated with the monoamine neurotransmitters. For example, in the rhesus monkey, DA D1, NA $\alpha$1 and $\alpha$2, and 5-HT1 receptors are found most densely in the more dorsal cortical layers (I, II and IIIA), whereas NA $\beta$1 and $\beta$2, and 5-HT2 receptors are found more predominantly in layers IIB and IV and DA D2 in layer V, whether within DL-PFC or OM-PFC regions (area 12) (Goldman-Rakic et al. 1990). It is important to note that the same pyramidal cell may be innervated by both ascending DA and 5-HT neurons, although the D1 and 5-HT2A receptors are located on different parts of the cell, the D1 receptors predominantly on the apical dendrites and spines, and the 5-HT2A receptors on more proximal regions of the pyramidal cell (Goldman-Rakic 1999). These observations suggest that the neuromodulation by these systems is distinct in its effects even for the same cell within the PFC.

The density of innervation does not, however, always correspond to the receptor density. For example, while the ascending 5-HT system has a diffuse cortical projection, 5-HT receptor subtypes show greater regional specificity. Thus, the 5-HT2A receptor is very densely located in the prefrontal cortex, particularly in area 11 in both humans (Pazos et al. 1987) and marmosets (Gebhard et al. 1995) as well as in the IL-PFC in rats (Santana et al. 2004). This suggests that 5-HT could have quite specific functions at those receptors within the region of the OM-PFC. Finally, neurochemical evidence shows variations in levels of the neurotransmitters and their metabolites in different PFC regions. Thus, several authors have reported on major differences between DA and 5-HT levels in the rat PFC. DA levels in the mPFC are greater than in ventrolateral regions of PFC (Berger et al. 1976; Emson and Koob 1978), whereas the opposite applies to 5-HT levels (Audet et al. 1987). However, although these differences are probably functionally significant, it can be dangerous to infer too much from them. For example, Winstanley et al. (2006) recently replicated these basal differences for the rat mPFC and OFC, but found that it was DA levels that changed most significantly within the OFC upon functional challenge (see also below).

The functions of these ascending systems are undoubtedly susceptible to developmental influence with likely functional implications for the adult animal. For example, in the rat, following isolation-rearing, there is increased 5-HT2A expression specifically in the medial and orbital sectors of the PFC (Preece et al. 2004). Early social deprivation also alters the pattern of serotoninergic and dopaminergic innervation of the OM-PFC in a regionally specific manner (Poeggel et al. 2003). This supports the hypothesis that these PFC sectors are recruited under different circumstances, in this case involving social stress, with differential functional consequences that depend on their pattern of neurochemical innervation.

Reward processes are also thought to be mediated in part by the ascending monoamine systems, in particular DA and 5-HT, although these systems have been implicated in other affective, but non-reward related processes such as error monitoring and stress. Whether reward or stress is simply a function of the level of activation of the system, or the context under which this occurs, or as seems plausible, both of these factors, remains to be determined. Environmental contingencies certainly affect the functioning of these monoamine systems, and the mediation of these effects is itself dependent in part on the descending projections to the brainstem from the PFC, including the OFC. For example, it has recently been shown that the effects of cognitive 'control' over instrumental responding on the functioning of the brainstem 5-HT systems are mediated by projections to the dorsal raphé neurons from the IL-PFC in rats. Thus, some of the effects of 'learned helplessness', a putative model of depression and post-traumatic stress disorder, on the ascending 5-HT system are blocked by inactivation of this region of the PFC in rats (Amat *et al.* 2005). This phenomenon of regulation of the ascending monoamine systems by environmental contingencies is of enormous importance, as it shows a mechanism by which the PFC can exert executive control over systems that exert modulatory effects on processing in widespread regions of the forebrain, quite apart from the PFC itself.

In summary, there is evidently specificity, as well as nonspecificity in the neurochemical modulation of PFC function. There are differences of innervation among the ascending systems that suggest they exert quite different forms of control on the neuronal networks within the PFC, even at the level of single pyramidal cells. There are also suggestions of regional differences or gradations that suggest these systems may be preferentially involved in certain PFC functions, including those mediated by the OM-PFC. The contribution of specific chemical systems will now be examined in that context.

## Role of indoleamine systems

Evidence of a specific involvement of the ascending 5-HT systems with the OFC has recently been provided by a series of studies of reversal learning, already known to depend on the integrity of the OFC, in marmoset monkeys. The neurotoxin 5,7 dihydroxytryptamine was used with appropriate pre-treatments to induce relatively specific and selective loss of 5-HT in the PFC, especially the OFC (Clarke *et al.* 2004, 2005). The effects of the neurotoxin were assessed by both *ex vivo* and *in vivo* (microdialysis) neurochemical assays, employing high performance liquid chromatography with electrochemical detection, and also by immunoreactive staining for tryptophan hydroxylase, an important stage in the biosynthesis of 5-HT. We trained marmosets pre-operatively on a visual discrimination task using computer-generated discriminanda presented on a touch-sensitive screen. Following surgery, we first examined the effects of 5-HT depletion on the retention of the original trained discrimination, and then on the learning of a novel visual discrimination. There were no significant effects at these two stages. The marmosets then received a series of reversals of this acquired discrimination, and errors were analysed according to the Jones and Mishkin (1972) concept of parcellating effects on perseveration of the previous stimulus-reward association from those of stimulus-reward learning, using a sensitive quantitative analytical method, based on signal detection theory.

The major effect observed in the first study was a large increase in perseverative responding, especially beginning from the second reversal of the series (see Fig. 16.2). By contrast, there was little effect on reversal after the animals had extinguished their perseverative tendency to respond to the previously reinforced discriminanda. While it is possible that this perseveration resulted from increased proactive interference of previous stages, further experiments have shown that the main effect is to reduce response inhibitory processes recruited presumably by the OFC in mediating the reversal. A follow-up study has examined the effects of 5-HT depletion limited to the OFC and has replicated the original deficit, although this was evident even on the initial reversal (Clarke *et al.* 2005). Moreover, the effects of OFC 5-HT depletion had no effect whatsoever on the performance of an extra-dimensional shift, consistent with human data to be reviewed below (Rogers *et al.* 1999c) and unlike the effects of manipulations of the catecholamine systems on this task (Roberts *et al.* 1994; Rogers *et al.* 1999c; Crofts *et al.* 2001). Although much remains to be done to define the nature of 5-HT involvement within the OFC in reversal learning, it does appear that it has a special role in modulating behavior in this paradigm.

One interesting speculation is that 5-HT mechanisms may contribute to different aspects of reversal, for example, response inhibition (Clarke *et al.* 2004), reward (Rogers *et al.* 2003) and the monitoring of negative feedback (Evers *et al.* 2005)—possibly in different regions of the PFC, which appear to be distinct from those regions involved in attentional-set shifting, spatial working memory and planning. By contrast, it appears that manipulation of PFC DA has little effect on either simple reversal or serial reversal learning (Clarke *et al.* 2004; Roberts *et al.* 1994). The effects of PFC 5-HT depletion do, however, somewhat resemble the effects of prefrontal cholinergic loss produced by excitotoxic lesions of the nucleus basalis on serial reversal learning (Roberts *et al.* 1992), and may, conceivably, reflect an interaction of cortical 5-HT and cholinergic mechanisms. Further work is required to investigate such interactions as well as the possible receptor sensitivity of these effects.

## Effects of tryptophan depletion on reversal learning in human volunteers

Converging evidence of involvement of 5-HT mechanisms in reversal learning has been provided by studies in human volunteers undergoing transient reductions of central 5-HT function using the tryptophan depletion technique. Food-deprived volunteers are given a drink replete in all of the essential amino-acids, except for tryptophan, the essential precursor of serotonin (Young *et al.* 1985). The amino acid load produces a rapid decrease in the synthesis and release of brain 5-HT, via processes of increased protein synthesis in the liver and increased competition for transport across the blood-brain barrier (Young *et al.* 1985, 1989). The tryptophan depletion procedure has been shown to precipitate a transient depressive relapse in remitted patients that had responded to SSRI treatment (Smith *et al.* 1999). This relapse is associated with altered rCBF in the OFC, as well as the middle frontal gyrus, anterior cingulate and caudate nucleus (Bremner *et al.* 1997). However, there is generally relatively little effect on mood in healthy volunteers with no personal or family history of mood disorder.

Tryptophan depletion in normal volunteers produced more obvious effects on reversal learning than extra-dimensional shifting (i.e. shifting attentional set between two perceptual dimensions) in two separate studies (Park *et al.* 1994; Rogers *et al.* 1999c). Visual discrimination reversal learning was impaired in these studies using two- and three-dimensional visual discrimination tasks, respectively. These effects were behaviourally selective, particularly in the study by Rogers *et al.* (1999c) where tryptophan depletion impaired a reversal of a discrimination involving stimuli with two dimensions but had no effect on attentional set-shifting between two perceptual dimensions ('extra-dimensional set-shifting'). Extra-dimensional shifting had been shown to be sensitive to lateral frontal lesions in marmosets rather than OFC lesions (Dias *et al.* 1996). By contrast, methylphenidate, which blocks the DA and NA transporters, with relatively little effect on 5-HT activity, affected performance on the intra- and extra-dimensional shifts, consistent with effects of other catecholamine manipulations, to be described below.

These striking effects of tryptophan depletion, together with the observations of impaired reversal learning in patients with OM-PFC lesions, provided the impetus for the recent investigations of the recent effects of OFC 5-HT depletion in marmosets described above (Clarke *et al.* 2004, 2005). The observations in humans urgently need to be consolidated by similar findings with other ways of influencing 5-HT function than tryptophan depletion. Moreover, their neuroanatomical basis needs to be elucidated: are they in fact due to a modulation of OFC function?

Two approaches have been used which have so far provided only a partial resolution of these issues. First, the SSRI paroxetine was employed as a treatment for the reversal deficits seen in patients with frontal lobe dementia, on the basis that this disorder has been linked with disruption of 5-HT function within the OFC that might be remediated by such treatment. However, paroxetine actually further *impaired* reversal performance, without having similar detrimental effects on other aspects of performance (Deakin *et al.* 2004). It is nonetheless significant that a 5-HT agent was again found to selectively affect reversal learning; however, to be consistent with the effects of tryptophan depletion, it is necessary to assume that the treatment somehow impaired 5-HT function. The second approach has been to employ a pharmacological fMRI procedure to investigate the effects of tryptophan depletion on performance of the probabilistic reversal task (Evers *et al.* 2005). A previous study had found some evidence of impaired probabilistic reversal following tryptophan depletion (Murphy *et al.* 2002). In the fMRI study of Evers *et al.* (2005) an altered response to both spurious and valid negative feedback produced by tryptophan depletion interacted with certain BOLD activations. However, the locus of this interaction was in the DM-PFC, not the OFC (where, however, the susceptibility artefacts may have prevented a full test of the hypothesis of OFC involvement). The interaction was selective in pharmacological terms, as another study has shown that methylphenidate interacted more specifically with the ventrolateral PFC locus and in terms of the reversal shift itself, rather than the negative feedback component of the task (see Fig. 16.3: Clark *et al.* 2004).

There is thus considerable evidence that manipulations of 5-HT in humans affects reversal learning, but the psychological nature of the effect requires further analysis. Comparison with other effects of tryptophan depletion suggests that it may reflect an effect on processing on the interface between cognition and the emotions. While there were deleterious effects on certain aspects of cognition such as visual recognition memory and visuo-spatial associative learning (Park *et al.* 1994), several aspects of processing associated with DL-PFC function, including self-ordered spatial working memory, spatial span and visuospatial planning (Tower of London task) were unimpaired, at least in male volunteers (Park *et al.* 1994). These tests use abstract stimuli and provide little in the way of performance feedback, which may help to determine whether the serotoninergic system is engaged in task performance.

Such lack of effects contrasts with the deficits produced by tryptophan depletion in tasks involving emotional stimuli or explicit feedback. For example, such depletion also impairs aspects of decision-making cognition, including the Cambridge Gamble Task

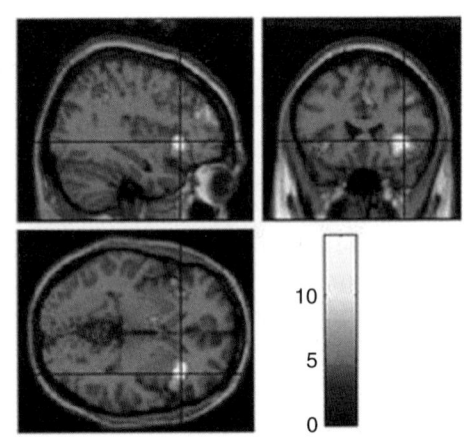

**Fig. 16.3** Reversal-related activity in an event-related fMRI task of probabilistic reversal learning Reproduced with permission from Cools *et al.* (2002). These sections show BOLD signal changes accompanying the shift from the previously-reinforced stimulus to the newly-reinforced stimulus, contrasted against a baseline of correct rule-adherent responding. The SPM99 maps are plotted at a reduced statistical threshold (P<.001 uncorrected for multiple comparisons) to illustrate response in the ventral striatum in addition to the ventrolateral prefrontal cortex, dorsomedial prefrontal cortex, and parietal cortex. In further studies using the same paradigm, methylphenidate was found to modulate the ventrolateral PFC locus, specifically for the reversal shift and independent of feedback. (Clark *et al.* 2004), whereas tryptophan depletion interacted with the dorsomedial PFC activation, specifically with respect to error feedback. Evers *et al.* (2005).

(Rogers et al. 1999a, 2003) as well as an 'affective go/no-go' task involving discrimination of happy and sad words (Murphy et al. 2002). This affective go/no-go task has proven sensitive in different ways to both depression and bipolar mania (Murphy et al. 1999), and depends on the activation of circuitry in the OFC (Elliott et al. 2002, 2004). Tryptophan depletion in healthy volunteers mimics the effects of depression in producing a relative speeding of responding to sad over happy words (Murphy et al. 2002). Tryptophan depletion also influences reinforcement sensitivity on a speeded reaction time task (Cools et al. 2005). This task uses an 'odd-one-out' display where rapid responses are disproportionately rewarded. A color cue on each trial signals the probability that reinforcements is available (there are three cues signalling 10%, 50%, or 90% probability). Under placebo conditions, healthy subjects increase their speed of responding at the higher probabilities, but this effect is abolished by tryptophan depletion, particularly in subjects reporting high questionnaire impulsivity.

Another line of work by Luciana et al. (2001) has contrasted effects of tryptophan depletion with tryptophan *loading* in normal volunteers. Relative to loading (no neutral control condition was included), tryptophan depletion had no effect on spatial working memory, but impaired verbal and affective working memory. Some of these data are consistent with previous data by Luciana et al. 1998 (using the indirect 5-HT agonist fenfluramine) that too much, as well as too little, 5-HT activity can be detrimental to PFC performance, thus obeying the ubiquitous Yerkes-Dodson relationship linking neurotransmitter activity to function. The sensitivity of the affective working memory task to this impairment is consistent with the hypothesis of selective involvement of 5-HT mechanisms within the OFC.

## Studies of OM-PFC 5-HT function in rats

There is evidence that pharmacological manipulation of central 5-HT systems in the rat affects reversal learning (Barnes et al. 1990; Domeney et al. 1991). Recently we have found preliminary evidence of deficits following selective depletion of 5-HT from the lateral OFC (O. Lehmann, J.W. Dalley and T.W. Robbins, unpublished findings 2005). However, there has been considerable evidence of involvement of PFC 5-HT on response inhibitory processes in other paradigms that focus on impulsivity, as distinct from the perseveration that can impede reversal learning.

For example, manipulation of 5-HT can affect certain measures of impulsivity in tests such as in the 5-Choice Serial Reaction Time Task (5CSRTT) in which rats are required to detect brief visual signals to earn food, and is analogous to the Continuous Performance Test used widely in human clinical studies (Robbins 2002). The 5CSRTT has measures of attentional selectivity in terms of accuracy of detecting the brief visual signals, as well as of inappropriate, premature responses which index impulsivity. Different aspects of performance on the 5CSRTT are sensitive to damage to distinct PFC loci (Muir et al. 1996; Chudasama et al. 2003). For example, damage to the IL-PFC and the postgenual anterior cingulate cortex selectively increases impulsive responding, whereas damage to the dorsal PL region predominantly affects accuracy and lateral OFC lesions increase perseverative responding. Pharmacological manipulations of the ascending neurotransmitter systems also have selective effects; for example, global depletion of 5-HT selectively increases impulsive responding without affecting accuracy (see Robbins and Everitt 1995).

Recent work has focused on the way in which these ascending systems interact with the PFC in determining 5CSRTT performance. Thus, we have used *in vivo* microdialysis to highlight a significant correlation between 5-HT efflux in the IL/PL regions of the mPFC and the number of premature, 'impulsive' responses shown in individual rats—but in no other aspects of performance (Dalley *et al.* 2002). There were no such correlations with measures of extracellular DA. Moreover, 5-HT2A receptor antagonists infused into the same region reduce impulsive responses often without affecting accuracy (Passetti *et al.* 2003; Winstanley *et al.* 2003). In fact, effects of DA and 5-HT manipulations of the rat mPFC on 5CSRTT performance can be distinguished to some extent, with DA particularly influencing attentional accuracy (Granon *et al.* 2000) and 5-HT, impulsive behavior. This is significant when taking into account the observations that different sub-regions of the rat PFC appear to be implicated in controlling attentional accuracy (dorsal prelimbic, Cg1); impulsive behavior (IL) and perseverative responding (lateral OFC) (Chudasama *et al.* 2003). The implication is that neuromodulation via 5-HT is of particular importance in the control of impulsive responding mediated by the IL-PFC, whereas DA D1 receptors are especially implicated in the process of response selection, mediated by area Cg1 of the dorsal PL-PFC (Granon *et al.* 2000). These data in the rat are thus consonant with those reviewed above in primates suggesting differential involvement of the catecholamine and indoleamine ascending neurotransmitter systems in the modulation of DL-PFC and OFC function. The IL-PFC in the rat is now generally regarded as homologous to certain regions of the primate anterior cingulate cortex (area 25), immediately adjacent to the OM-PFC (Vertes 2004).

However, under certain circumstances it is also clear that the catecholamine and 5-HT systems probably *interact* to determine performance, possibly directly within the same region of PFC and consistent with their possible functional interactions on the same pyramidal cells (Goldman-Rakic 1999). For example, in defined conditions, 5-HT2A receptor antagonists infused intra-mPFC can improve attentional accuracy (Winstanley *et al.* 2003). Moreover, the 5-HT indirect agonist fenfluramine has disruptive effects on human working memory that appear opposed to the beneficial effects of DA D2 receptor activation, via bromocriptine (Luciana *et al.* 1998: Mehta *et al.* 2001). Thus, it appears likely that the different modulatory influences can interact directly, affecting both pyramidal cell output (possibly via different receptors on the same pyramidal cells—Jakab and Goldman-Rakic 2000; Williams *et al.* 2002), and different *aspects* of behavioral performance. It is possible that the latter effects may also occur via modulatory actions on anatomically distinct systems in control of these different aspects of performance.

The delay discounting or delayed gratification paradigm described above provides another method of measuring impulsivity, in terms of impulsive choice. Several studies indicate that manipulation of 5-HT function using 5,7 dihydroxytryptamine or selective

5-HT receptor agents affects performance on this task (Winstanley *et al.* 2005). However, the involvement of 5-HT mechanisms within the OM-PFC has to date not been much studied. In an *in vivo* microdialysis study of delayed gratification performance, Winstanley *et al.* (2006) found no evidence for changes in 5-HT or 5-HIAA in the lateral OFC, although extracellular DA was specifically increased in this region in relation to instrumental choice rather than instrumental responding *per se* or food presentation. This observation argues against a simple view that the modulatory effects of DA and 5-HT respectively mainly affect the DL-PFC and OM-PFC; however, they are consistent with the hypothesis that these systems mediate different forms of neuromodulation in common regions of the PFC. In the mPFC, an almost opposite pattern of findings was found: selective increases in 5-HT in relation to instrumental choice, but no specific increases in mPFC DA. The most likely functional correlate of the changes in 5-HT in this region may be a contribution to timing functions, as mPFC lesions have the effect of flattening the delayed discounting function in rats (Cardinal *et al.* 2001). By contrast, the changes in mPFC DA were increased over control values in all conditions, suggesting a rather general neuromodulatory function, possibly linked to reward or stress (Winstanley *et al.* 2006).

## Catecholamine systems

### Dopamine

In monkeys, depletion of PFC DA but not NA or 5-HT, impairs spatial working memory processes dependent on the DL-PFC (Brozoski *et al.* 1979). These deficits that can be remediated by systemic treatment with DA agonists. Dopaminergic effects on spatial working memory are mainly mediated by DA D1-like receptors (Sawaguchi and Goldman-Rakic 1991). In humans, similar DA-dependent effects are seen on tests of spatial working memory and planning in patients with Parkinson's disease (for review, see Robbins 2000). Additionally, the catecholamine indirect agonist methylphenidate enhances spatial working memory performance while selectively reducing rCBF in a network including the DL-PFC and parietal cortex (Mehta *et al.* 2000). There is little or no evidence to date that this modulation of working memory depends on DA receptors in the OFC; rather the effect represents a specific interaction between the DL-PFC and the ascending DA system. Parallel work in rats suggests that the DA D1 receptor is implicated in the mPFC in similar functions of working memory, but also in tests of sustained visuospatial attention (Floresco and Phillips 2001; Granon *et al.* 2000; Chudasama and Robbins 2004). However, it is unclear what structures correspond in the primate brain to the rat mPFC, the anatomical homologies favouring either a relationship with the primate anterior cingulate cortex (Preuss 1995) or also including the DL-PFC (Ongur and Price 2000). A similar role for DA in attentional function is suggested by the effect of PFC catecholamine loss in marmosets induced by intra-PFC infusion of the neurotoxin 6-hydroxydopamine which impairs the 'learning set' implicated in intra-dimensional set-shifting across several shifts that normally occurs in control subjects (Fig. 16.2). Moreover, the behavior of marmosets with PFC DA depletion was more susceptible to disruption by background distractor cues during visual discrimination performance (Crofts *et al.* 2001). While there was no evidence of 5-HT PFC depletion in the study by Crofts (2001), there was substantial loss of PFC NA, which might therefore also be responsible for some of these effects. However, the marmosets with PFC DA loss, both in this study and an earlier one in which PFC NA loss was less evident (Roberts *et al.* 1994), showed normal reversal learning of such visual discriminations. Thus, it appeared that such behavior, although particularly dependent on the OFC, was relatively independent of DA despite quite substantial loss of this neurotransmitter in that region.

There is little doubt that DA-ergic manipulations at other brain sites can affect reversal learning. For example, intra-accumbens administration of d-amphetamine in the marmoset impairs reversal (Ridley *et al.* 1981) an effect that was blocked by systemic administration of the DA D2 receptor antagonist haloperidol. DA depletion of the nucleus accumbens also impaired reversal learning in the rat (Tagzhouti *et al.* 1985), presumably because both excess and depleted DA in this region produces behavioral impairments according to an inverted U-shaped, 'Yerkes-Dodson' relationship linking DA activity to efficiency of performance (Zahrt *et al.* 1997). Consistent with this theoretical formulation, the dopamine precursor medication, L-Dopa has been shown to impair probabilistic reversal in patients with Parkinson's disease (Swainson *et al.* 2000; Cools *et al.* 2001), probably via a direct interaction with the nucleus accumbens rather than the PFC (Cools *et al.* 2002, 2005). This contrasts with its beneficial effects on spatial working memory and task set switching in the same Parkinson's disease patients (Swainson *et al.* 2000). A subsequent observation that further strengthens the relationship of reversal learning to decision-making cognition is that medicated PD patients, like patients with OM-PFC damage, show increased betting on the Cambridge Gamble Task (Cools *et al.* 2003). These effects have been hypothesized to arise from the distinct interactions of L-Dopa with cortico-striatal loops depending on the dorsal striatum, which is heavily depleted of DA in Parkinson's disease, as compared with the ventral striatal loops that mediate reversal learning, which remain relatively unscathed in the early stages of this neurodegenerative disease (Kish *et al.* 1988; Cools *et al.* 2001; in press). The suggestion is that this leads to a relative 'overdosing' of DA in the ventral striatum, which indirectly impairs OM-PFC function. The data are also consistent with those of Volkow and Fowler (2000) showing a negative relationship between OFC metabolism and striatal D2 receptor binding in cocaine-dependent individuals.

There has been relatively little analysis to date of the behavioral functions of DA in the OFC to complement the single unit data on 'reward neurons'. A single study (Kheramin *et al.* 2004) has reported impaired discounting of reward in the delayed gratification paradigm following 6-OHDA-induced depletion of the catecholamines in this region—an effect probably dependent on OFC DA loss. These findings are supported by recent data from *in vivo* microdialysis showing a specific release of lateral OFC DA during delayed gratification performance that appears to depend specifically on the requirement for instrumental choice rather than the delay or the presentation of reward *per se* (Winstanley *et al.* 2006).

## Noradrenaline

The coeruleo-cortical noradrenergic system has been linked in monkeys to the modulation of spatial working memory dependent on the DL-PFC (Arnsten and Robbins 2002), and also appears to be important in the detection and evaluation of novel environmental contingencies (Aston-Jones *et al.* 1999; Dalley *et al.* 2001; Bouret and Sara 2004). There is evidence that the phasic and tonic firing of coeruleal NA neurons can influence performance in such tests as go/no-go discrimination, presumably via direct effects on the PFC (Usher *et al.* 1999; Bouret and Sara 2004). In the latter study, simultaneous recordings were made from the locus coeruleus (LC) and PL-PFC. LC activation was more tightly aligned to the behavioral response than to the CS+ in an odor-reward association task in rats. Such LC activation also preceded PL-PFC activation, when the response-reinforcer contingencies were changed, indicating that the LC plays an important role in modulation of PFC function, but also suggesting the possibility that it must be "instructed" about changing environmental contingencies by other forebrain structures, possibly including the OFC. Consistent with this hypothesis, Jodo *et al.* (1998) have reported that mPFC neurons exert an excitatory influence on LC neuron firing. In addition to affecting working memory function, one of the likely effects of such neuromodulation may be to enhance the later stages of memory consolidation (Tronel *et al.* 2004).

There are very few relevant human studies on the possible role of prefrontal adrenoceptors in cognition; however, drugs such as the mixed adrenoceptor-$\alpha$-1,2 agonist clonidine tend to affect attentional and working memory function rather than reversal learning (Coull *et al.* 1995; Rogers *et al.* 1999c). The $\alpha$-2 receptor antagonist idazoxan also produces substantial deficits in performance on these tests (Middleton *et al.* 1999), which preferentially tap DL-PFC rather than OFC function (Manes *et al.* 2002). Both idazoxan and clonidine tend to impair especially extra-dimensional attentional set-shifting (Middleton *et al.* 1999) rather than reversal learning, which is interesting given the recent demonstration of selective impairments at the extra-dimensional shift stage in rats following treatment within the PFC with saporin-DBH, which selectively destroys NA-containing neurons (Eichenbaum *et al.* 2003). Thus it is possible that the noradrenergic innervation of the PFC is especially associated with DL-PFC functions, although this hypothesis requires much more intensive testing.

## Cholinergic system

Manipulations of the ascending cholinergic system affect broad aspects of PFC function (Everitt and Robbins 1997 (Himmelhaber *et al.* 2001) including not only attention and working memory but also reversal learning. For example, selective lesions of the cholinergic basal forebrain cholinergic system impair not only attentional accuracy, but also other aspects of executive function of performance on the 5CSRTT. Recent evidence using either selective lesioning methods for the basal forebrain neurons, such as the immunotoxin IgG-192-saporin (McGaughy *et al.* 2002; Chudasama *et al.* 2004; Dalley *et al.* 2004), or the measurement of fluxes of acetylcholine in the mPFC during performance of the 5CSRTT (Passetti *et al.* 2000), suggest that the basal forebrain cholinergic system is engaged to optimize performance on the 5CSRTT or in working memory tasks. When saporin was infused directly into the mPFC, the attentional deficits were accompanied by other deficits reminiscent of those caused by IL-PFC and OFC lesions, i.e. premature, impulsive and perseverative responses, respectively (Dalley et al. 2004). In possible further support of a special relationship with OFC function, serial reversal learning is impaired by excitotoxic lesions of the nucleus basalis of Meynert in marmosets that lead to cholinergic loss within the PFC (Roberts et al. 1992). Thus, overall, it appears that acetylcholine modulates a broad range of PFC functions.

## 16.4. Conclusion

This chapter has provided considerable evidence of differential modulation of PFC function by the ascending neurotransmitter systems from experiments in animals and humans. A major conclusion is that these systems are recruited by diverse environmental demands and circumstances that implicate them in quite specific functions. There are also suggestions of differential regional involvement within the OFC—perhaps the most convincing evidence coming from data in monkeys that show a double dissociation of effects on tests, such as reversal learning and attentional set-shifting following selective 5-HT and DA depletion (Clarke et al. 2004, 2005). There is also evidence of interactions among these ascending systems in common functions within the same PFC region. An intriguing hypothesis is that the neuromodulation of PFC function has evolved to optimize fronto-executive functions required in a range of adaptive situations that include stress.

The distinct role for these chemical neuromodulatory functions is consistent with their probable contributions to different forms of psychopathology associated with PFC function. For example, ADHD is treated by drugs such as methylphenidate that appear to exert their major actions on the catcholamine systems, possibly associated with DL- and VL-PFC regions (Mehta et al. 2000) whereas OCD is treated most efficaciously with SSRIs, possibly acting within the OFC (Griest and Jefferson 1998; Saxena and Rauch 2000; Lucey 2001). One study has shown how the long-term treatment regimen with SSRIs customarily employed in OCD leads to increased 5-HT release in the OFC, associated with desensitization of the 5-HT terminal autoreceptor (Bergqvist et al. 1999). In this context, the finding that 5-HT depletion in the marmoset OFC is associated with increases in perseverative responding in reversal learning tasks is promising, especially as OM-PFC damage in humans has now been shown to be linked to impaired reversal learning.

Further challenges are posed by the need to understand how depression is ameliorated by chronic SSRI treatment—probably via quite different regional actions within the OFC than those hypothesized to underlie the successful medication of OCD. A major correlate of effective SSRI action is a *reduction* in metabolism in the subgenual cingulate (area 25) correlated with a normalisation of PFC hypometabolism in area 9/46 (Mayberg 2003). However, it is perhaps important to note that the OFC (area 11) appeared more sensitive to cognitive behavioral therapy than SSRI treatment.

Other important challenges are posed by the proposed involvement of ascending neuromodulatory systems to the OFC in frontal lobe dementia, the sequelae of drug addiction, including both cognitive impairment and craving and in understanding the effects of atypical anti-psychotic drugs in schizophrenia. The key question, whether the cognitive deficits associated with substance abuse are actually caused by the drugs themselves, possibly acting in a toxic manner on brain areas that may include the OFC, or whether both the drug-abuse and the cognitive deficits arise from another cause (e.g. specific genotype), which may lead in turn, for example, to OFC dysfunction and drug abuse. Thus, demonstrations of deficits in drug abusers on tests such as the Iowa (Bechara *et al.* 2002); Ernst *et al.* 2003) and Cambridge (Rogers *et al.* 1999a) Gambling tasks may not necessarily be indicative of drug-induced dysmodulation of OM-PFC function. One promising way of examining this question of cause and effect is via animal studies that can control the nature and extent of prior drug exposure. Jentsch *et al.* (2002) have elegantly shown how chronic cocaine administration to monkeys causes deficits in visual reversal learning, suggestive of OFC dysfunction. It will be useful in future studies to see whether chronic drug self-administration has such deleterious, and long-lasting, effects

We expect that future advances in specifying the roles of neuromodulatory systems in OM-PFC function will depend not only on conceptual advances about the possible computational effects of the modulation, but also on further evidence obtained with ever more sophisticated techniques, including the investigation of OM-PFC function with neurotransmitter ligands using PET, in animals and humans. Another approach for interrogating the neuromodulation of fronto-executive function will be via functional genomics, which can be used to resolve the mechanisms underlying individual differences in fronto-executive function and in response to drugs. This approach has already been exemplified by the study of cognitive and functional neuroimaging effects of polymorphisms of catechol-O-methyl-transferase (COMT) in tests of working memory and set-shifting that suggest the specific modulation of DL-PFC function (Goldman 2003; Mattay *et al.* 2003), consistent with other data described above. Investigation of functional polymorphisms affecting other modulatory neurotransmitter systems, as well as DA, (Neumister *et al.* 2002; Goldman 2003) will undoubtedly also yield fruit, especially if combined with the neuropsychological and psychopharmacological findings we have reviewed here.

## Acknowledgements

Much of this research was funded by a Wellcome Trust Programme Grant awarded to TWR, ACR, BJ Everitt and BJ Sahakian, and completed within the University of Cambridge Behavioral and Clinical Neuroscience Institute, supported by a joint award from the MRC and the Wellcome Trust. We thank our colleagues for their efforts in these studies.

## References

Amat, J., Baratta, M. V., Paul, E., Bland, S. T., Watkins, L. R., and Maier, S. F. (2005). The ventral medial prefrontal cortex determines how behavioral control over stress impacts behavior and dorsal raphé nucleus activity. *Nature Neuroscience* **8**:365–371.

Arnsten, A. F. T. and Goldman-Rakic, P. S. (1984). Selective prefrontal cortical projections to the region of the locus coeruleus and raphé nuclei in the rhesus monkey. *Brain Research* **306**:9–18.

Arnsten, A. F. T. and Robbins T. W. (2002). Neurochemical modulation of prefrontal cortical functions in humans and animals. In: D. Stuss, and R. Knight, (eds). *The Prefrontal Cortex*, pp. 51–84. New York:Oxford University Press.

Aston-Jones, G., Rajkowski, J., and Cohen, J. D. (1999). Role of locus coeruleus in attention and behavioral flexibility. *Biological Psychiatry* **46**:1309–1320.

Audet, M. A., Descarries, L., Doucet, G. (1989). Quantified regional and laminar distribution of the serotonin innervation in the anterior half of adult rat cerebral cortex. *Journal of Chemical Neuroanatomy* **B**: 29–44.

Azmitia, E. C., and Whitaker-Azmitia, P. M. (1995). Anatomy, cell biology and maturation of the serotonergic system. In: F. E. Bloom, and D. J. Kupfer, (eds). *Psychopharmacology, Fourth Generation of Progess*, New York:Raven Press.

Barnes, J. M., Costall, B., Coughlan, J., Domeney, A. M., Gerrard, P. A., Kelly, M., et al. (1990). The effects of ondansetron, a 5-HT3 receptor antagonist, on cognition in rodents and primates. *Pharmacology, Biochemistry and Behavior* **35**:955–962.

Bechara, A., and Damasio, H. (2002). Decision-making and addiction (part I): impaired activation of somatic states in substance dependent individuals when pondering decisions with negative future consequences. *Neuropsychologia* **40**:1675–1689.

Bechara, A., Damasio, A. R., Damasio, H., and Anderson, S. W. (1994). Insensitivity to future consequences following damage to human prefrontal cortex. *Cognition* **50**:7–15.

Bechara, A., Damasio, H., Damasio, A. R., and Lee, G. P. (1999). Different contributions of the human amygdala and ventromedial prefrontal cortex to decision-making. *Journal of Neuroscience* **19**:5473–5481.

Bechara, A., Damasio, H., Tranel, D., and Anderson, S. W. (1998). Dissociation of working memory from decision making within the human prefrontal cortex. *Journal of Neuroscience* **18**:428–437.

Bentivoglio, M., and Morelli, M. (2005). The organization and circuits of mesencephalic dopaminergic neurons and the distribution of dopamine receptors in the brain. In: S. B. Dunnett, M. Bentivoglio, A. Bjorklund, and T. Hokfelt, (eds). Chapter 1 in *Handbook of Chemical Neuroanatomy, Vol. 21. Dopamine.* pp. 1–107., Amsterdam:Elsevier.

Berger, B., Thierry, A. M., Moyne, M. A. (1976). Dopaminergic innervation of the rat prefrontal cortex: a fluorescence histochemical study. *Brain Research* **106**:133–145.

Bergqvist, P. B., Bouchard, C. and Blier, P. (1999). Effect of long-term administration of anti-depressant treatments on serotonin release in brain regions involved in obsessive-compulsive disorder. *Biological Psychiatry* **45**:164–174.

Berlin, H. A., Rolls, E. T., and Kischka, U. (2004). Impulsivity, time perception, emotuon and reinforcement sensitivty in patients with orbitofrontal cortex lesions. *Brain* **127**:1108–1126.

Bigl, V., Woolf, N. J., and Butcher, L. L. (1982). Cholinergic projections from the basal forebrain to frontal, parietal, temporal and cingulate cortices: a combined fluorescent tracer and acetylcholinesterase analysis. *Brain Research Bulletin* **8**:27–749.

Bouret, S. and Sara, S. J. (2004). Reward expectation, orientation of attention and locus coeruleus-medial frontal cortex interplay during learning. *European Journal of Neuroscience* **20**:791–802.

Bremner, J. D., Innis, R. B., Salomon, R. M., Staib, L. H., Ng, C. K., Miller, H. L., et al. (1997). Positron emission tomography measurement of cerebral metabolic correlates of tryptophan depletion-induced depressive relapse. *Archives of General Psychiatry* **54**:364–374.

Brody, A. L., Saxena, S., Silverman, D. H., Alborzian, S., Fairbanks, L. A., Phelps, M. E. et al. (1999). Brain metabolic changes in major depressive disorder from pre- to post-treatment with paroxetine. *Psychiatry Research* **91**:127–139.

Brozovski, T. J., Brown, R. M., Rosvold, H. E., and Goldman, P. (1979). Cognitive deficit caused by regional depletion of dopamine in prefrontal cortex of rhesus monkey. *Science* **205**:929–931.

Bussey, T. J., Muir, J. L., Everitt, B. J., and Robbins, T. W. (1997). Triple dissociation of anterior cingulate, posterior cingulate, and medial frontal cortices on visual discrimination tasks using a touchscreen testing procedure for the rat. *Behavioral Neuroscience* **111**:920–936.

Butter, C. (1969). Perseveration in extinction and in discrimination reversal tasks following selective frontal ablations in macaca mulatta. *Physiology and Behavior* **4**:163–171.

Cardinal, R. N., Pennicott, D. R., Sugathapala, C. L., Robbins, T. W., and Everitt, B. J. (2001). Impulsive choice induced in rats by lesions of the nucleus accumbens core. *Science* **292**:2499–2501.

Celada, P., Puig, M. V., Casanovas, J. M., Guillazo, G., and Artigas, F. (2001). Control of dorsal raphé serotonergic neurons by the medial prefrontal cortex: involvement of serotonin-1A, GABA(A), and glutamate receptors. *The Journal of Neuroscience* **21**:9917–9929.

Chudasama, Y. and Robbins, T. W. (2003). Dissociable contributions of the orbitofrontal and infralimbic cortex in pavlovian autoshaping and discrimination reversal learning: Further evidence for the functional heterogeneity of the rodent frontal cortex. *The Journal of Neuroscience* **23**:8771–8780.

Chudasama, Y., Passetti, F., Desai, A., Rhodes, S., Lopian, D., and Robbins, T. W. (2003). Dissociable aspects of performance on the 5 choice serial reaction time task following lesions of the dorsal anterior cingulate, infralimbic and orbitofrontal cortex in the rat: differential effects on selectivity, impulsivity and compulsivity. *Behavioral Brain Research* **146**:105–119.

Chudasama, Y., and Robbins, T. W. (2004). Dopaminergic modulation visual attention and working memory in the rodent prefrontal cortex. *Neuropsychopharmacology* **29**:1628–1636.

Chudasama, Y. Dalley, J. W., Nathwani, F., Bouger, P., and Robbins, T. W. (2004). Cholinergic modulation of visual attention and working memory: Dissociable effects of basal forebrain 192-IgG-saporin lesions and intraprefrontal infusions of scopolamine *Learning and Memory* **11**:78–86.

Clark, L., Manes, F., Antoun, N., Sahakian, B. J., et al. (2003). The contributions of lesion laterality and lesion volume to decision-making impairment following frontal lobe damage. *Neuropsychologia* **41**:1474–1483.

Clark, L., Cools, R., Evers, L. E., van der Veen, F., Jolles, J., Sahakian, B. J., et al. (2004). Neurochemical modulation of prefrontal cortex function. *FENS abstracts* **2**:A205(1).

Clarke, H., Dalley, J. F. W., Crofts, H. S., Robbins, T. W. and Roberts, A. C. (2004).Cognitive inflexibility following prefrontal serotonin depletion. *Science* **304**:878–880.

Clarke, H., Dalley, J. F. W., Crofts, H. S., Robbins, T. W., and Roberts, A. C. (2005). Perseveration in discrimination reversal learning following prefrontal serotonin depletion in the marmoset. *The Journal of Neuroscience* **12**:532–538.

Cools, R., Barker, R. A., Sahakian, B. J., and Robbins, T. W. (2001). Enhanced or impaired cognitive function in Parkinson's disease as a function of dopaminergic medication and task demands. *Cerebral Cortex* **11**:1136–1143.

Cools, R., Clark, L., Owen, A. M., and Robbins, T. W. (2002). Defining the neural mechanisms of probabilistic reversal learning using event-related functional magnetic resonance imaging. *The Journal of Neuroscience* **22**:4563–4567.

Cools, R., Barker, R. A., Sahakian, B. J., and Robbins, T. W. (2003). L-Dopa medication remediates cognitive inflexibility, but increases impulsivity in patients with Parkinson's disease. *Neuropsychologia* **41**:1431–1441.

Cools, R., Blackwell, A., Clark, L., Menzies, l, Cox, S., and Robbins, T. W. (2005). Tryptophan depletion disrupts the motivational guidance of goal-directed behavior as a function of trait impulsivity. *Neuropsychopharmacology* **30**:1362–1373.

Cools, R., Lewis, S. J. G., Clark, L., Barker, R., Robbins, T. W. (in press). L-Dopa disrupts activity in the nucleus accumbens during reversal learning in Parkinson's disease. *Neuropsychopharmacology*.

Coull, J.T., Middleton, H. C., Robbins, T. W., and Sahakian, B. J. (1995). Contrasting effects of clonidine and diazepam on tests of working memory and planning. *Psychopharmacology* **120**:311–321.

Crofts, H. S., Dalley, J. W., Collins, P., Van Denderen, J. C. M., Everitt, B. J., Robbins, T. W., and Roberts, A. C. (2001). Differential effects of 6-OHDA lesions of the prefrontal cortex and caudate nucleus on the ability to acquire an attentional set. *Cerebral Cortex* **11**:1015–1026.

Cusack, R., Russell, B., Cox, S. M. L., de Panfilis, C., Scharzbauer, C., Ansorge, R. ( 2005). An evaluation of passive shimming to improve frontal sensitivity in fMRI. *Neuroimage* **24**:82–91.

Dalley, J. W., McGaughy, J., O'Connell M. T., Cardinal, R., Levita, L. and Robbins, T. W. (2001). Distinct changes in cortical acetylcholine and noradrenaline efflux during contingent and non-continent performance of a visual attentional task. *The Journal of Neuroscience* **21**:4908–4914.

Dalley, J. W., Theobald, D. E., Eagle, D. M., Passetti, F., and Robbins, T. W. (2002). Deficits in impulse control associated with tonically elevated serotonergic function in rat prefrontal cortex. *Neuropsychopharmacology* **26**:716–728.

Dalley, J. W., Theobald, D. E., Bouger, P., Chudasama, Y., Cardinal, R. N., and Robbins, T. W. (2004). Cholinergic function and deficits in visual attentional performance in rats following 192 IgG-saporin-induced lesions of the medial prefrontal cortex. *Cerebral Cortex* **14**:922–932.

Damasio, A. (1994). *Descartes' error: Emotion, reason and the human brain*. G. P. Putnam. New York.

Damasio, A. R., Tranel, D., and Damasio, H. (1990). Individuals with sociopathic behavior caused by frontal damage fail to respond autonomically to social stimuli. *Behavioral Brain Research* **41**:81–94.

Daum, I., Schugens, M. M., Channon, S., Polkey, C. E., and Gray, J. A. (1991). T-maze, discrimination and reversal learning after unilateral temporal or frontal lobe lesions in man. *Cortex* **27**:613–622.

Deakin, J. B., Rahman, S., Nestor, P. J., Hodges, J. R., Sahakian, B. J. (2004). Paroxetine does not improve symptoms and impairs cognition in frontotemporal dementia: a double blind randomized controlled trial. *Psychopharmacology* **172**:400–408.

Dias, R., Robbins, T. W., and Roberts, A. C. (1996). Dissociation in prefrontal cortex of affective and attentional shifts. *Nature* **380**:69–72.

Dias, R., Robbins, T. W., and Roberts, A. C. (1997). Dissociable forms of inhibitory control within prefrontal cortex with an analogue of the Wisconsin card sort test: restriction to novel situations and independence from "on-line" processing. *The Journal of Neuroscience* **17**:9285–9297.

Domeney, A. M., Costall, B., Gerrard, P. A., Jones, D. N. C., Naylor, R. J., and Tyers, M. B. (1991). The effects of ondansetron on cognitive performance in the marmoset. *Pharmacology, Biochemistry and Behavior* **38**:169–175.

Drevets, W. C. (2001). Neuroimaging and neuropathological studies of depression: implications for the cognitive-emotional features of mood disorders. *Current Opinion in Neurobiology* **11**:240–249.

Eichenbaum, H. B., Ross, R., Raji, A., McGaughy, J. A. Noradrenergic, but not cholinergic, deafferentation of the infralimbic/prelimbic cortex impairs attentional set-shifting. *Society for Neuroscience Abstracts* **2003**:940.947.

Elliott, R., Rubinsztein, J. S., Sahakian, B. J., and Dolan, R. J. (2002). The neural basis of mood-congruent processing biases in depression. *Arch Gen Psychiatry* **59**:597–604.

Elliott, R., Ogilvie, A., Rubinsztein, J. S., Calderon, G., Dolan, R. J., and Sahakian, B. J. (2004). Abnormal ventral frontal response during performance of an affective go/no go task in patients with mania. *Biological Psychiatry* **55**:1163–1170.

Emson, P. C., and Koob, G. F. (1978) The origin and distribution of dopamine-containing afferents to the rat frontal cortex, *Brain Research* **142**:249–267.

Ernst, M., Grant, S. J., London, E. D., Contoreggi, C. S., Kimes, A. S., and Spurgeon, L. (2003). Decision making in adolescents with behavior disorders and adults with substance abuse. *American Journal of Psychiatry* **160**:33–40.

Ernst, M., Bolla, K., Mouratidis, M., Contoreggi, C., Matochik, J. A., Kurian, *et al.* (2002). Decision-making in a risk-taking task: a PET study. *Neuropsychopharmacology* **26**:682–691.

Evers, E. A. T., Cools, R., Clark, L., van der Veen, Jolles J., Sahakian, B. J., *et al.* (2005). Serotonergic modulation of prefrontal cortex during reversal learning. *Neuropsychopharmacology* **30**:1138–1147.

Fellows, L. K. and Farah, M. J. (2003). Ventromedial frontal cortex mediates affective shifting in humans: evidence from a reversal learning paradigm. *Brain* **126**:1830–1837.

Fellows, L. K. and Farah, M. J. (2005). Different underlying impairments in decision-making following ventromedial and dorsolateral frontal lobe damage in humans. *Cerebral Cortex* **15**:58–63.

Floresco, S. B. and Phillips, A. G. (2001). Delay-dependent modulation of memory retrieval by infusion of a dopamine D1 agonist into the rat medial prefrontal cortex. *Behavioral Neuroscience* **115**:934–939.

Gaykema, R. P. A., VanWeeghel, R., Hersh, L. B., and Luiten, P. G. M. (1991). Prefrontal cortical projections to the cholinergic neurons in the basal forebrain. *Journal of Comparative Neurology* **303**:563–583.

Gebhard, R., Zilles, K., Schleicher, A., Everitt, B. J., Robbins, T. W. (1995). Parcellation of the frontal cortex of the New World monkey *Callithrix jacchus* by eight neurotransmitter-binding sites. *Anatomy and Embryology* **191**:509–517.

Goldman, D. (2003). Genetics of human prefrontal function. *Brain Research Reviews* **43**:134–163.

Goldman-Rakic, P. S., Lidow, M. S., Gallagher, D. W. (1990). Overlap of dopaminergic, adrenergic, and serotoninergic receptors and complementarity of their sub-types in primate prefrontal cortex. *The Journal of Neuroscience* **10**:2125–2138.

Granon, S. Passetti, F., Thomas, K. L., Dalley, J. W., Everitt, B. J., and Robbins, T. W. (2000). Enhanced and impaired attentional performance after infusion of D1 dopaminergic receptor agents into rat prefrontal cortex. *The Journal of Neuroscience* **20**:1208–1215.

Griest, J. H. and Jefferson, J. W. (1998). Pharmacotherapy for obsessive-compulsive disorder. *British Journal of Psychiatry* **173**(suppl 135):64–70.

Hajos, M., Gartside, S. E., Varga, V., and Sharp, T. (1998). An electrophysiological and neuroanatomical study of the medial prefrontal cortical projection to the midbrain raphé nuclei in the rat. *Neuroscience* **87**:95–108.

Himmelhaber, A. M., Sarter, M., Bruno, J. P. (2001). The effects of manipulation of attentional demand on cortical acetylcholine release. *Cognitive Brain Research* **12**:353–370.

Holets, V. R. (1990). The anatomy and function of noradrenaline in the mammalian brain. In: D. J. Heal, and C, A. Marsden, (eds). *The Pharmacology of noradrenaline in the central nervous system.* pp. 1–40. Oxford: Oxford University Press.

Hollerman, J. R., Tremblay, L., Schultz, W. (2000). Involvement of basal ganglia and orbitofrontal cortex in goal-directed behavior. *Progress in Brain Research* **126**:193–215.

Hornak, J., O'Doherty, J., Bramham, J., Rolls, E. T., Morris, R. G., Bullock, P. R., Polkey, C. E. (2004). Reward-related reversal learning after surgical excisions in orbito-frontal or dorsolateral prefrontal cortex in humans. *Journal of Cognitive Neuroscience* **16**:463–478.

Hurd, Y. and Hall, H. (2005). Human forebrain dopamine systems: characterization of the normal brain and in relation to psychiatric disorders. Chapter 9 In: S. B. Dunnett, M. Bentivoglio, A Bjorklund, and T. Hokfelt, (eds). *Handbook of Chemical Neuroanatomy Vol. 21. Dopamine.* pp. 525–571. Amsterdam: Elsevier.

Insel, T. R. (1992). Towards a neuroanatomy of obsessive-compulsive disorder. *Current Opinions in Neurobiology* **28**:343–347.

Iversen, S. D. and Mishkin, M. (1970). Perseverative interference in monkeys following selective lesions of the inferior prefrontal convexity. *Experimental Brain Research* **11**:376–386.

Jakab, R. L., and Goldman-Rakic, P. S. (2000). Segregation of serotonin 5-HT$_{2A}$ and 5-HT$_3$ receptors in inhibitory circuits of the primate cerebral cortex. *Journal of Comparative Neurology* **417**:337–348.

Jentsch, J. D., Olausson, P., De La Garza, R., and Taylor, J. R. (2002). Impairments of reversal learning and response perseveration after repeated, intermittent cocaine administrations to monkeys. *Neuropsychopharmacology* **26**:183–190.

Jodo, E., Chiang, C., and Aston-Jones, (1998). Potent excitatory influence of prefrontal cortex activity on noradrenergic locus coeruleus neurons. *Neuroscience* **83**:63–79.

Jones, B. and Mishkin, M. (1972). Limbic lesions and the problem of stimulus—reinforcement associations. *Experimental Neurology* **36**:362–377.

Kheramin, S., Body, S., Ho, M. Y., Vlazquez-Martinez, D. N. Bradshaw, C. M., Szabadi, E., et al. (2004) Effects of orbital prefrontal cortex dopamine depletion on inter-temporal choice: a quantitative analysis. *Psychopharmacology* 175206–214.

Kish, S., Shannak, K., Hornykiewicz, O. (1988). Uneven patterns of dopamine loss in the striatum of patients with idiopathic Parkinson's disease. *New England Journal of Medicine* **318**:876–880.

Lewis, D. A. (2001). The catecholamine innervation of the preforntal cortex. Chapter 3 In: M. V. Solanto, A. F. T. Arnsten, and F. X. Castellanos, (eds). *Stimulant drugs and ADHD: Basic and clinical neuroscience*. pp. 77–103. New York:Oxford University Press.

Lucey, J. V. (2001). The neuroanatomy of OCD. In: N. Fineberg, D. Mazzin, and D. J. Stein, (eds). *Obsessive-compulsive disorder: A practical guide.* pp. 77–87. London: M. Donitz.

Luciana, M., Collins, P. F., and Depue, R. A. (1998). Opposing roles for dopamine and serotonin in the modulation of human spatial working memory functions. *Cerebral Cortex* **8**:218–226.

Luciana, M., Burgund, E. D., Berman, M., and Hanson, K. L. (2001). Effects of tryptophan loading on verbal, spatial and affective working memory functions in healthy adults. *Journal of Psychopharmacology* 15219–230.

Manes, F., Sahakian, B., Clark, L., Rogers, R., Antoun, N., Aitken, M., et al. (2002). Decision-making processes following damage to the prefrontal cortex. *Brain* **125**:624–639.

Mattay, V. S., Goldberg, T. E., Fear, F., Hariri, A. R., Tessitore, A., Egan, M. F., et al. (2003). Catechol O-methyltransferase val [158] met genotype and individual variation in the brain response to amphetamine. *Proceedings of the National Academy of Sciences of the United States of America* **100**:6186–6191

Mavaddat, N., Kirkpatrick, P. J., Rogers, R. D., and Sahakian, B. J. (2000). Deficits in decision-making in patients with aneurysms of the anterior communicating artery. *Brain* **123**:2109–2117.

Mayberg, H. S. (2003). Modulating dysfunctional limbic-cortical circuits in depression: towards development of brain-based algorithms for diagnosis and optimized treatment. *British Medical Bulletin* **65**:193–207.

Mazur, J. E. (1987) Choice, delay, probability and conditioned reinforcement. *Animal Learning and Behavior* **25**:131–147.

McGaughy, J., Dalley, J. W., Morrison, C., Everitt, B. J., and Robbins, T. W. (2002). Selective behavioral and neurochemical effects of cholinergic lesions produced by 192 IgG saporin on attentional performance in a 5 choice serial reaction time task. *The Journal of Neuroscience* **22**:1905–1913.

Mesulam, M. M. and Geula, C. (1992). Overlap between acetylcholinesterase-rich and choline acetyltransferase-positive (cholinergic) axons in human cerebral cortex. *Brain Research* **577**:112–120.

Mesulam, M. M., Mufson, E. J., Levey, A. I., and Wainer, B. H. (1983). Central cholinergic pathways in the rat: an overview based on alternative nomenclature (Ch-1-Ch-6). *Neuroscience* **10**:1185–1201.

Mehta, M. A., Owen, A. M., Sahakian, B. J., Mavaddat, N., Pickard, J. D., and Robbins, T. W. (2000). Methylphenidate enhances working memory by modulating discrete frontal and parietal lobe regions in the human brain. *The Journal of Neuroscience* **20**:RC65 1–6.

Mehta, M. A., Swainson, R., Ogilvie, A. D., Sahakian, J., and Robbins, T. W. (2001). Improved short-term spatial memory but impaired reversal learning following the dopamine D2 agonist bromocriptine in human volunteers. *Psychopharmacology* **159**:10–20.

Mesulam, M. M., and Mufson, E. J. (1984). Neural inputs into the nucleus basalis of the substantia innominata (Ch.4) of the rhesus monkey. *Journal of Comparative Neurology* 107:253–274.

Middleton, H. C., Sharma, A., Agouzoul, D., Sahakian, B. J., and Robbins, T. W. (1999). Idazoxan potentiates rather than antagonizes some of the cognitive effects of clonidine. *Psychopharmacology* 145:401–411.

Mobini, S., Body, S., Ho, M. Y., Bradshaw, C. M., Szabadi, E., Deakin, J. F., et al. (2002). Effects of lesions of the orbitofrontal cortex on sensitivity to delayed and probabilistic reinforcement. *Psychopharmacology* 160:290–298.

Mobini, S., Chiang, T. J., Ho, M. Y., Bradshaw, C. M., and Szabadi, E. (2000). Effects of central 5-hydroxytryptamine depletion on sensitivity to delayed and probabilistic reinforcement. *Psychopharmacology* 152:390–397.

Muir, J. L., Everitt, B. J., and Robbins, T.W. (1996). The cerebral cortex of the rat and visual attentional function: dissociable effects of mediofrontal, cingulate, anterior dorsolateral and parietal cortex lesions on a five choice serial reaction time task in rats. *Cerebral Cortex* 6:470–481.

Murphy, F. C., Sahakian, B. J., Rubinsztein, J. S., Michael, A., Rogers, R. D., Robbins, T. W., et al. (1999). Emotional bias and inhibitory control processes in mania and depression. *Psychological Medicine* 29:1307–21.

Murphy, F. C., Smith, K. A., Cowen, P. J., Robbins, T. W., and Sahakian, (2002). The effects of tryptophan depletion on cognitive and affective processing in healthy volunteers. *Psychopharmacology* 163:142–153.

Nagahama, Y., Okada, T., Katsumi, Y., Hayashi, T., Yamauchi, H., Oyanagi, C., et al. (2001). Dissociable mechanisms of attentional control within the human prefrontal cortex. *Cerebral Cortex* 1185–92.

Neumeister, A., Konstantinidis, A., and Stastny, J., (2002). Association between serotonin transporter gene promoter polymorphism (5HTTLPR) and behavioral responses to tryptophan depletion in healthy women with and without family history of depression. *Archives of General Psychiatry* 59:613–20.

O'Doherty, J., Kringelbach, M. L., Rolls, E. T., Hornak, J., and Andrews, C. (2001). Abstract reward and punishment representations in the human orbitofrontal cortex. *Nature Neuroscience* 4:95–102.

Ongur, D., and Price, J.L. (2000). The organization of networks within the orbital and medial prefrontal cortex of rats, monkeys and human. *Cerebral Cortex* 10206–219.

Owen, A. M., Roberts, A. C., Polkey, C. E., Sahakian, B. J., and Robbins, T. W. (1991) Extra-dimensional versus intradimensional set shifting performance following frontal lobe excision, temporal lobe excision or amygdalo-hippocampectomy in man. *Neuropsychologia* 29993–1006.

Park, S. B., Coull, J. T., McShane, R. H., Young, A. H., Sahakian, B. J., Robbins, T. W., and Cowen, P. J. (1994). Tryptophan depletion in normal volunteers produces selective impairments in learning and memory. *Neuropharmacology* 33:575–588.

Passetti, F., Dalley, J. W., O'Connell, M. T., Everitt, B. J., and Robbins, T. W. (2000). Increased acetylcholine release in the rat medial frontal cortex. *European Journal of Neuroscience* 123051–3058.

Passetti, F., Dalley, J. W., and Robbins, T. W. (2003). Double dissociation of serotoninergic and dopaminergic mechanisms in attentional performance using a rodent five choice reaction time task. *Psychopharmacology* 165:136–145.

Pazos, A., Probst, A., and Palacios, J. M. (1987). Serotonin receptors in the human brain. IV Autoradiographic mapping of serotonin-2-receptors. *Neuroscience* 21:123–139.

Poeggel, G., Nowicki, L., and Braun, K. (2003). Early social deprivation alters monoaminergic afferents in the orbital prefrontal cortex of *Octodon degus*. *Neuroscience* 166:617–620.

Porrino, L. J., and Goldman-Rakic, P. S. (1982). Brain stem innervation of prefrontal and anterior cingulate cortex in the rhesus monkey revealed by retrograde transport of HRP. *Journal of Comparative Neurology* 205:63–76.

Preece, M. A., Dalley, J. W., Theobald, D. E. H., Robbins, T, W., and Reynolds, G. P. (2004). Region specific changes in forebrain 5-hydroxytryptamine receptors in isolation-reared rats an *in vitro* autoradiography study. *Neuroscience* 123:725–732.

Preuss, T. M. (1995). Do rats have prefrontal cortex? The Rose-Woolsey-Akert Program reconsidered. *Journal of Cognitive Neuroscience* **7**:1–24.

Rahman, S., Sahakian, B. J., Hodges, J. R., Rogers, R. D., and Robbins, T. W. (1999). Specific cognitive deficits in mild frontal variant frontotemporal dementia. *Brain* **122**:1469–1493.

Ridley, R. M., Haystead, T. A., and Baker, H. F. (1981). An analysis of visual object reversal learning in the marmoset after amphetamine and haloperidol. *Pharmacology, Biochemistry and Behavior*, **14**:345–351.

Robbins, T. W. (2000). Chemical neuromodulation of frontal-executive function in humans and other animals. *Experimental Brain Research* **133**:130–138.

Robbins, T. W. (2002). The Five-Choice Serial Reaction Time Task: Behavioral pharmacology and functional neurochemistry. *Psychopharmacology* **163**:362–380.

Robbins, T. W. and Everitt, B. J. (1995). Arousal systems and attention. In: M. Gazzaniga, et al. (eds). *The Cognitive Neurosciences*, pp. 703–720. Cambridge MA: MIT press.

Roberts, A. C., Robbins, T. W., Everitt, B. J., Jones, G. H., Sirkia, T. E., Wilkinson, J., et al. (1990). The effects of excitotoxic lesions of the basal forebrain on retention, acquisition and reversal of visual discrimination in the common marmoset. *Neuroscience* **34**:311–329.

Roberts, A. C., De Salvia, M. A., Wilkinson, L. S., Collins, P., Muir, J. L., Everitt, B. J., et al. (1994). 6-hydroxydopamine lesions of the prefrontal cortex in monkeys enhance performance on an analogue of the Wisconsin Card Sorting test: Possible interactions with subcortical dopamine. *The Journal of Neuroscience* **14**:2531–2544.

Rogers, R. D., Everitt, B. J., Baldacchino, A., Blackshaw, A. J., Swainson, R., Wynne, K., et al. (1999a). Dissociable deficits in the decision-making cognition of chronic amphetamine abusers, opiate abusers, patients with focal damage to prefrontal cortex, and tryptophan-depleted normal volunteers: evidence for monoaminergic mechanisms. *Neuropsychopharmacology* **20**:322–339.

Rogers, R. D., Owen, A. M., Middleton, H. C., Williams, E. J., Pickard, J. D., Sahakian, B. J., et al. (1999b). Choosing between small, likely rewards and large, unlikely rewards activates inferior and orbital prefrontal cortex. *Journal of Neuroscience* 199029–9038.

Rogers, R. D., Blackshaw, A. J., Middleton, H. C., Matthews, K., Hawtin, K., Crowley, C., et al. (1999c). Tryptophan depletion impairs stimulus-reward learning while methylphenidate disrupts attentional control in healthy young adults: implications for the monoaminergic basis of impulsive behavior. *Psychopharmacology* **146**:482–491.

Rogers, R. D., Andrews, T. C., Grasby, P. M., Brooks, D. J., and Robbins, T. W. (2000). Contrasting cortical and subcortical activations produced by attentional-set shifting and reversal learning in humans. *Journal of Cognitive Neuroscience* **12**:142–162.

Rogers, R. D., Tunbridge, E. M., Bhagwagar, Z., Drevets, W. C., Sahakian, B. J., and Carter, C. S. (2003). Tryptophan depletion alters the decision-making of healthy volunteers through altered processing of reward cues. *Neuropsychopharmacology* **28**:153–162.

Rolls, E. T. (1999). *The Brain and Emotion*. Oxford: Oxford University Press.

Rolls, E. T., Hornak, J., Wade, D., and McGrath, J. (1994). Emotion-related learning in patients with social and emotional changes associated with frontal lobe damage. *Journal of Neurology, Neurosurgery and Psychiatry* **57**:1518–1524.

Salmon, E., Garraux, G., Delbeuck, X., Collette, F., Kalbe, E., Zuendorf, G., Perani, D., Fazio, F., and Herholz, K. (2003). Predominant ventromedial frontopolar metabolic impairment in frontotemporal dementia *Neuroimage* **20**:435–440.

Santana, N., Bortolozzi, A., Serrats, J., Mengod, G., and Artigas, F. (2004). Expression of serotonin$_{1A}$ and serotonin$_{2A}$ receptors in pyrmaidal and GABergic neurons of the rat preforntal cortex. *Cerebral Cortex* **14**:1100–1109.

Sawaguchi, T. and Goldman-Rakic, P. S. (1991). D1 Dopamine-receptors in prefrontal cortex—involvement in working memory. *Science* **251**:947–950.

Saxena, S., and Rauch, S. L. (2000). Functional neuroimaging and the neuroanatomy of obsessive - compulsive neurosis. *The Psychiatric Clinics of North America* **23**:563–586.

Sesack, S. R., Deutch, A. Y., Roth, R. H., and Bunney, B. S. (1989). Topographical organization of the efferent promections of the medial prefrontal cortex in the rat: an anterograde tract-tracing study with *Phaseolus vulgaris* leucoagglutinin. *Journal of Comparative Neurology* 290213–242.

Smith, K. A., Morris, J. S., Friston, K. J., Cowen, P. J., and Dolan, R. J. (1999). Brain mechanisms associated with depressive relapse and associated cognitive impairment following acute tryptophan depletion. *British Journal of Psychiatry* **174**:525–529.

Swainson, R., Rogers, R. D., Sahakian, B. J., Summers, B. A., Polkey, C. E., and Robbins, T. W. (2000). Probabilistic learning and reversal deficits in patients with Parkinson's disease or frontal or temporal lobe lesions: possible adverse effects of dopaminergic medication. *Neuropsychologia* **38**:596–612.

Taghzouti, K., Louilot, A., Herman, J. P., Le Moal, M., and Simon, H. (1985). Alternation behavior, spatial discrimination, and reversal disturbances following 6-hydroxydopamine lesions in the nucleus accumbens of the rat. *Behavioral and Neural Biology* **44**:354–363.

Thorpe, S. J., Rolls, E. T., and Maddison, S. (1983). The orbitofrontal cortex: neuronal activity in the behaving monkey. *Experimental Brain Research* **49**:93–115.

Tranel, D., Bechara, A., and Denburg, N. L (2002). Asymmetric functional roles of right and left ventromedial prefrontal cortices in social conduct, decision-making, and emotional processing. *Cortex* **38**:589–612.

Tronel, S., Feenstra, M. G. P., and Sara, S. J. (2004). Noradrenergic action in prefrontal cortex in the late stage of memory consolidation. *Learning and Memory* **11**453–458.

Usher, M., Cohen, J. D., Servan-Schreiber, D., Rajkowski, and Aston-Jones, G. (1999). The role of locus coeruleus in the regulation of cognitive performance. *Science* **283**:549–554.

Vertes, R. P. (2004). Differential projections of the infralimbic and prelimbic cortex in the rat. *Synapse* **51**:32–58.

Volkow, N. D. and Fowler, J. S. (2000). Addiction, a disease of compulsion and drive: involvement of the orbitofrontal cortex. *Cerebral Cortex* **10**:318–325.

Williams, G. V., Rao, S. G., and Goldman-Rakic, P. S. (2002). The physiological role of 5-HT$_{2A}$ receptors in working memory. *The Journal of Neuroscience* **22**:2843–2854.

Wilson, J. M., Kalasinsky, K. S., Levey, A. I., Bergeron, C., Reiber, G., Anthony, R. M., et al. (1996). Striatal dopamine nerve terminal markers in human, chronic methamphetamine users. *Nature Medicine* **2**:699–703.

Winstanley, C., Dalley, J. D., Chudasama, Y., Theobold, D., and Robbins, T. W. (2003). Intra-prefrontal 8-OH-DPAT and M100907 improve visuospatial attention and decrease impulsivity on the five choice serial reaction time task in rats. *Psychopharmacology* **167**:304–314.

Winstanley, C. A., Theobald, D. E. H., Cardinal, R. N. C., and Robbins, T. W. (2004). Contrasting roles of basolateral amygdala and orbitofrontal cortex in impulsive choice. *Journal of Neuroscience* **24**:4718–22.

Winstanley, C. A., Theobald, D. E. H., Dalley, J. W., Glennon, J. C., and Robbins, T. W. (2004). 5-HT2A and 5-HT2C receptor antagonists have opposing effects on a measure of impulsivity: interactions with global 5-HT depletion. *Psychopharmacology* **176**:376–385.

Winstanley, C. A., Theobald, D. E. H., Dalley, J. W., and Robbins, T. W. (2005). Interactions between serotonin and dopamine in the control of impulsive choice in rats.: therapeutic implications for impulse control disorders. *Neuropsychopharmacology*, **30**:669–682.

Winstanley, C. A., Theobald, D. E. H., Dalley, J. W., Cardinal, R. N. C., and Robbins, T. W. (2006). Double dissociation between serotonergic and dopaminergic modulation of medial prefrontal orbitofrontal cortex during a test of impulsive choice. *Cerebral Cortex* **16**:106–114.

Young, S. N., Smith, S. E., Pihl, R. O., Ervin, F. R. (1985). Tryptophan depletion causes a rapid lowering of mood in normal males. *Psychopharmacology* **87**:173–177.

Young, S. N., Smith, S. E., Pihl, R. O., Finn, P. (1989). Biochemical aspects of tryptophan depletion in primates *Psychopharmacology* **98**:508–511.

Zaborszky, L., Gaykema, R. P., Swanson, D. J., and Cullinan, W. E. (1997). Cortical input to the basal forebrain. *Neuroscience* **79**:1051–1078.

Zahrt, J., Taylor, J. R., Mathew, R. G., and Arnsten, A. F. (1997). Supranormal stimulation of D1 dopamine receptors in the rodent prefrontal cortex impairs spatial working memory performance. *The Journal of Neuroscience* **17**:8528–8535.

Zald, D. H., Kim, S. W. (2001). The orbitofrontal cortex. In: S. P. Salloway, P. F. Malloy, P. F., and J. D. Duffy, (eds). *The frontal lobes and neuropsychiatric Illness*, pp. 33–69. Washington:American Psychiatric Association Publications.

Chapter 17

# Technical considerations for BOLD fMRI of the orbitofrontal cortex

V. Andrew Stenger

## 17.1. Introduction

Neuroimaging plays a major role in enhancing our understanding of the orbitofrontal cortex (OFC). The purpose of this chapter is to present some of the technical challenges, limitations, and potential solutions with regard to using magnetic resonance imaging (MRI) to study the OFC. Although there are other imaging techniques which provide valuable information about the OFC, MRI is becoming one of the more common methods due to its non-invasive nature as well as its high spatial resolution and flexibility. One of the major challenges with MRI, however, is that it has low sensitivity or low signal to noise ratio (SNR). As a result, it often takes several minutes to acquire an image with a high resolution and brain coverage. One of the reasons that the reader may hear about MRI scanners going to higher field strengths, such as 3 Tesla (T), is that higher fields produce a larger MRI signal and improved SNR. A disadvantage of MRI at higher fields is an increased amount of artifact present in the images due to the physics of MRI in the human body. It is crucial to balance the tradeoffs between the improved imaging of brain function and the amount of artifact present in images as one goes to higher field strengths.

In this chapter we will focus on *magnetic susceptibility* in blood oxygen level dependent (BOLD) functional MRI (fMRI). The BOLD fMRI technique is gaining widespread use as one of the methods of choice for imaging brain function in neuropsychiatric research and clinical investigations. Magnetic susceptibility is important in fMRI because BOLD contrast arises from the difference in magnetic susceptibility between oxygenated and de-oxygenated blood. Magnetic susceptibility is a measure of how much a material (i.e. deoxyhemoglobin) is magnetized by an applied magnetic field (i.e. the scanner magnetic field) and becomes larger at higher fields. The use of high fields is desirable for fMRI due to increased BOLD contrast from the larger susceptibility differences between levels of blood oxygenation. A major concern, however, is that there are larger differences between the magnetic susceptibility of brain tissue and of the air in cavities in the head as well. These susceptibility differences produce distortions and large regions of no signal in the functional images, and are particularly severe in the OFC because of its proximity to the sinus regions. These distortions and signal voids are called *magnetic susceptibility artifacts* and are always present in fMRI data of inferior brain regions. Susceptibility artifacts become worse at higher field strengths. This 'catch-22' between increasing the BOLD contrast at high field and increasing image artifacts remains one of the major technical challenges in imaging the OFC with BOLD fMRI.

Susceptibility artifacts are such a common problem that the typical fMRI experiment is simply incapable of observing changes in the OFC unless specific steps are taken to rectify the problem. An absence of findings in the OFC (particularly the medial OFC) in many fMRI studies cannot be treated as a true negative finding because of susceptibility artifacts. There may also be bias, based on brain location as well. Susceptibility artifacts are greatest in the posterior-medial OFC, almost always encompassing areas 14, 13, and the subgenual cingulate regions (area 25). This may lead to a general bias in the literature towards observing more lateral or anterior foci over the more-difficult-to-image posterior-medial OFC. Furthermore, there may be a bias in fMRI studies of developing brain. As the brain matures in a developing adolescent and young adult, the degree of the susceptibility artifact will change as well. Differences in fMRI activation between age groups may be attributed to the changing geometry of the cortex, skull and sinuses, rather than the underlying brain circuitry.

Below we will present some of the basic physics of MRI and magnetic susceptibility artifacts, as well as show some methods that can be used to address this limitation. Many of the techniques, such as parallel imaging and field mapping, are becoming standard on many commercial MRI scanners, and the basic concepts presented should help the reader understand why investigators need to exploit these capabilities. Other methods, such as tailored RF pulses or gradient compensation, represent more sophisticated techniques that may or may not find their way to commercial systems but are indicative of the innovative research performed by physicists and engineers to address this important limitation in fMRI. This chapter is not meant to include every new technique developed to improve fMRI in regions with susceptibility artifacts. These techniques are continuously evolving, and are topics of intense research themselves. Furthermore there is no simple way to describe or deal with susceptibility artifacts without having to understand some MR physics. The goal of this chapter is to provide a neuroscientist with the concepts necessary to understand the principles underlying the problem and its potential solutions.

## BOLD contrast fMRI

The goals of neuroscience include understanding how the brain functions using fundamental sciences such as biology and chemistry, understanding how the brain produces complex human behavior, and understanding the origins of psychiatric or neurological illnesses and potential treatments. Recently imaging tools have become available for studying the function of the brain in alert, behaving humans. All methods are extremely valuable, and good neuroscience requires the use of all available technologies with the understanding of their relative strengths and weaknesses. Neural electromagnetic modalities, such as electro- and magneto-encephalography (EEG, MEG) (Naatanen 1992; Roland 1993), can be used to acquire important information about the time course of brain activity. Although these methods offer high temporal resolution, they provide poor and ambiguous spatial information and do not provide information about the anatomy of the brain. They are also limited by a noisy signal and only regions close to the skull can be sensed, leaving deeper brain structures such as the OFC difficult to image. Positron emission tomography (PET) and single photon emission computed tomography (SPECT) provide improved spatial resolution and greater coverage. PET studies of regional cerebral blood flow have been particularly important for advancing the understanding of OFC functions because the susceptibility issues that have often hindered fMRI investigations of the region do not affect PET. However, the limited temporal resolution and required use of radioactive isotopes provides a limit on the types of studies that can be performed with PET. Furthermore, these techniques are very costly and require significant infrastructure, such that relatively few sites can perform research studies with these techniques. Invasive optical imaging techniques (Grinvald *et al.* 1986, 1991) can achieve very high spatial and temporal resolution but have limited utility in human studies. Near Infrared (NIR) optical imaging techniques are non-invasive but have low spatial resolution (Gratton *et al.* 1997).

Methods have also been developed for the use of MRI to map functional brain activity. The first demonstration of brain activation with MRI (Belliveau *et al.* 1991) used a contrast agent to map cerebral blood volume during visual stimulation. Other functional MRI methods that do not require injection of contrast agents have been developed. Among these are several methods for imaging perfusion and blood flow (Detre *et al.* 1992; Kwong *et al.* 1992; Edelman *et al.* 1994; Kim 1995; Talagala and Noll 1998) that show great promise, but have not yet come into widespread use due to low signal strengths and lack of a standard approach. Furthermore, perfusion MRI methods are not available as part of commercial scanner software packages and require specific expertise for their successful implementation. The MRI method for functional imaging that has shown the most immediate promise is based on BOLD contrast (Bandettini *et al.* 1992; Kwong *et al.* 1992; Ogawa *et al.* 1992). The fMRI technique is capable of observing localized events in the whole brain at time scales that are useful for many cognitive paradigms. Hardware and software for performing fMRI is standard on all of the major MRI scanner platforms and are relatively straightforward to use. Many neuroscience and clinical psychiatry departments have dedicated MRI scanners for performing fMRI. Furthermore, the combination of fMRI with other neuroimaging methods such as PET and EEG is becoming more common as a means of answering more complex questions.

The premise behind BOLD fMRI is that the concentration of deoxyhemoglobin in blood affects the local magnetic field distribution around the blood vessels, changing it slightly from the global magnetic field produced by the magnet from the scanner. This perturbation of the magnetic environment arises from the strong field produced by the unpaired electrons used to bind oxygen to iron in hemoglobin. These extra local magnetic field gradients destroy the *coherence* of the proton spins near blood vessels, which results in a loss or decay of the MRI signal in the region. If the MRI acquisition is set such that there is a slight delay between the excitation and reception of the data from the spins (this delay is defined as the *time of echo* or TE), the loss of coherence is allowed to accumulate and the overall image intensity will be attenuated. The type of MRI scan, which is tuned for seeing susceptibility-related effects, is called a *gradient echo* acquisition, and the type of contrast is called T2* contrast. The parameter T2* is a phenomenological time constant that reflects the decay time for the gradient echo MRI signal due to a physical process. In the case of BOLD fMRI, the process is the interaction between the proton spins and magnetic fields produced by susceptibility of deoxyhemoglobin. One sets TE = T2* (approximately 25 ms at 3T) to sensitize the MRI scan to the deoxyhemoglobin concentration. In regions with increased blood oxygenation, such as during a neuronal event, there will be a reduced concentration of deoxyhemoglobin and the magnetic environment will be more homogenous. This increase in field homogeneity means it now takes longer for the MR signal to decay, and the image intensity near neuronal activity will be a little greater. BOLD contrast in fMRI is obtained from the statistically significant difference between images acquired during task and rest. This difference, or the BOLD response, reflects the increased magnitude of the signal within pixels in the images due to decreased deoxyhemoglobin concentration.

## Magnetic susceptibility

In order to understand how susceptibility artifacts occur in MRI, we will now present some mathematical definitions. One can define a physical property associated with hemoglobin or any material that indicates how well it is magnetized by an applied magnetic field. This property is called *magnetic susceptibility*. In mathematical terms, a simple linear model would make the induced magnetization $M(\mathbf{r})$ in the material proportional to the applied field $B_0$ (Schenck 1996):

$$M(\mathbf{r}) = \chi(\mathbf{r})B_0. \tag{1}$$

The $B_0$ field used in MRI is measured in units of Tesla and arises from the large magnet used in the scanner (i.e. 3T). The constant of proportionality $\chi(\mathbf{r})$ between the applied field and the magnetization is defined as the magnetic susceptibility; and for brain tissue (mainly water), it is typically on the order of $-10^{-6}$. Magnetic susceptibility is material dependent, and arises from different sources within a material. The magnetic susceptibility of water is a result of the orbital motion of electrons, which form small current loops creating dipoles, which oppose the applied field. Because it opposes the applied field it is *diamagnetic*. Deoxyhemoglobin is slightly *paramagnetic* ($+10^{-6}$) and results from the unpaired electrons used to bind oxygen to the iron atoms (minus the orbital shielding of the molecule). It is paramagnetic because the electron spins are aligned with the applied magnetic field $B_0$. Hemoglobin on the other hand is slightly more diamagnetic than water. The spatial variation (or gradients) in the magnetic susceptibilities in regions with vessels with greater concentrations of oxyhemoglobin and deoxyhemoglobin is the basic mechanism of BOLD contrast. These gradients produce more T2* decay in regions with greater deoxyhemoglobin concentration.

Ignoring contrast, an MR image is a map of the proton magnetization $M(\mathbf{r})$. Spin is a physical quantity that results from the intrinsic magnetic moment of the proton. The proton spins are very sensitive to the local magnetic field $B(\mathbf{r})$. In an MRI experiment, the protons precess about the local magnetic field at the Larmor frequency:

$$\begin{aligned}\omega(\mathbf{r}) &= \gamma B(\mathbf{r}) \\ &= \gamma(B_0 + \mathbf{G}(\mathbf{r})\cdot\mathbf{r} + \mathbf{G}_s(\mathbf{r})\cdot\mathbf{r})).\end{aligned} \quad [2]$$

Here $\gamma$ is the gyromagnetic ratio, which is a constant that is indicative of the strength of the magnetic moment of the proton. This precession is similar to the precession of a spinning top about the Earth's gravitational field. The local magnetic field is a superposition of the main static field $B_0$, the imaging gradients $\mathbf{G}(\mathbf{r})$ used for spatial encoding, and the gradients produced by susceptibility changes in the tissue $G_S(\mathbf{r})$. The application of imaging gradients as part of the local magnetic field gives rise to a spatial distribution of Larmor frequencies. This is fundamental to spatial localization of the proton spins in MRI. Image reconstruction in MRI is essentially the unscrambling of the frequencies of the protons from the raw signal and creating a spatial map or image based on an *a priori* knowledge of the applied gradients. As we will see below, if we are imaging a region that has large gradients from susceptibility variations, then the true mapping of frequency to location becomes unknown and the reconstructed image will be distorted. In severe cases, the gradients from the susceptibility variations will be strong enough that the signal in these voxels will be completely dephased due to T2* relaxation before data acquisition.

The signal detected in an MRI experiment contains information about the precession frequency of the nuclear spins. This information includes their spatial locations as a result of the imaging gradients as well as information about intrinsic fields and gradients in the material. The sensitivity of the nuclear spin system to its magnetic environment is what makes magnetic resonance experiments so powerful. In mathematical terms, the MRI signal is the sum or the 'integral' of the spin density $M(\mathbf{r})$ in the volume or 'slice' of interest. The signal is acquired as a function of time during which the spins precess about the local magnetic field at the Larmor frequency $\omega(\mathbf{r})$. In this manner the signal is encoded with the spatial information as well as the other magnetic field effects including BOLD contrast. The equation for the MRI signal can be written as:

$$s(t) = \int_{slice} M(\mathbf{r}) e^{-i\omega(r)t} d\mathbf{r}$$

$$\approx e^{-i\gamma B_0 t} \int_{slice} M(\mathbf{r}) e^{-i k(t)\cdot\mathbf{r} - t/T2^*(r)} d\mathbf{r}. \quad [3]$$

The second line in the above equation was obtained from Eq. [2] where we introduce the "k-space" vector $k(t)$ for representing the effect of the applied gradients and we approximated the effect of the susceptibility gradients $G_S(\mathbf{r})$ as a spatially dependent $T2^*(\mathbf{r})$

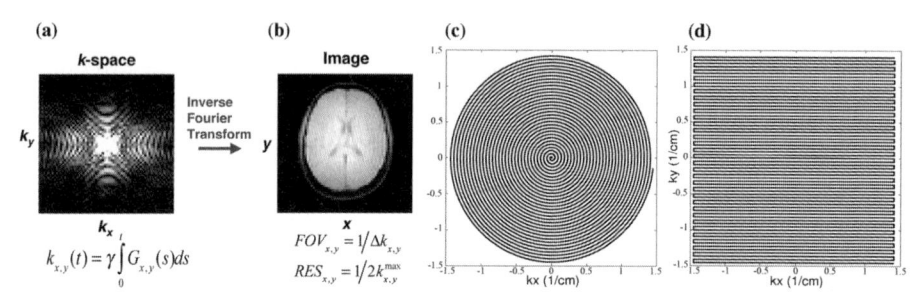

**Fig. 17.1** (a) The MRI k-space data can be placed on a Cartesian grid and (b) an inverse Fourier transform is used to reconstruct the image. The k-space locations are given by the integral of the applied gradients over time and the spacing and extent of the k-space data determine the FOV and resolution, respectively. (c) Spiral and (d) EPI k-space trajectories.

signal decay. The term k-space is a mathematical construct that allows one to understand the effects of the imaging gradients. The k-space location of the signal at a time $t$ is determined by the net effect of the imaging gradients at that time:

$$k(t) = \gamma \int_0^t G(s)ds$$
$$\approx \gamma G t. \quad [4]$$

If the data are acquired at the echo time TE, the reconstructed image will be equal to the true image $M_0(\mathbf{r})$ times a $T2^*(\mathbf{r})$ exponential decay term that scales the image intensity at different locations:

$$M(\mathbf{r}) = e^{-TE/T2^*(\mathbf{r})} M_0(\mathbf{r}). \quad [5]$$

Spins near blood vessels with lower concentrations of deoxyhemoglobin will have slightly longer (on the order of a few ms) $T2^*$ than spins with more deoxyhemoglobin due to the more uniform internal field. Therefore the pixel intensity in these image regions will be a little larger, giving rise to the BOLD effect. In regions where there are large susceptibility gradients from air-tissue interfaces, $T2^*$ can become very short and there will be little to no signal in the image in these regions. Fig. 17.1 (a) shows an example of the MRI signal displayed as an image of the k-space data and (b) shows the reconstructed image. The k-space is determined by the imaging gradients and defines the *field of view* (FOV) and *resolution* of the desired image. Higher resolutions or larger FOV's require more k-space sampling. Fig. 17.1 (c) and (d) show the two most common k-space trajectories used in fMRI: Spiral (Meyer et al. 1992; Noll et al. 1995) and echo planar imaging (EPI) (Mansfield 1977). The reconstruction of the image from the signal is performed using knowledge of the k-space trajectory. In the case of fMRI, one needs a snapshot of the brain and the spiral or EPI readout tends to be long (20–40 ms) as a result. Significant $T2^*$ decay will occur during these long readouts producing images that are distorted or have signal loss.

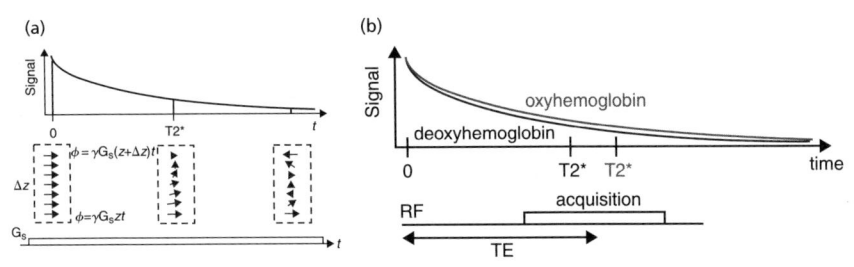

**Fig. 17.2** (a) Schematic of the T2* decay in a voxel with a susceptibility gradient along the z-direction. The phase difference across the voxel increases with time, producing a decrease in the vector sum of proton spins. (b) Diagram demonstrating the difference in T2* decay curves in a voxel containing vessels and capillaries with greater concentrations oxyhemoglobin or deoxyhemoglobin. The susceptibility gradients are larger in voxels with more deoxyhemoglobin concentration. This produces a faster T2* decay of the spins in the voxel and a decreased signal intensity. An fMRI sequence is weighted for BOLD contrast by putting TE where there is a large difference between the decay curves.

Fig. 17.2 (a) shows a schematic of signal loss through a voxel and how it can be parameterized by a T2* decay term. If there is a small background gradient $G_S$ along the z-direction due to a susceptibility change, for example, it will produce a phase difference through the voxel. As time progresses, the phase difference increases and the net magnetization, or the vector sum, of the proton spins in the voxel decreases. This is true for the x and y directions as well. In this manner, a spatially dependent susceptibility variation such as that produced by decreased deoxyhemoglobin will produce regions with less signal loss than others. The fMRI acquisition is sensitized to the BOLD T2* contrast by positioning the TE of the readout gradients to occur when there is the largest difference between the decay curves of tissue containing primarily deoxyhemoglobin or oxyhemoglobin. This time is typically 25 ms for 3 Tesla. Fig. 17.2 (b) shows a diagram of the decay curves for tissue with more deoxyhemoglobin or oxyhemoglobin and the positioning of the data acquisition at TE.

## Susceptibility artifacts

The objective of an fMRI acquisition is to use T2* contrast to maximize the differences in image intensity due to susceptibility variations between regions with varying degrees of deoxygenated blood. One major confound in fMRI is that the images are weighted to all susceptibility-related effects as well. Magnetic susceptibility variations in the brain due to differences in tissue types are much larger and have a greater spatial extent than those due to BOLD effects. These variations can produce two types of susceptibility artifacts: (1) image *distortions* and (2) *signal loss* due to strong T2* decay. In fMRI imaging modalities, such as EPI or spiral, the effects of these variations are magnified due to long readout times coupled with long echo times.

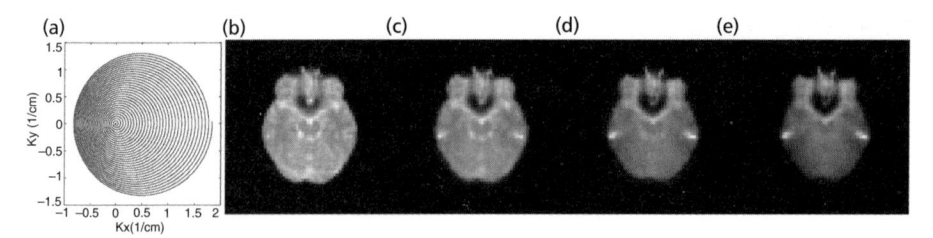

**Fig 17.3** (a) A simple diagram showing the shifting of a spiral k-space trajectory due to a linear susceptibility gradient along the x-direction. The 'true' k-space trajectory used for an image acquisition is typically unknown and is assumed to be ideal. (b–e) Simulations of the blurring artifact in spiral images (64 × 64 matrix size, 5 mm thick slice, 25 ms TE) with readout lengths of 5, 15, 25, and 35 ms, respectively. Note the loss of detail in the images as they become more blurred. The signal in the OFC is reduced as signal gets mapped to the wrong regions.

## Image distortion

The first type of artifact is image distortion or blurring. Image distortion is milder than complete signal loss and manifests itself not as a void but as a displacement of signal. In more mathematical terms, image distortion results from an improper mapping of the acquired data to the true k-space location during reconstruction. In other words, the gradients produced by magnetic susceptibility effects are indistinguishable from the gradients used for imaging and falsely spatially encode the MRI signal. For the susceptibility gradient $G_S(\mathbf{r})$, the k-space location of the acquired data can be defined as

$$k_{acquired}(t) = k(t) + \gamma G_S(\mathbf{r})t. \quad [6]$$

This equation can be inferred from the previous equations. This equation illustrates that as time progresses, the true k-space locations get more and more shifted from the ideal k-space. In the case of spiral imaging or EPI, where the readout times are very long (20–40 ms), the images can be very distorted and blurred compared to more conventional structural imaging modalities where readouts are shorter (5 ms). Due to the Cartesian or grid-like k-space trajectory of EPI, images will become stretched and skewed. Spiral images will become radially blurred because of the circular k-space trajectory. In both cases, the signal in a pixel is smeared into other pixels. Often one describes this smearing by a mathematical construct called the 'point spread function'. Resolution in MRI is not given by the pixel size, but by the characteristic width the point spread function. Knowledge of the k-space errors prior to reconstruction can allow for the correction of image distortion (using a field map for example). Also, the reduction of readout durations (with parallel imaging for example) can minimize image distortion as well. Figure 17.3 shows an example of the shifting of k-space due to a gradient along the x-direction as well as spiral images that contain blurring artifacts for increasingly longer readouts.

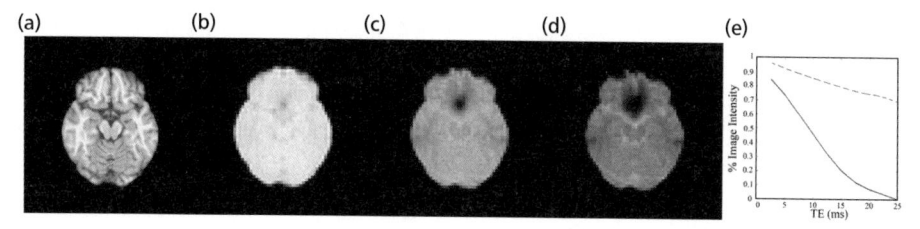

**Fig. 17.4** (a) T1-weighted 256 × 256 gradient echo image of a 5 mm thick axial slice at 3T. (b–d) Simulation of intravoxel dephasing artifact in 64x64 T2*-weighted images at echo times of 5, 15, and 25 ms, respectively. (e) Plot of the loss in image intensity in the OFC (solid line) compared to that to a region in the anterior of the slice (dashed line) where there is no susceptibility artifact as a function of TE. If TE is equal to 25 ms, there will be little to contribution to the MRI signal from the OFC.

## Signal loss

The second type of susceptibility artifact is signal loss or intravoxel dephasing. Complete signal loss will occur if the susceptibility gradient or variation across a voxel is so large that by the time the data are acquired the vector sum of the spins is zero. In other words, T2* is so short that the entire signal has decayed before the data acquisition even begins. The difficulty of the signal loss artifact is that it cannot be corrected post-acquisition nor can it be corrected completely by shortening readout durations. The acquisition needs to be modified to have a compensatory gradient or phase to cancel this effect. Gradient compensation and tailored RF pulses will be presented below as examples of modified data acquisitions. Fig. 17.4 (a–d) shows example images with signal loss artifact at 3T in the OFC. This Fig. shows that as TE increases there is an overall loss in image intensity, however, the T2* contrast between gray and white matter improves. Therefore, increasing TE is necessary to increasing tissue T2* contrast at the expense of overall image SNR. The T2* contrast due to BOLD is very small and is unobservable in a single image, which necessitates the use of statistical techniques and signal averaging in an fMRI experimental design. In the OFC regions, there is a complete loss of signal due to the close proximity of the sinus cavities. In this case the T2* is so short that all of the signal components from the OFC have decayed away before the signal was measured. The signal decay curves of the OFC and an anterior region are shown in the Fig. 17.4 (e). The plot indicates that in order to get reasonable BOLD contrasts at 3T using a TE of 25 ms there will be no MRI signal in the OFC.

## 17.2. Susceptibility artifact reduction methods
### Shimming

There are numerous ways for reducing susceptibility artifacts in fMRI. It is a difficult task to list every single technique, and we will only attempt here to cover the basic techniques, keeping in mind that most of the other methods are variants or combinations of these techniques. Probably the easiest way to reduce susceptibility artifacts is to simply shorten TE. This approach is impractical for fMRI because long TE's are required for functional contrast. Shortening TE to obtain functional contrast in a region with large susceptibility artifacts will eliminate the contrast in the rest of the brain. Another approach is to use automatic shimming with the shim coils on the MRI scanner (Reese et al. 1993, 1995). Shimming addresses both distortion and signal loss artifacts. Shim coils produce localized magnetic field sources that emanate from different positions in the MRI scanner. The currents in these coils can be adjusted to create small, compensatory fields to make the internal field more homogenous. Some form of shimming is a common procedure on all MRI scanners and does help with the bulk field inhomogeneity. Unfortunately, because the susceptibility effects which produce artifacts in fMRI have rapid spatial variations, a larger number of higher-order shims are required to truly fine-tune to the degree that is necessary. Although these shims are not commercially available, there are investigations into the use of passive and active oral shims for fMRI (Wilson and Jezzard 2003; Hsu and Glover 2005). The results from these studies look promising for the OFC, however these techniques may be slightly invasive of the subject, sometimes requiring the insertion of a coil or magnetic bar into the mouth. Nonetheless, as the technology develops, it is likely that there will be specialized head shim coils for fMRI in the future.

### Data post-processing

A common and practical approach for reducing image distortion is to perform a correction on the data during image reconstruction. These image *post-processing* methods can remove geometric distortions in EPI and blur in spiral images but not signal loss artifact. In a typical post-processing implementation, a *field map* is acquired along with the fMRI data set. This procedure is very quick, taking only a few seconds. In an MRI scan one typically only uses the magnitude of the reconstructed image, however, the image is really comprised of complex numbers with both a magnitude and a phase. The phase image is representative of the orientation of the spins in each pixel. The field map is usually constructed from the difference in the phases between two gradient echo images acquired at different echo times. Taking the difference between the phase images and dividing by the difference in time will give a frequency:

$$\omega_S(\mathbf{r}) \frac{\phi(TE_2) - \phi(TE_1)}{TE_2 - TE_1}. \qquad [7]$$

This technique is adequate as long as the two TEs are chosen such that there are no wraps in the phases and fat suppression is used. The first thing that can be done with $\omega_S(\mathbf{r})$ is to find the constant and linear terms through a numerical fitting routine (Irarrazabal et al. 1996). These terms are very easy to correct in the signal using Eq. [3], requiring a minimal computation time penalty. If a more complete correction is desired, several more computationally intensive approaches can be used for reconstruction including iterative and time-segmentation methods (Noll et al. 1991, 1992; Kadah and Hu 1997; Sutton

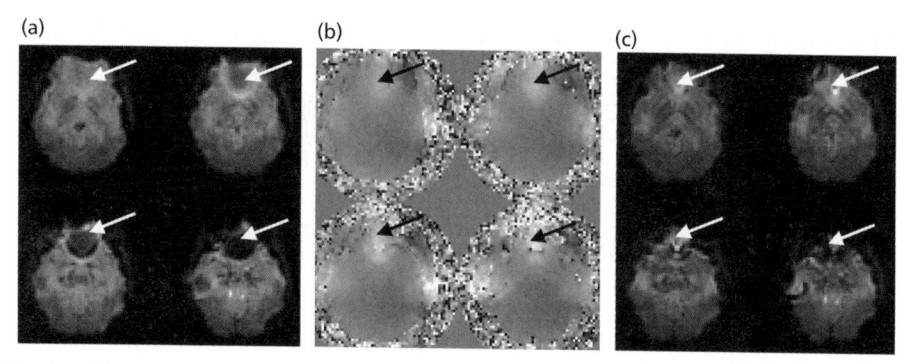

**Fig. 17.5** (a) Single-shot spiral images at 3T acquired with a 25 ms TE (3.2 mm thick slices and 64 × 64 matrix over a 20 cm FOV). (b) Corresponding field maps for each slice. The images are windowed between $-\pi/2$ and $\pi/2$ radians/ms. White regions have large frequency values due to susceptibility effects. The gray regions are homogeneous. (c) The same slices reconstructed using the field map. Note the recovered image magnitude in the areas where there are larger off-resonance frequencies.

*et al.* 2003). Although these involve more intensive post-processing, increased image reconstruction time is typically not a penalty in fMRI because it is performed off-line after the study is completed.

An advantage of post-processing techniques is that they only require a field map measurement, which is easily obtained in a few seconds prior to the fMRI scan, or even during the scan with no time penalty. There are several dynamic approaches for acquiring the field map that allow for a real-time correction to reduce confounds, such as head motion, and address other artifacts such as physiological noise (Sutton *et al.* 2004). Post-processing techniques, however, will be ineffective in severe cases of field inhomogeneity where there is significant intravoxel dephasing, because much of the signal will be lost. Nonetheless, a field map-based post-processing technique should be used in most fMRI studies and is often included as part of the scanner's commercial package for fMRI. Fig. 17.5 shows results of post-processing using a field map in spiral images containing the OFC. This correction used both a linear correction and a time-segmented method. Note that the field maps in (b) show larger frequencies (indicated by the white areas) in the OFC, which is also where the fMRI images in (a) show distortion. The corrected images are shown in (c).

## Reduced data methods and parallel imaging

Shortening the *k*-space trajectory and readout duration can also reduce image distortion. As discussed above, the resolution and FOV in MRI are determined by the extent as well as the spacing of the acquired data in *k*-space. Higher resolutions and larger FOVs require more *k*-space samples, which produces longer readouts creating more susceptibility artifacts. The easiest way to shorten readouts is to speed up the gradients. Unfortunately faster gradients can lead to patient safety concerns in the form of nerve stimulation. Another way of speeding up the acquisition is by acquiring only every other or every third line of *k*-space and reconstructing the under-sampled image using some *a priori* knowledge of the data. Shortening the readout length will reduce image distortion, however, there will be a penalty in overall image SNR. The loss in SNR is equal to the square root of the *k*-space reduction *R*:

$$SNR = \frac{SNR_0}{\sqrt{R}}.$$
[8]

Here $SNR_0$ is the optimal SNR. The tradeoff between lost SNR and improved image quality needs to be carefully considered in all reduced data methods. A second advantage of reduced data techniques is a throughput increase because shortened readouts lead to more slices per TR.

One reduced data technique is called a *partial k-space* acquisition, which relies on symmetry or redundancy in the acquired data (Stenger and Noll 1997; Jesmanowicz et al. 1998). This redundancy stems from the idea that the Fourier transform (the *k*-space data) of a real object (the image) has an underlying symmetry called *Hermitian symmetry*. Hermitian symmetry means that one half of the data is equal to the complex conjugate of the second half. If one acquires only half of the data, synthesizing the second half is a simple mathematical operation that can be performed after the data are acquired. Another reduced data approach uses the idea that fMRI data are acquired repeatedly over time such that the sampling domain is really *k-t-space*. By periodically varying the locations of the missing data during each successive image acquisition, the image aliases produced by the under-sampling will be periodic as well. The image aliases can be removed by filtering in the temporal frequency dimension (UNFOLD) (Madore et al. 1999). This idea relies on the assumption that fMRI paradigms are designed such that they produce activation with well-defined frequencies and the image aliases can be stored in the holes where there is no activation. Both partial *k*-space techniques and UNFOLD have the expected loss in SNR for the reduction *R* in the readout.

It is now becoming standard for MRI scanners to have *phased array* or multi-receiver head coils. Multi-receiver coils can be used for reducing data readouts using *parallel imaging* techniques (Huchinson and Raff 1988; Kelton et al. 1989; Kwait and Einav 1991; Ra and Rim 1991; Sodickson and Manning 1997; Pruessmann et al. 1999). The basic idea is that each receiver coil has a unique view or *sensitivity* to different brain regions depending on its location around the head. This provides a coarse spatial localization technique that provides additional information allowing for less *k*-space data to be acquired. There are several variants of parallel imaging that are currently being used, with the sensitivity encoding (SENSE) method being a popular approach, because it provides a generalized framework in which arbitrary coil configurations and *k*-space trajectories can be used. Like the other reduced data techniques the *k*-space reduction is parameterized by a reduction factor *R* and the readout is also reduced by a factor *R* in a single-shot imaging modality with fixed resolution. Although the reduction factor typically scales with the number of coils, the geometry of the coils and the image SNR limit the possible acceleration factors that can be achieved.

The advantages of parallel imaging have been explored for fMRI (de Zwart et al. 2001, 2002; deZwart et al. 2002). These results show that the loss in the activation sensitivity due to decreased SNR is compensated by the gain in SNR using the array coil compared to a standard volume head coil. The reason array coils can have more SNR than volume coils is that they tend to be closer to the brain. The decreased readouts allowed by parallel imaging have also been shown to reduce susceptibility artifacts in fMRI as well (Weiger et al. 2002). Parallel imaging can also be combined with other reduced data acquisition methods such as UNFOLD (Madore 2002). The disadvantage of parallel imaging include the lost SNR, increased artifact due to imperfect unwrapping of the aliases, the need for a reference scan for the coil sensitivity, and increased image reconstruction time. The reference scan problem can be addressed using acquisitions with intrinsic sensitivity map information (Griswold et al. 2002). Most commercial scanners have parallel imaging techniques for EPI, however spiral methods have yet to see widespread use due to image reconstruction complexity. Fig. 17.6 shows examples of spiral images acquired using a four-channel head coil at 3T.

## Acquisition trajectories

Due to the long readouts of spiral and EPI sequences, much of the data along the trajectory is weighted with different amounts of T2* decay. Typically, TE is defined as the time when the center of $k$-space gets acquired after slice selection and is usually set equal to the T2* for maximum BOLD contrast. In EPI the $k$-space center occurs at the center of the readout. This produces a very balanced $k$-space weighting because half of the $k$-space is acquired with a TE that is too short and the second half with a TE that is too long. The spiral sequence, however, starts at the center of $k$-space at TE and spirals 'out'. This makes all of the higher spatial frequencies acquired at a long TE. As a result, spiral will naturally show more signal loss than EPI. It has been shown that by reflecting the spiral gradients

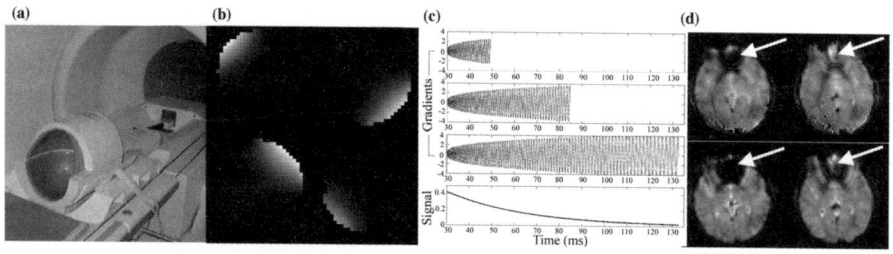

**Fig. 17.6** (a) Four-channel phased array coil for fMRI at 3T. (b) Coil sensitivities. Each coil provides spatial localization due to their different proximities. (c) Spiral readout gradients for images with different resolutions (3, 2, 1.6, and 1.0 mm from top to bottom) and signal T2* decay curve. Parallel imaging allows for images to be acquired with shorter readouts for reduced signal loss and image distortion. (d) Inferior 64 × 64 slices using $R = 2$ SENSE (top) and a standard head coil (bottom). Images acquired with parallel imaging techniques show more signal recovery in the OFC (see arrows). Data courtesy of Dr Douglas Noll, University of Michigan, and Dr Bradley Sutton, University of Illinois.

about TE, a 'reversed spiral' is generated (Bornert *et al.* 2000). The advantage of a reversed spiral is that the high spatial frequencies are acquired at a short TE. This property makes the reversed spiral acquisition less sensitive to signal loss artifact. The T2* weighting is obviously different for a reversed spiral due to the T2* decay curve being 'backwards,' however, there still remains substantial activation because the center of *k*-space is still acquired at TE. In order to alleviate this imbalance in T2* weighting and get the best of both worlds, the reversed spiral acquisition can also be inserted before a forward spiral acquisition generating an 'in-out' spiral (Glover and Law 2001). Combination of the two images from an in-out spiral provides an image with very balanced T2* weighting where one can combine the images or use whichever is more optimal. The in-out spiral has the added advantage of being able to dynamically acquire a field map for post-processing correction because each images has a slightly different TE (Sutton *et al.* 2004). Dynamic field map acquisitions are useful in cases where there is substantial head motion and physiological noise. It is also possible to alternate the phase encoding blips in an EPI sequence to generate an 'out-to-in' or *reversed* EPI sequence (Zhuo *et al.* 2003). Reversed EPI shows similar benefits as reversed spiral with regard to reduced susceptibility artifacts. An added advantage of acquiring all of the data before TE is that time is used more efficiently and the next scan can occur immediately after TE. This produces approximately a 30% increase in slice throughput, which translates to more coverage within a given TR. Fig. 17.7 compares images acquired with reversed spiral and forward spiral sequences.

Figure 17.8 shows results from an event-related cognitive fMRI study as an example of how reversed spiral techniques can be used to image the OFC. The task was an 'AX'

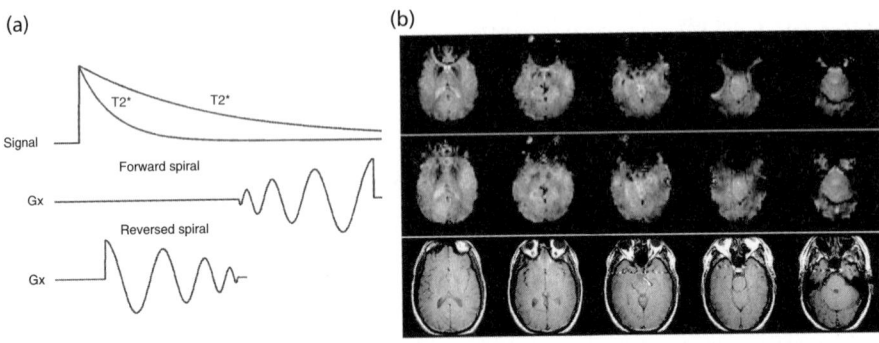

**Fig. 17.7** (a) Diagram of T2* decay and the position of one of the spiral readout gradients for forward and reversed spiral acquisitions. The dashed line shows the signal decay curve for a region with large susceptibility effects. The signal has decayed away to zero before the forward spiral readout begins. The reversed spiral readout begins immediately and can acquire some of the fast decaying signal. (b) Top row shows forward spiral images at 3T with a 25 ms TE. Note the signal loss in the OFC by the arrows. The second row shows the same slices acquired with a reversed spiral. There are less susceptibility artifacts compared to the forward spiral images. The bottom row shows spin-echo slices for comparison.

continuous performance decision-making task (AX-CPT) that is expected to produce activation in the OFC region due to decision-making (Carter *et al.* 1998). The subject sees either an 'A' or a 'B' cue followed by a 12-second wait. After this wait an 'X' or a 'Y' cue is presented. The subject responds by pressing one button for the 'AX' only and a second button for all other combinations. The slices were acquired along the coronal direction. There is a 2% activation-related change in the image magnitude in the OFC region from the images acquired with the reversed spiral following the 'A' cue.

## Gradient compensation

A simple approach to reducing signal loss artifact is to use higher resolutions or thinner slices (Haacke *et al.* 1989; Merboldt *et al.* 2000). By decreasing the voxel size or slice thickness, the susceptibility gradient across the voxel will be reduced. This will reduce the amount of signal loss. This approach is limited because as the voxels get smaller image SNR also becomes smaller. Furthermore, as discussed above, higher resolution requires longer readouts producing increased image distortion. A technique that is very similar to using high resolution or thin slice averaging is *gradient compensation* (Frahm *et al.* 1988, 1994; Constable 1995; Yang *et al.* 1997, 1998; Glover 1999; Yang *et al.* 1999). These methods implement an incremental re-phasing gradient $G_C$ after the slice selection pulse to refocus areas in the image where complete signal loss has occurred. Fig. 17.9 (a) shows a diagram of the signal loss as the result of a susceptibility gradient $G_S$. If a compensation gradient is applied with the opposite sign, the susceptibility gradient will be cancelled and the signal will be recovered. The gradient can be applied as a short pulse with high

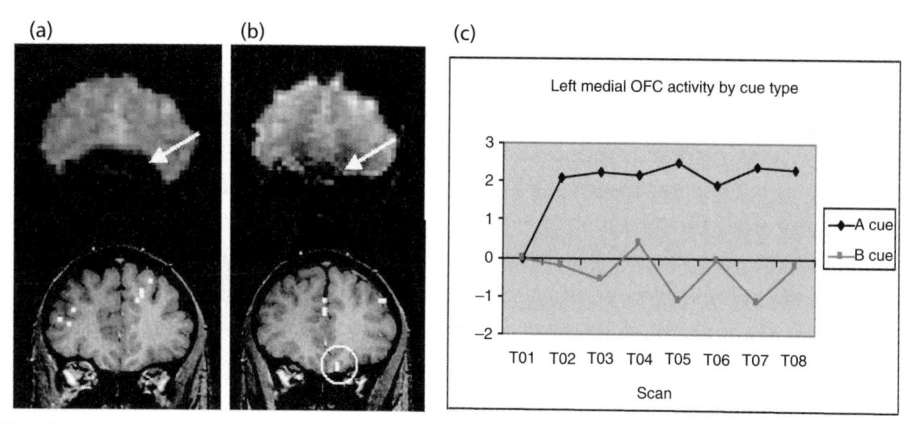

**Fig.17.8** Event-related fMRI data using the AX-CPT paradigm to activate the OFC, acquired using a reversed spiral sequence at 1.5T. (a) Coronal slice acquired with a forward spiral sequence (top) and corresponding activation map overlaid on the structural image (bottom). Note the large signal loss in the orbital frontal cortex. (b) Coronal slice acquired with a reversed spiral sequence (top) and corresponding activation (bottom). (c) Time course from the circled area in (b). Note in the 2% increase in signal form the 'A' cue in the orbital frontal cortex area. Data were collected with the assistance of Drs Stefan Ursu and Cameron Carter, University of California, Davis.

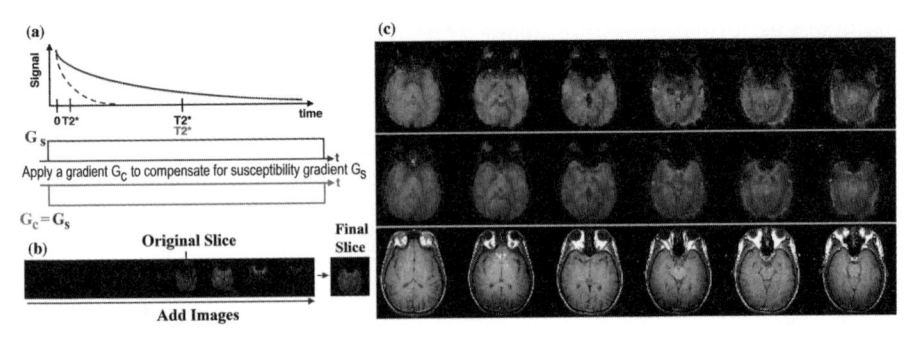

**Fig. 17.9** (Also see Color Plate 26.) (a) Diagram of T2* decay from a susceptibility gradient $G_s$. If an opposing compensation gradient $G_c$ is applied during the data acquisition, it will cancel out the effects of the susceptibility induced gradient. In practice several gradient compensation steps are needed for signal recovery in the whole slice. (b) Series of eight images of one slice each acquired with a different $G_c$. Each slice contains images with different amounts of correction and the final image is obtained by summation. (c) Slices acquired reversed spiral, gradient compensated reversed spiral, and spin echo acquisitions at 3T from top to bottom, respectively. Slice thickness was 5 mm, FOV = 20 cm, matrix = 64 × 64, TE = 25 ms, and four gradient compensation steps were used.

amplitude or a long pulse during the readout with low amplitude. Due to the distribution of susceptibility gradients in the brain at different positions, a series of images need to be acquired with varying degrees of refocusing and the final image is obtained by summation or Fourier transformation along the slice-select direction. Typically $G_C$ is applied along the z-direction and is often called *z-shim* as a result.

Fig. 17.9 (b) shows an axial slice acquired with a spiral sequence with different amplitudes of gradient compensation at 3T. The horizontal axis in (b) is the image as a function of compensation gradient area (the image in the fifth column has no compensation). Note, how changing the strength of the gradient not only refocuses the regions with no signal, but also defocuses the other parts of the image. The final image can be obtained by summing all of the respective images together. Fig. 17.9 (c) shows example images using gradient compensation and reversed spiral methods. The rows from top to bottom show reversed spiral images, reversed spiral images created by summing all of the images with varying $G_C$, and spin-echo images. By inspection, the images acquired with the gradient compensation have the least amount of signal loss and are more comparable to the spin echo images.

Fig. 17.10 (a) and (b) show results using gradient compensated reversed spiral at 1.5T for a Reward Contingent Decision (RCD) task developed to examine affective processes associated with decision-making (Rogers et al. 2003, 2004). The purpose of this paradigm is to be able to distinguish between affective processes associated with decision-making from affective processes associated with feedback such as reward and loss. The paradigm is presented to subjects as a game, in which subjects must choose between one of two sets of contingencies and either win or lose points, which translate into small financial rewards or losses. This decision phase lasts for 12 seconds and is followed by a 12-second feedback

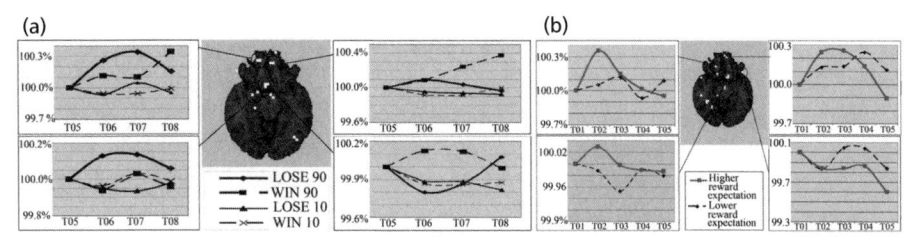

**Fig. 17.10** (a) Event-related fMRI data from fifteen normal subjects performing an RCD task. Activation in ventral striatum and medial OFC is highest for large rewards. Activation in amygdala and more lateral regions of OFC is highest for large losses. (b) Brain activity associated with expectation of reward during the decision phase of the RCD task. Medial OFC and ventral striatal activation are greatest for high probability high reward decisions, compared to all other decision types. Lateral OFC activation is lowest for these decisions compared to all others. Data were collected with the assistance of Drs Tim Lowry and Cameron Carter, University of California, Davis.

phase where the subject is informed of the outcome and their running total of points. Fifteen healthy adult subjects underwent fMRI during the RCD Paradigm. The scanning procedure implemented a reversed spiral pulse sequence with four-step gradient compensation sequence. This sequence allowed for thirteen 5 mm thick slices with a 64 × 64 matrix size and a 20 cm FOV to be acquired in 4 seconds.

The advantage of gradient compensation is that it is very simple to implement. The disadvantage is that it incurs a significant increase in the scan time. As many as four to eight gradient compensation steps are needed to recover the through-plane signal loss at 3T. Furthermore, the gradient compensation steps need to be applied to all of the slices in the acquisition, not just those that need correction, in order for the image steady state to be maintained during the scan. This penalty is particularly severe for fMRI because it produces decreased temporal resolution or decreased brain coverage, leaving these methods only effective for slower paradigms or for a few slices. Multi-slice gradient compensation methods have been proposed which can acquire 14 slices in 3.5 sec at 3 Tesla, however, these techniques give slices with sub-optimal and variable SNR which is undesirable for whole brain fMRI (Yang et al. 1999). A second disadvantage of gradient compensation is that it is applied only along the z-direction and does not correct for in-plane gradients (x-y directions). This is a valid assumption for many brain regions because the susceptibility-induced gradients vary primarily along the z-direction (such as the OFC for example). However, this assumption is not true for more inferior slices where gradient compensation needs to be applied along the x and y directions as well.

## Tailored RF pulses

Tailored RF (TRF) pulses are another technique that can be used to recover lost signal from intravoxel dephasing. The basic concept behind TRF pulses is that the phase at TE due susceptibility effects

$$\phi_S(\mathbf{r}, TE) = \omega_S(\mathbf{r}) TE \qquad [9]$$

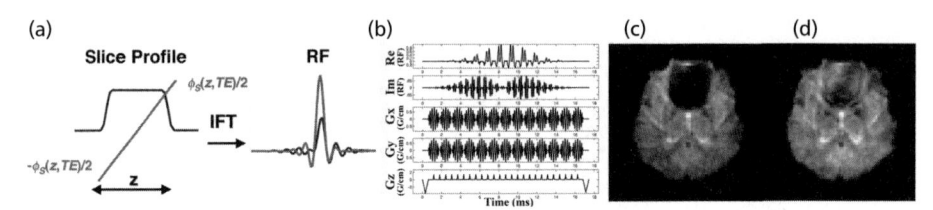

**Fig. 17.11** (Also see Color Plate 27.) (a) Phase correction can be put into a 1D tailored RF pulse by using the Fourier transform. (b) A 3D tailored RF pulse that applies the correction only in the OFC. (c) Image acquired without the 3D tailored RF pulse. (d) Same image acquired using the 3D tailored RF pulse.

can be cancelled by exciting a slice with the opposite of this phase:

$$M_{TE}(\mathbf{r}) = M(\mathbf{r})e^{-i\phi_s(\mathbf{r},\,TE)}. \qquad [10]$$

The idea is identical to gradient compensation only it is the phase of the RF excitation that produces the compensation and not a gradient. One-dimensional (1D) TRF pulses have been shown to be effective using a quadratic phase through the slice with an overall loss in SNR (Cho and Ro 1992). It has also been shown (Chen et al. 1998; Glover and Lai 1998; Chen and Wyrwicz 1999) that fMRI can be performed using 1D TRF pulses as well. Fig. 17.11 (a) shows a diagram of how a 1D TRF pulse can excite a slice with a phase difference through the slice. The slice profile with the phase correction is related to the RF pulse by the inverse Fourier transform. The disadvantage of using 1D TRF pulses is the same as that with gradient compensation: the correction is applied to the whole image, and not to the specific region where there is artifact. This requires the application of numerous 1D TRF pulses to the same slice for a single corrected image or a sacrifice of signal in normally homogenous regions.

Work has also been done with three-dimensional (3D) TRF pulses. The theoretical advantage of using a 3D TRF pulse is that the phase correction is applied only to the regions in the slice that need compensation. This can potentially eliminate the need to acquire a large number of slices. Furthermore, true anatomical 3D phase maps can be used providing more accurate corrections and removing the ambiguity associated with gradient compensation. The 3D TRF method has been shown to have great potential in fMRI acquisitions (Stenger et al. 2000). The major limitation of 3D TRF pulses is the necessity for high sampling resolution, which requires pulses that can be too long for a practical implementation. Multi-shot 3D TRF methods have been proposed that allow for susceptibility artifact reduction using two or four shots (Stenger et al. 2002). Fig. 17.11 (b) shows one shot from a four-shot 3D TRF pulse designed to reduce the signal loss artifact in the OFC. Images acquired without and with the 3D TRF pulse are shown in (c) and (d), respectively. Note the recovered signal in the OFC. Although this technique works well, a single-shot implementation of the 3D TRF technique is ultimately what is desirable for fMRI.

One approach to reducing 3D TRF pulse lengths is to use new pulse designs (Yip et al. 2005). Although the description of these designs is beyond the scope of this chapter,

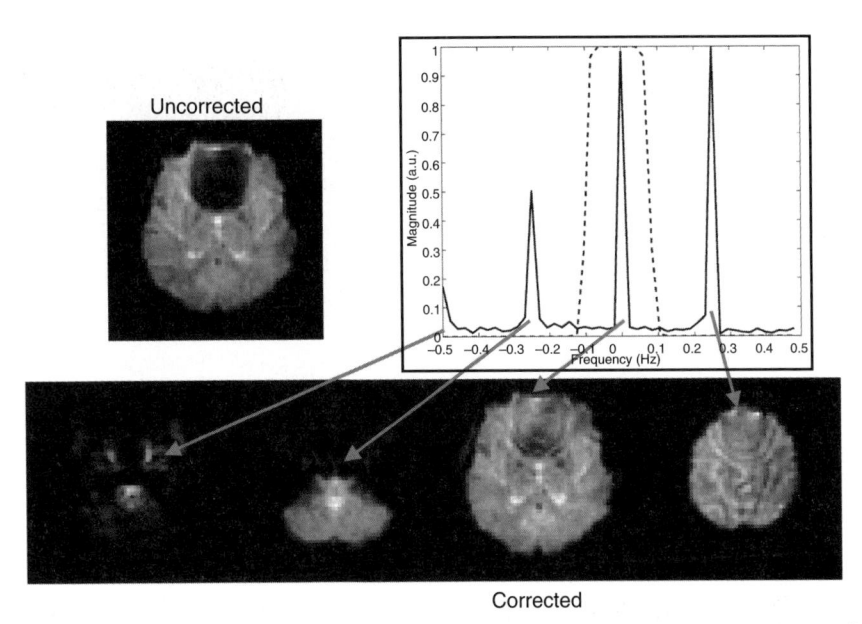

**Fig. 17.12** The use of the temporal direction can be an advantage in a dynamic imaging application such as fMRI. If a short 3D TRF pulse is of insufficient sampling, multiple slices will be excited in the head, which will all fold together during acquisition leaving a very undesirable image. However, if the shorter undersampled 3D TRF pulses are played out in a periodic manner as possible in an fMRI experiment, the extra slices can be filtered out in the temporal frequency domain using the UNFOLD method. The result will be the corrected slice acquired using a single-shot 3D TRF pulse

these methods look extremely promising. A second approach is to explore reduced data methods similar to those described above. Much of the future work in 3D TRF techniques will most likely involve reduced data schemes for shorter pulses. For example, because fMRI is a dynamic process, it is possible to use the temporal dimension to decrease 3D TRF pulses using a sliding window or temporal frequency filtering methods (Madore *et al.* 1999; Giurgi *et al.* 2003). Fig. 17.12 shows an example of how filtering in the temporal frequency domain in a dynamic fMRI experiment can be used to create a 'single-shot' 3D TRF pulse for the OFC. Recently, it has been proposed that parallel transmission can be used as a means of shortening multi-dimensional TRF pulses (Katscher *et al.* 2003; Zhu 2004). The concept is very similar to the acquisition method described above in that the transmit coil sensitivity will contain spatial information that can be used to replace missing excitation data. If multiple transmitters are used, then the lengths of the TRF pulses can be accordingly reduced. Although a lack of commercially available parallel transmitter hardware currently prohibits any applications in fMRI, the method looks extremely promising as a means of making 3D TRF pulses a practical and rapid solution to susceptibility artifact reduction. Fig. 17.13 shows a simulation example of how parallel transmission can create a single shot 3D TRF pulse for reducing intravoxel dephasing.

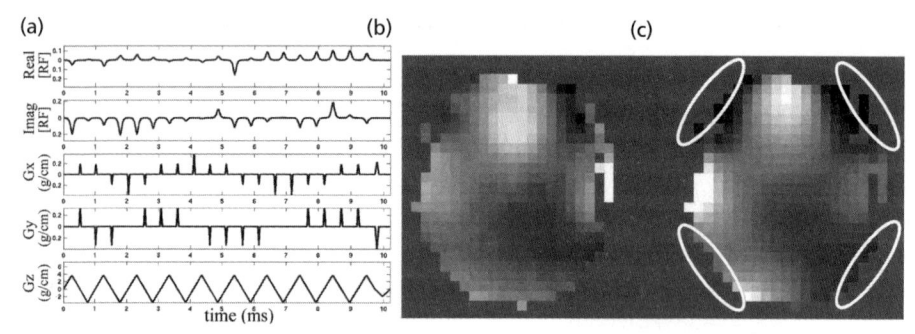

**Fig. 17.13** (a) Single-shot 3D TRF pulse that can be used with a parallel transmitter system. (b) Off-resonance map acquired at 3T used in the construction of the pulse shown in (a). (c) Phase in a slice produced by a simulation of the excitation using the pulse in (a) and multiple transmitters. The location of the transmitters is shown by the red ellipses. In this example we used a reduction factor of $R = 2$. The coil location and sensitivities were similar to those shown in Figure. 17.6.

## Conclusions

Functional MRI using BOLD contrast is one of the most effective techniques for imaging brain function in humans. The future points towards more psychiatry, psychology, and neuroscience departments operating their own fMRI scanners. Unfortunately, BOLD fMRI is still severely hampered by susceptibility artifacts that corrupt many inferior brain regions including the OFC. It is very important that great care be put into the design of fMRI experiments, and into the analysis of the results, when studying the OFC. It is also important that work continues by physicists and engineers to develop new scanning techniques to address this serious limitation. The development of techniques to address susceptibility artifacts remains an ongoing area of research with many breakthroughs; however, they need to be translated to more routine use. Interdisciplinary collaborations are a very important part of this translation of technology to research and the clinic.

This chapter focused on several methods that can be utilized to mitigate susceptibility artifacts. Many of these techniques require the assistance of an expert in pulse sequence programming and may not be useful to all neuroscientists performing fMRI. There are several simple approaches that anyone can try to reduce the effects of magnetic susceptibility artifacts. The use of field maps and parallel receivers is becoming a standard feature on commercial scanners, and it is prudent to take advantage of these features to reduce distortion. Investigators are encouraged to find out about these features and ask the vendor to assist with these protocols. One can perform a comparison with and without parallel imaging and see if the loss in SNR or increased aliasing artifact is worth the decrease in image distortion. A simple approach to reduce signal loss is to use thinner slices. Again there is a loss of SNR that needs to be considered. A portion of the lost SNR can be recovered by averaging the slices together. Slice orientation also plays a role. It is assumed that all of the slices are oriented close to axially and are above the air cavity when one uses thin slices or gradient compensation. Tilting the slices too much in the coronal direction will tend to decrease the effectiveness of these approaches. Coronal acquisitions will be more effective at reducing signal loss if the slice thickness is greater than the slice resolution. Slice orientation is something that is straightforward to test and optimize with the rule of thumb that the smallest voxel dimension needs to be along the direction perpendicular to the air cavity surface. In the case of EPI even the directions of the phase encoding and readout can improve the image quality (Deichmann *et al.* 2003). Distortion in EPI is worse along the phase encode direction. It is worth trying to switch their directions and see if the images look better. The use of passive oral shims is also very simple to implement and test.

In summary, it is most likely that a combination of methods or a hybrid technique will ultimately be needed to provide the most accurate fMRI data. For example, it is possible to combine in-out spirals or similar multi-echo EPI methods with gradient compensation (Guo and Song 2003). Gradient compensation has also been combined with parallel imaging and partial *k*-space techniques (Song 2001; Yang *et al.* 2004). The advantage of combining these methods is to decrease scan time, and to correct for both image distortion and signal loss artifacts in one sequence. Another possible combination could be to use multiple transmitters and multiple receivers such that single-shot TRF pulses can be used for refocusing the bulk susceptibility loss and shorter acquisitions will improve image distortion. The success in developing these methods will improve the reliability of fMRI data at high field such as 3T and may make fMRI at ultra-high field (i.e., 7T and up) a real possibility.

## References

Bandettini, P., Wong, A. E. C., Hinks, R. S., Tikofsky, R. S., and Hyde, J. S. (1992) Time course EPI of human brain function during task activation. *Magn. Reson. Med.* **25**:390–397.

Belliveau, J., Kennedy, D., McKinstry, R., Buchbinder, B., Weisskoff, R., Cohen M., Vevea, J., Brady, T., and Rosen, B. (1991) Functional mapping of the human visual cortex by magnetic resonance imaging. *Science* **254**:716–719.

Bornert, P., Aldefeld, B., and Eggers, H. (2000) Reversed spiral MR imaging. *Magn Reson Med* **44**:479–484.

Carter, C. S., Braver, T.S., Barch, D.M., Botvinick, M., Noll, D.C., and Cohen, J.D. (1998) Anterior cingulate cortex, error detection, and the on line monitoring of performance. *Science* **280**:747–749.

Chen, N., Li, L., and Wyrwicz, A. (1998) Optimized phase preparation and auto-shimming technique for gradient echo MRI. Int. Soc. of Magn. Reson. in Med., 5th Scientific Meeting.

Chen, N. and Wyrwicz, A. (1999) Removal of intravoxel dephasing artifact in gradient-echo images using a field-map based RF refocusing technique. *Magnetic Resonance in Medicine* **42**:807–812.

Cho, Z. and Ro, Y. (1992) Reduction of susceptibility artifact in gradient-echo imaging. *Magn Reson Med* **23**:193–200.

Constable, R. (1995) Functional MR imaging using gradient-echo echo-planar imaging in the presence of large static field inhomogeneities. *J Magn Reson Imag* **5(6)**:746–752.

Deichmann, R., Gottfried, J.A., Hutton, C., and Turner, R. (2003) Optimized EPI for fMRI studies of the orbitofrontal cortex. *Neuroimage* **19(2 Pt 1)**:430–41.

Detre, J. A., Leigh, J. S., Williams, D.S., and Koretsky, A.P. (1992) Perfusion imaging. *Magnetic Resonance in Medicine* **23**:37–45.

de Zwart, J. A., van Gelderen, P., Kellman, P., and Duyn, J. (2001) *Applicaiton of sensitivity-encoded EPI for BOLD fMRI*. Workshop on MInimum MR Data Acquisition Methods: Making More with Less, Marco Island, Florida, USA.

de Zwart, J., Ledden, P., Kellman, P., van Gelderen, P., and Duyn, J. (2002) Design of a SENSE-optimized high-sensitivity MRI receive coil for brain imaging. *Magn Reson Med* **47**:1218–1227.

de Zwart, J., van Gelderen, P., Kellman, P., and Duyn, J. (2002) Application of sensitivity-encoded echo-planar imaging for blood oxygen level-dependent functional brain imaging. *Magn Reson Med* **48**:1011–1020.

Edelman, R., Siewert, B., Darby, D., Thangaraj, V., Nobre, A., Mesulam, M., and Warach, S. (1994) Qualitative mapping of cerebral blood flow and functional localization with echo-planar MR imaging and signal targeting with alternating radio frequency. *Radiology* **192**:513–520.

Frahm, J., Merboldt, K. D., and Haenicke, W. (1988) Direct FLASH MR imaging of magnetic field inhomogeneities by gradient compensation. *Magnetic Resonance in Medicine* **6**:474–480.

Frahm, J., Merboldt, K.-D., and Hanicke, W. (1994) The Influence of the slice-selection gradient on functional MRI of human brain activation. *J Magn Reson B* **103**:91–93.

Giurgi, M. S., Boada, F. E., Noll, D. C., and Stenger, V. A. (2003) *Excitation UNFOLD using 3D Tailored RF Pulses*. In: Proc of the 11th Annual Meeting of the ISMRM, Toronto.

Glover, G. and Lai, S. (1998) *Reduction of susceptibility effects in BOLD fMRI using tailored RF pulses*. In: Proc 6th Annual Meeting ISMRM, Sydney.

Glover, G. (1999) 3D z-Shim method for reduction of susceptibility effects in BOLD fMRI. *Magnetic Resonance in Medicine* **42**:290–299.

Glover, G. H. and Law, C. S. (2001) Spiral-in/out BOLD fMRI for increased SNR and reduced susceptibility artifacts. *Magnetic Resonance in Medicine* **46**:512–522.

Gratton, G., Fabiani, M., Corballis, P. M., and Gratton, E. (1997) Noninvasive detection of fast signals from the cortex using frequency-domain optical methods. *Ann N Y Acad Sci* **820**:286–298; discussion 298–299.

Grinvald, A., Leike, E., Frostig, R. D., Gilbert, C. D., and Wiesel, T. N. (1986) Functional architecture of cortex revealed by optical imaging of intrinsic signals. *Nature* **324**:361–364.

Grinvald, A., Frostig, R. D., Siegel, R. M., and Bartfeld, R. M. (1991) High-resolution optical imaging of functional brain architeture in the awake monkey. *Proceedings of the National Academy of Science, USA* **88**:11559–11563.

Griswold, M. A., Jakob, P. M., Heidemann, R. M., Nittka, M., Jellus, V., Wang, J., Kiefer, B., and Haase, A. (2002) Generalized autocalibrating partially parallel acquisitions (GRAPPA). *Magn Reson Med* **47(6)**:1202–210.

Guo, H. and Song, A. W. (2003) Single-shot spiral image acquisition with embedded z-shimming for susceptibility signal recovery. *J Magn Reson Imaging* **18(3)**:389–395.

Haacke, E., Tkach, J., and Parrish, T. (1989) Reduction of T2* dephasing in gradient field-echo imaging. *Radiology* **170**:457–462.

Hsu, J. J. and Glover, G. H. (2005) Mitigation of susceptibility-induced signal loss in neuroimaging using localized shim coils. *Magn Reson Med* **53(2)**:243–248.

Huchinson, M. and Raff, U. (1988) Fast MRI data acquisition using multiple detectors. *Magnetic Resonance in Medicine* **6**:87–91.

Irarrazabal, P., Meyer, C.H., Nishmura, D.G., and Macovski, A. (1996) Inhomogeneity correction using an estimated linear field map. *Magn. Reson. Med.* **35**:278–282.

Jesmanowicz, A., Bandettini, P.A., and Hyde, J.S. (1998) Single-shot half k-space high-resolution gradient-recalled EPI for fMRI at 3 Tesla. *Magn Reson Med* **40(5)**:754–62.

Kadah, Y. M. and Hu, X. (1997) Simulated phase evolution rewinding (SPHERE): A technique for reducing B0 inhomogeneity effects in MR images. *Magn. Reson. Med.* **38**:615–627.

Katscher, U., Bornert, P., Leussler, C., and van den Brink, J. (2003) Transmit SENSE. *Magn Reson Med* **49**(1):144–150.

Kelton, J. R., Magin, R.L., and Wright, S.M. (1989) *An algorithm for rapid image acquisition using multiple receiver coils*. Proceedings of the SMRM 8th Annual Meeting, Amsterdam.

Kim, S. G. (1995) Quantification of relative cerebral blood flow change by flow-sensitive alternating inversion recovery (FAIR) technique: Application to functional mapping. *Magnetic Resonance in Medicine* **34**:293–301.

Kwait, D. and Einav, S. (1991) A decoupled coil detector array for fast image acquisition in magnetic resonance imaging. *Medical Physics* **18**:251–265.

Kwong, K. K., Belliveau, J. W., Chesler, D. A., Goldberg, I. E., Weisskoff, R. M., Poncelet, B. P., Kennedy, D. N., Hoppel, B.E., Cohen, M. S., Turner, R., Cheng, H-M., Brady, T. J., and Rosen, B. R. (1992) Dynamic magnetic resonance imaging of human brain activity during primary sensory stimulation. *Proc. Natl. Acad. Sci., USA* **89**:5675–5679.

Madore, B. (2002) Using UNFOLD to remove artifacts in parallel imaging and in partial-Fourier imaging. *Magn Reson Med* **48**(3):493–501.

Madore, B., Glover, G. H., and Pelc, N. J. (1999) Unaliasing by fourier encoding the overlaps using the temporal dimension (UNFOLD), applied to cardiac imaging and fMRI. *Magn Reson Med* **42**:813–828.

Mansfield, P. (1977) Multi-planar image formation using NMR spin echoes. *J Phys C* **10**:L55-L58.

Merboldt, K.-D., Finsterbusch, J., and Frahm, J. (2000) Reducing inhomogeneity artifacts in functional MRI of human brain activation-thin slices vs gradient compensation. *J Magn Reson* **145**:184–191.

Meyer, C., Hu, B., Nishimura, D., and Macovski, A. (1992) Fast Spiral Coronary Artery Imaging. *Magn. Reson. Med.* **28**:202–213.

Naatanen, T. (1992) *Attention and Brain Function*. Hillsdale, NJ: Erlbaum.

Noll, D., Pauly, J., Meyer, C., Nishimura, D., and Macovski, A. (1992) Deblurring for non-2D Fourier transform magnetic resonance imaging. *Magn Reson Med* **25**(2):319–333.

Noll D. C., Cohen, J. D., Meyer, C. H., and Schneider, W. (1995) Spiral K-space MR Imaging of Cortical Activation. *J. Magn. Reson. Imaging* **5**:49–56.

Noll, D. C., Meyer, C. H., Pauly, J. M., Nishimura, D. G., and Macovski, A. (1991) A homogeneity correction method for magnetic resonance imaging with time-varying gradients. *IEEE Trans. Med. Imaging* **10**(4):629–637.

Ogawa, S., Tank, D. W., Menon, R., Ellerman, J. M., Kim, S.-G., Merkle, H., and Ugurbil, K. (1992) Intrinsic signal changes accompanying sensory stimulation: Functional brain mapping with magnetic resonance imaging. *Proc Natl Acad Sci USA* **89**:5951–5955.

Pruessmann, K. P., Weiger, M., Scheidegger, M. B., and Boesiger, P. (1999) SENSE: Sensitivity encoding for fast MRI. *Magnetic Resonance in Medicine* **42**:952–962.

Ra, J. B. and Rim, C. Y. (1991) *Fast imaging using multiple receiver coils with subencoding data set*. Proceedings of the SMRM 10th Annual Meeting, San Fransisco.

Reese, T., Davis, T., and Weisskoff, R. (1993) *Automated shimming at 1.5T using echo-planar image frequency maps*. 12th Ann. Sci. Mtg., Soc. of Magn. Reson. in Med., New York, NY.

Reese T., Davis, T., and Weisskoff, R. (1995) Automated shimming at 1.5T using echo-planar image frequency maps. *J Magn Reson Imag* **5**(6):739–745.

Rogers, R. D., Tunbridge, E. M., Bhagwagar, Z., Drevets, W. C., Sahakian, B. J., and Carter, C. S. (2003) Tryptophan depletion alters the decision-making of healthy volunteers through altered processing of reward cues. *Neuropsychopharmacology* **28**(1):153–62.

Rogers, R. D., Ramnani, N., Mackay, C., Wilson, J. L., Jezzard, P., Carter, C. S., and Smith, S.M. (2004) Distinct portions of anterior cingulate cortex and medial prefrontal cortex are activated by reward processing in separable phases of decision-making cognition. *Biol Psychiatry* **55**(6):594–602.

Roland, P. E., ed. (1993) *General and Selective Attention*. Brain Activation. New York: Wiley-Liss.

Schenck, J. (1996) The role of magentic susceptibility in magnetic resonance imaging: MRI magnetic compatibility of the firs and second kinds. *Med Phys* **23(6)**:815–850.

Sodickson, D. K. and Manning, W. J. (1997) Simultaneous acquisition of spatial harmonics (SMASH). *Magnetic Resonance in Medicine* **38**:591–603.

Song, A. W. (2001) Single-Shot EPI with Signal Recovery from the Susceptibility-Induced Losses. *Magnetic Resonance in Medicine* **46**:407–411.

Stenger, V. A., Boada, F. E., and Noll, D. C. (2000) Three-dimensional tailored RF pulses for the reduction of susceptibility artifacts in T2*-weighted functional MRI. *Magn Reson Med* **44**:525–531.

Stenger, V. A., Boada, F. E., and Noll, D. C. (2002) Multishot 3D slice-select tailored RF pulses for MRI. *Magn Reson Med* **48**:157–165.

Stenger, V. A. and Noll, D. C. (1997) *Partial k-space reconstruction for 3D acquisitions in functional MRI*. Int. Soc. of Magn. Reson. in Med., 5th Scientific Meeting.

Sutton, B. P., Noll, D. C., and Fessler, J. A. (2003) Fast, iterative image reconstruction for MRI in the presence of field inhomogeneities. *IEEE Trans Med Imaging* **22(2)**:178–188.

Sutton, B. P., Noll, D. C., and Fessler, J. A. (2004) Dynamic field map estimation using a spiral-in/spiral-out acquisition. *Magn Reson Med* **51(6)**:1194–1204.

Talagala, S. L. and Noll, D. C. (1998) Functional MRI using steady-state arterial water labeling. *Magn. Reson. Med.* **39**:179–183.

Weiger, M., Pruessmann, K., Osterbauer, R., Bornert, P., and Jezzard, P. (2002) Sensitivity-encoded single-shot spiral imaging for reduced susceptibility artifacts in BOLD fMRI. *Magn Reson Med* **48**:860–866.

Wilson, J. L. and Jezzard, P. (2003) Utilization of an intra-oral diamagnetic passive shim in functional MRI of the inferior frontal cortex. *Magn Reson Med* **50(5)**:1089–1094.

Yang, Q., Dardzinski, B., Li, S., Eslinger, P., and Smith, M. (1997) Multi-gradient echo with susceptibility inhomogeneity compensation (MGESIC): demonstration of fMRI in the olfactory cortex at 3.0 T. *Magnetic Resonance in Medicine* **37**:331–335.

Yang, Q., Dardzinski, B., Li, S., Eslinger, P., and Smith, M. (1998) Removal of local field gradient artifacts in T2*-weighted images at high fields by gradient-echo slice excitation profile imaging. *Magn Reson Med* **39**:402–409.

Yang, Q. X., Demeure, R. J., and Smith, M. B. (1999) *Rapid whole-brain T2* echo planar imaging with removal of magnetic susceptibility Artifacts*. Int. Soc. of Magn. Reson. in Med., 6th Scientific Meeting, Philadelphia, PA, USA.

Yang, Q. X., Wang, J., Smith, M. B., Meadowcroft, M., Sun, X., Eslinger, P.J., and Golay, X. (2004) Reduction of magnetic field inhomogeneity artifacts in echo planar imaging with SENSE and GESEPI at high field. *Magn Reson Med* **52(6)**:1418–1423.

Yip, C. P., Fessler, J. A., and Noll, D. C. (2005) *A novel fast and adaptive trajectory in three-dimensional excitation k-Space*. In: Proc of the 13th Annual Meeting of the ISMRM, Miami, Florida, USA.

Zhu, Y. (2004) Parallel excitation with an array of transmit coils. *Magn Reson Med* **51(4)**:775–784.

Zhuo, J., Stenger, V. A., and Boada, F. E. (2003) *Reversed EPI data acquisition for improved fMRI in areas of large magnetic field susceptibility*. In: Proc 11th Annual Meeting of the ISMRM, Toronto.

Part 3

# Neuropsychiatry

Chapter 18

# Neuropsychological assessment of the orbitofrontal cortex

David H. Zald

## 18.1. Introduction

At least at a superficial level, individuals with orbitofrontal cortex (OFC) dysfunction often appear cognitively intact. Yet, the deficits associated with OFC damage can cause disastrous consequences, not infrequently leading to major interpersonal, occupational and legal problems. In such circumstances, it is not uncommon to have a request for neuropsychological evaluation in an attempt to substantiate dysfunction of the OFC. In some cases, the referral question may be directed at characterizing the functional consequences of a known lesion. Alternatively, the assessment may occur in the absence of a verified lesion, and the neuropsychologist is asked whether there is evidence of cognitive or behavioral dysfunction suggestive of a neurological etiology, and what circuits of the brain are most likely implicated in the dysfunction. Because the conclusions of these assessments may have an impact on both treatment decisions as well as a host of legal issues (such as liability, compensation, and disability decisions), it is important that we have a solid understanding of the evidence supporting any inferences or conclusions arising from the neuropsychological testing.

Neuropsychological tests, that are purported to be sensitive to OFC lesions, have also frequently been used in research studies of psychiatric and neurological patients. In this context, poor performance on such measures is often used to infer which neural systems underlie certain neuropsychiatric symptoms. Examples of studies taking this approach with 'OFC' measures include studies of patients with obsessive-compulsive disorder, schizophrenia, substance dependence, psychopathic personality disorder, and autism spectrum disorders. As in clinical situations, confidence in the inferences arising from 'OFC' measures is directly related to the tests' psychometric characteristics.

The psychometric properties of a neuropsychological test may be broadly considered in terms of the test's reliability (internal consistency, consistency over time), validity (how well it measures what it is intended to measure), and diagnostic utility (sensitivity and specificity). I will not discuss general issues of reliability and validity for standardized neuropsychological tests, since these issues are well-described by other sources (e.g., Lezak 1995), and are not specific to questions about the OFC. The issue of diagnostic utility represents a much more specific issue, since it is dependent upon the specific diagnostic question. In the present context, the question is, 'What is the accuracy of the tests in detecting OFC damage?'[1] In other words, are the tests sensitive enough to detect true cases of OFC damage and specific enough to reject cases with no OFC damage?

When examining neuropsychological tests, we may consider two levels of diagnostic utility. The first and more lenient level is the sensitivity and specificity of detecting patients with OFC lesions relative to healthy controls. However, if we want to make a specific inference about the location of a lesion, we ideally need to address a more stringent question regarding the ability of the test to discriminate between patients with lesions of the OFC versus patients with lesions elsewhere in the brain. This is a far more difficult question because it requires assessment of multiple groups of subjects with documented lesions of different parts of the brain. Given the specific nature of the question, it is not surprising that the ability to detect OFC lesions is not addressed by manuals for standard neuropsychological tests. Instead, its determination is largely dependent upon the extent and quality of published journal articles that have examined patients with OFC and other lesions. The present chapter reviews this literature, with a particular focus on experimental neuropsychological probes that have been implemented with a specific aim of detecting damage to the OFC.

## 18.2. Sources of OFC damage

Several types of neuropathology produce damage to the OFC and ventromedial prefrontal cortex (VMPFC) in humans. The term VMPFC, when used in the clinical literature, almost always encompasses both the ventral medial wall region as well as the medial portions of the orbital surface. Because this regional designation overlaps with the medial OFC, and pathologies affecting the region often affect both the OFC and overlying aspects of the ventromedial wall simultaneously, it is difficult to segregate these areas in the clinical literature. Indeed, papers reporting VMPFC damage almost always include patients with damage to the medial aspects of the OFC, and studies reporting on medial OFC lesions often include patients with damage to the overlying cortex along the medial wall. Given this lack of segregation, studies reporting patients with VMPFC damage are included in this review, and little attempt is made to distinguish these studies from those reporting OFC damage.[2] Sources of damage to these two regions range from closed head

---

[1] The word, dysfunction could be substituted for damage here. However, most of the data presented in this chapter comes from patients with damage to the OFC rather than dysfunction that is secondary to pathology elsewhere in the brain.

[2] This contrasts with many of the chapters in the book that highlight differences between the medial and orbital networks. At some point, the clinical literature may advance to a point where there is enough data to make similar distinctions in clinical practice, but at present such distinctions appear largely speculative.

injuries and penetrating head wounds to cerebrovascular accidents, neurosurgical excisions and neurodegenerative disorders. As such conditions vary widely in terms of their specificity for the OFC, and their specific pattern of pathological impact, it is useful to briefly consider the effects of different types of pathology affecting the OFC.

Patients with OFC damage as a consequence of closed head injury are relatively common in typical neuropsychology clinics. However, such patients are often not the most informative from a research standpoint, because the damage is rarely specific to the OFC, and frequently involves substantial damage to the frontal and temporal poles (Courville 1937), as well as more widespread axonal shearing (Pang 1989). Unfortunately, structural scanning is often insensitive to the extent of such damage (Richardson 2000), which means that reports of highly specific OFC lesions following traumatic brain injury must be treated with some caution. In contrast, patients with surgical excisions in the OFC (e.g., for removal of tumors or intractable epilepsy) often have quite restricted lesions. The size of the excisions and the degree of specificity to the OFC varies in such cases. In some reported cases the excision impinges upon overlying aspects of the frontal cortex, and even with smaller lesions, there may be continued disturbance in functioning of the remnant tissue surrounding the excised tissue or tumor. Yet, when the extent of the lesions are well-characterized, patients with surgical excisions provide perhaps the best opportunity to examine the effects of OFC lesions in isolation from comorbid neuropathology of other brain regions.

Cerebrovascular accidents are also a common source of injury to the OFC. The vascular supply of the OFC includes the orbital branch (or branches) of the anterior cerebral artery (which serves the gyrus rectus and medial orbital gyrus), the frontopolar branch of the anterior cerebral artery (in the anterior OFC) and the orbital branch of the middle cerebral artery (which serves the lateral aspects of the OFC). The posterior aspects of the OFC are also frequently injured in the course of ruptured aneurysm of the anterior communicating artery (ACoA: see Fig. 18.1). However, interpretation of such cases is complicated by concomitant injury to the basal forebrain (DeLuca 1993). This is particularly true for measures of memory and attention, which appear especially sensitive to damage to the cholinergic producing cells of the basal nucleus of Mynert in the basal forebrain.

Disruption of OFC functions also occurs as part of a number of neurodegenerative disorders. However, in most neurodegenerative diseases the neuropathological process is neither specific to, nor centered on, the OFC. Rather, the OFC dysfunction is part of a larger picture of neurobehavioral disruption. The primary exception is the frontal variant of frontotemporal dementia (fvFTD: also known as dementia of the frontal lobe type or the behavioral variant of frontotemporal dementia) in which disruption of the OFC may be a particularly prominent focus of both the pathology and the clinical presentation of the disorder (see Chapter 24, by Lu *et al.*). Because of this, deficits in the context of fvFTD are described in this chapter in support of the sensitivity of tasks to OFC dysfunction. However, in considering this interpretation, it must be stated that the

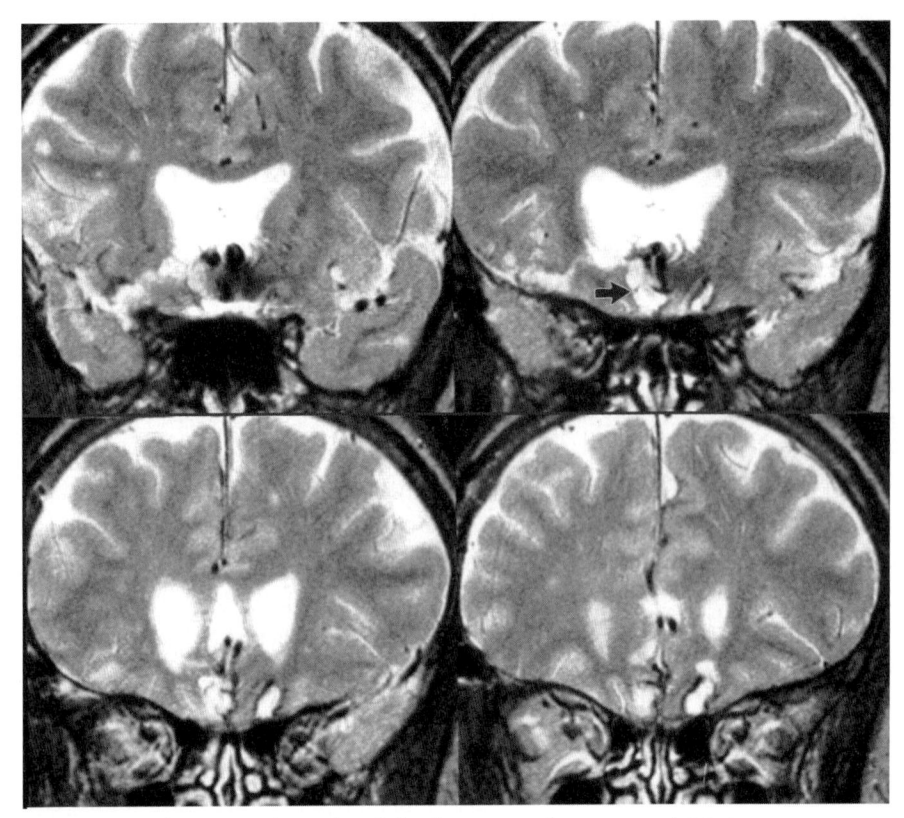

**Fig. 18.1** T2 weighted MRI of a patient following surgery for a ruptured ACoA aneurysm. Coronal sections show involvement of the basal forebrain (see arrow on right top row), bilateral lesions of the gyrus rectus and neighboring orbital gyri (bottom row), as well artifact from the aneurysm clip. Adapted from Stefanie et al. (1998) and reprinted with permission from BMJ Publishing.

pathology of fvFTD is by no means limited to the OFC or VMPFC, and frequently involves other frontal regions, particularly more dorsomedial regions (Franceshsi et al. 2005; Williams et al. 2004; Varrone et al. 2004). Broe et al. (2003) describe that atrophy in FTD (including fvFTD cases) arises first in the OFC, the superior medial frontal cortex, and the hippocampus, but then spreads to other areas (although the severity of atrophy in the OFC, superior medial frontal cortex and hippocampus appears to continue to progress ahead of other cortical and subcortical regions). Thus, while providing general support for OFC involvement in tasks, fvFTD certainly cannot be considered to reflect a 'pure' OFC pathology, and in some cases the deficits may very well reflect disruption of other frontal circuits. Furthermore, since neuropsychological studies of fvFTD do not restrict themselves to patients with verified pathology of the OFC (using neuroimaging, or subsequent autopsy data), the relative balance of OFC/VMPFC pathology to other frontal pathology is generally unknown in the samples studied. Thus, for the purpose of the present review, deficits in the context of fvFTD are treated as weaker evidence than studies of patients with verified OFC lesions.

## 18.3. Traditional neuropsychological measures

The most striking finding from neuropsychological studies of patients with circumscribed OFC lesions is the general sparing of functions on the vast majority of traditional neuropsychological and IQ measures (Anderson et al. 1992; Angrilli et al. 1999; Eslinger and Damasio 1985; Stuss et al. 1983). Indeed, part of the 'enigma' of the OFC for many years was the source of the real life problems experienced by OFC lesion patients, given their appearance of normality on most neuropsychological tests.

From a theoretical perspective, the most notable sparing of performance comes from the minimal effects of OFC lesions on the Wisconsin Card Sort Task (WCST). The WCST requires subjects to learn a sorting rule through trial and error learning, and then flexibly shift to new sorting rules when the experimenter changes to a new sorting rule. In his 1964 chapter on the OFC, Mortimer Mishkin put forth the hypothesis that animals with OFC lesions have a 'perseveration of central sets', in which they are unable to overcome or inhibit prepotent responses. Given that the most critical measure of the WCST is perseverative errors (continued responding with a sorting rule after the sorting rule has changed), it follows from Mishkin's hypothesis that patients with OFC lesions should have excessive perseverative errors on the WCST. However, this is not what is observed. Patients with OFC lesions show no increases in perseverative errors relative to controls (Stuss et al. 2000). They do, however, tend to complete fewer sorting categories, primarily due to a failure to maintain set (i.e., they fail to stick with a sorting rule that is working) (Stuss et al. 1983, 2000). The lack of perseverative errors on the WCST (and analogous tests in animals) indicates that the OFC is not broadly necessary to suppress all classes of perseverative responses, and has led to increasing specification as to the types of situations in which OFC-lesioned individuals perseverate (discussed more below).

While most traditional neuropsychological measures of memory (such as the Wechsler Memory Scale) appear normal in patients with OFC lesions, it must be noted that memory problems can arise in conjunction with OFC lesions (DeLuca 1993; Stuss et al. 1982). These memory issues are principally covered in the chapter by Brand and Markowitsch and will not be detailed here, other than to note that two questions are warranted when one sees memory problems in a patient with ventral frontal lesions. The first question is whether or not the lesions extend into the basal forebrain. This question is particularly pertinent for cases involving ruptured aneurysm of the ACoA, which often involve lesions in the basal forebrain resulting in amnestic symptoms with prominent confabulation (DeLuca 1993). Indeed, given the importance of basal forebrain lesions following ACoA aneurysm, it is somewhat surprising that studies of VMPFC damage (which are usually at least partially comprised of ACoA aneurysm patients) frequently fail to detail the extent of basal forebrain involvement. In such cases, information regarding the memory status of the patients might give increased confidence that the observed cognitive or emotional findings relate to OFC dysfunction, rather than a basal forebrain-related memory deficit. Unfortunately, such information is lacking in many studies in the literature.

The second important question to ask when observing memory deficits following OFC lesions is whether the deficits might be due to problems with distractibility rather than a mnemonic problem per se. In such cases, one generally expects to see uneven memory performance, with most tasks appearing normal, but tasks in which there are prominent distractor tasks showing evidence of impairment. The most notable example of such a finding comes from a study of schizophrenics who had undergone OFC leukotomies many years prior to being assessed (Stuss et al. 1982). The leukotomy patients demonstrated short-term memory problems only in situations in which they had to perform a distractor task (arithmetic problems).

## 18.4. Experimental tasks derived from the animal literature

Given the lack of sensitivity of most traditional neuropsychological measures to OFC lesions, researchers and clinical practitioners have increasingly employed cognitive tasks that are directly derived from paradigms that were originally developed for research with monkeys. The advantage of such tasks is that they are based on an animal literature that allows for analysis of the effects of selective focal lesions. The disadvantage is that normative data is often poor or completely lacking, which hinders the interpretation of quantitative results. Additionally, since the tests are often not standardised, they may be subject to variations in their implementation across laboratories, and the impact of these variations is often unknown. Nevertheless, such tasks are increasingly utilised in research settings, and may prove quite useful in specific clinical settings.

### Acquisition of alternation tasks

Object alternation (OA) tasks are increasingly used as behavioral probes of ventral prefrontal functions in humans (Abbruzzese et al. 1995; Abbruzzese et al. 1997; Cavedini et al. 1998; Faraone et al. 1999; Freedman 1990, 1994; Gansler et al. 1996; Good et al. 2002; Koenen et al. 2001; Marie et al. 1999; Pantelis and Brewer, 1995; Seidman et al. 1995; Zohar et al. 1999). In OA tasks, subjects view two objects, and on a trial-by-trial basis, must select whichever object they did not select on the previous trial (see Fig. 18.2). The relative position of the objects varies randomly, so that the subject must rely on object features in making their response selections. The task was originally developed as a complement to spatial alternation tasks in which the spatial location alternates from trial to trial (Pribram and Mishkin 1956). Lesion studies with monkeys demonstrated marked deficits in the reacquisition of OA following lesions to the lateral OFC or inferior convexity (Mishkin et al. 1969; Mishkin and Manning 1978; Pribram and Mishkin 1956). Strikingly, lesioned animals often showed no improvement in task performance even after thousands of trials. In essence, such animals appeared unable to re-learn the task rule. The effect appeared specific to more ventral frontal lesions, in that animals with lesions centered above the inferior convexity were largely unimpaired on the task.

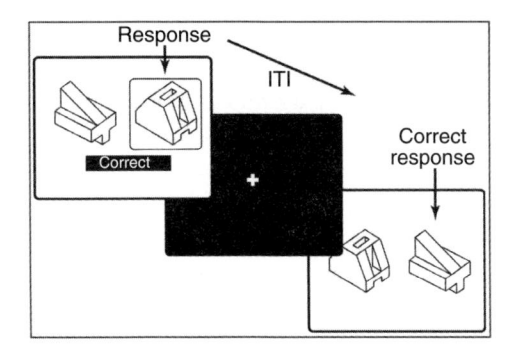

**Fig. 18.2** Schematic of the Object Alternation Paradigm. The subject must learn the task rule that the rewarded object alternates between each trial regardless of the position of the objects. The intertribal interval (ITI) is usually 1 or 5-s in duration.

The most direct evidence regarding the sensitivity of OA to ventral frontal lesions in humans comes from a small study by Freedman *et al.* (1998). These investigators studied 6 patients with bilateral frontal lesions and fifteen healthy controls on an OA acquisition task. In this paradigm, as in most of the human studies of OA, the subjects were not informed of the task rule and had to learn it through trial and error learning. The source of frontal damage ranged from anterior communicating artery haemorrhage to bilateral frontal leukotomy for schizophrenia. Consistent with the animal literature, the frontal lesion patients showed significantly more errors than the controls on OA acquisition. While the study supports the sensitivity of the OA acquisition task to ventral frontal lesions, it was not able to directly address the issue of specificity due to the large scope of the lesions, which in addition to ventrolateral cortical regions (areas 11 and 47/12) variably included the frontal pole and the medial prefrontal cortex extending into the dorsal anterior cingulate.

PET data also support the involvement of the OFC in OA. Ventral frontal activations have been observed during both the acquisition and the practiced performance of OA tasks (Curtis *et al.* 2000; Zald *et al.* 2002, 2005). However, the neuroimaging data also raises questions regarding the specificity of OA tasks to the OFC. As with many 'frontal lobe' tasks, the neuroimaging data indicate that multiple areas become active during both the acquisition and practiced performance of OA (Curtis *et al.* 2000; Zald *et al.* 2002, 2005). This finding itself is not concerning, since obviously a task of this sort must involve the coordinated activity of a network of brain areas. However, the most striking finding from the OA acquisition study (Zald *et al.* 2005) was not the engagement of the OFC (which was quite circumscribed), but the far more robust and large activation of the pre-supplementary motor area (pre-SMA) and dorsal anterior cingulate region. Moreover, activity in the pre-SMA/cingulate region showed an association with the ability to acquire the task. The involvement of the pre-SMA/cingulate region during task acquisition is of particular interest given recent evidence of its importance in other tasks requiring rule acquisition, such as the WCST (Stuss *et al.* 2000). This activation is of particular concern for claims of OFC specificity for OA acquisition because many of the patients in the Freedman *et al.* (1998) study had lesions that extended into the anterior cingulate. Unfortunately, no published studies have specifically examined the effects of pre-SMA/cingulate lesions on OA acquisition. Thus, although OA deficits may lead to the natural speculation of OFC dysfunction, dysfunction in other brain regions such as the dorsomedial frontal cortex cannot be ruled out as an alternative source of the impairment.

Lesions to the lateral OFC/inferior convexity region in monkeys have also been reported to impair *spatial* alternation (Mishkin *et al.* 1969). However, this effect appears more variable than the impairments in OA (Passingham 1975). Furthermore, spatial alternation is robustly impaired following more dorsally placed lesions (Butters and Pandya 1969; Goldman *et al.* 1971; Mishkin 1957; Mishkin *et al.* 1969; Passingham 1975), especially those centered on the principal sulcus. The added dorsolateral involvement in spatial alternation has generally been attributed to the spatial working memory demands of the task. Because of the involvement of dorsolateral regions, spatial alternation is not believed to have the same localizing specificity as OA. However, in a study, which included six fontal lobe lesion patients, particularly severe deficits in acquiring a delayed spatial alternation task emerged only in the two patients whose lesions included the OFC (Freedman and Oscar-Berman 1986). While these data raise the possibility that spatial alternation may be sensitive to OFC lesions in humans, the assessment of both object and spatial alternation acquisition is not recommended in the same subjects, because whichever task is performed first is likely to contaminate acquisition of the second alternation task. Given this limitation, and the present state of the animal data, the OA paradigm appears preferable to spatial alternation tasks when the primary focus of the assessment is on ventral frontal functioning.

## Reversal learning

A number of monkey lesion studies have demonstrated impairments in object reversal learning following OFC lesions (Butter 1969; Butter *et al.* 1973; Dias *et al.* 1996; Iversen and Mishkin 1970; Meuneir *et al.* 1997; Mishkin and Manning 1978; Voytko 1985; Passingham 1975). In object reversal learning tasks, an animal is rewarded with food for selecting an object. After a certain unpredicted amount of trials, the reward contingency reverses so that the previously rewarded object is no longer rewarded, and the previously non-rewarded stimulus becomes the rewarded stimulus. Depending upon the specific paradigm, the task may undergo a single reversal, or may undergo multiple reversals, with the reversals occurring every time the subject has reached a certain criterion performance level. The most prominent reversal deficits arise as a sequela of inferior convexity lesions, and appear to be part of a general problem with perseveration (Iversen and Mishkin 1970). In other cases, particularly with more medial OFC lesions, the deficit appears more related to an inability to update which item was rewarded on a trial-by-trial basis. In these cases, the deficit only arises after both stimuli have been associated with reward and non-reward. A simple associative process alone cannot be used for determining which stimulus is the correct one to respond to in this situation, because both stimuli have been associated with reward and non-reward. Rather, the animal must be able to hold on-line in working memory information about which object is currently rewarded, and which object is not rewarded. This demand resembles features of the OA task, in that information about which object is currently being rewarded must be maintained on-line. However, in reversal tasks, knowledge of when to switch objects can only be determined based on the receipt of error feedback. In contrast, in the OA task, once the rule is acquired and the subject is performing the task correctly, the task can be performed perfectly without ever receiving and processing another error.

In addition to deficits in object reversals, monkeys with OFC lesions also frequently show deficits in spatial reversal tasks (in which the location that is rewarded reverses). However, the effects of OFC lesions on spatial reversals have been much less consistent than the effects on object reversals (Butter 1969; Butters 1973; Passingham 1975; Goldman et al. 1971). Because of this inconsistency, spatial reversal tasks have not been commonly used as probes of OFC functioning in humans.

Human subjects with OFC damage have been found to exhibit deficient performance on reversal tasks. Several studies support the sensitivity of reversal learning to OFC lesions (Berlin et al. 2004; Fellows and Farah 2003; Hornak et al. 2004; Rolls et al. 1994). It must be noted that the specific tasks have varied substantially from study to study. For instance, Rolls et al. (1994) used a paradigm in which subjects only saw one stimulus (S+ or S−) on the screen at a time, and were scored for both errors of commission (responding to the S−) and errors of omission (failure to respond to the S+). Patients with ventral frontal lesions demonstrated striking impairments on this task. However, the errors may have arisen due to a problem with inhibitory control rather than a specific failure of reversal. This concern arises because the subjects demonstrated a marked inability to withhold responses (three-fourths of the errors were errors of commission), similar to the problems on no-go deficits discussed below. More recently, Fellows and Farah (2003) demonstrated impairments in patients with VMPFC lesions on a reversal task in which subjects had to select cards from one of two decks of cards, with the winning deck switching following eight consecutive correct responses. This study also provided evidence for the specificity of the lesion: patients with VMPFC lesions performed significantly worse than both healthy control participants and patients with dorsolateral frontal lesions, whereas patients with dorsolateral lesions performed no worse than healthy controls. This result suggests that the task may be useful in distinguishing between dorsolateral and VMPFC patients, and is consistent with data from monkeys in this regard (Dias et al. 1996).

Hornak et al. (2004) and Berlin et al. (2004) have studied OFC lesion patients using a more complex 'probabilistic reversal task' in which subjects had to select from two abstract figures (S+ and S−), one of which was advantageous (in terms of probability of reward and size of pretend monetary rewards relative to losses) and the other of which was disadvantageous. If the subject got 16 out of 18 consecutive responses correct, the contingencies incrementally reversed over a period of 10 trials. It may be noted that because subjects only completed 100 trials, the actual number of reversals is limited (average in healthy controls = 3.5), so the task taps the ability to make an initial two or three reversals. In their initial study (Hornak *et al.* 2004), patients with bilateral OFC lesions showed severely impaired performance on this task, losing money, whereas patients with lesions, that did not encroach on the OFC, and healthy controls earned money. The deficit only occurred in patients with bilateral OFC lesions, as patients with unilateral lesions performed normally. In terms of specificity, patients with medial prefrontal lesions (Brodmann areas 8 and 9) performed normally. Patients with dorsolateral prefrontal lesions were highly variable, with some showing normal performance and others showing performance as bad as the OFC patients. However, post-test screening revealed that the dorsolateral patients who performed poorly had all failed to attend to the feedback information, making it virtually impossible for them to perform the task properly. In contrast, the patients with bilateral OFC damage reported that they understood the need to pay attention to the feedback screen, but nevertheless were unable to perform the task properly. In a second paper (Berlin *et al.* 2004), the authors indicated that a combined group of patients with both unilateral and bilateral OFC lesions were significantly impaired on the task relative to both healthy controls and to patients with frontal lesions that did not encroach upon the OFC. In both studies, the OFC patients who were impaired on the task appeared relatively insensitive to the outcome of trials, failing to switch away from the S− on trials immediately following losses, and failing to reselect the S+ immediately following wins.

The Cambridge Neuropsychological Test Automated Battery (CANTAB: Cambridge Cognition, Cambridge, UK) contains a well-normed and standardized measure of reversals as part of the ID/ED (intradimensional/extradimensional shift) task. This task attempts to capture the dissociation observed in studies of marmosets in which more ventral frontal lesions lead to impaired affective (reward/non-reward) reversals within a dimension, whereas more dorsolateral lesions lead to problems with extradimensional shifting (changes in the type of stimulus features that need to be responded to) (Dias *et al.* 1996). Unfortunately, data specifically testing this dissociation in humans with well-defined frontal lesions remains limited. The strongest support at present comes from a study of fvFTD. Rahman *et al.* (1999) reported that these patients show a specific deficit in the reversal phase of the ID/ED task, while demonstrating intact performance on non-reversal components on the task. The patients further performed normally on spatial working memory and planning tasks, which are sensitive to dorsolateral lesions. These data are consistent with the idea of a relatively selective ventral frontal involvement in fvFTD.

The most relevant neuroimaging data on reversal learning derives from a study by Kringelbach and Rolls (2003), who utilized their probabilistic reversal paradigm with positive and negative facial expressions as reinforcers. The authors reported activations in the lateral OFC during reversal acquisitions relative to initial task acquisitions. However, they also observed significant activations in the anterior cingulate/paracingulate region. These data appear to parallel the OA data in suggesting involvement of both the OFC and cingulate in task performance. While specific data addressing the effects of cingulate lesions on reversal learning in humans remains lacking, data from a primate lesion study suggests a greater importance of the OFC than the cingulate for object discrimination reversals. Meunier, Bachevalier and Mishkin (1997) examined three monkeys with medial OFC lesions and three monkeys with cingulate lesions. On the initial several reversals, the OFC-lesioned and cingulate-lesioned monkeys did not show a significant difference in performance. However, following the 4th–9th reversals, the OFC animals averaged over four times as many errors as the cingulate-lesioned animals. In contrast, during spatial discrimination reversals, the cingulate-lesioned monkeys performed as poorly as the OFC-lesioned animals. This finding appears consistent with the alternation literature in indicating a greater ability to discriminatively detect OFC dysfunction in tasks requiring attention to object rather than spatial features of the stimuli.

Interestingly, failure of reversal learning appears related to a number of other deficits in patients with ventral frontal lobe lesions. For instance, Rolls *et al.* (1994) indicate a relationship between errors on their reversal task and behavioral ratings by staff members regarding the degree to which subjects were socially inappropriate and disinhibited. Similarly, Fellows and Farah (2003) observed an inverse association between reversal performance and ratings of independent activities of daily living. Thus there appear to be real-world correlates to these deficits.

As noted above, the lesion literature suggests two potential types of reversal problems: 1) a deficit that arises during an initial reversal and occurs secondary to a more general pattern of perseveration or failure to inhibit prepotent responses; and 2) a more specific failure to update stimulus-reward associations after multiple reversals. These two types of deficits are difficult to disentangle in the current human literature because data regarding initial reversals are usually not reported separately from data on subsequent reversals. Moreover, there are several cases in the literature where tasks that possess only a single reversal were utilized. Such tasks may prove useful in detecting a problem with perseveration, but cannot address the more specific form of reversal problems associated with medial OFC lesions in monkeys. Indeed, based on the monkey literature, we would predict that some OFC-lesioned patients would perform normally on tasks involving just one reversal, even though these same patients would be impaired in a task with a large number of reversals. Closer attention to the distinction between initial and subsequent reversals is necessary if reversal tasks are to meet their full potential in neuropsychological assessment.

## Go/no-go

Go/no-go tasks provide a measure of inhibitory control of responses. Such tasks are typically set up to provide a prepotent go response (for instance hitting a button), with the critical measure being whether the subject can withhold the response on trials in which they need to withhold the prepotent response (no-go trials). Many continuous performance tasks can also be viewed as go/no-go tasks when the proportion of go trials significantly outnumbers no-go trials. The relevance of the task to the OFC dates back to an early study by Brutkowski *et al.* (1963), which demonstrated that lesions of the OFC in monkeys produced impaired inhibition responses on no-go trials. Subsequent studies in monkeys with more focal lesions demonstrated that the deficit arose following lesions of the inferior convexity (lateral OFC and ventral principal sulcus region) (Butters *et al.* 1973; Iversen and Mishkin 1970) but not following lesions to more medial orbital areas (Butter 1969; Iversen and Mishkin 1970). Based on these animal data, some investigators have argued for the use of the go/no-go task as a measure of OFC functioning in humans. Potential support for the use of go/no-go tasks in the assessment of ventral frontal function comes from a study of healthy controls by Spinella (2002), who reported correlations between go/no-go performance and measures of delayed alternation and olfactory performance (both of which are discussed below as useful measures of OFC functions). The human lesion literature provides clear evidence that prefrontal lesions cause problems inhibiting responses on no-go trials (Black *et al.* 2000; Drewe 1975; Godefroy and Rousseaux 1996; Leimkuhler and Mesulam 1985; Salmaso and Denes 1982). Deficits have also been observed in patients with fvFTD (Slavchesky *et al.* 2004). However, the human lesion studies provide little evidence for a specific OFC source of the deficit. Perhaps the strongest support for inferior frontal involvement in the deficit comes from Black *et al.* (2000), who studied patients who had undergone prefrontal leukotomies (which typically cause particularly strong damage to the afferents and efferents of ventral frontal regions). The schizophrenic patients who had undergone prefrontal leukotomy were significantly impaired in go/no-go performance relative to age-matched schizophrenic controls. However, most of the other lesion data have emphasized the effects of medial frontal, rather than inferior frontal lesions on commission errors (Drewe 1975; Godefroy and Rousseaux 1996; Leimkuhler and Mesulam 1985). A relatively large neuroimaging literature has emerged in recent years on cognitive control tasks including response inhibition during go/no-go tasks. This literature provides compelling evidence for the involvement of the ventrolateral cortex (particularly in the right hemisphere), which is highly consistent with the early animal literature regarding the inferior convexity. However, the activations are generally superior to the OFC proper, falling for instance in BA 45 (pars triangularis), rather than the pars orbitalis sections of the inferior frontal gyrus (Aron and Poldrack 2005). These data suggest that to the extent that go/no-go deficits reflect frontal damage, they may be a better measure of ventrolateral or more dorsomedial functioning than that of the OFC proper (see chapter by Hooker and Knight for further discussion of the role of the ventrolateral prefrontal cortex in behavioral inhibition).

## 18.5. Decision-making (gambling) tasks

In recent years, intense interest has developed on the potential use of gambling tasks as probes for VMPFC dysfunction. Unlike the experimental tasks described already, gambling tasks have emerged strictly within the context of studies of human patients. The Iowa Gambling Task (IGT) is the most widely used of these tasks and is described in Chapter 13 in this volume by Naqvi, Bechara, and Tranel. The primary measure of interest for the IGT is the extent to which subjects select from riskier card decks despite their poorer outcome relative to other card decks over the course of the task. Learning is important in this task, since the only information about the reward or risk of a card selection is the subject's past experience of rewards or losses from prior card selections.

Evidence supports the sensitivity of the IGT to VMPFC lesions, at least when the lesions include the right VMPFC (Bechara et al. 1994, 1998; Tranel et al. 2002). Interestingly, lesions restricted to the left VMPFC region (or the left OFC more generally) do not appear to significantly lead to riskier deck selections on the IGT (Manes et al. 2002; Tranel et al. 2002). However, questions remain regarding the specificity of the IGT. Whereas early data from Bechara et al. (1998) suggested normal performance of patients with dorsolateral prefrontal lesions, data from studies by Manes et al. (2002) and Clark et al. (2003) indicate significant effects of lesions that include either dorsolateral or dorsomedial prefrontal regions. Again the laterality appears to relate more to the right hemisphere: Clark et al. (2003) report substantial effects only for right hemisphere lesion patients. Similarly, in studies of traumatic brain injury patients, Levine et al. (2005) report a general sensitivity to prefrontal lesions, particularly in the right hemisphere, but without evidence of specificity to the VMPFC.

Fellows and Farah (2004) have suggested that the disadvantageously risky card selections on the IGT shown by VMPFC lesion patients may relate not to a specific deficit in decision making, but from a more elemental deficit in reversal learning. Specifically, they note that the IGT has a reversal component in that the disadvantageous decks have an initial high reward associated with them before becoming disadvantageous. Support for this argument comes from data indicating that patients with VMPFC lesions show improved performance when the initial bias favoring the disadvantageous decks is removed by reordering the cards (see Fig. 18.3). Furthermore, the degree to which this task modification improved performance appeared directly correlated with the patient's level of reversal learning impairment. From a psychometric standpoint, this raises a question of whether the IGT provides incremental validity above and beyond that provided by reversal learning measures. At present the answer to this question has not been addressed.

The best-known alternative to the IGT is the Cambridge Gambling Test (Rogers et al. 1999). In contrast to the learning elements of the IGT, the CGT explicitly provides probabilities on each trial, and past trials do not have an impact on current probabilities. On each trial the subject views a number of colored squares and determines whether a token is under a blue square or a red square. The probability is reflected in the number of blue or red squares appearing for the given trial. On each trial the subject is first asked to make a probabilistic judgement of whether the token is under one of the blue or red squares, and then are asked to determine how much they want to bet that they are correct. In an initial study, Rogers et al. (1999) reported that patients with OFC lesions were impaired in their probabilistic judgements and made suboptimal bets compared to healthy controls. In contrast, subjects with dorsomedial frontal lesions appeared normal. However, subsequent patient studies have not provided consistent support for either the sensitivity or the specificity of the CGT, although other measures, such as deliberation

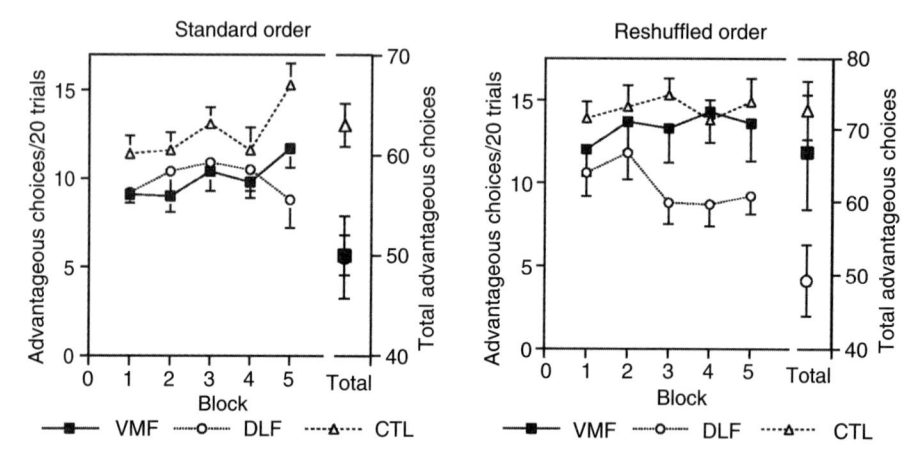

**Fig. 18.3** Results from Fellows and Farah (2005). The figure on the left shows impaired performance by both VMPFC patients (black squares) and DLPFC patients (open circles) using the standard order for the IGT. In contrast, the figure on the right shows normal performance on the task for VMPFC patients when the decks are reshuffled so that the unfavorable decks are not initially positive in value. Note: the scales differ slightly across figures (and the control groups differ across studies). Adapted from Fellows and Farah (2005). Reprinted with permission from Oxford University Press.

time for making the probabilistic judgement have suggested that these patients may have a subtle deficit in probabilistic reasoning that is not fully captured by the task's outcome measures (Clark et al. 2003; Mavaddat et al. 2000; Rahman et al. 1999). The inconsistency of the results so far do not provide strong support for using the CGT as a specific measure of OFC functioning.

## 18.6. Social processing and theory of mind

Facial expressions are a critical aspect of human non-verbal communication of emotion. A number of neuroimaging studies involving explicit judgements about faces implicate the OFC in aspects of facial judgements (see Chapter 15, by Dougherty et al., this volume for review). However, the importance of the OFC to emotional recognition of facial expressions is unresolved. Large lesions that include the OFC have been observed to cause deficits in facial emotion recognition (Hornak et al. 1996). Yet, the deficit is clearly not universal: patients with unilateral OFC excisions typically showed normal performance, and this occurred even in patients who showed impairments on a vocal emotion recognition task (Hornak et al. 2003). A more sizable literature indicates problems with emotional recognition in patients with fvFTD (Fernandez-Duque and Black 2005; Keane et al. 2002; Lavenu and Pasquier 2005; Lavenu et al. 1999; Lough et al. 2006; Rosen et al. 2004), although the effect does not appear to particularly differentiate between fvFTD and other dementias. Thus, at present its diagnostic utility remains questionable, although from a functional perspective, an assessment of this domain is useful when trying to understand the source of problematic social interactions.

Theory of mind (ToM) refers to the ability to make inferences regarding the mental state (knowledge, intentions, and beliefs) of others. Multiple neuroimaging and electrophysiological studies (see Gallagher and Frith 2003; Sabbagh 2004 for reviews) and studies of patients with frontal lobe dysfunction (Gregory et al. 2002; Lough et al. 2006; Rowe et al. 2001; Shamay-Tsoory et al. 2005a, 2005b; Stone et al. 1998; Stuss, et al. 2001) indicate a general association between ToM deficits and the frontal lobes. Of particular interest in the present context is the ability to detect faux pas. Faux pas refers to someone saying or doing something they should not have said, without knowing or realizing that they should not have said it. Recognition of faux pas involves an element of ToM in that it requires the subject to infer what the person did or did not know when they uttered the offending words. Baron-Cohen et al. (1999) developed the faux Pas Recognition Test as an advanced test of ToM. The test includes 10 faux pas and 10 matched control scenarios. The following is an example of one of the faux pas stories:

> Mike, a 9-year old boy, just started at a new school. He was in one of the cubicles in the toilets at school. Joe and Peter, two other boys at school, came in and were standing at the sinks talking. Joe said, 'You know that new guy in the class? His name's Mike. Doesn't he look weird? And he's so short!' Mike came out of the cubicles, and Joe and Peter saw him. Peter said, 'Oh, hi Mike! Are you going out to play football now?'

After each story, the subject is asked; 1) whether someone said something they should not have, 2) to identify what it was that they should not have said, 3) answer a comprehension question, and 4) recognize that the faux pas was a consequence of a false belief. In a pioneering study, Stone et al. (1998) reported that subjects with bilateral OFC lesions ($n = 5$) were impaired in recognizing faux pas. In contrast, patients with left dorsolateral lesions performed normally. In a second study Shamay-Tsoory et al. (2005b) examined patients with either VMPFC ($n = 12$) or dorsolateral ($n = 7$) lesions, and observed significantly greater impairment in the VMPFC patients compared to both healthy controls and patients with posterior cortical lesions. These deficits were particularly severe in subjects whose lesions involved the right VMPFC. In contrast, subjects with dorsolateral lesions did not reach statistical significance in comparison to either healthy controls or posterior cortical lesion patients. Gregory et al. (2002) has similarly observed deficits in faux pas detection in subjects with fvFTD. In that study, the number of ToM tasks that were compromised (including the faux pas task + other ToM tasks) was associated with a rating of the amount of VMPFC atrophy observable on MRI. It may be noted that one negative finding exists in this area. The study reported a patient with a large bilateral medial lesion that extended into VMPFC areas who showed normal performance on the Faux Pas Recognition task (Bird et al. 2004). However, substantial portions of the right OFC were intact in this subject, which might have contributed to the spared performance. To date, the only other lesion site that has been associated with deficits on the Faux Pas Recognition Task is the amygdala. Stone et al. (2003) reported impaired performance in two patients with bilateral amygdala damage. The similarity of this effect to the deficit arising from ventral frontal damage is consistent with tight anatomical and functional connections between the amygdala and OFC (Zald and Kim 2001). In considering the faux pas recognition deficit arising from ventral frontal lesions, it is important to note that these deficits generally do not occur in isolation, and may be accompanied by deficits in detecting irony, sarcasm, and deception (Stuss *et al.* 2001; Shamay-Tsoory *et al.* 2005a, 2005b). In each case these deficits appear worse following VMPFC (particularly right VMPFC) lesions relative to more dorsal (particularly dorsolateral) regions. These deficits are also likely accompanied by measurable deficits in scales tapping empathy (Shamay *et al.* 2002; Shamay-Tsoory *et al.* 2004; Rankin *et al.* 2005). However, present data indicate that empathy deficits in isolation are probably not particularly specific to the OFC or the VMPFC, as these deficits occur with reasonable frequency both in patients with right parietal lesions and patients with semantic dementias (Rankin *et al.* 2005; Shamay-Tsoory *et al.* 2004).

In summary, although the literature on faux pas recognition and the recognition of irony, sarcasm, and deception remains limited, the data so far are encouraging, and the inclusion of such measures may prove useful when augmenting a neuropsychological test battery to focus on ventral frontal functions. Furthermore, the assessment of these domains is directly relevant to understanding the real-life social problems exhibited by patients with ventral frontal lesions.

## 18.7. Olfactory testing

Although olfactory functioning has often been overlooked as a major focus of clinical neuropsychology, the existing data suggest that it may actually be among the most sensitive and selective measures of OFC dysfunction. In order for most olfactory processing tests to be informative about OFC dysfunction, it first must be demonstrated that the person has normal olfactory detection thresholds. Detection thresholds are typically tested by asking a patient to sniff from bottles containing different concentrations of an odorant dissolved in water, and determining at what concentration they reliably detect the odorant. Detection thresholds can be rapidly determined using a forced two-choice ascending staircase procedure, in which the patient is presented with two sample bottles per trial, one containing the odorant and one containing water, and the patient is asked to report which one smells stronger (Cain *et al.* 1983). Butanol is probably the most frequently used odorant in such trials, with steps increasing by a factor of three-fold each level, although other odorants and step factors can be used. It is often useful to perform testing with one nostril plugged, in order to test for laterality effects.

Deterioration in threshold detection is a common consequence of head trauma, and is often linked to damage to the olfactory bulb or nerve (Doty *et al.* 1997; Levin *et al.* 1985; Sumner 1964; Yousem *et al.* 1996, 1999; Zusho 1982). In such cases, impairments in other olfactory abilities must be considered secondary to the bulb or nerve damage rather than a consequence of OFC damage. However, because of their proximal location, head traumas that injure the olfactory nerves enough to cause anosmia frequently also produce damage to the OFC. For instance, Varney, Pinkston and Wu (2001) observed decreased OFC metabolism in head trauma patients with post-traumatic anosmia. Post-traumatic anosmia has also been associated with poor vocational outcomes and behaviors resembling those of patients with OFC damage (Varney 1988). Thus, although anosmia most likely reflects a primary impairment to the olfactory nerves or bulbs, it may nevertheless suggest the presence of concomitant damage to the OFC.

When olfactory threshold detection abilities are within reasonable levels, other tests of olfactory processing begin to have additional localizing significance. The most widely used standardized measure of olfactory functioning is the University of Pennsylvania Smell Identification Test (UPSIT: Doty et al. 1984). On each item of this test, participants scratch a microencapsulated odorant, and then indicate which of four choices the odozr smells like. The test includes 40 items, a few of which specifically engage the trigeminal nerve and hence should not be affected by olfactory nerve damage (thus providing a potential to detect malingering). The strongest data supporting the UPSIT's utility in regards to OFC functions derives from a study by Jones-Gotman and Zatorre (1988), who examined 120 patients with focal surgical brain lesions. Among frontal lobe lesion patients, impairments only emerged in subjects whose lesions invaded the OFC, suggesting an ability to discriminate between ventral frontal vs. other frontal lesions. Patients with unilateral temporal lobectomies also demonstrated impairments, although these were significantly weaker than patients with OFC lesions. Thus, although not completely specific relative to anterior temporal lesions, the data suggest that particularly severe deficits are indicative of OFC damage. Notably, this deficit occurs despite normal olfactory detection thresholds.

A second source of support for the sensitivity and specificity of the UPSIT comes from studies which examined the relationship between plaques detected with MRI in patients with multiple sclerosis (Doty et al. 1998, 1999). An initial study revealed a strong inverse association between the number of plaques in the inferior frontal lobes and temporal olfactory regions ($r = -0.94$), and performance on the UPSIT (Doty et al. 1998). In contrast, no relationship was seen between plaque numbers in other brain regions. In a second study, Doty and colleagues (Doty et al. 1999) observed that the waxing and waning of task performance over time correlated with change in plaque levels in the inferior frontal lobe and temporal olfactory regions. A similarly strong inverse association between plaques in the inferior frontal lobe and temporal olfactory regions has been reported with the Cross Cultural Smell Identification Test (Zorzon et al. 2000). Taken together, the current data strongly support the utility of smell identification tests in detecting inferior frontal damage, although they are unlikely to discriminate inferior frontal vs. temporal contributions to such dysfunction.

A number of other olfactory related tasks appear substantially impaired by OFC lesions (see review in chapter by Gottfried, Small and Zald, this volume), and some may be more sensitive to OFC vs. temporal lobe damage. However, because these tasks have not been standardized and lack normative data, they remain largely experimental in nature. The best studied of these tasks is olfactory discrimination in which patients have to determine if two odorants are different or the same. A number of studies have observed olfactory discrimination impairments following both right OFC and temporal lobe lesions (Hulshoff Pol *et al.* 2002; Potter and Butters 1980; Zatorre and Jones-Gotman 1991). Importantly, these studies indicate a critical difference between temporal lobe and frontal lesions on olfactory discrimination. Temporal lobectomies only impair discrimination on the nostril ipsilateral to the lesion site. In contrast, right OFC lesions cause a birhinal deficit in olfactory discrimination. This represents a clear-cut distinction between the effects of temporal lobe and OFC damage, and may be useful when trying to discriminate between these two potential sources of olfactory deficits.

As should be clear from the above discussions, the primary challenge to the specificity of olfactory identification and discrimination deficits involves the temporal cortices. This differs dramatically from object reversal and alternation tasks, where more concern was raised in terms of potential pre-SMA and cingulate contributions to deficits. While a number of neuroimaging studies have reported changes in cingulate activity during various olfactory tasks (Royet *et al.* 2001; Savic *et al.* 2000; Small *et al.* 2004), at present there is little evidence that the cingulate lesions produce any impairments in olfactory discrimination or identification tasks. This differential pattern of challenges to specificity is advantageous in that combined deficits in both olfactory processing and cognitive tasks such as object alternation and reversals are unlikely to be attributable to an isolated lesion in either the cingulate or the anterior-medial temporal cortices. Moreover, because of their markedly different locations in the brain and different vascular supplies, combined cingulate and medial temporal damage are unlikely to co-occur unless the person has extremely diffuse pathology, in which case a far greater range of cognitive problems will also be present. Thus, the combination of olfactory deficits and deficits on experimental OFC tasks should provide a dramatically greater level of confidence in interpretation than can be achieved when observing either deficit in isolation.

## 18.8. Interview and questionnaire data

Data obtained during interview can play an important role in developing hypotheses regarding OFC dysfunction. In fact quite often, interview data characterizing changes in personality, disinhibition, and poor decision-making provide the first inkling of OFC dysfunction. Moreover, because these symptoms can occur in the absence of gross deficits on traditional neuropsychological measures, they provide information that is often not assessed if taking a straight neuropsychological battery approach to assessment. Several interview/questionnaire schedules are now available that capture some of the symptoms of ventral frontal dysfunction. Such schedules are designed for administration to caregivers or family members who rate the patient on a number of characteristics.

The Frontal Systems Behavioral Scale (FrSBe: originally titled the Frontal Lobe Personality Scale (Grace *et al.* 1999; Grace and Malloy 2001)) is unique among frontal measures in that it attempts to distinguish a specific set of symptoms connected to OFC pathology. The measure contains three subscales—Apathy, Executive Dysfunction, and Disinhibition—that were developed to capture dysfunction of different prefrontal regions. Of particular importance in the current context is the Disinhibition scale, which was specifically designed to capture symptoms of ventral frontal dysfunction. The scale, although focused on disinhibitory behaviors, also includes a number of items outside of this dimension such as changes in taste and smell, and difficulties with the law. Factor analysis of the scale indicates that these latter items do not appear to actually load on the same factor as the disinhibition items, but their relevance to OFC pathology is clear (Stout *et al.* 2003). The manual for the FrSBe describes adequate reliability for this scale, and normative data are provided stratified by gender, age, and education, with an age range extending from 18–95 years. The most important validation data for the FrSBe comes from a study demonstrating its utility for detecting frontal lobe pathology relative to healthy controls and patients with nonfrontal neuropathology (Boyle *et al.* 2001; Grace *et al.* 1999). However, it remains to be seen whether the subscales discriminate between ventral vs. non-ventral frontal lesions.

The Iowa Rating Scales for Personality Change by Barrash and Anderson (1993) is aimed at capturing aspects of personality change following frontal injury, and emphasizes a number of personality features specifically related to ventral frontal lesions. The scale includes items related to both changes in emotional functioning and real-world competencies (see Chapter 23, by Koenigs and Tranel, who describe this measure in some detail). Although normative data remains lacking for this scale, the authors have demonstrated that it has utility in discriminating VMPFC patients from patients with nonfrontal lesions (Barrash *et al.* 2000; Anderson *et al.* in press). Importantly, in a recent study of patients with VMPFC lesions vs. other frontal lesions, Anderson *et al.* observed significantly higher reports of emotional changes in the VMPFC group relative to patients with other frontal lesions, and patients with nonfrontal lesions. The VMPFC patients were rated both as having increased emotional reactivity (poor frustration tolerance, lability, irritability) and hypo-emotionality (impoverished emotions, apathy, blunted affect). The presence of increased emotional reactivity, although frequent in the VMPFC sample (71%), was not entirely specific (25% of the other frontal group also showed disturbance in this area). However, the presence of both emotional reactivity and hypo-emotionality problems in the same person appears unique to the VMPFC population: 3 of 7 VMPFC patients showed both elements, whereas no patients with other frontal or nonfrontal lesions demonstrated a similar pattern (from $n = 14$, and $n = 36$ respectively). The real-world competencies part of the Iowa rating scales also appears quite sensitivity to VMPFC lesions. Of the 13 items included in the scale, 10 were rated as significantly higher in VMPFC patients than in patients with nonprefrontal lesions, and 8 were higher in VMPFC patients than in patients with other frontal lesions.

Other interview measures in current use do not attempt to specifically identify a ventral frontal syndrome, but include items or subscales that get at specific behaviors associated with ventral lesions as part of a broader measurement of frontal pathology. For instance, the 24-item Frontal Behavioral Inventory (Kertesz *et al.* 1997), which was developed to specifically assess aspects of fvFTD, contains a number of questions specifically related to OFC pathology, including inappropriateness, jocularity, impulsivity, aggression, and hypersexuality. The Neuropsychiatric Inventory (Cummings et al. 1994), which is one of the most widely used measures in the field, includes a number of items related to frontal pathology as part of a broader assessment of psychiatric symptoms arising in neurological conditions. For instance, items covering agitation, dysphoria, anxiety, apathy, irritability, euphoria, disinhibition, and appetite and eating abnormalities, may all be of relevance for frontal lobe patients. It may be noted that the Disinhibition subscale from the Neuropsychiatric Inventory is more conceptually limited to disinhibition phenomena than the corresponding scale in the FrSBe, although the two scales appear highly correlated (Norton et al. 2001).

The Dysexecutive Questionnaire is a 20-item questionnaire that is meant as a supplement to a larger battery of tests called the Behavioral Assessment of the Dysexecutive Syndrome (Wilson et al. 1996). The scale has items related to emotional, behavioral, cognitive, and personality changes. Unfortunately, although the authors have provided some modest normative data, published studies have not directly addressed the sensitivity or the specificity of the measure to frontal lobe lesions.

Two additional scales warrant mention, but are limited by a lack of adequate normative data, which hinders their implementation in clinical practice. Berlin et al. (2004), describe a 20-item, self-report Frontal Behavior Questionnaire, based on an original scale designed by Rolls et al. (1994), and indicate that OFC patients scored significantly higher than patients with prefrontal lesions that did not encroach upon the OFC. Lhermitte (Lhermitte 1986; Lhermitte et al. 1986) described a scale that has shown some utility in research with frontal lobe patients. Although not well-defined psychometrically, the scale is of interest in the specific range of symptoms that it captures. Like other frontal measures it includes symptoms such as apathy, restlessness, impulsiveness, indifference, euphoria, disinterestedness, cheerfulness, personality deficits, attention deficits, and loss of intellectual or emotional control. More uniquely, it also assesses dependence on social environment, indifference to rules, and dependence on stimuli from the physical environment, which may be relevant in certain cases of OFC dysfunction.

The biggest limitation of many of the behavioral scales described above lies in the relative dearth of specific data on patients with focal OFC lesions relative to other lesion patients (particularly other frontal lobe lesion patients). However, an increasing literature highlights their utility in patients with fvFTD. The FrSBe, Frontal Behavioral Inventory, and the Neuropsychiatric Inventory have each shown success in discriminating fvFTD from Alzheimer's (Boyle et al. 2001; Kertesz et al. 2003; Levy et al. 1996). Indeed, Kertesz et al. (2003) report that the Frontal Behavioral Inventory outperforms cognitive tests in discriminating fvFTD patients from patients with Alzheimer's, especially in milder cases of the disorders. Interestingly, however, the Frontal Behavioral Inventory appeared relatively insensitive to detection of the effects of frontal leukotomy relative to schizophrenic controls in a recent study by Black et al. (2000), but the fact that this study involved patients who started off with severe schizophrenia may limit the generalizability of this findings.

Specific correlates have also emerged between several of these measures and quantitative evidence of OFC pathology. Tekin *et al.* (2001) report a correlation between previous ratings of agitation on the Neuropsychiatric Inventory and levels of neurofibrillary tangles in the OFC at autopsy (strong correlations also emerged in the cingulate). Similarly, a number of items from Lhermitte's scale have shown associations with reduced OFC metabolism in patients with frontal lesions (Sarazin *et al.* 1998, 2003). Of particular interest was the finding that decreased metabolism in the right OFC was associated with indifference to rules (Sarazin *et al.* 2003), which appears to further highlight the value of assessing anti-social characteristics in this patient population.

In summary, the use of interview schedules shows substantial promise for capturing additional information relevant to ventral frontal pathology. While the lack of clearly defined psychometric properties or well-developed normative data sets limits the implementation of many of these measures in clinical practice, several of them have already demonstrated clear clinical utility. Additional data regarding specific populations of patients with focal frontal lesions would substantially improve the confidence in interpreting these scales, but they nevertheless should aid in assessment when added to a broader neuropsychological battery.

## 18.9. Network issues and the utility of a broad assessment approach

In summary, neuropsychological evaluation of the OFC is still a field in development. Traditional neuropsychological batteries, while useful in ruling out other potential pathologies, are generally insensitive to OFC lesions. A number of experimental measures such as reversal learning, object alternation learning, the IGT, faux pas recognition and behavioral questionnaires appear far more sensitive to OFC lesions than traditional measures. However, the psychometric properties of most of these measures, in particular their specificity, remain uncertain. When taken in isolation, researchers and practitioners alike would be wise to use appropriate levels of caution when using such measures to infer damage to the OFC. This concern is heightened by evidence that several of the tasks (or their variants) are also sensitive to lesions in brain regions connected to the ventral frontal lobe. Such studies were often not reviewed in this chapter if they did not include a sample of patients with frontal lobe damage. These studies were not reviewed for two reasons. First, variations in methods and sample selection frequently make it difficult to compare results from these studies with other studies using OFC-lesioned patients. Second, in some cases the observation of a positive effect of a lesion may not challenge the specificity of the task for detecting OFC dysfunction if the damage occurs in an area that is part of a network that is critical for OFC functioning. The clearest example of this type of situation involves patients with basal ganglia dysfunction. Disruption of normal basal ganglia functioning, either through lesions, or in disorders such as Parkinson's or Huntington's disease, disrupts frontal related functions (Stocchi and Brusa 2000; Zgaljardic *et al.* 2003). Indeed, alternation and reversal tasks have both been observed to be impaired in patients with basal ganglia related disorders (Hsieh *et al.* 1998;

Lawrence et al. 1999; Marie et al. 1999; Swainson et al. 2000). Because these effects likely occur through the dysfunction of the corticostriatal loops connecting the frontal cortex, striatum and thalamus (Alexander et al. 1986; Cummings 1993), they do not specifically challenge the hypothesis that impaired performance reflects disruption of OFC functioning. Indeed, depending upon the specific location of the basal ganglia disruption they may highlight the importance of OFC dysfunction in a task, even if the source of the dysfunction lies in the striatum rather than the OFC itself. Similar issues arise for the amygdala, whose input and interaction with the OFC appear essential for normal OFC/VMPFC functioning. Thus it is hardly a surprise that lesions affecting the amygdala often have impacts on tests that are sensitive to OFC lesions (e.g., Bar-On et al. 2003; Bechara et al. 1999; Stone et al. 2003).

Given the network issues, a single test is probably unlikely to provide sufficient specificity when taken in isolation. However, when the full constellation of symptoms associated with OFC dysfunction emerge, the confidence in interpreting the data increases dramatically. It is hard to find explanations other than ventral frontal dysfunction in patients demonstrating acquired emotional dysregulation and social processing deficits if they also show deficits on a combination of experimental 'OFC tasks' and olfactory identification and discrimination tasks. Based on this reasoning, evaluation of OFC dysfunction appears best undertaken with an integrative approach to assessment that incorporates measures that evaluate cognitive, behavioral and sensory domains that are not assessed by traditional neuropsychological batteries.

## 18.10. Functional assessment and treatment planning for patients with ventral frontal lesions

As noted at the outset of the chapter, a major goal of neuropsychological assessment is the characterization of the patient's functional deficits. In this regard, the neural specificity of a task may in many situations be far less important than its ability to reveal functional deficits. Framed in this context, the question becomes to what extent do tasks help reveal the fundamental cognitive or emotional problems that are likely to impact the patient's real world functioning. For instance, it may be more important to evaluate a task such as the IGT in terms of the extent to which it meets its authors' goal of simulating real-life decision-making. In other words, is the task ecologically valid? Similarly, it is essential to know whether a task provides clear enough information regarding the specific deficit that leads to failure on a task. In this regard, questions about the extent to which the IGT reflects a lack of risk aversion, a somatic marker deficit, or a problem with reversal learning become critical for understanding the roots of a patient's decision-making problems.

Understanding the specific functional deficits becomes particularly important when the neuropsychological data are used to make treatment recommendations. In this regard, knowledge of a patient's specific deficits is essential in designing a rehabilitation program for the patient. The more a neuropsychological assessment reveals a patient's strengths and core deficits, the greater the ability to tailor a rehabilitation program to their specific needs. That said, there are unfortunately few empirically validated rehabilitation programs that are specifically designed for OFC or VMPFC patients. Whereas there are well-developed and validated rehabilitation programs for language and other cognitive functions (Cicerone *et al.* 2005), the functions that are most disrupted by OFC dysfunction have received far less attention. For instance, studies examining methods to help patients with frontal related olfactory deficits are virtually nonexistent. While compared to a loss of other sensory processing in other sensory modalities, olfactory functioning would seem to be a relatively tolerable deficit. However, in a recent study of patients with anosmia (from any cause), only 50% of patients with persistent anosmia reported being satisfied with life (compared to 87% of patients who had recovered their sense of smell (Miwa *et al.* 2001). Safety problems related to difficulty recognizing spoiled food, and detection of smoke or natural gas were also noted in such patients.

ToM deficits have also received minimal attention in terms of rehabilitation. Treatment programs related to ToM issues have emerged in recent years for patients with Asperger's syndrome and high functioning autism. For instance, Solomon and colleagues (2004) have developed a program to teach autism spectrum children facial emotion recognition and ToM skills such as perspective taking. Although sample sizes are small, children enrolled in this program have demonstrated significant improvements in facial emotion recognition. Improvement on the Faux Pas Recognition Test were noted, although this did not reach significance relative to a waiting list control group that also showed improvement upon re-testing. Whether a similar course of treatment in patients with acquired ToM problems would be beneficial is not known.

Park and colleagues (2003) have described a tailored rehabilitation program for individuals who have problems utilizing information about risk in their decision making. The treatment program, called Strategic Evaluation of Alternatives, attempts to make patients explicitly retrieve negative attributes associated with potential actions prior to performing them. The core component of the program involves analyzing specific situations (for instance, job opportunities) with an explicit evaluation of the positive and negative attributes of each situation. In the early stages of training this process is modeled to the patient, with the patient taking more initiative in this process over the course of treatment. The goal of the training is to get the patient to utilize this controlled evaluation step, rather than relying on their automatic reactions that fail to utilize or appropriately weigh negative information. Unfortunately, direct data regarding the efficacy of the program is currently lacking. The authors describe its application to a single case, a formerly successful businessman who had, since a traumatic brain injury, been repeatedly involved with relatively risky start-up ventures from which he made little gain. The program appeared partially successful in that the patient did begin to receive direct financial compensation for his consulting with companies. However, a larger issue relating to his ambitious (and more risky) job search strategy did not show any alteration. Thus, at least in the exemplar given, the approach appeared to be more beneficial in regard to the patient's approach to specific, relatively concrete situations, but less successful in terms of more big-picture decisions.

A larger rehabilitation literature addresses techniques for reducing socially inappropriate or disinhibited behavior. Various behavioral modification strategies have been applied to deal with these sorts of problems (Alderman 2004). In recent years, behavioral modification has been complemented with strategies designed to make the patient aware of their deficits or problem behaviors, and to teach them self-monitoring strategies (Turner and Levine 2004). Alderman and colleagues (Alderman, Fry, and Youngson 1995; Knight, *et al.* 2002) have described the most formalized version of this approach, Self-Monitoring Training, which combines techniques aimed at improving the subject's ability to attend to their own behavior with operant (differential reinforcement) conditioning to increase appropriate behavior and reduce unwanted behaviors. This approach may provide more generalized results than simple differential reinforcement schedules, which often fail to generalize beyond the specific environment (e.g., rehabilitation facility) in which the schedule is in place. The issue of generalizability may be particular pertinent for ventral frontal patients who often have sufficient knowledge of social and cultural rules, but fail to use them in guiding their real-world behaviours.

The ability to self-monitor may also work best when there are clear cues reminding the patient to be aware of their behavior. For example, Cicerone and Tanenbaum (1997) describe severe disturbances in emotional regulation and social cognition in a patient with OFC/VMPFC damage. The rehabilitation focused on increasing the patient's awareness of problem areas and their impact on her ability to perform activities of daily living and social functioning. The intervention started with a videotaping of her behavior on several community trips, which captured on film her difficulties inhibiting socially inappropriate behavior, and her problematic social interactions. These examples were shown to the patient, who was able to acknowledge the problems and began to modify her behavior when the camera was present. However, she continued to display problems when the cue from the camera was not present. In summary, the ability to teach self-monitoring to ventral frontal lobe patients may prove quite helpful, but the challenge is to develop approaches that maximize the patient's ability to generalize these strategies across multiple real-world environments.

## Acknowledgments

I thank Amy Cooter for assistance in editing this chapter, and Natalie Denburg, Scott Rauch and an anonymous reviewer for their useful comments on an earlier version of this chapter.

## References

Abbruzzese, M., Bellodi, L., Ferri, S., and Scarone, S. (1995) Frontal-lobe dysfunction in schizophrenia and obsessive-compulsive disorder—A neuropsychological study. *Brain and Cognition* **27**:202–212.

Abbruzzese, M., Ferri, S., and Scarone, S. (1997) The selective breakdown of frontal functions in patients with obsessive-compulsive disorder and in patients with schizophrenia: a double dissociation experimental finding. *Neuropsychologia* **35**:907–912.

Alderman, N. (2004) Disorders of Behavior. In: *Cognitive and Behavioral Rehabilitation: From Neurobiology to Clinical Practice*, (J. Ponsford, ed.). New York: Guilford. pp. 269–298.

Alderman, N., Fry, R., and Youngson, H. A. (1995) Improvement of self-monitoring skills, reduction of behavioral disturbance and the dysexecutive syndrome: Comparison of response cost and a new programme of self-monitoring training. *Neuropsychological Rehabilitation* 5:193–221.

Alexander, G. E., DeLong, M.R., and Strick, P. L. (1986) Parallel organization of functionally segregated circuits linking basal ganglia and cortex. *Annual Review of Neuroscience* 9:357–381.

Anderson, S. W., Damasio, H., Tranel, D., and Damasio, A. R. (1992) Cognitive sequelae of focal lesions in ventromedial frontal lobe. Journal of Clinical and Experimental Neuropsychology 14:83.

Anderson, S. W., Barrash, J., Bechara, A., Tranel, D. (in press). Impairments of emotion and real-world complex behavior following childhood- or adult-onset damage to ventromedial prefrontal cortex. *Journal of International Neuropsychological Society.*

Angrilli, A., Palomba, D., Cantagallo, A., Maietti, A., and Stegagno, L. (1999) Emotional impairment after right orbitofrontal lesion in a patient without cognitive deficits. *Neuroreport* 10:1741–1746.

Aron, A. R., and Poldrack, R. A. (2005) The cognitive neuroscience of response inhibition: relevance for genetic research in attention-deficit/hyperactivity disorder. *Biological Psychiatry* 57:1285–1292.

Bardenhagen, F. J. and Bowden, S. C. (1998) Cognitive components in perseverative and non-perseverative errors on the object alternation task. *Brain and Cognition* 37:224–236.

Bar-On, R., Tranel, D., Denburg, N. L., and Bechara, A. (2003).Exploring the neurological substrate of emotional and social intelligence, *Brain* 126:1790–800.

Baron-Cohen, S., O'Riordan, M., Stone, V., Jones, R, and Plaisted, K. (1999) Recognition of faux pas by normally developing children and children with Asperger syndrome or high-functioning autism. *jouranal of Autism and Developmental Disorders* 29:407–418.

Barrash, J., Tranel, D., and Anderson, S. W. (2000) Acquired personality disturbances associated with bilateral damage to the ventromedial prefrontal region. *Developmental Neuropsychology* 18:355–381.

Bechara, A., Damasio, A. R., Damasio, H., and Anderson, S. W. (1994) Insensitivity to future consequences following damage to human prefrontal cortex. *Cognition* 50:7–15.

Bechara, A., Damasio, H., Tranel, D., and Anderson, S. W. (1998) Dissociation of working memory from decision making within the human prefrontal cortex. *Journal of Neuroscience* 18:428–437.

Bechara, A., Damasio, H., Damasio, A. R., Lee, and GP (1999) Different contributions of the human amygdala and ventromedial prefrontal cortex to decision-making. *Journal of Neuroscience* 19:5473–81.

Berlin, H. A., Rolls, E. T., and Kischka, U. (2004) Impulsivity, time perception, emotion and reinforcement sensitivity in patients with orbitofrontal cortex lesions. *Brain* 127: 1108–1126.

Bird, C. M., Castelli, F., Malik, O., Frith, U., and Husain, M. (2004) The impact of extensive medial frontal lobe damage on "Theory of Mind" and cognition. *Brain* 127:914–28.

Black, D. N., Stip, E., Bedard, M., Kabay, M., Paquette, I., and Bigras, M. J. (2000) Leukotomy revisited: late cognitive and behavioral effects in chronic institutionalized schizophrenics. *Schizophrenia Research* 43:57–64.

Boyle, P. A., Grace, J., Zawacki, T. M., Ott, B. R., and Stout, J. C. (2001) Frontal behavior in neurodegenerative and acute neurological disorders.

Broe, M., Hodges, J. R., Schofield, E., Shepherd, C. E., Kril, J. J., and Halliday, G. M. (2003) Staging disease severity in pathologically confirmed cases of frontotemporal dementia. *Neurology* 60:1005–11.

Brutkowski, S., and Davrowska, J. (1963) Disinhibition after prefrontal lesions as a function of duration of intertrial intervals. *Science* 139:505–506.

Butter, C. M. (1969) Perseveration in extinction and in discrimination reversal tasks following selective frontal ablations in Macaca mulatta. *Physiology and Behavior* 4:163–171.

Butters, N., Butter, C., Rosen, J., and Stein, D. (1973) Behavioral effects of sequential and one-stage ablations of orbital prefrontal cortex in the monkey. *Experimental Neurology* 39:204–214.

Cain, W. S., Gent, J., Catalanotto, F. A., and Goodspeed, R. B. (1983) Clinical evaluation of olfaction. *American Journal of Otolaryngology* 4:252–6.

Cavedini, P., Ferri, S., Scarone, S., and Bellodi, L. (1998) Frontal lobe dysfunction in obsessive-compulsive disorder and major depression: a clinical-neuropsychological study. *Psychiatry Research* **78**:21–28.

Cicerone, K. D., Dahlberg, C., Malec, J. F., Langenbahn, D. M., Felicetti, T., Kneipp, S., et al. (2005) Evidence-based cognitive rehabilitation: updated review of the literature from 1998 through 2002. *Archives of Physical Medicine Rehabilitation* 86: 1681–92.

Cicerone, K. D. and Tanenbaum, L. N. (1997) Disturbance of social cognition after traumatic orbitofrontal brain injury. *Archives of Clinical Neuropsychology* **12**:173–88.

Clark, L., Manes, F., Nagui, A., Sahakian, B. J., and Robbins, T. W. (2003) The contributions of lesion laterality and lesion volume to decision-making impairment following frontal lobe damage. *Neuropsychologia* **41**:1474–1483.

Courville, C. B. (1937) *Pathology of the Central Nervous System; a Study Based Upon a Survey of Lesions Found in a Series of 15,000 Autopsies.* Oxford, UK: Pacific Press Publishing Association.

Cummings, J. L. (1993) Frontal-subcortical circuits and human behavior. [Review] [71 refs], *Archives of Neurology* **50**:873–880.

Curtis, C. E., Zald, D. H., Lee, J. T., and Pardo, J. V. (2000) Object and spatial alternation tasks with minimal delays activate the right anterior hippocampus proper in humans. *Neuroreport* **11**:2203–2207.

DeLuca, J. (1993). Predicting neurobehavioral patterns following anterior Communicating Artery Aneurysm. *Cortex* **29**:639–647.

Dias, R., Robbins, T. W., and Roberts, A. C. (1996) Dissociation in prefrontal cortex of affective and attentional shifts. *Nature* **380**:69–72.

Doty, R. L., Li, C., Mannon, L. J., and Yousem, D. M. (1998) Olfactory dysfunction in multiple sclerosis. Relation to plaque load in inferior frontal and temporal lobes. *Annals of the New York Academy of Sciences* **855**:781–786.

Doty, R. L., Li, C., Mannon, L. J., and Yousem, D. M. (1999) Olfactory dysfunction in multiple sclerosis: relation to longitudinal changes in plaque numbers in central olfactory structures. *Neurology* **53**:880–882.

Doty, R. L., Shaman, P., Kimmelman, C. P., and Dann, M. S. (1984) University of Pennsylvania Smell Identification Test: a rapid quantitative olfactory function test for the clinic. *Laryngoscope* **94**:176–8.

Doty, R. L., Yousem, D. M., Pham, L. T., Kreshak, A. A., Geckle, R., and Lee, W. W. (1997) Olfactory dysfunction in patients with head trauma. *Archives of Neurology* **54**:1131–1140.

Drewe, E. A. (1975) Go—no go learning after frontal lobe lesions in humans. *Cortex* **11**:8–16.

Elliott, R., Frith, C. D., and Dolan, R. J. (1997) Differential neural response to positive and negative feedback in planning and guessing tasks. *Neuropsychologia* **35**:1395–1404.

Eslinger, P. J. and Damasio, A. R. (1985) Severe Disturbance of Higher Cognition After Bilateral Frontal-Lobe Ablation—Patient EVR, *Neurology* **35**:1731–1741.

Faraone, S. V., Seidman, L. J., Kremen, W. S., Toomey, R., Pepple, J. R., and Tsuang, M. T. (1999) Neuropsychological functioning among the non-psychotic relatives of schizophrenic patients: a 4-year follow-up study. *Journal of Abnormal Psychology* **108**:176–181.

Fellows, L. K. and Farah, M. J. (2003) Ventromedial frontal cortex mediates affective shifting in humans: evidence from a reversal learning paradigm. *Brain* **126**:1830–1837.

Fellows, L. K. and Farah, M. J. (2005) Different underlying impairments in decision-making following ventromedial and dorsolateral frontal lobe damage in humans. *Cerebral Cortex* **15**:58–63.

Fernandez-Duque, D. and Black, S. E. (2005) Impaired recognition of negative facial emotions in patients with frontotemporal dementia. *Neuropsychologia* **43**:1673–87.

Fletcher, P. C. and Henson, R. N. (2001) Frontal lobes and human memory: insights from functional neuroimaging. *Brain* **124**:849–881.

Franceschi, M., Anchisi, D., Pelati, O., Zuffi, M., Matarrese, M., Moresco, R. M., et al. (2005) Glucose metabolism and serotonin receptors in the frontotemporal lobe degeneration. *Annals of Neurology* **57**:216–225.

Freedman, M. (1990) Object alternation and orbitofrontal system dysfunction in Alzheimer's and Parkinson's disease. *Brain and Cognition* **14**:134–143.

Freedman, M. (1994) Frontal and parietal lobe dysfunction in depression: delayed alternation and tactile learning deficits. *Neuropsychologia* **32**:1015–1025.

Freedman, M., Black, S., Ebert, P., and Binns, M. (1998) Orbitofrontal function, object alternation and perseveration. *Cerebral Cortex* **8**:18–27.

Gallagher, H. L., and Frith, C. D. (2003) Functional imaging of "theory of mind". *Trends in Cognitive Sciences* **7**:77–83.

Gansler, D. A., Covall, S., McGrath, N., and Oscar-Berman, M. (1996) Measures of prefrontal dysfunction after closed head injury. *Brain and Cognition* **30**:194–204.

Godefroy, O., and Rousseaux, M. (1996) Divided and focused attention in patients with lesion of the prefrontal cortex. *Brain and Cognition* **30**:155–174.

Good, K. P., Kiss, I., Buiteman, C., Woodley, H., Rui, Q., Whitehorn, D., and Kopala, L. (2002) Improvement in cognitive functioning in patients with first-episode psychosis during treatment with quetiapine: an interim analysis. *British Journal of Psychiatry* **43**:(Suppl.):s45-s49.

Grace, J. and Malloy, P. F. (2001) *Frontal Systems Behavior Scale: Professional Manual.* Lutz, Florida: Psychological Assessment Resources.

Grace, J., Stout, J. C., and Malloy, P. F. (1999) Assessing frontal lobe behavioral syndromes with the Frontal Lobe Personality Scale *Assessment* **6**:269–284.

Gregory, C., Lough, S., Stone, V., Erzinclioglu, S., Martin, L., Baron-Cohen, S., and Hodges, J.R. (2002) Theory of mind in patients with frontal variant frontotemporal dementia and Alzheimer's disease: theoretical and practical implications. *Brain* **125**:752–764.

Hornak, J., Bramham, J., Rolls, E. T., Morris, R. G., O'Doherty, J., Bullock, P. R., et al. (2003) Changes in emotion after circumscribed surgical lesions of the orbitofrontal and cingulate cortices. *Brain* **126**:1691–712.

Hornak, J., O'Doherty, J., Bramham, J., Rolls, E. T., Morris, R. G., Bullock, P. R., et al. (2004) Reward-related reversal learning after surgical excisions in orbito-frontal or dorsolateral prefrontal cortex in humans. *Journal of Cognitive Neuroscience* **16**:463–478.

Hornak, J., Rolls, E. T., and Wade, D. (1996) Face and voice expression identification in patients with emotional and behavioral changes following ventral frontal lobe damage, *Neuropsychologia* **34**:247–61.

Hsieh, S., Chuang, Y. Y., Hwang, W. J., and Pai, M. C. (1998) A specific shifting deficit in Parkinson's disease: a reversal shift of consistent stimulus-response mappings. *Perceptual and Motor Skills* **87**:1107–1119.

Hulshoff Pol, H. E., Hijman, R., Tulleken, C. A., Heeren, T. J., Schneider, N., and van Ree, J. M. (2002) Odor discrimination in patients with frontal lobe damage and Korsakoff's syndrome. *Neuropsychologia* **40**:888–891.

Iversen, S. and Mishkin, M. (1970) Perseverative interference in monkeys following selective lesions of the inferior prefrontal convexity. *Experimental Brain Research* **11**:376–386.

Jones-Gotman, M. and Zatorre, R. J. (1988) Olfactory identification deficits in patients with focal cerebral excision. *Neuropsychologia* **26**:387–400.

Keane, J., Calder, A. J., Hodges, J. R., and Young, A. W. (2002) Face and emotion processing in frontal variant frontotemporal dementia. *Neuropsychologia* **40**:655–65.

Kertesz, A., Davidson, W., and Fox, H. (1997) Frontal behavioral inventory: diagnostic criteria for frontal lobe dementia. *Canadian Journal of Neurological Sciences* **24**:29–36.

Kertesz, A., Davidson, W., McCabe, P., and Munoz, D. (2003) Behavioral quantitation is more sensitive than cognitive testing in frontotemporal dementia. *Alzheimer Disease and Associated Disorders* **17**:223–229.

Knight, C., Rutterford, N. A., Alderman, N., and Swan, L. J. (2002) Is accurate self-monitoring necessary for people with acquired neurological problems to benefit from the use of differential reinforcement methods? *Brain Injury* **16**:75–87.

Koenen, K. C., Driver, K. L., Oscar-Berman, M., Wolfe, J., Folsom, S., Huang, M. T. and Schlesinger L. (2001) Measures of prefrontal system dysfunction in posttraumatic stress disorder. *Brain and Cognition* **45**:64–78.

Kringelbach, M. L., and Rolls, E. T. (2003) Neural correlates of rapid reversal learning in a simple model of human social interaction. *Neuroimage* **20**:1371–1383.

Lavenu, I. and Pasquier, F. (2005) Perception of emotion on faces in frontotemporal dementia and Alzheimer's disease: a longitudinal study. *Dementia and Geriatric Cognitive Disorders* **19**:37–41.

Lavenu, I., Pasquier, F., Lebert, F., Petit, H., and Van der, L. M. (1999) Perception of emotion in frontotemporal dementia and Alzheimer disease. *Alzheimer Disease and Associated Disorder* **13**:96–101.

Lawrence, A. D., Sahakian, B. J., Rogers, R. D., Hodges, J. R., and Robbins, T. W. (1999) Discrimination, reversal, and shift learning in Huntington's disease: mechanisms of impaired response selection. *Neuropsychologia* **37**:1359–1374.

Leimkuhler, M. E., Mesulam, M. M. (1985) Reversible go-no go deficits in a case of frontal lobe tumor. *Annals of Neurology* **18**:617–619.

Levin, H. S., High, W. M., and Eisenberg, H. M. (1985) Impairment of olfactory recognition after closed head injury. *Brain* 108 ( Pt 3):579–591.

Levine, B., Black, S. E., Cheung, G., Campbell, A., O'Toole, C. O., Schwartz, M. L. (2005) Gambling task performance in traumatic brain injury: Relationships to injury severity, atrophy, lesion location, and cognitive and psychosocial outcome. *Cognitive Behavioral Neurology* **18**:45–54.

Levy, M.L., Miller, B.L., Cummings, J.L., Fairbanks, L.A., and Craig, A. (1996) Alzheimer disease and frontotemporal dementias. Behavioral distinctions. *Archives of Neurology* **53**:687–690.

Lezak, M.D. (1995) *Neuropsychological Assessment*. London: Oxford University Press.

Lhermitte, F. (1986) Human autonomy and the frontal lobes. Part II: Patient behavior in complex and social situations: the "environmental dependency syndrome". *Annals of Neurology* **19**:335–343.

Lhermitte, F., Pillon, B., and Serdaru, M. (1986) Human autonomy and the frontal lobes. Part I: Imitation and utilization behavior: a neuropsychological study of 75 patients. *Annals of Neurology* **19**:326–334.

Lough, S., Kipps, C. M., Treise, C., Watson, P., Blair, J. R., and Hodges, J. R. (2006) Social reasoning, emotion and empathy in frontotemporal dementia. *Neuropsychologia* **44**:950–958.

Manes, F., Sahakian, B., Clark, L., Rogers, R., Antoun, N., Aitken, M., *et al.* (2002) Decision-making processes following damage to the prefrontal cortex. *Brain* **125**:624–639.

Marie, R. M., Barre, L., Dupuy, B., Viader, F., Defer, G., and Baron, J. C. (1999). Relationships between striatal dopamine denervation and frontal executive tests in Parkinson's disease. *Neuroscience Letters* **260**:77–80.

Mavaddat, N., Kirkpatrick, P. J., Rogers, R. D., and Sahakian, B. J. (2000) Deficits in decision-making in patients with aneurysms of the anterior communicating artery. *Brain* **123**:2109–2117.

Meunier, M., Bachevalier, J., and Mishkin, M. (1997) Effects of orbital frontal and anterior cingulate lesions on object and spatial memory in rhesus monkeys. *Neuropsychologia* **35**:999–1015.

Mishkin, M. (1964) Preservation of central sets after frontal lesions in monkeys. In *The frontal granular cortex and behavior* (J. M. Warren and K. Akert, eds). pp. 219–241. New York: McGraw-Hill.

Mishkin, M. and Manning, F. J. (1978) Non-spatial memory after selective prefrontal lesions in monkeys. *Brain Research* **143**:313–323.

Mishkin, M., Vest, B., Waxler, M., and Rosvold, H. E. (1969) A re-examination of the effects of frontal lesions on object alternation. *Neuropsychologia* **7**:357–363.

Miwa, T., Furukawa, M., Tsukatani, T., Costanzo, R. M., DiNardo, L. J., and Reiter, E. R. (2001) Impact of olfactory impairment on quality of life and disability. *Archives of Otolaryngololgy -Head Neck Surgery* **127**:497–503.

Norton, L. E., Malloy, P. F., and Salloway, S. (2001) The impact of behavioral symptoms on activities of daily living in patients with dementia. *American Journal of Geriatric Psychiatry* **9**:41–48.

Oscar-Berman, M. and Bardenhagen, F. (1998) Nonhuman animal models of memory dysfunction in neurodegenerative disease. In: *Memory in Neurodegenerative Disease* ( A. Tröster, ed.), pp. 3–20. New York: Cambridge University Press.

Pang, D. (1989) Physics and pathology of closed head injury. In: *Assessment of the behavioral consequences of head trauma* ( M. Lezak, ed.), pp 1–17. New York: Alan R. Liss.

Pantelis, C. and Brewer, W. (1995) Neuropsychological and olfactory dysfunction in schizophrenia: relationship of frontal syndromes to syndromes of schizophrenia. *Schizophrenia Research* **17**:35–45.

Park, N. W., Conrod, B., Hussain, Z., Murphy, K. J., Rewilak, D., and Black, S. E. (2003) A treatment program for individuals with deficient evaluative processing and consequent impaired social and risk judgement. *Neurocase* **9**:51–62.

Potter, H. and Butters, N. (1980) An assessment of olfactory deficits in patients with damage to prefrontal cortex. *Neuropsychologia* **18**:621–628.

Pribram, K. H. and Mishkin, M. (1956) Analysis of the effects of frontal lesions in monkey: III. Object alternation. *Journal of Comparative and Physiological Psychology* **49**:41–45.

Rahman, S., Robbins, T. W., and Sahakian, B. J. (1999) Comparative cognitive neuropsychological studies of frontal lobe function: Implications for therapeutic strategies in frontal variant frontotemporal dementia. *Dementia and Geriatric Cognitive Disorders* **10**:15–28.

Rankin, K. P., Kramer, J. H., and Miller, B. L. (2005) Patterns of cognitive and emotional empathy in frontotemporal lobar degeneration. *Cognitive and Behavioral Neurology* **18**:28–36.

Richardson, J. (2000) *Clinical and neruopsychological aspects of closed head injury*. Hove, UK: Psychology Press.

Rogers, R. D., Everitt, B. J., Baldacchino, A., Blackshaw, A. J., Swainson, R., Wynne, K., et al. (1999) Dissociable deficits in the decision-making cognition of chronic amphetamine abusers, opiate abusers, patients with focal damage to prefrontal cortex, and tryptophan-depleted normal volunteers: Evidence for monoaminergic mechanisms. *Neuropsychopharmacology* **20**:322–339.

Rolls, E. T., Hornak, J., Wade, D., and McGrath, J. (1994) Emotion-related learning in patients with social and emotional changes associated with frontal lobe damage. *Journal of Neurology, Neurosurgery and Psychiatry* **57**:1518–1524.

Rosen, H. J., Pace-Savitsky, K., Perry, R. J., Kramer, J. H., Miller, B. L., and Levenson, R. W. (2004). Recognition of emotion in the frontal and temporal variants of frontotemporal dementia, *Dementia and Geriatric Cognitive Disorders* **17**:277–81.

Rowe, A. D., Bullock, P .R., Polkey, C. E., and Morris, R. G. (2001) "Theory of mind" impairments and their relationship to executive functioning following frontal lobe excisions. *Brain* **124**:600–616.

Royet, J. P., Hudry, J., Zald, D. H., Godinot, D., Gregoire, M. C., Lavenne, F., et al. (2001) Functional neuroanatomy of different olfactory judgements, *Neuroimage* **13**:506–519.

Sabbagh, M. A., Moulson, M. C., and Harkness, K. L. (2004). Neural correlates of mental state decoding in human adults: an event-related potential study. *Journal of Cognitive Neuroscience* **16**:415–426.

Salmaso, D., Denes, G. (1982) Role of the frontal lobes on an attention task: a signal detection analysis. *Perceptual and Motor Skills* **54**:1147–1150.

Sarazin, M., Michon, A., Pillon, B., Samson, Y., Canuto, A., Gold, G., et al. (2003) Metabolic correlates of behavioral and affective disturbances in frontal lobe pathologies. *Journal of Neurology* **250**:827–833.

Sarazin, M., Pillon, B., Giannakopoulos, P., Rancurel, G., Samson, Y., and Dubois, B. (1998) Clinicometabolic dissociation of cognitive functions and social behavior in frontal lobe lesions. *Neurology*, **51**:142–148.

Savic, I., Gulyas, B., Larsson, M., and Roland, P. (2000) Olfactory functions are mediated by parallel and hierarchical processing. *Neuron* **26**:735–745.

Seidman, L. J., Oscar-Berman, M., Kalinowski, A. G., and Ajilore, O. (1995) Experimental and clinical neuropsychological measures of prefrontal dysfunction in schizophrenia. *Neuropsychologia* **9**:481–490.

Shamay, S. G., Tomer, R., and Aharon-Peretz, J. (2002) Deficit in understanding sarcasm in patients with prefronal lesion is related to impaired empathic ability. *Brain and Cognition* **48**:558–563.

Shamay-Tsoory, S. G., Tomer, R., Goldsher, D., Berger, B. D., and Aharon-Peretz, J. (2004) Impairment in cognitive and affective empathy in patients with brain lesions: anatomical and cognitive correlates. *Journal of Clinical and Experimental Neuropsychology* **26**:1113–1127.

Shamay-Tsoory, S. G., Tomer, R., and Aharon-Peretz, J. (2005a) The neuroanatomical basis of understanding sarcasm and its relationship to social cognition. *Neuropsychology* **19**:288–300.

Shamay-Tsoory, S. G., Tomer, R., Berger, B. D., Goldsher, D., and Aharon-Peretz, J. (2005b) Impaired "affective theory of mind" is associated with right ventromedial prefrontal damage. *Cognitive and Behavioral Neurology* **18**:55–67.

Slachevsky, A., Villalpando, J. M., Sarazin, M., Hahn-Barma, V., Pillon, B., and Dubois, B. (2004) Frontal assessment battery and differential diagnosis of frontotemporal dementia and Alzheimer disease. *Archives of Neurology* **61**:1104–1107.

Small, D. M., Voss, J., Mak, Y. E., Simmons, K. B., Parrish, T., and Gitelman, D. (2004) Experience-dependent neural integration of taste and smell in the human brain. *Journal of Neurophysiology* **92**:1892–1903.

Solomon, M., Goodlin-Jones, B. L., and Anders, T. F. (2004) A social adjustment enhancement intervention for high functioning autism, Asperger's syndrome, and pervasive developmental disorder NOS. *Journal of Autism and Developmental Disorders* **34**:649–68.

Spinella, M. (2002) Correlations among behavioral measures of orbitofrontal function. *International Journal of Neuroscience* **112**:1359–1369.

Stocchi, F. and Brusa, L. (2000) Cognition and emotion in different stages and subtypes of Parkinson's disease. *Journal of Neurology* 247, Suppl 2, **II**: 114-II121.

Stone, V. E., Baron-Cohen, S., and Knight, R. T. (1998) Frontal lobe contributions to theory of mind. *Journal of Cognitive Neuroscience* **10**:640–656.

Stone, V. E., Baron-Cohen, S., Calder, A., Keane, J., and Young, A. (2003) Acquired theory of mind impairments in individuals with bilateral amygdala lesions. *Neuropsychologia* **41**:209–220.

Stout, J. C., Ready, R. E., Grace, J., Malloy, P. F., and Paulsen, J. S. (2003) Factor analysis of the Frontal Systems Behavior Scale (FrSBe), *Assessment* **10**:79–85.

Stuss, D. T., Benson, D. F., Kaplan, E. F., Weir, W. S., Naeser, M. A., Lieberman, I., and Ferrill, D. (1983) The involvement of orbitofrontal cerebrum in cognitive tasks. *Neuropsychologia* **21**: 235–248.

Stuss, D. T., Kaplan, E. F., Benson, D. F., Weir, W. S., Chiulli, S., and Sarazin, F. F. (1982) Evidence for the involvement of orbitofrontal cortex in memory functions: an interference effect. *Journal Of Comparative And Physiological Psychology* **96**:913–925.

Stuss, D. T., Levine, B., Alexander, M. P., Hong, J., Palumbo, C., Hamer, L., Murphy, K. J., and Izukawa, D. (2000) Wisconsin Card Sorting Test performance in patients with focal frontal and posterior brain damage: effects of lesion location and test structure on separable cognitive processes. *Neuropsychologia* **38**:388–402.

Stuss, D. T., Gallup, G. G., Jr., and Alexander, M. P. (2001) The frontal lobes are necessary for "theory of mind". *Brain* **124**:279–286.

Sumner, D. (1964) Post-traumatic anosmia. *Brain* **87**:107–120.

Swainson, R., Rogers, R. D., Sahakian, B. J., Summers, B. A., Polkey, C. E., and Robbins, T. W. (2000) Probabilistic learning and reversal deficits in patients with Parkinson's disease or frontal or temporal lobe lesions: possible adverse effects of dopaminergic medication. *Neuropsychologia* **38**:596–612.

Tekin, S., Mega, M. S., Masterman, D. M., Chow, T., Garakian, J., Vinters, H. V., et al. (2001) Orbitofrontal and anterior cingulate cortex neurofibrillary tangle burden is associated with agitation in Alzheimer disease. *Annals of Neurology* **49**:355–361.

Turner, G. R. and Levine, B. (2004) Disorders of executive functioning and self-awareness. In: *Cognitive and Behavioral Rehabilitation: From Neurobiology to Clinical Practice* (J. Ponsford, ed.). New York: Guilford. pp. 224–268.

Tranel, D., Bechara, A., and Denburg, N. L. (2002) Asymmetric functional roles of right and left ventromedial prefrontal cortices in social conduct, decision-making, and emotional processing. *Cortex* **38**:589–612.

Varney, N. R. (1988) Prognostic significance of anosmia in patients with closed-head trauma. *Journal of Clinical and Experimental Neuropsychology* **10**:250–254.

Varney, N. R., Pinkston, J. B., and Wu, J. C. (2001) Quantitative PET findings in patients with posttraumatic anosmia. *Journal of Head Trauma Rehabilitation* **16**:253–259.

Varrone, A., Pappata, S., Caraco, C., Soricelli, A., Milan, G., Quarantelli, M., et al. (2002) Voxel-based comparison of rCBF SPET images in frontotemporal dementia and Alzheimer's disease highlights the involvement of different cortical networks. *European Journal of Nuclear Medicine and Molecular Imaging* **29**:1447–1454.

Williams, G.B., Nestor, P.J., and Hodges, J.R. (2005) Neural correlates of semantic and behavioral deficits in frontotemporal dementia. *Neuroimage* **24**:1042–1051.

Wilson, B, Alderman, N., Burgess, P., Emsley, H., Evans, J. (1996) *Behavioral assessment of the dysexecutive syndrome (BADS) manual*. Bury St. Edmunds: Thames Valley Test Company.

Yousem, D. M., Geckle, R. J., Bilker, W. B., Kroger, H., and Doty, R. L. (1999) Posttraumatic smell loss: Relationship of psychophysical tests and volumes of the olfactory bulbs and tracts and the temporal lobes. *Academic Radiology* **6**:264–272.

Yousem, D. M., Geckle, R. J., Bilker, W. B., McKeown, D. A., and Doty, R. L. (1996) Posttraumatic olfactory dysfunction: MR and clinical evaluation. *American Journal of Neuroradiology* **17**:1171–1179.

Zald, D. H., Curtis, C. E., Chernitsky, L. A., and Pardo, J. V. (2005) Frontal lobe activation during object alternation acquisition. *Neuropsychology* **19**:97–105.

Zald, D. H., Curtis, C., Folley, B. S., and Pardo, J. V. (2002). Prefrontal contributions to delayed spatial and object alternation: A positron emission tomography study. *Neuropsychology* **16**:182–189.

Zald, D. H. and Kim, S. W. (2001) The Orbitofrontal Cortex. In: *The Frontal Lobes and Neuropsychiatric Illness* ( S. Salloway, J.D. Duffy, and P.F. Malloy, eds). pp. 33–70. Washington DC: American Psychiatric Press.

Zatorre, R. J. and Jones-Gotman, M. (1991) Human olfactory discrimination after unilateral frontal or temporal lobectomy. *Brain* **114**:71–84.

Zgaljardic, D. J., Borod, J. C., Foldi, N. S., and Mattis, P. (2003) A review of the cognitive and behavioral sequelae of Parkinson's disease: relationship to frontostriatal circuitry. *Cognitve and Behavioral Neurology* **16**:193–210.

Zohar, J., Hermesh, H., Weizman, A., Voet, H., and Gross-Isseroff, R. (1999) Obitofrontal cortex dysfunction in obsessive-compulsive disorder? I. Alternation learning in obsessive-compulsive disorder: male-female comparisons. *European Neuropsychopharmacology* **9**:407–413.

Zorzon, M., Ukmar, M., Bragadin, L. M., Zanier, F., Antonello, R. M., Cazzato, G., and Zivadinov, R. (2000) Olfactory dysfunction and extent of white matter abnormalities in multiple sclerosis: a clinical and MR study. *Multiple Sclerosis* **6**:386–390.

Zusho, H. (1982) Posttraumatic anosmia. *Archives Otolaryngology* **108**:90–92.

# Chapter 19

# The orbitofrontal cortex in drug addiction

Rita Z. Goldstein, Nelly Alia-Klein, Lisa A. Cottone, and Nora D. Volkow

## 19.1. Introduction

### Why study the orbitofrontal cortex in the context of addiction?

As a paralimbic heterogeneous structure, the orbitofrontal cortex (OFC) is well localized to: (a) integrate sensory information (as it receives inputs from association areas of each sensory modality) with valence input from its amygdala connections; and (b) evaluate new inputs vis-à-vis previously processed information from its connections with structures subserving memory (London *et al.* 2000). Gathering, comparing, and attributing value to stimuli or objects are essential prerequisites of complex behaviors. Converging evidence from clinical human and primate studies have consistently implicated the OFC in these higher-order organizational and self-monitoring functions, such as social adjustment and the control of mood and drive, traits that crucially encompass personality composition. The case of Phineas Gage has provided an archetypical example as to how pervasive and profound damage to the ventral aspects of the prefrontal cortex (PFC) can be in undermining the integrity of the self (Steegmann 1962). From a well-mannered, stable working man, Phineas Gage became disrespectful, profane, obstinate, impatient and fitful. This case triggered studies of other patients with lesions to the frontal lobe that have corroborated (and were further validated by animal studies) the role of the OFC in personality and higher-order cognitive factors (e.g., Bechara 2004). This case has also triggered interest in the neuropsychological sequelae of drug addiction.

This chapter focuses on the role of the OFC in addiction. Defined as compulsive drug use despite considerable negative consequences, addiction is a disease characterized behaviorally by deficits in inhibitory control and drive, as well as the other neurocognitive functions classically attributed to the OFC (Goldstein and Volkow 2002). Since this book provides an in-depth account of the anatomical, biological and behavioral aspects of the OFC, in the current chapter we will focus our review on delineating this region's involvement specifically in the unique symptomatology of the drug addiction syndrome. A cautionary note: while the OFC appears to have unique and integrative functions and corresponding 'dysfunctions' in addiction, the reader is encouraged to conceive the OFC as part of a larger brain circuit that is compromised in the addicted individual. Before discussing neuronal mechanisms, however, we must first describe the core characteristics, deficits, and behavioral models that contribute to the definition of this complex disorder.

## 19.2. Addiction

### Definition and core neuropsychological deficits

Addiction is a chronic disease that is best understood behaviorally as a continuous cycle involving self-administration of a reinforcing substance with consequent intoxication, craving, bingeing and withdrawal compelling the organism back to self-administration of the substance (Fig. 19.1) (Goldstein and Volkow 2002). Despite serious negative consequences this pattern continues, whereby use of a reinforcing substance leads to an increased desire (or craving) for further consumption of that substance. After procuring and uncontrollably consuming the substance (culminating in bingeing for drugs such as cocaine or alcohol but not for nicotine, marijuana, or heroin), symptoms of withdrawal develop the intensity of which varies depending on the drug (very intense for heroin and alcohol and milder for cocaine). These withdrawal symptoms (including excessive fatigue, irritability, anxiety, dysphoria, and anhedonia (Gawin and Kleber 1986)) cease upon drug re-administration, even when the substance is no longer pleasurable (Fischman et al. 1985).

This vicious cycle, however, does not invariably ensue upon drug self-administration. According to a recent report by the National Institute on Drug Abuse (NIDA) (Services and Abuse 2004), up to 51% of adolescents experiment with various drugs (76% with alcohol); however only 2%–15% (depending on the substance; cocaine: 24.5% of adolescents and 16.7% for all age groups) (Anthony et al. 1994; APA 1994) become addicted. Interestingly, a similar percentage of rats (17% in the case of cocaine) trained to self-administer drugs exhibit behaviors that mimic the human disease of addiction

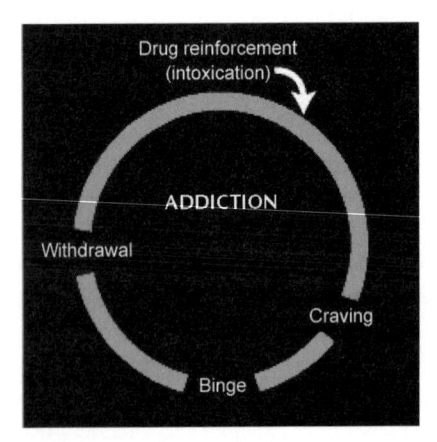

**Fig. 19.1** Core behavioral manifestations of the I-RISA (Impaired Response Inhibition and Salience Attribution) syndrome of drug addiction. Reproduced with permission from Goldstein and Volkow (2002).

(persistence in drug-seeking, resistance to punishment, motivation for the drug, and drug-induced reinstatement), while the majority of the rats do not develop these characteristics (Deroche-Gamonet et al. 2004).

This gap between experimentation with drug use or recreational consumption and the onset of the disorder of addiction has led investigators to generally agree that drug addiction is the result of an interplay between repeated drug self-administration, genes and other biological factors that predispose individuals to addiction, and the effects on these vulnerable individuals of environmental stressors at critical periods. An ongoing debate among researchers pertains to the etiology of these vulnerabilities and/or deficits; do they pre-date the development of addiction? Or do they develop as a result of chronic exposure to these substances? There is data to support both positions (Mackler and Eberwine 1991; Nestler 2000, 2001; Volkow et al. 2002), and it is progressively becoming clear that the onset and maintenance of the cycle of addiction finds fertile ground in individuals with certain vulnerabilities while concomitantly the continuous exposure to these substances imparts its own neuronal change. Keeping the aforementioned in mind, we will review findings on cognitive and emotional deficits in addicted individuals.

Cognitive deficits in drug addiction can be detected by using standard neuropsychological testing batteries tapping memory, attention and executive functions (Rogers and Robbins 2001; Goldstein et al. 2004). Taking advantage of the wealth of accumulated neuropsychological data from substance dependent individuals in our laboratory, we obtained a quantitative measure of deviation from the norm (our group of healthy controls) on several dimensions of cognitive functioning in cocaine-addicted and alcoholic subjects (Goldstein et al. 2004). We found that the severity of neuropsychological impairment in the substance-dependent individuals was significant across several cognitive dimensions including verbal and visual memory, verbal knowledge, attention and executive functioning. Furthermore, compared to cocaine addicts, alcoholics showed a greater impairment in attention and executive functioning, possibly pointing to the role of a specific substance in a particular deficit expression. Notwithstanding, the observed overall deviations as graphed in Fig. 19.2 were modest, with the addicted groups' means not exceeding one standard deviation below the means for healthy subjects. As a frame of reference, individuals with other brain disorders, such as schizophrenia, show deviations at about one and a half standard deviations below the norm on comparable cognitive measures (Bilder et al. 2000).

It is important to note here, however, that the impact of such statistically modest deviations in neuropsychological functioning on the quality of life and possibly on relapse in drug-addicted individuals may be quite substantial. This discrepancy between *subtle* cognitive deficits and *dramatic* life consequences in drug-addicted individuals may be responsible for the relatively slow acceptance of drug addiction as a *disease* among the health and public sectors. Accordingly, the conceptualization of addiction as a disease of the brain is a relatively recent one (Leshner 1997), developed in part from similar

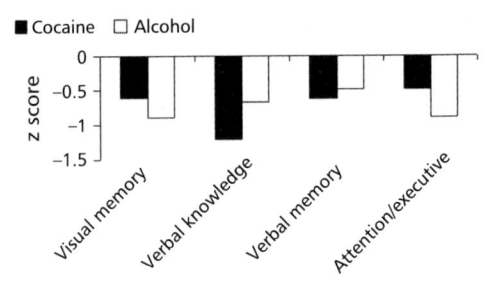

**Fig. 19.2** Deficits in neuropsychological functioning of cocaine (N=42) and alcohol (N=40) dependent subjects as compared to healthy controls (N=72; mean score of zero on each scale), Brookhaven National Laboratory 1988–1996. Reproduced with permission from Goldstein *et al.* (2004).

neuropsychological and imaging studies that have started to characterize the functional and neurochemical abnormalities in the brains of drug-addicted subjects (e.g., Volkow *et al.* 1990).

Increasingly, scientific efforts target bridging and converging neuropsychology and neuroimaging research to investigate how cognitive deficits map onto specific brain areas implicated in the core characteristics of addiction. As we dissect the syndrome into its cognitive-behavioral components, we focus on the brain areas and circuits subserving these behavioral manifestations. This iterative process of investigation from behavior to brain and back to behavior is essential as we take upon ourselves the task of associating (and of equal importance, dissociating) brain function to the development and chronicity of this disease. The overarching goal, of course, is to provide an integrative understanding of this disorder that may eventually aid in the advancement of effective treatment strategies (Rogers and Robbins 2001). Below, a brief (and by no means complete) synopsis of the theories of addiction and their integration into a unifying brain-behavior theoretical approach is provided.

## Theories of addiction: brain-behavioral models

There have been several theories of addiction, each highlighting different aspects of the disease, with more recent ones integrating the underlying brain function with behavioral and cognitive mechanisms. We will briefly review the theories with special emphasis on brain-behavior models.

A conceptual account of addiction is based on Solomon and Corbit's 'opponent-process' theory of motivation (Solomon and Corbit 1973) and has been further described by Robinson and Berridge (2003). This theory holds that each hedonic or pleasurable response is accompanied by an opposite (or opponent) unpleasant response. In addiction, the immediate 'rush' or 'high' is the pleasurable response, followed by the longer-lasting withdrawal acting as the 'opponent', unpleasant response (Koob and Bloom 1988; Koob *et al.* 1997). As addiction progresses, the aversive effects of the drug become pervasive and dominate its pleasurable effects. Indeed, it has been argued that the maintenance of drug abuse may be driven by negative reinforcement, namely, taking the drug to escape the negative effects of withdrawal. Critiques (Robinson and Berridge 2003) of the opponent-process theory of addiction argue that escaping withdrawal is not as potent a reinforcer as the positive reinforcing effects of drugs of abuse (Stewart and Wise 1992; Shalev et al. 2002) and that relapse occurs even after withdrawal is no longer experienced by drug-addicted individuals pointing to the limits of this theory in explaining the formidable power and chronicity addiction entails.

Robinson and Berridge (Robinson and Berridge 1993, 2001) proposed an alternative 'incentive-sensitization' theory. Drawing from brain research on drug addiction, this theory postulates that drugs of abuse cause neural changes in brain regions involved in incentive motivation and reward, such that the underlying brain circuit becomes hypersensitive to the reinforcing properties of the drugs of abuse and associated stimuli (drug cues). Hypersensitivity of this brain circuit increases the incentive salience or 'wanting' of the drug but does not affect or even dampens the euphoric effects or 'liking' of the drug, leading to compulsive drug-pursuing behavior and susceptibility for relapse. Indeed, basic research has shown that escalated cocaine use in rats produces drug reward threshold elevations as measured behaviorally (e.g., increased number of infusions and amount per infusion) (Ahmed and Koob 1998) or by intracranial self-stimulation (Ahmed et al. 2002).

These drug reward threshold elevations can be further explained by Koob and Le Moal's conception of drug addiction as a state of homeostatic dysregulation. They contend that organisms strive to maintain a state of equilibrium or homeostasis. Through the process of allostasis, the organism evaluates the demands placed upon it and must direct resources to retain homeostasis (e.g., changing neurotransmitter and receptor levels). Drugs of abuse can lead to such extreme levels of disequilibrium that allostasis cannot properly regulate balance, and hallmarks of addiction (e.g., loss of control, compulsive behavior, dysfunctional reward processing) ensue (Koob and Le Moal 2001).

Volkow and colleagues (Volkow and Fowler 2000; Volkow et al. 2003) suggested that four behavioral systems underlie addictive behavior: reward (saliency), motivation (drive), memory (conditioning), and control (disinhibition). According to this model, the initial experiences with the drug are exceedingly pleasurable such that overwhelming saliency (value) is attributed to the drug and a positive memory is formed. After repeated administrations, the drug itself or related cues trigger the recollection of this memory, thus leading to a decrease in the addicted individual's ability to control their drive to consume the substance excessively. These positive associations with the drug are easily formed (through habit learning and declarative memory processes) and retained for the addicted individual, while their extinction or replacement with the negative associations that ensue (e.g., withdrawal symptoms, adverse consequences) (Butter et al. 1963; Volkow and Fowler 2000) is more difficult.

Goldstein and Volkow (Goldstein and Volkow 2002) further elaborated on the interplay between these processes by proposing the I-RISA (Impaired Response Inhibition and Salience Attribution) model of drug addiction. This model, as depicted in Fig. 19.3, emphasizes the dysfunctional interplay between reward processing and control, such that

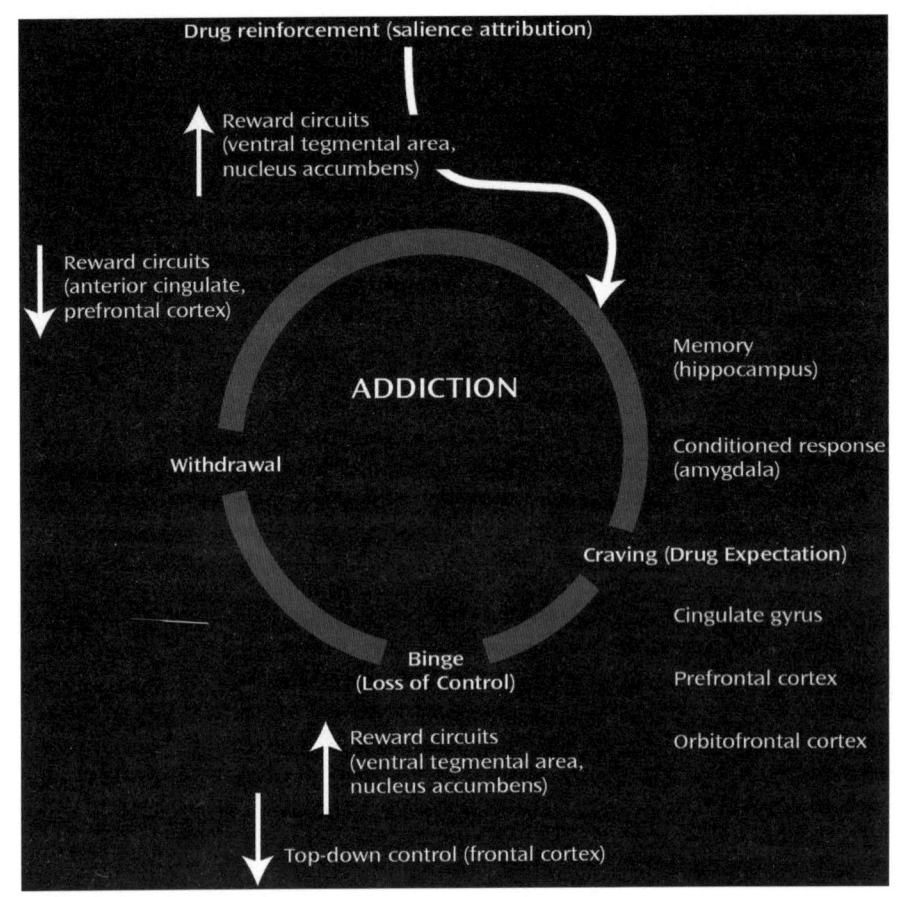

**Fig. 19.3** Integrative model of brain and behavior: the I-RISA (Impaired Response Inhibition and Salience Attribution) syndrome of drug addiction. Reproduced with permission from Goldstein and Volkow (2002).

an addicted individual attributes salience grossly biased toward the drug and at the expense of other potentially but no-longer rewarding stimuli, with a concomitant decrease in the ability to inhibit the maladaptive drug use (Goldstein and Volkow 2002). The interplay of these behavioral components, namely, response inhibition and salience attribution, were hypothesised to map onto frontocorticolimbic brain areas, advancing the idea that a systemic interaction between behavioral components can be mirrored by the interplay between the brain circuits subserving these behaviors in addiction. Specifically, the authors noted the circular nature of this interaction: the attribution of salience to a given stimulus (which is a function of the OFC) depends on the relative value of a reinforcer compared to simultaneously available reinforcers, which requires knowledge of the strength of the stimulus as a reinforcer (Schultz *et al.* 2000), a function of the hippocampus and amygdala. Consumption of the drug in turn further activates cortical circuits (the OFC and anterior cingulate gyrus, ACC) in proportion to the dopamine (DA) stimulation by favoring the target response and decreasing non-target-related background activity (Kiyatkin and Rebec 1996). The activation of these interacting circuits was suggested to be indispensable for maintaining the chronically relapsing nature of drug addiction (Goldstein and Volkow 2002).

## The OFC in the core clinical characteristics of drug addiction: intoxication, craving, bingeing, and withdrawal

Neuroanatomically, addiction takes root in a complex web of overlapping circuits comprised of multiple brain regions extending from the frontal cortices (OFC, dorsolateral PFC, ACC) to subcortical (nucleus accumbens, ventral tegmental area, striatum) and limbic (amygdala, hippocampus) structures. In particular, an OFC-striatum-globus pallidus-thalamus-OFC loop is linked with other circuits involved in sensation (Rolls and Baylis 1994) (sensory cortices), mediating reinforcement and motivation (London *et al.* 2000), reward (through the subcortical structures), memory (limbic), and inhibitory control (PFC) (Volkow *et al.* 2003). Accordingly, the OFC has been described by Elliott *et al.* (2000) as a 'convergence zone' (pp.308) integrating the sensory experience of a stimulus with mnemonic or non-mnemonic emotional responses, and generating an estimated 'value' of a potential reward, thus driving behavior.

In laboratory animals, damage to the OFC impairs reversal of stimulus-reinforcement associations, and leads to perseverations and resistance to extinction of reward-associated behaviors (for review, see Volkow (Volkow and Fowler 2000) and Chapters 5 and 16 by Rolls and by Robbins *et al.* in this volume). Studies of humans who sustained damage to ventromedial regions of the OFC show similar impairments, persistently displaying disadvant-ageous decision-making (for review, see Bechara (Bechara *et al.* 2000) and Chapters 13 and 18 by Naqvui *et al.* and by Zald in this volume). This is reminiscent of anecdotal clinical observations of addicts who frequently report the inability to stop self-administration of the no-longer pleasurable substance. Therefore, several investigators had initiated functional as well as neuroanatomical analyses that documented changes in the activity as well as the tissue composition of frontal regions in drug addiction. Specifically, decreased brain gluc-ose metabolism (marker of tissue function) activity at baseline in the OFC, ACC, PFC and were documented in abstinent cocaine abusers (Volkow *et al.* 1992, 1993), alcoholics (Volkow *et al.* 1992, 1993, 1997; Adams *et al.* 1993; Dao-Castellana *et al.* 1998) and methamphetamine abusers (Volkow *et al.* 2001; Kim *et al.* 2005). In addition, reductions in gray matter in frontal cortical regions (ACC, medial and lateral OFC), limbic and paralimbic structures (including anteroventral insula and superior temporal cortex) were documented in abstinent cocaine addicts (O'Neill *et al.* 2001; Fein *et al.* 2002; Franklin *et al.* 2002; Matochik *et al.* 2003), methamphetamine users (Thompson *et al.* 2004), alcoholics (Jernigan *et al.* 1991; Pfefferbaum *et al.* 1997; O'Neill *et al.* 2001; Fein *et al.* 2002), and polysubstance users (Liu *et al.* 1998) as compared to non-drug-using control subjects.

Similarly, evidence has been accumulating to demonstrate white matter abnormalities in drug-addicted individuals, encompassing white matter hyperintensities (Volkow *et al.* 1988), reduced white matter integrity (O'Neill et al. 2001; Lim et al. 2002), or overall white matter increased volumes (Thompson et al. 2004). In general, it appears that the frequency of white matter changes increases with age (Bartzokis et al. 1999) and longer duration of drug use (O'Neill et al. 2001). Recent studies have revealed changes specifically in *frontal* white matter integrity among stimulant-dependent subjects. For example, Lyoo and colleagues (Lyoo et al. 2004) have differentiated an OFC striatal tract by using a two-point tractography and found compromised OFC white matter tract integrity in 31 chronic methamphetamine users as compared to age-matched control subjects. This finding was correlated with the reported amount of methamphetamine abuse (Lyoo et al. 2004). Very recently, reduced anterior corpus callosum white matter integrity (poor crossing between axonal tracts in the PFC) was related to impulsivity (impaired inhibitory control) in cocaine-dependent subjects (Moeller et al. 2005).

The functioning of the PFC, including the OFC and the ACC, has emerged as a primary target in characterizing the *core* clinical characteristics of drug addiction as reviewed below.

## Intoxication

Intoxication occurs when a person consumes enough of a substance to produce behavioral, physical or cognitive impairments. Here we review the results from neuroimaging studies using Positron Emission Tomography (PET) that have assessed the effects of drug administration on functional measures, such as metabolism with 2-deoxy-2[$^{18}$F]fluoro-D-glucose (FDG) and cerebral blood flow (CBF). Few studies have measured regional brain activity during drug intoxication, and most of these studies have employed a single drug exposure (for exceptions, see recent animal studies (Febo et al. 2005)). Such studies have shown reduced glucose metabolism throughout the brain, particularly in the frontal cortex, during cocaine, morphine, or alcohol intoxication (de Wit et al. 1990; London et al. 1990; London et al. 1990; Volkow et al. 1990). In contrast, marijuana intoxication is associated with higher levels of glucose metabolism in the PFC, OFC, and striatum in marijuana-addicted individuals, but not in non-addicted individuals (Volkow et al. 1996). Similarly, increased metabolism in the PFC, ACC, OFC, and striatum has been reported in cocaine addicts after sequential administration of intravenous methylphenidate, a psychostimulant drug with DA agonist properties like cocaine, that cocaine addicts report to be similar to intravenous cocaine (Volkow et al. 1999).

Although individual differences exist (e.g., Nakamura et al. (2000)) and studies vary in measuring normalized vs. absolute CBF changes, evidence suggests that CBF in the PFC increases during intoxication with nicotine (Nakamura et al. 2000; Rose et al. 2003), marijuana (Mathew et al. 1992), and alcohol (Volkow et al. 1988; Tiihonen et al. 1994; Ingvar et al. 1998) as compared to a non-drug baseline. Moreover, the activation of the right PFC during alcohol intoxication has been shown to be associated with euphoria (Tiihonen et al. 1994), and during marijuana intoxication with the subjective sense of intoxication (Mathew et al. 1992). In contrast, cocaine lowered CBF throughout the brain, including the frontal cortex, an effect that could be attributed to cocaine's vasoconstricting actions (Wallace et al. 1996), but which could also be secondary to the decrease in glucose metabolism shown by London and colleagues (London et al. 1990a).

Mapping studies during drug intoxication with functional magnetic resonance imaging (fMRI) to measure the blood-oxygenation-level-dependent (BOLD) response have reported activation of the anterior PFC and ACC during cocaine intoxication, an effect that has been correlated with drug reinforcement properties (Breiter et al. 1997; Kufahl et al. 2005; Risinger et al. 2005). Studies of nicotine administration have also shown activation in the frontal cortex and ACC coinciding in time with the subjective experiences of 'rush' and 'high' (Stein et al. 1998). Here it is relevant to note that the suspicious absence of OFC involvement in intoxication especially as associated with subjective reports of 'high' may be due to methodological differences between PET and fMRI. Potential sources of these discrepancies in activation patterns could reflect: 1) that the OFC is more difficult to map in fMRI due to artifacts observed particularly in this area (just above the eye orbits; see Stenger (Chapter 17),. this volume); and 2) the difference in the temporal course of the processes measured (metabolism studies, 30 minutes; CBF-water studies, 60 seconds; BOLD studies, 3–5 seconds). Thus, all imaging methods present different methodological challenges. Since the metabolism and CBF-water studies are limited by their poor time resolution, the BOLD method may be better suited for assessing the relationship between regional functional changes and drug-induced fast behavioral effects, such as 'rush' and 'high'. On the other hand, the BOLD method is limited by its sensitivity to vasoactive changes and motion that may occur during drug administration and intoxication.

Nevertheless, most of the studies show activation of the PFC, including the OFC and ACC, during drug intoxication. Also, PFC activation appears to be associated with the subjective perception of intoxication, the reinforcing effects of the drug, or enhanced mood. It is also intriguing that in the case of marijuana or methylphenidate, the activation of the frontal regions, particularly the OFC, was predominantly observed in the users but not in the non-users (Volkow et al. 2005). This suggests that PFC regions and the ACC are involved in the intoxication process and that their response to drugs is in part related to previous drug experiences.

## Craving

Craving, a learned response which links the drug and its environment to a pleasurable or an intensely overpowering experience, refers to the urgent desire to consume the drug of abuse. Thus, acute drug administration is not necessary for the activation of the frontal cortex in individuals previously exposed to the drug of choice, in whom, because of prior exposure and consequent reinforcement learning, craving alone is possibly sufficient to activate frontolimbic circuits. Indeed, higher levels of brain activation (measured with CBF, FDG, or BOLD) in frontolimbic areas, primarily in the OFC and ACC, have been demonstrated in cocaine users exposed to videotapes depicting drug-related stimuli (Grant et al. 1996; Maas et al. 1998; Childress et al. 1999; Garavan et al. 2000; Wexler et al. 2001). Self-reports of craving significantly correlated with glucose metabolism changes in the dorsolateral PFC in one study (Grant et al. 1996) and with the spatial extent of activation in the dorsolateral PFC and ACC in another (Maas *et al.* 1998). In all five studies, the drug-related stimuli elicited craving only in the cocaine users and not in the comparison subjects, pointing again to the importance of prior drug experience.

The mechanism that underlies craving may entail *recall* of *emotionally*-laden previous experiences underscoring the critical role of the pathways between the OFC and mesolimbic subcortical regions. Indeed, craving correlated with activation of the amygdala in one study (Grant *et al.* 1996) and with the temporal insula in another (Wang *et al.* 1999); in the latter study, we asked cocaine users to recall and describe their own method of preparing cocaine; OFC activation was observed in this condition, but not when they described their own family genealogical tree (e.g., an active control condition) (Fig. 19.4) (Wang *et al.* 1999). A more recent study using fMRI BOLD showed correlations between craving and both amygdala and insula, but also the OFC and dorsolateral PFC, in 11 cocaine abusers (Bonson *et al.* 2002). An issue that remains to be resolved is the relative contribution of the lateral vs. medial OFC to cue-induced craving; for example, the medial OFC has been implicated in some studies (Grant *et al.* 1996, Garavan *et al.* 2000; Volkow *et al.* 1991), while the lateral OFC has been implicated in others (Bonson *et al.* 2002).

There is evidence pointing to a central role of the OFC in other aspects of craving. For example, craving may involve *anticipation* of a future drug reward. The dissociation of craving from anticipation and of both from the acute drug experience has not been thoroughly examined. Nevertheless, the actual drug experience may be related to more circumscribed activations than the anticipation phase (Grant *et al.* 1996), in line with evidence for a similar dissociation of anticipation from an actual sensory (tactile) experience (Ploghaus *et al.* 1999). Indeed, Volkow *et al.* (2003) examined the effect of expectation on regional brain metabolism in cocaine addicts and documented a greater (50% larger) cerebellar and thalamic and smaller lateral OFC metabolic response to methylphenidate when it was expected versus when it was not expected.

**Fig. 19.4** (Also see Color Plate 28.) OFC activation in active cocaine abusers during a cocaine theme interview and a neutral theme interview, as measured by FDG PET. Reproduced with permission from Goldstein and Volkow (2002).

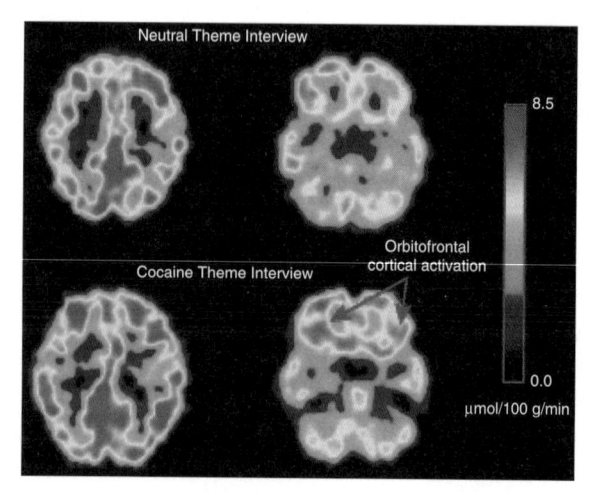

Another line of evidence supporting the role of the frontal cortex in craving derives from studies conducted shortly after the subjects' last drug use. For example, we have demonstrated higher regional brain glucose metabolism, including in the medial OFC and striatum, in cocaine users tested during early withdrawal (<1 week since last cocaine use) as compared to healthy subjects (Volkow et al. 1991). These higher levels were proportional to the reported intensity of craving, such that the higher the metabolism, the greater was the drug craving. Similarly, as we have mentioned above, enhanced metabolism in frontal and subcortical brain areas has been reported in cocaine users after sequential administration of intravenous methylphenidate. Interestingly, methylphenidate increased OFC metabolism only in subjects in whom it induced intense craving and in the PFC in the subjects in whom it enhanced mood (Volkow et al. 1999).

Of note, the observed link between craving and the OFC may not be specific to stimulant users. In a recent CBF study, 12 abstinent opiate-dependent subjects were exposed to recorded autobiographical scripts to induce opiate craving. In response to their opiate-craving memories, connectivity maps were identified involving the left OFC region, which correlated with activity in the right OFC and left parietal and posterior insular regions. There was also a positive association with the hippocampus and brainstem. The authors concluded that these OFC activations reflected the ability of drug-related stimuli to activate attentional and memory circuits (Daglish et al. 2003). Similarly, compared to 20 non-smoking controls, 20 heavy cigarette smokers had greater increases to a cigarette cue as compared to a neutral cue scan in relative glucose metabolism in the perigenual ACC. Significant positive correlations were found between intensity of craving and metabolism in the OFC, dorsolateral PFC, and anterior insula bilaterally (Brody et al. 2002).

## Bingeing

Drug or alcohol craving is not directly predictive of compulsive drug intake, and, in fact, the association of craving with drug use and relapse continues to be challenged (Tiffany 1990). Nevertheless, for certain drugs intense cravings may be followed by bingeing, where for a discrete period of time there is an extreme loss of control during which the individual engages in repeated consumption of the substance. Bingeing is associated with DA, serotonergic, and glutamatergic circuits (Loh and Roberts 1990; Cornish et al. 1999) and probably involves the activation of the thalamo-OFC circuit and the ACC.

Experimentally demonstrating the involvement of the OFC in human compulsive drug self-administration would require investigation of the drug-addicted individual during actual drug use in which drug supply is unrestricted or is at its peak (note that the OFC has been demonstrated to be critical in other disorders involving compulsive behaviors such as obsessive-compulsive disorder; see Chapter 20 by Milad and Rauch, this volume, and Zald and Kim (1996)). For understandable reasons, this is not immediately feasible in human subjects although elegant laboratory designs can circumvent these difficulties. For example, a recent fMRI study of non-treatment-seeking cocaine-dependent subjects, who chose both when and how often intravenous cocaine administration occurred within a medically supervised one hour self-administration procedure, showed that repeated drug-self-administration-induced-high correlated negatively with activity in limbic, paralimbic, and mesocortical regions including the OFC and ACC, while craving correlated positively with activity in these regions (Foltin et al. 2003; Risinger et al. 2005). To circumvent technical difficulties in such drug self-administration fMRI paradigms (e.g., effect of motion), paradigms that simulate compulsive behavior (such as gambling when it is clearly no longer beneficial) might offer invaluable insight into the neural circuits underlying loss of control in addiction. Such gambling paradigms and studies on inhibition are discussed in more detail in the sections on *decision-making* and *inhibition*.

Animal studies provide important clues into the processes involved in bingeing behaviors. For example, rat models of the clinical eating disorder bulimia nervosa have shown that binge-eating is associated with increased DA release and metabolism in the PFC and ventrolateral striatum (Inoue et al. 1998), implicating these regions specifically in compulsion. Moreover, two recent studies have documented addiction-like or compulsive behaviors in rats (Deroche-Gamonet et al. 2004; Vanderschuren and Everitt 2004) that closely mimic the clinical symptoms of addiction in humans (APA 1994), including the persistent seeking of cocaine despite its unavailability, intense motivation toward procuring the drug, and continued use of the substance despite adverse consequences. After three months of cocaine self-administration and 5 days of withdrawal, Deroche-Gamonet et al. were able to separate rats into two groups based on their response levels on a device previously associated with drug delivery: rats with high response levels (referred to as 'addicted' rats), and rats with low response levels ('non-addicted' rats). The authors showed that even when cocaine delivery was associated with punishment (electric shock), the 'addicted' rats continued to self-administer cocaine, while the 'non-addicted' rats did not. Moreover, the 'addicted' rats had a higher 'breaking point' (the maximum amount of effort the animal put toward procuring the reward before ceasing) than the 'non-addicted' rats and continued to seek the drug even when a cue indicated that the drug was not available (Deroche-Gamonet et al. 2004). As Robinson (2004) pointed out, 'this is reminiscent of the cocaine addict, who has run out of drug, compulsively searching the carpet for a few white crystals ("chasing ghosts") that they know will most likely be sugar' (p.951).

Similarly, Vanderschuren and Everitt (2004) showed that presentation of an aversive conditioned stimulus (a tone previously paired with a foot-shock) suppressed drug-seeking behavior in rats only after a short history of cocaine self-administration, but not after a more extended history, suggesting that addiction developed only after numerous repeated exposures. Further, the drug of abuse was more resistant to conditioned suppression (behavior suppressed by aversive stimuli) than natural reinforcers such as food. Thus, despite aversive circumstances, the *compulsion* to consume the drug of abuse persisted, while this was not true for other valued items (this may vary with the type addiction, as the compulsion to eat food might persist in the obese individual despite side effects that severely compromise health; see Wang et al. (2004)). The loss of control to the extent that vital survival actions (e.g., feeding) are overridden by drug bingeing is analogous to anecdotal accounts of heroin or cocaine addicts who binge for days until they either run out of the drug or collapse in exhaustion. It remains to be determined whether such compulsion involves the PFC (see recent studies in animals, e.g., (Bailey et al. 2005) suggesting that it may).

## Withdrawal

Withdrawal occurs upon terminating drug use, and is accompanied by a variety of symptoms including excessive fatigue, irritability, anxiety and anhedonia (Gawin and Kleber 1986). Withdrawal symptoms vary in intensity and quality depending on the type of drug and the amount of time that has passed since last drug intoxication and they are typically characterized as 'early' versus 'protracted'.

Abnormalities in the human cortex associated with withdrawal from cocaine in regular cocaine users were documented as early as 1988 in our laboratory (Volkow et al. 1988). We demonstrated that the relative CBF values for the PFC and the left lateral frontal cortex were significantly lower in the cocaine users than in healthy comparison subjects. A follow-up study in active cocaine users demonstrated differences in regional brain glucose metabolism between cocaine users tested within 1 week of last cocaine use and cocaine users tested 2–4 weeks after last cocaine use (Volkow et al. 1991). Of interest, glucose metabolism was higher in the OFC and in the striatum in the former group than in the healthy comparison subjects. During more protracted withdrawal (1–6 weeks since last use), brain metabolism was lower in cocaine users than in the comparison subjects, an effect that was most accentuated in the frontal cortex (Volkow et al. 1992).

Studies of alcohol users have provided similar evidence. For example, glucose metabolism abnormalities (including in the frontal cortex) were documented in otherwise healthy alcoholic subjects with mean duration of alcohol withdrawal of 11 days (Volkow et al. 1992). Persistent lower striatal and OFC metabolism has been shown in regional metabolism studies after more protracted alcohol withdrawal (Volkow et al. 1993; Volkow et al. 1994). In addition, alcoholic subjects have shown less sensitivity to the lower metabolism induced by lorazepam, a benzodiazepine that facilitates gamma-aminobutyric acid neurotransmission in the striatal-thalamo-OFC circuit during early (1–4 weeks) detoxification (Volkow et al. 1993) and in the OFC during protracted (8–11 weeks) detoxification (Volkow et al. 1997), suggesting long-lasting drug-related adjustments in these brain regions. Persistent abnormalities after alcohol detoxification were also documented for the ACC (Volkow et al. 1997). Lower activity in the PFC in alcoholic subjects during detoxification was also documented in other laboratories using slightly different study groups (Cloninger-type 2 alcoholics) and techniques (single photon emission computed tomography) (Catafau et al. 1999). Alcoholic subjects show less sensitivity in the striatal-thalamo-OFC circuit to the serotonin agonist $m$-chlorophenylpiperazine, which provides evidence for the relevance of serotonin in these abnormalities (Hommer et al. 1997). Studies from our laboratory also point to the relevance of DA in withdrawal as described in the next section.

## The underlying mechanism: role of dopamine

As demarcated in the I-RISA model, we have described how the functioning of the OFC and associated structures of the mesocorticolimbic system map onto the behavioral manifestations of addiction (intoxication, craving, bingeing, withdrawal). What could be a potential neurochemical mechanism of such brain-behavior associations? All addictive substances increase DA in mesolimbic brain regions such as the nucleus accumbens and the striatum (Volkow *et al.* 1997). Stimulant drugs are the most potent drugs of abuse in increasing DA in the nucleus accumbens by blocking DA transporters and in the case of amphetamine and methamphetamine by also releasing DA. Imaging studies have documented that these DA-enhancing effects are directly associated with the reinforcing effects of cocaine, methylphenidate and amphetamine (as evidence by self-reports of 'high', 'rush', and 'euphoria') in the human brain (Laruelle *et al.* 1995; Volkow *et al.* 2002). This is different from natural reinforcers for which increases in DA appear to occur mostly for reward prediction rather than to the reward itself (Koob and Bloom 1988; Pontieri *et al.* 1996, 1998; Schultz *et al.* 2000). Furthermore, these changes in DA by drugs of abuse also encode for motivation to procure the drug irrespective of whether the drug is pleasurable or not (McClure *et al.* 2003). In primates, this DA-motivational coding occurred at both the initial stage of learning as well as throughout the learning process (Satoh *et al.* 2003), suggesting that DA changes may be intimately involved in the development of addiction at multiple stages.

Indeed, we have reliably documented lower baseline striatal DA D2 receptor availability in cocaine, methamphetamine and heroin abusers and in alcoholic subjects (Volkow *et al.* 2002) (Fig. 19.5). Note that DA recovery with abstinence does not appear to be complete; we documented that in cocaine users during early (up to 1 month after last cocaine use) and more protracted (up to 4 months after last cocaine use) withdrawal periods, striatal DA receptor availability was still significantly lower than in healthy comparison subjects (Volkow and Fowler 2000). However, longer-term studies await completion before a conclusion can be reached about rate and amount of recovery of DA function in addicted individuals.

In general, it has been hypothesized that the low D2 receptor availability may underlie the reduced sensitivity in drug users to natural rewards and the perpetual need for the drug of abuse as the predominant reinforcer (Volkow *et al.* 2002, 2004). Not surprisingly, the OFC is directly and indirectly connected with brain regions dense with DA $D_2$ receptors such as the ventral tegmental area (Oades and Halliday 1987), nucleus accumbens (Koob and Bloom 1988), and striatum (for review, see Volkow *et al.* (2002)); it is therefore modulated by DA changes at multiple points within the corticomesolimbic pathway. In the human subject, Volkow *et al.* (1993) showed that regional glucose metabolism in the OFC (and ACC) is positively associated with DA D2 receptor availability in the striatum of detoxified cocaine addicts (Fig. 19.6). Similar findings (lower levels of striatal D2 receptors associated with lower metabolism in the OFC) were reported in methamphetamine addicts (Volkow *et al.* 2001). More recently, higher metabolism in the ventral ACC and in the medial OFC has been shown in response to methylphenidate (Volkow

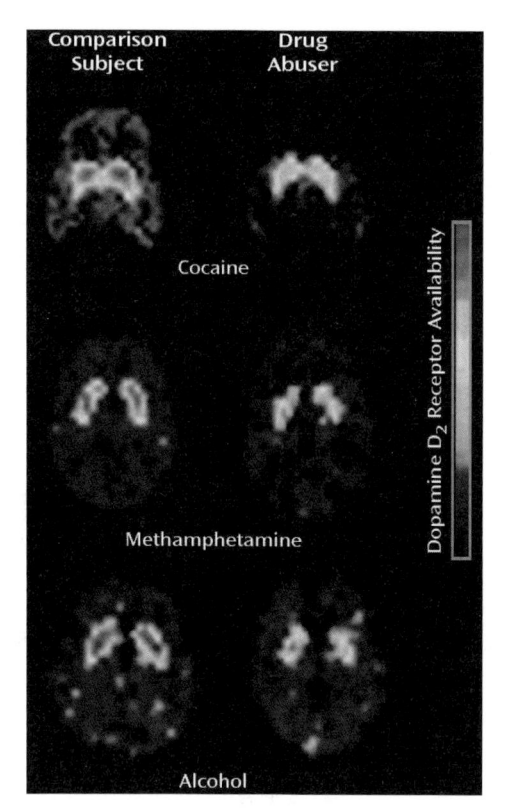

**Fig. 19.5** (Also see Color Plate 29.) Lower striatal dopamine D₂ receptor availability in drug users during withdrawal from cocaine, methamphetamine, and alcohol than in normal comparison subjects. Reproduced with permission from Goldstein and Volkow (2002).

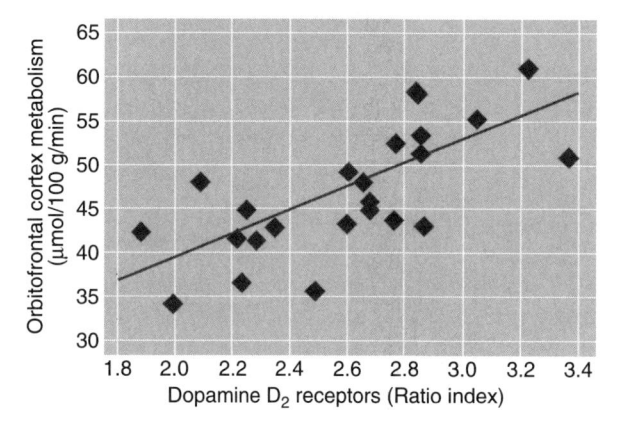

**Fig. 19.6** Relation of striatal dopamine D₂ receptor availability and OFC metabolism in cocaine users ($r = 0.70$, $p < 0.0001$). Reproduced with permission from Goldstein and Volkow (2002).

*et al.* 2005), which increases DA by blocking the DA transporter (Volkow *et al.* 1999), but only in cocaine addicted subjects and not controls providing further support for the role of DA in PFC function specifically associated with drug use experience. Moreover, in 11 detoxified male alcoholics, less availability of $D_2$-like receptors in the ventral striatum (as assessed with PET (18F)desmethoxyfallypride) was associated with alcohol craving severity and with greater cue-induced activation of the medial PFC and ACC (as assessed with fMRI) (Heinz *et al.* 2004). It is unclear, however, whether the primary dysfunction stems from striatal DA mechanisms or the OFC and ACC. Furthermore, as Volkow and Fowler (2000) note, while DA clearly affects OFC function, it is likely that other neurotransmitters also modulate OFC function (e.g., glutamate, serotonin, GABA).

## 19.3. The OFC in the neuropsychological substrates associated with drug addiction: the I-RISA syndrome

We reviewed above the extensive body of studies providing strong evidence that the dopaminergic striatal-thalamo-OFC circuit is dysfunctional specifically as relates to the core clinical symptoms of drug addiction. We now proceed to describe the role of this circuit in the neuropsychological (cognitive, behavioral, emotional, and personality) aspects that are thought to be integrally related to the development and maintenance of the drug addiction syndrome, as described in the I-RISA model of addiction (Goldstein and Volkow 2002).

### Reward, saliency, attribution, and motivation

A central role for the OFC has been implicated in the motivation to gain a reward. Monkeys with stimulating electrodes implanted in the OFC will perform for self-stimulation of this region as well as other regions that are neuroanatomically linked with it, suggesting that the mere stimulation of the OFC is reinforcing (Mora *et al.* 1979, 1980). Monkeys with OFC damage perform poorly on object-reversal reward tasks; the monkeys keep responding to objects that are no longer associated with reward (Butter 1969; Jones and Mishkin 1972). Moreover, neurons in the OFC have been shown to distinguish between reward and penalty (Thorpe *et al.* 1983), reflecting the size of an upcoming reward (Wallis and Miller 2003), the monkeys' reward preferences (Tremblay and Schultz 1999; Hikosaka and Watanabe 2004), and the degree of the animal's preference for a certain reward, that is, the animal's motivational state (Hikosaka and Watanabe 2000). In analogous human functional neuroimaging studies in healthy subjects, the OFC has been identified to respond maximally to extremes of a reward range (including best vs. worse outcome) (Elliott *et al.* 2003; O'Doherty *et al.* 2001), demonstrating an association with the relative magnitude of reinforcement (monetary, see (Breiter *et al.* 2001)) or the subjective pleasantness (i.e., value) of the rewarding stimulus (liquid food, (Kringelbach *et al.* 2003); see also association of the OFC with valence estimations of odorants (Anderson *et al.* 2003)).

In particular, Elliott and colleagues emphasized a functional distinction between the medial and lateral OFC (Elliott *et al.* 2000), whereby the former correlates with numerous measures of pleasantness such as monetary reinforcers (O'Doherty *et al.* 2001), pleasant odors (Gottfried *et al.* 2002; Anderson *et al.* 2003), and facial attractiveness (O'Doherty *et al.* 2001); while the latter is more aligned with the negative or punishing aspects of reinforcers (Gottfried *et al.* 2002) (e.g. unpleasant odors (Anderson *et al.* 2003)).

Thus, the authors suggested that the medial OFC has a role in making associ-ations between stimuli and correct or rewarded responses, while the lateral OFC has a role in the suppression of previously rewarded responses (see (Elliott *et al.* 2000) and (Kringelbach and Rolls 2004)). Indeed, the medial and lateral OFC subdivisions each have distinct inter-connections with other cortical regions (for review, see Cavada *et al.* (2000) and Price, this volume). Note that the role of these functionally distinct OFC sub-regions needs to be demarcated in drug addiction. One could postulate that the medial PFC/OFC will be associated with drug craving and cue-induced behavioral and brain activations (for example see (Heinz *et al.* 2004)) while a dysfunctional lateral OFC will be associated with impairments in the control of behavior as reviewed further below.

## Conditioning and extinction

Chronic drug administration in addiction was suggested to disrupt normal learning and memory systems, which in turn may underlie maladaptive (i.e., compulsive) drug-seeking habits (Everitt *et al.* 2001); indeed, individuals with cocaine use disorders are generally believed to be particularly prone to relapse (Washton and Stone-Washton 1990; Miller and Gold 1994) even after prolonged abstinence (O'Brien *et al.* 1998). We suggest that these powerful conditioned responses to cocaine-related stimuli (see O'Brien *et al.* (1998) for review) may also underlie the above discussed compromised reward processing, salience attribution, and motivation in drug-addicted individuals.

Multiple learning and memory systems have been proposed to be involved in the process of drug addiction (reviewed in White (1996)). For example, through conditioned incentive learning, neutral stimuli coupled with the drug of abuse acquire reinforcing properties and motivational salience even in the absence of the drug. Extinction learning and retention are other processes designating the ability of a conditioned stimulus to reinitiate drug-seeking behavior after a non-reinforced delay; these processes can be tested with extinction paradigms (identical to training sessions for drug self-administration, except that no drug is delivered after completion of the response requirement) that have predictive validity as a measure of drug relapse (Koob 2000).

These multiple learning and memory systems map uniquely with the functioning of the PFC. For example, employing an appetitive associative learning task, Chudasama and Robbins (Chudasama and Robbins 2003) have recently reported that damage to the rat infralimbic cortex (subgenual ACC) resulted in impaired new stimulus-reward learning (after reversal of stimulus-reward contingencies) but not in initial acquisition learning or in inhibitory control of previously reinforced responses (see also (Jones and Mishkin 1972)). In contrast, damage to the rat OFC was characterized by an inability to inhibit previously reinforced responses, causing excessive perseverative responding during the early stages of reversal learning. Similar results have been reported in human fMRI studies of aversive conditioning and extinction (Gottfried and Dolan 2004; Phelps *et al.* 2004): although active throughout all stages of learning, the ventromedial PFC (subgenual ACC) response was primarily linked to the expression of the conditioned response during the retention of extinction learning.

A question that remains to be explored is how these conditioned responses and their underlying neurobiological mechanisms contribute to the enhanced salience attributed to the drug or drug-related targets in drug addiction.

## Higher order executive function: Inhibitory control and decision-making

### Inhibitory control

Inhibitory control is a neuropsychological construct referring to the capacity to control the inhibition of harmful and/or inappropriate emotion, cognition or behavior. The use of psychoactive substances impairs mechanisms of inhibitory control, thus contributing to the initiation or escalation of undesirable behaviors. For example, the inability to suppress pre-potent drug-seeking behavior may be particularly significant during periods of relapse and drug bingeing. Other examples include the uninhibited over-expression of private emotions, thoughts, and behaviors during intoxication (regretful the next day); anger and aggressive behavior are examples that will be reviewed further below.

Impaired response inhibition is common to other psychopathological disorders of impulsivity and frontal lobe dysfunction; therefore it has been relatively well-studied in neuroimaging paradigms. Before we review the literature, it is important to note that the putative inhibitory control mechanisms are not unitary: functional neuroimaging and lesion studies suggest several substrates of inhibitory control (e.g., cognitive vs. emotional) that can be measured by different batteries of tests and linked to multiple regions of the PFC, including Brodmann area 9 and the OFC and ACC (see Bechara (2003) and Hooker and Knight, Chapter 12, this volume). Indeed, the nomenclature of brain regions found to be associated with inhibitory control is not consistent, indicative of the heterogeneity of the underlying neuronal circuits. Ongur and Price (2000) provided anatomical description of the ACC and OFC as situated in the orbital-medial-PFC, covering Brodmann areas 24, 25, 32, and 10 for the medial and 47 and 11 for the orbital areas. A recent review of cognitive control has implicated partially overlapping regions, naming them instead, 'poster-ior medial frontal cortex' (Ridderinkhof *et al.* 2004). It is important to consider this richness and heterogeneity in both the behavioral construct and the underlying brain function when studying the role of the PFC in inhibitory control.

The most commonly used inhibitory control tasks in neuroimaging studies, go/no-go paradigms, establish a prepotent response tendency by requiring button presses to rapidly presented stimuli immediately upon target detection. These 'go' stimuli are mixed with rare events in which inhibition is unexpectedly imposed by requiring the withholding of the prepotent response (i.e., 'no-go'). In healthy subjects, the OFC, ACC, and striatum were activated in go/no-go tasks in several fMRI studies (e.g., Casey *et al.* (1997); Vaidya *et al.* (1998)). Better response inhibition was associated with greater volume of activation in the OFC and a smaller magnitude of activation in the ACC, possibly implicating the OFC in the effort exerted when inhibiting a response and the ACC in error detection (Casey *et al.* 1997).

Garavan and colleagues tested cocaine users in go/no-go fMRI tasks and found the ACC and PFC to be less responsive in these subjects as compared to controls (Kaufman et al. 2003; Hester and Garavan 2004). Furthermore, unlike drug-naïve controls, and opposite to the ACC pattern, cocaine users showed an over-reliance on the left cerebellum, a compensatory pattern previously seen in alcohol addiction (Hester and Garavan 2004). Similar findings of diminished ACC activity were found in a group of opiate addicted subjects (Forman et al. 2004) and authors in all studies concluded that the hypoactivation in the ACC was primarily indicative of compromised detection and regulation of error responses, which may be particularly prominent when working-memory demands are increased (e.g., during cue-induced craving for the drug) (Hester and Garavan 2004).

There have been important critiques of using the go/no-go paradigm to test inhibitory control. These critiques highlight a potential confound—the 'odd-ball' effect—where attentional strategies that do not load exclusively on inhibitory control can develop. Indeed go/no-go paradigms may engage bottom-up striato-frontal pathways and may become automatic with practice (e.g., stopping to a red light). A task similar to go/no-go is the stop signal paradigm; it has been proposed to tap a higher load on inhibitory control than go/no-go tasks since the stop signal appears after a motor ('go') response has been triggered (e.g., stopping to an unexpected obstacle). Indeed results from a study using this task showed involvement of the inferior PFC when errors (unsuccessful inhibitions) were subtracted from correct inhibitory performance in healthy subjects (Rubia et al. 2003), adding support to the notion of dissociable inhibitory control roles within the PFC.

There are two important missing ingredients in go/no-go tasks that render their use limited in the assessment of inhibitory control in drug addiction. First, the 'go' and 'no-go' contingencies are not generally linked to emotional or conditioned stimuli. Second, task requirements occur in a setup where the rules are clearly understood. This is not what occurs in the case of control over emotionally salient and strongly conditioned 'no-go' situations such as drug intoxication and withdrawal. We suggest that more ecologically valid go/no-go paradigms will better characterize the inhibitory control impairments in drug addiction possibly by more specifically engaging the OFC.

Another classic paradigm widely used in neuroimaging studies of the attentional networks involved in the modulation of cognitive control is the Stroop color-word task. Here, latency (reaction time) to name the color of a word that is incongruent with the word's meaning is measured; it takes longer to name blue fonts if they spell the word 'red' as compared to the same blue color fonts that now spell the word 'blue' (Stroop 1935). The first incongruent condition (word 'red' in blue) conflicts and therefore interferes with the task's demand to name the color of the word (blue) since reading (red) is an automatic and faster response. This conflict engages inhibitory control processes thus producing slower reaction time and activations in areas subserving inhibitory control. Indeed, similarly to the go/no-go task, the classic Stroop task activates primarily the ACC and middle frontal gyrus along with the OFC and insula. In particular, results point to the importance of the ACC in guiding the execution of a correct response by error detection and performance monitoring, thus resolving the conflict inherent in this task

(Carter *et al.* 2000; MacDonald *et al.* 2000). Although not frequently discussed, studies also show the OFC to be among the most strongly activated regions during the Stroop effect (Leung *et al.* 2000; Milham *et al.* 2003).

While the neurobiology of the Stroop effect has been extensively studied in healthy groups, it has been less frequently explored in patient studies (see, for recent exceptions in subjects with bipolar disorder (Blumberg *et al.* 2003; Gruber *et al.* 2004); with post-traumatic stress disorder (Bremner *et al.* 2004) and with obsessive-compulsive disorder (Nakao *et al.* 2005)). Nevertheless, recent neuroimaging studies that examined the Stroop effect in drug-addicted subjects, provide preliminary data that suggest a change in the PFC functioning when solving the conflict inherent in this cognitive task. Thus, one fMRI study suggested that despite similar task performance compared to healthy controls, marijuana smokers exhibited different patterns of neural responsiveness (lower activity in focal areas of the ACC and a more diffuse pattern of dorsolateral PFC activation) during the Stroop interference (incongruent) condition (Gruber and Yurgelun-Todd 2005) (see also (Potenza *et al.* 2003) for decreased activity in the left ventromedial PFC in response to infrequent incongruent stimuli in pathological gamblers compared to controls).

A differential brain responsiveness to the Stroop task (including hypoactivity in the left ACC and lateral PFC) was also suggested in a PET (15)O study in another group of abstinent heavy marijuana users (Eldreth *et al.* 2004). Using similar methods, comparable results were reported in abstinent cocaine abusers; here the average amount of cocaine used per week was negatively correlated with activity in the ACC and PFC (Bolla *et al.* 2004). In an earlier study we provided evidence for the role of the OFC in the Stroop effect in drug addiction (Goldstein *et al.* 2001) (Fig. 19.7): higher levels of relative OFC metabolism at *baseline* (measured with PET FDG) were associated with lower conflict (higher Stroop interference score) for cocaine-addicted and alcoholic subjects, while the pattern was reversed for controls. In this latter study, cross-correlations were observed between brain function at a resting state with standard performance of the Stroop task during neuropsychological testing.

Taken together, the involvement of the PFC, including the OFC and ACC, in the Stroop task in drug-addicted subjects still remains to be reliably demonstrated, especially using simultaneous brain-behavior recordings and designs that utilize higher temporal resolution. Documenting differential brain responsivity to this classic cognitive conflict in drug-addicted subjects would provide additional support for a top-down regulatory deficit, which may be involved in the inability of drug-addicted subjects to abstain from drug taking. Further, integrating emotional processes in the classic Stroop effect may increase the task's ecological validity since what often creates interference in daily life is incongruence between current demands and emotionally salient responses (Compton *et al.* 2003).

Modified Stroop tasks have indeed begun to address this concern by replacing the color words (e.g., red) with emotionally salient words (e.g., murder); the latter can be specifically selected to tap core concerns within particular patient populations. For example, an emotional Stroop task was adapted specifically to war veterans with post-traumatic stress disorder by inclusion of negative combat-related words. Results of this fMRI study

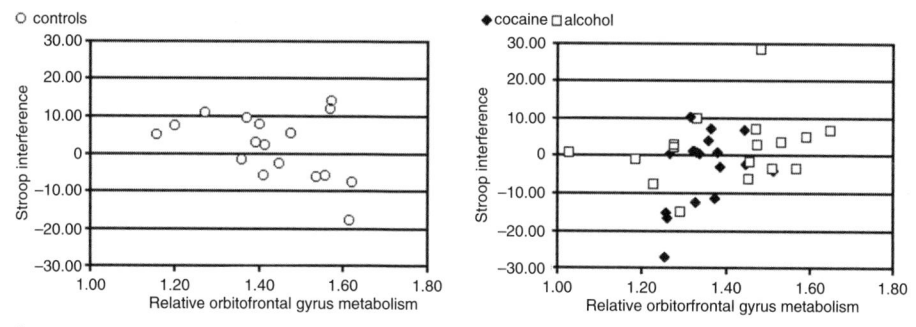

**Fig. 19.7** Association between Stroop interference and relative OFC metabolism in right-handed male 17 control subjects, 17 alcoholics, and 17 cocaine addicts matched for age, education, and estimates of IQ. Reproduced with permission from Goldstein et al. (2001).

revealed significant ACC signal increases to combat as compared to general negative words in Vietnam combat veterans without post-traumatic stress disorder but not in those with post-traumatic stress disorder (Shin et al. 2001). In recent behavioral studies, drug-related words (alcohol, heroin, or cocaine words, respectively, as compared to neutral words) were shown to bias attentional processes (i.e., slow reaction time) in alcoholic (Lusher et al. 2004), heroin-dependent (Franken et al. 2004), or cocaine (Copersino et al. 2004) subjects as compared to non-drug-using controls. Using similar drug Stroop designs in the fMRI environment will allow the study of the neurobiological underpinnings of this cue-induced attentional bias in drug addiction.

## Personality

Going beyond basic cognitive tasks that can be extended to include an emotional component, trait aspects of human psychological functioning, namely personality, can also measure behavioral control. For example, we examined the relationship between inhibitory control constructs on the Multidimensional Personality Questionnaire (MPQ, (Tellegen and Waller 1997)) and relative glucose metabolism in the OFC (Goldstein et al. 2002). Results showed a positive correlation between the MPQ harm avoidance scale and relative OFC metabolism in methamphetamine-dependent subjects but not controls (Fig. 19.8). Thus, in the drug-addicted group the higher the OFC glucose metabolism at rest, the higher were the self-reported avoidance of potentially harmful situations and inhibitory constraint on inappropriate approach behaviors (Depue and Collins 1999). These results are consistent with our previous Stroop study, where higher relative OFC glucose metabolism at rest correlated with better inhibitory control of pre-potent responses in drug-addicted subjects (Goldstein et al. 2001).

Note that impulsivity has been the most frequently implicated trait (or sustained emotional/behavioral disposition) of lack of control in substance abuse. Indeed, the inability to forgo immediate rewards for the benefit of gaining larger but delayed rewards has been classically implicated as core cognitive impairment in drug addiction (see for example

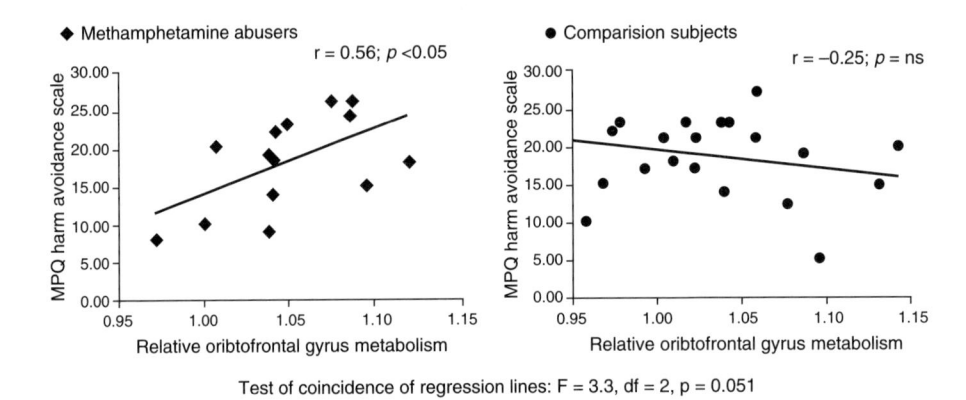

**Fig. 19.8** Association between Tellegen's Multidimensional Personality Questionnaire (MPQ) Harm Avoidance scale and relative OFC metabolism in 14 methamphetamine abusers and 22 comparison subjects. Reproduced with permission from Goldstein et al. (2002).

(McClure *et al.* 2004)). Thus, studies utilizing decision-making paradigms and skin conductance responses point to the existence of drug-addicted subgroups that are insensitive to future reinforcement or that are hypersensitive to reward (Bechara *et al.* 2002). Similarly, heroin and cocaine abusers report discounting the importance of delayed monetary reward more than non-drug-using controls (Kirby and Petry 2004), see also Petry (2003), and Bjork *et al.* (2004). A promising line of research is exemplified in a recent fMRI study that showed a disruptive dose-dependent alcohol effect in the OFC and ACC during simulated driving conditions among adolescents who report emerging symptoms of tolerance (Calhoun *et al.* 2004), possibly due to increased impulsivity and decreased self-control.

### Anger

Anger is an intense approach emotion that can mobilize behavioral change; when coupled with poor inhibitory control, anger can trigger an aggressive response (Davidson, Putnam *et al.* 2000); additionally, another person's anger in a social exchange can serve as a cue to cease current behavior (Elliott *et al.* 2000). In a recent study, we therefore chose the Anger content scale of the Minnesota Multiphasic Personality Inventory—2 (MMPI-2) as an emotionally based trait measure of inhibitory control; this scale assesses anger expression and anger control problems (Schill and Wang 1990; Greene 1991; Kawachi *et al.* 1996; Strassberg and Russell 2000). Deriving from our previous results (Goldstein *et al.* 2001; Goldstein *et al.* 2002) and from other studies that suggested anger to be specifically associated with the lateral OFC (see for example two PET $H_2^{15}O$ activation studies (Blair *et al.* 1999; Drexler *et al.* 2000); also for review see Murphy *et al.* (Murphy *et al.* 2003)), we hypothesized that better emotional and behavioral control (lower scores on the MMPI-2 Anger scale) will be associated with *greater* relative glucose metabolism

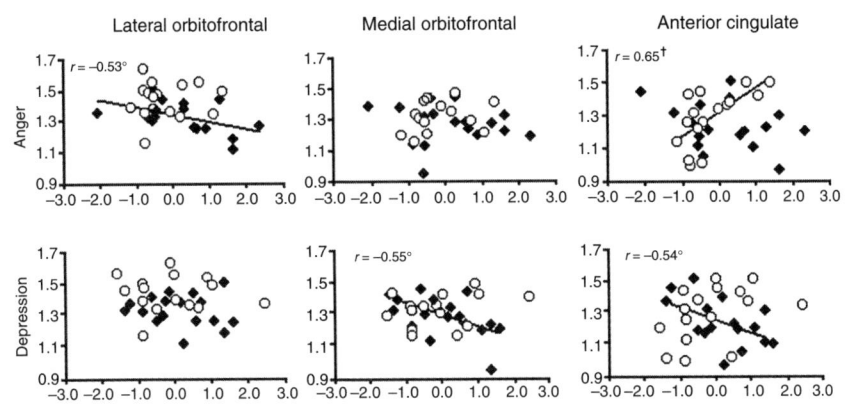

**Fig. 19.9** Association between MMPI-2 Anger content scale with relative resting glucose metabolism in lateral OFC, medial OFC, and ACC in 17 cocaine dependent (diamond) and 16 comparison subjects (circle). *Significant Pearson correlation at $p < 0.05$ (two-tailed). †Significant Pearson correlation at $p < 0.01$ (two-tailed). Reproduced with permission from Goldstein et al. (2005).

in the lateral OFC for drug-dependent subjects. Indeed, results showed an inverse association between reports of anger and the lateral OFC in the cocaine subjects; in control subjects anger instead was directly associated with the ACC (Fig. 19.9). This converges with rodent models where an increase in aggressiveness occurs after OFC lesions (Fuster 1997). It also converges with a reduction in self-reported anger after cingulotomy for intractable pain (Cohen et al. 2001).

## Aggression

Aggression and violence are common derivatives of anger and poor inhibitory control, particularly when associated with the abuse of psychoactive drugs (Krug et al. 2002). Indeed, 42%–78% of persons arrested for assault behaviors also test positive for illicit substance use (National Institute of Justice 1999). As for legal substances, the U.S. Department of Justice Report on Alcohol and Crime established that alcohol abuse was a contributing factor in 40% of violent crimes committed in the U.S. Similarly, about 4 in 10 criminal offenders report that they were using alcohol at the time of their offense (Greenfeld 1998).

It has also been reported that individuals with earlier age of onset of problem drinking and family history of alcoholism (i.e., a problem-drinking father or type 2-like alcohol dependence) show more impulsivity during neuropsychological assays of inhibitory control (e.g., continuous performance test) and report more aggression (which correlated with severity of impulsivity) (Bjork et al. 2004). Similarly, alcohol consumption was associated with an increased likelihood of non-severe intimate partner violence among men enrolled in a violence or alcoholism treatment programs (note that drinking was more strongly associated with a likelihood of severe violence among men with antisocial personality disorder) (Fals-Stewart et al. 2005).

Beyond reporting prevalence and describing trends, however, very few human studies have investigated the neurobiology of aggression with its co-occurring substance abuse (De La Rosa *et al.* 1990). An ideal starting point is the study of OFC function, because OFC lesions seem to increase aggressive behavior in both animals (Fuster, pp.87, (Fuster 1997)), and humans (Grafman *et al.* (1996); see Blake and Grafman, Chapter 22, this volume). Further, reminiscent of Phineas Gage and common to individuals with OFC lesions, aggressive behavior (or its most frequent subtype) can be characterized as impulsive, socially inappropriate, irresponsible, and of compromised insight. Indeed, similarly to substance-dependent individuals, aggressive individuals perform poorly on tasks that challenge inhibitory control and tasks that tap into other functions of the OFC, such as shifting a dominant behavior when contingencies have been reversed (Blair *et al.* 2001; Blair 2003). Overall, these behavioral observations led to the conceptualization of aggression as a behavior that stems from PFC-mediated dysregulation of cognitive and emotional control (Seguin 2004).

Nevertheless, evidence supporting these neuropsychological findings has yet to be established. Thus, although significant reductions (~10%) in prefrontal gray matter volumes were reported in persons with anti-social aggressive behavior when compared with psychiatric controls and substance-dependent individuals (Raine *et al.* 2000; Woermann *et al.* 2000), review of the studies assessing OFC function in anti-social personality disorder, for which aggression is a hallmark (Rolls *et al.* 1994), failed to establish a reliable account of OFC in aggression, mainly due to methodological difficulties and the characteristic heterogeneity in groups with anti-social behavioral disorders (for review see Seguin (2004)).

A cautionary note is in order here. Aggressive or criminal behavior in the drug-addicted subjects is frequently a means to procure the drug and does not necessarily indicate that these individuals are more aggression prone. Thus, although we draw similarities between drug addiction and aggression in the context of inhibitory dyscontrol and PFC impairment, we do not suggest that substance-dependent individuals are more aggression-prone than those not addicted. In fact, there is no data to support such a suggestion (certainly not for marijuana or heroin, with a possible exception for stimulants such as cocaine where intoxication may be associated with excitability, irritability and paranoia (Brody 1990)). On the contrary, it is more plausible that drugs of abuse evoke violent behavior in those who are already aggression-prone, as mentioned above in domestic abusers (Bjork *et al.* 2004) and Type-II alcoholics (Fals-Stewart *et al.* 2005), who collectively form a minority subgroup within the drug-addicted population. It is important to draw this distinction between aggression/violence and the clear lowering of self-control thresholds with drug intoxication—both for research and clinical purposes.

## Inhibitory control: summary

To summarise our review of inhibitory control, there appears to be a fairly consistent relationship between the lateral OFC and behavioral control, and it seems to vary as a function of drug addiction: we observed an association between higher metabolism in the lateral OFC and better control in drug-addicted but not healthy control subjects (Goldstein *et al.* 2002; Goldstein *et al.* 2005; Goldstein *et al.* 2001). This positive brain-behavior correlation in drug addiction may denote a pathological state, where higher OFC activity indicates an ineffective or over-exerted allocation of control resources possibly due to compensatory processes needed to maintain a steady baseline state. Thus, this positive brain-behavior association may point to a mechanism that possibly underlies greater vulnerability for impulsivity (including over-responsiveness to immediate rewards and under-avoidance of longer-term punishment) and bouts of severe lack of control, together culminating in compulsive drug intake reminiscent of perseverative behaviors.

In the healthy state, activity would reveal a compromised performance (i.e., a negative correlation), because the worse the performance the more adaptation required (e.g., greater effort). A similar mechanism was embedded in Miller and Cohen's description of the evaluative top-down processing role suggested for the PFC (Miller and Cohen 2001). Indeed, regional signal *decreases* in the OFC during focused attention seem to reflect a necessary interplay between continuous cognitive and emotional processes (Gusnard and Raichle 2001) while OFC *hyper-metabolism* may be a characteristic of individuals with obsessive-compulsive disorder (for review see Saxena *et al.* (1998)) and Milad and Rauch, Chapter 20, this volume)) and Tourette's syndrome (Chase *et al.* 1986), disorders that are characterized by perseverative behavior or the compromised ability to respond appropriately to changing contingencies in the environment.

If the positive correlation between inhibitory control and OFC metabolism at a resting *baseline* is indeed indicative of pathology in the drug-addicted population, OFC activation to a *challenge* should uncover such pathology and be associated with measures of *impaired* performance in the cocaine users. This concept of reactivity to a challenge (stress) to uncover subclinical or not easily detectable underlying pathologies is widely used in clinical cardiovascular and autoimmune research (see (Turner 1994)), but remains to be applied to the neurobiological study of reward processing and inhibitory control in drug addiction.

## Decision-making

While reward processing and inhibitory control are clearly compromised in addiction, so is the ability to make advantageous decisions. When an addicted individual is faced with the decision of whether or not to take the drug, is it really a choice? Evidence from neuropsychological and neuroimaging studies suggest that what may be a well-considered decision for the healthy individual, may not be a well-considered decision for the addicted or at-risk individual. This perspective lends further credence to the concept of addiction as a *disease*, not a choice (Kalivas 2004).

Executive functions include the ability to plan, solve problems, and make choices (Stern and Prohaska 1996) and are highly dependent on the frontal lobes (Stuss and Alexander 2000). A widely used task to examine decision-making, especially as it relates to the possibility to gain monetary reward, is the Iowa Gambling Task (Bechara *et al.* 1994). This task requires subjects to select a card from one of four decks of cards, and with each card the subject can win or lose money. Some of the decks include high gains but also high losses, while other decks result in low gains but also low losses. The subject is more likely to win if he or she selects repeatedly from the low gain/low loss deck(s). While healthy individuals tend to pick more from the low gain/low loss deck(s), drug-addicted individuals perform poorly on this task (cocaine (Grant *et al.* 2000; Bartzokis *et al.* 2000), see also (Bechara and Damasio 2002) and (Bechara *et al.* 2002); opiate (Petry *et al.* 1998); and alcohol (Mazas *et al.* 2000)), as do individuals with lesions to the ventromedial PFC (Bechara *et al.* 1994; Bechara *et al.* 1997; Bechara *et al.* 1999; Bechara *et al.* 2000; Bechara *et al.* 2000. In fact, many similarities exist between the performance on this task of drug-addicted individuals and patients with ventromedial PFC damage including denial or unawareness of deficit, and a tendency to choose short-term gains in the face of negative future consequences (Bechara 2003). Other common deficits on similar tasks include longer deliberation times and greater likelihood to make the wrong decisions (Rogers *et al.* 1999; Rogers and Robbins 2001).

The severity of impairment in decision-making may vary depending on the drug of abuse and length of abstinence. For example, Bechara *et al.* showed that methamphetamine-dependent subjects have worse performance on the Iowa Gambling Task than cocaine- or alcohol-dependent individuals (Bechara and Martin 2004). Bartzokis *et al.* compared cocaine-addicted subjects that were abstinent for less than four days with those abstinent for more than four days and found improved performance on this task with abstinence from cocaine ((Bartzokis *et al.* (2000); see also Bolla *et al.* (1999) for previously reported dose-related effects of cocaine on other tests of executive functioning).

In a PET study, Bolla and colleagues examined CBF changes in the OFC while abstinent cocaine users and healthy controls performed the Iowa Gambling Task or a control task and found that the right OFC was activated more in the cocaine users (while the dorsolateral PFC was activated less) (Bolla *et al.* 2003). In contrast, using a similar methodology these authors reported less activation in the right lateral OFC and the right dorsolateral PFC and greater activation in the left cerebellum in abstinent marijuana users than controls (Bolla *et al.* 2005). Clearly, more functional neuroimaging studies utilizing this or similar decision-making tasks in drug-addicted individuals are necessary before reliable conclusions can be reached. Note that recent evidence suggests a laterality effect and the contribution of prefrontal regions outside the ventromedial PFC region to the Iowa Gambling Task performance; alternative tasks (Cambridge Gamble Task and Risk Task) may be more selectively associated with ventral prefrontal damage (Clark *et al.* 2003).

A related field of study pertains to regret ('what could have been'), or counterfactual thinking, where the outcome of a decision is compared with alternative outcomes, and where autonomic signals are important internal guiding mechanisms. Based on self-report and skin conductance measures during a modified gambling task, it has been recently suggested that individuals with OFC lesions did not experience regret similarly to healthy controls (Camille *et al.* 2004). While there is no direct evidence as to the role of regret in the decision-making capacity of drug-addicted individuals, it is certainly an area that deserves further exploration.

## 19.4. Summary and future directions

We have reviewed the role of the OFC in the clinical characteristics of drug addiction and their putative neuropsychological mechanisms. We suggest that the OFC is intricately related to the core characteristics of drug addiction through impaired response Inhibition and salience attribution (I-RISA) contributing to the development and maintenance of this pervasive and chronically relapsing disorder. Its role in reward processing, saliency attribution and motivation is hypothesized to allow for a drug of abuse to assume such intense value that upon repeated administration, the drug-addicted individual experiences intense *wanting* (or *craving*) of this valuable commodity at the cost of a generalized relative indifference for all other (no-longer valuable) stimuli. We suggest that the OFC is not only involved in attributing value to the drug initially, but also in maintaining a representation of the drug as 'valuable', despite its numerous adverse consequences and inability over time to elicit 'pleasure'. With this impairment in salience attribution, loss of inhibitory control ensues, thus depicting the *bingeing* component of addiction. Note that the direction of association could also be reversed, where impaired control over drug intake (possibly due to deficient prefrontal control mechanisms) in association with impaired insight into difficulties may lead to justification (possibly through cognitive dissonance mechanisms) of continued drug use despite adverse consequences, increasing the value of the drug and drug taking behaviors. When the drug is no longer available, the individual experiences *withdrawal*, and OFC abnormalities, which continue to be observed during this stage, may feedback into relapse to drug use. Indeed, *relapse* can occur even after extended periods of abstinence, when the drug itself or merely a cue triggers the well-maintained association between the drug and its salient value; the inability of the individual to extinguish this association needs to be further explored especially vis-à-vis the functioning of the OFC.

### Sensitivity and specificity

The OFC has a clear role as an integrator of affective/emotional with cognitive information, establishing the salience value of stimuli in the environment thus driving motivation and approach behavior; yet it influences other functions such as inhibitory control and aggression through an intricate array of afferent and efferent connections with nigrostriatal and mesolimbic regions. Many of these OFC-associated neuropsychological functions are impaired in populations with substance use disorders. However, there is much work that remains to be accomplished in terms of dissociating brain and behavioral functions. For example, it is important to establish the brain regions and neuropsychological functions that are well preserved in drug-addicted individuals (this needs to be done after controlling for age, education, socioeconomic status and other possibly confounding variables that may unduly amplify the level of impairment in addicted individuals as compared to controls). Further, it is critical to dissociate functions of different OFC subregions (e.g., lateral vs. medial, rostral vs. caudal) and of the OFC from other associated regions (e.g., ACC, dorsolateral PFC). Indeed current research is engaged in developing more sensitive neuroimaging techniques and behavioral assays to tap into specific brain regions relevant to specific neuropsychological functions (e.g., responses to positive vs. negative or abstract vs. primary reinforcement) underlying a given complex disorder such as drug addiction.

## Clinical applications

Whenever studying patient populations, a long-term goal is to use reliable and valid scientific results to inform clinically relevant hypotheses for the development of effective intervention and prevention efforts. Neurobiological and neurobehavioural information about the drug-addicted individual's response inhibition and attribution of salience (I-RISA) can help guide the development of treatment interventions at the individual and program levels. For example, cognitive-behavioral rehabilitation efforts geared towards improving inhibitory control functions (i.e., gaining control over cognitive functions to suppress craving and other automatic processes, see (Beck 1993)), and lessening the value attributed to the drugs while increasing the value attributed to non-drug reinforcers, could prove promising in longer-term relapse prevention efforts. Promising approaches involve progressive cue-exposure, where subjective and objective (sometimes automatic) responses to drug-related stimuli are targeted for extinction (Childress *et al.* 1987; O'Brien *et al.* 1990; Franken *et al.* 1999). Combined with the improvement of learning and memory skills, these interventions could be tailored to ameliorate the cognitive dysfunction in drug-addicted individuals (or at least to consider patients' level of cognitive functioning (Aharonovich *et al.* 2003)) and have the potential of improving outcome of the classically used treatment approaches, such as cognitive behavioral therapy for relapse prevention and motivational interviewing. Such behavioral therapies may be enhanced through the addition of pharmacological interventions as extensively reviewed elsewhere (Volkow *et al.* 2004) and summarized briefly below.

Medications that affect the same neurotransmitter system as the drug of abuse but with slower pharmacokinetic properties can interfere with the drug's reinforcing effects, and have been remarkably successful in the management of heroin addiction and to a certain extent smoking (see review (Kreek *et al.* 2002)), but not for treating stimulant addiction (Grabowski *et al.* 1997; Shearer *et al.* 2003). Results from animal studies and preliminary clinical trials with human subjects highlight the potential therapeutic effects of GABA-enhancing drugs (Dewey *et al.* 1998; Johnson *et al.* 2003), selective antagonism of cannabinoid (CB1) receptors (De Vries *et al.* 2001), and MAO B inhibitors (George *et al.* 2003). These agents can successfully prevent cue- and drug-induced increases in brain DA, they modulate both DA cells and postsynaptic responses from DA stimulation, and they increase the amount of DA release in response to DA cell firing, respectively.

Brain imaging studies also suggest that strategies to interfere with conditioned responses may be effective in treating addiction. As mentioned above, this approach currently is based on behavioral desensitization of addicted subjects to responses linked with conditioned stimuli, but it is reasonable to predict that medications could facilitate these processes by impeding circuits linked to memory processes in the hippocampus and amygdala or affecting those linked with salience attribution and inhibitory control in the OFC and ACC. Specifically, beta-blockers have been shown to inhibit conditioned responses to natural reinforcers and to aversive stimuli (Miranda et al. 2003), but to our knowledge these have not been tested on drug-induced conditioned responses. GABAergic stimulation attenuates Pavlovian conditioned responses and impairs conditioned responses to drugs of abuse (Bailey et al. 2002; Dewey et al. 1998; Franklin and Druhan 2000; Hotsenpiller and Wolf 2003). Indeed this may be one of the mechanisms contributing to the therapeutic effectiveness in drug-addicted subjects of some of the GABA-enhancing drugs that have recently been reported. A pharmacological agent that would particularly target the OFC and ACC in drug addiction has yet to be developed.

It is important to recognize that more research is required to delineate the unique differences among the various drugs of abuse. This knowledge may help in the design of specific medications, but also help us understand the neurobiological processes that underlie the specificity for an addiction that occurs in most cases to a given drug but not to another, as well as preferences for a particular drug of abuse. Continuing research in drug addiction must also seek to understand genes involved in the predisposition for drug abuse, which may offer new targets for the development of pharmacological and—pharmacological interventions. Similarly, a better understanding of the interactions between genes, environment, and neurobiology will help develop behavioral interventions to counteract deleterious effects of stressor(s) on the brain that in turn facilitate drug abuse and addiction.

As we acquire basic knowledge about addiction-related brain circuits and how they are affected by environmental variables, dual approaches pairing behavioral interventions with medications will likely offer new and effective treatments for drug addiction and its resulting neurobiological changes. For example, one could conceive of interventions designed to "exercise" brain circuits by specific cognitive and behavioral interventions to remediate and strengthen circuits affected by chronic drug use in an analogous way to some interventions used for reading disabilities (Papanicolaou et al. 2003). Dual interventions to activate and strengthen circuits involved in inhibitory control and salience attribution may increase successful abstinence from drug-taking.

## Acknowledgements

This review was supported by grants from the National Institute on Drug Abuse (DA06891-06; 1K23 DA15517-01; and T32 DA07316-01A1); Laboratory Directed Research and Development from U.S. Department of Energy (OBER); National Institute on Alcohol Abuse and Alcoholism (AA/ODO9481-04); NARSAD Young Investigator Award and Stony Brook/Brookhaven National Laboratory seed grant (79/1025459); and General Clinical Research Center (5-MO1-RR-10710).

## References

Adams, K. M., Gilman, S., Koeppe, R. A., Kluin, K. J., Brunberg, J. A., Dede, D. et al. (1993) Neuropsychological deficits are correlated with frontal hypometabolism in positron emission tomography studies of older alcoholic patients. *Alcohol Clin Exp Res* **17**:205–210.

Aharonovich, E., Nunes, E., and Hasin, D. (2003) Cognitive impairment, retention and abstinence among cocaine abusers in cognitive-behavioral treatment. *Drug Alcohol Depend* **71**:207–211.

Ahmed, S. H. and Koob, G. F. (1998) Transition from moderate to excessive drug intake: change in hedonic set point. *Science* **282**:298–300.

Ahmed, S. H., Kenny, P. J., Koob, G. F., and Markou, A. (2002) Neurobiological evidence for hedonic allostasis associated with escalating cocaine use. *Nat Neurosci* **5**:625–626.

American Psychiatric Association (1994) *Diagnostic and Statistical Manual of Mental Disorders, Fourth Edition.* Washington, DC: American Psychiatric Association.

Anderson, A. K., Christoff, K., Stappen, I., Panitz, D., Ghahremani, D. G., Glover, G. *et al.* (2003) Dissociated neural representations of intensity and valence in human olfaction. *Nat Neurosci* **6**:196–202.

Anthony, J. C., Warner L. A., and Kessler, R. C. (1994) Comparative epidemiology of dependence on tobacco, alcohol, controlled substances, and inhalants: basic findings from the national comorbidity survey. *Experimental and Clinical Psychopharmacology* **2**: 244–268.

Bailey, A., Yuferov, V., Bendor, J., Schlussman, S. D., Zhou, Y., Ho, A. *et al.* (2005) Immediate withdrawal from chronic 'binge' cocaine administration increases mu-opioid receptor mRNA levels in rat frontal cortex. *Brain Res Mol Brain Res* **137**:258–262.

Bailey, D. J., Tetzlaff, J. E., Cook, J. M., He, X., and Helmstetter, F. J. (2002) Effects of hippocampal injections of a novel ligand selective for the alpha 5 beta 2 gamma 2 subunits of the GABA/benzodiazepine receptor on Pavlovian conditioning. *Neurobiol Learn Mem* **78**:1–10.

Bartzokis, G., Beckson, M., Hance, D. B., Lu, P. H., Foster, J. A., Mintz, J. *et al.* (1999) Magnetic resonance imaging evidence of 'silent' cerebrovascular toxicity in cocaine dependence. *Biol Psychiatry* **45**:1203–1211.

Bartzokis, G., Lu, P. H., Beckson, M., Rapoport, R., Grant, S., Wiseman, E. J. *et al.* (2000) Abstinence from cocaine reduces high-risk responses on a gambling task. *Neuropsychopharmacology* **22**:102–103.

Bechara, A. (2003) Risky business: emotion, decision-making, and addiction. *J Gambl Stud* **19**:23–51.

Bechara, A. (2004) The role of emotion in decision-making: evidence from neurological patients with orbitofrontal damage. *Brain Cogn* **55**:30–40.

Bechara, A. and Damasio, H. (2002) Decision-making and addiction (part I): impaired activation of somatic states in substance dependent individuals when pondering decisions with negative future consequences. *Neuropsychologia* **40**:1675–1689.

Bechara, A. and Martin, E. M. (2004) Impaired decision-making related to working memory deficits in individuals with substance addictions. *Neuropsychology* **18**:152–162.

Bechara, A., Tranel, D., and Damasio, H. (2000) Characterization of the decision-making deficit of patients with ventromedial prefrontal cortex lesions. *Brain* **123** ( Pt 11): 2189–2202.

Bechara, A., Damasio, A. R., Damasio, H., and Anderson, S. W. (1994) Insensitivity to future consequences following damage to human prefrontal cortex. *Cognition* **50**:7–15.

Bechara, A., Damasio, H., and Damasio, A. R. (2000) Emotion, decision-making and the orbitofrontal cortex. *Cereb Cortex* **10**:295–307.

Bechara, A., Damasio, H., Damasio, A. R., and Lee, G. P. (1999) Different contributions of the human amygdala and ventromedial prefrontal cortex to decision-making. *J Neurosci* **19**:5473–5481.

Bechara, A., Damasio, H., Tranel, D., and Damasio, A. R. (1997) Deciding advantageously before knowing the advantageous strategy. *Science* **275**:1293–1295.

Bechara, A., Dolan, S., and Hindes, A. (2002) Decision-making and addiction (part II): myopia for the future or hypersensitivity to reward? *Neuropsychologia* **40**:1690–1705.

Beck, A. T. (1993) Cognitive therapy: past, present, and future. *J Consult Clin Psychol* **61**:194–198.

Bilder, R. M., Goldman, R. S., Robinson, D., Reiter, G., Bell, L., Bates, J. A. et al. (2000) Neuropsychology of first-episode schizophrenia: initial characterization and clinical correlates. *Am J Psychiatry* **157**: 549–559.

Bjork, J. M., Hommer, D. W., Grant, S. J., and Danube, C. (2004) Impulsivity in abstinent alcohol-dependent patients: relation to control subjects and type 1-/type 2-like traits. *Alcohol* **34**:133–150.

Blair, R. J. (2003) Facial expressions, their communicatory functions and neuro-cognitive substrates. *Philos Trans R Soc Lond B Biol Sci* **358**:561–572.

Blair, R. J., Colledge, E., and Mitchell, D. G. (2001) Somatic markers and response reversal: is there orbitofrontal cortex dysfunction in boys with psychopathic tendencies? *J Abnorm Child Psychol* **29**:499–511.

Blair, R. J., Morris, J. S., Frith, C. D., Perrett, D. I., and Dolan, R. J. (1999) Dissociable neural responses to facial expressions of sadness and anger. *Brain* **122 (Pt 5)**: 883–893.

Blumberg, H. P., Leung, H. C., Skudlarski, P., Lacadie, C. M., Fredericks, C. A., Harris, B. C. et al. (2003) A functional magnetic resonance imaging study of bipolar disorder: state- and trait-related dysfunction in ventral prefrontal cortices. *Arch Gen Psychiatry* **60**:601–609.

Bolla, K. I., Eldreth, D. A., London, E. D., Kiehl, K. A., Mouratidis, M., Contoreggi, C. et al. (2003) Orbitofrontal cortex dysfunction in abstinent cocaine abusers performing a decision-making task. *Neuroimage* **19**:1085–1094.

Bolla, K. I., Eldreth, D. A., Matochik, J. A., and Cadet, J. L. (2005) Neural substrates of faulty decision-making in abstinent marijuana users. *Neuroimage* **26**:480–492.

Bolla, K., Ernst, M., Kiehl, K., Mouratidis, M., Eldreth, D., Contoreggi, C. et al. (2004) Prefrontal cortical dysfunction in abstinent cocaine abusers. *J Neuropsychiatry Clin Neurosci* **16**:456–464.

Bolla, K. I., Rothman, R., and Cadet, J. L. (1999) Dose-related neurobehavioral effects of chronic cocaine use. *J Neuropsychiatry Clin Neurosci* **11**:361–369.

Bonson, K. R., Grant, S. J., Contoreggi, C. S., Links, J. M., Metcalfe, J., Weyl, H. L. et al. (2002) Neural systems and cue-induced cocaine craving. *Neuropsychopharmacology* **26**: 376–386.

Breiter, H. C., Aharon, I., Kahneman, D., Dale, A., and Shizgal, P. (2001) Functional imaging of neural responses to expectancy and experience of monetary gains and losses. *Neuron* **30**:619–639.

Breiter, H. C., Gollub, R. L., Weisskoff, R. M., Kennedy, D. N., Makris, N., Berke, J. D. et al. (1997) Acute effects of cocaine on human brain activity and emotion. *Neuron* **19**:591–611.

Bremner, J. D., Vermetten, E., Vythilingam, M., Afzal, N., Schmahl, C., Elzinga, B. et al. (2004) Neural correlates of the classic color and emotional stroop in women with abuse-related posttraumatic stress disorder. *Biol Psychiatry* **55**:612–620.

Brody, A. L., Mandelkern, M. A., London, E. D., Childress, A. R., Lee, G. S., Bota, R. G. et al. (2002) Brain metabolic changes during cigarette craving. *Arch Gen Psychiatry* **59**:1162–1172.

Brody, S. L. (1990) Violence associated with acute cocaine use in patients admitted to a medical emergency department. *NIDA Res Monogr* **103**:44–59.

Butter, C. M. (1969) Perseveration in extinction and in discrimination reversal tasks following selective frontal ablations in Macaca mulatta. *Physiology and Behavior* **4**:163–171.

Butter, C. M., Mishkin, M., and Rosvold, H. E. (1963) Conditioning and extinction of a food-rewarded response after selective ablations of frontal cortex in rhesus monkeys. *Exp Neurol* **7**:65–75.

Calhoun, V. D., Pekar, J. J., and Pearlson, G. D. (2004) Alcohol intoxication effects on simulated driving: exploring alcohol-dose effects on brain activation using functional MRI. *Neuropsychopharmacology* **29**:2097–2017.

Camille, N., Coricelli, G., Sallet, J., Pradat-Diehl, P., Duhamel, J. R., and Sirigu, A. (2004) The involvement of the orbitofrontal cortex in the experience of regret. *Science* **304**:1167–1170.

Carter, C. S., Macdonald, A. M., Botvinick, M., Ross, L. L., Stenger, V. A., Noll, D. *et al.* (2000) Parsing executive processes: strategic vs. evaluative functions of the anterior cingulate cortex. *Proc Natl Acad Sci USA* **97**:1944–1948.

Casey, B. J., Trainor, R., Giedd, J., Vauss, Y., Vaituzis, C. K., Hamburger, S. *et al.* (1997) The role of the anterior cingulate in automatic and controlled processes: a developmental neuroanatomical study. *Dev Psychobiol* **30**:61–69.

Catafau, A. M., Etcheberrigaray, A., Perez de los, C. J., Estorch, M., Guardia, J., Flotats, A. *et al.* (1999) Regional cerebral blood flow changes in chronic alcoholic patients induced by naltrexone challenge during detoxification. *J Nucl Med* **40**:19–24.

Cavada, C., Company, T., Tejedor, J., Cruz-Rizzolo, R. J., and Reinoso-Suarez, F. (2000) The anatomical connections of the macaque monkey orbitofrontal cortex. A review. *Cereb Cortex* **10**:220–242.

Chase, T. N., Geoffrey, V., Gillespie, M., and Burrows, G. H. (1986) Structural and functional studies of Gilles de la Tourette syndrome. *Rev Neurol (Paris)* **142**:851–855.

Childress, A. R., McLellan, A. T., Ehrman, R. N., and O'Brien, C. P. (1987) Extinction of conditioned responses in abstinent cocaine or opioid users. *NIDA Res Monogr* **76**:189–195.

Childress, A. R., Mozley, P. D., McElgin, W., Fitzgerald, J., Reivich, M., and O'Brien, C. P. (1999) Limbic activation during cue-induced cocaine craving. *Am J Psychiatry* **156**:11–18.

Chudasama, Y. and Robbins, T. W. (2003) Dissociable contributions of the orbitofrontal and infralimbic cortex to pavlovian autoshaping and discrimination reversal learning: further evidence for the functional heterogeneity of the rodent frontal cortex. *J Neurosci* **23**:8771–8780.

Clark, L., Manes, F., Antoun, N., Sahakian, B. J., and Robbins, T. W. (2003) The contributions of lesion laterality and lesion volume to decision-making impairment following frontal lobe damage. *Neuropsychologia* **41**:1474–1483.

Cohen, R. A., Paul, R., Zawacki, T. M., Moser, D. J., Sweet, L., and Wilkinson, H. (2001) Emotional and personality changes following cingulotomy. *Emotion* **1**:38–50.

Compton, R. J., Banich, M. T., Mohanty, A., Milham, M. P., Herrington, J., Miller, G. A. *et al.* (2003) Paying attention to emotion: an fMRI investigation of cognitive and emotional stroop tasks. *Cogn Affect Behav Neurosci* **3**:81–96.

Copersino, M. L., Serper, M. R., Vadhan, N., Goldberg, B. R., Richarme, D., Chou, J. C. *et al.* (2004) Cocaine craving and attentional bias in cocaine-dependent schizophrenic patients. *Psychiatry Res* **128**:209–218.

Cornish, J. L., Duffy, P., and Kalivas, P. W. (1999) A role for nucleus accumbens glutamate transmission in the relapse to cocaine-seeking behavior. *Neuroscience* **93**:1359–1367.

Daglish, M. R., Weinstein, A., Malizia, A. L., Wilson, S., Melichar, J. K., Lingford-Hughes, A. *et al.* (2003) Functional connectivity analysis of the neural circuits of opiate craving: 'more' rather than 'different'? *Neuroimage* **20**:1964–1970.

Dao-Castellana, M. H., Samson, Y., Legault, F., Martinot, J. L., Aubin, H. J., Crouzel, C. *et al.* (1998) Frontal dysfunction in neurologically normal chronic alcoholic subjects: metabolic and neuropsychological findings. *Psychol Med* **28**:1039–1048.

Davidson, R. J., Putnam, K. M., and Larson, C. L. (2000) Dysfunction in the neural circuitry of emotion regulation—a possible prelude to violence. *Science* **289**:591–594.

De La Rosa, M., Lambert, E. Y., and Gropper, B. (eds). (1990) *Drugs and Violence: Causes, Correlates, and Consequences*. Rockville, MD: National Institute on Drug Abuse.

De Vries, T. J., Shaham, Y., Homberg, J. R., Crombag, H., Schuurman, K., Dieben, J. *et al.* (2001) A cannabinoid mechanism in relapse to cocaine seeking. *Nat Med* **7**:1151–1154.

de Wit, H., Metz, J., Wagner, N., and Cooper, M. (1990) Behavioral and subjective effects of ethanol: relationship to cerebral metabolism using PET. *Alcohol Clin Exp Res* **14**: 482–489.

Depue, R. A. and Collins, P. F. (1999) Neurobiology of the structure of personality: dopamine, facilitation of incentive motivation, and extraversion. *Behav Brain Sci* **22**:491–517.

Deroche-Gamonet, V., Belin, D., and Piazza, P. V. (2004) Evidence for addiction-like behavior in the rat. *Science* **305**:1014–1017.

Dewey, S. L., Morgan, A. E., Ashby, C. R., Jr., Horan, B., Kushner, S. A., Logan, J. et al. (1998) A novel strategy for the treatment of cocaine addiction. *Synapse* **30**:119–129.

Drexler, K., Schweitzer, J. B., Quinn, C. K., Gross, R., Ely, T. D., Muhammad, F. et al. (2000) Neural activity related to anger in cocaine-dependent men: a possible link to violence and relapse. *Am J Addict* **9**:331–339.

Eldreth, D. A., Matochik, J. A., Cadet, J. L., and Bolla, K. I. (2004) Abnormal brain activity in prefrontal brain regions in abstinent marijuana users. *Neuroimage* **23**:914–920.

Elliott, R., Dolan, R. J., and Frith, C. D. (2000) Dissociable functions in the medial and lateral orbitofrontal cortex: evidence from human neuroimaging studies. *Cereb Cortex* **10**:308–317.

Elliott, R., Friston, K. J., and Dolan, R. J. (2000) Dissociable neural responses in human reward systems. *J Neurosci* **20**:6159–6165.

Elliott, R., Newman, J. L., Longe, O. A., and Deakin, J. F. (2003) Differential response patterns in the striatum and orbitofrontal cortex to financial reward in humans: a parametric functional magnetic resonance imaging study. *J Neurosci* **23**:303–307.

Elliott, R., Rubinsztein, J. S., Sahakian, B. J., and Dolan, R. J. (2000) Selective attention to emotional stimuli in a verbal go/no-go task: an fMRI study. *Neuroreport* **11**:1739–1744.

Everitt, B. J., Dickinson, A., and Robbins, T. W. (2001) The neuropsychological basis of addictive behavior. *Brain Res Brain Res Rev* **36**:129–138.

Fals-Stewart, W., Leonard, K. E., and Birchler, G. R. (2005) The occurrence of male-to-female intimate partner violence on days of men's drinking: the moderating effects of antisocial personality disorder. *J Consult Clin Psychol* **73**:239–248.

Febo, M., Segarra, A. C., Nair, G., Schmidt, K., Duong, T. Q., and Ferris, C. F. (2005) The neural consequences of repeated cocaine exposure revealed by functional MRI in awake rats. *Neuropsychopharmacology* **30**: 936–943.

Fein, G., Di, S. V., Cardenas, V. A., Goldmann, H., Tolou-Shams, M., and Meyerhoff, D. J. (2002) Cortical gray matter loss in treatment-naive alcohol dependent individuals. *Alcohol Clin Exp Res* **26**:558–564.

Fein, G., Di, S. V., and Meyerhoff, D. J. (2002) Prefrontal cortical volume reduction associated with frontal cortex function deficit in 6-week abstinent crack-cocaine dependent men. *Drug Alcohol Depend* **68**:87–93.

Fischman, M. W., Schuster, C. R., Javaid, J., Hatano, Y., and Davis, J. (1985) Acute tolerance development to the cardiovascular and subjective effects of cocaine. *J Pharmacol Exp Ther* **235**:677–682.

Foltin, R. W., Ward, A. S., Haney, M., Hart, C. L., and Collins, E. D. (2003) The effects of escalating doses of smoked cocaine in humans. *Drug Alcohol Depend* **70**:149–157.

Forman, S. D., Dougherty, G. G., Casey, B. J., Siegle, G. J., Braver, T. S., Barch, D. M. et al. (2004) Opiate addicts lack error-dependent activation of rostral anterior cingulate. *Biol Psychiatry* **55**:531–537.

Franken, I. H., de Haan, H. A., van der Meer, C. W., Haffmans, P. M., and Hendriks, V. M. (1999) Cue reactivity and effects of cue exposure in abstinent posttreatment drug users. *J Subst Abuse Treat* **16**:81–85.

Franken, I. H., Hendriks, V. M., Stam, C. J., and Van den, B. W. (2004) A role for dopamine in the processing of drug cues in heroin dependent patients. *Eur Neuropsychopharmacol* **14**:503–508.

Franklin, T. R., Acton, P. D., Maldjian, J. A., Gray, J. D., Croft, J. R., Dackis, C. A. et al. (2002) Decreased gray matter concentration in the insular, orbitofrontal, cingulate, and temporal cortices of cocaine patients. *Biol Psychiatry* **51**:134–142.

Franklin, T. R. and Druhan, J. P. (2000) Involvement of the nucleus accumbens and medial prefrontal cortex in the expression of conditioned hyperactivity to a cocaine-associated environment in rats. *Neuropsychopharmacology* **23**:633–644.

Fuster, J. M. (1997) *The prefrontal cortex: Anatomy, physiology, and neuropsychology of the frontal lobe.* Lippinicott-Raven Publishers.

Garavan, H., Pankiewicz, J., Bloom, A., Cho, J. K., Sperry, L., Ross, T. J. *et al.* (2000) Cue-induced cocaine craving: neuroanatomical specificity for drug users and drug stimuli. *Am J Psychiatry* **157**:1789–1798.

Gawin, F. H. and Kleber, H. D. (1986) Abstinence symptomatology and psychiatric diagnosis in cocaine abusers. Clinical observations. *Arch Gen Psychiatry* **43**:107–113.

George, T. P., Vessicchio, J. C., Termine, A., Jatlow, P. I., Kosten, T. R., and O'Malley, S. S. (2003) A preliminary placebo-controlled trial of selegiline hydrochloride for smoking cessation. *Biol Psychiatry* **53**:136–143.

Goldstein, R. Z. and Volkow, N. D. (2002) Drug addiction and its underlying neurobiological basis: neuroimaging evidence for the involvement of the frontal cortex. *Am J Psychiatry* **159**: 642–1652.

Goldstein, R. Z., Volkow, N. D., Wang, G. J., Fowler, J. S., and Rajaram, S. (2001) Addiction changes orbitofrontal gyrus function: involvement in response inhibition. *Neuroreport* **12**:2595–2599.

Goldstein, R. Z., Volkow, N. D., Chang, L., Wang, G. J., Fowler, J. S., Depue, R. A. *et al.* (2002) The orbitofrontal cortex in methamphetamine addiction: involvement in fear. *Neuroreport* **13**:2253–2257.

Goldstein, R. Z., Leskovjan, A. C., Hoff, A. L., Hitzemann, R., Bashan, F., Khalsa, S. S. *et al.* (2004) Severity of neuropsychological impairment in cocaine and alcohol addiction: association with metabolism in the prefrontal cortex. *Neuropsychologia* **42**:1447–1458.

Goldstein, R. Z., Alia-Klein, N., Leskovjan, A. C., Fowler, J. S., Wang, G. J., Gur, R. C. *et al.* (2005) Anger and depression in cocaine addiction: association with the orbitofrontal cortex. *Psychiatry Res* **138**:13–22.

Gottfried, J. A., Deichmann, R., Winston, J. S., and Dolan, R. J. (2002) Functional heterogeneity in human olfactory cortex: an event-related functional magnetic resonance imaging study. *J Neurosci* **22**:10819–10828.

Gottfried, J. A. and Dolan, R. J. (2004) Human orbitofrontal cortex mediates extinction learning while accessing conditioned representations of value. *Nat Neurosci* **7**:1144–1152.

Grabowski, J., Roache, J. D., Schmitz, J. M., Rhoades, H., Creson, D., and Korszun, A. (1997) Replacement medication for cocaine dependence: methylphenidate. *J Clin Psycho Pharmacol* **17**:485–488.

Grafman, J., Schwab, K., Warden, D., Pridgen, A., Brown, H. R., and Salazar, A. M. (1996) Frontal lobe injuries, violence, and aggression: a report of the Vietnam Head Injury Study. *Neurology* **46**:1231–1238.

Grant, S., Contoreggi, C., and London, E. D. (2000) Drug abusers show impaired performance in a laboratory test of decision-making. *Neuropsychologia* **38**:1180–1187.

Grant, S., London, E. D., Newlin, D. B., Villemagne, V. L., Liu, X., Contoreggi, C. *et al.* (1996) Activation of memory circuits during cue-elicited cocaine craving. *Proc Natl Acad Sci USA* **93**:12040–12045.

Greene, R. L. (1991) *MMPI-2/MMPI: An Interpretive Manual.* Boston, MA: Allyn and Bacon.

Greenfeld, L. A. (1998) *Alcohol and crime. National Symposium on Alcohol Abuse and Crime.* B. o. J. Statistics. Washington, D.C.: U.S. Department of Justice.

Gruber, S. A., Rogowska, J., and Yurgelun-Todd, D. A. (2004) Decreased activation of the anterior cingulate in bipolar patients: an fMRI study. *J Affect Disord* **82**:191–201.

Gruber, S. A. and Yurgelun-Todd, D. A. (2005) Neuroimaging of marijuana smokers during inhibitory processing: a pilot investigation. *Brain Res Cogn Brain Res* **23**:107–118.

Gusnard, D. A., Raichle, M. E., and Raichle, M. E. (2001) Searching for a baseline: functional imaging and the resting human brain. *Nat Rev Neurosci* **2**:685–694.

Heinz, A., Siessmeier, T., Wrase, J., Hermann, D., Klein, S., Grusser, S. M. *et al.* (2004) Correlation between dopamine D(2) receptors in the ventral striatum and central processing of alcohol cues and craving. *Am J Psychiatry* **161**:1783–1789.

Hester, R. and Garavan, H. (2004) Executive dysfunction in cocaine addiction: evidence for discordant frontal, cingulate, and cerebellar activity. *J Neurosci* **24**:11017–11022.

Hikosaka, K. and Watanabe, M. (2000) Delay activity of orbital and lateral prefrontal neurons of the monkey varying with different rewards. *Cereb Cortex* **10**:263–271.

Hikosaka, K. and Watanabe, M. (2004) Long- and short-range reward expectancy in the primate orbitofrontal cortex. *Eur J Neurosci* **19**:1046–1054.

Hommer, D., Andreasen, P., Rio, D., Williams, W., Ruttimann, U., Momenan, R. *et al.* (1997) Effects of m-chlorophenylpiperazine on regional brain glucose utilization: a positron emission tomographic comparison of alcoholic and control subjects. *J Neurosci* **17**:2796–2806.

Hotsenpiller, G. and Wolf, M. E. (2003) Baclofen attenuates conditioned locomotion to cues associated with cocaine administration and stabilizes extracellular glutamate levels in rat nucleus accumbens. *Neuroscience* **118**:123–134.

Ingvar, M., Ghatan, P. H., Wirsen-Meurling, A., Risberg, J., Von Heijne, G., Stone-Elander, S. *et al.* (1998) Alcohol activates the cerebral reward system in man. *J Stud Alcohol* **59**:258–269.

Inoue, K., Kiriike, N., Okuno, M., Fujisaki, Y., Kurioka, M., Iwasaki, S. *et al.* (1998) Prefrontal and striatal dopamine metabolism during enhanced rebound hyperphagia induced by space restriction—a rat model of binge eating. *Biol Psychiatry* **44**:1329–1336.

Jernigan, T. L., Butters, N., DiTraglia, G., Schafer, K., Smith, T., Irwin, M. *et al.* (1991) Reduced cerebral gray matter observed in alcoholics using magnetic resonance imaging. *Alcohol Clin Exp Res* **15**:418–427.

Johnson, B. A., Ait-Daoud, N., Bowden, C. L., DiClemente, C. C., Roache, J. D., Lawson, K. *et al.* (2003) Oral topiramate for treatment of alcohol dependence: a randomised controlled trial. *Lancet* **361**:1677–1685.

Jones, B. and Mishkin, M. (1972) Limbic lesions and the problem of stimulus—reinforcement associations. *Exp Neurol* **36**:362–377.

Kalivas, P. W. (2004) Choose to study choice in addiction. *Am J Psychiatry* **161**:193–194.

Kaufman, J. N., Ross, T. J., Stein, E. A., and Garavan, H. (2003) Cingulate hypoactivity in cocaine users during a GO-NOGO task as revealed by event-related functional magnetic resonance imaging. *J Neurosci* **23**:7839–7843.

Kawachi, I., Sparrow, D., Spiro, A., III, Vokonas, P., and Weiss, S. T. (1996) A prospective study of anger and coronary heart disease. The Normative Aging Study. *Circulation* **94**:2090–2095.

Kim, S. J., Lyoo, I. K., Hwang, J., Sung, Y. H., Lee, H. Y., Lee, D. S. *et al.* (2005) Frontal glucose hypometabolism in abstinent methamphetamine users. *Neuropsychopharmacology* **30**:1383–1391.

Kirby, K. N. and Petry, N. M. (2004) Heroin and cocaine abusers have higher discount rates for delayed rewards than alcoholics or -drug-using controls. *Addiction* **99**:461–471.

Kiyatkin, E. A. and Rebec, G. V. (1996) Dopaminergic modulation of glutamate-induced excitations of neurons in the neostriatum and nucleus accumbens of awake, unrestrained rats. *J Neurophysiol* **75**:142–153.

Koob, G. F. (2000) Animal models of craving for ethanol. *Addiction* 95 Suppl 2: S73–S81.

Koob, G. F. and Bloom, F. E. (1988) Cellular and molecular mechanisms of drug dependence. *Science* **242**:715–723.

Koob, G. F., Caine, S. B., Parsons, L., Markou, A., and Weiss, F. (1997) Opponent process model and psychostimulant addiction. *Pharmacol Biochem Behav* **57**:513–521.

Koob, G. F. and Le Moal, M. (2001) Drug addiction, dysregulation of reward, and allostasis. *Neuropsychopharmacology* **24**:97–129.

Kreek, M. J., LaForge, K. S., and Butelman, E. (2002) Pharmacotherapy of addictions. *Nat Rev Drug Discov* **1**:710–726.

Kringelbach, M. L., O'Doherty, J., Rolls, E. T., and Andrews, C. (2003) Activation of the human orbitofrontal cortex to a liquid food stimulus is correlated with its subjective pleasantness. *Cereb Cortex* **13**:1064–1071.

Kringelbach, M. L. and Rolls, E. T. (2004) The functional neuroanatomy of the human orbitofrontal cortex: evidence from neuroimaging and neuropsychology. *Prog Neuro Biol* **72**:341–372.

Krug, E. G., Mercy, J. A., Dahlberg, L. L., and Zwi, A. B. (2002) The world report on violence and health. *Lancet* **360**:1083–1088.

Kufahl, P. R., Li, Z., Risinger, R. C., Rainey, C. J., Wu, G., Bloom, A. S. *et al.* (2005) Neural responses to acute cocaine administration in the human brain detected by fMRI. *Neuroimage* **28**:904–914.

Laruelle, M., Abi-Dargham, A., van Dyck, C. H., Rosenblatt, W., Zea-Ponce, Y., Zoghbi, S. S. *et al.* (1995) SPECT imaging of striatal dopamine release after amphetamine challenge. *J Nucl Med* **36**:1182–1190.

Leshner, A. I. (1997) Addiction is a brain disease, and it matters. *Science* **278**:45–47.

Leung, H. C., Skudlarski, P., Gatenby, J. C., Peterson, B. S., and Gore, J. C. (2000) An event-related functional MRI study of the stroop color word interference task. *Cereb Cortex* **10**:552–560.

Lim, K. O., Choi, S. J., Pomara, N., Wolkin, A., and Rotrosen, J. P. (2002) Reduced frontal white matter integrity in cocaine dependence: a controlled diffusion tensor imaging study. *Biol Psychiatry* **51**:890–895.

Liu, X., Matochik, J. A., Cadet, J. L., and London, E. D. (1998) Smaller volume of prefrontal lobe in polysubstance abusers: a magnetic resonance imaging study. *Neuropsychopharmacology* **18**:243–252.

Loh, E. A. and Roberts, D. C. (1990) Break-points on a progressive ratio schedule reinforced by intravenous cocaine increase following depletion of forebrain serotonin. *Psychopharmacology (Berl)* **101**:262–266.

London, E. D., Broussolle, E. P., Links, J. M., Wong, D. F., Cascella, N. G., Dannals, R. F. *et al.* (1990a) Morphine-induced metabolic changes in human brain. Studies with positron emission tomography and [fluorine 18]fluorodeoxyglucose. *Arch Gen Psychiatry* **47**:73–81.

London, E. D., Cascella, N. G., Wong, D. F., Phillips, R. L., Dannals, R. F., Links, J. M. *et al.* (1990b) Cocaine-induced reduction of glucose utilization in human brain. A study using positron emission tomography and [fluorine 18]-fluorodeoxyglucose. *Arch Gen Psychiatry* **47**:567–574.

London, E. D., Ernst, M., Grant, S., Bonson, K., and Weinstein, A. (2000) Orbitofrontal cortex and human drug abuse: functional imaging. *Cereb Cortex* **10**:334–342.

Lusher, J., Chandler, C., and Ball, D. (2004) Alcohol dependence and the alcohol Stroop paradigm: evidence and issues. *Drug Alcohol Depend* **75**:225–231.

Lyoo, I. K. and Chung, A. I. (2004) Compromised orbitofrontal white matter tract integrity in chronic intravenous methamphetamine users: the consecutively varying threshold tractography study. *College of Problems of Drug Dependence*. Puerto Rico.

Maas, L. C., Lukas, S. E., Kaufman, M. J., Weiss, R. D., Daniels, S. L., Rogers, V. W. *et al.* (1998) Functional magnetic resonance imaging of human brain activation during cue-induced cocaine craving. *Am J Psychiatry* **155**:124–126.

MacDonald, A. W., III, Cohen, J. D., Stenger, V. A., and Carter, C. S. (2000) Dissociating the role of the dorsolateral prefrontal and anterior cingulate cortex in cognitive control. *Science* **288**:1835–1838.

Mackler, S. A. and Eberwine, J. H. (1991) The molecular biology of addictive drugs. *Mol Neuro Biol* **5**:45–58.

Mathew, R. J., Wilson, W. H., Humphreys, D. F., Lowe, J. V., and Wiethe, K. E. (1992) Regional cerebral blood flow after marijuana smoking. *J Cereb Blood Flow Metab* **12**:750–758.

Matochik, J. A., London, E. D., Eldreth, D. A., Cadet, J. L., and Bolla, K. I. (2003) Frontal cortical tissue composition in abstinent cocaine abusers: a magnetic resonance imaging study. *Neuroimage* **19**:1095–1102.

Mazas, C. A., Finn, P. R., and Steinmetz, J. E. (2000) Decision-making biases, antisocial personality, and early-onset alcoholism. *Alcohol Clin Exp Res* **24**:1036–1040.

McClure, S. M., Daw, N. D., and Montague, P. R. (2003) A computational substrate for incentive salience. *Trends Neurosci* **26**:423–428.

McClure, S. M., Laibson, D. I., Loewenstein, G., and Cohen, J. D. (2004) Separate neural systems value immediate and delayed monetary rewards. *Science* **306**:503–507.

Milham, M. P., Banich, M. T., and Barad, V. (2003) Competition for priority in processing increases prefrontal cortex's involvement in top-down control: an event-related fMRI study of the stroop task. *Brain Res Cogn Brain Res* **17**:212–222.

Miller, E. K. and Cohen, J. D. (2001) An integrative theory of prefrontal cortex function. *Annu Rev Neurosci* **24**:167–202.

Miller, N. S. and Gold, M. S. (1994) Dissociation of 'conscious desire' (craving) from and relapse in alcohol and cocaine dependence. *Ann Clin Psychiatry* **6**:99–106.

Miranda, M. I., LaLumiere, R. T., Buen, T. V., Bermudez-Rattoni, F., and McGaugh, J. L. (2003) Blockade of noradrenergic receptors in the basolateral amygdala impairs taste memory. *Eur J Neurosci* **18**:2605–2610.

Moeller, F. G., Hasan, K. M., Steinberg, J. L., Kramer, L. A., Dougherty, D. M., Santos, R. M. *et al.* (2005) Reduced anterior corpus callosum white matter integrity is related to increased impulsivity and reduced discriminability in cocaine-dependent subjects: diffusion tensor imaging. *Neuropsychopharmacology* **30**:610–617.

Mora, F., Avrith, D. B., Phillips, A. G., and Rolls, E. T. (1979) Effects of satiety on self-stimulation of the orbitofrontal cortex in the rhesus monkey. *Neurosci Lett* **13**:141–145.

Mora, F., Avrith, D. B., and Rolls, E. T. (1980) An electrophysiological and behavioral study of self-stimulation in the orbitofrontal cortex of the rhesus monkey. *Brain Res Bull* **5**:111–115.

Murphy, F. C., Nimmo-Smith, I., and Lawrence, A. D. (2003) Functional neuroanatomy of emotions: a meta-analysis. *Cogn Affect Behav Neurosci* **3**:207–233.

Nakamura, H., Tanaka, A., Nomoto, Y., Ueno, Y., and Nakayama, Y. (2000) Activation of fronto-limbic system in the human brain by cigarette smoking: evaluated by a CBF measurement. *Keio J Med* 49 Suppl **1**: A122–A124.

Nakao, T., Nakagawa, A., Yoshiura, T., Nakatani, E., Nabeyama, M., Yoshizato, C. *et al.* (2005) Brain activation of patients with obsessive-compulsive disorder during neuropsychological and symptom provocation tasks before and after symptom improvement: a functional magnetic resonance imaging study. *Biol Psychiatry* **57**:901–910.

National Institute of Justice, (1999) *1998 ADAM: Annual Report on Drug Use Among Adult and Juvenile Arrestees.*

Nestler, E. J. (2000) Genes and addiction. *Nat Genet* **26**:277–281.

Nestler, E. J. (2001) Psychogenomics: opportunities for understanding addiction. *J Neurosci* **21**:8324–8327.

O'Brien, C. P., Childress, A. R., Ehrman, R., and Robbins, S. J. (1998) Conditioning factors in drug abuse: can they explain compulsion? *J Psycho Pharmacol* **12**:15–22.

O'Brien, C. P., Childress, A. R., McLellan, T., and Ehrman, R. (1990) Integrating systemic cue exposure with standard treatment in recovering drug dependent patients. *Addict Behav* **15**:355–365.

O'Doherty, J., Kringelbach, M. L., Rolls, E. T., Hornak, J., and Andrews, C. (2001) Abstract reward and punishment representations in the human orbitofrontal cortex. *Nat Neurosci* **4**:95–102.

O'Neill, J., Cardenas, V. A., and Meyerhoff, D. J. (2001) Separate and interactive effects of cocaine and alcohol dependence on brain structures and metabolites: quantitative MRI and proton MR spectroscopic imaging. *Addict Biol* **6**:347–361.

Oades, R. D. and Halliday, G. M. (1987) Ventral tegmental (A10) system: neurobiology. 1. Anatomy and connectivity. *Brain Res* **434**:117–165.

Ongur, D. and Price, J. L. (2000) The organization of networks within the orbital and medial prefrontal cortex of rats, monkeys and humans. *Cereb Cortex* **10**:206–219.

Papanicolaou, A. C., Simos, P. G., Breier, J. I., Fletcher, J. M., Foorman, B. R., Francis, D. *et al.* (2003) Brain mechanisms for reading in children with and without dyslexia: a review of studies of normal development and plasticity. *Dev Neuropsychol* **24**:593–612.

Petry, N. M. (2003) Discounting of money, health, and freedom in substance abusers and controls. *Drug Alcohol Depend* **71**:133–141.

Petry, N. M., Bickel, W. K., and Arnett, M. (1998) Shortened time horizons and insensitivity to future consequences in heroin addicts. *Addiction* **93**:729–738.

Pfefferbaum, A., Sullivan, E. V., Mathalon, D. H., and Lim, K. O. (1997) Frontal lobe volume loss observed with magnetic resonance imaging in older chronic alcoholics. *Alcohol Clin Exp Res* **21**:521–529.

Phelps, E. A., Delgado, M. R., Nearing, K. I., and LeDoux, J. E. (2004) Extinction learning in humans: role of the amygdala and vmPFC. *Neuron* **43**:897–905.

Ploghaus, A., Tracey, I., Gati, J. S., Clare, S., Menon, R. S., Matthews, P. M. *et al.* (1999) Dissociating pain from its anticipation in the human brain. *Science* **284**:1979–1981.

Pontieri, F. E., Tanda, G., Orzi, F., and Di Chiara, G. (1996) Effects of nicotine on the nucleus accumbens and similarity to those of addictive drugs. *Nature* **382**:255–257.

Pontieri, F. E., Passarelli, F., Calo, L., and Caronti, B. (1998) Functional correlates of nicotine administration: similarity with drugs of abuse. *J Mol Med* **76**:193–201.

Potenza, M. N., Leung, H. C., Blumberg, H. P., Peterson, B. S., Fulbright, R. K., Lacadie, C. M. *et al.* (2003) An FMRI Stroop task study of ventromedial prefrontal cortical function in pathological gamblers. *Am J Psychiatry* **160**:1990–1994.

Raine, A., Lencz, T., Bihrle, S., LaCasse, L., and Colletti, P. (2000) Reduced prefrontal gray matter volume and reduced autonomic activity in antisocial personality disorder. *Arch Gen Psychiatry* **57**:119–127.

Ridderinkhof, K. R., Ullsperger, M., Crone, E. A., and Nieuwenhuis, S. (2004) The role of the medial frontal cortex in cognitive control. *Science* **306**:443–447.

Risinger, R. C., Salmeron, B. J., Ross, T. J., Amen, S. L., Sanfilipo, M., Hoffmann, R. G. *et al.* (2005) Neural correlates of high and craving during cocaine self-administration using BOLD fMRI. *Neuroimage* **26**:1097–1108.

Robinson, T. E. (2004) Neuroscience. Addicted rats. *Science* **305**:951–953.

Robinson, T. E. and Berridge, K. C. (1993) The neural basis of drug craving: an incentive-sensitization theory of addiction. *Brain Res Brain Res Rev* **18**:247–291.

Robinson, T. E. and Berridge, K. C. (2001) Incentive-sensitization and addiction. *Addiction* **96**:103–114.

Robinson, T. E. and Berridge, K. C. (2003) Addiction. *Annu Rev Psychol* **54**:25–53.

Rogers, R. D., Owen, A. M., Middleton, H. C., Williams, E. J., Pickard, J. D., Sahakian, B. J. *et al.* (1999) Choosing between small, likely rewards and large, unlikely rewards activates inferior and orbital prefrontal cortex. *J Neurosci* **19**:9029–9038.

Rogers, R. D. and Robbins, T. W. (2001) Investigating the neurocognitive deficits associated with chronic drug misuse. *Curr Opin Neuro Biol* **11**:250–257.

Rolls, E. T. and Baylis, L. L. (1994) Gustatory, olfactory, and visual convergence within the primate orbitofrontal cortex. *J Neurosci* **14**:5437–5452.

Rolls, E. T., Hornak, J., Wade, D., and McGrath, J. (1994) Emotion-related learning in patients with social and emotional changes associated with frontal lobe damage. *J Neurol Neurosurg Psychiatry* **57**:1518–1524.

Rose, J. E., Behm, F. M., Westman, E. C., Mathew, R. J., London, E. D., Hawk, T. C. *et al.* (2003) PET studies of the influences of nicotine on neural systems in cigarette smokers. *Am J Psychiatry* **160**:323–333.

Rubia, K., Smith, A. B., Brammer, M. J., and Taylor, E. (2003) Right inferior prefrontal cortex mediates response inhibition while mesial prefrontal cortex is responsible for error detection. *Neuroimage* **20**:351–358.

Satoh, T., Nakai, S., Sato, T., and Kimura, M. (2003) Correlated coding of motivation and outcome of decision by dopamine neurons. *J Neurosci* **23**:9913–9923.

Saxena, S., Brody, A. L., Schwartz, J. M., and Baxter, L. R. (1998) Neuroimaging and frontal-subcortical circuitry in obsessive-compulsive disorder. *Br J Psychiatry Suppl*: 26–37.

Schill, T. and Wang, S. (1990) Correlates of the MMPI-2 anger content scale. *Psychol Rep* **67**:800–802.

Schultz, W., Tremblay, L., and Hollerman, J. R. (2000) Reward processing in primate orbitofrontal cortex and basal ganglia. *Cereb Cortex* **10**:272–284.

Seguin, J. R. (2004) Neurocognitive elements of antisocial behavior: Relevance of an orbitofrontal cortex account. *Brain Cogn* **55**:185–197.

Services, U. S. D. o. H. a. and N. I. o. D. Abuse (2004) http://www.drugabuse.gov/Infofax/HSYouthtrends.html. 2004.

Shalev, U., Grimm, J. W., and Shaham, Y. (2002) Neurobiology of relapse to heroin and cocaine seeking: a review. *Pharmacol Rev* **54**:1–42.

Shearer, J., Wodak, A., van, B. I., Mattick, R. P., and Lewis, J. (2003) Pilot randomized double blind placebo-controlled study of dexamphetamine for cocaine dependence. *Addiction* **98**:1137–1141.

Shin, L. M., Whalen, P. J., Pitman, R. K., Bush, G., Macklin, M. L., Lasko, N. B. *et al.* (2001) An fMRI study of anterior cingulate function in posttraumatic stress disorder. *Biol Psychiatry* **50**:932–942.

Solomon, R. L. and Corbit, J. D. (1973) An opponent-process theory of motivation. II. Cigarette addiction. *J Abnorm Psychol* **81**:158–171.

Steegmann, A. T. (1962) Dr. Harlow's famous case: the "impossible" accident of Phineas P. Gage. *Surgery* **52**:952–958.

Stein, E. A., Pankiewicz, J., Harsch, H. H., Cho, J. K., Fuller, S. A., Hoffmann, R. G. *et al.* (1998) Nicotine-induced limbic cortical activation in the human brain: a functional MRI study. *Am J Psychiatry* **155**:1009–1015.

Stern, R. A. and Prohaska, M. L. (1996) Neuropsychological Evaluation of Executive Functioning. *American Psychiatric Press Review of Psychiatry*. In: L. J. Dickenstein, M. B. Riba, and J. M. Oldham(eds). Washington, DC:American Psychiatric Press. 15: 243–266.

Stewart, J. and Wise, R. A. (1992) Reinstatement of heroin self-administration habits: morphine prompts and naltrexone discourages renewed responding after extinction. *Psychopharmacology (Berl)* **108**:79–84.

Strassberg, D. S. and Russell, S. W. (2000) MMPI-2 content scale validity within a sample of chronic pain patients. *Journal of Pyschopathology and Behavioural Assessment* **22**:47–60.

Stroop, J. R. (1935) Studies of interference in serial verbal reactions. *Journal of Experimental Psychology* **18**:643–662.

Stuss, D. T. and Alexander, M. P. (2000) Executive functions and the frontal lobes: a conceptual view. *Psychol Res* **63**:289–298.

Tellegen, A. and Waller, N. G. (1997) Exploring personality through test construction: development of the multidimensional personality questionnaire. *Personality measures: development and evaluation*. S. R. Briggs and J. M. Cheek (eds). Greenwich: JAI Press. 1.

Thompson, P. M., Hayashi, K. M., Simon, S. L., Geaga, J. A., Hong, M. S., Sui, Y. *et al.* (2004) Structural abnormalities in the brains of human subjects who use methamphetamine. *J Neurosci* **24**:6028–6036.

Thorpe, S. J., Rolls, E. T., and Maddison, S. (1983) The orbitofrontal cortex: neuronal activity in the behaving monkey. *Exp Brain Res* **49**:93–115.

Tiffany, S. T. (1990) A cognitive model of drug urges and drug-use behavior: role of automatic and nonautomatic processes. *Psychol Rev* **97**:147–168.

Tiihonen, J., Kuikka, J., Hakola, P., Paanila, J., Airaksinen, J., Eronen, M. *et al.* (1994) Acute ethanol-induced changes in cerebral blood flow. *Am J Psychiatry* **151**:1505–1508.

Tremblay, L. and Schultz, W. (1999) Relative reward preference in primate orbitofrontal cortex. *Nature* **398**:704–708.

Turner, J. (1994) *Cardiovascular reactivity and stress: Patterns of physiological response.* New York: Plenum Press.

Vaidya, C. J., Austin, G., Kirkorian, G., Ridlehuber, H. W., Desmond, J. E., Glover, G. H. *et al.* (1998) Selective effects of methylphenidate in attention deficit hyperactivity disorder: a functional magnetic resonance study. *Proc Natl Acad Sci USA* **95**:14494–14499.

Vanderschuren, L. J. and Everitt, B. J. (2004) Drug seeking becomes compulsive after prolonged cocaine self-administration. *Science* **305**:1017–1019.

Volkow, N. D., Valentine, A., and Kulkarni, M. (1988) Radiological and neurological changes in the drug abuse patient: a study with MRI. *J Neuroradiol* **15**:288–293.

Volkow, N. D., Mullani, N., Gould, K. L., Adler, S., and Krajewski, K. (1988) Cerebral blood flow in chronic cocaine users: a study with positron emission tomography. *Br J Psychiatry* **152**:641–648.

Volkow, N. D., Mullani, N., Gould, L., Adler, S. S., Guynn, R. W., Overall, J. E. *et al.* (1988) Effects of acute alcohol intoxication on cerebral blood flow measured with PET. *Psychiatry Res* **24**:201–209.

Volkow, N. D. and Fowler, J. S. (2000) Addiction, a disease of compulsion and drive: involvement of the orbitofrontal cortex. *Cereb Cortex* **10**:318–325.

Volkow, N. D., Chang, L., Wang, G. J., Fowler, J. S., Ding, Y. S., Sedler, M. *et al.* (2001) Low level of brain dopamine D2 receptors in methamphetamine abusers: association with metabolism in the orbitofrontal cortex. *Am J Psychiatry* **158**:2015–2021.

Volkow, N. D., Fowler, J. S., Wang, G. J., Ding, Y. S., and Gatley, S. J. (2002) Role of dopamine in the therapeutic and reinforcing effects of methylphenidate in humans: results from imaging studies. *Eur Neuropsycho Pharmacol* **12**:557–566.

Volkow, N. D., Fowler, J. S., Wang, G. J., and Goldstein, R. Z. (2002) Role of dopamine, the frontal cortex and memory circuits in drug addiction: insight from imaging studies. *Neuro BiolLearn Mem* **78**:610–624.

Volkow, N. D., Fowler, J. S., and Wang, G. J. (2003) The addicted human brain: insights from imaging studies. *J Clin Invest* **111**:1444–1451.

Volkow, N. D., Fowler, J. S., and Wang, G. J. (2004) The addicted human brain viewed in the light of imaging studies: brain circuits and treatment strategies. *Neuropharmacology* 47 Suppl **1**: 3–13.

Volkow, N. D., Fowler, J. S., Wolf, A. P., Schlyer, D., Shiue, C. Y., Alpert, R. *et al.* (1990) Effects of chronic cocaine abuse on postsynaptic dopamine receptors. *Am J Psychiatry* **147**: 719–724.

Volkow, N. D., Hitzemann, R., Wolf, A. P., Logan, J., Fowler, J. S., Christman, D. *et al.* (1990) Acute effects of ethanol on regional brain glucose metabolism and transport. *Psychiatry Res* **35**:39–48.

Volkow, N. D., Fowler, J. S., Wolf, A. P., Hitzemann, R., Dewey, S., Bendriem, B. *et al.* (1991) Changes in brain glucose metabolism in cocaine dependence and withdrawal. *Am J Psychiatry* **148**:621–626.

Volkow, N. D., Hitzemann, R., Wang, G. J., Fowler, J. S., Burr, G., Pascani, K. *et al.* (1992) Decreased brain metabolism in neurologically intact healthy alcoholics. *Am J Psychiatry* **149**:1016–1022.

Volkow, N. D., Hitzemann, R., Wang, G. J., Fowler, J. S., Wolf, A. P., Dewey, S. L. et al. (1992) Long-term frontal brain metabolic changes in cocaine abusers. *Synapse* **11**:184–190.

Volkow, N. D., Fowler, J. S., Wang, G. J., Hitzemann, R., Logan, J., Schlyer, D. J. et al. (1993) Decreased dopamine D2 receptor availability is associated with reduced frontal metabolism in cocaine abusers. *Synapse* **14**:169–177.

Volkow, N. D., Wang, G. J., Hitzemann, R., Fowler, J. S., Wolf, A. P., Pappas, N. et al. (1993) Decreased cerebral response to inhibitory neurotransmission in alcoholics. *Am J Psychiatry* **150**:417–422.

Volkow, N. D., Wang, G. J., Hitzemann, R., Fowler, J. S., Overall, J. E., Burr, G. et al. (1994) Recovery of brain glucose metabolism in detoxified alcoholics. *Am J Psychiatry* **151**:178–183.

Volkow, N. D., Gillespie, H., Mullani, N., Tancredi, L., Grant, C., Valentine, A. et al. (1996) Brain glucose metabolism in chronic marijuana users at baseline and during marijuana intoxication. *Psychiatry Res* **67**:29–38.

Volkow, N. D., Wang, G. J., Fowler, J. S., Logan, J., Gatley, S. J., Hitzemann, R. et al. (1997) Decreased striatal dopaminergic responsiveness in detoxified cocaine-dependent subjects. *Nature* **386**:830–833.

Volkow, N. D., Wang, G. J., Overall, J. E., Hitzemann, R., Fowler, J. S., Pappas, N. et al. (1997) Regional brain metabolic response to lorazepam in alcoholics during early and late alcohol detoxification. *Alcohol Clin Exp Res* **21**:1278–1284.

Volkow, N. D., Wang, G. J., Fowler, J. S., Hitzemann, R., Angrist, B., Gatley, S. J. et al. (1999) Association of methylphenidate-induced craving with changes in right striato-orbitofrontal metabolism in cocaine abusers: implications in addiction. *Am J Psychiatry* **156**:19–26.

Volkow, N. D., Wang, G. J., Fowler, J. S., Logan, J., Gatley, S. J., Gifford, A. et al. (1999) Prediction of reinforcing responses to psychostimulants in humans by brain dopamine D2 receptor levels. *Am J Psychiatry* **156**:1440–1443.

Volkow, N. D., Wang, G. J., Ma, Y., Fowler, J. S., Zhu, W., Maynard, L. et al. (2003) Expectation enhances the regional brain metabolic and the reinforcing effects of stimulants in cocaine abusers. *J Neurosci* **23**:11461–11468.

Volkow, N. D., Fowler, J. S., Wang, G. J., and Swanson, J. M. (2004) Dopamine in drug abuse and addiction: results from imaging studies and treatment implications. *Mol Psychiatry* **9**:557–569.

Volkow, N. D., Wang, G. J., Ma, Y., Fowler, J. S., Wong, C., Ding, Y. S., et al. (2005) Activation of orbital and medial prefrontal cortex by methylphenidate in cocaine addicted subjects but not in controls: relevance to addiction. *Journal of Neuroscience* **25**:3932–3939

Wallace, E. A., Wisniewski, G., Zubal, G., vanDyck, C. H., Pfau, S. E., Smith, E. O. et al. (1996) Acute cocaine effects on absolute cerebral blood flow. *Psychopharmacology (Berl)* **128**:17–20.

Wallis, J. D. and Miller, E. K. (2003) Neuronal activity in primate dorsolateral and orbital prefrontal cortex during performance of a reward preference task. *Eur J Neurosci* **18**:2069–2081.

Wang, G. J., Volkow, N. D., Fowler, J. S., Cervany, P., Hitzemann, R. J., Pappas, N. R. et al. (1999) Regional brain metabolic activation during craving elicited by recall of previous drug experiences. *Life Sci* **64**:775–784.

Wang, G. J., Volkow, N. D., Thanos, P. K., and Fowler, J. S. (2004) Similarity between obesity and drug addiction as assessed by neurofunctional imaging: a concept review. *J Addict Dis* **23**:39–53.

Washton, A. M. and Stone-Washton, N. (1990) Abstinence and relapse in outpatient cocaine addicts. *J Psychoactive Drugs* **22**:135–147.

Wexler, B. E., Gottschalk, C. H., Fulbright, R. K., Prohovnik, I., Lacadie, C. M., Rounsaville, B. J. et al. (2001) Functional magnetic resonance imaging of cocaine craving. *Am J Psychiatry* **158**:86–95.

White, N. M. (1996) Addictive drugs as reinforcers: multiple partial actions on memory systems. *Addiction* **91**:921–949.

Woermann, F. G., van Elst, L. T., Koepp, M. J., Free, S. L., Thompson, P. J., Trimble, M. R. *et al.* (2000) Reduction of frontal neocortical gray matter associated with affective aggression in patients with temporal lobe epilepsy: an objective voxel by voxel analysis of automatically segmented MRI. *J Neurol Neurosurg Psychiatry* **68**:162–169.

Zald, D. H. and Kim, S. W. (1996) Anatomy and function of the orbital frontal cortex, II: Function and relevance to obsessive-compulsive disorder. *J Neuropsychiatry Clin Neurosci* **8**:249–261.

Chapter 20

# The orbitofrontal cortex and anxiety disorders

Mohammed R. Milad and Scott L. Rauch

## 20.1. Introduction

In this chapter, we discuss evolving neurocircuitry models of anxiety disorders. Though much of the data comes from human neuroimaging experiments, we seek to integrate results from other modes of inquiry as well, including animal studies and human psychophysiological research. In so doing, we highlight the commonalities and differences among anxiety disorders with respect to clinical presentation as well as purported mediating anatomy. Though prior such reviews have typically been amygdalocentric, we focus here on the role of the orbitofrontal cortex (OFC), while duly considering the complex interplay among various interconnected brain regions. Further, we discuss schemes for conceptualizing the distinct functions of OFC sub-territories and their respective contributions to the etiology, pathophysiology and treatment of various anxiety disorders.

## 20.2. Clinical characteristics of anxiety disorders

Anxiety disorders represent a major category of psychiatric syndromes (American Psychiatric Association 1994), and are characterized by exaggerated anxiety and inappropriate fear responses to relatively innocuous stimuli. The etiology of anxiety disorders remains mostly unclear with the exception of *post-traumatic stress disorder (PTSD)*. In PTSD, individuals develop a constellation of cardinal symptoms in the aftermath of a severe emotionally traumatic event. These include: re-experiencing phenomena (e.g., flashbacks—which can occur spontaneously or in response to reminders of the traumatic event), avoidance (e.g., avoiding situations that remind the individual of the traumatic event), and hyperarousal (e.g., exaggerated startle response). It has been hypothesized that hyper-conditionability (the ability to show abnormally-strong conditioned responses after conditioning) (Orr et al. 2000, 2002) along with failure to extinguish conditioned fear responses in PTSD may underlie the chronicity of this condition, as well as the emotional perseveration commonly observed in PTSD patients (Rauch et al. 1998; Pitman et al. 2001; Rothbaum and Davis 2003). Thus, the fear conditioning paradigm has been considered a valuable model for developing and testing hypotheses about the pathophysiology of this disorder, as well as anxiety disorders more generally (Charney 2004; Milad et al. 2005; Dolan et al. Chapter 10, this volume).

Panic disorder (PD) is characterized by recurrent panic episodes, typically occurring spontaneously, without overt precipitants. Panic attacks entail a rapid escalation to extreme anxiety, accompanied by physical symptoms, such as rapid breathing, palpitations, sweating, and dizziness, as well as emotional and cognitive symptoms, such as the feeling that something catastrophic is about to happen, or that immediate escape is necessary. Interestingly, though the experience of having an isolated panic attack is very common, it is only a fraction of individuals that go on to develop PD. Individuals with PD typically develop anticipatory anxiety, repeated panic episodes, and avoidance of places or situations where they believe panic attacks are more likely to occur (including where such attacks have occurred in the past), or where escape may be difficult (e.g., crowded or confined places). In this context, it is interesting to consider how the evolution from a panic attack to PD resembles PTSD; where the emotionally traumatic event in PD is the initial panic attack(s). To date, models of PD have tended to focus on the neural substrates that confer risk for or mediate a panic attack. However, we submit that the mechanisms that underlie the progression from panic attack to PD may be equally pertinent; we propose here that these may be similar or even identical to the factors that confer risk for developing PTSD.

Phobias are characterized by reliable, exaggerated anxiety responses to innocuous stimuli or situations. Such phobic responses include both the psychic experience of anxiety and peripheral manifestations of anxiety as well as the tendency to avoid the feared object or situation. Social phobia (also known as social anxiety disorder; SAD) entails phobic responses to any of a range of different social situations (e.g., from more circumscribed abnormal performance anxiety, to anxiety about more generalized exposure to social situations). Specific phobias (SP) entail phobic responses to any of various particular stimuli or situations (e.g., small animals, heights, water, enclosed spaces). Though there is some appeal to models based on learned fear for these disorders (as for PTSD), existing data about phobias do not provide support for the hypothesis that these are typically learned phenomena. On the contrary, current evidence suggests that there may be a strong familial component to such conditions, perhaps based on inherited aspects of temperament (Rosenbaum *et al.* 1991; Kendler *et al.* 2002). The critical phenomenologic features across the phobias are that: A) the cause of fear reactions are generally quite circumscribed and specific; and B) the objects of classic phobias tend to be representations of archetypal stimuli or settings that do actually pose some degree of threat, at least from an evolutionary perspective, when in an extreme form, or in a particular setting.

Obsessive compulsive disorder (OCD) is characterized by intrusive, unwanted thoughts (i.e., obsessions) and ritualized, repetitive behaviors (i.e., compulsions). In OCD, the obsessions are typically accompanied by anxiety that drives the compulsions; the compulsions are performed to neutralize the obsessions and attendant anxiety. Interestingly, OCD is the one anxiety disorder, which does not actually include anxiety as a defining feature. Practically, the sine qua non of this disorder is intrusive bothersome thoughts, and a significant subgroup of patients with OCD do not suffer from pathological anxiety per se.

Given that patients with anxiety disorders generally either suffer exaggerated fear responses to relatively innocuous stimuli (e.g., phobias) or inappropriate learned fear responses to reminders of past emotionally evocative events (e.g., PTSD and PD) (Grillon et al. 1994; Pole et al. 2003), contemporary models of these conditions have, to varying degrees, emphasized dysfunction within the amygdala and/or failure of cortical structures to provide top-down governance over the output of the amygdala (Rauch et al. 1998; Quirk 2004; Milad et al. 2005).

## 20.3. Relevant functions and functional anatomy of the OFC

As detailed elsewhere in this volume, the OFC subserves a wide variety of functions, in concert with other brain regions via their connections. It has been challenging to parse out the unique contributions of the OFC. However, in general, the OFC is believed to play a central role in performance monitoring and detection of performance errors, decision-making, and representation as well as anticipation of reward (Elliott et al. 2000), evaluating the impact of negative outcomes (Ursu and Carter 2005), and mediating autonomic responses (Critchley et al. 2000) (see Naqvi et al. Chapter 13, this volume). Given this wide range of functions, attempts have been made to develop schemes for topographically dissecting the OFC into its corresponding functional subterritories. The prevailing schemes have emphasized the functional distinctions between the medial and lateral OFC (mOFC, lOFC) (Elliott et al. 2000). For instance, Kringelbach and Rolls (2004) proposed that the mOFC mediates monitoring reward value, whereas the lOFC plays a role in evaluating punishers to alter behavior. Similarly, Markowitsch et al. (2003) showed that the mOFC is preferentially activated during the recall of positive affects, whereas the lOFC is preferentially activated during recall of negative affects. Extending this theme, Ursu and Carter (2005) suggested that the lOFC is responsible for evaluating the impact of negative outcomes, whereas the mOFC represents the impact of positive outcomes. Elliott et al. (2000) have suggested that the mOFC is involved in the selection of responses on the basis of their reward value whereas the lOFC is required when the response selected requires suppression of previous rewarded responses. Fig. 20.1 represents some of these distinct functions of the lateral and medial OFC in the human brain.

Perhaps most germane to anxiety disorders, the OFC, as well as the ventromedial prefrontal cortex (VMPFC), have also been implicated in the extinction of conditioned fear responding (see Dolan et al. Chapter 10, this volume). Cumulating evidence from rodents has shown that the mOFC is involved in the extinction of conditioned responding. For example, Butter et al (1963) found the lesions to the mOFC in monkeys retarded extinction of operant conditioned responses. Compulsive-like behavior in bar pressing for food was also observed in rats with lesions of the mOFC (Joel et al. 2005). Several other animal studies have directly implicated the VMPFC in rats in the recall of extinction memory (Milad and Quirk 2002; Morgan et al. 2003; Herry and Mons 2004; Lebron et al. 2004; Santini et al. 2004). For example, in rats, neuronal firing in the VMPFC is

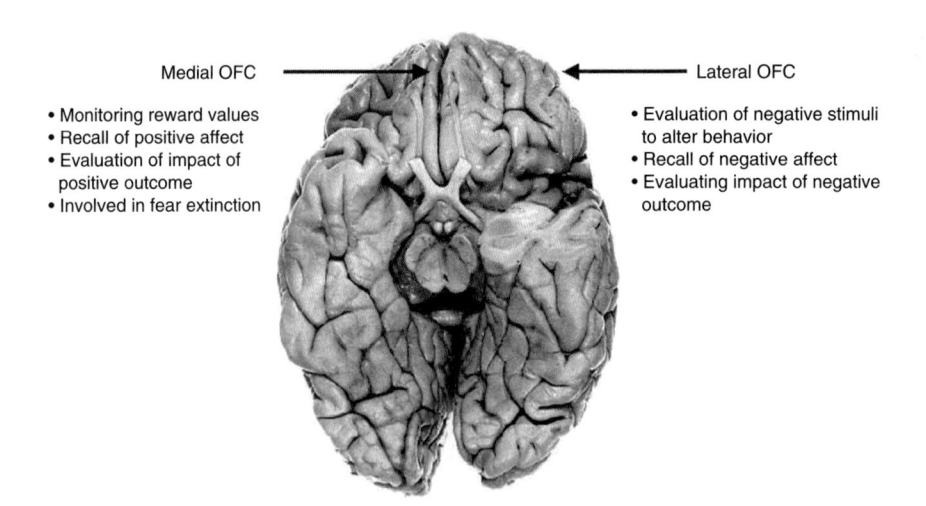

**Fig. 20.1** Representation of the various functions of the lateral and medial orbitofrontal cortex.

Medial OFC
- Monitoring reward values
- Recall of positive affect
- Evaluation of impact of positive outcome
- Involved in fear extinction

Lateral OFC
- Evaluation of negative stimuli to alter behavior
- Recall of negative affect
- Evaluating impact of negative outcome

increased during extinction recall, and inversely correlated with freezing behavior (Milad and Quirk 2002). This inverse correlation between VMPFC activity and conditioned fear responses has also been observed in several other animal studies (Herry and Garcia 2002; Barrett *et al*. 2003). Additional recent animal studies also show that protein synthesis and other cellular changes in the VMPFC are necessary during extinction training and recall (Herry and Mons 2004; Santini *et al*. 2004; Baretta *et al*. 2005).

The role of the human OFC in extinction is becoming increasingly evident. Recently, it has been shown that the mOFC is directly implicated in the extinction of olfactory aversive conditioning (Gottfried and Dolan 2004). Reversal of conditioning contingency during classical conditioning (i.e., stimulus 'X' previously paired with shock is now safe, whereas stimulus 'Y' previously safe is now paired with a shock) has been shown to activate the OFC (Morris and Dolan 2004). Activity of the VMPFC has recently been shown to be correlated with extinction recall (Phelps *et al*. 2004). Specifically, Phelps *et al*. (2004) showed that during extinction recall, the activity of the VMPFC is inversely correlated with conditioned fear responses as measured by skin conductance. In parallel, we have recently shown that thickness of the mOFC was also directly correlated with skin conductance responses during extinction recall (Milad *et al*. 2005). We found that subjects with the thickest mOFC showed the strongest memory for extinction (see Fig. 20.2). These recent human neuroimaging data strongly support the role of the mOFC and adjacent VMPFC in fear inhibition.

The amygdala and the hippocampus are two key structures that appear to interact with the OFC to modulate emotional learning and memory. The role of the amygdala in mediating 'appropriate' fear responses and the modulation of emotional memory consolidation is well-documented in the literature (e.g. Cahill *et al*. 1994; Davis and Whalen 2001; Maren and Quirk 2004; Rodrigues *et al*. 2004). The amygdala is clearly

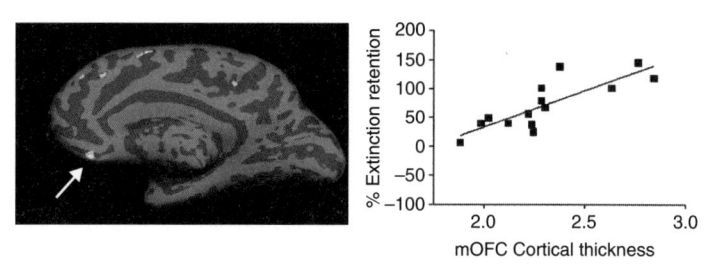

**Fig. 20.2** (Also see Color Plate 30.) Regions with positive correlations between cortical thickness and extinction retention (indicated by arrow) and regression plots for the correlations between percent extinction retention and cortical thickness in the mOFC. Threshold is set at $p < 0.01$ (light grey) to $p < 0.001$ (white). Adapted with permission from Milad et al. (2005).

implicated in 'tagging' stimuli encountered in our daily lives with emotional significance, much like forming an association between a tone and an electric shock in rats undergoing Pavlovian conditioning. In this way, the amygdala engraves our memories of emotional events in our synapses (McGaugh 2000; McGaugh and Roozendaal 2002). In addition to forming associations between conditioned and unconditioned stimuli during learning, the amygdala, through its connections with various brainstem structures as well as the hypothalamus, is critical for mediating the conditioned responses commonly seen during learning, such as changes in heart rate, blood pressure, and freezing (for reviews, see Ledoux 2000; Davis and Whalen 2001; Maren and Quirk 2004). The hippocampus appears to play critical role in the contextual modulation of memories. For example, several studies have shown that the hippocampus is involved in contextual conditioning as well as in gating the expression of extinction memory via its anatomical connections with the mOFC (Corcoran and Maren 2004; Maren and Holt 2004).

While the anatomical connections of the OFC are described in detail elsewhere in this volume (see Price and Barbas et al., this volume), we briefly highlight the connections between the OFC and other cortical and subcortical regions that correspond to and support the topographic functional scheme of the OFC described above. For example, the mOFC projects massively to the ventral striatum (Ferry et al. 2000) and such projections are thought to subserve the modulation of behaviors related to reward. Furthermore, the mOFC projects to the amygdala (Ongur and Price 2000; Barbas et al. 2003) as well as to the lateral hypothalamus (Floyd et al. 2000), the periacquaductal gray (PAG) (Floyd et al. 2000), and the hippocampus (Cavada et al. 2000). These projections are thought to mediate and/ or influence emotional expression (Barbas et al. 2003; Gusnard et al. 2003). Specifically with respect to fear conditioning and extinction, the mOFC along with VMPFC regions may be involved in inhibiting conditioned fear during extinction via their connections to the amygdala and to the PAG (Quirk et al. 2003). The lateral OFC, on the other hand, projects to more central and dorsal regions of the striatum (Ferry *et al.* 2000) which may influence behaviors in response to punishment, as well as automated or ritualized behaviors evolved to escape or otherwise mitigate danger.

## 20.4. Current hypotheses regarding pathophysiology of anxiety disorders

The symptomatology of anxiety disorders, together with a basic knowledge of functional neuroanatomy, have led investigators to various neurocircuitry models of pathophysiology. For example, for PTSD, the prevailing model hypothesizes hyper-responsivity within the amygdala to threat-related stimuli, with inadequate top-down governance over the amygdala and hippocampus by the medial prefrontal cortex (including mOFC and rostral anterior cingulate cortex (rACC)) (e.g., see Milad *et al.* 2005).

With respect to PD, several leading theories, such as the false suffocation alarm model (Klein 1993), have focused on a decreased threshold for panic attacks. For instance, one hypothesis holds that panic attacks evolve in the context of what should be minor anxiety episodes, because of failures in the systems responsible for limiting such normal responses—similar to the aforementioned model of PTSD (Rauch *et al.* 2000). Alternatively, panic episodes could reflect anxiety responses to stimuli that are not processed at the conscious (i.e., explicit) level, but instead, recruit anxiety circuitry without awareness (i.e., implicitly). In fact, there is strong evidence that the amygdala can be recruited into action in the absence of awareness that a threat-related stimulus has been presented (Whalen *et al.* 1998). This model might characterize PD by fundamental amygdala hyper-responsivity to subtle environmental cues, triggering full-scale threat-related responses in the absence of conscious awareness of the initial triggers. Regardless of the etiology of initial attacks, however, we propose that the basis for the evolution from panic attacks to PD parallels the pathogenesis of PTSD, as detailed above.

For phobias, given that the stimuli tend to be more circumscribed and of a characteristic nature, the clinical features may reflect hypersensitivity in the pathways that mediate evolutionarily honed fear responses to prepared stimuli[1]. This would most likely involve exaggerated activity of the amygdala and sensory or interoceptive regions, as well as possible dysfunction within the frontal cortex (Aouizerate *et al.* 2004b). Further, such dysfunction might only be evident during exposure to specific phobic stimuli.

As noted, OCD appears to be distinct from the other anxiety disorders in several respects. In general, pathophysiological models of this disorder have focused on cortico-striato-thalamo-cortical circuitry (Baxter, Jr. *et al.* 1996; Saxena *et al.* 2001; Cannistraro and Rauch 2003). One hypothesis suggests that primary pathology within the striatum (specifically, the caudate nucleus) leads to inefficient gating at the level of the thalamus, which results in hyperactivity within the OFC (corresponding to the intrusive thoughts) and hyperactivity within anterior cingulate cortex (corresponding to anxiety, in a non-specific manner) (Zald and Kim 1996a, b; Graybiel and Rauch 2000; Cannistraro and

---

[1] Prepared stimuli are objects or stimuli that evolution has made sources of innate fear (Ohman and Mineka 2001). Examples of such stimuli are snakes and spiders.

Rauch 2003; Aouizerate *et al.* 2004a). Further, it is noteworthy that OCD is a highly heterogeneous condition, such that its sub-types may be characterized by somewhat different pathophysiological profiles. To the extent that OCD patients can exhibit phobic fear of specific stimuli, such as perceived contaminants, the amygdala may be involved in this context.

Cognitive behavioral therapy is considered effective for essentially all anxiety disorders. Given that behavioral therapy in particular relies on extinction-based processes (Rothbaum and Davis 2003), this highlights the importance of VMPFC—amygdala interactions across all anxiety disorders. As with fear extinction in rats, during behavioral therapy patients with anxiety disorders are exposed to the anxiety-inducing stimuli in the absence of any negative reinforcement (Wald and Taylor 2003). After several sessions of exposure therapy, the majority of patients learn to inhibit their fear responses (Foa *et al.* 1991). However, some patients with anxiety disorders fail to respond to exposure therapy (Foa 2000; van Minnen and Hagenaars 2002), consistent with the hypothesis that extinction learning may be especially deficient in these treatment resistant patients. Thus, the OFC, as one element of the VMPFC, along with the amygdala are implicated across anxiety disorders, at least with respect to their role in extinction and extinction-based therapies. As we will now elaborate, a more thorough examination of the relevant data and the precise anatomy will reveal commonalities and differences among the disorders with respect to relevant neurocircuitry.

## 20.5. Excessive 'bottom-up' influence of the amygdala

The amygdala projects to various cortical regions including the VMPFC and OFC (McDonallads *et al.* 1996; Barbas 2000). These connections may provide a way for the amygdala to influence cortical responses to emotional events and stimuli. Shoenbaum and colleagues have shown that lesions of the amygdala disrupt OFC-dependent goal directed behaviors (Shoenbaum *et al.* 2003). Activation of the amygdala has been shown to modulate frontal cortical responses to aversive sensory stimuli (Dringenberg *et al.* 2001). Most relevant is the work of Garcia *et al.* (1999) showing that during fear conditioning, OFC activity is inhibited, and that such inhibition is completely abolished when the amygdala is lesioned, providing direct evidence of the amygdala's ability to inhibit the OFC. Amygdala modulation of prefrontal activity has also been observed in human neuroimaging studies. For instance, Kilpatrick and Cahill (2003) showed that during storage of emotional memory in healthy humans, the amygdala appears to modulate activity in the hippocampus and the ventrolateral prefrontal cortex.

Amygdala hyperactivity to threat-related stimuli has been hypothesized to play a key role in the pathophysiology of several anxiety disorders. Patients with PTSD exhibit exaggerated amygdala responses when viewing masked-fearful faces (Rauch *et al.* 2000a) and during the processing of auditory neutral stimuli (Semple *et al.* 2000) as well as during exposure to reminders of traumatic events (Shin *et al.* 1997; Liberzon *et al.* 1999; Rauch *et al.* 2000; Pissiota *et al.* 2002; Hendler *et al.* 2003; Shin *et al.* 2004). A recent fMRI study using words also showed exaggerated amygdala activation in PTSD patients, and further indicated that they failed to show normal rates of habituation relative to normal controls when exposed to negative stimuli (Protopopescu *et al.* 2005). Furthermore, Bremner *et al.* (2005) have recently shown abnormal activation in amygdala, as well as the rostral anterior cingulate cortex, during the acquisition and extinction of conditioned fear in PTSD.

With respect to PD, exaggerated amygdala responses have not been widely reported, though pertinent studies are few to date. Boshuisen *et al.* (2002) reported decreased amygdala responses during anticipatory anxiety in PD patients and smaller amygdala volumes have also been observed (Massana *et al.* 2003). However, the findings of exaggerated amygdala responses to masked-fearful vs. masked-happy faces noted in PTSD (Rauch *et al.* 2000) have now been ostensibly replicated for PD as well (Whalen, Rauch and colleagues, unpublished data), underscoring the similarities between these two disorders.

Phobias are characterized by exaggerated amygdala responses to the objects of the phobias, but not to general threat-related stimuli. For example, amygdala hyperactivation has been observed when SAD patients are exposed to social events that trigger anxiety responses such as in anticipation (Lorberbaum *et al.* 2004) or during delivery of public speeches (Tillfors *et al.* 2001). SAD subjects have also shown exaggerated amygdala responses to disorder relevant (angry or contemptuous) faces (Stein *et al.* 2002). In SP, amygdala hyperactivation was also observed when patients were exposed to phobia-related stimuli (Dilger *et al.* 2003). However, excessive amygdala responses were not observed when SP patients were exposed to disorder-irrelevant emotionally valenced human faces (Wright *et al.* 2003) or even when exposed to phobia-related words (Straube *et al.* 2004).

In OCD, few studies have shown exaggerated amygdala responses to disorder specific stimuli, and this may reflect amygdala hyper-responsivity limited to the sub-type of OCD that includes phobic elements (e.g., contamination fears). For example, van den Heuvel *et al.* (2004) exposed OCD patients with fear of contamination to pictures of dirty and clean surroundings. They found exaggerated amygdala activity in OCD patients in response to contamination stimuli. Similarly, exposure to highly provocative stimuli in OCD patients with fear of contamination resulted in amygdala activation, not seen in control subjects exposed to analogous stimuli (Breiter *et al.* 1996). Studies of OCD using general threat-related stimuli (e.g., faces) have not found amygdala hyperresponses (Shapira *et al.* 2003; Cannistraro *et al.* 2004); to the contrary, the amygdala appears to be generally hyporesponsive in OCD (e.g., Cannistraro *et al.* 2004). Likewise, among anxiety disorders, OCD is the only disorder where peripheral anxiety measures have been reported to be reduced. For example, while startle responses have been shown to be increased in OCD (Kumari *et al.* 2001), decreases in SCR and autonomic reactivity have been reported in children and adolescents with OCD (Zahn *et al.* 1996) as well as in adult OCD patients undergoing psychological stress (Hoehn-Saric *et al.* 1995).

## 20.6. Failure of 'top-down' control by the medial OFC

Given that anxiety disorders are generally accompanied by uncontrolled fear and inappropriate anxiety, it is not surprising that dysfunction in various subregions of the mPFC and the OFC have been noted for essentially all of the anxiety disorders. For example, the majority of studies comparing PTSD to non-PTSD control groups have consistently shown diminished mPFC activation. Specifically, symptom provocation studies indicate that when exposed to reminders of traumatic events, PTSD patients exhibit attenuated responses within mPFC areas (Shin *et al.* 1999, 2001, 2004; Lanius *et al.* 2001, 2002; Liberzon *et al.* 2003). Relative to non-PTSD subjects, PTSD patients also exhibited decreased rACC activation while performing emotional Stroop tasks (Bremner *et al.* 2004; Shin *et al.* 2001). Recently, Shin *et al.* (2004) showed that observed decreases in mPFC activity were inversely correlated with increases in amygdala activity in PTSD. Additional support for the involvement of prefrontal areas in PTSD comes from a recent morphometric MRI study (Rauch *et al.* 2003), where PTSD patients showed smaller rACC and subcallosal cortical volumes in comparison with non-PTSD controls. While many of the frontal structural and functional abnormalities in PTSD extend beyond the OFC, several studies have reported such findings that are referable to the mOFC regions as well as adjacent areas. For example, PTSD patients exhibit attenuated mOFC and subcallosal cortical activation during retrieval of emotionally valenced words (Bremner *et al.* 2003), and during exposure to reminders of traumatic events (Bremner *et al.* 1999a, b).

As for PD, a recent fMRI experiment in subjects with PD essentially replicated the earlier findings in PTSD with the emotional counting Stroop task (Whalen, Rauch and colleagues, unpublished data). Specifically, subjects with PD vs. matched healthy controls exhibited deficient rACC responses to panic-related vs. control words. Several other studies have used symptom provocation technique in which the panic episode was pharmacologically induced. Woods *et al.* (1988) employed SPECT and yohimbine infusions and found that, in comparison to the normal control subjects, PD subjects experienced increased anxiety and exhibited decreased rCBF in bilateral frontal cortex, though the precise anatomical subregions were not specified. Interestingly, a spontaneous panic attack was captured during PET acquisition in a single case; the data showed decreased rCBF in the right OFC, anterior cingulate (area 25), and anterior temporal cortex during the acute event (Fischer *et al.* 1998). Neutral state studies have found increased OFC activity at rest in PD, whereas successful treatment with imipramine was accompanied by an attenuation of this OFC hyperactivity (Nordahl *et al.* 1998). However, inverse correlations were observed between the regional glucose metabolism in OFC in the treated PD patients and their anxiety levels, suggesting again that the OFC may be recruited to dampen anxiety in a compensatory fashion. Nonetheless, additional investigations are needed to clarify the role of the OFC in PD.

OFC dysfunction has also been implicated in SAD as well as SP; such frontal cortical differences in phobias tend to emerge only in the setting of exposure to the phobic object.

For example, one PET symptom provocation study contrasting public vs. private speaking conditions found decreases in the OFC and insular cortex in the SAD vs. the non-phobic group (Tillfors *et al.* 2001). Another PET symptom provocation study recently reported deactivation of mOFC in SAD during exposure to phobia-inducing events (Van Ameringen *et al.* 2004). Analogous studies have found increased activation in the lOFC regions when spider phobic subjects viewed phobia-related pictures (Dilger *et al.* 2003).

## 20.7. Excessive lateral OFC activity in concert with the striatum

Hyperactivity in the OFC has been one of the most consistent findings from neuroimaging studies in OCD (Zald and Kim 1996a, b; Saxena *et al.* 1998; Graybiel and Rauch 2000; Saxena and Rauch 2000; Cannistraro and Rauch 2003). Pre/post-treatment studies have likewise indicated attenuation of abnormal regional brain activity within the OFC, anterior cingulate cortex, and caudate nucleus associated with successful treatment (Baxter, Jr. 1992; Swedo *et al.* 1992; Perani *et al.* 1995; Saxena *et al.* 2002). Moreover, similar changes have been observed for both pharmacological and behavioral therapies (Baxter, Jr. 1992; Brody *et al.* 1998). Treatment studies have also consistently indicated that the magnitude of OFC activity prior to pharmacotherapy predicts subsequent response to serotonergic reuptake inhibitors (Swedo *et al.* 1992; Brody *et al.* 1998; Saxena *et al.* 2002; Rauch *et al.* 2002); specifically, lesser magnitude of OFC hyperactivity is predictive of superior treatment response. It is important to note that most early studies of OCD used region of interest-based methods of data analysis that did not distinguish between medial vs. lateral subdivisions of the OFC.

Symptom provocation studies employing PET (McGuire *et al.* 1994; Rauch *et al.* 1994), as well as functional MRI (Breiter and Rauch 1996; Adler *et al.* 2000), have also most consistently shown increased brain activation within the OFC, anterior cingulate cortex, and caudate nucleus associated with the OCD symptomatic state. As these studies have employed voxelwise brain mapping methods of analysis, they have enabled investigators to localize the involvement of the OFC to anterior and/or lateral subdivisions. Further, in comparison with other groups, the anterolateral OFC and caudate activation appears to be somewhat specific to OCD, whereas activation of the mOFC and anterior cingulate cortex has been observed nonspecifically across other anxiety disorders and normal anxiety states (Rauch and Baxter 1998; Rauch *et al.* 1998). Note that in a symptom provocation study, Rauch *et al.* (1994) showed that whereas OCD symptom severity was directly correlated with anterolateral OFC activation, severity of anxiety in OCD was inversely correlated with posteromedial OFC activity (Fig. 20.3).

It has also been shown that activity within the OFC correlated directly with subsequent response to cognitive behavior therapy (Brody *et al.* 1998), consistent with the concept that magnitude of mOFC function may be associated with the capacity for patients to benefit from an extinction-based therapy.

There are numerous theories regarding the relationship between OFC hyperactivity and the phenomenology of OCD. For instance, Zald and Kim (1996b) proposed several

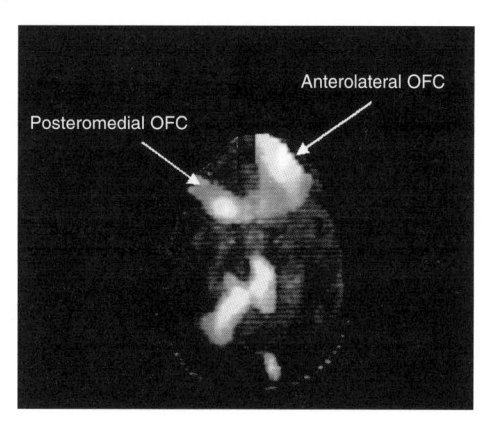

**Fig. 20.3** (Also see Color Plate 31.) A PET image showing increases in regional cerebral blood flow during symptom provocation (versus a control state) in subjects with OCD. Note that extracted values showed that OCD symptom severity was directly correlated with anterolateral OFC activation, where as severity of anxiety in OCD was inversely correlated with posteromedial OFC activation. Adapted with permission from Rauch et al. (1994).

domains of possible relationships in this regard including the following: (1) aversive representations—given that the OFC is involved in the processing of negative emotions, increased activity in OCD may therefore reflect increased anticipatory anxiety; (2) mnemonic characteristics—given that the OFC holds representations of value reinforcement in working memory, increased OFC activity in OCD may reflect heightened working memory of internally generated representations thus leading to obsessions; (3) erroneous error detection—since the OFC contains cells that are involved in signaling errors, increased OFC activity in OCD might mediate false error signals and cause acts aimed to correct those false errors; (4) and social-affiliative behavior—given that OFC lesions often result in pseudopsychopathy (see Chapter 23 by Koenigs and Tranel, this volume) with symptoms that include a lack of concern about the consequences that their actions have on others, which appears antithetical to the overconcern about the effects of their actions on others that often characterize patients with OCD. More recently, Elliott *et al.* (2000) suggested that the OFC is engaged in ambiguous situations where the solution requires taking into account the reward value of stimuli or responses; they proposed that familiarity and feeling of 'rightness' are included as types of reward-based decisions. The lateral OFC, in particular, is engaged when the action selected requires the suppression of previously rewarded responses (i.e., priority/response shifts).

Thus, we propose that the two extremes of OFC dysfunction are manifested in anxiety disorders. On the one hand, hypoactive mOFC is detected in disorders where there is a failure to inhibit inappropriate fear and anxiety responses (e.g., in PTSD). On the other hand, hyperactive lOFC is most commonly detected when there are constant anxiety-laden cognitions and drive to perform repetitive behavior (i.e., OCD).

**Table 20.1** Summary of the most commonly observed findings from neuroimaging studies regarding increased activation (+), decreased activation (−), both (±) or not prominently reported (ø) across the different anxiety disorders: Orbitofrontal cortex (OFC), Ventromedial prefrontal cortex (VMPFC). Results from psychophysiological studies measuring indices of arousal are also summarized.

| Region | Disorder | | | | |
|---|---|---|---|---|---|
|  | PTSD | PD | SP | SAD | OCD |
| Medial OFC/VMPFC | − | − | − | ø | + |
| Lateral OFC | ø | ø | ø | ø | + |
| Amygdala (specific threat) | + | + | + | + | + |
| Amygdala (general threat) | + | + | ø | ø | − |
| Hippocampus/PHP | − | +/− | ø | ø | + |
| Peripheral Indices of Anxiety | + | + | + | + | − |

## 20.8. Peripheral physiology and the OFC in anxiety disorders

Peripheral measures of anxiety are consistent with this scheme of OFC dysfunction, given the well-documented role of the OFC in influencing autonomic responses (Critchley *et al.* 2000; Patterson *et al.* 2002; Critchley *et al.* 2003). As reviewed above, whereas autonomic responses appear to be decreased in OCD patients, increased peripheral responses are reported across the rest of the anxiety disorders. For example, psychophysiological responses to fear memory imagery have been reported in PTSD, SP, SAD, and PD (Cuthbert *et al.* 2003). Fear-potentiated startle has been shown to increase in PD (Grillon *et al.* 1994) as well as PTSD (Shalev *et al.* 1997; Orr *et al.* 2003; Pole *et al.* 2003; Guthrie and Bryant 2005). Increased startle response along with reduced habituation and decreased pre-pulse inhibition has been reported in PD patients (Ludewig *et al.* 2002, 2005). Increased autonomic reactivity during $CO_2$ inhalation has been reported in PD (Sigmon *et al.* 2000). Psychological stress has been reported to increase heart rate and blood pressure and decrease SCR in PD patients (Hoehn-Saric *et al.* 1991). Thus, OCD appears to be associated with elevated OFC activity and reduced peripheral physiological responses, whereas other anxiety disorders are principally associated with deficient mOFC activity and exaggerated peripheral physiological responses (see Table 20.1).

## 20.9. The role of the OFC in extinction: implications for anxiety disorders

As we have reviewed: (1) the mOFC and VMPFC are involved in fear extinction in humans; (2) these prefrontal regions are dysfunctional in anxiety disorders; and (3) extinction forms the basis for exposure therapy commonly used to treat anxiety disorders. We propose that these three points provide a compelling rationale for testing the integrity of the OFC in anxiety disorders in the context of fear extinction and extinction recall paradigms. The neuroimaging studies reviewed above have undoubtedly provided enormously valuable information regarding the pattern of brain activity in patients with anxiety disorders. However, while most of the studies have employed neutral (e.g., resting) state, neurocognitive or symptom-provocation paradigms, direct tests of the functional integrity of the OFC in anxiety disorders during fear extinction remain to be conducted. Understanding the neural mechanisms of fear extinction may be fundamental to elucidating the pathophysiology of anxiety disorders; moreover such a line of inquiry could also enhance our understanding of the mechanism of action by which extinction-based treatments confer their therapeutic effects (e.g., see Rauch 2003; Milad et al. 2005).

## 20.10. Conclusion

Taken together, the data on anxiety disorders implicate amygdalo-cortical circuitry in the pathophysiology of PTSD, PD, SAD, and SP. For all of these disorders, it appears that the amygdala is hyper-responsive to threat-related stimuli, while key cortical regions, including the mOFC, are deficient in their top-down governance over the amygdala's response. One manifestation of this imbalance between amygdala and cortex, across these four anxiety disorders, is exaggerated peripheral physiological responses. This is in accord with the purported role of the mOFC in modulating autonomic functions including skin conductance. Despite the substantially similar features across these four anxiety disorders, there are also important distinguishing characteristics as well. For example, whereas the phobias (SAD and SP) manifest abnormal amygdalo-cortical responses only to the circumscribed disorder-specific stimuli that represent the objects of their phobic fears, PTSD and PD seem to exhibit abnormal amygdalo-cortical responses to general threat-related as well as disorder-specific stimuli.

OCD, in particular, seems quite distinct from the other four anxiety disorders reviewed. Firstly, OCD is characterized by pan-OFC hyperactivity. Hyperactivity of the mOFC, via its top-down influences, may be responsible for both the generalized hypoactivity of the amygdala and the reduced peripheral physiological responding in OCD. Interestingly, however, exaggerated amygdala activation has been observed in OCD, but only in response to provocative stimuli that target the phobia-like elements of OCD subtypes (e.g., contamination phobia). In contrast to the mOFC, which is purportedly essentially anxiolytic in its action, hyperactivity of the lOFC may more directly mediate the obsessions of OCD and the subjective experience of anxiety when present in this disorder. Thus, in OCD, current theories emphasize the role of lOFC-striato-thalamo-cortical circuitry, rather than the amygdala-mOFC pathway, in the pathophysiology of this disorder. Nonetheless, the findings on OCD underscore the importance of individual differences among subjects who share the same diagnosis; the magnitude of OFC hyperactivity was shown to predict subsequent response to behavioral (positive correlation) or medication (inverse correlation) treatment. Based on the above scheme, behavioral therapy response may be most closely related to mOFC function.

Extinction processes are germane to all of the anxiety disorders. Deficiencies in extinction function may play a role in the pathogenesis, maintenance, and treatment resistance of anxiety disorders. Conversely, extinction-based mechanisms are doubtless at the heart of behavioral therapies. Therefore, as the OFC (and perhaps, especially the mOFC) is implicated in human extinction retention and recall, understanding the neuroscience of this brain region is critical to advancing models of anxiety disorder etiology, pathophysiology and treatment.

As we reflect, anxiety disorders provide an apt vehicle for raising key themes regarding the OFC in humans. From this chapter it is evident that the OFC has meaningful functional subdivisions—with several schemes emphasizing distinctions between medial and lateral territories. These subregions of the OFC are distinguished by their connections, which in turn influence their participation in specified circuits as well as more widely distributed networks. In turn, the connections of the OFC subregions govern their functional impact, which then provide the substrates for disease through dysfunction. In the context of anxiety, the interactions between the mOFC and the amygdala as well as the lOFC and the striatum have been emphasized. Future research should help to advance our understanding of the OFC subregions and their respective functions. We foresee imminent progress with respect to the role of amygdalo-OFC circuits in extinction as well as OFC-striatal pathways in implicit information processing and automated behaviors.

## Acknowledgment

The authors of this chapter were supported in part by the Massachusetts General Hospital Tosteson Fellowship, and NIMH grants R21MH072156 and RO1MH60219.

## References

Adler, C. M., McDonough-Ryan, P., Sax, K. W., Holland, S. K., Arndt, S., and Strakowski S. M. (2000) fMRI of neuronal activation with symptom provocation in unmedicated patients with obsessive compulsive disorder. *J Psychiatr Res* **34**:317–324.

American Psychiatric Association (1994) *Diagnostic and Statistical Manual of Mental Disorders* 4th Edition. Washington, DC:APA.

Aouizerate, B., Guehl, D., Cuny, E., Rougier, A., Bioulac, B., Tignol, J., and Burbaud P. (2004a) Pathophysiology of obsessive-compulsive disorder:a necessary link between phenomenology, neuropsychology, imagery and physiology. *Prog Neurobiol* **72**:195–221.

Aouizerate, B., Martin-Guehl, C., and Tignol J. (2004b) Neurobiology and pharmacotherapy of social phobia. *Encephale* **30**:301–313.

Barbas, H. (2000) Connections underlying the synthesis of cognition, memory, and emotion in primate prefrontal cortices. *Brain Res Bull* **52**:319–330.

Barbas, H., Saha, S., Rempel-Clower, N., and Ghashghaei, T. (2003) Serial pathways from primate prefrontal cortex to autonomic areas may influence emotional expression. *BMC Neurosci* **4**:25.

Barrett, D., Shumake, J., Jones, D., and Gonzalez-Lima, F. (2003) Metabolic mapping of mouse brain activity after extinction of a conditioned emotional response. *J Neurosci* **23**:5740–5749.

Berretta, S., Pantazopoulos, H., Caldera, M., Pantazopoulos, P., and Pare, D. (2005) Infralimbic cortex activation increases c-Fos expression in intercalated neurons of the amygdala. *Neuroscience* **132**:943–953.

Baxter, L. R., Jr. (1992) Neuroimaging studies of obsessive compulsive disorder. *Psychiatr Clin North Am* **15**:871–884.

Baxter, L. R., Jr., Saxena, S., Brody, A. L., Ackermann, R. F., Colgan, M., Schwartz, J. M., Allen-Martinez, Z., Fuster, J. M., and Phelps, M. E. (1996) Brain mediation of obsessive-compulsive disorder symptoms: evidence from functional brain imaging studies in the human and nonhuman primate. *Semin Clin Neuropsychiatry* **1**:32–47.

Boshuisen, M. L., Ter Horst, G. J., Paans, A. M., Reinders, A. A., and den Boer J. A. (2002) rCBF differences between panic disorder patients and control subjects during anticipatory anxiety and rest. *Biol Psychiatry* **52**:126–135.

Breiter, H. C., and Rauch S. L. (1996) Functional MRI and the study of OCD:from symptom provocation to cognitive-behavioral probes of cortico-striatal systems and the amygdala. *Neuroimage* **4**:S127–S138.

Breiter, H. C., Rauch, S. L., Kwong, K. K., Baker, J. R., Weisskoff, R. M., Kennedy, D. N., *et al.* (1996) Functional magnetic resonance imaging of symptom provocation in obsessive-compulsive disorder. *Arch Gen Psychiatry* **53**:595–606.

Bremner, J. D., Narayan, M., Staib, L. H., Southwick, S. M., McGlashan, T., and Charney, D. S. (1999a) Neural correlates of memories of childhood sexual abuse in women with and without posttraumatic stress disorder. *Am J Psychiatry* **156**:1787–1795.

Bremner, J. D., Staib, L. H., Kaloupek, D., Southwick, S. M., Soufer, R., and Charney, D. S. (1999b) Neural correlates of exposure to traumatic pictures and sound in Vietnam combat veterans with and without posttraumatic stress disorder:a positron emission tomography study. *Biol Psychiatry* **45**:806–816.

Bremner, J. D., Vythilingam, M., Vermetten, E., Southwick, S. M., McGlashan, T., Staib, L. H., *et al.* (2003) Neural correlates of declarative memory for emotionally valenced words in women with posttraumatic stress disorder related to early childhood sexual abuse. *Biol Psychiatry* **53**:879–889.

Bremner, J. D., Vythilingam, M., Vermetten, E., Vaccarino, V., and Charney, D. S. (2004) Deficits in hippocampal and anterior cingulate functioning during verbal declarative memory encoding in midlife major depression. *Am J Psychiatry* **161**:637–645.

Bremner, J. D., Vermetten, E., Schmahl, C., Vaccarino, V., Vythilingam, M., Afzal, N., Grillon, C., and Charney D. S. (2005) Positron emission tomographic imaging of neural correlates of a fear acquisition and extinction paradigm in women with childhood sexual-abuse-related post-traumatic stress disorder. *Psychol Med* **35**:791–806.

Brody, A. L., Saxena, S., Schwartz, J. M., Stoessel, P. W., Maidment, K., Phelps, M. E., *et al.* (1998) FDG-PET predictors of response to behavioral therapy and pharmacotherapy in obsessive compulsive disorder. *Psychiatry Res* **84**:1–6.

Butter, C. M., Mishkin, M., and Rosvold, H. E. (1963) Conditioning and extinction of a food-rewarded response after selective ablations of frontal cortex in rhesus monkeys. *Exp Neurol* **7**:65–75.

Cahill, L., Prins, B., Weber, M., and McGaugh, J. L. (1994) Beta-adrenergic activation and memory for emotional events. *Nature* **371**:702–704.

Cannistraro, P. A. and Rauch, S. L. (2003) Neural circuitry of anxiety:evidence from structural and functional neuroimaging studies. *Psychopharmacol Bull* **37**:8–25.

Cannistraro, P. A., Wright, C. I., Wedig, M. M., Martis, B., Shin, L. M., Wilhelm, S., *et al.* (2004) Amygdala responses to human faces in obsessive-compulsive disorder. *Biol Psychiatry* **56**:916–920.

Cavada, C., Company, T., Tejedor, J., Cruz-Rizzolo, R. J., and Reinoso-Suarez, F. (2000) The anatomical connections of the macaque monkey orbitofrontal cortex. A review. *Cereb Cortex* **10**:220–242.

Charney, D. S. (2004) Discovering the neural basis of human social anxiety:a diagnostic and therapeutic imperative. *Am J Psychiatry* **161**:1–2.

Corcoran, K. A. and Maren, S. (2004) Factors regulating the effects of hippocampal inactivation on renewal of conditional fear after extinction. *Learn Mem* **11**:598–603.

Critchley, H. D., Elliott, R., Mathias, C. J., and Dolan, R. J. (2000) Neural activity relating to generation and representation of galvanic skin conductance responses:a functional magnetic resonance imaging study. *J Neurosci* **20**:3033–3040.

Critchley, H. D., Mathias, C. J., Josephs, O., O'Doherty, J., Zanini, S., Dewar, B. K., et al. (2003) Human cingulate cortex and autonomic control:converging neuroimaging and clinical evidence. *Brain* **126**:2139–2152.

Cuthbert, B. N., Lang, P. J., Strauss, C., Drobes, D., Patrick, C. J., and Bradley, M. M. (2003) The psychophysiology of anxiety disorder:fear memory imagery. *Psychophysiology* **40**:407–422.

Davis, M., and Whalen, P. J. (2001) The amygdala:vigilance and emotion. *Mol Psychiatry* **6**:13–34.

Dilger, S., Straube, T., Mentzel, H. J., Fitzek, C., Reichenbach, J. R., Hecht, H., et al. (2003) Brain activation to phobia-related pictures in spider phobic humans:an event-related functional magnetic resonance imaging study. *Neurosci Lett* **348**:29–32.

Dringenberg, H. C., Saber, A. J., and Cahill L. (2001) Enhanced frontal cortex activation in rats by convergent amygdaloid and noxious sensory signals. *Neuroreport* **12**:2395–2398.

Elliott, R., Friston, K. J., and Dolan, R. J. (2000) Dissociable neural responses in human reward systems. *J Neurosci* **20**:6159–6165.

Ferry, A. T., Ongur, D., An, X., and Price J. L. (2000) Prefrontal cortical projections to the striatum in macaque monkeys:evidence for an organization related to prefrontal networks. *J Comp Neurol* **425**:447–470.

Fischer, H., Andersson, J. L., Furmark, T., and Fredrikson, M. (1998) Brain correlates of an unexpected panic attack:a human positron emission tomographic study. *Neurosci Lett* **251**:137–140.

Floyd, N. S., Price, J. L., Ferry, A. T., Keay, K. A., and Bandler, R. (2000) Orbitomedial prefrontal cortical projections to distinct longitudinal columns of the periaqueductal gray in the rat. *J Comp Neurol* **422**:556–578.

Foa, E. B. (2000) Psychosocial treatment of posttraumatic stress disorder. *J Clin Psychiatry 61 Suppl* 5:43–48.

Foa, E. B., Rothbaum, B. O., Riggs, D. S., and Murdock, T. B. (1991) Treatment of posttraumatic stress disorder in rape victims:a comparison between cognitive-behavioral procedures and counseling. *J Consult Clin Psychol* **59**:715–723.

Garcia, R., Vouimba, R. M., Baudry, M., and Thompson, R. F. (1999) The amygdala modulates prefrontal cortex activity relative to conditioned fear. *Nature* **402**:294–296.

Gottfried, J. A., and Dolan, R. J. (2004) Human orbitofrontal cortex mediates extinction learning while accessing conditioned representations of value. *Nat Neurosci* **7**:1144–1152.

Graybiel, A. M., and Rauch, S. L. (2000) Toward a neurobiology of obsessive-compulsive disorder. *Neuron* **28**:343–347.

Grillon, C., Ameli, R., Goddard, A., Woods, S. W., and Davis, M. (1994) Baseline and fear-potentiated startle in panic disorder patients. *Biol Psychiatry* **35**:431–439.

Gusnard, D. A., Ollinger, J. M., Shulman, G. L., Cloninger, C. R., Price, J. L., Van Essen, D. C., et al. (2003) Persistence and brain circuitry. *Proc Natl Acad Sci U S A* **100**:3479–3484.

Guthrie, R. M. and Bryant, R. A. (2005) Auditory startle response in firefighters before and after trauma exposure. *Am J Psychiatry* **162**:283–290.

Hendler, T., Rotshtein, P., Yeshurun, Y., Weizmann, T., Kahn, I., Ben Bashat, D., et al. (2003) Sensing the invisible:differential sensitivity of visual cortex and amygdala to traumatic context. *Neuroimage* **19**:587–600.

Herry, C. and Mons, N. (2004) Resistance to extinction is associated with impaired immediate early gene induction in medial prefrontal cortex and amygdala. *Eur J Neurosci* **20**:781–790.

Hoehn-Saric, R., McLeod, D. R., and Hipsley P. (1995) Is hyperarousal essential to obsessive-compulsive disorder? Diminished physiologic flexibility, but not hyperarousal, characterizes patients with obsessive-compulsive disorder. *Arch Gen Psychiatry* **52**:688–693.

Hoehn-Saric, R., McLeod, D. R., and Zimmerli, W. D. (1991) Psychophysiological response patterns in panic disorder. *Acta Psychiatr Scand* **83**:4–11.

Joel, D., Zohar, O., Afek, M., Hermesh, H., Lerner, L., Kuperman, R., et al. (2005) Impaired procedural learning in obsessive-compulsive disorder and Parkinson's disease, but not in major depressive disorder. *Behav Brain Res* **157**:253–263.

Kendler, K. S., Myers, J., and Prescott, C. A. (2002) The etiology of phobias:an evaluation of the stress-diathesis model. *Arch Gen Psychiatry* **59**:242–248.

Kilpatrick, L. and Cahill, L. (2003) Amygdala modulation of parahippocampal and frontal regions during emotionally influenced memory storage. *Neuroimage* **20**:2091–2099.

Klein, D. F. (1993) False suffocation alarms, spontaneous panics, and related conditions. An integrative hypothesis. *Arch Gen Psychiatry* **50**:306–317.

Kringelbach, M. L. and Rolls E. T. (2004) The functional neuroanatomy of the human orbitofrontal cortex:evidence from neuroimaging and neuropsychology. *Prog Neurobiol* **72**:341–372.

Kumari, V., Kaviani, H., Raven, P. W., Gray, J. A., and Checkley, S. A. (2001) Enhanced startle reactions to acoustic stimuli in patients with obsessive-compulsive disorder. *Am J Psychiatry* **158**:134–136.

Lanius, R. A., Williamson, P. C., Boksman, K., Densmore, M., Gupta, M., Neufeld, R. W., et al. (2002) Brain activation during script-driven imagery induced dissociative responses in PTSD:a functional magnetic resonance imaging investigation. *Biol Psychiatry* **52**:305–311.

Lanius, R. A., Williamson, P. C., Densmore, M., Boksman, K., Gupta, M. A., Neufeld, R. W., et al. (2001) Neural correlates of traumatic memories in posttraumatic stress disorder:a functional MRI investigation. *Am J Psychiatry* **158**:1920–1922.

Lebron, K., Milad, M. R., and Quirk, G. J. (2004) Delayed recall of fear extinction in rats with lesions of ventral medial prefrontal cortex. *Learn Mem* **11**:544–548.

Ledoux, J. E. (2000) Emotion circuits in the brain. *Annu Rev Neurosci* **23**:155–184.

Liberzon, I., Britton, J. C., and Phan, K. L. (2003) Neural correlates of traumatic recall in posttraumatic stress disorder. *Stress* **6**:151–156.

Liberzon, I., Taylor, S. F., Amdur, R., Jung, T. D., Chamberlain, K. R., Minoshima, S., et al. (1999) Brain activation in PTSD in response to trauma-related stimuli. *Biol Psychiatry* **45**:817–826.

Lorberbaum, J. P., Kose, S., Johnson, M. R., Arana, G. W., Sullivan, L. K., Hamner, M. B., et al. (2004) Neural correlates of speech anticipatory anxiety in generalized social phobia. *Neuroreport* **15**:2701–2705.

Ludewig S., Geyer M. A., Ramseier M., Vollenweider F. X., Rechsteiner E., Cattapan-Ludewig K. (2005) Information-processing deficits and cognitive dysfunction in panic disorder. *J Psychiatry Neurosci* **30**:37–43.

Ludewig, S., Ludewig, K., and Geyer, M. A., Hell, D., and Vollenweider, F. X. (2002) Prepulse inhibition deficits in patients with panic disorder. *Depress Anxiety* **15**:55–60.

Maren, S., and Holt, W. G. (2004) Hippocampus and Pavlovian fear conditioning in rats:muscimol infusions into the ventral, but not dorsal, hippocampus impair the acquisition of conditional freezing to an auditory conditional stimulus. *Behav Neurosci* **118**:97–110.

Maren, S. and Quirk, G. J. (2004) Neuronal signalling of fear memory. *Nat Rev Neurosci* **5**:844–852.

Markowitsch, H. J., Vandekerckhovel, M. M., Lanfermann, H., and Russ M. O. (2003) Engagement of lateral and medial prefrontal areas in the ecphory of sad and happy autobiographical memories. *Cortex* **39**:643–665.

Massana, G., Serra-Grabulosa, J. M., Salgado-Pineda, P., Gasto, C., Junque, C., Massana, J., et al. (2003) Amygdalar atrophy in panic disorder patients detected by volumetric magnetic resonance imaging. *Neuroimage* **19**:80–90.

McDonald, A. J. (1996) Glutamate and aspartate immunoreactive neurons of the rat basolateral amygdala:colocalization of excitatory amino acids and projections to the limbic circuit. *J Comp Neurol* **365**:367–379.

McGaugh, J. L. (2000) Memory—a century of consolidation. *Science* **287**:248–251.

McGaugh, J. L., and Roozendaal B. (2002) Role of adrenal stress hormones in forming lasting memories in the brain. *Curr Opin Neurobiol* **12**:205–210.

McGuire, P. K., Bench, C. J., Frith, C. D., Marks, I. M., Frackowiak, R. S., and Dolan, R. J. (1994) Functional anatomy of obsessive-compulsive phenomena. *Br J Psychiatry* **164**:459–468.

Milad, M. R. Rauch, S. L., Pitman, R. K., and Quirk, G. J. (2005). Fear extinction in rats:implications for human brain imaging and anxiety disorders. *Biol.Psychol.* In Press.

Milad, M. R., Quinn, B. T., Pitman, R. K., Orr, S. P., Fischl, B., and Rauch, S. L. (2005) Thickness of ventromedial prefrontal cortex in humans is correlated with extinction memory. *Proc Natl Acad Sci U S A*.

Milad, M. R. and Quirk, G. J. (2002) Neurons in medial prefrontal cortex signal memory for fear extinction. *Nature* **420**:70–74.

Morgan, M. A., Schulkin, J., and Ledoux, J. E. (2003) Ventral medial prefrontal cortex and emotional perseveration:the memory for prior extinction training. *Behav Brain Res* **146**:121–130.

Morris, J. S. and Dolan, R. J. (2004) Dissociable amygdala and orbitofrontal responses during reversal fear conditioning. *Neuroimage* **22**:372–380.

Nordahl, T. E., Stein, M. B., Benkelfat, C., Semple, W. E., Andreason, P., Zametkin, A., et al. (1998) Regional cerebral metabolic asymmetries replicated in an independent group of patients with panic disorders. *Biol Psychiatry* **44**:998–1006.

Ohman, A. and Mineka, S. (2001) Fears, phobias, and preparedness:toward an evolved module of fear and fear learning. *Psychol Rev* **108**:483–522.

Ongur, D. and Price J. L. (2000) The organization of networks within the orbital and medial prefrontal cortex of rats, monkeys and humans. *Cereb Cortex* **10**:206–219.

Orr, S. P., Metzger, L. J., Lasko, N. B., Macklin, M. L., Peri, T., and Pitman, R. K. (2000) De novo conditioning in trauma-exposed individuals with and without posttraumatic stress disorder. *J Abnorm Psychol* **109**:290–298.

Orr, S. P., Metzger, L. J., and Pitman, R. K. (2002) Psychophysiology of post-traumatic stress disorder. *Psychiatr Clin North Am* **25**:271–293.

Orr, S. P., Metzger, L. J., Lasko, N. B., Macklin, M. L., Hu, F. B., Shalev, A. Y., et al. (2003) Physiologic responses to sudden, loud tones in monozygotic twins discordant for combat exposure:association with posttraumatic stress disorder. *Arch Gen Psychiatry* **60**:283–288.

Patterson, J. C., Ungerleider, L. G., and Bandettini, P. A. (2002) Task-independent functional brain activity correlation with skin conductance changes:an fMRI study. *Neuroimage* **17**:1797–1806.

Perani, D., Colombo, C., Bressi, S., Bonfanti, A., Grassi, F., Scarone, S., et al. (1995) [18F]FDG PET study in obsessive-compulsive disorder. A clinical/metabolic correlation study after treatment. *Br J Psychiatry* **166**:244–250.

Phelps, E. A., Delgado, M. R., Nearing, K. I., and Ledoux, J. E. (2004) Extinction learning in humans:role of the amygdala and VMPFC. *Neuron* **43**:897–905.

Pissiota, A., Frans, O., Fernandez, M., von Knorring, L., Fischer, H., and Fredrikson M. (2002) Neurofunctional correlates of posttraumatic stress disorder:a PET symptom provocation study. *Eur Arch Psychiatry Clin Neurosci* **252**:68–75.

Pitman, R. K., Shin, L. M., and Rauch, S. L. (2001) Investigating the pathogenesis of posttraumatic stress disorder with neuroimaging. *J Clin Psychiatry 62 Suppl* **17**:47–54.

Pole, N., Neylan, T. C., Best, S. R., Orr, S. P., and Marmar, C. R. (2003) Fear-potentiated startle and posttraumatic stress symptoms in urban police officers. *J Trauma Stress* **16**:471–479.

Protopopescu, X., Pan, H., Tuescher, O., Cloitre, M., Goldstein, M., Engelien, W., Epstein, J., Yang, Y., Gorman, J., Ledoux, J., Silbersweig, D., and Stern, E. (2005) Differential time courses and specificity of amygdala activity in posttraumatic stress disorder subjects and normal control subjects. *Biol Psychiatry* **57**:464–473.

Quirk, G. J. (2004) Learning not to fear, faster. *Learn Mem* **11**:125–126.

Quirk, G. J., Likhtik, E., Pelletier, J. G., and Pare, D. (2003) Stimulation of medial prefrontal cortex decreases the responsiveness of central amygdala output neurons. *J Neurosci* **23**:8800–8807.

Rauch, S. L. (2003) Neuroimaging and neurocircuitry models pertaining to the neurosurgical treatment of psychiatric disorders. *Neurosurg Clin N Am* **14**:213–224.

Rauch, S. L., and Baxter, L. R. (1998) Neuroimaging of OCD and related disorders. In:Jenike MA, Baer L, Minichiello WE (eds.) Obsessive-Compulsive Disorders:Practical Management. Boston:Mosby 289–317.

Rauch, S. L., Jenike, M. A., Alpert, N. M., Baer, L., Breiter, H. C., Savage, C. R., *et al.* (1994) Regional cerebral blood flow measured during symptom provocation in obsessive-compulsive disorder using oxygen 15-labeled carbon dioxide and positron emission tomography. *Arch Gen Psychiatry* **51**:62–70.

Rauch, S. L., Shin, L. M., Dougherty, D. D., Alpert, N. M., Fischman, A. J., and Jenike, M. A. (2002) Predictors of fluvoxamine response in contamination-related obsessive compulsive disorder:a PET symptom provocation study. *Neuropsychopharmacology* **27**:782–791.

Rauch, S. L., Shin, L. M., Segal, E., Pitman, R. K., Carson, M. A., McMullin, K., Whalen, P. J., and Makris, N. (2003) Selectively reduced regional cortical volumes in post-traumatic stress disorder. *Neuroreport* **14**:913–916.

Rauch, S. L., Shin, L. M., Whalen, P. J., and Pitman, R. K. (1998) Neuroimaging and the neuroanatomy of PTSD. *CNS Spectr* **3**:30–41.

Rauch, S. L., Whalen, P. J., Dougherty, D. D., and Jenike, M. A. (1998) Neurobiological models of obsessive compulsive disorders. In:Jenike MA, Baer L, Minichiello WE (eds.) Obsessive-Compulsive Disorders:Practical Management. Boston:Mosby. 222–253.

Rauch, S. L., Whalen, P. J., Shin, L. M., McInerney, S. C., Macklin, M. L., Lasko, NB, Orr, SP, and Pitman, RK (2000) Exaggerated amygdala response to masked facial stimuli in posttraumatic stress disorder:a functional MRI study. *Biol Psychiatry* **47**:769–776.

Rodrigues, S. M., Schafe, G. E., and Ledoux, J. E. (2004) Molecular mechanisms underlying emotional learning and memory in the lateral amygdala. *Neuron* **44**:75–91.

Rolls, E. T., Hornak, J., Wade, D., and McGrath, J. (1994) Emotion-related learning in patients with social and emotional changes associated with frontal lobe damage. *J. eurol Neurosurg Psychiatry* **57**:1518–1524.

Rosenbaum, J. F., Biederman, J., Hirshfeld, D. R., Bolduc, E. A., Faraone, S. V., Kagan, J., *et al.* (1991) Further evidence of an association between behavioral inhibition and anxiety disorders:results from a family study of children from a non-clinical sample. *J Psychiatr Res* **25**:49–65.

Rothbaum, B. O. and Davis, M. (2003) Applying learning principles to the treatment of post-trauma reactions. *Ann N Y Acad Sci* **1008**:112–121.

Santini, E., Ge, H., Ren, K., Pena, D. O., and Quirk, G. J. (2004) Consolidation of fear extinction requires protein synthesis in the medial prefrontal cortex. *J Neurosci* **24**:5704–5710.

Saxena, S., Bota, R. G., and Brody, A. L. (2001) Brain-behavior relationships in obsessive-compulsive disorder. *Semin Clin Neuropsychiatry* **6**:82–101.

Saxena, S., Brody, A. L., Ho, M. L., Alborzian, S., Maidment, K. M., Zohrabi, N., *et al.* (2002) Differential cerebral metabolic changes with paroxetine treatment of obsessive-compulsive disorder vs major depression. *Arch Gen Psychiatry* **59**:250–261.

Saxena, S., Brody, A. L., Schwartz, J. M., and Baxter, L. R. (1998) Neuroimaging and frontal-subcortical circuitry in obsessive-compulsive disorder. *Br J Psychiatry Suppl* 26–37.

Saxena, S. and Rauch, S. L. (2000) Functional neuroimaging and the neuroanatomy of obsessive-compulsive disorder. *Psychiatr Clin North Am* **23**:563–586.

Schoenbaum, G., Setlow, B., Saddoris, M. P., and Gallagher, M. (2003) Encoding predicted outcome and acquired value in orbitofrontal cortex during cue sampling depends upon input from basolateral amygdala. *Neuron* **39**:855–867.

Semple, W. E., Goyer, P. F., McCormick, R., Donovan, B., Muzic, R. F., Jr., Rugle, L., et al. (2000) Higher brain blood flow at amygdala and lower frontal cortex blood flow in PTSD patients with comorbid cocaine and alcohol abuse compared with normals. *Psychiatry* **63**:65–74.

Shalev, A. Y., Peri, T., Gelpin, E., Orr, S. P., and Pitman, R. K. (1997) Psychophysiologic assessment of mental imagery of stressful events in Israeli civilian posttraumatic stress disorder patients. *Compr Psychiatry* **38**:269–273.

Shapira, N. A., Liu, Y., He, A. G., Bradley, M. M., Lessig, M. C., James, G. A., et al. (2003) Brain activation by disgust-inducing pictures in obsessive-compulsive disorder. Biol Psychiatry **54**:751–756.

Shin, L. M., McNally, R. J., Kosslyn, S. M., Thompson, W. L., Rauch, S. L., Alpert, N. M., et al. (1997) A positron emission tomographic study of symptom provocation in PTSD. *Ann N Y Acad Sci* **821**:521–523.

Shin, L. M., McNally, R. J., Kosslyn, S. M., Thompson, W. L., Rauch, S. L., Alpert, N. M., et al. (1999) Regional cerebral blood flow during script-driven imagery in childhood sexual abuse-related PTSD:A PET investigation. *Am J Psychiatry* **156**:575–584.

Shin, L. M., Whalen, P. J., Pitman, R. K., Bush, G., Macklin, M. L., Lasko, N. B., et al. (2001a) An fMRI study of anterior cingulate function in posttraumatic stress disorder. *Biol Psychiatry* **50**:932–942.

Shin, L. M., Whalen, P. J., Pitman, R. K., Bush, G., Macklin, M. L., Lasko, N. B., et al. (2001b) An fMRI study of anterior cingulate function in posttraumatic stress disorder. *Biol Psychiatry* **50**:932–942.

Shin, L. M., Orr, S. P., Carson, M. A., Rauch, S. L., Macklin, M. L., Lasko, N. B., et al. (2004) Regional cerebral blood flow in the amygdala and medial prefrontal cortex during traumatic imagery in male and female Vietnam veterans with PTSD. *Arch Gen Psychiatry* **61**:168–176.

Sigmon, S. T., Dorhofer, D. M., Rohan, K. J., Hotovy, L. A., Boulard, N. E., and Fink, C. M. (2000) Psychophysiological, somatic, and affective changes across the menstrual cycle in women with panic disorder. *J Consult Clin Psychol* **68**:425–431.

Stein, M. B., Goldin, P. R., Sareen, J., Zorrilla, L. T., and Brown, G. G. (2002) Increased amygdala activation to angry and contemptuous faces in generalized social phobia. *Arch Gen Psychiatry* **59**:1027–1034.

Straube, T., Mentzel, H. J., Glauer, M., and Miltner, W. H. (2004) Brain activation to phobia-related words in phobic subjects. *Neurosci Lett* **372**:204–208.

Swedo, S. E., Leonard, H. L., and Rapoport, J. L. (1992) Childhood-onset obsessive compulsive disorder. *Psychiatr Clin North Am* **15**:767–775.

Tillfors, M., Furmark, T., Marteinsdottir, I., Fischer, H., Pissiota, A., Langstrom, B., et al. (2001) Cerebral blood flow in subjects with social phobia during stressful speaking tasks:a PET study. *Am J Psychiatry* **158**:1220–1226.

Ursu, S. and Carter, C. S. (2005) Outcome representations, counterfactual comparisons and the human orbitofrontal cortex:implications for neuroimaging studies of decision-making. *Brain Res Cogn Brain Res* **23**:51–60.

van Ameringen, M., Mancini, C., Szechtman, H., Nahmias, C., Oakman, J. M., Hall, G. B., et al. (2004) A PET provocation study of generalized social phobia. *Psychiatry Res* **132**:13–18.

van den Heuvel, O. A., Veltman, D. J., Groenewegen, H. J., Dolan, R. J., Cath, D. C., Boellaard, R., et al. (2004) Amygdala activity in obsessive-compulsive disorder with contamination fear:a study with oxygen-15 water positron emission tomography. *Psychiatry Res* **132**:225–237.

van Minnen, A., and Hagenaars, M. (2002) Fear activation and habituation patterns as early process predictors of response to prolonged exposure treatment in PTSD. *J Trauma Stress* **15**:359–367.

Wald, J. and Taylor, S. (2003) Preliminary research on the efficacy of virtual reality exposure therapy to treat driving phobia. *Cyberpsychol Behav* **6**:459–465.

Whalen, P. J., Rauch, S. L., Etcoff, N. L., McInerney, S. C., Lee, M. B., and Jenike, M. A. (1998) Masked presentations of emotional facial expressions modulate amygdala activity without explicit knowledge. *J Neurosci* **18**:411–418.

Woods, S. W., Charney, D. S., Goodman, W. K., and Heninger, G. R. (1988) Carbon dioxide-induced anxiety. Behavioral, physiologic, and biochemical effects of carbon dioxide in patients with panic disorders and healthy subjects. *Arch Gen Psychiatry* **45**:43–52.

Wright, C. I., Martis, B., McMullin, K., Shin, L. M., and Rauch, S. L. (2003) Amygdala and insular responses to emotionally valenced human faces in small animal specific phobia. *Biol Psychiatry* **54**:1067–1076.

Zald, D. H. and Kim, S. W. (1996a) Anatomy and function of the orbital frontal cortex, I:anatomy, neurocircuitry; and obsessive-compulsive disorder. *J Neuropsychiatry Clin Neurosci* **8**:125–138.

Zald, D. H. and Kim, S. W. (1996b) Anatomy and function of the orbital frontal cortex, II:Function and relevance to obsessive-compulsive disorder. *J Neuropsychiatry Clin Neurosci* **8**:249–261.

Zahn, T. P., Leonard, H. L., Swedo, S. E., and Rapoport, J. L. (1996) Autonomic activity in children and adolescents with obsessive-compulsive disorder. *Psychiatry Res* **60**:67–76.

# Chapter 21

# The role of the ventral prefrontal cortex in mood disorders

Carolyn A. Fredericks, Jessica H. Kalmar, and Hilary P. Blumberg

## 21.1. Introduction

In this chapter we review evidence supporting a prominent role for the ventral prefrontal cortex (vPFC) in the neuropathophysiology of mood disorders. Findings from multiple studies of mood disorders, using a variety of research methods, converge on a broad vPFC region that encompasses the medial and lateral orbitofrontal cortex (OFC), the subgenual and pregenual anterior cingulate, as well as the medial and rostral frontal cortices. Despite heterogeneity in the specific subregions of the vPFC reported in these studies, the vPFC region emerges as a crucial node within a distributed cortico-limbic neural system disrupted in mood disorders. Research that has elucidated the functions of the vPFC and brain structures with significant connectivity to the vPFC, as reviewed here and in other chapters of this volume, strongly implicates the vPFC and its connected circuitry in the deficits in emotional and behavioral regulation characteristic of mood disorders. We review the data related to this more general vPFC region, and point out, where possible, the specific subregions that may contribute to certain aspects of mood disorders.

The hallmark of mood disorders is the presence of the dramatic and sustained emotional changes of acute depressive or manic episodes. During these acute mood episodes, patients characteristically have disruptions not only in their subjective affective experiences, but also in their ability to respond adaptively to shifts in the positive and negative reinforcing values of internal and external stimuli (Lewinsohn and Graf 1973). Sadness, amotivation and anhedonic symptoms are common in depression, wherein patients no longer take interest in activities they previously found pleasurable, and behavior becomes constricted and isolative. Conversely, euphoria and excessive drive towards hedonic behaviors are common in mania, and can result in expansive and risky behaviors. The severe emotional changes are invariably accompanied by a broad array of abnormalities ranging from disruptions in the more basic, circadian rhythms to deficits in higher order cognition. Impairments in more primitive neurovegetative functions, such as sleep and appetite, often herald the onset of acute mood episodes. Disturbances in attention and higher cognitive processes can impair school and work performance. Thus, a neural model for mood disorders must take into account the core abnormalities in the regulation of emotion and motivation, as well as the broad range of accompanying symptoms. The vPFC is especially implicated in mood disorders as it has a central role in regulating adaptive emotional and social responses to changing reinforcement contingencies. Moreover, its connectivity with the hypothalamus, the amygdala, and the paralimbic and multimodal cortices (Amaral and Price 1984; Morecraft *et al.* 1992; Carmichael and Price 1995a, b; Zald and Kim 1996; Rempel-Clower and Barbas 1998) places it at a nexus between the subcortical limbic and heteromodal cognitive areas, and thus in a position to integrate and modulate the broader range of functions that are disrupted in primary mood disorders.

In the following sections we provide a brief introduction to mood disorders, and consider the history of vPFC lesions that have resulted in mood disorder-like symptoms, as well as the neurobehavioral profile of individuals with mood disorders, including their performance on tests of emotional and cognitive functioning. We provide an overview of evidence from structural and functional neuroimaging studies of persons with mood disorders, both at rest and during performance of emotional and cognitive tasks. Cellular-level postmortem studies of those who suffered from mood disorders provide further evidence for vPFC abnormalities in mood disorders, and inform emerging research at the level of neurons and glia. We consider hypotheses about the neurodevelopmental course of mood disorders, and about the ways in which treatment interventions for mood disorders may affect vPFC neural systems. Finally, we consider the role of the vPFC in the neural circuitry implicated in two of the major mood disorders, major depressive disorder (MDD) and bipolar disorder (BD, also known as manic-depressive illness), in summary.

## 21.2. Mood disorders: prevalence and symptomatology

Mood disorders are highly common, and often devastating medical problems. About 16% of Americans experience a major depressive episode at some point during their lifetime (Kessler *et al.* 2003), and approximately 1.2% may suffer from BD (Weissman *et al.* 1988). In the past few decades, mood disorders have been increasingly recognized as national and global health problems. The World Health Organization predicts that by 2020, depression will be second only to ischemic heart disease in contributing to global disease burden (Murray and Lopez 1996). Mood disorders carry a large economic burden that has been estimated at approximately $45 billion per year for the U.S., both in direct costs of treatment and in indirect costs of lost wages and disability (Wyatt and Henter 1995). Importantly, mood disorders carry a significant risk of mortality. Completed suicide occurs most commonly in persons with mood disorders (Institute of Medicine 2002). For those with a diagnosis of BD, it is estimated that approximately 19% will die from suicide (Goodwin and Jamison 1990).

### Depression

Depressive episodes are the defining features of MDD and are common elements of BD. A depressive episode is characterized by an abnormally low mood that persists for at least two weeks (APA 1994). Individuals who are depressed are typically anhedonic, i.e., they lose interest or pleasure in activities that they previously enjoyed, and their behavior becomes constricted. Loss of motivation may also lead to social withdrawal and isolation. Depression is often associated with a range of neurovegetative symptoms, including changes in sleep ranging from insomnia to sleeping excessive hours, changes in appetite patterns accompanied by weight loss or gain, as well as diminished sexual drive. Energy changes and psychomotor abnormalities may be present, ranging from an inability to remain still to marked slowing. Feelings of hopelessness, excessive guilt, worthlessness, and irritability are common. Cognitive dysfunction may be present, including difficulty sustaining attention, as well as memory impairment. Individuals who are depressed may become ruminative, report decreased flexibility in thinking, and have trouble making decisions. When depression is severe, individuals may have disordered thinking processes, and may experience psychotic symptoms, including mood-congruent delusions or hallucinations. For instance, they may believe that God has a particular hatred for them, believe their body is decaying, or feel that they are personally responsible for world problems. Thoughts of death and suicide may develop, and suicide may be attempted.

## Mania

In contrast to unipolar MDD, for which acute depressive episodes are the defining feature, the defining features of BD are the presence of manic episodes that may alternate with depressive episodes. Manic episodes are characterized by extreme mood elevation, which can take the form of euphoria or irritability. In contrast to individuals who are depressed, individuals who are manic report feelings of high energy and display excessive motivational drive. Persons experiencing mania often have difficulty acting in an adaptive manner in negotiating conflicting rewards and punishments, and participate excessively in pleasurable activities that may have painful consequences. In brief periods, persons who are manic may spend into bankruptcy, lose jobs due to argumentativeness, and have sexual indiscretions that disrupt important relationships. Reduced sleep often marks the onset of manic episodes (Bunney *et al.* 1972; Wehr *et al.* 1987). Persons who are manic may report decreases in sleep to a few hours a night, or even several nights without sleep, yet they may report feeling energized. Mania is often associated with psychomotor agitation, as well as pressured, rapid speech that contains a flight of ideas and is difficult to interpret. Distractibility is common, in that attention is easily drawn to irrelevant stimuli. Grandiosity is also common, and in severe cases, as with persons who are depressed, abnormalities of thought may reach psychotic levels in the form of mood-congruent delusions and hallucinations. For example, a person who is manic may believe that they have a special relationship with celebrities or with God, that they should advise world leaders or other influential figures on how things should be done, or that they should write a novel or a symphony when they have had no previous experience in doing so. Dr. Kay Redfield Jamison's book *An Unquiet Mind* includes the following vivid description of her experiences with mania:

> There is a particular kind of pain, elation, loneliness, and terror involved in this kind of madness. When you're high it's tremendous. The ideas and feelings are fast and frequent like shooting stars

and you follow them until you find better and brighter ones. Shyness goes, the right words and gestures are suddenly there, the power to seduce and captivate others a felt certainty. . . . Feelings of ease, intensity, power, well-being, financial omnipotence and euphoria now pervade one's marrow. But, somewhere, this changes. The fast ideas are far too fast, and there are far too many; overwhelming confusion replaces clarity. Memory goes. Humor and absorption on friends' faces are replaced by fear and concern. Everything previously moving with the grain is now against—you are irritable, angry, frightened, uncontrollable, and enmeshed totally in the blackest caves of the mind. You never knew those caves were there. It will never end, for madness carves its own reality.—(Jamison 1995, page 67)

### Euthymia

Between extreme mood episodes, individuals who have primary mood disorders are generally euthymic, i.e., they have emotional variations that are within the normal range, and may exhibit a high level of adaptive functioning. However, a subset of individuals with mood disorders may show subtle symptoms that persist into euthymia (Altshuler *et al.* 2004; Strakowski *et al.* 2004). MDD and BD can be quite heterogeneous in clinical presentation in terms of the frequency and severity of mood episodes, and in responsiveness to treatment. But, in general, the goal of treatment is to decrease the severity and frequency of depressive or manic episodes, lengthen euthymic periods, and prevent future episodes.

## 21.3. Brain lesions and secondary mood symptoms

The study of patients with vPFC lesions provided the first evidence that the vPFC was a key area of the brain for emotional and social functioning. Such lesions disrupt the neural systems subserving emotional and reward-related behaviors, and highlight a subset of the symptoms associated with mood disorders. The effect of vPFC lesions on mood and aggression is discussed in detail by Blake and Grafman (Chapter 22, this volume); here we review the history of this association as it relates to our understanding of mood disorders. Furthermore, we discuss the effects of lesions in other brain structures with significant connectivity to the vPFC, representing components of a distributed vPFC neural system implicated in mood disorders.

### Frontal lesions

Since at least the late 1800s, insult to the prefrontal cortex (PFC) has had a recognized association with mood abnormalities, such as inappropriate euphoria (Jastrowitz 1888; Oppenheim 1889). In the first years of the 20$^{th}$ century Starr described lesions in the frontal lobe and subadjacent white matter that resulted in disinhibited emotional states (Starr 1903). Such lesions, he reported, could lead to 'undue excitement, causeless laughter, unusual crying, [and] great depression'. In his discussion, Starr also raised the idea that frontal cortex is a key area for integrating information from the external environment with internal emotional or motivational states. Damage to this brain area, in his words, could yield 'a lack of harmony between the association of ideas and the states of feeling which they should awaken'. Starr's work, too, pointed towards the importance of considering the functioning of frontal cortex in the context of its connectivity within frontal-subcortical neural systems. He described subsets of mood symptoms resulting from lesions to the PFC and the thalamus, as well as the corpus callosum and frontothalamic white matter.

Case descriptions, from the past few decades, of patients who have sustained vPFC lesions illustrate behavioral changes that are similar to those observed in primary mood disorders. Cummings (1985) described the symptoms of patients with vPFC lesions that are especially reminiscent of symptoms observed in patients with manic episodes, such as inappropriate euphoric or irritable affect, impaired judgment with decreased tact and behavioral restraint, distractibility, hyperkinetic behavior, sexual promiscuity, and sometimes paranoia or grandiosity. A case study published by Eslinger and Damasio (1985) highlights the distinction between the ability of individuals with vPFC lesions to discriminate reasonably in theoretical ethical and social situations presented in the laboratory setting, and their difficulty in carrying over information about the motivational relevance of situations into directed, adaptive responses in real-world settings. Eslinger and Damasio's patient, E.V.R., demonstrated striking disinhibition and inappropriate social behavior after the bilateral ablation of orbital and lower mesial frontal cortices. E.V.R. consistently gave thoughtful and well-reasoned responses to hypothetical ethical scenarios posed to him by experimenters. He was unable, however, to integrate the same reasoning into his life. Soon after his injury, he entered a business partnership with a 'man of questionable reputation' and, against the advice of his family, put all his assets into the venture, which quickly failed and left him bankrupt. His marriage fell apart, and he was fired from various jobs because of his disorganization and tardiness. The Damasio group has characterised numerous other vPFC lesion patients whose symptoms are similar to those of E.V.R., and of Phineas Gage (whose story is told elsewhere in this volume): their 'ability to make rational decisions in personal and social matters is invariably compromised and so is their processing of emotion', although other higher-order cognitive functions are intact (Damasio *et al.* 1994).

## Right-left lesion laterality and mood

Evidence suggests that the valence of mood symptoms is related to the right-left hemispheric laterality of the lesion. Electroencephalographic measures have yielded evidence for a right-left hemispheric lateralization of emotional functions. The left hemisphere tends to show a greater association with the processing of stimuli that convey positive affective or motivational significance and elicit approach behaviors, whereas the right hemisphere is associated with the processing of negative stimuli and withdrawal behaviors (Sobotka *et al.* 1992; Davidson 2004). For example, Wheeler *et al.* (1993) reported that left frontal activation correlated with the intensity of positive affect, while right frontal activation correlated with the intensity of negative affect. Lesions in one hemisphere are theorised to result in the disinhibition of the affect associated with the contralateral hemisphere. Indeed, whereas insult to the left hemisphere is associated with negative, depressive states, lesions in the right hemisphere are associated with manic-type symptoms (Wexler 1980; Sackeim *et al.* 1982; Starkstein *et al.* 1991).

The emotional dysregulation exhibited in mood disorders may be related to abnormalities in the balance of function between the two hemispheres. Studies of functional connectivity in mood disorders suggest that impaired coordination of activity between the right and left frontal cortices may contribute to mood symptoms. Research has shown decreased prefrontal coupling (the degree of correlation between activity levels in the left and right PFC) in individuals experiencing major depressive episodes, and a return to normal coupling levels when depression remits (Mallet et al. 1998). In a related study, Pettigrew and Miller report findings that suggest that inter-hemispheric switching is abnormally slow or "sticky" in BD (Pettigrew and Miller 1998). Emotional lateralization and prefrontal-uncoupling studies have tended to focus on the PFC more generally, or on dorsal rather than vPFC structures. Starkstein et al. (1991) have, however, noted that when brain lesions are associated with mania (a finding less common than that of brain lesions associated with depression) they tend to be right-sided and involve the OFC and basotemporal cortex.

## Mood changes related to subcortical lesions

Damage to the brain regions with substantial connectivity to the vPFC may also result in mood symptomatology, implicating the involvement of a distributed vPFC neural system in mood disorders. For instance, the vPFC has major reciprocal interconnections with the amygdala (Amaral and Price 1984). Mood symptoms have long been described in association with seizures with mesial temporal lobe foci that encompass the amygdala (Flor-Henry 1969). Observations of parallels between the vulnerability of mesial temporal structures to sensitization and sensitization-like increases in the frequency and intensity of mood episodes with accumulation of untreated episodes contributed to the development of a sensitization model for BD and the successful implementation of anti-convulsant medications as treatments for BD (Post et al. 1982; Bowden et al. 2000). Evidence for vPFC deficits in mood disorders raises the possibility that aberrant amygdala activity could result from a disruption of the normal inhibitory vPFC inputs to the amygdala. However, it is also possible that vPFC abnormalities in mood disorders relate to abnormalities in amygdala inputs.

Lesions in other connected subcortical and posterior fossa regions, such as striatum, thalamus, pons, and cerebellum, have also been associated with secondary mood disturbances (Sackeim et al. 1982; Cummings and Mendez 1984; Drake et al. 1990; Lauterbach 1996). Associations between the hemispheric laterality of these lesions and the valence of resulting mood symptoms are generally similar to lesions in frontal regions (Cummings and Mendez 1984). The presence of subcortical involvement may relate to disturbances in the regulation of more primitive functions. Pontine lesions, for example, have been shown to result in sleeplessness (Drake et al. 1990), a neurovegetative symptom present in mood disorders.

Individuals with neurodegenerative diseases of subcortical origin often present with mood symptoms. Dr. Helen Mayberg has noted the common presence of secondary depression in patients with neurological disorders of the basal ganglia, such as Parkinson's disease and Huntington's disease. Her seminal neuroimaging studies of patients with these disorders demonstrated that the presence of depression was associated with bilateral hypometabolism in the orbito-inferior prefrontal cortex (Mayberg 1994, 2002). This important work suggests that although lesions in subcortical regions may contribute to mood disorders, disruption of their input to vPFC and thus in vPFC function may be a necessary component for the full expression of the range of symptoms typical of mood disorders.

## Conclusions

The lesion literature highlights the role of the vPFC in adaptive emotional and motivational functioning and the similarity between behavioral changes exhibited following vPFC lesions and primary mood disorders, although discrete vPFC lesions may approximate only a subset of mood symptomatology. Disruptions at other nodes within vPFC neural systems can produce symptoms of mood disorders, but vPFC functional abnormalities seem to be an essential component for the appearance of major emotional disturbances. Therefore, dysfunction in the vPFC may be necessary but not sufficient for the expression of the broad range of symptoms characteristic of primary mood disorder phenotypes, which may relate to abnormalities in a more widely distributed vPFC neural system. The lesion literature also introduces the idea that the laterality of the lesion may play a significant role in the valence of the affective abnormality.

## 21.4. Cognitive and behavioral findings in mood disorders

The performance of individuals with mood disorders on a wide range of neuropsychological tasks has been studied in order to better characterize the neurobehavioral profile of the disorders and to contribute towards an understanding of neural circuitry implicated in the disorders. Research has focused on the types of cognitive and behavioral difficulties that are frequently reported by people suffering from mood disorders, as well as functions subserved by the neural systems implicated in the disorders. Results suggest that both MDD and BD are associated with deficits in emotional processing, as well as on tests that target attentional, mnemonic, and executive functions. Perhaps particularly salient to the involvement of the vPFC in mood disorders are the results of studies that probe patients' ability to respond to changing reward and motivational contingencies. An in-depth discussion of neuropsychological tests sensitive to vPFC functions is presented in the chapter by Zald (Chapter 18, this volume).

### Processing facial affect

The processing of facial affect is a critical component of adaptive emotional and social behavior. The vPFC and the amygdala both play important roles in this processing. They receive inputs from posterior association cortices that process face stimuli (Amaral and Price 1984; Morecraft *et al.* 1992; Zald and Kim 1996), and they also contain populations of cells that respond to faces, and seem particularly specialized to respond to facial emotion

(Leonard et al. 1985; Adolphs et al. 1998; Rolls 1999; Kreiman et al. 2000; Fried et al. 2002; Frey and Petrides 2003). The amygdala may be especially important in the rapid signaling of the presence of aversive or threatening stimuli (LeDoux 2000; Whalen et al. 2001; Rauch et al. 2003; Zald 2003). While the OFC has been implicated in negative emotional states, it may also play an important role in processing the positive reinforcement value of a face. In healthy individuals, the degree of OFC activation correlates with a person's judgment about the relative happiness of facial expressions viewed, as well as the positive affect associated with viewing the face (Gorno-Tempini et al. 2001; Nitschke et al. 2004).

Studies in persons experiencing symptoms of either MDD or BD have shown deficits in facial affect recognition and discrimination (Gur et al. 1992; Rubinow and Post 1992; Addington and Addington 1998). These deficits appear to remit during periods of euthymia in BD, as euthymic BD patients have not shown impairments on measures of facial affect recognition (Harmer et al. 2002; Venn et al. 2004). An emerging literature also suggests that MDD and BD patients show mood-congruent biases in the processing of emotional stimuli. Individuals who are depressed exhibit a negative bias when processing stimuli. They are more likely to interpret facial affect as sadness (Gur et al. 1992), to selectively attend to negative facial expressions (Suslow et al. 2001), or to rate negative expressions as more intense than control subjects (Hale 1998). In contrast, when individuals are manic, they are less adept than comparison participants at recognizing negative facial emotion (Lembke and Ketter 2002).

## Cognitive dysfunction

A wide variety of neuropsychological measures has been employed to study cognition in mood disorders with varied results. This may be due, at least in part, to the heterogeneity in the tasks used, as well as in the clinical characteristics and mood state at testing of study participants. Across the literature, the cognitive constructs of attention, memory and executive functions are most consistently reported to be impaired in persons with mood disorders. These cognitive domains may reflect vPFC involvement (Savage et al. 2001; Zald et al. 2005), and are also subserved by other components of the neural circuitry implicated in mood disorders. Depression and mania have been associated with impairments in attentional functions, including spatial and sustained attention (Sweeney et al. 2000; Clark et al. 2001; Clark and Goodwin 2004; Weiland-Fiedler et al. 2004). In the domain of memory, depression in MDD or BD appears to be especially associated with vulnerability to abnormalities of episodic memory (Sweeney et al. 2000). On the other hand, while cognitive difficulties in the area of executive functions, such as planning and set-shifting, may be present in MDD (Elliott et al. 1997; Purcell et al. 1997), they may be more marked in BD depressed individuals than in MDD depressed individuals. Investigations of performance on the Stroop color-word task, and measures of verbal fluency, have also supported the idea that executive dysfunction is more pronounced in BD than MDD (Borkowska and Rybakowski 2001).

Cognitive dysfunction in individuals with mood disorders may be especially pronounced during the processing of stimuli that carry emotional and motivational salience, further implicating the vPFC. As in the processing of facial affect, individuals with mood disorders tend to show mood-congruent biases reflective of their current emotional state, when performing tests of executive function and in decision-making. For example, individuals who are depressed tend to be biased towards negative stimuli in an emotional Stroop task (Williams *et al.* 1996), and towards the recall of negative information in memory tasks (Bradley *et al.* 1995). While the mood-congruent bias for individuals in a depressed state is towards negative stimuli, manic patients display a bias towards positive stimuli. On an affective shifting task, relative to healthy comparison participants, depressed individuals responded slower to happy stimuli but not to sad targets, while manic patients reacted slower to sad stimuli but not to positive targets (Murphy *et al.* 1999). On one task that involved probability-based reward assessment and required adaptation to changing reward contingencies, persons who were manic performed worse than healthy comparison participants, and showed more marked deficits than depressed persons with MDD (Murphy *et al.* 2001). Mania in BD was also associated with impaired performance on the Iowa Gambling Task, in which participants are asked to assess probability-based outcomes in choosing from four decks of cards in order to win the most money possible (Clark *et al.* 2001). On average, the participants who were manic selected more cards from the risky decks than the comparison participants.

## Trait deficits and distribution

Investigators have raised the possibility of trait-related cognitive deficits that persist into remitted euthymic states of MDD or BD. While several studies reviewed by Clark and Goodwin (2004) have reported that remitted MDD patients do not show deficits in sustained attention, a recent study conducted by Weiland-Fiedler *et al.* (2004) suggests that persons with MDD in remission continue to display deficits in sustained attention, relative to healthy controls. Clark and Goodwin (2004) note that individuals with BD are impaired in target detection on a sustained attention task across mood states, including during euthymia. Euthymia in BD is also associated with impairments in working memory (Adler *et al.* 2004), and episodic memory (van Gorp *et al.* 1999; Altshuler *et al.* 2004). Deckersbach *et al.* (2004a, b, 2005) suggest that the trait deficit in BD may be in the initiation and implementation of organizational strategies used to encode verbal, as well as non-verbal material. Some individuals with BD have a very high level of functioning, but for others, cognitive complaints may persist across mood states, and the disorder may be more disabling. Altshuler *et al.* (2004) noted a bimodal distribution to executive functioning in BD euthymia, supporting the idea that there may be a subset of individuals with BD who are more vulnerable to developing persistent cognitive dysfunction.

The origin of the cognitive deficits that appear to persist into euthymia in adults with mood disorders is unclear. The possibility that they represent 'trait' abnormalities or expressions of genetic vulnerability to the disorders is intriguing. It is, however, possible that they represent residual effects of previous mood episodes. It is also possible that they are neurodegenerative features of the disorders that present in some individuals, or that they represent 'toxic' effects of acute mood episodes. Supporting the latter hypothesis, Basso and

Bornstein (1999), for example, have shown that individuals with recurrent MDD depressive episodes perform significantly worse on a word-list memory task than first-episode depressed individuals. However, emerging data suggest that cognitive deficits similar to those in adult euthymic BD appear early in the course of illness. Adolescents diagnosed with BD who are euthymic show impaired performance on executive and visual memory tasks (Dickstein et al. 2004). Unaffected first-degree relatives of individuals with BD also show a pattern of cognitive deficits on measures of memory and executive functions similar to that of BD euthymia (Keri et al. 2001; Ferrier et al. 2004; Zalla et al. 2004). These findings suggest that certain deficits may be vulnerability markers, present before or at least near BD onset, and they may also be present in relatives of individuals with BD. It is also possible that, in line with Dr. Altshuler's bimodal distribution hypothesis, the neuropsychological deficits may be present only in a subgroup of persons vulnerable to the disorder.

## Conclusions

The range of reported neuropsychological deficits in MDD and BD is wide. Nonetheless, the general pattern that emerges is one of difficulties in emotional processing, attention, memory and executive functions. Although dysfunction in multiple brain regions likely contributes to these difficulties, many of the deficits point towards dysfunction in the vPFC and the associated cortico-limbic circuitry. Classic vPFC-mediated abilities, such as reversal learning, set-shifting, and response to changing reinforcement contingencies, have been cited as disrupted in mood disorder patients, as have frontotemporal-dependent emotional processes, such as facial affect recognition and emotional bias. Some cognitive and emotional disturbances are potential trait deficits which may persist into euthymia, and may be present in adolescents with mood disorders and the first-degree relatives of persons with mood disorders. Taken together, these data imply that, at least in a subset of individuals with mood disorders, vPFC-related behavioral deficits may represent markers of vulnerability towards development of the disorder.

## 21.5. Structural neuroimaging in mood disorders

### Frontal abnormalities

Early structural imaging studies suggested that individuals with MDD, and those with BD, show decreases in various measures of frontal lobe size relative to individuals without a mood disorder. Studies of MDD have generally shown small but significant decreases in frontal lobe volume and frontal width, and increases in the width of prefrontal sulci (Krishnan et al. 1992; Coffey et al. 1993; Elkis et al. 1996). Similar findings have been reported in individuals with BD. Coffman et al. (1990) reported a trend towards decreased mean frontal area in persons with BD, and that smaller frontal volume in people with BD may correlate with the degree of neuropsychological impairment. Sax et al. (1999) found that small prefrontal volume in individuals with mania correlated with poor performance on a test of attentional function.

More recent structural neuroimaging studies have focused specifically on frontal subregions, and have repeatedly found significant volume differences in the vPFC between individuals with MDD or BD and individuals without these disorders. The seminal work of Drevets *et al.* (1997) showed decreased volume in left subgenual PFC in both MDD and BD (Fig. 21.1): a 48% reduction in MDD and a 39% reduction in BD. Bremner *et al.* (2002) found significantly smaller bilateral medial OFC (gyrus recti) volumes in individuals with remitted MDD as compared to participants without the disorder. A study in BD by Lopez-Larson *et al.* (2002) showed significantly smaller volume in right inferior frontal gray matter, as well as volume decreases bilaterally in other PFC regions.

## vPFC measures across the lifespan

There is some evidence that vPFC volume abnormalities may develop during adolescence in MDD or BD. A study by Botteron *et al.* (2002) detected smaller subgenual PFC volumes in female adolescents with MDD relative to subgenual PFC volumes in comparison participants without the disorder. In another study, adolescents with MDD also showed smaller frontal white matter but larger frontal gray matter volumes relative to healthy comparison participants (Steingard *et al.* 2002). With regard to BD, one study found that adolescent BD patients displayed reduced gray matter volume in both the subgenual anterior cingulate and the OFC relative to healthy comparison participants (Wilke *et al.* 2004). In a recent investigation, Blumberg *et al.* (2006) suggest that vPFC volume abnormalities in BD may progress over the course of adolescence to become

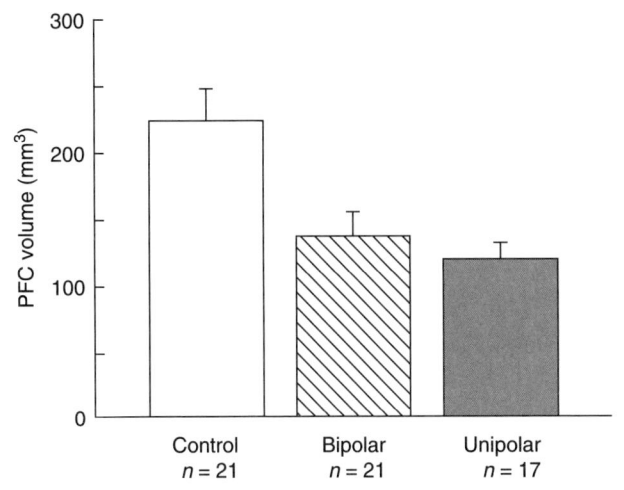

**Fig. 21.1** This graph of MRI-based volumes of the left subgenual prefrontal cortex (PFC) demonstrates the 48% volume reduction in a group of patients with MDD and the 39% volume reduction in a group of patients with BD, as compared to a group of healthy comparison subjects. Reproduced with permission from revets *et al.* (1997).

especially significant in early adulthood at the time when the more typical adult clinical phenotype for BD is usually expressed. This work implies that an abnormality in the maturation of the vPFC in adolescents vulnerable to BD may contribute to the development of the adult phenotype of BD.

Findings reported in elderly adults with mood disorders converge in describing significant volume deficits in the vPFC. Lai and colleagues showed smaller bilateral OFC in two studies of older adults with mood disorders compared to healthy age-matched comparison adults (Lai et al. 2000; Lee et al. 2003). Taylor et al. (2003) replicated the finding with a group of 41 elderly adults with mood disorders. Additionally, a study by Kumar et al. (2000) confirmed that the magnitude of this reduction was associated with the severity of the symptoms. Similarities between findings in mood disorders in young adult and geriatric samples suggest that vPFC volume abnormalities may develop in adolescence and may persist throughout adulthood in persons with these disorders.

### Measures in other brain regions

Volume abnormalities have been found in mesiotemporal limbic structures in both MDD and BD. In MDD, studies have generally shown decreases in the amygdala volume in adults (Sheline et al. 1998; Siegle et al. 2003), and small amygdala volume has also been reported in children and adolescents with MDD (Rosso et al. 2005). In BD, reports of amygdala size in adults are inconsistent at present (e.g. Swayze et al. 1992; Pearlson et al. 1997; Strakowski et al. 1999; Altshuler et al. 2000; Blumberg et al. 2003b) with increases, decreases, or no differences in volume seen relative to healthy comparison participants. In contrast, the few studies conducted in adolescents with BD concur and report smaller amygdala volumes in adolescents with BD relative to adolescents without BD (Blumberg et al. 2003b; DelBello et al. 2004; Chen et al. 2004). The presence of amygdala volume abnormalities in adolescents with mood disorders suggests that these may already be present at the time that vPFC abnormalities are developing. This raises the question of whether early amygdala abnormalities might influence the development of vPFC deficits in mood disorders.

Volume abnormalities in other cortico-limbic structures have been reported in mood disorders. The presence of episodic memory disturbances in mood disorders implicates abnormalities in the hippocampus. In MDD, hippocampal volume deficits have been reported, and may be associated with exposure to trauma and associated changes in stress hormones, as well as with recurrent depressive episodes (Sheline et al. 1999). Reports of hippocampal volumes in BD have been more variable, although decreases have been noted in adolescents with BD (Blumberg et al. 2003b; Frazier et al. 2005). Caudate and cerebellar vermis volume abnormalities have also been implicated (Aylward et al. 1994; DelBello et al. 1999; Brambilla et al. 2001; Noga et al. 2001) in mood disorders.

### Conclusions

Structural neuroimaging studies implicate reductions in volume of the vPFC in both MDD and BD. These volume abnormalities may emerge in adolescence, progress over this developmental epoch, and persist into young and older adulthood. This theorized progression parallels the emergence of the typical, clinical phenotypes of these disorders that are recognized most commonly in early adulthood. Volume abnormalities in subcortical structures with significant vPFC connectivity have also been found in individuals with mood disorders, may manifest early, and may influence the development of vPFC abnormalities. Future research that helps to elucidate the development of the vPFC in humans, and models of abnormalities in vPFC development, may contribute to understanding of the neurodevelopment of mood disorders.

## 21.6. Functional neuroimaging

While structural neuroimaging studies have revealed differences in regional brain volumes between individuals with mood disorders and healthy comparison participants, positron emission tomography (PET) and functional magnetic resonance imaging (fMRI) techniques can provide *in vivo* measures of regional brain function in individuals experiencing mood symptoms. Studies of resting activity, as well as of task-related activation in MDD and BD, have also led to converging findings in vPFC neural systems.

### Resting state studies

Studies of resting regional cerebral blood flow and metabolism in mood disorders reveal different mood state-related patterns of hyper- and hypo-activity in the vPFC for depression and mania. In MDD, the observed pattern of resting activity consists of hyperactivity in the OFC (Drevets 2000), but hypoactivity in the dorsolateral prefrontal cortex (DLPFC) (Baxter *et al.* 1989; Bench *et al.* 1993; Mayberg *et al.* 1999). Hyperactivity has also been demonstrated in the amygdala during major depressive episodes (Drevets *et al.* 1992, 1995, 2001; Abercrombie *et al.* 1998), and correlates positively with depressive severity and with cortisol levels (Drevets *et al.* 2002). The pattern of abnormal resting activity in the depressed phase of BD is similar to that of MDD. The OFC (Drevets 2000), as well as the associated subcortical regions, like the thalamus and the ventral striatum (Ketter *et al.* 2001), show significantly increased resting activity in BD depression. While one early study of BD showed decreased right amygdala activity (Post *et al.* 1987), more recent studies have generally reported increased amygdala activity in both the right (Ketter *et al.* 2001) and left (Drevets *et al.* 2002) amygdala. The DLPFC, as in MDD, is abnormally hypoactive in individuals with BD depression, particularly in the left hemisphere (Baxter *et al.* 1989).

Overall, bipolar mania presents a pattern of resting state activity that is inverse to that seen in depression. Studies have reported decreased resting state activity in the OFC during mania (Rubin *et al.* 1995; Blumberg *et al.* 1999), while dorsal frontal areas, including the dorsal anterior cingulate, show increased resting activity (Goodwin *et al.* 1997; Blumberg *et al.* 2000). Thus, studies of resting brain activity suggest mood state-related abnormalities in the vPFC with increased activity associated with depression and decreased activity associated with mania, as well as abnormalities in the regions with substantial connectivity to the vPFC.

## Task-related activation: PET and fMRI studies

Studies of task-related brain activation using PET and blood-oxygen-level-dependent (BOLD) fMRI in MDD and BD have implicated functional brain abnormalities in regions similar to those areas highlighted by studies of resting activity. While structural and resting-state functional imaging studies implicate volume and functional abnormalities in vPFC neural systems in mood disorders, task-related activation studies can reveal abnormalities in the on-line functioning of brain regions within the context of an established cognitive or behavioral task. Recent research has used both emotional and cognitive activation tasks, targeting the vPFC and the related cortico-limbic structures, in the study of MDD and BD.

## Brain activation abnormalities during emotional processing tasks

Functional neuroimaging studies in which participants process emotional stimuli (such as faces, emotionally-valenced words and autobiographical scripts) have pointed to abnormalities in vPFC functioning in mood disorders. The direction of findings (relative increases or decreases in activation) differs across the studies. Authors have noted the possibility of interesting relationships between the behavioral task studies, the mood disorder subtype, the mood state at scanning, or the presence of a specific subset of symptoms, that may help to explain what appear to be heterogeneous findings.

During performance of an emotional go/no-go task, individuals with MDD who were depressed displayed elevated response in the rostral anterior cingulate and the anterior medial prefrontal cortex to sad stimuli but decreased response in these regions to happy emotional stimuli, raising the possibility of mood-congruent bias in vPFC recruitment (Elliott *et al.* 2002). Dougherty *et al.* (2004) provided the interesting perspective of subdividing MDD by the presence or absence of anger attacks. Employing angry and neutral autobiographical scripts, this group found deficits in vPFC recruitment that were specific to the subset of individuals with MDD who also had anger attacks when they processed anger-related stimuli. In the study of BD depression and euthymia, viewing of both positive and negative emotional face expressions was associated with vPFC hyperactivation (Lawrence *et al.* 2004). Ventral and medial PFC hyperactivation was also observed in BD mania during an emotional go/no-go task, especially in association with happy stimuli (Elliott *et al.* 2004).

Functional neuroimaging during emotional processing has also revealed abnormalities in other components of the cortico-limbic circuitry implicated in mood disorders, including the amygdala. Depressed individuals with MDD viewing masked emotional faces show abnormal increases in left amygdala response (Sheline *et al.* 2001). Siegle *et al.* (2002) reported that depression in MDD was associated with a sustained amygdala response to emotionally-valenced words that persisted beyond the amygdala response of healthy comparison participants. The sustained response was moderately related to self-report of rumination, a common symptom of MDD (Siegle *et al.* 2002). Two fMRI studies of emotional face viewing revealed that individuals with BD exhibited hyperactivation of the amygdala, whether in a mixed, depressed or euthymic state (Yurgelun-Todd *et al.* 2000; Lawrence *et al.* 2004) and in association with faces depicting both positive and negative emotions.

These studies suggest potential differences between MDD and BD in the processing of emotional stimuli. In MDD, abnormal amygdala and vPFC responses were observed in response to stimuli of negative emotional valence. However, in BD, abnormal amygdala and vPFC responses were observed in response to both positive as well as negative stimuli. This suggests that BD may be characterized by an enhanced perception of salience of both positive and negative emotional stimuli (Lawrence *et al.* 2004).

Activation abnormalities in additional cortico-limbic structures have been reported in mood disorders. For example, BD participants who were depressed displayed hyperactivation of the ventral striatum, thalamus and hippocampus when viewing emotional facial expressions (Lawrence *et al.* 2004) and BD participants who were hypomanic showed abnormal activation of caudate and thalamus (Malhi *et al.* 2004). Malhi *et al.* (2004) suggest that BD patients may draw more heavily on these subcortical regions when performing emotional tasks.

Taken together, these studies suggest the presence of abnormalities in the response of the vPFC, the amygdala, and other connected subcortical structures in persons with mood disorders during the processing of emotional stimuli. The various directions of findings suggest the presence of complex interactions between diagnosis, mood state, and the valence of the emotional stimuli.

## Brain activation abnormalities during cognitive tasks

Studies of brain function in individuals with mood disorders during performance of tasks that target attentional, mnemonic, and executive processes demonstrate a distributed pattern of brain abnormalities that generally echoes the pattern found in the lesion literature and structural and resting imaging studies, wherein depression is associated with left-hemisphere dysfunction, and with increased vPFC and decreased dorsal frontal activation, and mania is associated with right-hemisphere dysfunction, and with decreased vPFC and increased dorsal frontal activity.

Blumberg and colleagues found manic state-related deficits in activation in BD in right caudal OFC and rostral vPFC during word generation (Blumberg *et al.* 1999). A related study, in which individuals who were manic performed a decision-making task, yielded similar results, finding decreased activation in right vPFC as well as increased activation in dorsal anterior cingulate (Rubinsztein *et al.* 2001). In a later fMRI study, Blumberg *et al.* (2003c) employed a color-naming Stroop paradigm. Results showed deficits in task-related activation in right caudal OFC in association with BD mania consistent with the previous work. An inverse activation pattern was exhibited in BD depression, with increased task-related activation in left caudal OFC relative to euthymic participants (Fig. 21.2). In contrast to the increased dorsal anterior cingulate activation in manic participants, reported by Rubinsztein *et al.* (2001), several studies have found decreased dorsal anterior cingulate activity in depressed individuals during the performance of cognitive tasks. George and colleagues found decreased dorsal anterior cingulate

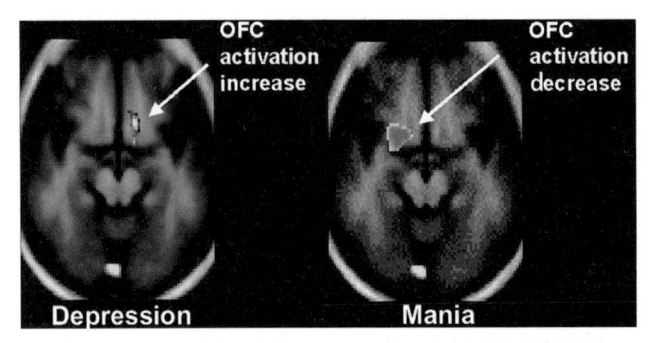

**Fig. 21.2** (Also see Color Plate 32.) The left image demonstrates the relative increases in left caudal OFC activation in a group of BD participants who were depressed at the time of scanning as compared to BD participants who were euthymic at scanning. The right image demonstrates relative decreases in right caudal OFC in BD mania as compared to euthymia. Images are presented in radiological convention, that is the right side of the image displays the left side of the brain. Reproduced with permission from Blumberg et al. (2003c).

activation in a mixed group of depressed MDD and BD subjects on a Stroop task (George et al. 1997). In a geriatric MDD sample, de Asis et al. (2001) found decreased dorsal anterior cingulate, as well as decreased hippocampal activity, during a word generation task.

Functional abnormalities associated with the BD-trait have been suggested by studies that demonstrate the presence of vPFC abnormalities during euthymic periods of the disorder. For example, in addition to the mood state-related abnormalities in caudal OFC noted by Blumberg et al. this group also noted the presence of activation abnormalities in a more rostral region of left vPFC that was independent of mood state (i.e., present in depressed, manic or euthymic states), perhaps indicating an vPFC functional deficit related to the BD trait (Blumberg et al. 2003c). However, Strakowski et al. found that when persons with BD were euthymic they demonstrated higher activation in the vPFC and the amygdala relative to healthy comparison participants during a test of sustained attention (Strakowski et al. 2004). This is reminiscent of the results of the functional neuroimaging studies designed to test emotional processing more directly. The authors suggest that while the participants with BD were performing a putatively non-emotional, attentional task, they may have displayed a trait vulnerability in attaching a heightened emotional valence to the task, suggesting a potential vulnerability to hyperarousal of the vPFC and the amygdala in BD. The differing results of these studies again implicate abnormalities in the vPFC in mood disorders, and also again highlight the complex interactions that may occur between specific mood disorder diagnoses, mood state at the time of scanning, and the specific features of the paradigms employed during scanning.

## Conclusions

Taken together, resting state and task-related functional neuroimaging studies implicate similar vPFC neural system abnormalities in mood disorders. Depression tends to involve abnormally increased activity in the left vPFC, while mania often appears to involve abnormally decreased activity in the right vPFC. Some studies hint at trait-related abnormalities that may be present across mood states, especially in BD. The task-related activation literature again implicates subcortical structures, such as the amygdala, in MDD and BD, and suggests the possibility that frontal regulation of amygdala activity may be abnormal in both disorders. Abnormal responsiveness of these structures may be especially elicited by emotionally-charged stimuli in BD. Further elucidation of the relationship between the PFC and the amygdala (Zald et al. 1998), and of how this relationship goes awry in mood disorders, may be especially informative for understanding mood disorder pathophysiology.

## 21.7. Postmortem studies of mood disorders: neuronal and glial involvement

While imaging studies have consistently pointed to the vPFC as a site of structural and functional abnormalities in mood disorders, the types of cellular abnormalities that contribute to these findings are not known. In the last decade, however, several post-mortem studies of MDD and BD have shown cellular-level abnormalities in the vPFC that may contribute to the neuropathophysiology of mood disorders, with particular emphasis on findings in glial cells and, most recently, of oligodendrocyte abnormalities.

The work of Rajkowska and colleagues has provided evidence of vPFC cellular abnormalities in MDD. They found decreases in cortical thickness, neuronal size, and neuronal and glial density in the upper cortical layers of the rostral vPFC, and decreases in neuronal size, as well as marked reductions in glial densities, in the caudal vPFC in MDD (Rajkowska et al. 1999). Ongur et al. (1998) showed that glial cell counts are lower in the subgenual PFC in MDD, especially in persons with a family history of mood disorders. Post-mortem studies in MDD have also focused on areas with strong connectivity to the vPFC, including the DLPFC and the amygdala. Glial and neuronal size and density were decreased in the DLPFC in MDD (Rajkowska et al. 1999). Glial density and the glia:neuron ratio were significantly reduced in the amygdala in MDD, particularly on the left side (Bowley et al. 2002); this reduction was primarily due to a decreased density of oligodendrocytes in patients with MDD (Hamidi et al. 2004).

Ongur et al. (1998) found reduced glial cell counts in BD comparable to those found in subgenual PFC in MDD. This group also found preliminary evidence of glial decreases in posterior-medial OFC (Ongur et al. 1998). Preliminary observations of neuronal and glial decreases in the vPFC have also been observed by Dr. Grazyna Rajkowska and others (Rajkowska 2002; Selemon and Rajkowska 2003), including a lateralized finding of 15–20% decreased neuronal density in the left subgenual PFC (Bouras et al. 2001). Cellular abnormalities have been demonstrated in the brain areas with strong connectivity to the vPFC in BD: for instance, glial and neuronal densities were reduced in the DLPFC (Rajkowska et al. 2001). Studies of glial cells in the amygdala in BD have been scant but suggestive. Bowley et al. (2002) note that the two BD subjects in their study, who were not treated with lithium or valproic acid, had significant reduction of glia in the amygdala, but that BD subjects being treated with medications did not; this raises the possibility of protective effects of lithium and other medications (see Molecular Abnormalities and Pharmacotherapy below). Finally, preliminary evidence of oligodendrocyte abnormalities in BD has been reported, with decreased oligodendrocyte density in BD in Brodmann areas 9 and 10, and in caudate (Uranova *et al.* 2001, 2004). Data are also emerging from studies of gene expression, with Tkachev *et al.* (2003) reporting that the expression of multiple oligodendrocyte- and myelination-related genes is down-regulated in the frontal cortex in BD. However, this tissue was sampled from the DLPFC. Direct studies of the vPFC are needed.

## Conclusions

Findings from postmortem studies suggest that glial and neuronal pathology contribute to macroscopic structural and functional vPFC abnormalities in mood disorders. Future studies that more specifically investigate cell pathology in the vPFC in BD will be especially valuable, such as investigations into the specific subpopulations of glia involved, and into mechanisms related to the development of cellular pathology in mood disorders.

## 21.8. Development

A better understanding of the developmental pathophysiology of mood disorders has the potential to aid in our ability to detect these disorders at a younger age, and to intervene earlier with the hope of preventing progression and improving prognosis. A correct identification of mood disorders at an early age is crucial. Currently, many youth with BD are misdiagnosed with attention-deficit/hyperactivity disorder (ADHD) or MDD (e.g., Isaac 1992; Wozniak *et al.* 2001; Cicero *et al.* 2003), and treated with stimulants or antidepressants that may hasten the onset of manic episodes and worsen the clinical course for the patient (Faedda *et al.* 1995; DelBello *et al.* 2001; Cicero *et al.* 2003; Coyle *et al.* 2003). The neurodevelopmental trajectories of cortico-limbic circuitry may provide hints as to why MDD, BD and ADHD are difficult to differentiate in youth, as well as shed light on the role of the vPFC in mood disorders in particular.

### Normal developmental trajectories of PFC and subcortical structures

Evidence from both human and nonhuman primate studies suggests the presence of dynamic developmental changes in the PFC in adolescence (Yakovlev and Lecours 1967; Machado and Bachevalier 2003). Nonhuman primate and human studies have shown that, after a series of progressive synaptic developmental steps that occur before adolescence, adolescence brings substantial regression of synaptic processes in the PFC thought to refine higher order behavioral responses (Alexander and Goldman 1978; Huttenlocher 1984; Bourgeois *et al.* 1994; Lewis 1997; Erickson *et al.* 1998). Structural MRI studies of healthy humans are consistent with this result, showing increased gray matter in the PFC as adolescence approaches, and decreased gray matter thereafter (Pfefferbaum et al. 1994; Passe et al. 1997; Giedd et al. 1999). Changes in brain function and behavior accompany this sequence of developmental changes in structure. Studies in nonhuman primates have shown that a shift occurs over the course of adolescence from the recruitment of subcortical structures to the recruitment of the PFC, when engaging in goal-directed behavior. For instance, during adolescence, nonhuman primates increasingly recruit the OFC relative to the amygdala as adult social behavior emerges (Machado and Bachevalier 2003). FMRI studies in humans also suggest that as adolescence progresses, dependence on the prefrontal cortex increases when inhibiting motor and prepotent responses (Rubia et al. 2000; Luna et al. 2001; Blumberg et al. 2003a). Likewise, when processing emotional faces, subjects rely progressively more heavily on the PFC relative to the amygdala in adulthood (Killgore et al. 2001). The neuroimaging studies have primarily noted increases in DLPFC function with age, and have provided less data regarding VPFC development. VPFC development warrants further study.

## Evidence for abnormal trajectories in pediatric BD

Preliminary evidence suggests an abnormal trajectory for vPFC development in BD. In the structural neuroimaging section above we noted preliminary evidence in support of an abnormal trajectory of vPFC structural maturation in BD. With regard to vPFC function, differences were observed in age-correlated vPFC activation (during a color-naming Stroop task) between adolescents with and without BD. While differences between healthy and BD adolescents in PFC function may be evolving during this developmental epoch, differences in activation are already apparent in the left putamen and thalamus (Blumberg et al. 2003a). Another study of pediatric BD demonstrated increased subcortical (including thalamic) activation during tests of visuospatial working memory and the viewing of pictures with positive valence (Chang et al. 2004). Combined, these observations suggest that adolescents with BD may already differ from adolescents without BD in the recruitment of striatum and thalamus; however, group differences in the task-related response of the vPFC may be continuing to progress during adolescence. Disorders that are difficult to distinguish from BD in youth may have subcortical abnormalities in the striatum and the thalamus in common. Striatal and thalamic abnormalities have been reported in neuroimaging studies of Tourette syndrome, obsessive compulsive disorder, and ADHD as well as in BD (Leckman et al. 1997; Shafritz et al. 2004; Spessot et al. 2004). The similarities in subcortical abnormalities in youth across these disorders may contribute to some of their common behavioral features. As vPFC abnormalities may be continuing to progress in their expression over the adolescent period, behavioral abnormalities related to the vPFC and characteristic of the prototypical adult BD phenotype may become increasingly apparent (Blumberg et al. 2004, 2006).

## Conclusions

Preliminary evidence suggests a developmental sequence for the expression of regional brain abnormalities in mood disorders, such that subcortical abnormalities may be expressed earlier than vPFC abnormalities, which continue to progress over adolescence. Understanding neurodevelopmental trajectories in mood disorders may help to identify early abnormalities and provide clues as to the pathophysiological mechanisms underlying the development of the disorders, and thus lead to improvements in early treatment interventions.

## 21.9. Molecular abnormalities and pharmacotherapy

While lesion, neuropsychological, structural and functional neuroimaging studies have provided converging evidence for the involvement of vPFC neural systems in mood disorders, the molecular contributions to these abnormalities are not known. 'Challenge' studies can provide inroads into understanding contributions of particular neurotransmitter systems to regional brain abnormalities. A range of treatments are effective for mood disorders including psychotherapy, medication classes including antidepressants, lithium, anticonvulsants, and antipsychotic agents, as well as electroconvulsive therapy. The mechanisms of action underlying their therapeutic effects are unclear, but emerging evidence for the influence of treatments on regional brain abnormalities has begun to shed light in this area.

### Pharmacological challenge studies

Serotonin is a neurotransmitter that has been especially implicated in mood disorders (Goodwin and Jamison 1990) and thus pharmacological 'challenge' studies conducted in an attempt to explain the neurochemical mechanism underlying structural and functional abnormalities in mood disorders have often manipulated the serotonergic system. An example of such a challenge study is the tryptophan-depletion paradigm, in which patients are given an amino acid cocktail including all essential amino acids except the serotonin precursor tryptophan. The resulting transient depletion of tryptophan is associated with a temporary relapse of depressive symptoms in individuals with remitted MDD (Delgado *et al.* 1990; Leyton *et al.* 2000). Tryptophan-depletion was also noted to be associated with changes in OFC metabolism in individuals with remitted depression (Bremner *et al.* 1997; Neumeister *et al.* 2004). In one report, OFC changes were associated with relapse of depressive symptoms (Bremner *et al.* 1997). This evidence, in addition to a recent study that demonstrated disruption of vPFC-mediated reward processing by tryptophan depletion (Rogers *et al.* 2003), suggests that serotonin plays an important role in OFC-related abnormalities in mood disorders.

### Modulation of frontolimbic circuitry by treatment

The work of Mayberg (2002) suggests that serotonergic pharmacotherapy (e.g. selective serotonin reuptake inhibitors, SSRIs) can modulate the functioning of vPFC neural systems. This group has reported that individuals with MDD who respond to the SSRI fluoxetine have pre-treatment elevations in resting activity in ventral anterior cingulate. Reductions in these abnormalities are associated with treatment response. Moreover, the time course of this effect is similar to that of clinical response to fluoxetine, i.e. reductions in subgenual PFC metabolism can be detected after six weeks of treatment (Mayberg *et al.* 2000) (Fig. 21.3). Decreases in ventral anterior cingulate activity are also sustained during an 18-month period of remission (Mayberg 2002). Most recently, the group has demonstrated that chronic deep brain stimulation of white matter tracts adjacent to the subgenual cingulate can lead to marked improvement in individuals with treatment-resistant MDD. Antidepressant effects of this treatment were also associated with reductions in abnormal metabolism in vPFC circuitry: subgenual cingulate and OFC activity decreased, while dorsal PFC and brainstem activity increased (Mayberg *et al.* 2005). Similar findings have been noted by numerous research groups, including those of Brody and colleagues, who demonstrated that antidepressant response to the SSRI paroxetine is associated with return of vPFC activation to healthy levels (Brody *et al.* 1999, 2001a,b). Interestingly, vPFC changes were also found in association with positive response to cognitive behavioral and interpersonal psychotherapies, suggesting that the effect of these non-pharmacological treatments might also be mediated by their influence on vPFC function (Brody *et al.* 2001b; Goldapple *et al.* 2004). Preliminary evidence suggests that similar treatment effects may exist in BD. Mood stabilisers may help to normalize the functional abnormalities found in BD in vPFC and amygdala (Baxter *et al.* 1989; Goodwin *et al.* 1997; Drevets *et al.* 2002; Ketter and Wang 2002; Lawrence *et al.* 2004; Blumberg *et al.* 2005).

## Neuroprotective effects

Recently, some medications used to treat mood disorders have been associated with neurotrophic and neuroprotective effects. For example, lithium, a long-standing first-line

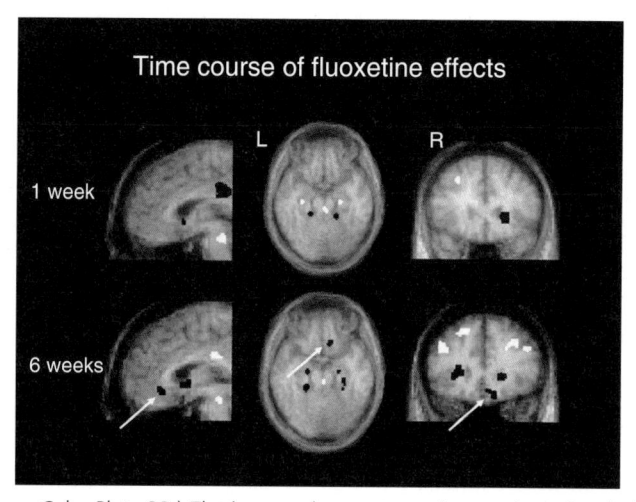

**Fig. 21.3** (Also see Color Plate 33.) The images demonstrate changes in regional glucose metabolism after 1 week and 6 weeks of fluoxetine treatment for MDD. Areas depicting decreases in metabolism are shown in green and areas of increased metabolism are shown in red (black and white respectively in the black and white figure ). Decreases in subgenual prefrontal cortex metabolism (white arrows) are detected after 6 weeks of treatment. Reproduced with permission from Mayberg *et al.* (2000).

treatment for BD, may upregulate the neuroprotective protein bcl-2, (Manji *et al.* 1999) and is associated with increases in *N*-acetyl-aspartate, a putative marker for neuronal viability (Moore *et al.* 2000b, Silverstone *et al.* 2003). Lithium treatment is also associated with increased gray matter volume in persons with BD (Moore *et al.* 2000a). Preliminary evidence suggests that lithium and valproic acid are associated with normalization of gray matter volume in vPFC in individuals with BD (Drevets 2001, Blumberg *et al.* 2005). This raises the possibility that treatments for mood disorders have the potential not only to normalize functioning within vPFC neural systems, but also to help reverse structural abnormalities in vPFC.

## Conclusions

Current research is beginning to unravel the neural systems implicated in medication response in patients with mood disorder. Serotonin may have an important role in vPFC-related regulation of emotional and reward processing. Evidence suggests that the key to successful treatment may be a modulation or normalization of functional abnormalities in frontolimbic circuitry and possibly an ability to reverse structural abnormalities in vPFC.

## 21.10. Overall conclusions and future directions

Evidence from a variety of research modalities has implicated vPFC as a key region in the pathophysiology of mood disorders. Given its primary functions as well as its connectivity, the vPFC has the potential to be either directly or indirectly involved in the broad range of symptoms characteristic of MDD and BD. The central role of the vPFC in the adaptive regulation of behavior, based on its emotional and motivational significance (Schoenbaum *et al.* 1998; Bechara *et al.* 1999; Rolls 2004), implicates it in the core emotional, motivational and social disturbances of primary mood disorders. VPFC is also tightly interconnected with both subcortical limbic structures important to neurovegetative functioning (Amaral and Price 1984; Carmichael and Price 1995a) as well as to dorsal frontal regions associated with attention and higher-order cognitive processes (Barbas and Pandya 1989; Carmichael and Price 1995b), and thus vPFC is implicated in the broader range of symptoms also characteristic of mood disorders.

Lesion studies first demonstrated that insult to vPFC could cause emotional and social difficulties, but highlighted only a subset of the complex variety of symptoms characteristic of MDD and BD. Neuropsychological work has further characterized the emotional and cognitive deficits associated with mood disorders, some of which may develop early on in the course of illness. Preliminary evidence suggests that some cognitive deficits associated with mood disorders may be present in individuals who are vulnerable to development of MDD or BD. Future longitudinal studies that specifically target vPFC-related tasks could be especially valuable in identifying trait features of the disorders.

Research on vPFC structure, in both volumetric MRI and postmortem cellular studies, has provided evidence that mood disorders are associated with pathology in vPFC volume and in measures of neuronal and especially of glial density. Future work on the particular cell types affected and clarification of their roles will help to illuminate the neuropathology associated with mood disorders.

Functional neuroimaging work has contributed important information about regional brain functioning in individuals with mood disorders during the resting state, as well as during emotional processing and the performance of cognitive tasks. This literature has advanced our understanding of the intra- as well as inter-hemispheric abnormalities present in mood disorders, including the role of hemispheric laterality in the illnesses, where mania is often associated with right-sided deficits and depression with left-sided dysfunction. Furthermore, a pattern has emerged with respect to dorsal and ventral frontal regions: while depression is associated with increased vPFC activity, but decreases in dorsal frontal activity, mania shows the opposite pattern. Functional findings support the presence of abnormalities in the interactions between vPFC and amygdala in mood disorders that may stem from abnormal modulation of amygdala by vPFC.

Emerging evidence, including that gleaned from neuroimaging studies, suggests that the normal pattern of development of cortico-limbic circuitry may be disrupted in individuals with mood disorders. Ongoing research in this area may help to contribute to new methods to identify the disorders early in their course and design early treatment interventions that could have a salutary effect on prognosis. Finally, investigations of the neurochemical mechanisms involved in mood disorders suggest that normalization of vPFC function, and perhaps structure, may be important components of successful treatment of mood disorders.

Studies of mood disorders in general are limited by several factors. Mood disorders show a great deal of heterogeneity, including differences in the frequency, severity and type of mood episodes, history of trauma, and presence of psychosis. In particular, studies including manic subjects are limited in that patients with the most severe symptoms might exhibit psychomotor and agitated behaviors that would make research participation in many study modalities difficult, e.g., neuroimaging and neuropsychological testing. Studies of both MDD and BD are also complicated by common comorbid psychiatric and medical conditions, e.g., many individuals with mood disorders may also abuse alcohol or illicit substances, and by the range of medications prescribed to patients with these disorders. These issues are especially salient in neuroimaging studies that have generally used fairly small patient samples, increasing the likelihood that sources of heterogeneity might confound results. However, the future is hopeful: studies with increased numbers of subjects are becoming more common, as are studies of mood disorders in pediatric samples as well as of unaffected first-degree relatives of mood disorder patients. Better methods for studying these heterogeneous patient populations are also being actively investigated, including the subdivision of individuals with mood disorders by certain symptom constellations or by specific genetic variations. Additionally, technological advancements are allowing the development of tasks and methodologies that target vPFC function more directly and could be used in the study of mood disorders.

The future promises many exciting advances in our understanding of mood disorders and vPFC's role within them. Diffusion tensor MRI (DT-MRI) is an imaging technique that provides measures of the structural integrity of white matter connections. DT-MRI has already been used to aid in the study of schizophrenia (Buchsbaum *et al.* 1998, Lim *et al.* 1999) and has great potential as a method to investigate possible abnormalities in white matter tracts within cortico-limbic circuits in mood disorders. A better understanding of the functional connections implicated in mood disorders, combined with an increased understanding of relevant molecular and neurotransmitter-level mechanisms, will be extremely valuable. At present, striking genetic findings in schizophrenia have emerged (Hariri *et al.* 2003, Egan *et al.* 2004), with particular genotypes relating not only to risk of developing the disorder but also to specific behavioral and neural profiles. Research on the influence of genetic variations within mood disorders on regional brain abnormalities holds great promise to advance our understanding of molecular contributions to the neuropathophysiology of mood disorders and to point towards novel treatment strategies.

## References

**Abercrombie, H. C., Schaefer, S. M., Larson, C. L. *et al.*** (1998) Metabolic rate in the right amygdala predicts negative affect in depressed patients. *Neuroreport* **9**:3301–3307.

**Addington, J. and Addington, D.** (1998) Facial affect recognition and information processing in schizophrenia and bipolar disorder. *Schizophrenia Research* **32**:171–181.

**Adler, C. M., Holland, S. K., Schmithorst, V., Tuchfarber, M. J., and Strakowski, S. M.** (2004) Changes in neuronal activation in patients with bipolar disorder during performance of a working memory task. *Bipolar Disord* **6**:540–549.

**Adolphs, R., Tranel, D., and Damasio, A. R.** (1998) The human amygdala in social judgment. *Nature* **393**:470–4.

**Alexander, G. E. and Goldman, P. S.** (1978) Functional development of the dorsolateral prefrontal cortex: an analysis utilizing reversible cryogenic depression. *Brain Res* **143**:233–249.

**Altshuler, L. L., Bartzokis, G., Grieder, T. *et al.*** (2000) An MRI study of temporal lobe structures in men with bipolar disorder and schizophrenia. *Biol Psychiatry* **48**:147–162.

**Altshuler, L. L., Ventura, J., van Gorp, W. G., Green, M. F., Theberge, D. C., and Mintz, J.** (2004) Neurocognitive function in clinically stable men with bipolar I disorder or schizophrenia and normal control subjects. *Biol Psychiatry* **56**:560–569.

**Amaral, D. G., and, Price, J. L.** (1984) Amygdalo-cortical projections in the Monkey (Macaca fascicularis). *J Comp Neurology* **230**:465–96.

**APA** (1994) *Diagnostic and Statistical Manual of Mental Disorders, 4th Edition.* Washington, DC: American Psychiatric Association.

**Aylward, E. H., Roberts-Twillie, J. V., Barta, P. E. *et al.*** (1994) Basal ganglia volumes and white matter hyperintensities in patients with bipolar disorder. *Am J Psychiatry* **151**:687–693.

**Barbas, H. and Pandya, D. N.** (1989) Architecture and intrinsic connections of the prefrontal cortex in the rhesus monkey. *J Comp Neurol* **286**:353–375.

**Basso, M. R. and Bornstein, R. A.** (1999) Relative memory deficits in recurrent versus first-episode major depression on a word-list learning task. *Neuropsychology* **13**:557–563.

**Baxter, L. R., Schwartz, J. M., Phelps, M. E. *et al.*** (1989) Reduction of prefrontal glucose metabolism common to three types of depression. *Arch Gen Psychiatry* **46**:243–250.

**Bechara, A., Damasio, H., Damasio, A. R., and Lee, G. P.** (1999) Different contributions of the human amygdala and ventromedial prefrontal cortex to decision-making. *J Neurosci* **19**:5473–5481.

Bench, C. J., Friston, K. J., Brown, R. G., Frackowiak, R. S., and Dolan, R. J. (1993) Regional cerebral blood flow in depression measured by positron emission tomography: the relationship with clinical dimensions. *Psychol Med* **23**:579–590.

Blumberg, H. P., Stern, E., Ricketts, S. *et al.* (1999) Rostral and orbital prefrontal cortex dysfunction in the manic state of bipolar disorder. *Am J Psychiatry* **156**:1986–1988.

Blumberg, H. P., Stern, E., Martinez, D. *et al.* (2000) Increased anterior cingulate and caudate activity in bipolar mania. *Biol Psychiatry* **48**:1045–52.

Blumberg, H. P., Martin, A., Kaufman, J. *et al.* (2003a) Frontostriatal abnormalities in adolescents with bipolar disorder: preliminary observations from functional MRI. *Am J Psychiatry* **160**:1345–1347.

Blumberg, H. P., Kaufman, J., Martin, A. *et al.* (2003b) Amygdala and hippocampus volumes in adolescents and adults with bipolar disorder. *Arch Gen Psychiatry* **60**:1201–1208.

Blumberg, H. P., Leung, H. C., Skudlarski, P. *et al.* (2003c) A functional magnetic resonance imaging study of bipolar disorder. *Arch Gen Psychiatry* **60**:601–609.

Blumberg, H. P., Kaufman, J., Martin, A., Charney, D. S., Krystal, J. H., and Peterson, B. S. (2004) Significance of adolescent neurodevelopment for the neural circuitry of bipolar disorder. *Ann NY Acad Sci* **1021**:376–383.

Blumberg, H. P., Donegan, N. H., Sanislow, C. A., *et al.* (2005) Preliminary evidence for medication effects on functional abnormalities in the amygdala and anterior cingulate in bipolar disorder. *Psychopharmacology*, **183**: 308–313.

Blumberg, H. P., Krystal, J. H., Bansal, R. *et al.* (2006) Age, rapid-cycling, and pharmacotherapy effects on ventral prefrontal cortex in bipolar disorder: a cross-sectional study. *Biol Psychiatry* **59**:611–618.

Borkowska, A. and Rybakowski, J. K. (2001) Neuropsychological frontal lobe tests indicate that bipolar depressed patients are more impaired than unipolar. *Bipolar Disord* **3**:88–94.

Botteron, K. N., Raichle, M. E., Drevets, W. C., Heath, A. C., and Todd, R. D. (2002) Volumetric reduction in left subgenual prefrontal cortex in early onset depression. *Biol Psychiatry* **51**:342–344.

Bouras, C., Kovari, E., Hof, P. R., Riederer, B. M., and Giannakopoulos, P. (2001) Anterior cingulate cortex pathology in schizophrenia and bipolar disorder. *Acta Neuropathol* **102**:373–379.

Bourgeois, J. P., Goldman-Rakic, P. S., and Rakic, P. (1994) Synaptogenesis in the prefrontal cortex of rhesus monkeys. *Cerebral Cortex* **4**:78–96.

Bowden, C. L., Calabrese, J. R., and McElroy, S. L. (2000) A randomized, placebo-controlled 12-month trial of divalproex and lithium in treatment of outpatients with bipolar I disorder. *Arch Gen Psychiatry* **57**:481–9.

Bowley, M. P., Drevets, W. C., Ongur, D., and Price, J. L. (2002) Low glial numbers in the amygdala in major depressive disorder. *Biol Psychiatry* **52**:404–412.

Bradley, B. P., Mogg, K., and Williams, R. (1995) Implicit and explicit memory for emotion-congruent information in clinical depression and anxiety. *Behav Res Ther* **33**:755–770.

Brambilla, P., Harneski, K., Nicoletti, M. *et al.* (2001) MRI study of posterior fossa structures and brain ventricles in bipolar disorder. *J Psychiatry Res* **35**:313–22.

Bremner, J. D., Innis, R. B., Salomon, R. M. *et al.* (1997) Positron emission tomography measurement of cerebral metabolic correlates of tryptophan depletion-induced depressive relapse. *Arch Gen Psychiatry* **54**:364–374.

Bremner, J. D., Vythilingam, M., Vermetten, E. *et al.* (2002) Reduced volume of orbitofrontal cortex in major depression. *Biol Psychiatry* **51**:273–279.

Brody, A. L., Saxena, S., Silverman, D. H. *et al.* (1999) Brain metabolic changes in major depressive disorder from pre- to post-treatment with paroxetine. *Psychiatry Res* **91**:127–139.

Brody, A. L., Saxena, S., Mandelkern, M. A., Fairbanks, L. A., Ho, M. L., and Baxter, L. R. (2001a) Brain metabolic changes associated with symptom factor improvement in major depressive disorder. *Biol Psychiatry* **50**:171–178.

Brody, A. L., Saxena, S., Stoessel, P. *et al.* (2001b) Regional brain metabolic changes in patients with major depression treated with either paroxetine or interpersonal therapy: preliminary findings. *Arch Gen Psychiatry* **58**:631–640.

Buchsbaum, M. S., Tang, C. Y., Peled, S. *et al.* (1998) MRI white matter diffusion anisotropy and PET metabolic rate in schizophrenia. *Neuroreport* **9**:425–430.

Bunney, W. E. Jr., Goodwin, F. K., Murphy, D. L. *et al.* (1972) The 'switch process' in manic-depressive illness: II: relationship to catecholoamines, REM sleep, and drugs. *Arch Gen Psychiatry* **27**:312–317.

Carmichael, S. T. and Price, J. L. (1995a) Limbic connections of the orbital and medial prefrontal cortex in macaque monkeys. *J Comp Neurol* **363**:615–641.

Carmichael, S. T. and Price J. L. (1995b) Sensory and premotor connections of the orbital and medial prefrontal cortex of macaque monkeys. *J Comp Neurol* **363**:642–664.

Chang, K., Adleman, N. E., Dienes, K, Simeonova, D. I., Menon, V., and Reiss, A. (2004) Anomalous prefrontal-subcortical activation in familial pediatric bipolar disorder. *Arch Gen Psychiatry*, **61**:781–792.

Chen, B. K., Sassi, R, Axelson, D., *et al.* (2004) Cross-sectional study of abnormal amygdala development in adolescents and young adults with bipolar disorder. *Biol Psychiatry* **56**:399–405.

Cicero, D., El-Mallakh, R. S., Holman, J., and Robertson, J. (2003) Antidepressant exposure in bipolar children. *Psychiatry* **66**:317–322.

Clark, L. and Goodwin, G. M. (2004) State- and trait-related deficits in sustained attention in bipolar disorder. *Eur Arch Psychiatry Clin Neurosci* **254**:61–68.

Clark, L., Iversen, S. D., and Goodwin, G. M. (2001) A neuropsychological investigation of prefrontal cortex involvement in acute mania. *Am J Psychiatry* **158**:1605–1611.

Coffey, C. E., Wilkinson, W. E., and Weiner, R. D. (1993) Quantitative cerebral anatomy in depression. A controlled magnetic resonance imaging study. *Arch Gen Psychiatry* **50**:7–16.

Coffman, J. A., Bornstein, R. A., Olson, S. C., Schwarzkopf, S. B., and Nasrallah, H. A. (1990) Cognitive impairment and cerebral structure by MRI in bipolar disorder. *Biol Psychiatry* **27**: 1188–1196.

Coyle, J. T., Pine, D. S., Charney, D. S. *et al.* (2003) Depression and bipolar support alliance consensus statement on the unmet needs in diagnosis and treatment of mood disorders in children and adolescents. *J Am Acad Child Adolesc Psychiatry* **42**:1494–1503.

Cummings, J. L. (1985) *Clinical neuropsychiatry*. New York: Grune and Stratton.

Cummings, J. L. and Mendez, M. F. (1984) Secondary mania with focal cerebrovascular lesions. *Am J Psychiatry* **141**:1084–1087.

Damasio, H., Grabowski, T., Frank, R., Galaburda, A. M., and Damasio, A. R. (1994) The return of Phineas Gage: the skull of a famous patient yields clues about the brain. *Science* **264**:1102–1105.

Davidson, R. J. (2004) What does the prefrontal cortex 'do' in affect: perspectives on frontal EEG asymmetry research. *Biol Psychiatry* **67**:219–233.

de Asis, J. M., Stern, E., Alexopoulos, G. S. *et al.* (2001) Hippocampal and anterior cingulate activation deficits in patients with geriatric depression. *Am J Psychiatry* **158**:1321–1323.

Deckersbach, T., Savage, C. R., Reilly-Harrington, N. *et al.* (2004a) Episodic memory impairment in bipolar disorder and obsessive-compulsive disorder: the role of memory strategies. *Bipolar Disord* **6**:233–244.

Deckersbach, T., McMurrich, S., Ogutha, J. *et al.* (2004b) Characteristics of non-verbal memory impairment in bipolar disorder: the role of encoding strategies. *Psychological Med* **34**:823–32.

Deckersbach, T., Savage, C. R., Dougherty, D. D. *et al.* (2005) Spontaneous and directed application of verbal learning strategies in bipolar disorder and obsessive-compulsive disorder. *Bipolar Disord* **7**:166–75.

DelBello, M. P., Soutullo, C. A., Hendricks, W. *et al.* (2001) Prior stimulant treatment in adolescents with bipolar disorder: association with age at onset. *Bipolar Disord* **3**:53–57.

DelBello, M. P., Strakowski, S. M., Zimmerman, M. E. *et al.* (1999) MRI analysis of the cerebellum in bipolar disorder: a pilot study. *Neuropsychopharmacology* **21**:63–8.

DelBello, M. P., Zimmerman, M. E., Mills, N. P., Getz, G. E., and Strakowski, S. M. (2004) Magnetic resonance imaging analysis of amygdala and other subcortical brain regions in adolescents with bipolar disorder. *Bipolar Disord* **6**:43–52.

Delgado, P. L., Charney, D. S., Price, L. H., Aghajanian, G. K., Landis, H., and Heninger, G. R. (1990) Serotonin function and the mechanism of antidepressant action. Reversal of antidepressant-induced remission by rapid depletion of plasma tryptophan. *Arch Gen Psychiatry* **47**:411–418.

Dickstein, D. P., Treland, J. F., Snow, J. *et al.* (2004) Neuropsychological performance in pediatric bipolar disorder. *Biol Psychiatry* **55**:32–39.

Dougherty, D. D., Rauch, S. L., Deckersbach, T. *et al.* (2004) Ventromedial prefrontal cortex and amygdala dysfunction during an anger induction positron emission tomography study in patients with major depressive disorder with anger attacks. *Arch Gen Psychiatry* **61**:795–804.

Dougherty, D. D., Shin, L. M., Alpert, N. M. *et al.* (1999) Anger in healthy men: a PET study using script-driven imagery. *Biol Psychiatry* **46**:466–72.

Drake, M. E., Pakalnis, A., and Phillips, B. (1990) Secondary mania after ventral pontine infarction. *J Neuropsychiatry Clin Neurosci* **2**:322–325.

Drevets, W. C. (2000) Neuroimaging studies of mood disorders. *Biol Psychiatry* **48**:813–29.

Drevets, W. C. (2001) Neuroimaging and neuropathological studies of depression: implications for the cognitive-emotional features of mood disorders. *Curr Opin Neurobiol* **11**:240–249.

Drevets, W. C., Videen, T. O., Price, J. L., Preskorn, S. H., Carmichael, S. T., and Raichle, M. E. (1992) A functional anatomical study of unipolar depression. *J Neurosci* **12**:3628–3641.

Drevets, W. C., Spitznagel, E., and Raichle, M. (1995) Functional anatomical differences between major depressive subtypes. *Neuroreport* **9**:3301–3307.

Drevets, W. C., Price, J. L., Simpson, J. R. Jr. *et al.* (1997) Subgenual prefrontal cortex abnormalities in mood disorders. *Nature* **386**:824–827.

Drevets, W. C., Price, J. L., Bardgett, M. E., Reich, T., Todd, R. D., and Raichle, M. E. (2002) Glucose metabolism in the amygdala in depression: relationship to diagnostic subtype and plasma cortisol levels. *Pharmacol Biochem Behav* **71**:431–447.

Egan, M. F., Straub, R. E., Goldberg, T. E. *et al.* (2004) Variation in GRM3 affects cognition, prefrontal glutamate, and risk for schizophrenia. *Proc Natl Acad Sci USA* **101**:12604–12609.

Elkis, H., Friedman, L., Buckley, P. F. *et al.* (1996) Increased prefrontal sulcal prominence in relatively young patients with unipolar major depression. *Psychiatry Res* **67**:123–134.

Elliott, R., Baker, S. C., and Rogers, R. D. (1997) Prefrontal dysfunction in depressed patients performing a complex planning task: a study using positron emission tomography. *Psychol Med* **27**:931–942.

Elliott, R., Rubinsztein, J. S., Sahakian, B. J., and Dolan, R. J. (2002) The neural basis of mood-congruent processing biases in depression. *Arch Gen Psychiatry* **59**:597–604.

Elliott, R., Ogilvie, A., Rubinsztein, J. S., Calderon, G., Dolan, R. J., and Sahakian, B. J. (2004) Abnormal ventral frontal response during performance of an affective go/no go task in patients with mania. *Biol Psychiatry* **55**:1163–1170.

Erickson, S. L., Akil, M., Levey, A. I., and Lewis, D. A. (1998) Postnatal development of tyrosine hydroxylase- and dopamine transporter-immunoreactive axons in monkey rostral entorhinal cortex. *Cereb Cortex* **8**:415–427.

Eslinger, P. J. and Damasio, A. R. (1985) Severe disturbance of higher cognition after bilateral frontal lobe ablation: Patient EVR. *Neurology* **35**:1731–1741.

Faedda, G. L., Baldessarini, R. J., Suppes, T., Tondo, L., Becker, I., and Lipschitz, D. S. (1995) Pediatric-onset bipolar disorder: a neglected clinical and public health problem. *Harv Rev Psychiatry* **3**:171–195.

Ferrier, I. N., Chowdhury, R., Thompson, J. M., Watson, S., and Young, A. H. (2004) Neurocognitive function in unaffected first-degree relatives of patients with bipolar disorder: a preliminary report. *Bipolar Disord* **6**:319–322.

Flor-Henry, P. (1969) Schizophrenia-like reactions and affective psychoses associated with temporal lobe epilepsy: etiological factors. *Am J Psychiatry* **126**:400–404.

Frazier, J. A., Chiu, S., Breeze, J. L. *et al.* (2005) Structural brain magnetic resonance imaging of limbic and thalamic volumes in pediatric bipolar disorder. *Am J Psychiatry* **162**:1256–65.

Frey, S., and Petrides, M. (2003) Greater orbitofrontal activity predicts better memory for faces. *Eur J Neurosci* **17**: 2755–8.

Fried, I., Cameron, K. A., Yashar, S., Fong, R., and Morrow, J. W. (2002) Inhibitory and excitatory responses of single neurons in the human medial temporal lobe during recognition of faces and objects. *Cereb Cortex* **12**:575–584.

George, M. S., Ketter, T. A., Parekh, P. I. *et al.* (1997) Blunted left cingulate activation in mood disorder subjects during a response interference task (the Stroop). *J Neuropsychiatr* **9**:55–63.

Giedd, J. N., Blumenthal, J., Jeffries, N. O. *et al.* (1999) Brain development during childhood and adolescence: a longitudinal MRI study. *Nat Neurosci* **2**:861–863.

Goldapple, K., Segal, Z., Garson, C. *et al.* (2004) Modulation of cortical-limbic pathways in major depression: treatment specific effects of cognitive behavioral therapy. *Arch Gen Psychiatry* **61**:34–41.

Goodwin, F. K., and Jamison, K. R. (1990) *Manic-Depressive Illness*. New York: Oxford University Press.

Goodwin, G. M., Cavanagh, J. T. O., Glabus, M. F. *et al.* (1997) Uptake of 99mTc-exametazime shown by single photon emission computed tomography before and after lithium withdrawal in bipolar patients: associations with mania. *Brit J Psychiatry* **170**:426–30.

Gorno-Tempini, M. L., Pradelli, S., Serafini, M. *et al.* (2001) Explicit and incidental facial expression processing: an fMRI study. *NeuroImage* **14**:465–73.

Gur, R. C., Erwin, R. J., Gur, R. E., Zwil, A. S., Heimberg, C., and Kraemer, H. C. (1992) Facial emotion discrimination: II. behavioral findings in depression. *Psychiatry Res* **42**:241–251.

Hale, W. W. (1998) Judgment of facial expressions and depression persistence. *Psychiatry Res* **80**:265–274.

Hamidi, M., Drevets, W. C., and Price, J. L. (2004) Glial reduction in amygdala in major depressive disorder is due to oligodendrocytes. *Biol Psychiatry* **55**:563–569.

Hariri, A. R., Goldberg, T. E., Mattay, V. S. *et al.* (2003) Brain-derived neurotrophic factor val66met polymorphism affects human memory-related hippocampal activity and predicts memory performance. *J Neurosci* **23**:6690–6694.

Harmer, C. J., Grayson, L., and Goodwin, G. M. (2002) Enhanced recognition of disgust in bipolar illness. *Biol Psychiatry* **51**:298–304.

Harmon-Jones, E. (2003) Clarifying the emotive functions of asymmetrical frontal cortical activity. *Pathophysiology* **40**:838–48.

Huttenlocher, P. R. (1984) Synapse elimination and plasticity in developing human cerebral cortex. *Am J Ment Defic* **88**:488–496.

Institute of Medicine (2002) *Reducing Suicide. A National Imperative*. Goldsmith, S. K., Pellmar, T. C., Kleinman, A. M., Bunney, W. E. (eds). National Academies Press.

Isaac, G. (1992) Misdiagnosed bipolar disorder in adolescents in a special educational school and treatment program. *J Clin Psychiatry* **53**:133–136.

Jamison, K. R. (1995) *An Unquiet Mind*. New York: Alfred A. Knopf.

Jastrowitz, M. (1888) Beitrage zur localisation im grosshirn and uber deren praktische verwerthung. *Dtsch Med Wochenshr* **14**:81.

Keri, S., Kelemen, O., Benedek, G., and Janka, Z. (2001) Different trait markers for schizophrenia and bipolar disorder: a neurocognitive approach. *Psychol Med* **31**:915–922.

Kessler, R. C., Berglund, P., Demler, O. et al. (2003) The epidemiology of major depressive disorder: results from the national comorbidity survey replication (NCS-R). *J Am Med Assoc* **289**:3095–3105.

Ketter, T. A., Kimbrell, T. A., George, M. S. et al. (2001) Effects of mood and subtype on cerebral glucose metabolism in treatment-resistant bipolar disorder. *Biol Psychiatry* **49**:97–109.

Ketter, T. A. and Wang, P. W. (2002) Predictors of treatment response in bipolar disorders: evidence from clinical and brain imaging studies. *J Clin Psychiatry* **63**(Suppl 3):21–25.

Killgore, W. D., Oki, M., and Yurgelun-Todd, D. A. (2001) Sex-specific developmental changes in amygdala responses to affective faces. *Neuroreport* **12**:427–433.

Kreiman, G., Koch, C., and Fried, I. (2000) Category-specific visual responses of single neurons in the human medial temporal lobe. *Nat Neurosci* **3**:946–953.

Krishnan, K. R., McDonald, W. M., Escalona, P. R. et al. (1992) Magnetic resonance imaging of the caudate nuclei in depression. Preliminary observations. *Arch Gen Psychiatry* **49**:553–557.

Kumar, A., Bilker, W., Jin, Z., and Udupa, J. (2000) Atrophy and high intensity lesions: complementary neurobiological mechanisms in late-life major depression. *Neuropsychopharmacology* **22**:264–274.

Lai, T., Payne, M. E., Byrum, C. E., Steffens, D. C., and Krishnan, K. R. (2000) Reduction of orbital frontal cortex volume in geriatric depression. *Biol Psychiatry* **48**:971–975.

Lauterbach, E. C. (1996) Bipolar disorders, dystonia, and compulsion after dysfunction of the cerebellum, dentatorubrothalamic tract, and substantia nigra. *Biol Psychiatry* **40**:726–730.

Lawrence, N. S., Williams, A. M., Surguladze, S. et al. (2004) Subcortical and ventral prefrontal cortical neural responses to facial expressions distinguish patients with bipolar disorder and major depression. *Biol Psychiatry* **55**:578–587.

Leckman, J. F., Peterson, B. S., Anderson, G. M., Arnsten, A. F., Pauls, D. L., and Cohen, D. J. (1997) Pathogenesis of Tourette's syndrome. *J Child Psychol Psychiatry* **38**:119–142.

LeDoux, J. E. (2000) Emotion circuits in the brain. *Ann Rev Neurosci* **23**:155–84

Lee, S. H., Payne, M. E., Steffens, D. C. et al. (2003) Subcortical lesion severity and orbitofrontal cortex volume in geriatric depression. *Biol Psychiatry* **54**:529–533.

Lembke, A. and Ketter, T. A. (2002) Impaired recognition of facial emotion in mania. *Am J Psychiatry* **159**:302–304.

Leonard, C. M., Rolls, E. T., Wilson, F. A., and Baylis, G. C. (1985) Neurons in the amygdala of the monkey with responses selective for faces. *Behav Brain Res* **15**:159–176.

Lewinsohn, P. M. and Graf, M. (1973) Pleasant activities and depression. *J Consult Clin Psychol* **41**:261–268.

Lewis, D. A. (1997) Development of the prefrontal cortex during adolescence: insights into vulnerable neural circuits in schizophrenia. *Neuropsychopharmacology* **16**:385–398.

Leyton, M., Ghadirian, A. M., Young, S. N. et al. (2000) Depressive relapse following acute tryptophan depletion in patients with major depressive disorder. *J Psychopharmacol* **14**:284–287.

Lim, K. O., Hedehus, M., Moseley, M. et al. (1999) Compromised white matter tract integrity in schizophrenia inferred from diffusion tensor imaging. *Arch Gen Psychiatry* **56**:367–374.

Lopez-Larson, M. P., DelBello, M. P., Zimmerman, M. E., Schwiers, M. L., and Strakowski, S. M. (2002) Regional prefrontal gray and white matter abnormalities in bipolar disorder. *Biol Psychiatry* **52**:93–100.

Luna, B., Thulborn, K. R., Munoz, D. P. et al. (2001) Maturation of widely distributed brain function subserves cognitive development. *Neuroimage* **13**:786–793.

Machado, C. J., and Bachevalier, J. (2003) Non-human primate models of childhood psychopathology: the promise and the limitations. *J Child Psychol Psychiatry* **44**:64–87.

Malhi, G. S., Lagopoulos, J., Sachdev, P., Mitchell, P. B., Ivanovski, B., and Parker, G. B. (2004) Cognitive generation of affect in hypomania: an fMRI study. *Bipolar Disord* **6**:271–285.

Mallet, L., Mazoyer, B., and Martinot, J. L. (1998) Functional connectivity in depressive, obsessive-compulsive, and schizophrenic disorders: an explorative correlational analysis of regional cerebral metabolism. *Psychiatry Res* **82**:83–93.

Manji, H. K., Moore, G. J., and Chen, G. (1999) Lithium at 50: have the neuroprotective effects of this unique action been overlooked? *Biol Psychiatry* **46**:929–940.

Mayberg, H. S. (1994) Frontal lobe dysfunction in secondary depression. *J Neuropsychiatry Clin Neurosci* **6**:428–442.

Mayberg, H. S. (2002) Modulating limbic-cortical circuits in depression: targets of antidepressant treatments. *Semin Clin Neuropsychiatry* **7**:255–268.

Mayberg, H. S., Brannan, S. K., Mahurin, R. K. *et al.* (2000) Regional metabolic effects of fluoxetine in major depression: serial changes and relationship to clinical response. *Biol Psychiatry* **48**:830–843.

Mayberg, H. S., Liotti, M., Brannan, S. K. *et al.* (1999) Reciprocal limbic-cortical function and negative mood: converging PET findings in depression and normal sadness. *Am J Psychiatry* **156**:675–682.

Mayberg, H. S., Lozano, A. M., Voon, V. *et al.* (2005) Deep brain stimulation for treatment-resistant depression. *Neuron* **45**:651–660.

Moore, G. J., Bebchuk, J. M., Hasanat, K. *et al.* (2000b) Lithium increases N-acetyl-aspartate in the human brain: in vivo evidence in support of bcl-2's neurotrophic effects? *Biol Psychiatry* **48**: 1–8.

Moore, G. J., Bebchuk, J. M., Wilds, I. B., Chen, G., and Manji, H. K. (2000a) Lithium-induced increase in human brain grey matter. *Lancet* **356**:1241–1242.

Morecraft, R. J., Geula, C., and Mesulam, M. M. (1992) Cytoarchitecture and neural afferents of orbitofrontal cortex in the brain of the monkey. *J Comp Neurol* **323**:341–58.

Murphy, F. C., Rubinzstein, J. S., Michael, A. *et al.* (2001) Decision-making cognition in mania and depression. *Psychol Med* **31**:679–693.

Murphy, F. C., Sahakian, B. J., Rubinsztein, J. S. *et al.* (1999) Emotional bias and inhibitory control processes in mania and depression. *Psychol Med* **29**:1307–1321.

Murray, C. J. L. and Lopez, A. D., eds. (1996) *The global burden of disease and injury series, volume 1: a comprehensive assessment of mortality and disability from diseases, injuries, and risk factors in 1990 and projected to 2020.* Harvard School of Public Health on behalf of the World Health Organization and the World Bank. Cambridge, MA: Harvard University Press.

Neumeister, A., Nugent, A. C., and Waldeck, T. (2004) Neural and behavioral responses to tryptophan depletion in unmedicated patients with remitted major depressive disorder and controls. *Arch Gen Psychiatry* **61**:765–773.

Nitschke, J. B., Nelson, E. E., Rusch, B. D. *et al.* (2004) Orbitofrontal cortex tracks positive mood in mothers viewing pictures of their newborn infants. *NeuroImage* **21**:583–592.

Noga, J. T., Vladar, K., and Torrey, E. F. (2001) A volumetric magnetic resonance imaging study of monozygotic twins discordant for bipolar disorder. *Psychiatry Res* **106**:25–34.

Ongur, D., Drevets, W. C., and Price, J. L. (1998) Glial reduction in the subgenual prefrontal cortex in mood disorders. *Proc Natl Acad Sci USA* **95**:13290–13295.

Opppenheim, H. (1889) Zur pathologie der grosshirngeschwulste. *Arch Psychiatry* **21**:560.

Passe, T. J., Rajagopalan, P., Tupler, L. A. *et al.* (1997) Age and sex effects on brain morphology. *Prog Neuro-Psychopharmacol Biol Psychiatry* **21**:1231–1237.

Pearlson, G. D., Barta, P. E., Powers, R. E. *et al.* (1997) Medial and superior temporal gyral volumes and cerebral asymmetry in schizophrenia versus bipolar disorder. *Biol Psychiatry* **41**:1–14.

Pettigrew, J. D. and Miller, S. M. (1998) A 'sticky' interhemispheric switch in bipolar disorder? *Proc Biol Sci* **265**:2141–2148.

Pfefferbaum, A., Mathalon, D., Sullivan, E. V. *et al.* (1994) A quantitative magnetic resonance imaging study of changes in brain morphology from infancy to late adulthood. *Arch Neurol* **51**: 874–887.

Post, R. M., Uhde, T. W., Putnam, F. W. et al. (1982) Kindling and carbamazepine in affective illness. *J Nerv Ment Dis* **170**:717–31.

Post, R. M., DeLisi, L. E., Holcomb, H. H. et al. (1987) Glucose utilization in the temporal cortex of affectively ill patients: positron emission tomography. *Biol Psychiatry* **22**:545–553.

Purcell, R., Maruff, P., Kyrios, M., and Pantelis, C. (1997) Neuropsychological function in young patients with unipolar major depression. *Psychol Med* **27**:1277–1285.

Rajkowska, G. (2002) Cell pathology in mood disorders. *Sem Clin Neuropsychiatry* **7**:281–292.

Rajkowska, G., Miguel-Hidalgo, J. J., Wei, J. et al. (1999) Morphometric evidence for neuronal and glial prefrontal cell pathology in major depression. *Biol Psychiatry* **45**:1085–1098.

Rajkowska, G., Halaris, A., and Selemon, L. D. (2001) Reductions in neuronal and glial density characterise the dorsolateral prefrontal cortex in bipolar disorder. *Biol Psychiatry* **49**:741–752.

Rauch, S. L., Shin, L. M., and Wright, C. I. (2003) Neuroimaging studies of amygdala function in anxiety disorders. *Annals NY Acad Sci* **985**: 389–410.

Rempel-Clower, N. L., and Barbas, H. (1998) Topographic organization of connections between the hypothalamus and prefrontal cortex in the rhesus monkey. *J Comp Neurol* **398**:393–419.

Rogers, R. D., Tunbridge, E. M., Bhagwagar, Z., Drevets, W. C., Sahakian, B. J., and Carter, C. S. (2003) Tryptophan depletion alters the decision-making of healthy volunteers through altered processing of reward cues. *Neuropsychopharmacology* **28**:153–162.

Rolls, E. T. (1999) *The Brain and Emotion*. New York: Oxford University Press.

Rolls, E. T. (2004) The functions of the orbitofrontal cortex. *Brain and Cognition* **55**:11–29.

Rosso, I. M., Cintron, C. M., Steingard, R. J., Renshaw, P. F., Young, A. D., and Yurgelun-Todd, D. A. (2005) Amygdala and hippocampus volumes in pediatric major depression. *Biol Psychiatry* **57**:21–26.

Rubia, K., Overmeyer, S., Taylor, E. et al. (2000) Functional frontalisation with age: mapping neurodevelopmental trajectories with fMRI. *Neurosci Biobeh Rev* **24**:13–19.

Rubin, E., Sackeim, H. A., Prohovnik, I. et al. (1995) Regional cerebral blood flow in mood disorders: IV. Comparison of mania and depression. *Psychiatry Res* **61**:1–10.

Rubinow, D. R. and Post, R. M. (1992) Impaired recognition of affect in facial expression in depressed patients. *Biol Psychiatry* **31**:947–953.

Rubinsztein, J. S., Fletcher, P. C., Rogers, R. D. et al. (2001) Decision-making in mania: a PET study. *Brain* **124**:2550–63.

Sackeim, H. A., Greenberg, M. S., Weiman, A. L. et al. (1982) Hemisphere asymmetry in the expression of positive and negative emotions. *Arch Neurol* **39**:210–218.

Savage, C. R., Deckersbach, T., Heckers, S. et al. (2001) Prefrontal regions supporting spontaneous and directed application of verbal learning strategies: evidence from PET. *Brain* **124**:219–31.

Sax, K. W., Strakowski, S. M., Zimmerman, M. E., DelBello, M. P., Keck, P. E. Jr, and Hawkins, J. M. (1999) Frontosubcortical neuroanatomy and the continuous performance test in mania. *Am J Psychiatry* **156**:139–141.

Schoenbaum, G., Chiba, A. A., and Gallagher, M. (1998) Orbitofrontal cortex and basolateral amygdala encode expected outcomes during learning. *Nature Neurosci* **1**:155–159.

Selemon, L. D. and Rajkowska, G. (2003) Cellular pathology in the dorsolateral prefrontal cortex distinguishes schizophrenia from bipolar disorder. *Curr Mol Med* **3**:427–436.

Shafritz, K. M., Marchione, K. E., Gore, J. C., Shaywitz, S. E., and Shaywitz, B. A. (2004) The effects of methylphenidate on neural systems of attention in attention deficit hyperactivity disorder. *Am J Psychiatry* **161**:1990–1997.

Sheline, Y. I., Gado, M. H., and Price, J. L. (1998) Amygdala core nuclei volumes are decreased in recurrent major depression. *Neuroreport* **9**:2023–2028.

Sheline, Y. I., Sanghavi, M., Mintun, M. A., and Gado, M. H. (1999) Depression duration but not age predicts hippocampal volume loss in medically healthy women with recurrent major depression. *J Neurosci* **19**:5034–43.

Sheline, Y. I., Barch, D. M., Donnelly, J. M., Ollinger, J. M., Snyder, A. Z., and Mintun, M. A. (2001) Increased amygdala response to masked emotional faces in depressed subjects resolves with antidepressant treatment: an fMRI study. *Biol Psychiatry* **50**:651–8.

Siegle, G. J., Konecky, R. O., Thase, M. E., and Carter, C. S. (2003) Relationships between amygdala volume and activity during emotional information processing tasks in depressed and never-depressed individuals: an fMRI investigation. *Ann N Y Acad Sci* **985**:481–484.

Siegle, G. J., Steinhauer, S. R., Thase, M. E., Stenger, A., and Carter, C. S. (2002) Can't shake that feeling: event-related fMRI assessment of sustained amygdala activity in response to emotional information in depressed individuals. *Biol Psychiatry* **51**:693–707.

Silverstone, P. H., Wu, R. H., O'Donnell, T., Ulrich, M., Asghar, S. J., and Hanstock, C. C. (2003) Chronic treatment with lithium, but not sodium valproate, increases cortical N-acetyl-aspartate concentrations in euthymic bipolar patients. *Int Clin Psychopharmacol* **18**:73–79.

Sobotka, S. S., Davidson, R. J., and Senulis, J. A. (1992) Anterior brain electrical asymmetries in response to reward and punishment. *Electroencephalogr Clin Neurophysiol* **83**:236–247.

Spessot, A. L., Plessen, K. J., and Peterson, B. S. (2004) Neuroimaging of developmental psychopathologies: the importance of self-regulatory and neuroplastic processes in adolescence. *Ann N Y Acad Sci* **1021**:86–104.

Starkstein, S. E., Fedoroff, P., Berthier, M. L., and Robinson, R. G. (1991) Manic-depressive and pure manic states after brain lesions. *Biol Psychiatry* **29**:149–58.

Starr, M. A. (1903) *Organic Nervous Diseases.* New York: Lea Brothers and Co.

Steingard, R. J., Renshaw, P. F., Hennen, J. et al. (2002) Smaller frontal lobe white matter volumes in depressed adolescents. *Biol Psychiatry* **52**:413–417.

Strakowski, S. M., DelBello, M. P., Sax, K. W. et al. (1999) Brain magnetic resonance imaging of structural abnormalities in bipolar disorder. *Arch Gen Psychiatry* **56**:254–260.

Strakowski, S. M., Adler, C. M., Holland, S. K., Mills, N., and DelBello, M. P. (2004) A preliminary fMRI study of sustained attention in euthymic unmedicated bipolar disorder. *Neuropsychopharmacology* **29**:1734–1740.

Suslow, T., Junghanns, K., and Arolt, V. (2001) Detection of facial expressions of emotions in depression. *Percept Mot Skills* **92**:857–868.

Swayze, I. I. V. W., Andreasen, N. C., Alliner, R. J., Yuh, W. T., and Ehrhardt, J. C. (1992) Subcortical and temporal structures in affective disorder and schizophrenia: a magnetic resonance imaging study. *Biol Psychiatry* **31**:221–240.

Sweeney, J. A., Kmiec, J. A., and Kupfer, D. J. (2000) Neuropsychologic impairments in bipolar and unipolar mood disorders on the CANTAB neurocognitive battery. *Biol Psychiatry* **48**:674–685.

Taylor, W. D., Steffens, D. C., McQuoid, D.R., et al. (2003) Smaller orbital frontal cortex volumes associated with functional disability in depressed elders. *Biol Psychiatry* **53**:144–149.

Tkachev, D., Mimmack, M. L., Ryan, M. M. et al. (2003) Oligodendrocyte dysfunction in schizophrenia and bipolar disorder. *Lancet* **362**:798–805.

Uranova, N., Orlovskaya, D., Vikhreva, O. et al. (2001) Electron microscopy of oligodendroglia in severe mental illness. *Brain Res Bull* **55**:597–610.

Uranova, N. A., Vostrikov, V. M., Orlovskaya, D. D., and Rachmanova, V. I. (2004) Oligodendroglial density in the prefrontal cortex in schizophrenia and mood disorders: a study from the Stanley Neuropathology Consortium. *Schizophr Res* **67**:269–275.

van Gorp, W. G., Altshuler, L., Theberge, D. C., and Mintz, J. (1999) Declarative and procedural memory in bipolar disorder. *Biol Psychiatry* **46**:525–531.

Venn, H. R., Gray, J. M., Montagne, B. *et al.* (2004) Perception of facial expressions of emotion in bipolar disorder. *Bipolar Disord* **6**:286–293.

Wehr, T. A., Sack, D. A., and Rosenthal, N. E. (1987) Sleep reduction as a final common pathway in the genesis of mania. *Am J Psychiatry* **144**:201–204.

Weiland-Fiedler, P., Erickson, K., Waldeck, T. *et al.* (2004) Evidence for continuing neuropsychological impairments in depression. *J Affect Disord* **82**:253–258.

Weissman, M. M., Leaf, P. J., Tischler, G. L. *et al.* (1988) Affective disorders in five United States communities. *Psychol Med* **18**:141–153.

Wexler, B. E. (1980) Cerebral laterality and psychiatry: a review of the literature. *Am J Psychiatry* **137**:279–91.

Whalen, P. J., Shin, L. M., McInerney, S. C., Fischer, H., Wright, C. I., and Rauch, S. L. (2001) A functional MRI study of human amygdala responses to facial expressions of fear versus anger. *Emotion* **1**:70–83.

Wheeler, R. E., Davidson, R. J., and Tomarken, A. J. (1993) Frontal brain asymmetry and emotional reactivity: a biological substrate of affective style. *Psychophysiology* **30**:82–89.

Wilke, M., Kowatch, R. A., DelBello, M. P. *et al.* (2004) Voxel-based morphometry in adolescents with bipolar disorder: first results. *Psychiatry Res* **131**:57–69.

Williams, J. M. G., Mathews, A., and McLeod, C. (1996) The emotional stroop task and psychopathology. *Psychol Bull* **120**:3–24.

Wozniak, J., Biederman, J., and Richards, J. A. (2001) Diagnostic and therapeutic dilemmas in the management of pediatric-onset bipolar disorder. *J Clin Psychiatry* **14** (suppl):10–15.

Wyatt, R. J. and Henter, I. (1995) An economic evaluation of manic-depressive illness—1991. *Soc Psychiatry Psychiatr Epidemiol* **30**:213–219.

Yakovlev, P. I. and Lecours, A. (1967) The myelogenetic cycles of regional maturation in the brain. In: Minowski, A. (ed.). *Regional Development of the Brain in Early Life*. Oxford: Blackwell.

Yurgelun-Todd, D. A., Gruber, S. A., Kanayama, G., Killgore, W. D., Baird, A. A., and Young, A. D. (2000) fMRI during affect discrimination in bipolar affective disorder. *Bipolar Disord* **2**: 237–248.

Zald, D. H. (2003) The human amygdala and the emotional evaluation of sensory stimuli. *Brain Research—Brain Res Rev* **41**:88–123.

Zald, D. H. and Kim, S. W. (1996) The anatomy and function of the orbital frontal cortex, I: anatomy, neurocircuitry and obsessive-compulsive disorder. *J Neuropsychiatry Clin Neurosci* **8**:125–38.

Zald, D. H., Donndelinger, M. J., and Pardo, J. V. (1998) Elucidating dynamic brain interactions with across-subjects correlational analyses of positron emission tomographic data: the functional connectivity of the amygdala and orbitofrontal cortex during olfactory tasks. *J Cereb Blood Flow Metab* **18**:896–905.

Zald, D. H., Curtis, C., Chernitsky, L. A., and Pardo, J. V. (2005) Frontal lobe activation during object alternation acquisition. *Neuropsychology* **19**:97–105.

Zalla, T., Joyce, C., Szoke, A. *et al.* (2004) Executive dysfunctions as potential markers of familial vulnerability to bipolar disorder and schizophrenia. *Psychiatry Res* **121**:207–217.

Chapter 22

# Effect of orbitofrontal lesions on mood and aggression

Pamela Blake and Jordan Grafman

## 22.1. Introduction

The role of the orbitofrontal cortex (OFC), defined in this chapter as Brodmann areas 10, 11, 12, 13, 14, and 47, in the maintenance of normal social and emotional behavior is well described (Phillips *et al.* 2002). The rich interconnections of the OFC with other cortical and subcortical areas allow for the unique function of this cortical site in integrating inputs necessary for decision-making, behavioral modulation, impulse control, personality and social comportment, among others. Since Harlow's paragon description of Phineas Gage and the dramatic change in his personality following a penetrating injury to the OFC (Harlow 1868), lesions of the OFC have been known to result in a variety of symptoms including alterations in mood, personality, social comportment, and behavior. Similarly, the orbitofrontal cortex plays a key role in the inhibition of aggressive actions that occurs under normal circumstances; impairment of this inhibitory action by the OFC in a diseased state may be associated with recurrent aggression. In some areas there is general consensus as to the effect of OFC lesions on these functions, but in other areas there is considerable controversy.

In this chapter we will review the effects of lesions in the OFC on mood states and on aggressive behavior. Some of the lesion types to be reviewed, such as traumatic brain injury, are well-researched and described, with an abundant literature. Other lesion types such as stroke are not as well studied with respect to their effect on mood or aggression, and there is, therefore, a much more limited literature available. Furthermore, there are several limitations of the available literature: most of the cited literature does not include specific references as to the exact cortical (Brodmann) areas involved, and several of the studies employ CT scanning which, by virtue of its reduced spatial resolution, is less sensitive for determining involved cortical areas. Many studies report lesions or injury to the 'frontal lobe' without better specificity. None of the studies of patients with dementia have histopathology to support the suggestion of preferential involvement of certain cortical areas. When the authors indicate specific details such as Brodmann areas, we will note it.

## 22.2. Effect of OFC lesions on mood states

### The presentation and assessment of mood disorders in individuals with orbitofrontal lesions

A mood disorder may be either an accompanying symptom of a primary neurological process, such as a head injury, or the initial or presenting symptom of a hitherto undiagnosed condition such as dementia. The patient with an indolent frontal lobe-related mood disorder may present to a practitioner in the primary care setting or to an acute setting if the symptoms are pronounced or unmanageable. Not uncommonly, poor insight on the part of the patient results in the evaluation being prompted by a family member or an employer, as along with the mood disturbance there is frequently an accompanying deterioration of social skills or occupational performance.

The role of the prefrontal cortex in primary depression has been suggested by imaging studies that reported structural and metabolic abnormalities in the orbitofrontal cortex (Bremner et al. 2002), dorsolateral cortex (Biver et al. 1994), and the anterior cingulate cortex (Drevets et al. 1997) in patients with depression (Fredericks et al., Chapter 21, this volume).

Abnormal mood states are typically assessed by psychiatrists utilizing interview techniques based on the Diagnostic and Statistical Manual of Mental Disorders diagnoses (APA 2000) or standard scales such as the Beck Depression Inventory (Beck et al. 1961) or the Hamilton Depression Rating Scale (Hamilton 1960). The patient, who presents with significant anxiety, mania or depression in the absence of a known primary neurological process, will usually generate a psychiatric referral with a resulting primary psychiatric diagnosis, and thus the recognition that the symptoms are the result of an orbitofrontal lesion may escape notice. For example, the apathy associated with indolent prefrontal tumors may be diagnosed as depression, as may the cognitive and motor slowing associated with gradually worsening hydrocephalus. Lesions of the dorsolateral or orbitofrontal cortex typically spare motor systems, so there may be no other neurological symptom, such as weakness or a gait disorder, present to alert the examiner that the underlying condition is in fact 'neurological' as opposed to 'psychiatric'. A high index of suspicion and familiarity with neuropsychiatric syndromes is necessary to alert the examiner to the possibility of a structural brain disorder underlying mood state changes in some patients.

As early as the 1970s, investigators recognized that lesions of the prefrontal cortex resulted in differing clinical presentations (Blumer and Benson 1975). Lesions of the dorsolateral cortex resulted in apathy and reduced motivation; orbitofrontal lesions produced a decrease in empathy and a general disinhibition of behavior. These original observations are born out and elaborated upon with newer, more sensitive data derived from both lesion studies and studies of the pathological state in the intact brain.

We will consider in this section the most typical mood disorders that may be associated with frontal lesions, namely, depression, anxiety and mania.

## Traumatic brain injury

**Depression.** It is generally accepted that traumatic brain injury (TBI) is associated with an increased prevalence of depression. The rate of the development of depression following traumatic brain injury has been reported to be as low as 18.5% in World War II veterans (Holsinger et al. 2002) to as high as 61% in a study of 100 subjects who sustained TBI as adults (Hibbard et al. 1998). Perhaps fluctuating individual definitions of depression and changing societal expectations are in part responsible for this broad variation in prevalence rates. Nevertheless, most studies demonstrate significantly elevated rates of depression, typically around 40% (Jorge et al. 1993; Kreutzer et al. 2001), following a brain injury.

The neuropathological correlates of the site of injury and the development of depression have been investigated in some published reports, but the findings reported in the many studies of individuals with frontal lobe, or even more specifically orbitofrontal lesions, is surprisingly inconsistent, and at times conflicting. One of the challenges of assigning mood state changes to lesion site is the commonly widespread nature of the injury, involving diffuse frontal as well as non-frontal brain areas. Careful imaging is a critical factor in the success of such studies.

Grafman et al. (1986) studied the mood effects of frontal and non-frontal lesions in 52 Vietnam veterans evaluated approximately 15–20 years following their injury. They utilized CT scanning to localize the lesions site and performed a battery of neuropsychological with measures of both cognitive function and mood. The authors noted that the head-injured Vietnam veterans with orbitofrontal lesions, based on self-report and observation, were more anxious and dorsolateral lesion patients were more likely to be 'sluggish' and introverted. They compared brain-injured subjects to control subjects, and also compared left-brain to right-brain lesions. They further compared the patients with unilateral lesions to control subjects with isolated ipsilateral, non-frontal lesions to determine if there were within-hemisphere differences on measures of mood state.

Patients with isolated right or bilateral orbitofrontal lesions were more likely to have received psychiatric care following their injuries than left lesion or control patients; 50% of patients with right orbitofrontal lesions and 47% of patients with bilateral orbitofrontal lesions sought psychiatric care, compared to 20% of left orbitofrontal lesioned patients and 21% of controls. The authors did not list a specific psychiatric diagnosis, nor was a structured psychiatric interview conducted to determine psychiatric conditions. The patients with right orbitofrontal lesions were more likely to experience symptoms of depression, and anxiety than were those patients with left frontal lesions, while those with left frontal lesions experienced increased anger and hostility. In general, the within-hemisphere comparisons indicated that both left and right orbitofrontal lesions had the greatest adverse effect upon the regulation of mood and behavior. As would be expected, greater impairment in cognitive function and attention were seen in the patients with dorsolateral lesions. The authors concluded that the lateralization of the lesion was indeed an important factor in the pattern of clinical symptomatology and the regulation of mood, with the right OFC more involved in the control or inhibition of anxiety, so that lesions in this area resulted in increased levels of anxiety. In general the patients with right-sided lesions experienced more psychological distress; in contrast, patients with left-sided lesions were relatively undisturbed by psychological symptoms. Importantly, depression did not stand out as a prominent symptom in this group of patients with orbitofrontal lesions.

Another large series of patients with a post-traumatic mood disorder found results that in part conflict with the previous study. Jorge et al. (2004) examined the psychiatric status, the neuropsychological function and the location and severity of the brain lesion in a group of 91 consecutive patients with TBI and compared this group to 27 patients with multiple trauma that spared the central nervous system. The severity of injury in the TBI group was assessed using a combination of the 24-hour Glasgow Coma Scale Score, the need for an intracranial surgical procedure, and the presence of focal lesions greater than 15mL. The severity of injury was mild in 44.3%, moderate in 32.5%, and severe in 23.2%. Consistent with previously reported studies, the incidence of mood disorders in the TBI group was 51.6%, while the incidence of mood disorders in the non-TBI group was 22.2%. The mood disorders in the TBI group were classified as major depression disorder in 30 (33%), Depression without major depressive features in 9 (9.9%) and manic or mixed features in 8 (8.8%). Anxiety was co-morbid in 23 (76.7%) of the 30 patients with isolated major depressive disorder, encompassing a variety of forms of anxiety including generalized anxiety, panic attacks and post-traumatic stress disorder.

The authors studied the 30 patients with TBI and isolated major depressive disorder in detail, and compared them to the 44 patients with TBI without major depressive disorder. The frequency of mild, moderate and severe TBIs was not significantly different between the two groups. The subjects were evaluated at 3, 6, and 12 months, and underwent a neuropsychological evaluation and 1.5T MRI with volumetric analysis at the three-month follow-up visit. The comparison of background variables of the depressed and non-depressed groups revealed some interesting findings. The subjects in both the depressed and non-depressed groups were matched with respect to socioeconomic status, sex, marital status, educational status, and annual income. They differed, however, in two important respects. The depression group had higher rates of pre-injury unemployment, and also a significantly higher frequency of a personal history of mood disorders. There were no significant differences in the rates of drug or alcohol use among the two groups.

Measures of cognitive function such as the Mini-Mental Status Examination were identical in the two groups. Neuropsychological testing including tests sensitive for prefrontal cortex function, such as the Wisconsin Card-Sorting Test (Heaton et al. 1993) and Trail-Making Test (Reitan 1971), was also conducted. The patients with depression scored lower across the battery of neuropsychological tests, despite similar scores of global cognitive function.

Radiological studies with volumetric analysis were completed at the 3-month follow-up visit. A group of 17 patients with TBI and major depressive disorder were age- and sex-matched with 17 patients with TBI without major depressive disorder; the patients were also matched with respect to the severity of the TBI. The pattern of injury (diffuse versus focal), the location of the lesion and the volume of focal mass lesions were similar in the two groups. The volumetric analysis revealed no difference between the depressed and non-depressed groups in the total brain volume, total gray matter or total white matter volume. However, patients with major depression had significantly decreased frontal gray matter volume compared with the non-depressed TBI subjects. In addition, the volume of *left* frontal gray matter was reduced in the depressed group, owing to smaller volumes of the left inferior, superior and middle gyri. Interestingly, the volume of the orbitofrontal gray matter volume was identical. Thus, the depression present in these patients was not associated with alterations in the volume of the orbitofrontal cortex. The authors pointed out that since, as a group, the depressed patients had higher rates of mood disorder and unemployment, it was possible that the observed volumetric differences constituted a pre-existing trait associated with depression. It is possible that this pre-existing trait, or predisposition for, depression was 'uncovered' by the superimposed traumatic injury.

The findings in this carefully-done study are inconclusive with respect to the role of the OFC in the development of depression following head injury, but they suggest that depression following a head injury is more closely associated with dorsolateral injury, particularly on the left side, than with orbitofrontal injury. It is interesting to note that in the Grafman study, the presence of a right-sided lesion of the orbitofrontal cortex lesion was more likely to result in a psychiatric condition, but that anxiety disorders appeared to be more common than depression. These findings can both be contrasted to a recent report that described the volumetric measurements of the OFC in 31 patients with untreated MDD compared 34 controlled subjects (Lacerda *et al.* 2004). The authors found a reduction in the gray matter volume of the right medial OFC and the left lateral OFC. The exact role of the OFC, and the effect of lesions on mood state, therefore remains elusive.

**Anxiety.** As already noted above, one large study found an association between injury of the right OFC and anxiety (Grafman *et al.* 1986). As with the findings concerning the development of depression following a head injury, however, the published data regarding anxiety after orbitofrontal injury are similarly conflicting. Neuroimaging studies usually implicate the OFC as relevant in anxiety, but the direction of the effect varies. Two functional imaging studies in the intact brains of adult volunteers with anxiety disorders showed similar but not identical results:

Rauch *et al.* (1997) employed a symptom provocation paradigm to elicit anxiety in 23 patients with obsessive compulsive disorder, simple phobia, or post-traumatic stress disorder, and measured cerebral blood flow with positron emission tomography (PET) scanning. The subjects displayed activation in the right inferior frontal cortex, the right posterior medial orbitofrontal cortex, bilateral insular cortex, bilateral lenticulate nuclei and bilateral brainstem foci. A similar study by Chen et al. (2004) found that the provocation of symptoms in 10 unmedicated patients with obsessive-compulsive disorder was associated with a significant increase in regional cerebral blood flow in the head of the right caudate nucleus, the thalamus and the bilateral OFCs. These two studies therefore concluded that variants of anxiety disorders are associated with increased blood flow, that is, activation, of brain areas involving the OFC and caudate nucleus. In contrast to these studies, which suggest that OFC activation is associated with the acute symptom state of anxiety disorders, other studies link lesioned, and therefore presumably hypoactive, OFC with increased anxiety in head-injured (Grafman et al. 1986) and stroke patients (Castillo et al. 1993).

One large study in a pediatric population attempted to answer in a more definitive manner the question of whether damage to specific brain areas, particularly the OFC and the temporal lobe, relates to childhood anxiety. Vasa and colleagues enrolled 97 subjects, aged 4–19, following a head injury that was classified as severe according to the Glasgow Coma Scale (Vasa et al. 2004). An MRI was performed at three months post-injury; lesion size was noted with particular attention to the OFC, the dorsolateral cortex, the mesial cortex, the temporal lobe, and the non-frontal, non-temporal brain regions. Anxiety was assessed using the DICA, a structured interview (Reich et al. 1995). In addition, the subjects were screened for the presence of a total of 48 anxiety symptoms that spanned the spectrum of anxiety disorders. The subjects were evaluated with respect to pre-injury and post-injury characteristics. The five categories of anxiety that were assigned, based on DSM-III-R criteria available at the time of the study, include overanxious disorder (OAD), separation anxiety disorder (SAD), obsessive-compulsive disorder (OCD), simple phobia (SP), and avoidant disorder (AD). A measure of psychosocial adversity was also assessed.

Seventy seven of the total 97 children (81.1%) enrolled had neither a pre-injury nor post-injury anxiety disorder. Nine children (9.5%) had developed a new post-injury disorder. Five children (5.3%) had remission of pre-injury anxiety, and four (4.2%) had anxiety disorders both pre-injury and post-injury.

The main finding of the study was an inverse correlation between lesion volume of the orbitofrontal cortex and the degree of post-injury anxiety. OFC variables that were most strongly associated with significantly decreased risk for anxiety included the total lesion volume, the greater volume of either right- or left-sided lesions, the presence of any lesion, and the presence of a right lesion. Of further note, an increased number of OFC lesions was strongly associated with a reduction in the number of anxiety symptoms experienced by the subjects, including those who did not meet diagnostic criteria for an anxiety disorder. These findings stood even when correcting for the presence of pre-injury anxiety. Interestingly, no association was found between anxiety and temporal lobe injury. In summary, this study demonstrated that the more damage there is to the OFC in children with TBI, the less likely the subject is to develop anxiety. These findings are consistent with the outcomes of newer surgical approaches to treating refractory OCD by disrupting fronto-temporal connections with bilateral anterior capsulotomy (Oliver et al. 2003) or bilateral anterior cingulotomy (Kim et al. 2003).

The data indicating reduction in anxiety symptoms in patients with OFC lesions makes sense from an anatomical standpoint. As the OFC is important for comportment of behavior and the online manipulation of data and social inputs, one may theorise that these functions are associated with a heightened state of arousal in the individual with an awareness of the need for proper function. If the OFC is lesioned, the prompts to maintain adequate behavioral controls may be lacking, thus resulting in decreased awareness and care for these functions, and thus decreased levels of anxiety. These impairments in function following OFC lesion that affect one's ability to regulate emotional state are relevant to our next area of interest, the theory of mind (ToM).

**Empathy and impaired theory of mind.** The most compelling and classic symptoms that develop following a lesion of the OFC is the loss of the functions that are necessary for normal social interaction (Damasio et al. 1991). These functions include, among others, the ability to perceive and respond to socially relevant information, to recognize facial expression, and to understand the perspective and emotions of another. Loss of inhibition of social responses is also commonly reported. Acquired damage to the OFC results in impairment of these functions in a manner that may be dramatic or more insidious, but one consistent deficit that is observed is the failure of the OFC-lesioned individual to experience empathy. One feature of empathy is the ability to infer the mental state of another. This process has been termed the theory of mind (Premack and Woodruff 1978), and it also encompasses the ability to understand irony and the concept of a *faux pas*. ToM impairment localizes to the OFC according to lesion studies (Stone et al. 1998); dorsolateral lesions do not affect ToM or empathic abilities. Damage to the right ventromedial cortex was more strongly associated with ToM impairment (Shamay-Tsoory et al. 2005). Another study assessed the ability of brain-injured subjects to interpret nonverbal social cues (Mah et al. 2004). Subjects with lesions of the OFC demonstrated significantly impaired social perception.

As a marker for this impairment in function, OFC-lesioned subjects have also been noted to have reduced autonomic activity. Hopkins et al. (2002) found reductions in the normal expected increase in electrodermal activity in orbit frontal-lesioned patients upon viewing negative facial expressions. Taken together, these data suggest that OFC lesions are associated with a reduction in the ability to generate expectations of another's emotional response, perhaps particularly negative responses. Impairment of empathetic responses and ToM function is not typically associated with an outward mood disorder, however, there are some case reports of overly or altered emotional responses in individuals with OFC damage (Angrilli et al. 1999).

**Mania.** Mania is the least frequently described mood disorder to develop as a result of an OFC lesion, and is a relatively infrequent abnormal mood state following head injury. One must be careful to distinguish the disinhibition that may be seen with OFC lesions with true mania. Clark and colleagues examined the neuropsychological function of 15 acutely manic inpatients with 30 non-psychiatric subjects (Clark et al. 2001). A full battery of testing directed at assessing frontal lobe function, such as the Iowa Gambling Task and the Stroop test, were completed. The manic patients resembled the normal control subjects in their performance on the frontal tasks. The authors concluded that mania is not mediated via the prefrontal cortex. Most studies assign the neuropathologic focus of mania to the temporal lobes (Jorge et al. 1993).

### Degenerative disorders

**Depression.** Mood disorders can occur in the setting of primary degenerative disorders such as Alzheimer's Disease (AD) and frontotemporal dementia (FTD). It stands to reason that the involvement of the frontal lobes, a known site of impairment in depression (Bremner et al. 2002), by a neurodegenerative disorder may result in a clinical manifestation of depression in the same way that involvement of the visual cortex may lead to visual loss or to positive visual phenomena. However, depression that develops early in the course of the disease, while cognitive functions remain relatively intact, may be reactive; the later development of depression may represent just one manifestation of the known psychiatric symptomatology that occurs in dementing illnesses.

The rate of depression in AD varies in clinical studies from 25–50%. Zubenko et al. (2003) described a collaborative study of the rates of depression in patients with Alzheimer's disease recruited from five sites around the United States. The study compared the findings on the Clinical Assessment of Depression in Dementia, a structured diagnostic interview (Zubenko et al. 2003), in 243 AD patients to 151 non-demented elderly patients. The incidence of depression in the AD patients varied across sites from 22.5% to 54.4%. The highest rates of depression were present in the most severely demented patients. Bozeat et al. (2000) studied the neuropsychiatric and behavioral features of patients with AD ($n = 37$) or the frontal ($n = 13$) or temporal ($n = 20$) variants of FTD. The information was obtained from questionnaires that were administered to caregivers. The rate of depression in patients with AD was 24%; the rate of depression is patients with the temporal variant of FTD was 45% and the in the frontal variant of FTD 7%. None of these studies used neuroimaging to attempt to demonstrate increased involvement of frontal areas in the depressed patients versus the non-depressed patients. Similarly, no histopathological studies of postmortem brains have been conducted to demonstrate altered or increased involvement of the frontal lobe in depressed, demented patients.

**Apathy.** Several authors have made a distinction between depression and apathy, as recent research has challenged the traditional notion that apathy is simply one aspect of depression. Apathy is defined as the lack of motivation in behavior, cognition and affect. Apathy may co-exist with depression or may be one of the symptoms of depression, but some aspects of apathy are unique to that condition and may be distinguished from depression with the objective assessments of clinical scales including the Neuropsychiatric Interview (Cummings et al. 1994), the Structured Interview for Apathy

(Starkstein et al., 2005), and the Apathy Inventory (Robert et al. 2002), which can distinguish the three dimensions of apathy, including lack of initiative, lack of interest, and emotional blunting. Clinical studies have distinguished the rates of apathy from those of depression in patients with a variety of degenerative diseases, including AD, FTD, Parkinson's Disease, Huntington's Disease, and progressive supranuclear palsy (Levy et al. 1998). Apathy did not correlate with depression and was associated with a greater level of cognitive impairment as demonstrated by performance on the Mini Mental Status Examination. Starkstein and colleagues (2005) found varying rates of apathy and depression in a sample of 150 patients with AD. Twelve per cent of the patients met criteria for both apathy and depression, 31% met the criteria for depression only and 7% met the criteria for apathy only. As noted by Levy, the presence of apathy was associated with a greater degree of cognitive dysfunction. Imaging studies have demonstrated the neuroanatomic correlate of apathy. Benoit et al performed single photon emission tomography scans in 30 non-depressed patients with AD who were assessed for apathy with the Apathy Inventory (Benoit et al. 2004). There was a negative correlation between the severity of overall apathy and brain perfusion in the left and right superior orbitofrontal gyrus and less so in the left middle frontal gyrus (BA 10). Lack of initiative was most associated with reduction in perfusion of the right anterior cingulate cortex. Lack of interest was associated with reduced bloodflow in the right middle orbitofrontal gyrus, and emotional blunting was negatively correlated with reduced perfusion in the left superior dorsolateral cortex.

## Stroke

Several studies have shown that there is an increased incidence of depression after stroke that is on average 25% in the acute setting and 13% in the chronic setting (Robinson 1998), and lesions affecting either the right (Eastwood et al., 1989) or the left (Robinson and Starkstein 1990) hemisphere have been more strongly implicated. Studies have also indicated that the association of post-stroke depression and lesion location may change over time. One study examined 60 patients with respect to lesion location and size, degree of neurological impairment, and the presence and severity of depression in the acute setting and also at short-term (3–6 months) and long-term (1–2 years) follow-up (Shimoda and Robinson 1999). The authors found that in the acute, hospitalized setting, lesions of the left anterior hemisphere only were associated with a higher degree of depression. At the short-term follow-up visit this effect was lost, however, and the presence of depression was associated with proximity of the lesion to the frontal pole and lesion volume. By the 1–2 year follow-up evaluation, depression was significantly associated with strokes of the right hemisphere and more posterior placement, close to the occipital pole.

Other than these studies which involve rather general anatomic descriptions, there are no careful imaging studies of frontal lobe lesion location in stroke and the development of a classic orbitofrontal cortex clinical syndrome. Perhaps this reflects the vascular supply to the OFC; isolated infarcts of the OFC or the dorsolateral cortex are uncommon.

Furthermore, strokes involving the left hemisphere may involve language or sensory association areas that may complicate the patient's ability to participate in testing due to aphasia or neglect syndromes. There is not a clear explanation for the time-related changes in the demographics of depression and stroke, however, even with these reports; the important role of impairment of the frontal lobe in the generation of post-depression stroke has been suggested via other methods which study other symptoms of frontal lobe involvement. Vataja and et al. (2005) examined 158 patients who had suffered a stroke; the subjects were tested with a neuropsychological battery for impairments of executive function, which is more typically associated with dorsolateral prefrontal cortex. Twenty-one patients had executive dysfunction, one hundred and thirty-seven had none. Depression was significantly more commonly present in the patients with executive dysfunction, and these subjects were much more likely to have strokes affecting frontal subcortical structures than the 137 patients who did not show executive dysfunction. This association of depression with dorsolateral lesions is consistent with the data of head-injured patients who develop depression.

## 22.3. The effect of OFC lesions on aggression

### Definition and assessment of aggression

Aggression is a voluntary or involuntary action that results in, or is intended to result in, injury to an individual, an object or an animal. Aggression is usually manifested as a series of behaviors rather than a single isolated event. Aggression may be expressed in verbal, physical or sexual behaviors. Scales may be used to measure and quantify aggressive actions, such as the Modified Overt Aggression Scale (MOAS) (Kay et al. 1988), and aggressive thoughts and tendencies, such as the Buss-Perry Aggression Questionnaire (Buss and Perry 1992). These scales have limitations; for instance, the MOAS does not query sexual aggression.

Aggression may present in a persistent pattern of behaviors with onset in childhood, or may arise as a secondary phenomenon following or related to disorders of the brain. Based on the earliest description of Phineas Gage, who experienced an altered personality with antisocial features following injury to the ventromedial cortex (Harlow 1868) and subsequent supporting research, there is a general consensus that aggressive and antisocial behaviors are associated with impairment of the function of the prefrontal cortex. This theory represents a shift of thought over the last three decades, away from original theories of amygdala or limbic system structure damage underling aggressive behavior.

The association of ventromedial prefrontal cortex damage in association with aggression has been supported in a number of ways. Pietrini et al. (2000) found in a PET study that imagined physical aggression was associated with a reduction in perfusion of the ventromedial prefrontal cortex in normal volunteers. Furthermore, studies of aggressive individuals have shown impairment in prefrontal cortical functioning through a variety of methods. A study of 31 murderers (Blake et al. 1995) found prefrontal cortical and motor abnormalities on the neurological examination, a relative insensitive tool, in 20 subjects. Similarly, a study of 18 men committed to death row in Texas for crimes committed as juveniles found evidence of PFC impairment on the neurological examination in 13, and on neuropsychological testing in all 18 (Lewis *et al.* 2004). Further studies employing neuropsychological testing (Best *et al.* 2002) and neuroimaging (Raine *et al.* 2000) support the presence of PFC impairment in the repeatedly aggressive person. The etiology of PFC impairment is unclear in the majority of chronically aggressive subjects. The differential diagnosis includes a number of causes including congenital disorders, head injury, toxic effects of alcohol or illicit drugs, or structural acquired lesions. Given the prominent role of PFC dysfunction in the chronically aggressive, it follows that damage to the ventromedial PFC is associated with an increase in aggressive behavior, and such an increase has indeed been well-documented in subjects who have experienced damage to the PFC through mechanisms including a head injury, structural lesion, or degenerative disorders, as described below.

## Aggression following acquired lesions of the OFC

### Traumatic brain injury

The Vietnam Head Injury Study was a pivotal study which examined the association of increased aggressive behaviors with damage to the PFC in a group of Vietnam veterans. A cohort of 279 veterans who sustained penetrating frontal lobe brain injuries was compared to a matched control group of 57 veterans who sustained penetrating injury to other areas of the brain. The veterans were examined approximately 15 years after the injury; they underwent neurological and neuropsychological testing, electroencephalography, evoked potentials, and CT scanning. A close family member, usually the veteran's wife, completed a number of questionnaires regarding the subject's aggressive or violent behavior. Veterans with lesions restricted to the mediofrontal or OFC had a significantly greater likelihood of engaging in aggressive behavior. Verbal aggression was the most common form of aggression, and was present at higher rates in the orbitofrontally-injured patients in a variety of forms: arguing was present in 48% compared to 19% of controls, swearing in 29% versus 12% of controls, making threats to injure others in 20% versus 1%, and threatening to 'tell people off' in 30% versus 8%. Physical aggression against others was less common, occurring in about 16% of OFC-lesioned veterans compared to 4% of controls. Physical aggression against objects (breaking things) also occurred in about 15% of subjects compared to 4% of controls. The authors did not report the presence of criminal charges or legal involvement resulting from aggressive behavior, nor did they report the incidence of sexual aggression in the group studies. The findings presented in this study were the first careful assessment of the role of frontal lobe involvement in the generation of aggressive behavior following brain injury. Other, earlier studies of German veterans of World Wars I and II had shown a link between orbitofrontal injuries and antisocial behaviors (Blumer and Benson 1975) but had not provided a detailed assessment of aggression per se.

Subsequent studies have confirmed the higher rates of aggressive behavior in patients who sustained injury to the frontal lobes following a closed head injury. Tateno and colleagues examined 89 consecutive patients with closed head injury (Tateno et al. 2003). Aggressive behavior was assessed with the Overt Aggression Scale (Yudofsky et al. 1986), a precursor to the Modified Overt Aggression Scale, and which also measures verbal aggression or aggression aimed at objects, the self, or others. The authors also gathered data regarding criminal charges or legal involvement. The location of the head injury was determined with CT or MRI imaging. Lesions were described as being frontal or non-frontal; frontal lesions were not further classified as being orbitofrontal or dorsolateral. The patients also underwent a psychiatric assessment. Comparison was made to a group of patients who had been admitted with multiple traumas but without clinical or radiological evidence of head injury.

Aggressive behavior occurred in 30 (33.7%) of the head-injured subjects, five of whom had been arrested, compared to one subject of 59 in the non-aggressive group. The patients with post-traumatic aggression differed from the head-injured patients without post-traumatic aggression in one important aspect—they were more likely to have had a history of a mood disorder, alcohol abuse or drug abuse. The frequency of frontal lobe lesions was significantly higher among patients who developed post-traumatic aggression than in those who did not. Furthermore, patients with focal frontal lobe lesions had significantly higher scores on the Overt Aggression Scale. Depression was significantly more common in the aggressive group than in the non-aggressive group; the authors, however, did not make a distinction in the subjects as to the location of the injury within the frontal lobe in the depressed vs. non-depressed group. The findings of this important study, which demonstrate the increased rates of both aggression and depression in subjects who have sustained injury to the frontal lobes, are clearly limited by the lack of more specific anatomic delineation of the lesion site.

Not all studies implicate injury to the OFC as correlating most closely with behavioral changes after a head injury. Simpson et al. (2001) studied the characteristics of a group of head-injured patients who exhibited sexual aggression following injury to a matched group of head-injured patients who did not exhibit such behavior. They found no significant differences on neuroimaging reflecting the site of the injury between the case and control groups. Furthermore, neuropsychological testing of the two groups revealed similar findings; in general, the rates of impairment of executive function were similar in the two groups. The patients who exhibited sexually aberrant behavior were also more likely to engage in non-sexual criminal activity and were less likely to return to work, thus, the sexual aggression was considered part of a larger pattern of behavioral dyscontrol. However, in the absence of findings that reflected higher levels of frontal lobe impairment in the control group, the authors caution against automatically associating sexual aggression following a head injury with damage to the frontal lobes.

### Degenerative disorders

Aggression or agitation is common behavioral disturbance in patients with dementia (Levy et al. 1996; Patel and Hope 1992), ranging in incidence from 12% to 65%. Clinical reports and anatomic studies of aggressive behavior in subjects with dementia have shown conflicting results. Aggressive behavior has been show to occur at a high rate in subjects with frontotemporal dementia (FTD), who usually have impairment of dorsolateral frontal regions (Miller et al. 1997). A better understanding of the clinical and pathologic features of dementia is important, as aggressive behavior in the affected patient not only poses a risk to the patient, increasing the likelihood of harm and institutionalization, but also to the caregivers. As with the literature in patients with dementia and depression, however, there are no studies that contain histopathological correlation of greater degree of involvement of any cortical area in demented patients with aggression.

Hirono and colleagues (Hirono et al. 1998) studied two groups of ten patients each with dementia with and without aggression/agitation, as measured by the Neuropsychiatry Inventory. The subjects included patients with AD, vascular dementia, or probable dementia with lewy bodies. The subjects were free of pre-morbid psychiatric disorders or a history of head injury. A thorough behavioral assessment, caregiver interview, and assessment of cognitive skills were obtained. The two groups were comparable with respect to cognitive function and also with respect to medication use. The patients then underwent single photon emission computed tomography. Patients with aggression had statistically significantly reduced perfusion in the left anterior temporal cortex; the orbitofrontal perfusion was not statistically different in the aggressive patients compared to the non-aggressive patients. There was hypoprefusion in the right and left superior frontal cortices (Brodmann area 9) and the right superior parietal cortex (Brodmann area 7). Thus, in contrast to most studies of aggression which implicate OFC involvement, this study suggested that dorsolateral cortex impairment was more relevant to the development of aggressive behavior.

Lateralization has been suggested in some patients with dementia-associated aggression. Mychack and colleagues studied 41 patients with frontotemporal dementia (Mychack et al. 2001). SPECT scanning was used to determine whether the disease process was more severe in the right or left hemisphere. The presence of socially undesirable behaviors, such as aggression, criminal behavior, loss of a job, estrangement from family or friends, sexually or financially risky behavior, and abnormal response to spousal crisis, were recorded. Of the 12 patients with right-sided FTD, 11 (91.6%) engaged in such behaviors; of the 19 patients with left-sided FTD, 2 (10.5%) had. The authors concluded that right-sided FTD is associated with socially undesirable behavior, and that the presence of such clinical symptoms in an afflicted person may suggest localization.

## Stroke

There are limited studies of aggression or the development of socially unacceptable behaviors in patients with stroke, a relatively uncommon clinical finding. Paradiso and colleagues examined patients who had experienced violent outbursts after stroke and compared them to patients with stroke who had not developed these clinical syndromes (Paradiso et al. 1996). They found that the percentage of patients with violent outbursts were more likely to have lesions close to the frontal pole, were more likely to have left hemisphere lesions, and were more likely to have cognitive impairment. Even after controlling for the presence of depression, the higher rates of violent outbursts in the patients with left and anterior lesions was sustained.

## 22.4. Summary and future directions

Lesion studies of the OFC with respect to their effects on mood states and aggression are generally consistent with the widely held principles regarding the function of this area of the brain, but there is also considerable conflict in important areas. It is accepted that orbitofrontal lesions are not as closely associated with depression as are dorsolateral lesions; but the effect of OFC lesions in the development of anxiety are frankly contradictory (also see Chapter 20, this volume). The effect on personality may be more widely accepted, but there is little data in the lesioned populations. The data regarding aggression as a result of OFC lesions is a little more consistent and a general principal may be described; lesions of the OFC cause a disinhibited, impulsive state that results in an increase in impulsive aggression. There may be contributions from the amygdala temporal lobe and the dorsolateral cortex, but the OFC appears to be instrumental to the development of the disinhibition syndrome.

Research in this area, particularly with respect to lesion studies, is hampered by some common methodological factors. First, CT scanning was used in many of the older studies reported in our chapter but does not offer the same high degree of spatial resolution as MRI; this factor is especially important when considering that the OFC, positioned directly above bone and air-filled sinus spaces, is prone to artifact. Second, there is commonly a lack of anatomic specificity in described affected areas of the OFC or even the frontal lobes in general. Third, many studies offer limited clinical evaluation and information regarding the behavior under study. For instance most of the studies in the literature employ one or perhaps two methods of quantifying mood or aggression; a more thorough investigation (using both self-report and observer ratings/interviews) will provide much more detailed description of the effects of lesions on these conditions.

The future direction of this area of study follows naturally from the limitations of the current literature. The common application of more precise imaging methods will demand far more sensitive tools for detecting and quantifying the presence of behavioral irregularities. Also, the wider use of complementary functional imaging in addition to structural imaging studies will be helpful. Finally, the effects of therapeutic interventions and their impact on functional, and even structural, changes in the brain will not only greatly enhance our understanding of the results of OFC lesions, but will also provide greater translation of this body of knowledge from the laboratory to the clinic. The directions we have articulated here indicate that a multidisciplinary approach to evaluating and studying mood disorders and aggression after brain lesions is optimal with cognitive neuroscientists, psychiatrists, neurologists, and molecular biologists combining their unique talents and knowledge towards a better understanding of the role of different regions in the prefrontal cortex in setting our moods and mediating our social behavior.

# References

Angrilli, A., Palomba, D., Cantagallo, A., Maietti, A., and Stegagno, L. (1999) Emotional impairment after right orbitofrontal lesion in a patient without cognitive deficits. *Neuroreport* **10**:1741–1746.

American Psychiatric Association, (2000) *Diagnostic and statistical manual of mental disorders.* Washington, DC: American Psychiatric Association.

Beck, A. T., Ward, C. H. M. M, Mock, J., and Erbaugh, J. (1961) An inventory for measuring depression. *Arch Gen Psychiatry* **4**:561–571.

Benoit, M., Clairet, S., Koulibaly, P. M., Darcourt, J., and Robert, P. H. (2004) Brain perfusion correlates of the apathy inventory dimensions of Alzheimer's disease. *Int J Geriatr Psychiatry* **19**:864–869.

Best, M., Williams, J. M., Coccaro, E. F. (2002) Evidence for a dysfunctional prefrontal circuit in patients with an impulsive aggressive disorder. *Proc Natl Acad Sci* U S A **99**:8448–8453.

Biver, F., Goldman, S., Delvenne, V., Luxen, A., de Maertelaer, V., and Hubain, P. (1994) Frontal and parietal metabolic disturbances in unipolar depression. *Biol Psychiatry* **36**:381–388.

Blake, P. Y., Pincus, J. H., and Buckner, C. (1995) Neurologic abnormalities in murderers. *Neurology* **45**:1641–1647.

Blumer, D. and Benson, D. F. (1975) Personality changes with frontal and temporal lobe lesions. In: Benson, D. F. and Blumer, D., (eds.) *Psychiatric aspects of neurological disease.* New York: Grune and Stratton, pp. 151–170.

Bozeat, S., Gregory, C. A., Ralph, M. A., and Hodges, J. R. (2000) Which neuropsychiatric and behavioural features distinguish frontal and temporal variants of frontotemporal dementia from Alzheimer's disease? *J Neurol Neurosurg Psychiatry* **69**:178–186.

Bremner, J. D., Vythilingam, M., Vermetten, E., Nazeer, A., Adil, J. and Khan, S. (2002) Reduced volume of orbitofrontal cortex in major depression. *Biol Psychiatry* **51**:273–279.

Buss, A. H. and Perry, M. (1992) The aggression questionnaire. *J Pers Soc Psychol* **63**:452–459.

Castillo, C. S., Starkstein, S. E., Fedoroff, J. P., and Price, T. R., and Robinson, R. G. (1993) Generaliszed anxiety disorder after stroke. *J Nerv Ment Dis* **181**:100–106.

Chen, X. L., Xie, J. X., Han, H. B., Cui, Y. H., and Zhang, B. Q. (2004) MR perfusion-weighted imaging and quantitative analysis of cerebral hemodynamics with symptom provocation in unmedicated patients with obsessive-compulsive disorder. *Neurosci Lett* **370**:206–211.

Clark, L., Iversen, S. D., and Goodwin, G. M. (2001) A neuropsychological investigation of prefrontal cortex involvement in acute mania. *Am J Psychiatry* **158**:1605–1611.

Cummings, J. L., Mega, M., Gray, K., Rosenberg-Thompson, S., Carusi, D. A., and Gornbein, J. (1994) The Neuropsychiatric Inventory: comprehensive assessment of psychopathology in dementia. *Neurology* **44**:2308–2314.

Damasio, A. R. D. T. and Damasio, H. (1991) Somatic markers and guidance of behavior: Theory and preliminary testing. In: Levin, H. S., Eisenberg, H. M. and Benton, A. L. (eds.) *Frontal lobe function and dysfunction.* New York: Oxford University Press, pp. 217–229.

Drevets, W. C., Price, J. L., Simpson, J. R., jr., Todd, R. D., Reich, T., Vannier, M., *et al.* (1997) Subgenual prefrontal cortex abnormalities in mood disorders. *Nature* **386**:824–827.

Eastwood, M. R., Rifat, S. L., Nobbs, H., and Ruderman, J. ( 1989) Mood disorder following cerebrovascular accident. *Br J Psychiatry* **154**:195–200.

Grafman, J., Vance, S. C., Weingartner, H., Salazar, A. M., and Amin, D. (1986) The effects of lateralised frontal lesions on mood regulation. *Brain* **109** (Pt 6):1127–1148.

Hamilton, M. (1960) A rating scale for depression. *J Neurol Neurosurg Psychiatry* **23**:56–62.

Harlow, J. M. (1868) Recovery from the passage of an iron bar through the head. *Massachusetts Medical Society Publication* **12**:327–346.

Heaton, R. K., Chelune, G. J., and Tailey, J. L. ( 1993) The Wisconsin card sorting test. Odessa, FL, USA: Psychological Assessment Resources.

Hibbard, M. R., Uysal, S., Kepler, K., Bogdany, J., and Silver, J. (1998) Axis I psychopathology in individuals with traumatic brain injury. *J Head Trauma Rehabil* **13**:24–39.

Hirono, N., Mori, E., Ishii, K., Ikejiri, Y., Imamura, T., Shimomura, T., *et al.* (1998) Frontal lobe hypometabolism and depression in Alzheimer's disease. *Neurology* **50**:380–383.

Holsinger, T., Steffens, D. C., Phillips, C., Helms, M. J., Havlik, R. J., Breitner, J. C., *et al.* (2002) Head injury in early adulthood and the lifetime risk of depression. *Arch Gen Psychiatry* **59**:17–22.

Hopkins, M. J., Dywan, J., and Segalowitz, S. J. (2002) Altered electrodermal response to facial expression after closed head injury. *Brain Inj* **16**:245–257.

Jorge, R. E., Robinson, R. G., Starkstein, S. E., Arndt, S. V., Forrester, A. W., and Geisler, F. H. (1993) Secondary mania following traumatic *brain* injury. *Am J Psychiatry* **150**:916–921.

Jorge, R. E., Robinson, R. G., Moser, D., Tateno, A., Crespo-Facorro, B., and Arndt, S. (2004) Major depression following traumatic brain injury. *Arch Gen Psychiatry* **61**:42–50.

Kay, S. R., Wolkenfeld, F., and Murrill, L. M. (1988) Profiles of aggression among psychiatric patients. I. Nature and prevalence. *J Nerv Ment Dis* **176**:539–546.

Kim, C. H., Chang, J. W., Koo, M. S., Kim, J. W., Suh, H. S., Park, I. H., *et al.* (2003) Anterior cingulotomy for refractory obsessive-compulsive disorder. *Acta Psychiatr Scand* **107**:283–290.

Kreutzer, J. S., Seel, R. T., and Gourley, E. (2001) The prevalence and symptom rates of depression after traumatic brain injury: a comprehensive examination. *Brain Inj* **15**:563–576.

Lacerda, A. L., Keshavan, M. S., Hardan, A. Y., Yorbik, O., Brambilla, P., Sassi, R. B., *et al.* (2004) Anatomic evaluation of the orbitofrontal cortex in major depressive disorder. *Biol Psychiatry* **55**:353–358.

Levy, M. L., Cummings, J. L., Fairbanks, L. A., Masterman, D., Miller, B. L., Craig, A. H., *et al.* Apathy is not depression. *J Neuropsychiatry Clin Neurosci* **10**:314–319.

Levy, M. L., Miller, B. L., Cummings, J. L., Fairbanks, L. A., and Craig, A. (1996) Alzheimer disease and frontotemporal dementias. Behavioral distinctions. *Arch Neurol* **53**:687–690.

Lewis, D. O., Yeager, C. A., Blake, P., Bard, B., and Strenziok, M. (2004) Ethics questions raised by the neuropsychiatric, neuropsychological, educational, developmental, and family characteristics of 18 juveniles awaiting execution in Texas. *J Am Acad Psychiatry Law* **32**:408–429.

Mah, L., Arnold, M. C., and Grafman, J. (2004) Impairment of social perception associated with lesions of the prefrontal cortex. *Am J Psychiatry* **161**:1247–1255.

Miller, B. L., Darby, A., Benson, D. F., Cummings, J. L., and Miller, M. H. (1997) Aggressive, socially disruptive and antisocial behaviour associated with fronto-temporal dementia. *Br J Psychiatry* **170**:150–154.

Mychack, P., Kramer, J. H., Boone, K. B., and Miller, B. L. (2001) The influence of right frontotemporal dysfunction on social behavior in frontotemporal dementia. *Neurology* **56**:S11–15.

Oliver, B., Gascon, J., Aparicio, A., Ayats, E., Rodriguez, R., Maestro, de Leon, J. L., *et al.* (2003) Bilateral anterior capsulotomy for refractory obsessive-compulsive disorders. *Stereotact Funct Neurosurg* **81**:90–95.

Paradiso, S., Robinson, R. G., and Arndt, S. (1996) Self-reported aggressive behavior in patients with stroke. *J Nerv Ment Dis* **184**:746–753.

Patel, V. and Hope, R. A. (1992) Aggressive behaviour in elderly psychiatric inpatients. *Acta Psychiatr Scand* **85**:131–135.

Phillips, L. H., MacPherson, S. E. S., and Della Sala, S. (2002) Age, cognition, and emotion: the role of anatomical segregation in the frontal lobes. In: **Boller, F.** and **Grafman, J.**, eds. *Handbook of neuropsychology.* New York: Elsevier pp. 73–97.

Pietrini, P., Guazzelli, M., Basso, G., Jaffe, K., and Grafman, J. (2000) Neural correlates of imaginal aggressive behavior assessed by positron emission tomography in healthy subjects. *Am J Psychiatry* **157**:1772–1781.

Premack, D. and Woodruff, G. (1978) Chimpanzee problem-solving: a test for comprehension. *Science* **202**:532–35.

Raine, A., Lencz, T., Bihrle, S., LaCasse, L., and Colletti, P. (2000) Reduced prefrontal grey matter volume and reduced autonomic activity in antisocial personality disorder. *Arch Gen Psychiatry* **57**:119–127; discussion 128–129.

Rauch, S. L., Savage, C. R., Alpert, N. M., Fischman, A. J., and Jenike, M. A. (1997) The functional neuroanatomy of anxiety: a study of three disorders using positron emission tomography and symptom provocation. *Biol Psychiatry* **42**:446–452.

Reich, W., Leacock, N., and Shanfeld, K. (1995) *Diagnostic Interview for Children and Adolescents*. St. Luis, MO: Washington University, Division of Child Psychiatry.

Reitan, R. M. (1971) Trail making test results for normal and *brain*-damaged children. *Percept Mot Skills* **33**:575–581.

Robert, P. H., Clairet, S., Benoit, M., Koutaich, J., Bertogliati, C., Tible, O., *et al.* (2002) The apathy inventory: assessment of apathy and awareness in Alzheimer's disease, Parkinson's disease and mild cognitive impairment. *Int J Geriatr Psychiatry* **17**:1099–1105.

Robinson, R. G. (1998) *The clinical neuropsychiatry of stroke*. Cambridge: University press.

Robinson, R. G. and Starkstein, S. E. (1990) Current research in affective disorders following stroke. *J Neuropsychiatry Clin Neurosci* **2**:1–14.

Shamay-Tsoory, S. G., Tomer, R., Berger, B. D., Goldsher, D., and Aharon-Peretz, J. (2005) Impaired 'affective theory of mind' is associated with right ventromedial prefrontal damage. *Cogn Behav Neurol* **18**:55–67.

Shimoda, K. and Robinson, R. G. (1999) The relationship between post-stroke depression and lesion location in long-term follow-up. *Biol Psychiatry* **45**:187–192.

Simpson, G., Tate, R., Ferry, K., Hodgkinson, A., and Blaszczynski, A. (2001) Social, neuroradiologic, medical, and neuropsychologic correlates of sexually aberrant behaviour after traumatic brain injury: a controlled study. *J Head Trauma Rehabil* **16**:556–572.

Starkstein, S. E., Ingram, L., Garau, M. L., and Mizrahi, R. (2005) On the overlap between apathy and depression in dementia. *J Neurol Neurosurg Psychiatry* **76**:1070–1074.

Stone, V. E., Baron-Cohen, S., and Knight, R. T. (1998) Frontal lobe contributions to theory of mind. *J Cogn Neurosci* **10**:640–656.

Tateno, A., Jorge, R. E., and Robinson, R. G. (2003) Clinical correlates of aggressive behaviour after traumatic brain injury. *J Neuropsychiatry Clin Neurosci* **15**:155–160.

Vasa, R. A., Grados, M., Slomine, B., Herskovits, E. H., Thompson, R. E., Salorio, C., *et al.* (2004) Neuroimaging correlates of anxiety after pediatric traumatic brain injury. *Biol Psychiatry* **55**:208–216.

Vataja, R., Pohjasvaara, T., Mantyla, R., Ylikoski, R., Leskela, M., Kalska, H., *et al.* (2005) Depression-executive dysfunction syndrome in stroke patients. *Am J Geriatr Psychiatry* **13**:99–107.

Yudofsky, S. C., Silver, J. M., Jackson, W., Endicott, J., and Williams, D. (1986) The Overt Aggression Scale for the objective rating of verbal and physical aggression. *Am J Psychiatry* **143**:35–39.

Zubenko, G. S., Zubenko, W. N., McPherson, S., Spoor, E., Marin, D. B., Farlow, M. R, *et al.* (2003) A collaborative study of the emergence and clinical features of the major depressive syndrome of Alzheimer's disease. *Am J Psychiatry* **160**:857–866.

Chapter 23

# Pseudopsychopathy: a perspective from cognitive neuroscience

Michael Koenigs and Daniel Tranel

## 23.1. Introduction

Some of the most significant breakthroughs in cognitive neuroscience have resulted from the study of brain-damaged patients. Individual landmark cases have often provided the initial footing for mapping complex cognitive abilities onto discrete brain areas. For example, Broca's (1861) description of a language production defect in a patient ('Tan') with left frontal operculum damage sparked the formulation of the left hemisphere language circuit. Similarly, Scoville and Milner's (1957) description of a declarative memory acquisition defect in a patient (H.M.) who underwent bilateral medial temporal lobe resection established that region's role in memory. These initial findings continue to be extended through systematic study of groups of lesion patients. But can all cognitive phenomena be mapped this way? Certainly some constructs, such as intelligence, would be less localizable; intelligence deficits are not associated with a single discrete lesion. *A priori*, one might assume that social and moral behavior would be similarly difficult to associate with discrete neural structures, but there is in fact evidence from lesion patients that particular brain areas are critical for social and moral behavior. Just as neuroanatomical and behavioral study of patients with 'acquired disturbance of language' (aphasia) or 'acquired disturbance of memory' (amnesia) have yielded breakthroughs in identifying the neural substrates of language and memory, respectively, so, too have patients with acquired disturbances of personality yielded breakthroughs in identifying the neural substrates of social and moral behavior.

## 23.2. Psychopathy

Undoubtedly the greatest disturbances of social and moral behavior, absent a macroscopic brain lesion, are those observed in psychopaths. In his pioneering profile of the psychopathic personality, Cleckley (1976) emphasizes the pervasive disregard for others as a primary feature. To facilitate the diagnosis of psychopathy, Hare (1991) created a specific and operationalized instrument: the Psychopathy Checklist-Revised (PCL-R; Fig. 23.1). The Hare PCL-R is a 20-item measure completed on the basis of a semi-structured interview and detailed collateral or file information. Each item is scored on a 3-point scale and the total score reflects the extent to which an individual is psychopathic. Cutoff scores divide subjects into one of three groups: non-psychopathic, moderately psychopathic, and

| Factor 1: Interpersonal/affective | Factor 2: Social deviance |
|---|---|
| • Glibness/superficial charm<br>• Grandiose sense of self worth<br>• Pathological lying<br>• Conning/manipulative<br>• Lack of remorse or guilt<br>• Shallow affect<br>• Callous/lack of empathy<br>• Failure to accept responsibility for own actions | • Need for stimulation/proneness to boredom<br>• Parasitic lifestyle<br>• Poor behavioral controls<br>• Early behavioral problems<br>• Lack of realistic, long-term goals<br>• Impulsivity<br>• Irresponsibility<br>• Juvenile delinquency<br>• Revocation of conditional release |

**Additional items**

| | |
|---|---|
| • Promiscuous sexual behavior<br>• Many short-term marital relationships | • Criminal versatility |

**Fig. 23.1** Items in the Hare Psychopathy Checklist-Revised (PCL-R). The 20 items fall into two clusters of correlated traits. Factor 1 consists of affective/interpersonal features, while Factor 2 consists of items that describe an impulsive, antisocial and unstable lifestyle, or social deviance. Three additional items do not load on either factor. The Hare PCL-R is completed on the basis of a semi-structured interview and detailed collateral or file information. Each item is scored on a three-point scale. Adapted from Hare (1991).

psychopathic. The PCL-R highlights the fact that psychopathy is a syndrome; a cluster of behaviors defines the condition. The behavioral profile of the psychopath outlined in the PCL-R offers a useful reference point for assessing disturbances in social and moral behavior in brain-damaged patients. As will be elaborated below, there are remarkable overlaps between the psychopathic traits noted in Hare's nomenclature and the personality manifestations shown by patients following brain damage.

## 23.3. Pseudopsychopathy

The landmark case in this respect is that of Phineas Gage. The notable enduring effect of Gage's head injury was a change in personality. As characterized by Harlow (1868), Gage changed from shrewd, persistent, and respectable to profane, capricious, and unreliable. Over the next century, reports of Gage-like personality changes following selected cases of frontal injury slowly accumulated. In 1975, Blumer and Benson coined the term 'pseudopsychopathy' to describe the personalities of a subset of frontal lobe patients. Blumer and Benson described such patients as 'best characterized by the lack of adult tact and restraints.' In contrast to conventional, or developmental, psychopathy, in which psychopathic traits emerge in childhood and adolescence with no gross structural brain lesion, pseudopsychopathic behaviors 'apparently follow injury to the orbital frontal lobe or pathways traversing this region' (Blumer and Benson 1975).

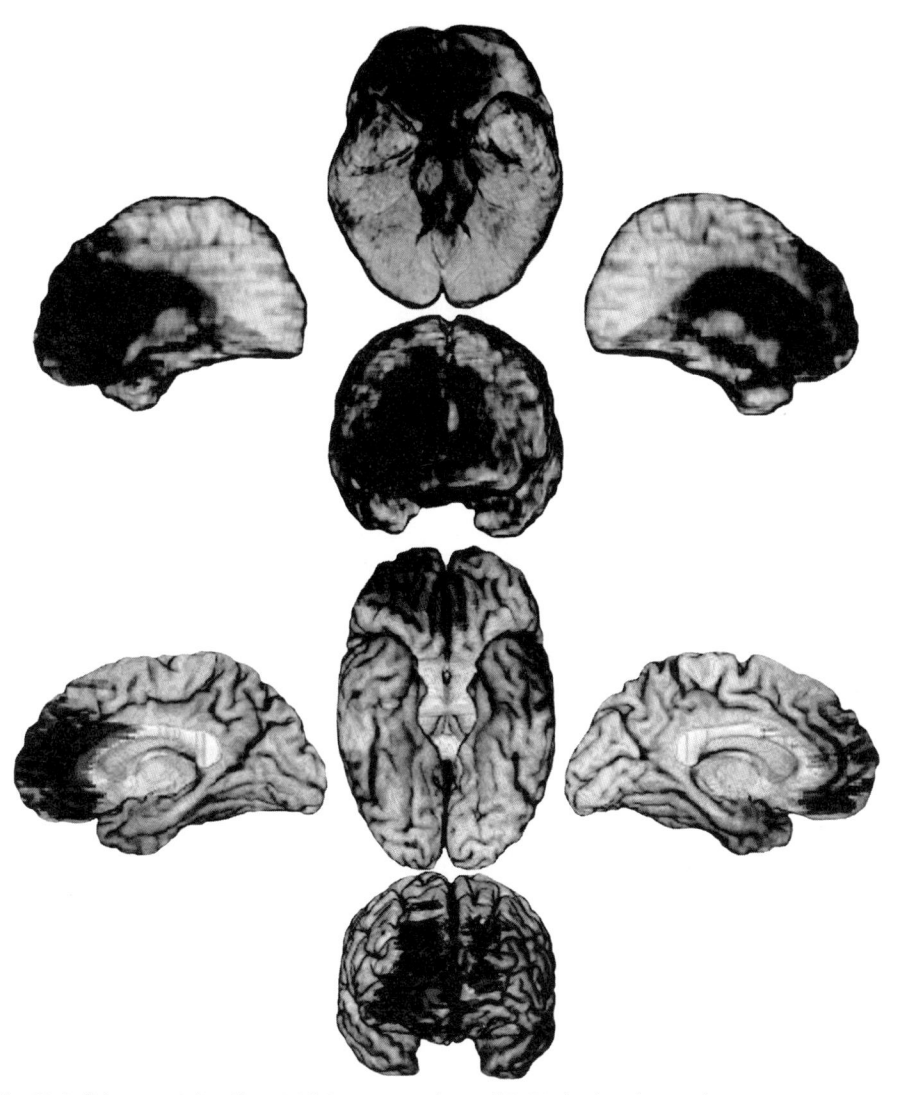

**Fig. 23.2** (Also see Color Plate 34.) Reconstructions of EVR's brain. The top images are 3-D reconstructions of EVR's brain from a computerized tomography (CT) scan. The bottom images depict EVR's lesion (in red) on a standard reference brain. Prefrontal damage includes most of the orbital cortex on the right and part of the orbital cortex on the left, most of the mesial cortex on the right and part of the mesial cortex on the left, part of the dorsolateral cortex on the right, and the frontal poles. Adapted with permission from Tranel (2002).

The association of psychopathic behavior with orbitofrontal cortex (OFC) damage has since been investigated in much greater detail. Perhaps the most illustrative example of this phenomenon is patient EVR, first reported by Eslinger and Damasio in 1985. At age 35, EVR underwent resection of a bilateral orbitofrontal meningioma. Orbital and lower mesial frontal cortices (collectively referred to as ventromedial prefrontal cortex; VMPFC) were excised with the tumor (Fig. 23.2). Profound personality changes ensued. Although his marriage had previously been stable and successful (EVR is the father of two well-adjusted children), he divorced his wife of many years, and within months entered into a second, short-lived marriage. Whereas formerly he had had a keen business sense with considerable financial success, post-surgery he entered into a series of disastrous business ventures over a brief period of time. One of these, a highly speculative venture with a disreputable partner, led to bankruptcy. Formerly he was an accomplished professional, working hard at a skilled job and securing promotions for good performance, but since the tumor resection he has never been able to maintain employment, and now lives in a sheltered environment. Although he is skilled and intelligent enough to hold a job, he cannot report to work promptly or regularly, and he is unacceptably unreliable in completion of various intermediate job steps. EVR's ability to plan his activities, both on a daily and long-term basis, is severely impaired. He fails to take into account his long-term interests and often pursues peripheral interests of no value. Unlike his former self, he is a poor judge of character and is often socially inappropriate. Despite all this, he is an avid conversationalist and often comes across as intelligent, charming, and witty.

Neuropsychological testing indicates EVR's intellectual and cognitive abilities remain superior. Administration of the WAIS-R yielded scores of Verbal IQ=132 and Performance IQ = 135. Speech, language, vocabulary, and memory are fully intact. Even his executive functioning abilities, insofar as conventional laboratory measures are concerned, are largely preserved—he performs flawlessly on the Wisconsin Card Sorting Task, the Category Test, and Word and Design Fluency tests. Clearly intellectual deficiency cannot explain the decline in his personal and professional life. It is quite notable that most of EVR's most salient post-operative changes (short-term marriages, poor long-term planning, unreliability/irresponsibility, dependence) appear nearly verbatim on the PCL-R. While striking, EVR is not an isolated case. A number of group studies with multiple VMPFC patients have been conducted. The main findings related to pseudopsychopathy (or 'acquired sociopathy' as it has been more commonly known; see Eslinger and Damasio, 1985) are summarized in the following sections. The focus of the chapter is on human lesion studies performed in our laboratory at the University of Iowa, which have taken advantage of a unique resource of neurological patients with focal lesions who have been fully characterized with neuropsychological and neuroanatomical studies.

Before turning to these experimental findings, we would like to clarify our terminology. First, it should be noted that the terms 'psychopathy' and 'sociopathy' have been used interchangeably in the abnormal psychology literature to refer to individuals with 'developmental' antisocial behavior, and criteria for these conditions overlap with the DSM-IV diagnosis of antisocial personality disorder. We will adhere to this usage. Second, the term 'pseudopsychopathy' apparently was first coined by Blumer and Benson (1975). A decade later, Eslinger and Damasio (1985) introduced the term 'acquired sociopathy' to refer to the same set of phenomena noted by Blumer and Benson. We will use pseudopsychopathy and acquired sociopathy interchangeably, to refer to the set of psychopathic personality manifestations that is reminiscent of the Hare depiction, but that follows an acquired brain lesion rather than occurring in the course of development. When the context may lack certainty, we will qualify psychopathy as being 'developmental' psychopathy to denote the Hare version.

## 23.4. Experimental findings

### Personality

Since the initial descriptions of psychopathic personality changes following VMPFC damage were largely based on clinical observations, Barrash et al. (1994, 2000) investigated the personality changes following PFC damage more systematically. Barrash et al. (1994) tested 6 PFC-damaged patients (all of whom had acquired blatant socially maladaptive behaviors after brain damage) with well-established psychometric instruments for detecting psychopathological features, including the Minnesota Multiphasic Personality Inventory (MMPI; Hathaway and McKinley 1983), MMPI-2 (Butcher et al. 1989), Eysenck Personality Questionnaire (Eysenck and Eysenck 1975), the Structured Interview for the revised third edition of the Diagnostic and Statistical Manual of Mental Disorders Personality (Pfohl et al. 1989), and Hare's PCL-R. Surprisingly, none of these instruments yielded an accurate reflection of the patients' real-world personalities. On these scales the patients' personality profiles were generally normal or significant misrepresentations of their actual selves. This result can be explained by a combination of factors: 1) these instruments generally rely on self-report, 2) the VMPFC patients had normal premorbid personalities, 3) the VMPFC patients are generally not aware of changes in their personalities (Barrash et al. 2000), 4) 'off-line' assessment of social situations and ethical dilemmas (verbally posed rather than real-life action) can be intact following VMPFC damage (Saver and Damasio 1991), and 5) the instruments do not explicitly detect *changes* in functioning; rather, they measure *level* of functioning.

To address these issues and provide an instrument for more accurate assessment of acquired psychopathic behavior following brain damage, Barrash and Anderson (1993) developed the Iowa Rating Scale of Personality Change (IRSPC). IRSPC is a standardised assessment of 30 specific personality changes obtained from a collateral (usually spouse or family member) that has observed the patient's real-world behavior before and after brain damage (see Fig. 23.3). Barrash et al. (2000) used the IRSPC to assess 7 patients with bilateral VMPFC damage, as well as 14 patients with PFC lesions outside of VMPFC and 36 patients with non-prefrontal lesions. Compared to the other two groups, all 7 VMPFC patients exhibited the following changes: dampening of emotional experience, poorly modulated emotional reactions, defective decision-making, impaired goal-directed behavior, and marked lack of insight into these changes. As expected, these features comprise the core of a syndrome that bears significant resemblance to psychopathy. Individual traits characteristic of pseudopsychopathy (acquired sociopathy) and developmental psychopathy have been further explored experimentally.

- Lack of insight
- Lack of initiative
- Irritability
- Social inappropriateness
- Poor judgment
- Lack of persistence
- Indecisiveness
- Lability
- Blunted emotional experience
- Apathy
- Inappropriate affect
- Lack of planning
- Poor frustration tolerance
- Inflexibility
- Behavioral rigidity
- Low emotional expressivity
- Insensitivity
- Excitability
- Anxiety
- Impulsivity
- Self-absorption
- Obsessiveness
- Social withdrawal
- Disorganization
- Manipulativeness
- Dependency
- Depression
- Type A behavior
- Delusional thinking
- Frugality

**Fig. 23.3** Personality traits assessed on the Iowa Rating Scale of Personality Change (IRSPC). IRSPC is a standardized assessment of 30 specific personality changes obtained from a collateral (usually spouse or family member) that has observed the patient's real-world behavior before and after brain damage. Adapted from Barrash and Anderson (1993).

## Autonomic function and emotion

The apparent non-emotionality observed in developmental psychopaths and pseudopsychopaths has been investigated psychophysiologically. The electrodermal skin conductance response (SCR) is a very sensitive index of psychosomatic arousal that is reliably elicited by stimuli with emotional or social valence. Damasio *et al.* (1990) employed this measure to investigate the hypothesis that pseudopsychopaths might have abnormal autonomic responses to emotionally charged stimuli. Three groups of subjects were tested: 1) 5 patients with bilateral VMPFC damage and acquired defects in social conduct, judgment, and planning; 2) 6 brain-damaged subjects with focal lesions outside of VMPFC; and 3) 5 neurologically and psychiatrically normal subjects. Subjects observed two types of pictures: 1) emotionally charged pictures such as mutilations, nudes, and social disasters ('targets'); and 2) neutral pictures such as farm scenes or abstract patterns ('non-targets'). Normal subjects and brain-damaged controls generated similar patterns of SCRs: little or no response to non-targets, but large SCRs to targets. The VMPFC patients, on the other hand, generated little or no response to both targets and non-targets (Fig. 23.4). Note that the unresponsiveness of the pseudopsychopaths is not simply an epiphenomenon of a global autonomic defect; the VMPFC patients generated normal SCRs to orienting stimuli such as an unexpected loud noise or a deep breath.

In an analogous study of developmental psychopaths, Patrick *et al.* (1994) recorded skin conductance levels (SCL) while subjects listened to neutral sentences (e.g. 'I'm looking out the window on a sunny autumn day'.) or fearful sentences (e.g. 'Alone in the house, I hear the sound of someone forcing the door and I panic'.). Normal subjects had greater SCLs to the fearful sentences than to the neutral sentences, but psychopaths (as identified

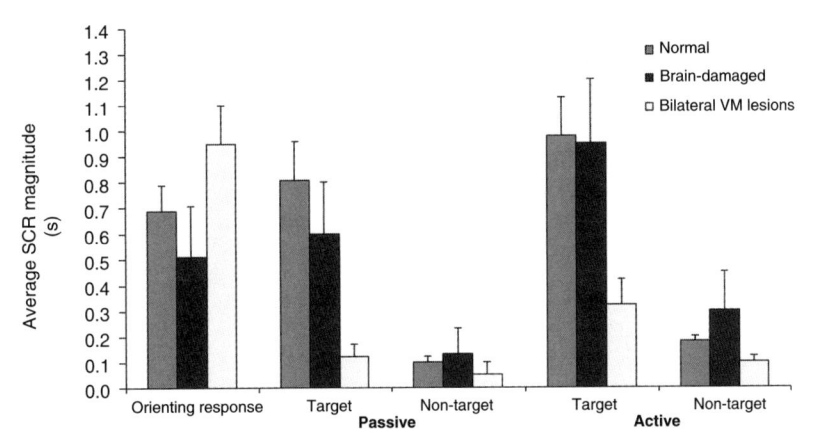

**Fig. 23.4** Skin conductance response (SCR) magnitudes to orienting stimuli, neutral pictures (non-targets), and emotionally charged pictures (targets) in normal controls, brain-damaged controls, and bilateral ventromedial (VM) prefrontal patients. In the VM group, SCR magnitude for the emotionally charged pictures is reduced. Interestingly, when the pictures were viewed 'actively' (i.e. when the patients had to describe the content verbally), the SCR magnitude from the VM patients was increased. Adapted from Tranel et al. (in press).

with PCL-R) showed no significant differentiation. The blink startle reflex has also been used to show diminished emotional reactivity in psychopaths. Whereas normal subjects exhibit large startle responses when they view unpleasant slides (mutilations, assault, threat) relative to neutral or pleasant slides, psychopaths exhibit no such difference (Patrick et al. 1993).

These studies indicate that both psychopaths and VMPFC patients manifest diminished physiological arousal when perceiving emotional external stimuli. Additional studies suggest that the same conclusion holds for internally generated percepts. In a study by Patrick et al. (1994), prison inmates memorized and read aloud a series of sentences with either neutral or emotional content. They were then instructed to imagine the content of each sentence for six seconds, during which time SCL was recorded. Non-psychopathic inmates (low PCL-R scores) exhibited larger increases in SCL for emotional content than for neutral content. By contrast, SCLs of psychopathic inmates (high PCL-R scores) were not significantly different for emotional and neutral content.

Tranel et al. (1998) obtained similar results from acquired sociopaths. In this study, VMPFC patients and normal subjects were asked to recall and re-experience emotional situations from their lives. Subjects recalled situations in which they felt happiness, sadness, fear, or anger. As a control, subjects also recalled a neutral, non-emotional event. First, subjects provided brief descriptions of each memory. Next, subjects imagined and re-experienced the individual emotional or neutral events while SCR and heart rate were recorded. Following this emotional imagery task, subjects rated the intensity of emotion they felt during recall on a scale of 0 ('none') to 4 ('intense'). The results are presented in Fig. 23.5. Both normal subjects and VMPFC patients were able to recall situations of

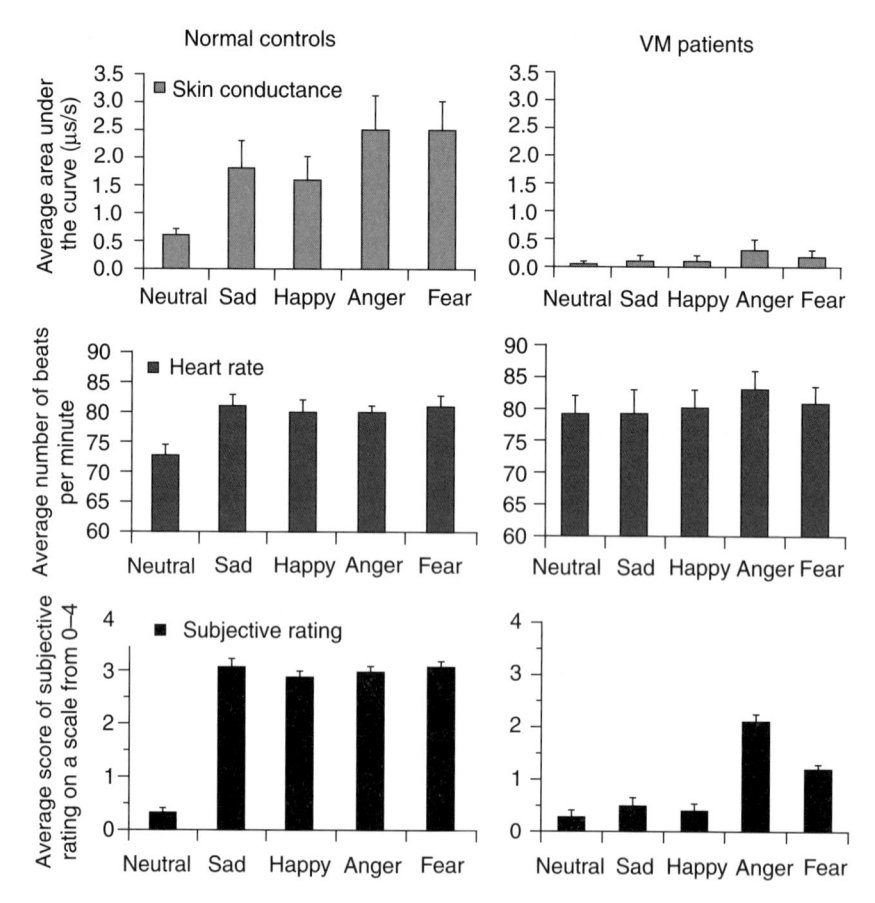

**Fig. 23.5** Skin conductance response (SCR), heart rate, and subjective rating of the feeling associated with the imagery of a personal emotional event experienced by individual normal control subjects versus vmPFC patients. In the vmPFC group, the SCRs generated during the imagery procedure were abnormally low, the changes in heart rate were not significant, and the subject ratings of feeling the target emotion were low, despite the fact that the patients had vivid memory and recall of the imagined personal emotional events. Adapted from Tranel et al. (in press).

happiness, sadness, fear, and anger (events like weddings, funerals, car accidents, family disputes). But compared to normal subjects, VMPFC patients experienced diminished physiological arousal and less subjective intensity of feelings during the emotional imagery.

The experience of diminished emotionality common to developmental psychopathy and pseudopsychopathy is described in the words of the subjects themselves. Compare the comments of EVR (as described above, a well-documented pseudopsychopath with bilateral VMPFC damage), to those of convicted killer and psychopath Jack Abbott. '(EVR) noted that he had not experienced the kind of "feeling" that he thought he ought to have in relation to some stimuli, given their content.' (Damasio et al. 1990).

Jack Abbott (1981) states, 'There are emotions—a whole spectrum of them—that I know only through words, through reading and in my imagination. I can imagine I feel these emotions but I do not.'

While the parallel findings in pseudopsychopathy and developmental psychopathy related to emotional unresponsiveness are indeed remarkable, the relationship should be qualified with mention of several findings. For one, VMPFC damage does not completely abolish all emotional experience. In Damasio *et al.*'s study (1990) in which subjects observed emotionally charged scenes, VMPFC patients exhibited increased physiological arousal when they actively verbalised the content of the emotional stimuli, compared to passive viewing (Fig. 23.4). Also, in Tranel *et al.*'s emotional imagery task (1998), VMPFC patients experienced greater physiological and subjective responses to memories of anger than to memories of other emotions (Fig. 23.5). In both cases the increases in arousal and feeling observed in VMPFC patients were still below those of non-brain-damaged subjects. In addition, at least some VMPFC patients have normal SCRs to loud noises (Damasio *et al.* 1990), and some VMPFC patients (apparently those with lesions sparing anterior cingulate cortex and/or basal forebrain) acquire normal conditioned SCRs to aversive loud noises (Bechara *et al.* 1999). By contrast, developmental psychopaths exhibit abnormally small SCRs in response to loud noises (Hare 1978) and reduced conditioned SCRs (Lykken 1957). Despite these qualifications, there remains strong evidence that VMPFC patients and psychopaths share the trait of emotional unresponsivenss, in both behavioral and psychophysiological terms.

## Decision-making

Poor decision-making is another hallmark psychopathic trait that has been extensively investigated in both VMPFC patients and psychopaths. One of the primary experimental paradigms used to investigate decision-making behavior is the Iowa Gambling Task (IGT) (Bechara *et al.* 1994). The IGT was developed as a laboratory probe to detect and measure the decision-making impairments that are so blatant in the day-to-day lives of VMPFC patients. Accordingly, the IGT factors in reward and punishment (winning and losing money) in such a way as to create a conflict between immediate reward and delayed punishment, although the precise contingencies cannot be calculated. (Tranel, in press) So, through variable experiences of reward and punishment, the subject must formulate a plan of action that is advantageous in the long run (i.e. pick mostly from the good decks) (for a complete description of the IGT, see Naqvi *et al.*, Chapter 13, this volume). Over the course of the task, the normal pattern of behavior is that subjects learn to avoid the bad decks and choose predominantly from the good decks. VMPFC patients, on the other hand, continue to select from the bad decks throughout the task (Fig. 23.6). Thus, the IGT captures the real-world defects in decision-making behavior observed in pseudopsychopaths.

Interestingly, the IGT has been administered to developmental psychopaths, who exhibit real-life decision-making impairments similar to VMPFC patients. Mitchell *et al.* (2002) found that psychopathic subjects, like VMPFC patients, continue to select from

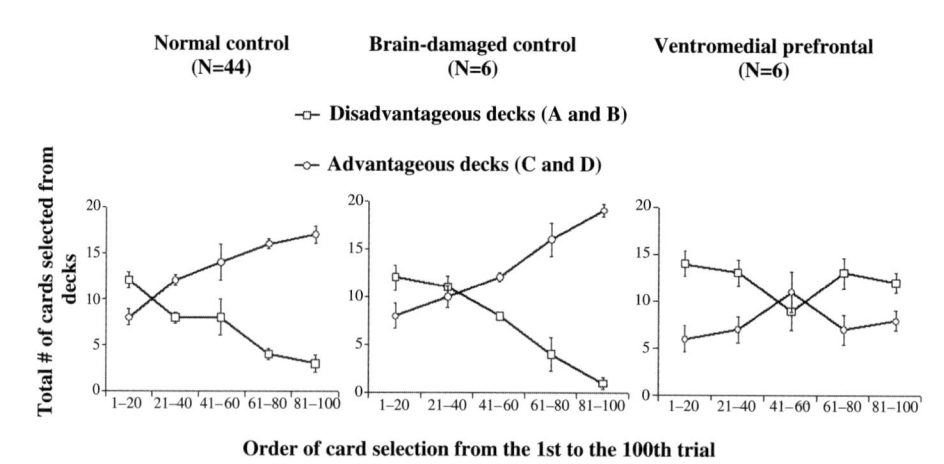

**Fig. 23.6** Card selections on the Iowa Gambling Task as a function of group (normal control, brain-damaged control, ventromedial prefrontal damage), deck type (disadvantageous versus advantageous), and trial block. The two control groups gradually shifted their response selections toward the advantageous decks, and this tendency became stronger as the game went on. The vmPFC subjects did not make a reliable shift, and continued to opt for the disadvantageous decks even during the latter stages of the game, when controls had almost completely abandoned choosing from the disadvantageous decks. Adapted from Tranel (2002).

the disadvantageous decks over the course of the task. In another study, Blair et al. (2001) found that boys (age 9–17) with psychopathic tendencies were impaired on the IGT compared to boys without psychopathic traits. Similarly, Van Honk et al. (2002) found that adults scoring the highest on a scale of psychopathic traits performed worse on the IGT than adults reporting the lowest degree of psychopathic behavior.

To investigate the pathophysiological basis of the decision-making impairment in VMPFC patients, Bechara et al. (1996) administered the IGT while recording SCRs. Both normal subjects and VMPFC patients generated SCRs after picking a card and learning they won or lost money (again indicating the absence of a global autonomic defect in the VMPFC patients). However, as normal subjects progressed through the task they began to generate SCRs prior to the selection of cards, during the time in which subjects consider which deck to choose from. Furthermore, these anticipatory SCRs were larger before picking from a bad deck than picking from a good deck. By contrast, VMPFC patients generated no SCRs prior to card selection.

In another instantiation of the IGT, Bechara et al. (1997) recorded SCRs but also intermittently interrupted the subject to ask what the subject knew about the game. Card selection and anticipatory SCRs were compared for different periods of the task (Fig. 23.7). In normal subjects, anticipatory SCRs began to develop as subjects began to choose advantageously and express a 'hunch' that some decks were better than others. VMPFC patients never generated anticipatory SCRs and continued to choose disadvantageously throughout the task. Remarkably, this pattern of poor choices even persisted

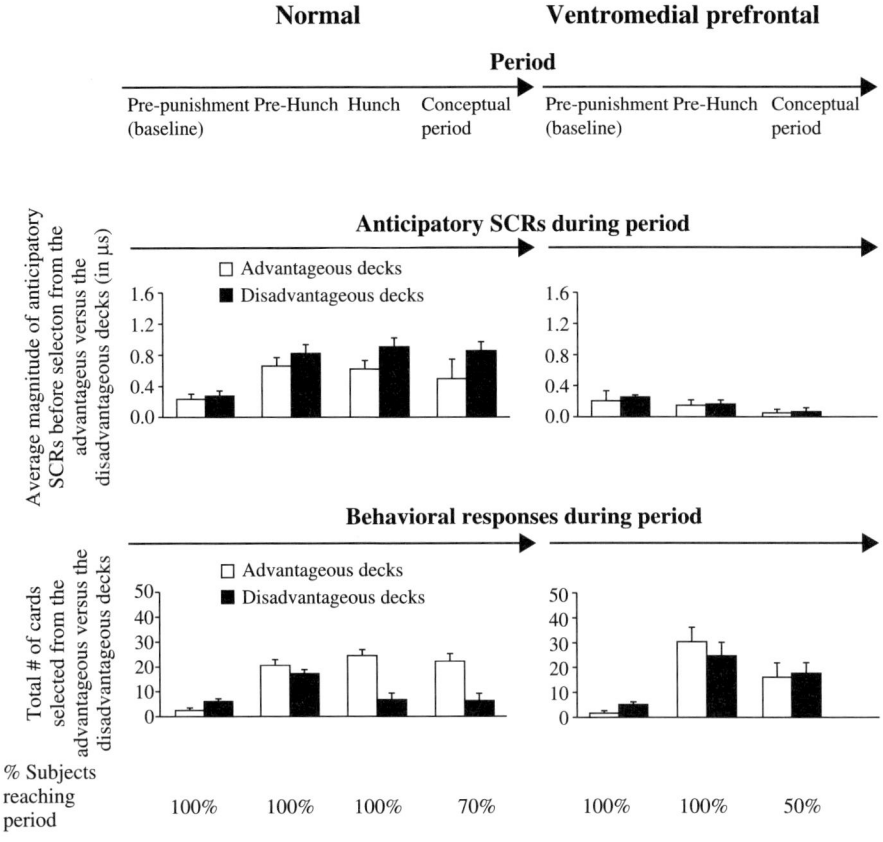

**Fig. 23.7** Psychophysiological (anticipatory skin conductance responses) and behavioral (card selection) data for control subjects (n=10) and subjects with ventromedial prefrontal lesions (n=6), as a function of four 'knowledge periods'. Control subjects, even before they knew anything consciously about how the game worked (pre-hunch period), began to generate anticipatory SCRs and shift their selections away from the bad decks. In the controls, anticipatory SCRs, especially to the bad decks, became more pronounced as the game progressed, and the subjects shifted almost exclusively to the good decks. The vmPFC subjects never produced anticipatory SCRs, and they also continued to opt more frequently for the bad decks. The pattern occurred even in vmPFC subjects (n=3) who knew at a conscious level (conceptual period) how the game worked, and that some decks were good and some were bad. Adapted from Tranel (2002).

in VMPFC patients who were able to articulate the relative long-term reward and punishment contingencies of each deck. Conversely, some normal subjects were never able to describe the long-term outcomes of each deck, but still generated anticipatory SCRs and chose advantageously. These data suggest that the critical determinant of successful decision-making is not declarative knowledge of likely outcomes, but a somatic signal mediated by VMPFC and derived from prior experiences with reward and punishment. Furthermore, these data provide an experimental model of pseudopsychopathic and psychopathic behavior; specifically, the ability of these individuals to say the right thing, but do the wrong thing.

The question of whether the decision-making impairments in developmental psychopathy and VMPFC damage share the same pathophysiological mechanism remains to be answered, but there is evidence of a similar psychophysiological impairment. One highly replicated finding in psychopathy has been made with 'countdown' experiments. In these studies, an aversive stimulus (like a loud noise or shock) is delivered at an explicit time point that is counted down to, so the subject knows when the aversive stimulus will occur. Reduced SCRs in anticipation of the punishment were observed in psychopaths compared to controls. (Hare *et al.* 1978; Tharp *et al.* 1980 Ogloff and Wong 1990). A second highly replicated finding in psychopathy is poor passive avoidance on Lykken's mental maze (Lykken 1957). In this task the subject must learn to navigate a maze with 20 choice points. At each point one of four choices is correct, advancing the subject to the next choice point, while one of the three incorrect choices is punished with a shock. A passive avoidance error occurs when the subject chooses a previously punished incorrect choice. Thus, like the IGT, Lykken's mental maze incorporates elements of learning, reward, punishment, and uncertainty in real time. While psychopaths are comparable to controls in learning the correct choice, they are impaired in learning to avoid the punished incorrect choice. Intriguingly, in one study in which SCR was recorded, psychopaths, unlike normal subjects, exhibited no SCRs prior to passive avoidance errors. (Schmauk 1970) In a study of normal subjects, Waid (1976) found a negative correlation between anticipatory SCRs (SCRs generated before making the choice) and passive avoidance errors. Waid obtained identical results when the punishment was shock or loss of money. These studies suggest decision-making defects in psychopathy, as in pseudopsychopathy, may be due to the absence of somatic arousal that signals impending or potential adverse consequences.

The preceding discussion of acquired sociopathy was based on neuropsychological studies of patients with *adult-onset, bilateral* VMPFC damage. Hence, these studies bear no information on two fundamental considerations of brain-behavior relationships: laterality and development. The following two sections, respectively, discuss acquired sociopathy in these terms.

## Laterality

To parse the relative contributions of right and left VMPFC damage to acquired sociopathy, Tranel *et al.* (2002) studied a rare series of patients with focal, unilateral lesions to VMPFC. The rarity of such patients can be attributed to the predominant etiologies of VMPFC damage. Ruptures of aneurysms in the anterior cerebral or anterior communicating arteries, resections of orbitofrontal meningiomas, and head injury all tend to inflict bilateral damage. Nonetheless, Tranel *et al.* assembled a group of 4 subjects with unilateral right VMPFC damage (RVMPFC group) and 3 subjects with unilateral left VMPFC damage (LVMPFC group). Neuropsychological assessment revealed no systematic differences between the two groups with respect to orientation, attention/concentration, intelligence, memory, speech/linguistic function, visuospatial processing, visuoconstructional function, or executive function. Experimental measures were used to assess three broad domains of function: social conduct, decision-making, and emotional processing/personality. Structured clinician-based ratings, ratings by family members, and employment status were used to measure changes in social conduct (Fig. 23.8) In stark contrast to the RVMPFC group, acquired deficits in social conduct were virtually nonexistent in the LVMPFC group. No RVMPFC patient was able to sustain gainful employment following brain damage, and severe disturbances in interpersonal functioning and social status were common in this group. The Iowa Gambling Task (IGT, described above and in Naqvi, Tranel and Bechara this volume) was used to assess decision-making ability (Fig. 23.9). As the task progressed, LVMPFC patients tended to shift their card selection preference from the bad decks towards to good decks, in a manner similar to normal subjects. RVMPFC patients, on the other hand, continued to select predominantly from the bad decks, in a manner similar to bilateral VMPFC patients. Despite the difference between RVMPFC and LVMPFC patients on the IGT, note that LVMPFC patients did not perform quite as well as normal subjects, nor did RVMPFC patients perform quite as poorly as bilateral VMPFC subjects. Accordingly, LVMPFC subjects developed anticipatory SCRs (albeit not as great in magnitude as controls) over the course of the task, while RVMPFC and bilateral VMPFC subjects failed to generate anticipatory SCRs (Fig. 23.10). To index changes in personality and emotional processing, the Iowa Rating Scales of Personality Change (IRSPC; described above) was administered. All three LVMPFC

*Social conduct*

| Group status | Social status | Employment functioning | Interpersonal rating | Clinician |
|---|---|---|---|---|
| Left VM | | | | |
| RF 297 | 1 | 1 | 1 | 1 |
| CG 1652 | 1 | 1 | 1 | 1 |
| CP 2183 | 1 | 1 | 1 | 1 |
| Right VM | | | | |
| CB 2310 | 1 | 2 | 2 | 2 |
| SB 2046 | 3 | 3 | 3 | 3 |
| RW 1768 | 3 | 3 | 2 | 3 |
| DV 1589 | 3 | 3 | 3 | 3 |

**Fig. 23.8** Extent of change or impairment in various aspects of social conduct was rated on a three-point scale (1 = no change; 2 = moderate change or impairment; 3 = severe change or impairment). A postmorbid change or impairment in Social status was defined as an alteration in the patient's financial security and/or peers' judgment of social attainment. A postmorbid change or impairment in Employment status was defined as an alteration in the patient's procurement or level of occupation (e.g. inability to sustain gainful employment). A postmorbid change in Interpersonal functioning was defined as an alteration in the patient's ability to maintain normal social relationships with significant others, such as friends or family. Clinician rating of change or impairment, was based on multiple sources of evidence (e.g. interview data, neuropsychological performance, information gathered from collaterals). Note that for subject SB 2046, who has an early-onset lesion, the ratings pertain to level of impairment, and not to change. Adapted from Tranel *et al.* (2002).

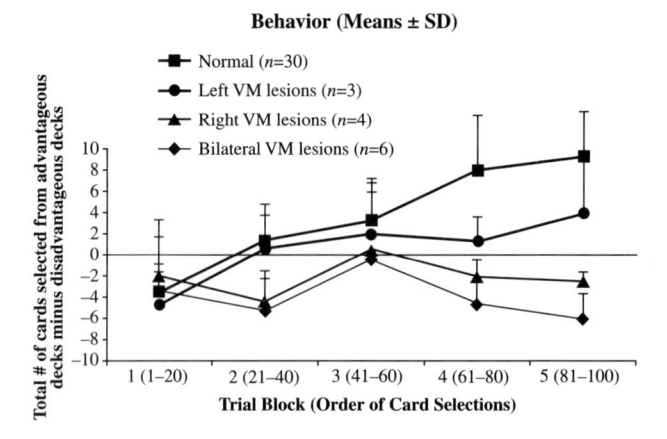

**Fig. 23.9** Iowa Gambling Task behavior, expressed as the total number of cards selected from the advantageous decks minus the total number of cards selected from the disadvantageous decks (positive scores indicate advantageous responding). Performances are graphed as a function of Group (Normal, Left VMPFC, Right VMPFC, Bilateral VMPFC) and Trial Block (1 through 5; each block corresponds to 20 card selections, in order). The Normal and Left VMPFC groups show advantageous responding over the course of the Iowa Gambling Task; the Right VMPFC and Bilateral VMPFC groups do not. Adapted from Tranel et al. (2002).

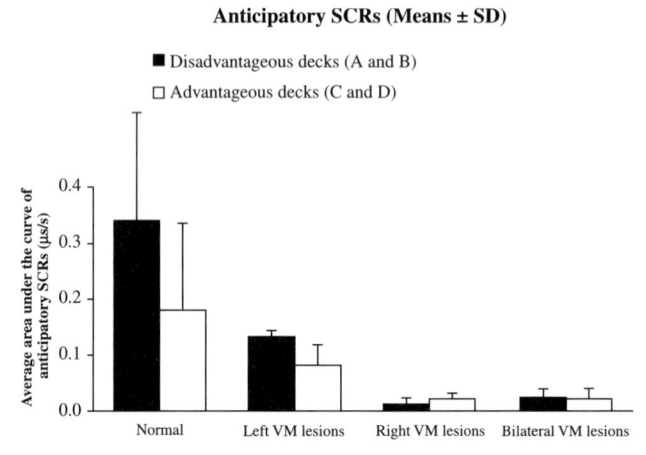

**Fig. 23.10** Anticipatory SCRs during the Iowa Gambling Task, expressed as average area under the curve. SCRs are graphed as a function of Group (Normal, Left VMPFC, Right VMPFC, Bilateral vmPFC) and Deck type (disadvantageous (A and B) versus advantageous (C and D)). The Normal and Left VMPFC groups showed higher anticipatory SCRs to the disadvantageous decks (although not statistically so in the Left VMPFC group); the Right VMPFC and Bilateral VMPFC groups did not. Adapted from Tranel et al. (2002).

patients were normal, but all three adult-onset RVMPFC patients exhibited significant decrements post-morbidly and met the researchers' criteria for 'acquired sociopathy'. In sum, these results support the notion that the right VMPFC is critically involved in social conduct, decision-making, and personality, whereas the contribution of the left VMPFC is less important. This conclusion is supported by the work of Manes *et al.* (2002), who showed that four patients with left OFC lesions perform normally on the IGT, and Angrilli *et al.* (1999), who demonstrated a pronounced impairment in psycho-physiological response to emotionally charged stimuli in a patient with unilateral right OFC damage.

Recent work in our laboratory has introduced another wrinkle into this story, namely, the possibility that gender acts as a potent modulator of VMPFC asymmetry. Specifically, our findings have suggested that in men, the right-sided VMPFC sector is more critical for social conduct, emotional processing, and decision-making, whereas in women, the left-sided VMPFC sector is more important (Denburg *et al.* 2004). It is possible that men and women use different approaches and strategies to solve similar problems—e.g., men may use a more holistic, gestalt-type strategy, and women may use a more analytic, verbally-mediated strategy. These findings are preliminary, but they have some interesting parallels with recent work on gender-related differences in how the amygdala subserves emotion, such as the evidence that the right amygdala is more critical in men, and the left amygdala is more critical in women (Cahill *et al.* 2004).

## Development

Like focal unilateral VMPFC lesions, early-onset VMPFC lesions are exceedingly rare. However, a handful of cases have been described and some common features emerge. Anderson and colleagues have assembled the most complete clinical and neuroanatomical descriptions of such subjects, including two cases published in previous articles (Anderson *et al.* 1999, 2000).[1] The first subject, FD, had a normal infancy until her head was run over by a motor vehicle at age 15 months. She appeared to fully recover from the accident within days, but behavioral abnormalities became apparent around age 3. Later neuroanatomical analysis at age 20 revealed bilateral anterior orbital damage (Fig. 23.11A). As a toddler, FD was unresponsive to verbal or physical punishment, and compulsively collected others' belongings for herself. In elementary school FD was considered intelligent and academically capable by her teachers, but she rarely completed assigned tasks and was extremely disruptive in the classroom and at home. Her adolescence was marked by worsening social behavior: repeated shoplifting and thievery, chronic lying, forgery, tardiness and absenteeism at school, verbal and physical aggression, failing classes, skipping assignments, sexual promiscuity starting at age 14, and lack of friends. FD became pregnant at age 18, and was a dangerously negligent and insensitive mother. She became entirely dependent on her family and social agencies.

---

[1] A larger series of cases has now been studied, and comparable results have been obtained; see Anderson *et al.* (in press).

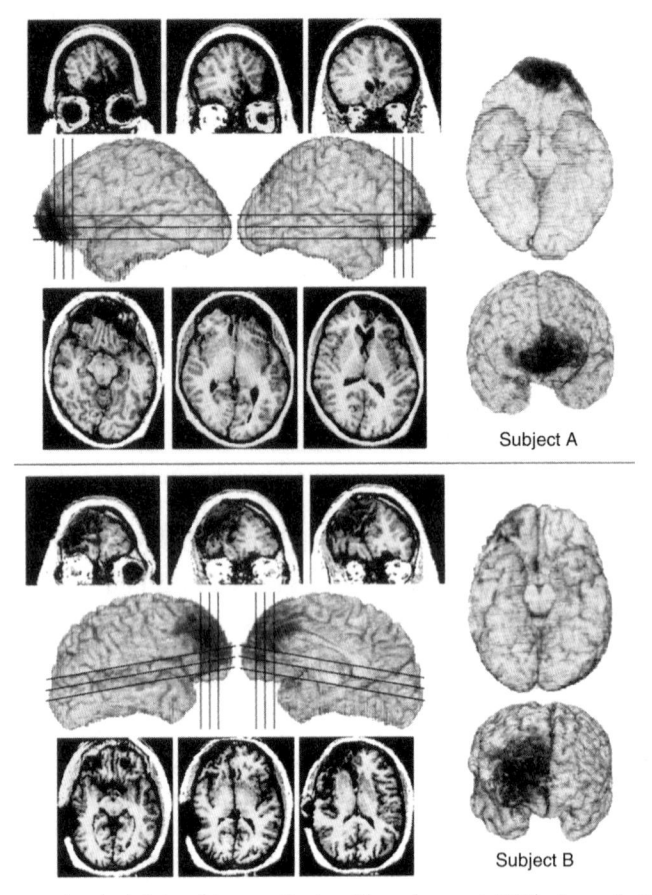

**Fig. 23.11** Neuroanatomical data of two patients with early-onset PFC lesions. A: FD has a bilateral lesion the prefrontal region, involving the anterior orbital sector, the right mesial orbital sector, and the left polar cortices. B: ML has extensive damage in the right frontal lobe, involving mesial, polar, and lateral prefrontal sectors. Adapted from Tranel (2002).

FD formulated no plans for her future and sought no employment. When employment was arranged for her, she was promptly dismissed for lack of dependability and rules violations. Her emotional reactions were described by family and caregivers as labile, fleeting, and often poorly matched to the situation. FD denied or underrepresented any behavioral problems, and she expressed no guilt or remorse for any of her actions. Nonetheless, it was repeatedly noted that she was often friendly, pleasant, and charming towards adults in authority positions, but also obviously manipulative.

The second subject, ML, underwent resection of a right frontal tumor at age 3 months. Later neuroanatomical analysis at age 23 revealed extensive right frontal damage including mesial, polar, and lateral PFC (Fig. 23.11B). Throughout school ML received normal scores on standardized tests of IQ and academic achievement, but he was often disruptive, impulsive, and failed to complete assignments. ML graduated high school after repeating his senior year, at which time his behavioral problems escalated. ML quit or was fired from multiple jobs. He repeatedly bought items on credit and failed to make payments, resulting in substantial debt. ML fathered a child in a casual relationship, but had little contact with the child or mother. He engaged in poorly planned thievery and frequent lying. He had no plans for the future. His parents described his affect as generally neutral or unconcerned, with occasional brief outbursts of anger that entailed profanity, threats, and sometimes physical assault. ML exhibited no worry, guilt, empathy, or remorse.

Despite being raised in stable, supportive families with well-adjusted siblings, the social and moral behavioral impairments of these two early-onset patients tend to exceed those of the adult-onset VMPFC patients. The cluster of psychopathic traits common to these two early-onset cases (superficial charm, lying, manipulative, lack of remorse or guilt, shallow affect, lack of empathy, failure to accept responsibility, parasitic, poor behavioral controls, early behavioral problems, lack of realistic, long-term goals, irresponsibility, juvenile delinquency, promiscuity) surpasses that of adult-onset cases in scope and severity, and clearly more closely approximates the profile of true psychopaths (see PCL-R, Fig. 23.1). Neuropsychological assessment of FD and ML indicates both have intact basic mental abilities (like the adult-onset cases), but significant impairments in executive function, social/moral reasoning, and verbal generation of responses to social situations (unlike adult-onset cases). As expected, FD and ML exhibited defective decision-making on the IGT and failed to generate anticipatory SCRs.

Anderson *et al.*'s findings of severely impaired social conduct and decision-making in early-onset VMPFC damage are in agreement with the small number of related cases. Patients JP (Ackerly and Benton 1948) and GK (Price *et al.* 1990) both suffered extensive bilateral damage to prefrontal cortex within the first week of life. Clinical observations of JP and GK echo those of FD and ML: preserved general intelligence, memory, and language; extreme interpersonal impairments; conspicuous lack of friends; stealing; lack of foresight; manipulative and shallow courteousness to adults; lack of concern or remorse; insensitivity to punishment; lewd and irresponsible sexual behavior; and egocentrism. In addition, several cases of severe social behavior impairment resulting from prefrontal damage occurring between ages 3 and 7 have been reported: MH (Price *et al.* 1990), DT (Eslinger *et al.* 1992) and PL (Marlowe 1992). Conversely, in several alternate cases of prefrontal damage with onset between ages 3 and 7 moral and social behavior is relatively preserved: SC2 (Daigneault *et al.* 1997), MJ (Eslinger *et al.* 1997) and JC (Eslinger *et al.* 1999). Damage in these latter three cases was restricted to either right dorsolateral PFC (MJ and JC) or left PFC (SC2), providing further support for the critical role of right VMPFC in moral and social behavior.[2]

Even though the sample of reported early-onset damage cases is so far small, the available data suggest that VMPFC is critically involved in the acquisition of social and moral knowledge during development. Adult-onset VMPFC patients, who presumably underwent normal social development, retain declarative access to social facts and

---

[2] For a review of early-onset PFC damage cases see Eslinger *et al.* (2004).

'off-line' moral reasoning, but they lose access to emotional signals that are necessary to guide appropriate social and decision-making behavior in real-life, real-time situations. Early-onset VMPFC patients seem to have never acquired factual social knowledge, nor do they have access to normal 'on-line' emotional signals, resulting in an even greater level of impairment.[3]

## Summary

In sum, the experimental evidence implicates the orbitofrontal and especially lower mesial prefrontal cortices as crucial neural substrates of social and moral behavior. Damage to this region results in a characteristic acquired cluster of traits that resembles psychopathy. The diminished emotionality of VMPFC-damaged patients, like developmental psychopaths, is manifest in reduced physiological arousal and subjective feeling to external stimuli as well as internally generated images. The marked real-life decision-making impairments common to VMPFC patients and psychopaths have been captured in the laboratory with the Iowa Gambling Task. Psychophysiological data from this task indicate that somatic signals, abolished by VMPFC damage, are the critical determinant of successful decision-making in real-time situations of uncertain reward and punishment. The relative contributions of right and left VMPFC damage to the acquisition of psychopathic traits are not equivalent; damage to the right VMPFC is generally necessary and sufficient for the condition. The observation of severely more psychopathic behavior in early-onset cases of VMPFC damage indicates a role for VMPFC in the acquisition, as well as execution, of social and moral knowledge.

It has not escaped the attention of a number of researchers, including ourselves, that the parallels between developmental psychopathy and acquired sociopathy must be more than a coincidence—for example, the similarities suggest a common pathophysiologic mechanism. Dysfunction in lower mesial and orbital prefrontal cortices appears to be a likely culprit, as the extensive evidence reviewed in this chapter suggests. However, it has been more elusive to flesh out the details of this story: clearly, neuronal damage in the critical regions (VMPFC generally) can lead to behavioral problems, but the mechanisms by which this occurs have not been identified. Nonetheless, it is tempting to speculate that in developmental psychopathy, there may very well be a genetically driven vulnerability that entails dysfunction of neurotransmitter systems in VMPFC, that may be exacerbated by impoverished learning experiences and the addition of factors such as

---

[3] As a noteworthy aside, the result of *greater impairment* following *early damage* to VMPFC is contrary to a long-held heuristic of brain-behavior relationships: the Kennard Principle, which holds that functional recovery from brain damage is greater when the injury occurs early in life. The initial basis for this idea was that primates lesioned at an early age recovered motor function to a greater extent than primates lesioned later in life. (Kennard, 1938) The Kennard principle was apparently corroborated by language studies, which showed that patients incurring left hemisphere damage early in life, even entire left hemisphere removal, exhibited less severe language deficits than patients incurring damage as adults. But the reverse finding, greater impairment following early damage, as in VMPFC patients, is not without precedent. In a large sample of patients with various cortical lesions, Duval et al. (2002) found greater recovery of intellectual function in patients sustaining damage after age seven than before age seven.

alcohol and substance abuse. The experimental evidence for this possibility is scant, to date, but this seems like a path that warrants careful scientific exploration.

At the same time, there are some notable differences between developmental psychopathy and the acquired variety. Perhaps most important is the proclivity for violent behavior and blatant acts of aggression against society, which typifies developmental psychopaths but is rare in the adult-onset patients whose problems follow the occurrence of brain damage. In our experience, the patients with acquired sociopathy do not tend to accumulate a trail of legal entanglements and alcohol- and drug-related offenses, which are common in developmental cases. The patients are more likely to harm themselves, especially through poor decision-making and more 'passive' types of behavioral problems such as lack of initiative, aspontaneity, and indecisiveness. And the strong reward-driven and thrill-seeking behavior of developmental psychopaths is not very typical of patients with acquired sociopathy.

One other parameter that warrants comment is the fact that in both developmental psychopathy and the acquired variety, the condition or 'syndrome' is not an all-or-none phenomenon. Rather, there are clear degrees of features and characteristics, and clear gradients in the manifestation of signs and symptoms. This is surely true of developmental psychopathy, where milder versions of the condition are not uncommon, and in fact fairly adaptive for certain activities, such as pursuing a career in law enforcement or serving in the military. We have observed considerable variability as well in patients with acquired sociopathy, who tend to run a gamut of impairment from mild to severe.

As a final comment in this summary section, we should note that our work on acquired sociopathy has been driven by and situated in a theoretical framework known as the 'somatic marker hypothesis' (Damasio 1994, 1999). An explication of this framework is well outside the scope of this chapter, but the importance of the theoretical developments outlined in the somatic marker hypothesis to the work summarized herein should be fully acknowledged. The somatic marker hypothesis has triggered some lively debates in the recent scientific literature (e.g., Maia and McClelland 2004; Bechara *et al.* 2005), which promise to spur even more empirical exploration which will help to address the many lingering questions in this field. We turn now to some of those questions.

## 23.5. **Future directions**

While the corpus of data relating to psychopathic behavior and VMPFC damage is indeed considerable, (and by no means exhausted in this chapter), there remain a number of empirical issues that warrant resolution. In general, experimental paradigms originally used in psychopaths have been applied to pseudopsychopaths, and vice versa, but for some measures comparative analyses are not complete. For example, no study to date has reported the psychophysiological response of developmental psychopaths while they perform the Iowa Gambling Task. These data could indicate that the decision-making impairments in psychopaths, like VMPFC patients, are due to a lack of appropriate somatic signals. Also, it would be informative to administer the Hare PCL-R to the early-onset VMPFC damage cases. Even though adult-onset cases are not accurately assessed with this measure (see 'Personality' section above), early-onset cases, who seem to develop more severely psychopathic behavior than adult-onset cases (and who develop at least some of this behavior before adulthood), may actually meet the clinical criteria for psychopathy.

Another topic that has begun to be explored in PFC-damaged patients is empathy. Lack of empathy is one of the most notable features of pseudopsychopathy and developmental psychopathy, but the pathophysiological basis of this egregious impairment has not been established. Comparing self-reported empathic behavior among patients with a variety of brain lesions, Grattan *et al.* (1994) found the greatest empathy impairments in patients with damage to orbitofrontal cortex. Shamay-Tsoory *et al.* (2003, 2004) found the greatest impairments in patients with damage to right VMPFC. However, as discussed above and reported in Barrash *et al.* (2000), self-report measures of personality traits by VMPFC patients are generally not sensitive in detecting acquired social and moral behavioral deficits. A more effective laboratory paradigm for characterizing empathic ability in VMPFC patients (and developmental psychopaths for that matter) might incorporate the following elements: 1) real-time empathy elicitation as opposed to 'off-line' self-report; 2) psychophysiological recording (like SCR) and subjective report of feeling during online task performance to index emotional arousal and valence; 3) distinction between empathy for social distress and empathy for physical pain; 4) distinction between empathy elicitation by external stimuli and internally generated images; 5) distinction between emotional unresponsiveness in general and empathy in particular; and 6) distinction between emotional/psychophysiological empathic response and cognitive apprehension of another's situation and mental state.

## 23.6. Conclusion

Clinical and laboratory exploration of pseudopsychopaths has linked social and moral behavior with lower mesial and orbital prefrontal cortices. Neuropsychological investigations of pseudopsychopaths and developmental psychopaths, as well as functional imaging studies in normal participants, have illuminated the interaction of emotion, cognition, and behavior. As in the cases of language in aphasics and memory in amnesics, continued study of the impairments in pseudopsychopaths promises to yield further insight into the neural substrates of social and moral behavior. This line of work will undoubtedly be complemented by functional imaging studies in normal participants. This is arguably one of the last and most important frontiers of cognitive neuroscience. The global demise of proper social and moral behavior has been widely decried, and has attracted increasingly urgent attention from many corners of the world. When coupled with the increasing complexity of modern society, which only adds to the acute and long-term challenges of charting social and moral function, the enormity of the issue can hardly be overemphasized. Regressive, rigid, narrow-minded approaches, such as those offered by war and religion, are not viable long-term solutions. Brain science has much to offer.

## Acknowledgment

This research was supported by the NINDS Program Project Grant P01 NS19632.

## References

Abbott, J. (1981) *In the Belly of the Beast: Letters from Prison.* New York: Random House.

Ackerly, S. S. and Benton, A. L. (1948) Report of a case of bilateral frontal lobe defect. *Proceedings of the Association for Research in Nervous and Mental Disease* 27:.479–504.

American Psychiatric Association. (1994) *Diagnostic and Statistical Manual of Mental Disorders* (Fourth Edition). Washington D.C.: American Psychiatric Association.

Anderson, S. W., Barrash, J., Bechara, A., and Tranel, D. Impairments of emotion and real world complex behavior following childhood- or adult-onset focal lesions in prefrontal cortex, *Journal of the International Neuropsychological Society* (in press)

Anderson, S. W., Bechara, A., Damasio, H., Tranel, D., and Damasio, A. R. (1999) Impairment of social and moral behavior related to early damage in human prefrontal cortex. *Nature Neuroscience* 2, 11:1032–1037.

Anderson, S. W., Damasio, H., Tranel, D., and Damasio, A. R. (2000) Long-term sequelae of prefrontal cortex damage acquired in early childhood. *Developmental Neuropsychology* 18, 3:281–296

Angrilli, A, Palomba, D., Cantagallo, A., Maietti, A., and Stegagno, L. (1999) Emotional impairment after right orbitofrontal lesion in a patient without cognitive deficits. *NeuroReport* 10:1741–1746.

Barrash, J. and Anderson, S. W. (1993) *The Iowa Rating Scales of Personality Change.* Iowa City: University of Iowa, Department of Neurology.

Barrash, J., Tranel, D., and Anderson, S. W. (1994) Assessment of dramatic personality changes after ventromedial frontal lesions. *Journal of Clinical and Experimental Neuropsychology* 16:66.

Barrash, J., Tranel, D., and Anderson, S. W. (2000) Acquired personality disturbances associated with bilateral damage to the ventromedial prefrontal region. *Developmental Neuropsychology* 18, 3:355–381.

Bechara, A., Damasio, A. R., Damasio, H., and Anderson, S. W. (1994) Insensitivity to future consequences following damage to human prefrontal cortex. *Cognition* 50:7–15.

Bechara, A., Tranel, D., Damasio, H., and Damasio, A. R. (1996).Failure to respond autonomically to anticipated future outcomes following damage to prefrontal cortex. *Cerebral Cortex* 6:215–225.

Bechara, A., Damasio, H., Tranel, D., and Damasio, A. R. (1997) Deciding advantageously before knowing the advantageous strategy. *Science* 275:1293–1295

Bechara, A., Damasio, H., Damasio, A. R., and Lee, G. P. (1999) Different contributions of the human amygdala and ventromedial prefrontal cortex to decision-making. *Journal of Neuroscience* 19:5473–5481.

Bechara, A., Damasio, H., Tranel, D., and Damasio, A. R. (2005) The Iowa Gambling Task and the Somatic Marker Hypothesis: Some questions and answers. *Trends in Cognitive Neurosciences* 4:159–162.

Blair, R. J., Colledge, E., and Mitchell, D. G. (2001) Somatic markers and response reversal: is there orbitofrontal cortex dysfunction in boys with psychopathic tendencies? *Journal of Abnormal Child Psychology* 6:499–511.

Blumer, D. and Benson, D. F. (1975) Personality changes with frontal and temporal lobe lesions. In: Benson, D.F. and Blumer, D. (eds) *Psychiatric Aspects of Neurological Disease.* New York: Grune and Stratton.

Broca, P. (1861) Remarques sur le siege de la faculte du langage articule, suivies d'une observation d'aphemie (perte de la parole). *Bulletin Societe Anatomique de Paris.* 6:330–357.

Butcher, J., Dahlstrom, W., Graham, J., Tellegen, A., and Kaemmer, B. (1989) *Manual for administering and scoring the MMPI-2*. Minneapolis, USA: University of Minnesota Press.

Cahill, L., Uncapher, M., Kilpatrick, L., Alkire, M., and Turner, J. (2004) Sex-related hemispheric lateralisation of amygdala function in emotionally influenced memory: An fMRI investigation. *Learning and Memory* 11:261–266.

Cleckly, H. (1976). *The Mask of Sanity*. St Louis, USA: C.V. Mosby.

Daigneault, S., Braun, C. M. J., and Montes, J. L. (1997) Pseudodepressive personality and mental inertia in a child with a focal left-frontal lesion. *Developmental Neuropsychology* 13:1–22.

Damasio, A. R. (1994) *Descartes' error: Emotion, reason and the human brain*. New York: Grosset/Putnam.

Damasio, A. R. (1999) *The feeling of what happens*. New York: Harcourt Brace and Co.

Damasio, A. R., Tranel, D., and Damasio, H. (1990). Individuals with sociopathic behavior caused by frontal damage fail to respond autonomically to social stimuli. *Behavioral Brain Research* 41:81–94.

Denburg, N. L., Tranel, D., and Bechara, A. (2004) Gender modulates functional asymmetry of the ventromedial prefrontal cortex. *Society for Neuroscience Abstracts* 548.3.

Duval, J., Dumont, M., Braun, C. M. J., and Montour-Proulx, I. (2002) Recovery of intellectual function after a brain injury: A comparison of longitudinal and cross-sectional approaches. *Brain and Cognition* 48:337–342.

Eslinger, P. J. and Damasio, A. R. (1985) Severe disturbance of higher cognition after bilateral frontal lobe ablation: patient EVR. *Neurology* 35:1731–1741.

Eslinger, P. J., Grattan, L. M., and Damasio, A. R. (1992) Developmental consequences of childhood frontal lobe damage. *Archives of Neurology* 49:764–769.

Eslinger, P. J., Biddle, K. R., and Grattan, L. M. (1997) Cognitive and social development in children with prefrontal cortex lesions, in: Krasnegor, A., Lyon, G.R., and Goldman-Rakic, P.S. (eds) *Development of the prefrontal cortex: Evolution, neurobiology, and behavior*. Baltimore, USA: Paul H. Brookes Publishing Co., Inc.

Eslinger, P. J., Biddle, K., Pennington, B, and Page, R. B. (1999) Cognitive and behavioral development up to 4 years after early right frontal lobe lesion. *Developmental Neuropsychology* 15:157–191.

Eslinger, P. J., Flaherty-Craig C. V., and Benton, A. L. (2004) Developmental outcomes after early prefrontal cortex damage. *Brain and Cognition* 55:84–103.

Eysenck, H. J, and Eysenck, S. B. G. (1975) *Eysenck Personality Questionnaire*. San Diego, USA: Educational Testing Service.

Grattan, L. M., Bloomer, R. H., Archambault, F. X., and Eslinger, P. J. (1994) Cognitive flexibility and empathy after frontal lobe lesion. *Neuropsychiatry, Neuropsychology, and Behavioral Neurology* 7, 4:251–259.

Hare, R. D. (1978) 'Psychopathy and electrodermal responses to non-signal stimulation', *Biological Psychology* 6: 237–246.

Hare, R. D. (1991) *The Hare Psychopathy Checklist-Revised*. Toronto: Multi-Health Systems.

Hare, R. D., Frazelle, J., and Cox, D. N. (1978) Psychopathy and physiological responses to threat of an aversive stimulus. *Psychophysiology* 15:165–172.

Harlow, J. M. (1868) Recovery from the passage of an iron bar through the head. *Publications of the Massachusetts Medical Society* 2:327–347.

Hathaway, S. R. and McKinley, J. C. (1983) *The Minnesota Multiphasic Personality Inventory Manual*. New York: Psychological Corporation.

Kennard, M. A. (1938) Reorganization of motor function in the cerebral cortex of monkeys deprived of motor and premotor areas in infancy. *Journal of Neuropsychology* 1:477–496.

Lykken, D. T. (1957) A study of anxiety in the sociopathic personality. *Journal of Abnormal and Social Psychology* **55**:6–10.

Maia, T. V. and McClelland, J. L. (2004) A reexamination of the evidence for the somatic marker hypothesis: What participants really know in the Iowa gambling Task. *Proceeding of the National Academy of Sciences* **101**:16075–16080.

Manes, F., Sahakian, B., Clark, L., Rogers, R., Antoun, N., Aitken, M., and Robbins, T. (2002) Decision-making processes following damage to the prefrontal cortex. *Brain* **125**:624–639.

Marlowe, W. (1992) The impact of right prefrontal lesion on the developing brain. *Brain and Cognition* **20**:205–213.

Mitchell, D. G. V., Colledge, E., Leonard, A., and Blair, R. J. R. (2002) Risky decisions and response reversal: is there evidence of orbitofrontal cortex dysfunction in psychopathic individuals? *Neuropsychologia* **40**:2013–2022.

Ogloff, J. R. and Wong, S. (1990) Electrodermal and cardiovascular evidence of a coping response in psychopaths. *Criminal Justice and Behavior* **17**:231–245.

Patrick, C. J., Bradley, M. M., and Lang, P. J. (1993) Emotion in the criminal psychopath: Startle reflex modulation. *Journal of Abnormal Psychology* **102**:82–92.

Patrick, C. J., Cuthbert, B. N., and Lang, P. J. (1994) Emotion in the criminal psychopath: Fear image processing. *Journal of Abnormal Psychology* **103**:523–534.

Pfohl, B., Blum, N., Zimmerman, M., and Stangl, D. (1989) *Structured Interview for DSM-III-R Personality: SIDP-R*. Iowa City: University of Iowa, Department of Psychiatry.

Saver, J. L. and Damasio, A. R. (1991) Preserved access and processing of social knowledge in a patient with acquired sociopathy due to ventromedial frontal damage. *Neuropsychologia* **29**:1241–1249.

Schmauk, F. J. (1970) Punishment, arousal, and avoidance learning in sociopaths. *Journal of Abnormal Psychology* **76**:325–335.

Scoville, W. B. and Milner, B. (1957) Loss of recent memory after bilateral hippocampal lesions. *Journal of Neuropsychiatry and Clinical Neuroscience* **12**:103–113.

Shamay-Tsoory, S. G., Tomer, R., Berger, B. D., and Aharon-Peretz, J. (2003) Characterisation of empathy deficits following prefrontal brain damage: The role of the right ventromedial prefrontal cortex. *Journal of Cognitive Neuroscience* **15**:324–337.

Shamay-Tsoory, S. G., Tomer, R., Berger, B. D., and Aharon-Peretz, J. (2004) Impairment in cognitive and affective empathy in patients with brain lesions: anatomical and cognitive correlates. *Journal of Clinical and Experimental Neuropsychology* **8**:1113–1127.

Tharp, V. K., Maltzman, I., Syndulko, K., and Ziskind, E. (1980) Autonomic activity during anticipation of an aversive tone in non-institutionalized sociopaths. *Psychophysiology* **17**:123–128.

Tranel, D. (2002) Emotion, decision making, and the ventromedial prefrontal cortex. In: Stuss, D.T. and Knight R.T. (eds) *Principles of frontal lobe function*. New York: Oxford University Press.

Tranel, D., Bechara, A., Damasio, H., and Damasio, A. R. (1998) Neural correlates of emotional imagery. *International Journal of Psychophysiology* **30**:107.

Tranel, D., Bechara, A., and Denburg, N. L. (2002) Asymmetric functional roles of right and left ventromedial prefrontal cortices in social conduct, decision-making, and emotional processing. *Cortex* **38**:589–612.

Tranel, D., Bechara, A., and Damaiso, A. R. (in press) Acquired sociopathy. In: Barch, D. (ed) *Cognitive and affective neuroscience of psychopathology*. New York: Oxford University Press.

van Honk, J., Hermans, E. J., Putnam, P., Montagne, B., and Schutter, D. J. (2002) Defective somatic markers in sub-clinical psychopathy. *Neuroreport* **13**:1025–1027.

Waid, W. M. (1976) Skin conductance response to punishment as a predictor and correlate of learning to avoid two classes of punishment. *Journal of Abnormal Psychology* **85**:498–504.

# Chapter 24

# Frontotemporal dementia and the orbitofrontal cortex

Po H. Lu, Negar Khanlou, and Jeffrey L. Cummings

## 24.1. Introduction

Neuropsychiatric illness associated with damage to the orbitofrontal cortex (OFC) was dramatically documented by the classic case of Phineas Gage, the railroad construction worker who accidentally blasted an iron rod through his head (Harlow 1848; Macmillan 2000). Frontal lobe injury produced marked personality changes. Similarly, patients with frontotemporal dementia (FTD) manifest severe behavioral and personality alterations associated with OFC dysfunction. Neuropathology and neuroimaging studies have identified this brain region to be prominently involved in the frontal-variant of FTD. Because of the strong relationship between altered social behavior and brain atrophy in FTD, this is an important population for studying the neurobiological basis of emotion and social behavior. This chapter will provide a review of the clinical features, neuropathology, neuroimaging, genetics, and neuropsychology of FTD as well as present two prototypical cases that provide a clinical picture of this disorder. The neuroanatomy, circuitry, and functions of the OFC will be summarized, emphasizing its role in emotional and social cognition. Theories involving deficits in recognition of emotional expression, decision-making, and theory of mind have been proposed to explain the mechanism underlying the clinical expression of FTD, and the OFC is intimately involved in studies examining the neural basis underlying these deficits.

## 24.2. Frontotemporal dementia

Frontotemporal dementia, characterized by profound personality and behavioral disturbance, is one of three major clinical manifestations of frontotemporal lobar degeneration (FTLD), a spectrum of dementing disorders resulting from degeneration of the frontal lobes, anterior temporal lobes, or both (Neary *et al.* 1998). Two other clinical syndromes of FTLD involve prominent aphasia. Progressive non-fluent aphasia (PNFA) is a disorder of expressive language characterized by hesitant and effortful speech production, phonologic and grammatical errors, and difficulties in word retrieval with relative preservation of comprehension. Semantic dementia (SD) involves severe loss of meaning for both verbal and non-verbal concepts resulting in impaired naming and word comprehension in the context of fluent, effortless speech output. There can be substantial overlap among the three clinical syndromes. Furthermore, other clinical conditions such as corticobasal degeneration (CBD) and progressive supranuclear palsy (PSP) share pathological features with FTLD and are often considered part of a common spectrum of disorders.

Of the clinical phenotypes of FTLD, FTD is the most common in occurrence, accounting for at least 70% of the patients with FTLD (Snowden *et al.* 2002), and most dramatic in appearance. FTD is the third most common neurodegenerative dementia syndrome following Alzheimer's disease (AD) and dementia with Lewy bodies, accounting for 2 to 5 per cent of all dementing cases (Neary *et al.* 1998). Onset of the disease typically occurs between the ages of 45 and 65 and averages around 57 years; therefore, it can represent up to 20% of presenile (age of onset less than 65 years) dementia cases (Knopman *et al.* 1990; Snowden *et al.* 2002). Occasional late-onset cases have been observed and postmortem confirmed FTD has been documented in patients as young as the early twenties (Jacob *et al.* 1999; Coleman *et al.* 2002; Snowden *et al.* 2004). Survival after symptom onset for FTD is usually 8 to 11 years but it can be highly variable, ranging from 2 to 20 years. Pathological variants such as those with motor neuron disease may have a shorter, more malignant course (Mendez *et al.* 1993; Snowden *et al.* 2002). A recent report on the demographic characteristics of a large cohort of patients with FTD documented a predominance of men (Johnson *et al.* 2005); however, conclusions regarding gender differences in FTD remain equivocal since some studies have reported a higher prevalence of women (Stevens *et al.* 1998) and others found an equal gender distribution (Chow *et al.* 1999; Hodges *et al.* 2003; Rosso *et al.* 2003).

## Clinical characteristics

Consensus criteria for the diagnosis of FTD were published by Neary and colleagues in 1998. The most common and early symptom of FTD is a profound alteration in social interpersonal conduct with relative preservation of instrumental functions of perception, spatial skills, praxis, and memory (Neary *et al.* 1998). Patients display breaches in social etiquette and decline in manners (e.g., violation of interpersonal space, tactlessness) that differs from the patient's pre-morbid behavior. Disinhibition and impulsiveness are characteristic features as patients often act out or say the first thought that comes to mind without reflecting on the potential consequences (e.g., make unrestrained lewd or sexual remarks, engage in inappropriate behaviors, make jokes at others' expense, be inappropriately jocular). Current consensus criteria (Neary *et al.* 1998) also describe quantitative departures from customary behavior ranging from purposeless overactivity (e.g., pacing, wandering, and increased verbal and motor activity) to passivity, inertia, and inactivity (e.g., reduced behavioral initiation and spontaneity, loss of interest). There is early loss of insight as evidenced by frank denial of symptoms with no appreciation of the changes in behavior or social, occupational, and financial consequences of behavioral changes.

Emotionally, FTD patients experience emotional blunting early in the disease and may lose the ability to demonstrate the basic emotions such as happiness, sadness, fear, and anger (Neary *et al.* 1998). They lack emotional warmth, empathy, and sympathy, and they are unconcerned or unaware of the needs of others. Some patients (approximately

40%; Mendez *et al.* 1997) may present with compulsion-like behaviors in the form of perseverative and stereotyped behaviors that encompass simple repetitive acts or motor mannerisms such as lip smacking, hand rubbing or clapping, counting aloud, and humming, as well as complex behavioral routines such as clock watching, avoiding cracks on pavements, or repeatedly checking doors (Mendez *et al.* 1997; Neary *et al.* 1998; Snowden *et al.* 2002). Neglect of personal hygiene and grooming (including disheveled appearance, change in taste in clothing, inappropriate clothing combinations), hyperorality and dietary changes (e.g., overeating, bingeing, altered food preferences, and food fads, particularly craving for sweets), distractibility and impersistence, mental rigidity and inflexibility (e.g., inability to adapt to novel situations, insistence in getting his or her own way, failure to take on another person's point of view, strict adherence to routine), and utilization behavior (grasping and repeatedly using objects in their visual field) are additional behaviors commonly displayed by patients with FTD (Neary *et al.* 1998).

## Neuropathology and neurochemistry of FTD

At autopsy, the brains of patients with FTD show marked bilateral atrophy of the frontal and temporal lobes with relative preservation of posterior gyri. Asymmetric changes also can be observed. In addition to the cortex, the adjacent white matter, the amygdala, and several basal ganglia nuclei are affected (Hodges and Miller 2001; Munoz *et al.* 2003). The OFC and the anterior and medial temporal areas show the most severe atrophic changes (Hodges and Miller 2001; Mendez and Cummings, 2003). The current consensus on the pathological classification of FTLD is based on the morphology and histochemical characterization of neuronal inclusions or their absence (Munoz *et al.* 2003).

Three major neuropathologic variants have been identified in FTD (McKhann *et al.* 2001). The most common pathology, accounting for approximately 60% of cases (Snowden *et al.* 2002), is a nonspecific frontotemporal atrophy 'lacking distinctive histology' (FTD-ldh) in which there is neuronal loss and astrogliosis with spongiosis (microvacuolation) affecting chiefly the upper lamina of the cortex with variable involvement of the subcortical and limbic structures (Schmitt *et al.* 1995; Hooten and Lyketsos 1996) (Fig. 24.10). This form has neither *tau* nor *ubiquitin* positive inclusions. The amount of neuronal synaptic loss and the degree of astrocytosis correlate with the clinical behavioral changes (Lipton *et al.* 2001; Martin *et al.* 2001).

A second histological pattern involves pathognomonic changes of Pick's disease, which include severe, 'knife-edged' frontotemporal atrophy, ballooned neurons (or Pick cells), and tau-positive, ubiquitin-positive Pick bodies (Figs. 24.2, 24.3 and 24.4). Tau-positive inclusions also are present in corticobasal degeneration (CBD), progressive supranuclear palsy (PSP), and neurofibrillary tangle dementia (McKhann *et al.* 2001), but the different isoforms of the insoluble deposits of post-translationally modified tau proteins differentiates FTD-Pick's type (three repeat tau) from CBD and PSP (four repeat tau) and neurofibrillary tangle predominant dementia (McKhann *et al.* 2001; Munoz *et al.* 2003).

In the third neuropathologic syndrome, FTD with motor neuron disease (FTD-MND) is characterized by neuronal loss and gliosis with ubiquitin-positive, tau-negative inclusions (Schmitt *et al.* 1995; McKhann *et al.* 2001). Small numbers of FTD patients have

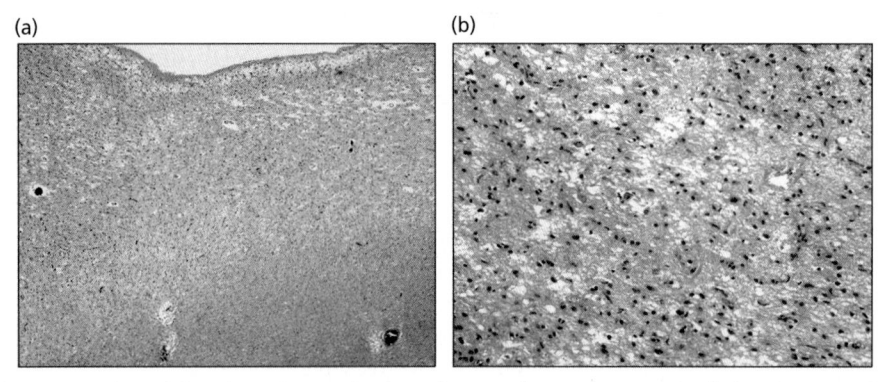

**Fig, 24.1** Section of dorsolateral posterior frontal cortex (from an FTD patient) showing spongiform change and astrocytic gliosis of the superficial cortical layers at low magnification (panel a) and high magnification (panel b). In both panels, pial surface is at top of the photomicrograph. Astrocytic gliosis is better appreciated at high magnification.

**Fig. 24.2** (Also see Color Plate 6.) Section of cingulate cortex from a patient with the Pick disease variant of FTD. Note a single 'balloon cell' (arrow) among smaller phenotypically normal neurons (hematoxylin and eosin-stained section).

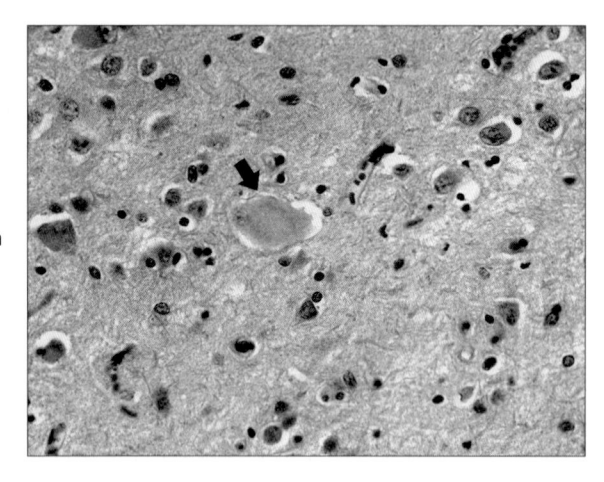

involvement of substantia nigra, striatopallidum, parietal cortex, thalamus, and other structures (Sam et al. 1991; Neary et al. 1998). In all forms, linkage to the tau gene in chromosome 17 or to other genes is possible. Additionally, FTD-ldh and FTD-MND have been recently linked to chromosome 3 and chromosome 9, respectively (Munoz et al. 2003).

The specific pathology has implications regarding its behavioral manifestation. The disinhibited form of FTD characterized by social misconduct is associated with pathological changes confined to the OFC and anterior temporal regions whereas in patients who exhibit predominantly symptoms of apathy, the pathological change extends into the dorsolateral prefrontal cortex (Snowden et al. 2002). A stereotypic form of FTD with compulsion-like behaviors occurs in association with marked striatal changes and variable cortical involvement, often with emphasis on temporal rather than frontal lobe pathology (Snowden et al. 2002). Neurochemical studies reveal preserved neocortical cholinergic function but decreased serotoninergic function in patients with FTD (Francis et al. 1993; Sparks et al. 1994; Mann 1998; Rinne et al. 2002).

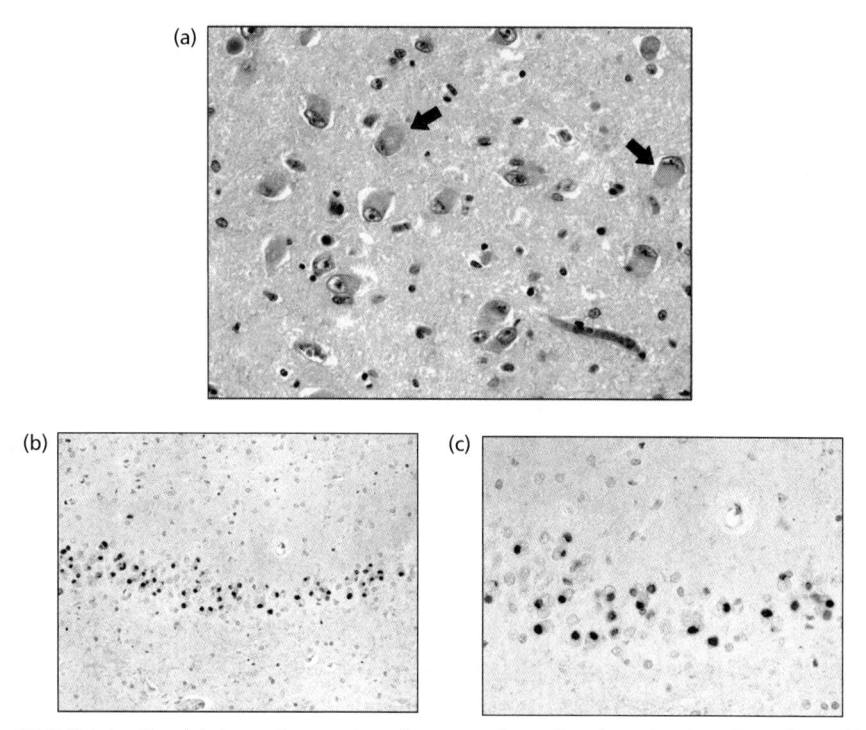

**Fig. 24.3** Pick bodies. (a). In routine sections (hematoxylin and eosin-stained section of entorhinal cortex), Pick bodies appear as faintly basophilic spherical structures in the neuronal cytoplasm (arrows). (b and c). Their presence can be highlighted by silver stains (e.g., modified Bielschowsky) or tau immunohistochemistry. These two panels show tau-immunoreactive Pick bodies in granule cell layer of the hippocampus.

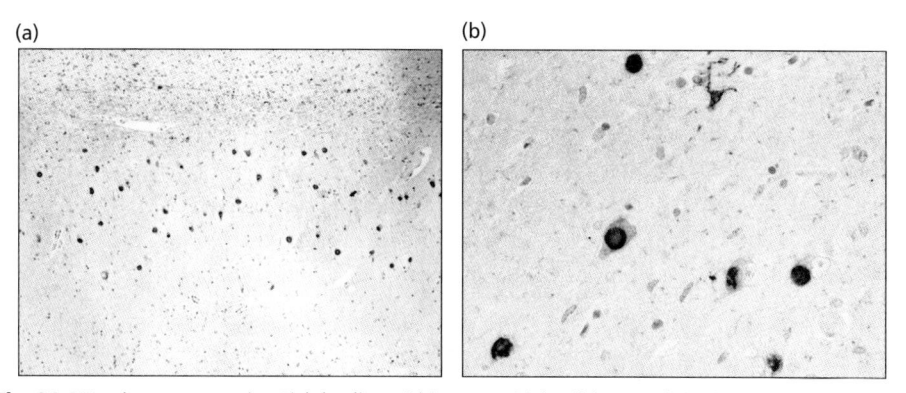

**Fig. 24.4** Tau-immunoreactive Pick bodies within pyramidal cell layers of the hippocampus photographed at low magnification (panel a) and high magnification (panel b).

## Genetics

The underlying cause of FTD is unknown but genetic factors are strongly implicated. A positive family history of a similar dementia is present in as many as 38 to 50 % of patients with FTD (Stevens *et al.* 1998). The first-degree relatives of FTD patients are 3.5 times more at risk for developing dementia before age 80 than relatives of control subjects (Stevens *et al.* 1998; Chow *et al.* 1999). About half of familial FTD patients have genetic forms mapped to chromosome 17 (Foster *et al.* 1997; Reed *et al.* 2001). Rare cases of familial motor neuron disease with FTD have been linked to chromosome 9, and rare cases of familial FTD have been linked to chromosome 3 (Brown 1998; Hosler *et al.* 2000). Several mutations of the tau gene have been identified, which may result in neurodegeneration from hyperphosphorylation of tau and disruption of its microtubular binding and stabilization properties. Pathological tau phenotypes lead to tau aggregation in neurons and glia and interference of axonal transport, thus resulting in symptoms. Not all FTDP-17 families have tau mutations or tau pathology. Within families, there is heterogeneity in clinical presentation, presence of associated motor abnormalities, patterns of neurodegeneration, and pathology, suggesting that other genetic or environmental effects can significantly modify the expression of the neurodegeneration.

## Neuroimaging

Structural and functional neuroimaging can be useful in the diagnosis of FTD and aid in differentiating between clinical subtypes of FTLD. Studies using computerized tomography (CT) demonstrate that patients with FTD may show marked symmetric or asymmetric frontal and temporal atrophy (Knopman *et al.* 1989). Magnetic resonance imaging (MRI) reveals greater atrophy of right dorsolateral prefrontal cortex in patients with FTD while those with SD have more left anterior temporal atrophy (Rosen *et al.* 2002a). Sagittal MRI sections reveal that the anterior corpus callosum is more atrophic in patients with FTLD than in patients with AD or normal elderly controls (Kitagaki *et al.* 1997). Bilateral temporal lobe atrophy is often evident in patients with PNFA (Fig. 24.5).

Functional neuroimaging may reveal abnormalities even when structural imaging does not reveal apparent atrophy. Studies of cerebral perfusion using single photon emission computed tomography (SPECT) reveal fairly widespread anterior frontal and temporal reductions in cerebral blood flow; moreover, the intensity of the reduction generally reflects the clinical syndrome. Patients with predominant language abnormalities (PNFA and SD) have asymmetrical cerebral perfusion with the greatest reductions in the left frontal or temporal regions (Snowden *et al.* 1992; Turner *et al.* 1996; Grossman *et al.* 1998). Those with the FTD syndrome evidence bilateral or right predominant frontal hypoperfusion. Within the FTD syndrome, patients with disinhibition as the predominant presentation have reduced perfusion in the ventral prefrontal cortex (including OFC), while those with apathetic symptoms exhibit reduced perfusion involving the dorsomedial prefrontal cortex (Didic *et al.* 1998). Positron emission tomography using fluorodeoxyglucose (FDG-PET) tends to confirm the observations from studies using SPECT (Chawluk *et al.* 1986; Kamo *et al.* 1987) (Fig. 24.6). One recent prospective FDG-PET study of patients with FTD showed a significant progression of metabolic deficits exclusively in the OFC (Grimmer *et al.* 2004). Using a voxel-based analysis of SPECT data, researchers were able to demonstrate that FTD patients, compared with AD, showed significantly reduced regional cerebral blood flow in the medial frontal, anterior

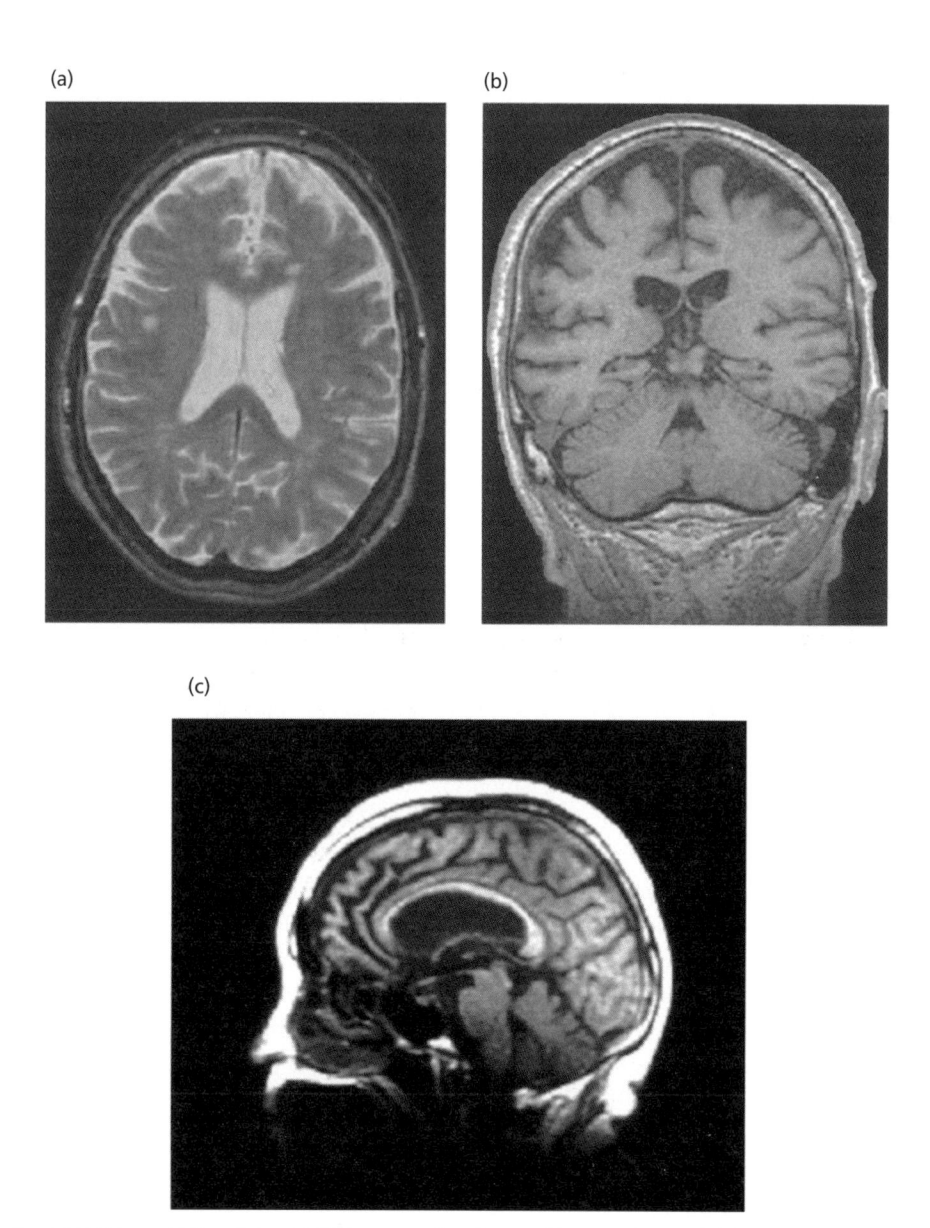

**Fig. 24.5** T1- and T2-weighted MRI images of patients with FTD. The axial (a), coronal (b), and sagittal (c) images demonstrate disproportionate frontotemporal atrophy.

cingulate, ventrolateral prefrontal (BA 47), and OFC (BA 11), with greater reduction in the right hemisphere (Varrone *et al.* 2002).

## Neuropsychology

Early investigations attempting to characterize the neuropsychological characteristics of FTD and distinguish this population from AD failed to demonstrate significant differences in cognitive impairment between the two patient groups (Jagust *et al.*

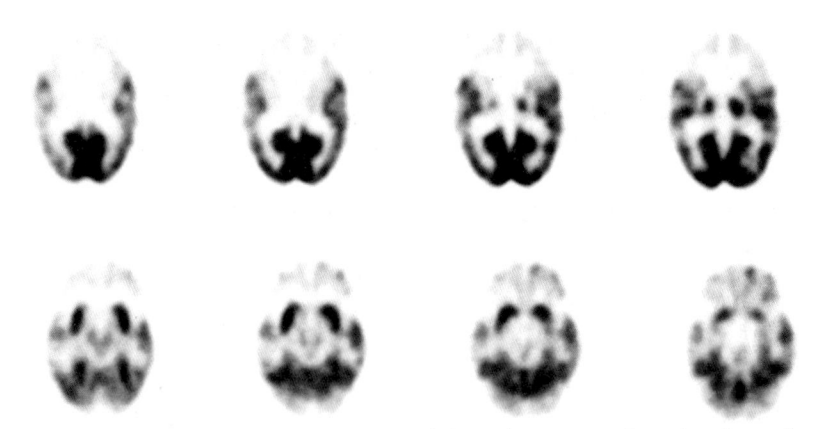

**Fig. 24.6** FDG-PET imaging of a FTD patient. Axial views demonstrate frontal and anterior temporal hypometabolism.

1989; Knopman *et al.* 1989). Another study reported that FTD patients outperformed the AD patients on non-verbal memory tasks, but the two groups did not differ in other cognitive domains, including executive functions (Pachana *et al.* 1996). The minimal cognitive differences, particularly in the executive domain, between the two groups in the early studies are likely attributed to small sample sizes and heterogeneity in the FTD group as they contained patients from all three FTLD sub-types.

Recent studies addressed the methodological limitations of the previous studies and compared the neurocognitive performance of the different FTLD phenotypes. Comparative studies indicated that FTD patients display better recall and recognition memory performances relative to the AD patients (Pasquier *et al.* 2001; Glosser *et al.* 2002), but executive dysfunction, as reflected by disorganisation, increased perseverations, or reduced phonemic fluency, was more apparent in FTD than in AD (Gregory *et al.* 1997; Mathuranath *et al.* 2000; Glosser *et al.* 2002). Kramer and colleagues (2003) found distinct neuropsychological patterns that appeared to differentiate between AD, FTD, and the SD subtype of FTLD; AD patients displayed severe impairment in verbal and visual memory, whereas executive dysfunction was the most prominent cognitive feature in the FTD group, and naming deficits were most apparent in the SD group. Discriminant function analysis identified five neuropsychological variables (naming, visual recall, word-list recall, category fluency, and executive errors) that discriminated the patient groups with 89.2% classification accuracy (Kramer *et al.* 2003). Boone and colleagues (1999) separated FTD patients based on left- or right-anterior hypoperfusion on SPECT and found that right-FTD patients exhibited relatively worse performance on visuoperceptual/visuospatial tasks and non-verbal executive tasks compared to their verbal analogs, with the left-FTD patients showing the opposite pattern manifested by worse performance on measures of language and verbal executive ability.

Most of the traditional neuropsychological tests of frontal/executive functioning are sensitive to dorsolateral prefrontal cortex changes and several studies suggest that patients in early stages of FTD may perform normally on these tasks (Gregory *et al.* 1999; Rahman *et al.* 1999). Given current evidence from functional neuroimaging and postmortem neuropathological studies, future research should employ a neuropsychological battery comprised of both classic executive tasks as well as frontal measures sensitive to the OFC such as Delayed Match to Sample Task (Elliott and Dolan 1999) and the Gambling Task (Bechara *et al.* 1994; see Zald, Chapter 18, this volume for a review of neuropsychological assessment of OFC functions).

## 24.3. **Two case examples of patients with frontotemporal dementia**

### Case 1: Mr. A

Mr. A is a 59-year-old, right-handed, Mexican-American male with 12 years of education, who presented with gradual onset and progression of cognitive symptoms for approximately one year. Specifically, he manifested short-term memory loss, poor reasoning ability, and word-finding difficulties for about a year; however, his wife had noticed significant behavioral changes for about four years. According to his wife, Mr. A would sometimes leave home and disappear for several hours; inquiries regarding his whereabouts would only elicit denials and confabulations. The patient was involved in two motor vehicle accidents over a six-month period, but more remarkably, after both incidents, he returned home and nonchalantly sat down to watch television. His wife did not know about the accidents until she discovered the damaged car in the garage. Additional behavioral alterations observed by his wife included a newly acquired craving for sweets, resulting in a 20-pound weight gain over four months, and newly developed interest in pornography. He was not embarrassed about this latter interest and would frequently leave pornographic material around the house. He had become apathetic and chronically inert, watching television and listening to music all day, though he would occasionally take a walk around the neighborhood.

During the examination, Mr. A's mood was euthymic with appropriate affect, although he frequently laughed out aloud at stimuli that amused him. Speech characteristics were normal and his thought processes were linear, organized, and relevant. He was generally socially appropriate, though odd behaviors were observed during the course of the examination. For example, he stood very close to the examiner when introducing himself, and he walked up to a campus security officer and inappropriately started a conversation. While crossing a street, he spontaneously performed a little dance in front of the car that had stopped for pedestrians.

Neuropsychological test scores revealed marked impairment in frontal/executive tasks (mental flexibility, set-shifting, logical reasoning, verbal and design fluency, picture sequencing) with mild weaknesses in basic attention, constructional ability, and language skills. Variable performance was noted on measures of mental speed with mild weaknesses observed on visuomotor speed tasks but impaired ability on tests involving language processing speed. In contrast, semantic knowledge (confrontation naming, fund of knowledge, knowledge of vocabulary words) and memory for both verbal and visual material remained relatively spared. Qualitatively, when assessing capacity for emotional expression, he began to growl like a lion, and simulated clawing with his hands when asked to make an angry face. On a measure of gesture fluency, he extended his middle finger as one of the gestures and made another gesture that simulated sexual intercourse. His SPECT scan revealed bilateral, symmetrical hypoperfusion in the OFC and inferior frontal lobes, areas adjacent to the Sylvian fissures (affecting the anteromedial temporal lobes), the hippocampi, areas adjacent to the anterior hemispheric fissure and the anterior cingulate gyrus.

## Case 2: Mrs. B

Mrs. B is a 54-year-old, right-handed woman who presented with memory decline, but her husband described marked behavioral and personality alterations approximately four-to-five years prior to presentation. He first noticed signs of disinhibition when Mrs. B saw a friend across the street and yelled, 'That's my lover!' During a cruise in the Greek isles, she demanded to be let off the ship and chose to sleep in a separate room from her husband. Mrs. B had decided to change her career to become a music businesswoman or producer, and she spent a great deal of money trying to secure a record deal for her children's piano teacher's nephew. She tried to push her children into careers in show business. She had been a victim of numerous scams and had spent over $250,000 over several years without any concern for her family's finances. Despite her husband's attempt to intervene, she was cunning about obtaining money and would forge her husband's signature on checks, or she would wait until he was on a business trip before withdrawing money.

She developed difficulty interacting with her family and became aggressive and hostile. She stopped cooking and house cleaning because she 'doesn't feel like it'. She developed hyper-religiosity and narrowed her range of reading to religious writings. Her preference in attire also changed from conservative dress to more ornate and colorful garments. Mrs. B displayed marked alteration in eating habits, restricting her diet to bacon, beef, lobster, and shrimp, and rarely ate fruits or vegetables. She ate more and developed an attraction to sweets, resulting in a 50-pound weight gain.

On neuropsychological evaluation, her Mini-Mental State Examination (MMSE) score was 26/30. She demonstrated poor verbal learning and retention; she was able to commit four of nine words to memory after four exposure trials, and remembered only two after a 30-second delay. Recognition memory was relatively preserved as she correctly identified seven of nine words with no false positive errors. She performed poorly on measures of executive abilities, including divided attention, response inhibition, both verbal and design fluency, and abstraction. Confrontation naming was also impaired. A brain MRI obtained one year prior to evaluation revealed bilateral frontal and temporal atrophy with greater involvement of the right hemisphere. High-resolution cerebral SPECT brain imaging revealed decreased perfusion of both frontal and temporal lobes, again with the right-side more affected.

## 24.4. Neural mechanisms underlying clinical symptoms of FTD and connection to OFC

### Neurocircuitry of OFC

The neuroanatomy and circuitry of the OFC are described in substantial detail in chapters 1–4 of this book and will only be summarized here. Of particular importance for the FTD, the OFC receives inputs from all the sensory modalities and processes gustatory, olfactory, somatosensory, auditory, and visual information (Rolls 1999). The OFC serves multimodal stimulus-reinforcement association learning, processing taste, olfactory, somatosensory, auditory, visual, and visceral information. In addition, the nuclei of the amygdala project extensively to widespread areas of the OFC (Kringelbach and Rolls 2004). The innervation by multiple sensory modalities implicates the OFC as a critical convergence zone for exteroceptive sensory association cortices, interoceptive information, limbic regions involved in emotional processing and memory, and subcortical regions involved in the control of autonomic and motor pathways (Zald and Kim 2001). The OFC projects back to temporal lobe areas such as the amygdala, entorhinal cortex, hippocampus, and the inferior temporal cortex. In terms of motor output, the OFC's contribution to cortico-striato-thalamocortical loops through the basal ganglia are of particular note. The lateral orbitofrontal circuit sends fibers to the ventromedial caudate nucleus. Neurons from this region of the caudate project to the medial part of the mediodorsal globus pallidus interna and to the rostromedial substantia nigra pars reticulata. Fibers from substantia nigra and globus pallidus connect to the ventral anterior and mediodorsal thalamus. The circuit then is closed by fibers projecting back to the OFC from the thalamus (Tekin and Cummings 2002). Other efferent projections include the cingulate cortex, preoptic region, lateral hypothalamus, and ventral tegmental area (Rolls 1999).

### Functions of the OFC

In humans, major changes in personality, emotion, behavior, and social conduct can result from direct OFC lesions or indirectly via disruption of the orbitofrontal circuitry. Patients are often described as socially disinhibited, coarse, tactless, impulsive, and generally lacking in empathy and social restraint (Eslinger and Damasio 1985; Stuss and Benson 1986; Damasio and Anderson 1993). They are prone to make risky business decisions, often with disastrous consequences. Perseveration on pleasure-seeking behaviors, especially sexual behaviors are common. It is useful to note that FTD patients exhibit clinical symptoms that resemble those with defined neurosurgical lesions of the OFC.

Though OFC is known to have a major role in emotion and social cognition, the exact role of this brain region is still a topic of investigation. A number of brain structures, including the hypothalamus, insula, nucleus accumbens, and various brainstem nuclei, such as the periaqueductal gray, have been found to be associated with emotional processing in primates. These brain regions are closely linked with the OFC as well as the amygdala and anterior cingulate cortex and are crucial for emotional processing. In early animal studies, ablations of bilateral frontal lobes resulted in behaviors described as hyperactive, impulsive, lacking in affection and socialization, and apathy (Ferrier 1876). Subsequent studies on nonhuman primates demonstrated that OFC lesion produced robust changes in emotional behaviors, including increased aggressive responses and increased social withdrawal (Raleigh et al. 1979). These research studies led to conclusions that the OFC-mediated relative reward value of many different primary reinforcers (i.e., taste and somatosensory stimuli) and secondary reinforcers (i.e., olfactory and visual stimuli). It mediates learning and rapid reversal of associations between secondary and primary reinforcers, that is, it mediates implementation of stimulus-reinforcement association learning, the principal type of learning involved in emotion. Analyses of the effects of lesions to the human OFC show that they impair patients in ways related to emotion, stimulus-reinforcement association and reversal, and decision-making.

## Deficits in emotional processing and social cognition

Deficits in emotional processing have been proposed as one mechanism underlying the behavioral disturbances of FTD as patients might misinterpret emotional cues that would normally help guide their behavior. For example, an inability of patients with FTD to recognize the emotional state of their caretakers could in turn result in inappropriate actions and responses in emotional situations, which could be interpreted as apathy or emotional blunting. Hence, the expression of neuropsychiatric symptoms in FTD may reflect deficits in emotional affect processing and recognition. A number of studies have investigated emotional affect processing and recognition in FTD patients, primarily assessing the ability to discriminate between different facial emotions, affective prosody, or emotional situations. They demonstrated that FTD patients are impaired in identifying facial expressions of emotions (Lavenu et al. 1999; Keane et al. 2002; Rosen et al. 2002b), particularly negative emotions (e.g., anger). In a recent study that stratified the FTD patients into frontal- versus temporal-variant sub-groups (Rosen et al. 2004), all FTD patients demonstrated difficulty recognizing negative emotions, but those with greater frontal involvement also displayed reduced ability in identifying positive emotions. A longitudinal study over a 3-year period (Lavenu and Pasquier 2004) observed similar decrements in perception of happiness in their sample of FTD patients. Happiness is usually easier to recognize than negative emotions; however, comprehension of happiness is affected in FTD patients with extensive OFC and amygdala damage (Rosen et al. 2002b, 2004), suggesting that processing of positive emotion may be subserved by these brain regions. Furthermore, functional MRI (fMRI) studies demonstrate an activation of OFC during explicit processing of happy faces (Gorno-Tempini et al. 2001).

Impairments in identification of facial as well as vocal emotional expression, particularly anger, have been demonstrated in patients with OFC lesions (Hornak et al. 1996; Blair and Cipolotti 2000). Hornak et al. (2003) reported that patients with bilateral lesions of the OFC, which were surgical and circumscribed, displayed impairment in all aspects of emotion relative to patients with lesions in dorsolateral prefrontal cortex or other medial areas. Neurophysiologic and lesion studies in monkeys indicate that OFC is important in representing rewards and punishments and in learning stimulus-reinforcement associations, and thus in emotion and social behavior (Rolls 1999). Direct projections to OFC from temporal lobe regions (including amygdala), where faces and face expressions are represented have been demonstrated (Rolls 1999), and Iidaka and colleagues (2001) suggest that the right OFC, by interacting with the amygdala and temporal structures, is involved in the interpretation of facial emotions. Thus, damage to the amygdala and the right temporal lobe results in impaired recognition of negative emotions, particularly fear (Scott *et al.* 1997; Adolphs 1999; Anderson *et al.* 2000).

Angry expressions are known to curtail the behavior of others in situations where social rules or expectations have been violated. The OFC receives inputs from several sensory modalities and are concerned with the analysis of the reward value or affective significance of the incoming stimuli, such as facial or vocal expressions of emotion (Barbas and Pandya 1989; Rolls 2000, 2002). The ventrolateral OFC is activated by negative emotional expressions, especially anger, but also fear and disgust, and it has been suggested that OFC activity, in interpreting and imagining the punitive emotional expression of others, inhibits one from violating social norms (Berthoz *et al.* 2002). In experimental studies, patients with OFC lesions exhibit impairment in appropriately attributing anger and embarrassment to story protagonists but not other emotions (Berthoz *et al.* 2002), and they also have a deficit in identifying violations of social norms (Stone *et al.* 1998; Blair and Cipolotti 2000). Berthoz *et al.* (2002) showed substantial activation of the left OFC (Brodmann's area (BA) 10 and 47) associated with both intentional and unintentional (embarrassment) violations of social norms stories; this region was previously found to be activated by angry expressions (Sprengelmeyer *et al.* 1998; Kesler-West *et al.* 2001). Therefore, activation of BA 47 might result in modulation of current behavioral responding to prevent individuals from engaging in inappropriate behavior. When representing a social norm violation or embarrassment situation, the individual forms an expectation of other's anger/social disapproval that reverses this response option in favor of another (Blair and Cipolotti 2000; Blair 2001). Consequently, damage to this system, as found in FTD and OFC- lesioned patients, will lead to socially inappropriate behavior as the patient considering an inappropriate action will not reverse their decision on the basis of the expected social disapproval.

Recent studies have developed the novel approach of using personality inventories to characterize the characterologic changes in FTD patients. Investigation into the neuroanatomic correlates of the personality variables again implicates the OFC as playing an important role in maintaining social skills. As reported in a recent study, Rankin and colleagues (2004) attempted to elicit the relationship between biological structures underlying reduced personal warmth and agreeableness (defined as the ability to maintain positive social relationships via cooperation and attention to social norms; Rankin *et al.* 2003). Using the NEO Personality Inventory, which measures five major domains of personality (Neuroticism, Extroversion, Openness to Experience, Agreeableness, and Conscientiousness), they found that OFC volumes were significantly related to agreeableness scores in FTD patients. More specifically, the right and left OFC were associated with opposite agreeableness score—smaller right OFC volume predicting lower agreeableness and smaller left OFC volume predicting higher agreeableness in FTD patients (Rankin et al. 2004). The investigators concluded that the right OFC contributes significantly to better interpersonal sensitivity and perspective-taking aspects of cognitive empathy leading to more effective social functioning. The right OFC may mediate cooperative and socially appropriate behavior, particularly behaviors requiring a complex weighing of personal needs and desires against social norms and needs of others. This is consistent with several recent fMRI discoveries surrounding a variety of higher social functions such as self-reflection, empathy, embarrassment (Berthoz et al. 2002), moral reasoning (Moll et al. 2002), and theory of mind (Castelli et al. 2000; Gallagher et al. 2000).

## Decision-making (somatic-marker hypothesis)

Bechara and colleagues (Bechara et al. 2000; see Naqvi et al., Chapter 13, this volume) have studied numerous patients with lesions of the ventromedial prefrontal region of the OFC. These patients, as exemplified by patient E.V.R., often considered a modern counterpart to Phineas Gage, develop severe impairment in real-life decision-making, especially in the personal and social realm. As described by Eslinger and Damasio (1985), E.V.R. was successfully employed, happily married and the father of two children. He underwent resection of an orbitofrontal meningioma involving bilateral excision of the orbital and mesial cortices. Following the surgery, EVR was unable to meet his personal and professional responsibilities and his marriage eventually deteriorated. Other patients with similar OFC injury fail to organize and plan future activity, make decisions contrary to their best interest, are unable to learn from past mistakes, choose friends poorly, and make decisions that repeatedly lead to negative consequences (Bechara et al. 1994). These behavioral descriptions are relevant to patients with FTD as they manifest similar disability in 'decision-making'.

Bechara et al. (1994) developed the Iowa Gambling Task to capture aspects of the decision-making defect demonstrated by these patients. The task models the uncertainty of premises and outcomes and the differential reward and punishments that characterize real-life decisions. Patients with damage to the ventromedial prefrontal cortex persist in selecting cards from the 'high-risk' decks (Bechara et al. 1994). Physiological measures indicate that healthy control subjects generated anticipatory skin-conductance responses (SCR) while they ponder risky choices, which were absent in the lesioned patients (Bechara et al. 1997).

A 'somatic marker hypothesis' (Damasio, 1996; Bechara et al. 2002) was proposed to provide a neural explanation for the real-life decision-making deficits characteristic of patients with OFC lesions, as reflected by their poor performance on the gambling task. The main point of the hypothesis postulates that decision-making is a process guided by emotions. In other words, abnormalities in emotion and feeling are associated with impaired judgment and decision-making in real life (Bechara et al. 2000, 2002). The theory proposes that these patients are unable to represent choice bias in the form of an emotional hunch; thus, they are unable to trigger normal emotional responses to socially relevant stimuli (Bechara et al. 1997). Such an emotional hunch or 'gut feeling' will influence behavior, which is experienced as a feeling that one would prefer to choose one action over another, prior to having overt knowledge of the reason underlying that particular choice. The somatic marker hypothesis suggests that the prefrontal cortex, specifically the OFC, participates in implementing a particular mechanism by which we acquire, represent, and retrieve that value of our actions. This mechanism relies on generating somatic states, or representations of somatic states, that correspond to the anticipated future outcome of decisions. Such 'somatic markers' guide the decision-making process toward those outcomes that are advantageous for the individual, on the basis of the individual's past experiences with a similar situation. Damasio speculates that the neural network required for such markers to operate involves the ventromedial prefrontal cortices (including part of the OFC), the amygdala, and somatosensory cortices (Damasio 1996). Thus, injury to the OFC disrupts this mechanism, resulting in abnormalities in emotion and feeling and leading to a compromise in judgment and decision-making process. Given the prominent involvement of the OFC in FTD patients, it would be reasonable to expect similar abnormalities in emotion and decision-making, resulting in the behavioral abnormalities that characterize this patient population. To date, no studies investigating the performance of FTD patients on the Iowa Gambling Task are available. However, on the Cambridge Gamble Task (Rogers *et al.* 1999), which involves making a probabilistic judgment on the location of a hidden token and then placing a bet on the confidence in the decision, Rahman and colleagues (1999) demonstrated increased risk-taking behavior in a group of patients with frontal variant FTD (fvFTD). A recent study reports preliminary, but encouraging, findings of attenuated risk-taking behavior on this task in fvFTD patients treated with methylphenidate (Rahman *et al.* 2006).

## Theory of mind

Common clinical manifestations of FTD patients include lack of empathy, paucity of emotional warmth, and inability to take on another person's point of view. Some researchers have observed the resemblance of this cluster of symptoms to children with autism and Asperger's syndrome (Baron-Cohen 1995, 1999). Borrowing from developmental studies on these populations, it has been hypothesised that deficits in social reasoning may explain the behavioral disturbance in FTD, focusing on the concept of 'theory of mind' (ToM), which refers to the ability to infer other people's mental states, thoughts, or feelings, as a principal construct underlying human social functioning. Children with autism and Asperger's syndrome perform poorly on ToM tasks relative to their mental age (Baron-Cohen 1995, 1999). ToM tasks place heavy demands on executive function but research have found a dissociation between ToM and other cognitive abilities, specifically executive functioning (Baron-Cohen 1995; Baron-Cohen *et al.* 1997). These studies triggered investigation into whether ToM deficits might be involved in patients with FTD given their social impairments.

Gregory *et al.* (2002) reported that FTD patients with mild cognitive impairment on a battery of neuropsychological tests were found to have significant deficits on ToM tests, most prominently displayed on a 'faux pas' test (Stone *et al.* 1998; Baron-Cohen *et al.* 1999) in which patients were read stories that either contain a social faux pas or involve a minor conflict but without a faux pas. They were asked to correctly identify the stories with a faux pas which required them to understand two mental states; namely that the person making the faux pas does not realise his/her error and that the person hearing it would be upset or hurt as a consequence. Analysis of the responses revealed that a high proportion of FTD patients showed deficits on this test, failing to detect when something hurtful or inappropriate had been said, which are analogous to lack of empathy (Gregory *et al.* 2002). The patients, as a group, displayed mild executive impairment though performance on ToM tasks was largely independent of the scores on the frontal measures. The dissociation between ToM and executive functioning was further illustrated by a case study presented by Lough and Hodges (2002). A patient with frontal variant FTD with gross antisocial behavior showed normal performance on a range of traditional neuropsychological measures assessing executive functioning but was markedly impaired on ToM and social faux pas tasks which were routinely passed by children aged 3–4 years (Lough and Hodges 2002).

Evidence for a neural basis of ToM is limited but existing research suggests certain regions within the prefrontal cortex may be critical. For example, patients with OFC lesions were significantly impaired on tests of faux pas recognition while patients with unilateral dorsolateral prefrontal lesions performed normally on the same task (Stone *et al.* 1998). The literature on impaired detection of emotional expression in patients of OFC lesions also led to the speculation that ToM abilities, and the ability to decode the emotional mental states of others, are subserved by a distributed system involving multiple regions of the prefrontal cortex, including OFC, and the limbic system, including the amygdala (Baron-Cohen *et al.* 1994; Baron-Cohen 1995; Rolls 2004; Stone *et al.* 1998). Neuroimaging studies also implicate the frontal lobes in the performance of ToM tasks. ToM appears to complement the somatic marker hypothesis in explaining different aspects of OFC-related behaviors observed in patients with FTD.

## Hyperorality

FTD causes profound changes in eating habits characterized by escalated desire for sweet food and reduced satiety, which is often followed by substantial weight gain. Lesions of the OFC also cause marked alterations in food preferences and impair the ability to discriminate food from non-food objects (Baylis and Gaffan 1991; Butter *et al.* 1969). Furthermore, OFC- lesioned monkeys sometimes demonstrate hyperorality and/or a failure to satiate (Butter *et al.* 1963, 1969). These monkeys will place non-food objects in their mouth, and readily eat foods deemed to be unpalatable by non-lesioned monkeys. Thus, the role of the OFC in gustatory reward and satiety may underlie the altered dietary preference seen in FTD patients.

Taste information is received from the ventral posteromedial nucleus of the thalamus by the primary taste cortex in the anterior insular and adjoining frontal operculum, which then projects to the caudolateral OFC containing the secondary taste cortex (Kringelbach and Rolls 2004; see Gottfried *et al.* Chapter 6, this volume). The gustatory projections to the OFC provide specific information about the physical properties of food, and projections from other sensory areas provide information about other stimuli (e.g., odor, sight) associated with food (Rolls and Baylis, 1994). The OFC integrates the sensory information and provides the ability to make cross-modal associations related to food reward. The recognition of the associates of specific foods could stimulate the craving for these foods. Therefore, the OFC plays an active role in modulating reward-related behaviors, which may be the mechanism that underlies other impulse-control and compulsive behaviors exhibited by FTD patients in which individuals fail to stop behaviors once started, or experience impulses when exposed to cues related to these rewarding actions (Zald and Kim 2001).

## 24.5. Summary

Classic cases, such as Phineas Gage and EVR, point to the OFC as an important brain structure in social conduct, personality, interpersonal judgment, and behavior. An important example of dysfunction of the OFC is FTD, a progressive neurodegenerative disorder affecting the frontal lobes and producing major compromise in personality and social functioning. Postmortem and neuroimaging studies have implicated the OFC as the structure with the greatest atrophic changes in FTD. Research into the functions of OFC and the neural mechanisms underlying the cognitive, social, emotional, and decision-making impairments are crucial in understanding the disturbances in socioemotional behavior manifest by FTD patients. There is a clear clinical need for the development of neuropsychological tasks to provide sensitive measures of OFC or ventromedial prefrontal lobe dysfunction as well as continued efforts for broad assessment of emotional functioning. These studies should lead to a greater understanding of the neuroanatomic basis of emotion as well as the relationship between emotional processing problems and behavioral dysfunction in FTD.

## Acknowledgment

This work was supported by a National Institute of Aging grant to the UCLA Alzheimer's Disease Research Center (P50 AG 16570) and the UCLA Alzheimer's Disease Research Center of California.

## References

Adolphs, R. (1999) Social cognition and the human brain. *Trends Cogn Sci.* **3**:469–479.

Anderson, A. K., Spencer, D. D., Fulbright, R. K., and Phelps, E. A. (2000) Contribution of the anteromedial temporal lobes to the evaluation of facial emotion. *Neuropsychology* **14**:526–536.

Baron-Cohen, S. (1995) *Mindblindness: an essay on autism and theory of mind*. Cambridge: MIT Press.

Baron-Cohen, S., Ring, H., Moriarty, J., Schmitz, B., Costa, D., and Ell, P. (1994) Recognition of mental state terms. Clinical findings in children with autism and a functional neuroimaging study of normal adults. *Br J Psychiatry* **165**:640–649.

Baron-Cohen, S., Jolliffe, T., Mortimore, C., and Robertson, M. (1997) Another advanced test of theory of mind: evidence from very high functioning adults with autism or asperger syndrome. *J Child Psychol Psychiatry* **38**:813–822.

Baron-Cohen, S., O'Riordan, M., Stone, V., Jones, R., Plaisted, K. (1999) Recognition of faux pas by normally developing children and children with Asperger syndrome or high-functioning autism. *J Autism Dev Disord* **29**:407–418.

Baylis, L. L., and Gaffan, D. (1991) Amygdalectomy and ventromedial prefrontal ablation produce similar deficits in food choice and in simple object discrimination learning for an unseen reward. *Exp Brain Res* **86**:617–622.

Bechara, A., Damasio, A. R., Damasio, H., and Anderson, S. W. (1994) Insensitivity to future consequences following damage to human prefrontal cortex. *Cognition.* **50**:7–15.

Bechara, A., Damasio, H., Tranel, D., and Damasio, A. R. (1997) Deciding advantageously before knowing the advantageous strategy. *Science.* **275**:1293–1295.

Bechara, A., Tranel, D., and Damasio, H. (2000) Characterisation of the decision-making deficit of patients with ventromedial prefrontal cortex lesions. *Brain* **123 (Pt 11)**:2189–2202.

Bechara, A., Tranel, D., and Damasio, H. (2002) The somatic marker hypothesis and decision-making. In: Boller, F., and Grafman, J. (eds.) *Handbook of Neuropsychology: Frontal Lobes*, Vol. 7, 2nd ed. Amsterdam: Elsevier pp. 117–143.

Berthoz, S., Armony, J. L., Blair, R. J., and Dolan, R. J. (2002) An fMRI study of intentional and unintentional (embarrassing) violations of social norms. *Brain.* **125**:1696–1708.

Blair, R. J. (2001) Neurocognitive models of aggression, the antisocial personality disorders, and psychopathy. *J Neurol Neurosurg Psychiatry* **71**:727–731.

Blair, R. J. and Cipolotti, L. (2000) Impaired social response reversal. A case of 'acquired sociopathy'. *Brain* **123 (Pt 6)**:1122–1141.

Boone, K. B., Miller, B. L., Lee, A., Berman, N., Sherman, D., and Stuss, D. T. (1999) Neuropsychological patterns in right versus left frontotemporal dementia. *J Int Neuropsychol Soc* **5**:616–622.

Brown, J. (1998) Chromosome 3-linked frontotemporal dementia. *Cell Mol Life Sci* **54**:925–927.

Butter, C. M., Mishkin, M., and Rosvold, H. E. (1963) Conditioning and extinction of a food-rewarded response after selective ablations of frontal cortex in rhesus monkeys. *Exp Neurol* **7**:65–75.

Butter, C. M., McDonald, J. A., and Snyder, D. R. (1969) Orality, preference behavior, and reinforcement value of non-food object in monkeys with orbital frontal lesions. *Science* **164**:1306–1307.

Castelli, F., Happe, F., Frith, U., and Frith C. (2000) Movement and mind: a functional imaging study of perception and interpretation of complex intentional movement patterns. *Neuroimage* **12**:314–325.

Chawluk, J. B., Mesulam, M. M., Hurtig, H., Kushner, M., Weintraub, S., Saykin, A., Rubin, N., Alavi, A., and Reivich, M. (1986) Slowly progressive aphasia without generalized dementia: studies with positron emission tomography. *Ann Neurol* **19**:68–74.

Chow, T. W., Miller, B. L., Hayashi, V. N., and Geschwind, D. H. (1999) Inheritance of frontotemporal dementia. *Arch Neurol* **56**:817–822.

Coleman, L. W., Digre, K. B., Stephenson, G. M., and Townsend, J. J. (2002) Autopsy-proven, sporadic pick disease with onset at age 25 years. *Arch Neurol* **59**:856–859.

Damasio, A. and Anderson, S. (1993) The frontal lobes. In: Heilman, K. M., and Valenstein, E. (eds.) *Clinical Neuropsychology*. New York: Oxford University Press pp. 409–460.

Damasio, A. R. (1996) The somatic marker hypothesis and the possible functions of the prefrontal cortex. *Philos Trans R Soc Lond B Biol Sci* **351**:1413–1420.

Damasio, H., Grabowski, T., Frank, R., Galaburda, A. M., and Damasio, A. R. (1994) The return of Phineas Gage: clues about the brain from the skull of a famous patient. *Science* **264**:1102–1105.

Elliott, R. and Dolan, R. J. (1999) Differential neural responses during performance of matching and non-matching to sample tasks at two delay intervals. *J Neurosci* **19**:5066–5073.

Eslinger, P. J. and Damasio, A. R. (1985) Severe disturbance of higher cognition after bilateral frontal lobe ablation: patient EVR. *Neurology* **35**:1731–1741.

Foster, N. L., Wilhelmsen, K., Sima, A. A., Jones, M. Z., D'Amato, C. J., and Gilman, S. (1997) Frontotemporal dementia and parkinsonism linked to chromosome 17: a consensus conference. Conference Participants. *Ann Neurol* **41**:706–715.

Francis, P, T., Holmes, C., Webster, M. T., Stratmann, G. C., Procter, A. W., and Bowen, D. M. (1993) Preliminary neurochemical findings in non-Alzheimer dementia due to lobar atrophy. *Dementia* **4**:172–177.

Gallagher, H. L., Happe, F., Brunswick, N., Fletcher, P. C., Frith, U., and Frith, C. D. (2000) Reading the mind in cartoons and stories: an fMRI study of 'theory of mind' in verbal and non-verbal tasks. *Neuropsychologia* **38**:11–21.

Glosser, G., Gallo, J. L., Clark, C. M., and Grossman, M. (2002), Memory encoding and retrieval in frontotemporal dementia and Alzheimer's disease. *Neuropsychology* **16**:190–196.

Gorno-Tempini, M. L., Pradelli, S., Serafini, M., Pagnoni, G., Baraldi, P., Porro, C., et al. (2001) Explicit and incidental facial expression processing: an fMRI study. *Neuroimage* **14**:465–473.

Gregory, C., Lough, S., Stone, V., Erzinclioglu, S., Martin, L., Baron-Cohen, S., et al. (2002) Theory of mind in patients with frontal variant frontotemporal dementia and Alzheimer's disease: theoretical and practical implications. *Brain* **125**:752–764.

Gregory, C. A., Orrell, M., Sahakian, B., and Hodges, J. R. (1997) Can frontotemporal dementia and Alzheimer's disease be differentiated using a brief battery of tests? *Int J Geriatr Psychiatry* **12**:375–383.

Gregory, C. A., Serra-Mestres, J., Hodges, J. R. (1999) Early diagnosis of the frontal variant of frontotemporal dementia: how sensitive are standard neuroimaging and neuropsychologic tests? *Neuropsychiatry Neuropsychol Behav Neurol* **12**:128–135.

Grimmer, T., Diehl, J., Drzezga, A., Forstl, H., and Kurz, A. (2004) Region-specific decline of cerebral glucose metabolism in patients with frontotemporal dementia: a prospective 18F-FDG-PET study. *Dement Geriatr Cogn Disord* **18**:32–36.

Grossman, M., Payer, F., Onishi, K., D'Esposito, M., Morrison, D., Sadek, A., et al. (1998) Language comprehension and regional cerebral defects in frontotemporal degeneration and Alzheimer's disease. *Neurology* **50**:157–163.

Harlow, J. M. (1868) Recovery from the passage of an iron bar through the head. *Publications of the Massachusetts Medical Society* **2**:327–347.

Hodges, J. R. and Miller, B. (2001) The classification, genetics and neuropathology of frontotemporal dementia. Introduction to the special topic papers: Part I. *Neurocase* **7**:31–35.

Hooten, W. M. and Lyketsos, C. G. (1006) Frontotemporal dementia: a clinicopathological review of four postmortem studies. *J Neuropsychiatry Clin Neurosci* **8**:10–19.

Hornak, J., Bramham, J., Rolls, E. T., Morris, R. G., O'Doherty, J., Bullock, P. R., et al. (2003) Changes in emotion after circumscribed surgical lesions of the orbitofrontal and cingulate cortices. *Brain* **126**:1691–1712.

Hornak, J., Rolls, E. T., and Wade, D. (1996) Face and voice expression identification in patients with emotional and behavioral changes following ventral frontal lobe damage. *Neuropsychologia* **34**:247–261.

Hosler, B. A., Siddique, T., Sapp, P. C., Sailor, W., Huang, M. C., Hossain, A., et al. (2000) Linkage of familial amyotrophic lateral sclerosis with frontotemporal dementia to chromosome 9q21-q22. *JAMA* **284**:1664–1669.

Iidaka, T., Omori, M., Murata, T., Kosaka, H., Yonekura, Y., Okada, T., et al. (2001) Neural interaction of the amygdala with the prefrontal and temporal cortices in the processing of facial expressions as revealed by fMRI. *J Cogn Neurosci* **13**:1035–1047.

Jacob. J., Revesz, T., Thom, M., and Rossor, M. N. (1999) A case of sporadic Pick disease with onset at 27 years. *Arch Neurol* **56**:1289–1291.

Jagust, W. J., Reed, B. R., Seab, J. P., Kramer, J. H., and Budinger, T. F. (1989) Clinical-physiologic correlates of Alzheimer's disease and frontal lobe dementia. *Am J Physiol Imaging* **4**:89–96.

Johnson, J. K., Diehl, J., Mendez, M. F., Neuhaus, J., Shapira, J. S., Forman, M., et al. (2005) Frontotemporal lobar degeneration: demographic characteristics of 353 patients. *Arch Neurol* **62**:925–930.

Kamo, H., McGeer, P. L., Harrop, R., McGeer, E. G., Calne, D. B., Martin, W. R., et al. (1987) Positron emission tomography and histopathology in Pick's disease. *Neurology* **37**:439–445.

Keane, J., Calder, A. J., Hodges, J. R., and Young, A. W. (2002) Face and emotion processing in frontal variant frontotemporal dementia. *Neuropsychologia* **40**:655–665.

Kesler-West, M. L., Andersen, A. H., Smith, C. D., Avison, M. J., Davis, C. E., Kryscio, R. J., et al. (2001) Neural substrates of facial emotion processing using fMRI. *Brain Res Cogn Brain Res* **11**:213–226.

Kitagaki, H., Mori, E., Hirono, N., Ikejiri, Y., Ishii, K., Imamura, T., et al. (1997) Alteration of white matter MR signal intensity in frontotemporal dementia. *AJNR Am J Neuroradiol* **18**:367–378.

Knopman, D. S., Christensen, K. J., Schut, L. J., Harbaugh, R. E., Reeder, T., Ngo, T., et al. (1989) The spectrum of imaging and neuropsychological findings in Pick's disease. *Neurology* **39**:362–368.

Knopman, D. S., Mastri, A. R., Frey, W. H., 2nd, Sung, J. H., and Rustan, T. (1990) Dementia lacking distinctive histologic features: a common non-Alzheimer degenerative dementia. *Neurology* **40**:251–256.

Kramer, J. H., Jurik, J., Sha, S. J., Rankin, K. P., Rosen, H. J., Johnson, J. K., et al. (2003) Distinctive neuropsychological patterns in frontotemporal dementia, semantic dementia, and Alzheimer disease. *Cogn Behav Neurol* **16**:211–218.

Kringelbach, M. L. and Rolls, E. T. (2004) The functional neuroanatomy of the human orbitofrontal cortex: evidence from neuroimaging and neuropsychology. *Prog Neurobiol* **72**:341–372.

Lavenu, I. and Pasquier, F. (2005) Perception of emotion on faces in frontotemporal dementia and Alzheimer's disease: a longitudinal study. *Dement Geriatr Cogn Disord* **19**:37–41.

Lavenu, I., Pasquier, F., Lebert, F., Petit, H., and Van der Linden, M. (1999) Perception of emotion in frontotemporal dementia and Alzheimer disease. *Alzheimer Dis Assoc Disord* **13**:96–101.

Lipton, A. M., Cullum, C. M., Satumtira, S., Sontag, E., Hynan, L. S., White, C. L., et al. (2001) Contribution of asymmetric synapse loss to lateralizing clinical deficits in frontotemporal dementias. *Arch Neurol* **58**:1233–1239.

Lough, S. and Hodges, J. R. (2002) Measuring and modifying abnormal social cognition in frontal variant frontotemporal dementia. *J Psychosom Res* **53**:639–646.

Macmillan, M. (2000) Restoring Phineas Gage: a 150th retrospective. *J Hist Neurosci* **9**:46–66.

Mann, D. M. (1998) Dementia of frontal type and dementias with subcortical gliosis. *Brain Pathol* **8**:325–338.

Martin, J. A., Craft, D. K., Su, J. H., Kim, R. C., and Cotman, C. W. (2001) Astrocytes degenerate in frontotemporal dementia: possible relation to hypoperfusion. *Neurobiol Aging* **22**:195–207.

Mathuranath, P. S., Nestor, P. J., Berrios, G. E., Rakowicz, W., and Hodges, J. R. (2000) A brief cognitive test battery to differentiate Alzheimer's disease and frontotemporal dementia. *Neurology* **55**:1613–1620.

McKhann, G. M., Albert, M. S., Grossman, M., Miller, B., Dickson, D., and Trojanowski, J. Q. (2001) Clinical and pathological diagnosis of frontotemporal dementia: report of the Work Group on Frontotemporal Dementia and Pick's Disease. *Arch Neurol* **58**:1803–1809.

Mendez. M. F. and Cummings, J. L. (2003) *Dementia: a clinical approach*. Philadelphia, Butterworth Heinemann.

Mendez, M. F., Perryman, K. M., Miller, B. L., Swartz, J. R., Cummings, J. L. (1997) Compulsive behaviors as presenting symptoms of frontotemporal dementia. *J Geriatr Psychiatry Neurol* **10**:154–157.

Mendez, M. F., Selwood, A., Mastri, A. R., and Frey, W. H., 2nd. ( 1993) Pick's disease versus Alzheimer's disease: a comparison of clinical characteristics. *Neurology* **43**:289–292.

Moll, J., de Oliveira-Souza, R., Eslinger, P. J., Bramati, I. E., Mourao-Miranda, J., Andreiuolo, P. A., et al. (2002) The neural correlates of moral sensitivity: a functional magnetic resonance imaging investigation of basic and moral emotions. *J Neurosci* **22**:2730–2736.

Munoz, D. G., Dickson, D. W., Bergeron, C., Mackenzie, I. R., Delacourte, A., and Zhukareva, V. (2003) The neuropathology and biochemistry of frontotemporal dementia. *Ann Neurol* **54** Suppl **5**:S24–28.

Neary, D., Snowden, J. S., Gustafson, L., Passant, U., Stuss, D., Black, S., et al. (1998) Frontotemporal lobar degeneration: a consensus on clinical diagnostic criteria. *Neurology* **51**:1546–1554.

Pachana, N. A., Boone, K. B., Miller, B. L., Cummings, J. L., and Berman, N. (1996) Comparison of neuropsychological functioning in Alzheimer's disease and frontotemporal dementia. *J Int Neuropsychol Soc* **2**:505–510.

Pasquier, F., Grymonprez, L., Lebert, F., and Van der Linden, M. (2001) Memory impairment differs in frontotemporal dementia and Alzheimer's disease. *Neurocase* **7**:161–171.

Rahman, S., Robbins, T. W., and Sahakian, B. J. (1999a) Comparative cognitive neuropsychological studies of frontal lobe function: implications for therapeutic strategies in frontal variant frontotemporal dementia. *Dement Geriatr Cogn Disord* **10** Suppl **1**:15–28.

Rahman, S., Sahakin, B. J., Hodges, J.R. Rogers, R.D., Robbins, T.W. (1999b) Specific cognitive deficits in mild frontal variant frontotemporal dementia. *Brain* **122**:1469–1493.

Rahman, S., Robbins, T. W., Hodges, J.R. Mehta, M.A., Nestor, P.J., Clark, L., et al. (2006) Methylphenidate ('Ritalin') can ameliorate abnormal risk-taking behavior in the frontal variant of frontotemporal dementia. *Neuropsychopharmacology* **13**:651–658.

Raleigh, M. J., Steklis, H. D., Ervin, F. R., Kling, A. S., and McGuire, M. T. (1979) The effects of orbitofrontal lesions on the aggressive behavior of vervet monkeys (Cercopithecus aethiops sabaeus). *Exp Neurol* **66**:158–168.

Rankin, K. P., Kramer, J. H., Mychack, P., and Miller, B. L. (2003) Double dissociation of social functioning in frontotemporal dementia. *Neurology* **60**:266–271.

Rankin, K. P., Rosen, H. J., Kramer, J. H., Schauer, G. F., Weiner, M. W., Schuff, N., et al. (2004) Right and left medial orbitofrontal volumes show an opposite relationship to agreeableness in FTD. *Dement Geriatr Cogn Disord* **17**:328–332.

Reed, L. A., Wszolek, Z. K., and Hutton, M. (2001) Phenotypic correlations in FTDP-17. *Neurobiol Aging* **22**:89–107.

Rinne, J. O., Laine, M., Kaasinen, V., Norvasuo-Heila, M. K., Nagren, K., and Helenius, H. (2002) Striatal dopamine transporter and extrapyramidal symptoms in frontotemporal dementia. *Neurology* **58**:1489–1493.

Rogers, R.D., Everitt, B.J., Baldacchino, A. Blackshaw, A.J. Swainson, R. Wynne, K., et al. (1999) Dissociable deficits in the decison-making cognition of chronic amphetamine abusers, opiate abusers, patients with focal damage to prefrontal cortex, and tryptophan-depleted normal volunteers: evidence for monoaminergic mechanisms. *Neuropsychopharmacology* **20**:322–339.

Rolls, E. T. and Baylis, L. L. (1994) Gustatory, olfactory, and visual convergence within the primate orbitofrontal cortex. *J Neurosci* **14**:5437–5452.

Rolls, E. T. (1999) Spatial view cells and the representation of place in the primate hippocampus. *Hippocampus* **9**:467–480.

Rolls, E. T. (2004) The functions of the orbitofrontal cortex. *Brain Cogn* **55**:11–29.

Rosen, H. J., Gorno-Tempini, M. L., Goldman, W. P., Perry, R. J., Schuff, N., Weiner, M., et al. (2002a) Patterns of brain atrophy in frontotemporal dementia and semantic dementia. *Neurology* **58**:198–208.

Rosen, H. J., Perry, R. J., Murphy, J., Kramer, J. H., Mychack, P., Schuff, N., et al. (2002b) Emotion comprehension in the temporal variant of frontotemporal dementia. *Brain* **125**:2286–2295.

Rosen, H. J., Pace-Savitsky, K., Perry, R. J., Kramer, J. H., Miller, B. L., and Levenson, R. W. (2004) Recognition of emotion in the frontal and temporal variants of frontotemporal dementia. *Dement Geriatr Cogn Disord* **17**:277–281.

Schmitt, H. P., Yang, Y., and Forstl, H. (1995) Frontal lobe degeneration of non-Alzheimer type and Pick's atrophy: lumping or splitting? *Eur Arch Psychiatry Clin Neurosci* **245**:299–305.

Scott, S. K., Young, A. W., Calder, A. J., Hellawell, D. J., Aggleton, J. P., and Johnson, M. (1997) Impaired auditory recognition of fear and anger following bilateral amygdala lesions. *Nature* **385**:254–257.

Snowden, J. S., Neary, D., Mann, D. M., Goulding, P. J., and Testa, H. J. (1992) Progressive language disorder due to lobar atrophy. *Ann Neurol* **31**:174–183.

Snowden, J. S., Neary, D., and Mann, D. M. (2002) Frontotemporal dementia. *Br J Psychiatry* **180**:140–143.

Snowden, J. S., Neary, D., and Mann, D. M. (2004) Autopsy proven sporadic frontotemporal dementia due to microvacuolar-type histology, with onset at 21 years of age. *J Neurol Neurosurg Psychiatry* **75**:1337–1339.

Sparks, D. L., Danner, F. W., Davis, D. G., Hackney, C., Landers, T., and Coyne, C. M. (1994) Neurochemical and histopathologic alterations characteristic of Pick's disease in a non-demented individual. *J Neuropathol Exp Neurol* **53**:37–42.

Sprengelmeyer, R., Rausch, M., Eysel, U. T., and Przuntek, H. (1998) Neural structures associated with recognition of facial expressions of basic emotions. *Proc Biol Sci* **265**:1927–1931.

Stevens, M., van Duijn, C. M., Kamphorst, W., de Knijff, P., Heutink, P., van Gool, W. A., *et al.* (1998) Familial aggregation in frontotemporal dementia. *Neurology* **50**:1541–1545.

Stone, V. E., Baron-Cohen, S., and Knight, R. T. (1998) Frontal lobe contributions to theory of mind. *J Cogn Neurosci* **10**:640–656.

Stuss, D.T. and Benson, D. F. (1986) *The frontal lobes*. New York: Raven.

Tekin, S. and Cummings, J. L. (2002) Frontal-subcortical neuronal circuits and clinical neuropsychiatry: an update. *J Psychosom Res* **53**:647–654.

Turner, R. S., Kenyon, L. C., Trojanowski, J. Q., Gonatas, N., and Grossman, M. (1996) Clinical, neuroimaging, and pathologic features of progressive nonfluent aphasia. *Ann Neurol* **39**:166–173.

Varrone, A., Pappata, S., Caraco, C., Soricelli, A., Milan, G., Quarantelli, M., *et al.* (2002) Voxel-based comparison of rCBF SPET images in frontotemporal
dementia and Alzheimer's disease highlights the involvement of different cortical networks. *Eur J Nucl Med Mol Imaging* **29**:1447–1454.

Zald, D. H. and Kim, S. W. (2001) The orbitofrontal cortex. In: Salloway, S. P., Malloy, P. F., and Duffy, J. D. (eds). *The Frontal Lobes and Neuropsychiatric Illness*. Washington D.C.: American Psychiatric Publishing. pp. 33–69.

# Index

acquired sociopathy 600, 611
acquisition of alternation tasks 454–6
affective information, regulation of 312–13
affective state
  production of 379–84
  regulation of 384–5, 385–6
aggression 503–4
  definition and assessment 588–9
  degenerative disorders 590–1
  frontotemporal dementia 591
  orbitofrontal lesions 588–92
  stroke 591–2
  traumatic brain injury 589–90
agranular cortex 29
agranular insula
  animals 6–7
  humans 11–13
alcohol abuse 484
alliesthesia 157
Alzheimer's disease 622
  aggression 591
  mood disorders 586
amygdala 46–47, 113–5, 219–26
  affective properties of stimuli 245–51
  anxiety disorders 528–9
  autonomic arousal 254–5, 257–8
  damage to 463–4
  decision-making functions 341–3
  discrimination learning 238–40
  fear responses 526–7
  goal selection 246–51
  hyperactivity 528–9
  input-output zones 65–6
  representational memory 219–22
    differentiation from orbitofrontal contribution 222–6
  reversal and extinction 242–3
  *see also* connections
anger 502–3
anorexia 161
anosmia 125
  post-traumatic 465
anterior cingulate 9–10, 14, 58, 67–8, 79–81, 378, 379, 393, 405, 407–9, 455–9, 466, 469, 486–503, 507, 509, 528, 530–2, 545, 557–60, 564–5, 630–1
anterior communicating artery aneurysm syndrome 299–300
anterior medial nucleus 71–2, 75
anterior olfactory nucleus 43
anterior orbital gyrus 21, 24
anteromedial nucleus 51
anticipation 200, 206–15
anxiety disorders 523–43

  clinical characteristics 523–5
  excessive influence of amygdala 528–9
  excessive lateral OFC activity 532–4
  failure of 'top-down' control 531–2
  functional anatomy of OFC in 525–8
  OFC and extinction 534–5
  pathophysiology 528–9
  peripheral physiology 534–5
  traumatic brain injury 583–5
apathy 586–7
Apathy Inventory 587
aphasia, progressive non-fluent 621
appetitive system 269–70
architectonic organization 3–15, 26–34
  human 10–5, 26–34
  nonhuman primate 4–9, 26–34
  rat 9–10
arcuate orbital sulcus 20
associative learning 144, 160, 199
attention-deficit/hyperactivity disorder 562
attitude regulation, inhibitory control 316–18
attribution 496–7
audition 45–6
autoassociation networks 115
autobiographical memory 293–9
  healthy subjects 293–5
  orbitofrontal damage 295–6
  psychiatric patients 296–9
autonomic arousal 254–7
autonomic function 602–5
autonomic regulation 386
aversive system 270

basal forebrain 291
basoventral trend 41
Behavioral Assessment of the Dysexecutive Syndrome 468
behavioral choice 276–7
behavioral deficits 215–19
behavioral disinhibition *see* disinhibition
behavioral tasks 173–4
  reward vs no reward 175–7
bingeing 491–3
bipolar disorder 546, 552, 553, 562
  pediatric 563
blood-oxygenation-level-dependent *see* BOLD
BOLD fMRI 423–46
  image distortion 430–1
  magnetic susceptibility 426–9
  signal loss 431
  susceptibility artifact reduction 431–42
    data post-processing 432–3
    gradient compensation 437–9
anxiety disorders 523–43

BOLD fMRI (*cont.*)
    reduced data methods and parallel imaging 433–5
    shimming 431–2
    tailored RF pulses 439–42
  susceptibility artifacts 429–30
bottleneck structures 285
brain-behavioral models 484–7
brainstem reward centres 64–5
Broca's region 297
Brodmann areas 4, 5, 26–8
  area 4 59
  area 9 402, 591
  area 10 4
  area 10m 15, 45
  area 10p 15
  area 10r 15
  area 11 31, 32, 289, 290
  area 11l 15, 45
  area 11m 32
  area 12 11, 59, 60
  area 12/47 238
  area 12l 44
  area 12o 45, 49
  area 13 11, 30, 31, 33, 59, 60
  area 13a 13, 49
  area 13l 45
  area 13m 45
  area 14 30, 31, 32, 33
  area 14c 14
  area 24 4, 327
  area 25 4, 14, 45, 49
  area 32 4, 14, 45, 49, 327
  area 32ac 14
  area 45 26, 44
  area 47 26
  area 47/12 30, 31, 33
  area FF 27
  area FFa 27
  area Iai 45, 49
  area Ial 44
  area Iapm 44
  TEav 44
  TGv 44
bulimia nervosa 161
Buss-Perry Aggression Questionnaire 588

calbindin 59
Cambridge Gamble Task 395–7, 406–7
Cambridge Neuropsychological Test Automated Battery (CANTAB) 458
Capgras syndrome 300
catechol-O-methyl-transferase 413
catecholamine systems 409–11
  dopamine 401, 403, 409–10
  noradrenaline 401, 411
caudal medial wall 14
centroposterior orbitofrontal area 129–31, 135
cerebrovascular accident *see* stroke
chemical senses 125–71
  flavor 101, 153–6

  olfaction 42, 43–4, 102–4, 125–45
  taste 98–101, 145–52
chemosensory integration 153–6
  humans 154–6
  nonhuman primates 153
5-Choice Serial Reaction Time Task 407
cholinergic system 411–12
circumplex model of emotion 377
cingulate *see* anterior cingulate and subgenual cingulate
Clinical Assessment of Depression in Dementia 586
cocaine abuse 394, 484, 506
cognitive behavioral therapy 529
cognitive deficit 483
cognitive dysfunction in mood disorders 552–3
cognitive tasks, functional neuroimaging 559–60
computational modelling of reward learning 114–7
conditional reward 107
conditioned reinforcement 240–2
conditioned stimulus 240, 271, 273
conditioning 485, 497–8
confabulations 299–300
connections 39–55
  amygdala 46–7, 62–4, 201–3
  audition 45–6
  external environment 60–2
  internal environment 62–2
  intrinsic 39–42
  limbic 46–9
  olfaction 43–4
  outputs to hypothalamus and midbrain 49–50, 66–8
  sensory inputs 42–3
  somatic sensation 45
  striatum 64
  taste/visceral afferents 44
  thalamic 9, 51–2, 62, 68, 71–3, 200–2
  striatum 51–2, 73–5, 201–2
  vision 44–5
contamination fears 530
corticobasal degeneration 622
craving 489–91
Cross Cultural Smell Identification Test 465
cue-sampling 211
cue-selectivity 206, 212–13, 222–3

data post-processing 432–3
decision and action 75–9
decision-making 505–6
  frontotemporal dementia 634–5
  in pseudopsychopathy 605–8
decision-making cognition 394–400
  functional imaging studies 396–7
  human lesion studies 395–6
  reversal learning 399–400
decision-making (gambling) tasks 270, 460–2
  Cambridge gamble task 395–7, 406–7
  Iowa Gambling Task 216, 227, 368–71, 395–7, 398–9, 505, 605, 605–6, 610
degenerative disorders
  aggression 590–1

mood 586–7
  apathy 586–7
  depression 586
delayed gratification 398, 408
depression 546–7
  degenerative disorders 586
  traumatic brain injury 581–3
  see also mood disorders
development
  mood disorders 562–4
  pseudopsychopathy 611–14
Diagnostic and Statistical Manual of Mental Disorders 580
Diagnostic and Statistical Manual of Mental Disorders Personality 601
diamagnetism 426
discrimination learning 238–9
  Pavlovian associations 239–40
disinhibition 238, 485
disruption theory of inhibition 308
distraction theory of inhibition 318
dopamine 113, 184, 192, 195, 276, 401, 403, 409–10
  in drug addiction 487, 494–6
drive 485
drug addiction 481–521
  bingeing 491–3
  clinical applications 508–9
  craving 489–91
  decision-making 505–6
  definition 482–4
  dopamine in 494–6
  I-RISA syndrome 496–506
    aggression 503–4
    anger 502–3
    conditioning and extinction 497–8
    inhibitory control 498–501
    personality 501–2
    reward, saliency, attribution and motivation 496–7
  intoxication 487–9
  neuropsychological deficits in 482–4
  sensitivity and specificity 507–8
  theories of 484–7
  withdrawal 493
dynamic filtering theory of inhibition 308, 311, 317
Dysexecutive Questionnaire 468
dysgranular areas 59

echo planar imaging 428
electrodermal skin conductance response 14, 227, 311, 331, 335, 362, 364, 372, 386, 502, 506, 526, 535, 602–4, 607
emotion
  circumplex model 377
  regulation 313–15
  theories of 377–8
emotion induction, functional neuroimaging 377–91
  amygdala-orbitofrontal cortex relationship 385
  autonomic regulation 386
  identification of emotional stimulus 378–9
  production of affective state 379–84
  regulation of affective state 384–5, 385–6

emotional inhibition see inhibition
emotional lateralization 449–50
emotional memory 69–72
  connections with medial temporal cortices 69–71
  connections with mediodorsal and anterior thalamic nuclei 71–2
emotional processing 57–91
  frontotemporal dementia 632–4
  functional neuroimaging 558–9
  lesions affecting 97
  medial prefrontal vs orbitofrontal areas 79–81
  orbitofrontal-striato-thalamic circuits 73–5
emotional visual stimuli 360–8
  faces 362–3
  scenes 363–8
empathy 585
encoding
  associated outcome 204–5
  memory 287–90
entorhinal cortex 46, 47
environment
  conscious awareness of significance of 68–9
  signals from 60–2
epilepsy surgery 356, 357
episodic memory 159–60, 286
error neurons 105, 107, 108, 110, 112, 115
errors of commission 457
errors of omission 457
euthymia 548
expectancy 200, 206–15
  generation of 207
expected reward value 271–2
extinction 242–3, 252–3, 497–8, 536
  role of OFC in 534–5
extra-dimensional set-shifting 405
  see also Intradimensional-Extradimensional (ID/ED) Shift Task
Eysenck Personality Questionnaire 601

face discrimination reversal 110
face expression identification impairment 110
faces, responses to 362–3
facial affect processing 551–2
familiarity processing 292
Faux Pas Recognition Test 463, 471
feedback 76
feedforward 76, 77
feeding behavior
  food selection 156–7
  internal state 157–61
  regulation of 156–61
  relative reward 156–7
  see also chemical senses
flavor 101, 153–6
food selection 156–7
Frontal Behavior Inventory 467
frontal lesions 97
  mood disorders 548–9
frontal lobe dementia 451
frontal pole 14–15
frontal sparing 141

Frontal Systems Behavioral Scale  466
frontotemporal dementia  396, 621–42
　aggression  591
　case histories  629–30
　clinical characteristics  622–3
　decision-making  634–5
　emotional processing deficit  632–4
　genetics  625–6
　hyperorality  636–7
　mood disorders  586
　neural mechanisms  631–7
　neuroimaging  626–7
　neuropathology and neurochemistry  623–5
　neuropsychology  627–9
　OFC functions  631–2
　OFC neurocircuitry  631
　social cognition deficits  632–4
　theory of mind  635–6
functional neuroimaging
　addiction  488–491, 493–496, 499–502
　decision-making cognition  396–7
　emotion induction  377–91
　　amygdala-orbitofrontal cortex relationship  385
　　autonomic regulation  386
　　facial expression  379–84
　　identification of emotional stimulus  378–9
　　production of affective state  379–84
　　regulation of affective state  384–5, 385–6
　　theories of emotion  377–8
　frontotemporal dementia  626–7
　mood disorders  557–61
　　cognitive tasks  559–60
　　emotional processing tasks  558–9
　　resting state studies  557
　　task-related activation  558
　reversal learning  400
　technical considerations  423–46
　see also BOLD fMRI
functional segregation  267–8

Gage, Phineas  v, 481, 579, 588, 598, 621
globus pallidus  68, 73
go/no-go tasks  175–7, 238–9, 459–60
goal selection  246–51
Goldman-Rakic, Patricia  199, 200
gradient compensation  437–9
granular (eulaminate) areas  59
gustation see taste
gyromagnetic ratio  427
gyrus rectus  25

Hamilton Depression Rating Scale  580
Hare Psychopathy Checklist-Revised  598
hedonic value
　olfaction
　　humans  137–9
　　nonhuman primates  136
　taste  149–52
Hermitian symmetry  434
hippocampal primordium  4
hippocampus  4, 527

autobiographical memory  293–4
　emotional memory  70–1
hunches  331, 336, 339, 340, 345
hunger  159, 160
Huntington's disease  587
hyper-metabolism  505
hyperorality  636–7
hypothalamus  68
　output to  49–50

I-RISA syndrome  496–506
　aggression  503–4
　anger  502–3
　conditioning and extinction  497–8
　core symptoms  482
　inhibitory control  498–501
　personality  501–2
　reward, saliency, attribution and motivation  496–7
image distortion  430–1
Impaired Response Inhibition and Salience Attribution see I-RISA syndrome
impulsive response  226
incentive-sensitization theory  485
indoleamine systems  403–4
inhibition  307–24
　anatomical considerations  308–9
　attitude regulation and memory  316–18
　emotion regulation  313–15
　mechanism  307–8
　mood congruent bias  315–16
　regulation of sensation  310–11
　selective attention  312–13
　self-regulation in social contexts  318–19
　suppression and modulation of pain  311–12
inhibitory control  498–501
instrumental conditioning  240–2
insular cortex  331
intercalated masses  66
intermediate orbital sulci  21, 22
International Affective Picture System  363, 384
interoceptive agnosia  334
intoxication  487–9
intracranial electrophysiology  355–76
　analysis of responses  359–60
　emotional visual stimuli  360–8
　　faces  362–3
　　scenes  363–8
　expectation of reward or punishment  368–71
　neuroanatomical localization  357–9
　technique  356–7
Intradimensional-Extradimensional (ID/ED) Shift Task  400–401
intrinsic connections  39–42
Iowa Gambling Task  216, 227, 368–71, 395–7, 398–9, 505, 605, 605–6, 610
　lesions impairing  335–9
Iowa Rating Scale of Personality Change  467, 601, 602

James, William  325
Jamison, Kay Redfield, *An Unquiet Mind*  547

*k*-space trajectory  428, 430

lateral orbital cortex  13
lateral orbital gyrus  21
lateral posterior orbitofrontal area  129–31, 135
laterality
  anxiety disorders  532–4
  emotional  449–50
  pseudopsychopathy  608–11
learned helplessness  403
learning  183–9
  associative  144, 160, 199
  behavior  184–5
  discrimination  238–9
  neural activity  186–8
  reversal *see* reversal learning
  striatum  192–3
lesions of orbitofrontal cortex  97–8
Lewy body dementia  622
limbic connections  46–9
limbic system  291
limbic-prefrontal gateway  202–6

magnetic susceptibility  423, 426–9
  artifacts *see* susceptibility artifacts
major depressive disorder  546, 552, 553, 562
mania  547–8
  traumatic brain injury  585–6
Mayberg, Helen  550
medial orbital gyrus  21
medial orbital sulcus  25
medial prefrontal cortex  8–9, 40, 41, 46
  as effector of emotions  80–1
  emotional processing  79–81
mediodorsal nucleus  51
mediodorsal thalamic nucleus  9
mediodorsal trend  41
memory  285–306
  autobiographical  293–9
    healthy subjects  293–5
    orbitofrontal damage  295–6
    psychiatric patients  296–9
  confabulations  299–300
  content-based classification  285–7
  emotional  69–72
  encoding  287–90
  episodic  159–60, 286
  inhibitory control  316–18
  lesions affecting  98
  olfactory  143–5
  recognition  144
  representational  207
    role of amygdala in  219–26
  retrieval  290–3
  semantic  286
  working  144
methamphetamine abuse  394
midbrain, output to  49–50
mindreading  298
Mini Mental Status Examination  587
Minnesota Multiphasic Personality Inventory  502, 601

mitral cells  126
Modified Overt Aggression Scale  588, 590
monkeys *see* nonhuman primates
mood congruent bias  315–16
mood disorders  545–77
  brain lesions and secondary mood
      symptoms  548–51
    frontal lesions  548–9
    right-left lesion laterality and mood  549–50
    subcortical lesions  550–1
  cognitive and behavioral findings  551–4
    cognitive dysfunction  552–3
    facial affect processing  551–2
    trait deficits and distribution  553–4
  development  562–4
  functional neuroimaging  557–61
    cognitive tasks  559–60
    emotional processing tasks  558–9
    resting state studies  557
    task-related activation  558
  future directions  566–8
  molecular abnormalities and
      pharmacotherapy  564–6
    modulation of frontolimbic circuitry  564–5
    neuroprotective effects  565–6
    pharmacological challenge studies  564
  and orbitofrontal lesions  580–8
  postmortem studies  561–2
    degenerative disorders  586–7
    presentation and assessment  580
    stroke  587–8
    traumatic brain injury  581–6
  prevalence and symptomatology  546–8
    depression  546–7
    euthymia  548
    mania  547–8
  structural neuroimaging  554–7
    frontal abnormalities  554–5
    ventral PFC measures across lifespan  555–6
  *see also individual disorders*
motivation  496–7
  incentive-sensitization theory  485
  opponent-process theory  484
motor neurone disease  623
Multidimensional Personality Questionnaire  501

NEO Personality Inventory  633
neocortical primordium  4
neural pathways  95–7
neural relationships  173–4
neurochemical modulation  393–422
  decision-making cognition and reversal
      learning  394–400
  prefrontal cortex  400–12
neuroimaging  265–83
  behavioral choice  276–7
  causality  279–80
  components of reward network  277–9
  expectation and receipt of reward  273–5
  expected reward value  271–2
  functional segregation  267–8

neuroimaging (*cont.*)
  prediction error responses 275–6
  reward value representations 266–7
  valence 268–71
neurophysiology 95–124
  and behavior 226–8
  flavor 101
  olfaction 102–4
  taste 98–101
  vision 104–10
neuroprotection 565–6
Neuropsychiatric Interview 586
Neuropsychiatric Inventory 468
neuropsychological assessment 449–79
  decision-making (gambling) tasks 460–2
  experimental tasks in animals 454–60
    acquisition of alternation tasks 454–6
    go/no-go tasks 459–60
    reversal learning 456–9
  functional assessment and treatment planning 470–2
  interview and questionnaire data 466–9
  network issues and broad assessment 469–70
  olfactory testing 464–6
  social processing and theory of mind 462–4
  sources of OFC damage 450–3
  traditional methods 453–4
nonhuman primates
  architectonics 4–9, 26–34, 57–60
  connections 39–52
  chemosensory integration 153
  gustatory areas 145–6
  neurophysiology and behaviour 95–117
  olfaction 127–31
  secondary olfactory cortex 134–6
    odor identity and hedonic value 136
    stimulus specificity 134–6
  topography 25–6
noradrenaline 401, 411
nucleus accumbens 64
nucleus magnocellularis 394

object alternation tasks 454–6
object discrimination 239–40
obsessive compulsive disorder 524, 528–9, 563
odor discrimination task 208
  effect of lesions on performance 225
odor hedonics
  humans 137–9
  nonhuman primates 136
odor identify 136
odor quality 139–41
olfaction 42, 43–4, 102–4, 125–45
  cognitive judgements 141–2
  humans 131–4
  nonhuman primates 127–31
olfaction-taste association 103
olfactory cortex 4
  primary 126
  secondary 126–7, 134–41

olfactory input 126–7
olfactory learning 143–5
olfactory memory 143–5
olfactory sulcus 25
olfactory testing 464–6
opponent-process theory 484
orbital and medial prefrontal areas
  agranular insula 6–7
  central orbital areas 8
  medial prefrontal cortex 8–9
  rostral orbital areas 8
orbital prefrontal network 40, 41
orbitofrontal cortex
  as effector of emotions 80–1
  transfer of information to lateral prefrontal cortices 75–9
orbitofrontal lesions 97–8, 215–19, 579–95
  effect on aggression 588–92
  effect on mood states 580–8
orbitofrontal-striato-thalamic circuits 73–5
orthonasal sensation 155, 156
outcome expectancies 200, 206–15
  behavioral deficits following loss of 215–19
outcome-expectant activity 209–10
  rate of development 211
Overt Aggression Scale 590

pain, suppression and modulation of 311–12
panic disorder 524
Papez circuit 287, 288
parabrachial nucleus 329
parallel imaging 433–5
paramagnetism 426
Parkinson's disease 587
partial *k*-space acquisition 434
parvalbumin 59
patient EVR 333–4, 335, 599–600, 604, 634, 637
Pavlovian conditioning 239–40
periallocortical areas 41
periamygdaloid cortex 43
periaqueductal grey 49, 50, 329
perirhinal cortex 48
perseveration 226, 398
personality 501–2
  in pseudopsychopathy 601–2
PFC *see* prefrontal cortex
pharmacological challenge 564
phobias 524, 528
Pick bodies 625
Pick disease 623, 624
piriform cortex 43
point spread function 430
positron emission tomography 110, 111, 131–2, 137, 142, 143, 146–7, 157, 289–90, 292, 396–7, 425, 455, 489–90, 500, 502, 506
post-traumatic stress disorder 296, 523
posterior cingulate cortex 48
posterior orbital gyrus 21, 24
posterior orbital sulci 22
posterior orbitofrontal cortex 60–8
  emotional processing 62–4

implications of orbitofrontal circuits in behavior 66–8
input-output zones of amygdala 65–6
relationship to striatal and brainstem reward centres 64–5
signals from external environment 60–2
signals from interal environment 62
posterior parahippocampal cortex 46, 48
prediction error responses 275–6
prefrontal networks 40
primary gustatory cortex 145
primary inducers 343
primary reinforcers 332
progressive non-fluent aphasia 621
progressive supranuclear palsy 587
pseudopsychopathy 597–619
  experimental findings
    autonomic function and emotion 602–5
    decision-making 605–8
    development 611–14
    laterality 608–11
    personality 601–2
  future directions 615–16
psychiatric disease
  autobiographical memory 296–9
  role of orbitoprefrontal cortex in 81
psychopathy 597–8
Psychopathy Checklist-Revised 597
punishment 268–71
  expectation of 368–71

rat
  architecture 9–10, 14
  central 5-HT systems in 407–9
  connections 200–2
  neuronal activity 206–15
reappraisal 314
recognition memory 144
reinforcer devaluation 217–8, 220–1, 228, 241, 245
reinforcer value 267
representational memory 207
  role of amygdala in 219–26
Rescorla-Wagner learning model 183
retrieval 290–3
retronasal sensation 155, 156
retrosplenial cortex 48
reversal 242–3, 252–3
reversal learning 112–7, 399–400
  animal lesion studies 456–7
  functional imaging studies 400
  human lesion studies 457–9
  tryptophan depletion 404–7
reward 268–71, 496–7
  expectation of 368–71
  prediction 271–2
reward centres 64–5
reward contingent decision tasks 438
reward discrimination 175–9
  behavioral task 177–8
  neural activity 178–9
  reward vs no reward 175–7

reward expectation 273–5
reward learning 333
reward network 277–9
reward preference and uncertainty 179–82
  behavior 180–1
  neural activity 181–2
  rationale 179–80
  striatum 191–2
reward receipt 273–5
reward reversal 115, 116
reward value representations 266–7
reward-prediction error 368
rodent *see* rat
rostral orbital cortex 8, 13–14

salience 485, 496–7
satiety 99, 159
scenes, response to 363–8
schizophrenia 297–8, 300
scissure en H 20
seasonal affective disorder 531
secondary inducers 343
secondary olfactory cortex 126–7, 134–41
  humans 137–41
    odor hedonics 137–9
    odor quality 139–41
  nonhuman primates 134–6
    odor identity and hedonic value 136
    stimulus specificity 134–6
secondary taste area 146, 147
selective attention 312–13
self-monitoring 472
semantic dementia 621
semantic memory 286
sensation, regulation of 310–11
sensory inputs 42–3
serotonin 401, 403–4
  mood disorders 564
shimming 431–2
signal loss 431
smell *see* olfaction
social behavior 318–19
social cognition deficits 632–4
social processing 462–4
social response reversal 319
somatic markers 333–4, 341, 343–7, 386, 393
  as executive processes 343–5
  function of 346–7
  role of feedback in decision-making 345–6
somatic marker hypothesis 333–4, 341, 343–7, 386, 615
  frontotemporal dementia 634–5
somatic sensation 42
somatosensory input 111
stimulus
  affective properties 243–6
  discrimination 251
  reward-value 266
  valence 268–71
stimulus-reinforcer association 112–17
Strategic Evaluation of Alternatives 471

striatal connections 51–2
striatal reward centres 64–5
striatum 189–94
  different reward magnitudes 191
  different rewards 191
  learning 192–3
  reward preference 191–2
  reward vs no reward 190–1
stroke 451, 452
  aggression 591–2
  mood disorders 587–8
structural neuroimaging
  mood disorders 554–7
    frontal abnormalities 554–5
    ventral PFC measures across lifespan 555–6
Structured Interview for Apathy 586
subcortical lesions 550–1
subgenual cingulate 4, 14, 294, 328, 363, 366, 412, 497, 545, 555, 561, 565
  *see also* anterior cingulate
sulcul and gyral morphology 19–37
sulci fragmentosi 22
sulcus orbitalis medius 20
sulcus orbitalis transversus 20
superior temporal gyrus 45
superior temporal sulcus 45
susceptibility artifacts 424, 429–30
  reduction of 431–42
    acquisition trajectories 435–7
    data post-processing 432–3
    gradient compensation 437–9
    reduced data methods and parallel imaging 433–5
    shimming 431–2
    tailored RF pulses 439–42
Sylvian fissure 146

tailored RF pulses 439–42
Talairach space 19, 24, 132
task-related activation 558
taste 98–101, 145–52
  anticipation of 152
  chemosensory integration and flavor assembly 153–6
  gustatory areas
    humans 146–7
    nonhuman primates 145–6
  orbital taste region 148–52
    anticipation of taste 152
    stimulus identity and specificity 148–52
  pleasant 150
  regulation of feeding behavior 156–61
  unpleasant 150
taste intensity 151
taste neurons, breadth of tuning 148, 149
taste/visceral afferents 42, 4412
thalamic connections 51–2, 200–2
thalamus 291
theories of emotion 377–8
theory of mind 298, 462–4, 471
  frontotemporal dementia 635–6
  impairment of 585

time of echo 426
Tourette syndrome 505, 563
Tower of London task 406
trait deficits 553–4
transreinforcer blocking 241
transverse orbital sulcus 24
traumatic brain injury
  aggression 589–90
  mood disorders 581–6
    anxiety 583–5
    depression 581–3
    empathy and impaired theory of mind 585
    mania 585–6
tryptophan depletion, and reversal learning 404–7
tufted cells 126

umami taste 98, 100, 145
uncinate fascicle, ventral division 291
unconditioned stimulus 271, 273
University of Pennsylvania Smell Identification Test 141, 465

valence 268–71
ventral anterior nucleus 73
ventral pallidum 51
ventral posterior medial nucleus 145
ventral prefrontal cortex 378, 555–6
ventral striatum 210, 211, 378
ventral tegmental area 64
ventral temporal pole 44
ventrolateral prefrontal cortex 309, 311–19
ventromedial prefrontal cortex 325–53, 450, 461
  decision-making functions 333–43
  dysfunction 237
  explicit awareness 339–41
  extinction of conditioned fear responding 525
  lesions of
    changes in decision-making 333
    impairment of visceral responses to emotional stimul 334–5
    impairments in Iowa gambling task 335–9
  somatic marker hypothesis 333–4
  visceral connectivity 328–30
  visceral functions 326–33
  visceral motor functions 330–1
  visceral sensory functions 331–3
  vs orbitofrontal cortex 326–8
vision 42, 44–5, 104–10
visual discrimination reversal 106
visual-olfactory associative learning 143
voice expression identification impairment 110

Walker's areas 4, 6, 28, 29–30
Wechsler Memory Scale 453
Wisconsin Card Sort Task 453
Wisconsin General Test Apparatus 251
withdrawal 482, 493
working memory 144

z-shim 438